D1285248

# DINOSAURS
# *of the* AIR

# DINOSAURS *of the* AIR

## *The Evolution and Loss of Flight in Dinosaurs and Birds*

Gregory S. Paul

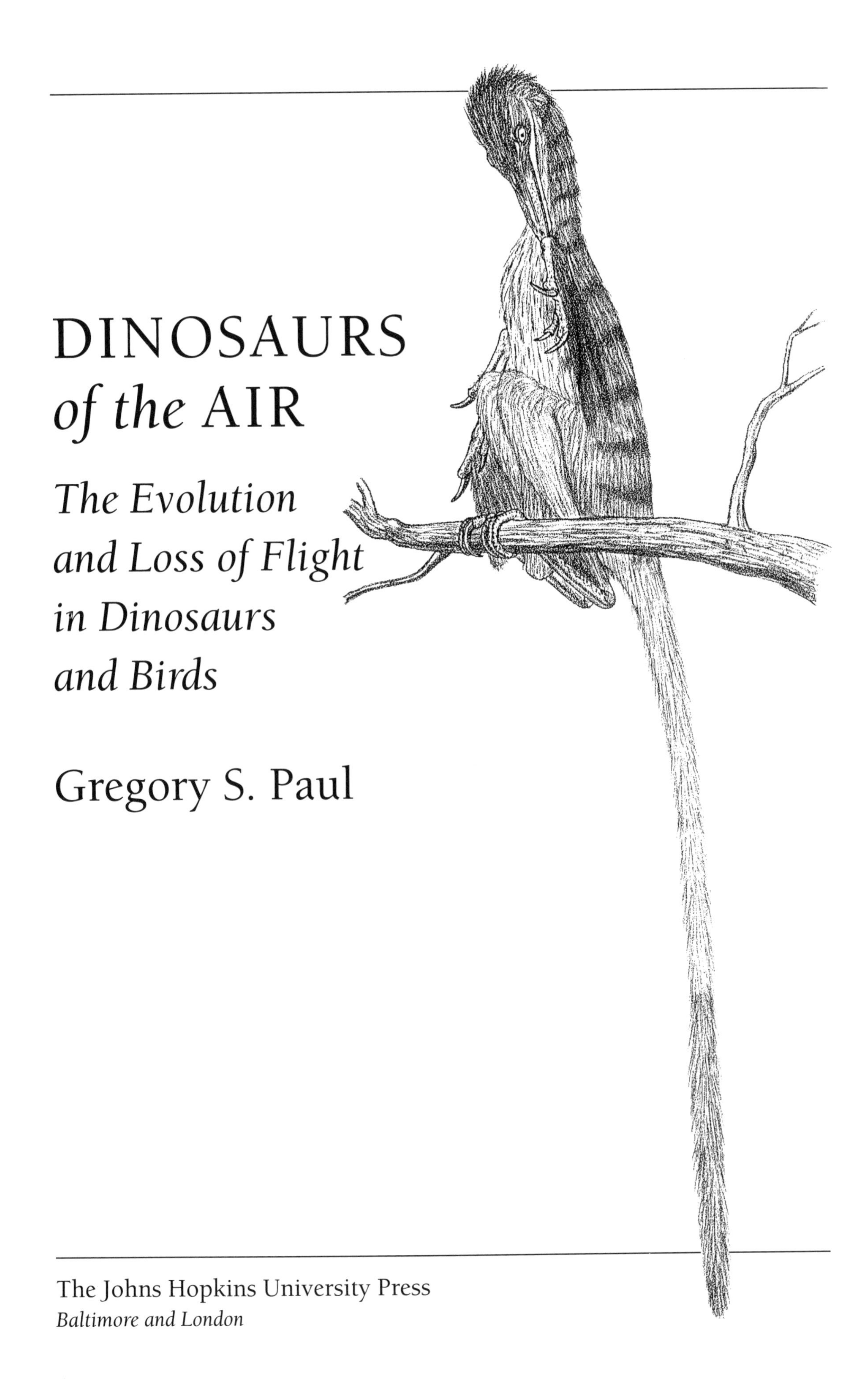

The Johns Hopkins University Press
*Baltimore and London*

Printed in the United States of America on acid-free paper
9 8 7 6 5 4 3 2 1

The Johns Hopkins University Press
2715 North Charles Street
Baltimore, Maryland 21218-4363
www.press.jhu.edu

Library of Congress Cataloging-in-Publication Data

Paul, Gregory S.
Dinosaurs of the air : the evolution and loss of flight in dinosaurs and birds / Gregory S. Paul.
p. cm.
Includes bibliographical references and index.
ISBN 0-8018-6763-0 (hardcover : acid-free paper)
1. Birds, Fossil. 2. Birds—Evolution. 3. Birds—Flight. I. Title.
QE871 .P38 2001
568—dc21 2001000242

A catalog record for this book is available from the British Library.

*Title page illustration:* The newly discovered Chinese dino-bird *Microraptor,* the smallest known dinosaur, tells us a lot about the origin of birds and their flight. That it was adapted for climbing adds support to the hypothesis that avian flight evolved among arboreal dinosaurs. Its possession of flight-related features that are in some respects better developed than those of *Archaeopteryx* and later dromaeosaurs suggests that the sickle claws were secondarily flightless.

To fellow paleoartist and researcher of bird origins Gerhard Heilmann

# *Contents*

# *Preface and Acknowledgments*

Having grown up reading that birds were at most only distantly related to dinosaurs, I was both startled and fascinated when a growing number of researchers began to argue that birds were the direct descendants of the terrible lizards. At the end of the 1970s, I was becoming involved in the argument both as an artist and a scientist. Art and science could not be separated; the science determined whether I should adorn the small predatory dinosaurs with scales or with feathers. Alas, there was no simple answer at the time, because there was no fossil evidence for either possibility. What I found disturbing was that many paleontologists were totally biased against illustrating dinosaurs with feathers, even though the available data was neutral on the subject. This bias only boosted my interest in researching bird origins.

While restoring the skeletons of dromaeosaurs and other close bird relatives, I noticed that they often had avian features lacking even in *Archaeopteryx,* features otherwise seen in flightless birds. This set me to wondering if some dinosaurs were themselves secondarily flightless, and closer to modern birds than the first bird was.

Since I presented this notion in *Predatory Dinosaurs of the World* in 1988, our knowledge of near-avian dinosaurs has exploded, and the hypothesis that some dinosaurs were birds has gained some support. At the same time, however, a few researchers have continued to deny that birds evolved from dinosaurs. I have long intended to present the totality of my research concerning my argument that birds are dinosaurs and that some dinosaurs may have been birds, and to cover other aspects of the evolution of birds from dinosaurs. The complexity of the argument precludes its presentation in a journal article. *Dinosaurs of the Air* follows in the tradition of fellow scientist's Gerhard Heilmann's *The Origin of Birds* in being both a technical and popular study.

There are many people to thank for their assistance, in one form or another, with this book. Indeed, their help extends back over decades of research, and I have especially benefited from discussions with and data provided by generous colleagues. Among them are Phil Currie and Larry Witmer, who especially encouraged this project; Andrzej Elzanowski, who provided valuable information on birds; Tom Holtz, Ken Carpenter, Alex Downs, Dan Chure, Guy Leahy, David Carrier, and Geoffrey Birchard, who provided helpful criticism on the energetics section; Steven Perry and John Brackenbury, who sent information on tetrapod respiration; John Ostrom, Peter Wellnhofer, Halzska Osmólska, Teresa Maryanska, Mark Norell, Peter Makovicky, Leon Claessens, Silvio Renesto, Alick Walker, Sankar Chatterjee, Nick Longrich, Fernando Novas, Julia Clarke, Ji Qiang, Xing Xu, Kevin Padian, Jacques Gauthier, Luis Chiappe, Alan Brush, Cristiano Dal Sasso, Marco Signore, Chris Brochu, John Brackenbury, Brian Cooley, Michael Raath, George Olshevsky, Steve and Sylvia Czerkas, Cathy Forster, Kristina Curry Rogers, John Horner, Jim Kirkland, Jim Farlow, Mike Brett-Surman, Storrs Olson, and Richard Zusi; Kathy Broder, who showed me the marvelous kiwis in her care; David Rimlinger; Trouper Walsh and his oras; Chris Sloan; Charles Martin; and many others.

I much appreciated the hospitality extended by many colleagues when I visited their institutions. Among them were Gunter Viohl at the marvelous Jura-Museum in Eichstätt, Bavaria, where Luftwaffe F-104s thundered overhead as I examined their *Archaeopteryx;* Herman Jaeger at the Humboldt Museum in the then East Berlin; and the border guard who cheerily passed me through Checkpoint Charlie at 6:00 A.M. I thank those who allowed me to use photographs they kindly provided. This book would not have been possible without the interest and help of Ginger Berman, and was greatly .assisted by the work of Sam Schmidt, both of the Johns Hopkins University Press. I am grateful to the staff of Princeton Editorial Associates for their fine work in the editing and production of the book.

# DINOSAURS
# *of the* AIR

PART 1

# GETTING STARTED

One of the wonderful coincidences of science is that immediately after Charles Darwin published *On the Origin of Species,* his famous explication of the mechanism behind evolution, dramatic support for his hypothesis appeared in Bavaria. In 1860, a feather and, in 1861, the skeleton of a Mesozoic vertebrate obviously intermediate in form between modern birds and their reptilian ancestors were uncovered in lithographic slate quarries. This vertebrate was, of course, the urvogel (original bird) *Archaeopteryx.* As our knowledge of fossil birds has expanded in the subsequent fourteen decades, the question of how birds arose has become ever more fascinating.

Most paleontologists now agree that birds—always popular with the public—happily happen to be the direct descendants of the best-liked group of extinct creatures, the dinosaurs. Of course, public opinion has no relevance to scientific debate, but the broad appeal of a dinosaur-bird link vexes the shrinking minority of researchers who dispute the link. A real problem with the science behind the dinosaur-bird hypothesis is the lack of detailed descriptions and illustrations of the specimens that support the hypothesis. At conferences on bird origins, speaker after speaker has presented cladogram after cladogram in support of the dinosaur-bird link, leaving the supporting anatomical information poorly covered. In addition, the detailed data that is available has yet to be presented in one volume. Therefore, it is not surprising that the skeptics, who are not dinosaur paleontologists, sometimes seem unfamiliar with the evidence showing how very similar certain dinosaurs are to birds. Thus, one of my goals in this work is to thoroughly present in textual form and especially in visual form the data that supports an ancestor-descendent relationship between the two groups.

However, this book is not another general purpose work that attempts to show that birds are descended from dinosaurs. The evidence favoring an ancestor-descendent relationship is so strong—especially since the recent and long-awaited discovery of feathered dinosaurs—that the validity of the relationship is no longer the center of the debate. The important questions now are how, why, and when did birds evolve from small predatory dinosaurs in Mesozoic times.

In the past, paleontologists generally tended to assume that bird evolution was a straightforward process by which some reptiles evolved increasingly sophisticated airfoils that allowed improvements in flight performance. Because few fossils were known, it was easy to construct such simple scenarios. In addition, some paleontologists unfortunately tended to simplify evolution so that it involved a system of progressive lineages that inevitably led to the advanced creatures we live among today. We now understand that bioevolution is an inefficient information-processing complex that generates intricate branching patterns of descent in which any increases in complexity are incidental, albeit important, side effects (Levy 1992, Paul and Cox 1996, Ridley 1996). Evolutionary parallelism and even major reversals are not only possible, but also common.

The fossil record is beginning to show, in ways still dimly illuminated and poorly understood, that the evolution of birds was yet another example of intricately branching evolutionary complexity that generated bewildering parallelism, dramatic reversals, and creatures that were in some cases simply bizarre. For example, this book presents evidence that some of the most birdlike of small predatory dinosaurs were not prefliers near to birds but rather the secondarily flightless descendants of ancestors that could fly about as well as if not better than *Archaeopteryx*. This book also shows that some of these dino-birds may have been more closely related to modern birds than the urvogel was, a possibility that raises interesting questions about what is and is not a bird.



In that *Dinosaurs of the Air* combines a large amount of technical data, some of it new, in a semipopular format that includes numerous artistic renderings, it follows in the tradition of a classic book produced three-quarters of a century ago by another artist-scientist, Gerhard Heilmann. His *Origin of Birds* was popular as well as influential, and I hope that my book will likewise appeal to nonprofessional enthusiasts, whether their interest be dinosaurs, birds, or evolution.

*Dinosaurs of the Air* is organized so that chapters covering closely related subjects are gathered together into parts that examine a common theme. The first five appendices contain technical data and cover other subjects related to bird origins, including the energetics of the origin of avian flight. The sixth appendix covers the latest and perhaps most important information to date on the loss of flight in dinosaurs.

CHAPTER 1

# A History

The history of avian flight began far back in the Mesozoic Era and continues today as humans research the complex story and add new fossils to fuel debate. This chapter is intended to bring nonprofessionals up to speed on the cast of often extinct characters discussed throughout the book and on the mysteries of geological time.

## Who Was Who in the Mesozoic Era, and About Time

Vertebrates, or backboned animals, are the enormous group that includes everything from fish to birds to humans. Tetrapods are vertebrates that have legs suitable for walking on land, or are descended from animals with such legs. This group includes amphibians, as well as whales, birds, and people. Amniotes are tetrapods that either lay shelled eggs or are descended from egg-laying ancestors. Reptiles are in this group, birds obviously are, and mammals are too. Most amniotes fall into two major categories: the synapsids, which include mammals and their ancestors, and the diapsids, the tetrapods this book focuses upon.

Diapsids include a wide array of largely extinct, usually lizardlike reptiles, of which the sole surviving example is the rare tuatara of New Zealand. The multitudes of true lizards and the snakes that evolved from them are also diapsids. The largest and most successful diapsid group is the archosaurs, which include the famous extinct thecodont, pterosaur, and dinosaur subgroups. The only living archosaurs are crocodilians and birds. Therefore, the closest living relatives of birds are crocodilians, followed by lizards and tuataras. Mammals and birds are only distant relations.

The category dinosaur includes a number of groups whose interrelationships are not always well understood. The herbivorous dinosaurs included the early, small-headed prosauropods, as well as sauropods, which were also small-headed but gigantic in body. The bird-hipped ornithischians were a very diverse group of beaked herbivores that included small to gigantic ornithopods that ran semibipedally; various armored types including stegosaurs and ankylosaurs; dome-headed pachycephalosaurs; and rhinolike horned ceratopsians. The organization of predatory dinosaurs is detailed in at the end of the next chapter and in Part 5.

Dinosaurs did not all live at the same time; their history was spread out over 100 million years (Myr), late in the history of our universe. So let us start at the beginning. The Big Bang occurred some 12 to 13 billion years before the present, and our solar system came into being a little more than 4.5 billion years ago. The planets then experienced an intense meteoritic bombardment—you can see the results by looking at the cratered surface of the moon—that may have hindered the development of life until 3.8 billion years ago. Ironically, the mass of incoming comets may have supplied the water and organic raw material for dinosaurs, birds, and the people who like them. Life seems to have remained single celled for billions of years, through most of the long stretch of time called the Precambrian. The Precambrian was followed by the Paleozoic Era, which began 540 Myr ago, when Earth was already fast approaching 90 percent of its current age.

The Paleozoic began with the Cambrian Explosion, a sudden—in geological terms—radiation of complex, often hard-shelled sea creatures such as trilobites. At about this time the first primitive vertebrates appeared in small numbers. Later in the Paleozoic, fishes became common, and plants and then animals colonized the land. Swamps, which would form the first of a long series of coal beds, and insects (including gigantic dragonfly-like forms), amphibians, and advanced mammal-like reptiles were abundant. The presence of the latter shows that the big split between synapsids and diapsids had already occurred. The synapsids were so successful that the closing periods of the

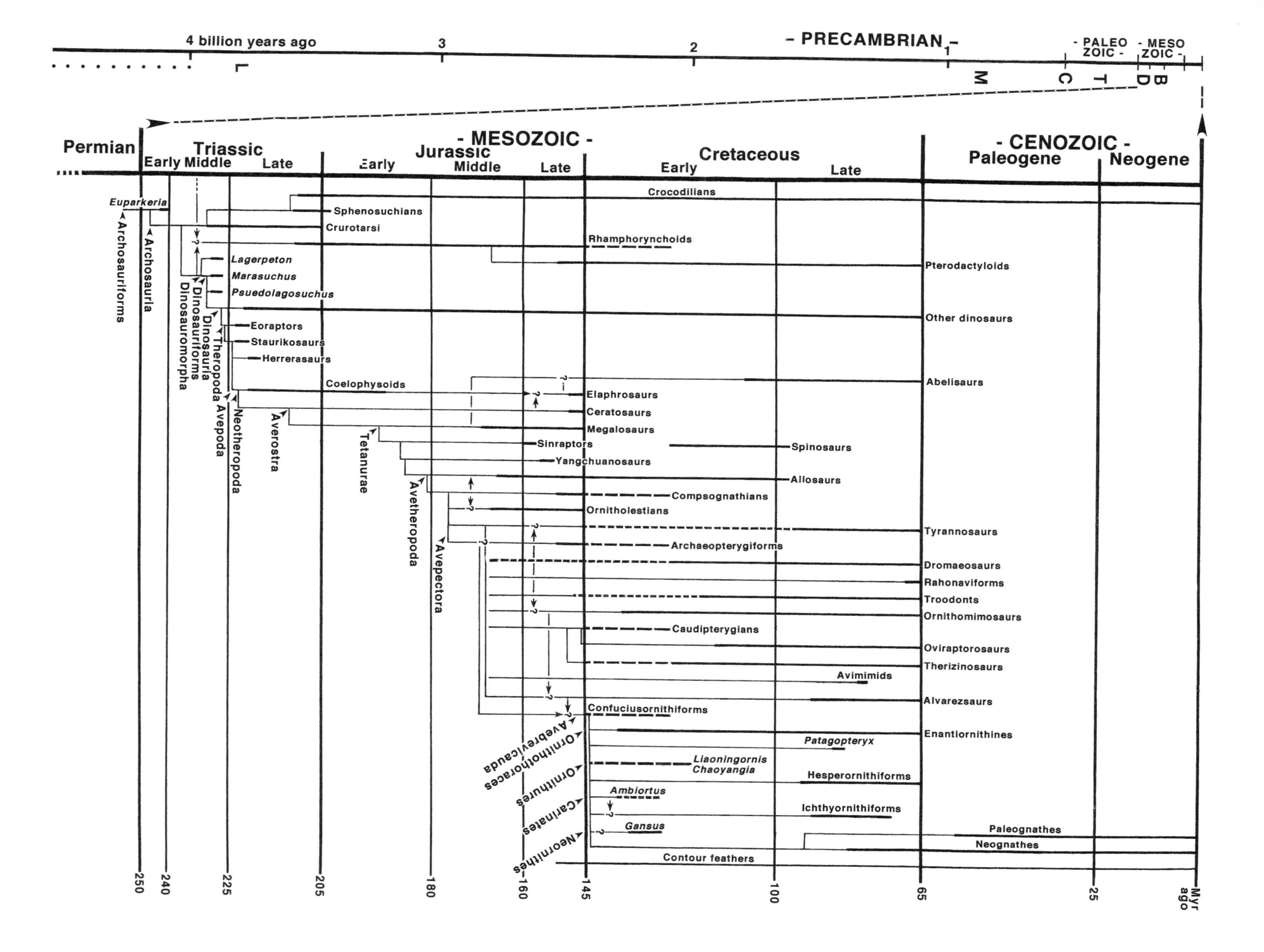
4 billion years ago
3
2
- PRECAMBRIAN -
1
- PALEO ZOIC -
- MESO ZOIC -
L
M
C
T
D
B
Permian
Triassic
Early Middle
Late
- MESOZOIC -
Jurassic
Early
Middle
Late
Cretaceous
Early
Late
- CENOZOIC -
Paleogene
Neogene
Crocodilians
Euparkeria
Sphenosuchians
Crurotarsi
Rhamphorynchoids
Lagerpeton
Marasuchus
Psuedolagosuchus
Pterodactyloids
Other dinosaurs
Eoraptors
Staurikosaurs
Herrerasaurs
Abelisaurs
Coelophysoids
Elaphrosaurs
Ceratosaurs
Megalosaurs
Sinraptors
Spinosaurs
Yangchuanosaurs
Allosaurs
Compsognathians
Ornitholestians
Tyrannosaurs
Archaeopterygiforms
Dromaeosaurs
Rahonaviforms
Troodonts
Ornithomimosaurs
Caudipterygians
Oviraptorosaurs
Therizinosaurs
Avimimids
Alvarezsaurs
Confuciusornithiforms
Enantiornithines
Patagopteryx
Liaoningornis
Chaoyangia
Hesperornithiforms
Ambiortus
Ichthyornithiforms
Gansus
Paleognathes
Neognathes
Contour feathers
Archosauriforms
Archosauria
Dinosauromorpha
Dinosauriforms
Dinosauria
Theropoda
Avepoda
Neotheropoda
Averostra
Tetanurae
Avetheropoda
Avepectora
Avebrevicauda
Ornithothoraces
Ornithures
Carinates
Neornithes
250
240
225
205
180
160
145
100
65
25
Myr ago

Paleozoic could be called the first age of synapsids; in comparison, the diapsids were doing only moderately well. The last period of the Paleozoic, the Permian, ended 250 Myr ago, when the planet was 94 percent of its current age. A few researchers have suggested that birds first evolved in the Permian, but this is very unlikely. What did appear at that time was a new group of diapsids, the then lizardlike archosaurs.

At the end of the Permian, a mass extinction, the worst in Earth's history, occurred. Despite the severity of the event, most of the major tetrapod groups survived—if just barely—into the Mesozoic Era. The new era's first period, the Triassic, saw both a reradiation of tetrapod groups (including advanced mammal-like therapsids) and the appearance of new groups (including the first small mammals near the end of the period). The Triassic can be called the "age of diversity" because both synapsids and diapsids were out in force. Among the latter, tuatara relatives and lizards started to become common. Of more interest to us, however, are the archosaurs, which began to flourish and diversify. The rather croclike thecodonts—some quite large—and the first true crocodilians, which were rather small, became the chief predators on land and in fresh waters. The very sophisticated pterosaurs also appeared, perhaps preempting the evolution of a new set of giant flying insects. A number of poorly known, small, and possibly arboreal archosaurs were also present. Some paleontologists think that birds evolved from these small archosaurs. In the last third of the Triassic, dinosaurs first evolved and became moderately successful predators and herbivores; none was gigantic. These included four-toed predatory dinosaurs that, although rather archaic in form, were like birds in being largely bipedal. Perhaps most important to the future of birds was the emergence of the first bird-footed, gracile, flesh-eating dinosaurs, such as *Coelophysis*. The end of the Triassic, about 205 Myr, was marked by a modest extinction that saw the loss of the thecodonts, as well as the four-toed predatory dinosaurs.

The dinosaurs survived in good order into the middle period of the Mesozoic, the Jurassic. With tuatara relatives, lizards, crocodilians, pterosaurs, dinosaurs, and eventually birds being common and even dominant in the air, on the land, and at sea—synapsids were limited to wee little mammals—the last three-quarters of the Mesozoic can be called "the great age of the diapsids." The Jurassic certainly was a global park for dinosaurs, which were the only large land predators and herbivores. Sauropods were the land equivalent of whales in terms of size, as well as the most important herbivores of the day. They, and a number of more modest-sized ornithischian herbivores including armored stegosaurs, were preyed upon by blade-toothed dinosaurs as big as rhinos and elephants. In the opening stages of the Jurassic, the bird-footed predatory dinosaurs differed little from those of the Triassic. Unfortunately, our knowledge of the small predatory dinosaurs of the Jurassic is poor. They must have been there, but only near the end of the period do we find reasonably good fossils for a few small, moderately advanced forms such as *Compsognathus* and *Ornitholestes*. Some tantalizing teeth hint that dinosaurs that were even more birdlike had evolved by that time.

Our knowledge base for birds is even poorer over most of the Jurassic. Nary a bird bone or feather has been found from the Early and Middle Jurassic, even though large numbers of pterosaurs and flying insects are known from a number of fine-grained, fossil-rich *lagerstätten* sediments. The absence of such feathers and bones suggests that flying birds were absent, rare, or confined to limited habitats. Birds and their feathers start to show up, most notably in the form of the famous *Archaeopteryx*, toward the end of the period, a mere 150 Myr ago, when 96.7 percent of Earth's

*Figure 1.1. (opposite) Left margin,* time line of Earth's history from 4.5 billion years before present to today. The series of dots on the left marks the period during which meteoritic bombardment precluded the development of life. *L,* the appearance of life; *M,* the appearance of multicellular life; *C,* the Cambrian Explosion; *T,* the appearance of tetrapods; *D,* the appearance of dinosaurs, crocodilians, pterosaurs, and mammals; *B,* the appearance of birds. Observe that complex terrestrial life has only been around for a tiny percentage of Earth's history. *Right,* chart of the Mesozoic and Cenozoic Eras showing the evolution of Archosauria over time (also see Fig. 2.1). Thick solid bars mark the approximate known spans of specific taxa; thick dashed bars indicate possible spans based on tentative fossil remains or uncertain dating of fossils (especially the Early Cretaceous lower Yixian fossils). The thin solid lines indicate possible relationships, with alternatives often presented. Notice the sparsity of avepod remains over much of the Jurassic and the absence of unambiguous bird remains, including contour feathers, before the Late Jurassic. Also notice that of all the many archosaurs, only crocodilians and neornithine birds have survived to today.

history had passed. There may have been an extinction event involving some dinosaurs and pterosaurs at the end of the Jurassic, but the details are murky.

During the Cretaceous, the last and longest Mesozoic period, reptiles continued to rule the seas, pterosaurs remained successful and grew to extraordinary sizes, and mammals stayed small even as they, like dinosaurs, diversified and became more sophisticated. Giant sauropods were somewhat pushed aside by an array of new large ornithischian herbivores, including heavily armored ankylosaurs, big semibipedal iguanodonts, duckbilled hadrosaurs, and, last of all, the rhino-like horned dinosaurs. All were the food supply for giant predatory dinosaurs. Some of these great flesh eaters were similar to those seen in the Jurassic, but toward the end of the Cretaceous, the advanced, small-armed tyrannosaurs became important.

It is the small predatory dinosaurs of these times that are most interesting, for their skeletons were quite different from those of their earlier relatives—*Compsognathus*-like and feathery *Sinosauropteryx* being an exception. The birdlike dinosaurs included sickle-clawed troodonts, dromaeosaurs (the misnamed raptors of the movie *Jurassic Park*), flying *Rahonavis,* beaked caudipterygians and oviraptorosaurs with misshapen heads, ostrichlike toothless ornithomimosaurs and avimimids, and tube-billed alvarezsaurs with short digging hands. What is most important about these Cretaceous forms is how extraordinarily birdlike they were and how smart, by dinosaurian standards: their brains were as large as those of smaller-brained birds.

As for birds themselves, their skeletons and feathers make a big appearance in the opening stages of this period. They were essentially modern-looking birds, although their skeletons retained numerous reptilian characteristics, including teeth in many cases. By the end of Cretaceous times, truly modern birds were beginning to appear. Indeed, by the end of the Mesozoic, the world was in many ways surprisingly modern. Birds filled the skies, and the dominant plants were flowering herbs, shrubs, and trees that would blend right into many contemporary landscapes.

As is well known, the end of the Cretaceous, 65 Myr ago, saw another severe extinction, one that was harder on large land animals than the one at the end of the Permian had been. All of the classic dinosaurs, as well as the last pterosaurs, went belly up at the time. This extraordinary event remains poorly explained. Many Mesozoic birds were also lost. Yet enough birds survived to spawn a new and even more spectacular radiation of their kind in the Cenozoic, the Age of Mammals, which could also be called "the second age of synapsids." True enough, mammals rule the land, are abundant in the seas, and even dominate the night skies, but diapsid archosaurs still own the daylight skies.

So next time you are in the park watching the squirrels and birds hop across the grass, you might pause to reflect upon the fact that you and the rodents share a history that diverged from that of our feathered diapsid friends some 300 Myr ago.

## A Brief History of the Study of Bird Origins

The development of ideas about the origin of birds over the past fourteen decades has been covered in detail by Witmer (1991), Feduccia (1996), and Shipman (1998). The brief summary presented here as a primer to those new to the subject emphasizes the issues that are particularly relevant to this study.

### *The Early Years*

Dinosaurs begin to be recognized as such in the early 1800s, but they were largely gigantic reptilian forms known only from fragmentary remains. By 1861, Lyell's geology and Darwin's evolutionary biology made birds obvious candidates for being "glorified reptiles" that, over vast stretches of time, developed feathers as they learned to fly. In 1860, the first avian fossil, a single feather, was discovered in the Solnhofen lithographic limestones of southern Germany. This feather was the initial, tantalizing evidence that something avian had lived in the middle of the Mesozoic, when giant dinosaurs and pterosaurs dominated the land and sky. The year 1861 was an exceptional one in that two truly spectacular items were found in the very same Jurassic sediments: a semi-articulated skeleton of *Archaeopteryx,* adorned with feathers, and an even more complete skull and articulated skeleton of the small predatory dinosaur *Compsognathus.* It is unfortunate that feathers were missing on the latter specimen; had they been present, views on bird origins would have developed differently. Even without feather evidence, E. D. Cope (1867) and especially Thomas "Darwin's bulldog" Huxley (1868) in a few years were noting the similarities between dinosaurs and birds and suggesting

*Figure 1.2. Top,* the London specimen of the oldest known and most celebrated dino-bird, *Archaeopteryx; bottom,* the first complete skeleton of a predatory dinosaur, *Compsognathus.* Feathers accompany the former, and their absence on the latter, which is probably an accident of nonpreservation, has hindered full appreciation of avian descent from small "terrible lizards." Drawn to same scale.

a close relationship. Thus the dinosaur-bird hypothesis was the initial, albeit tentative, explanation for how birds arose. On the basis of the birdlike features seen in *Compsognathus,* Huxley even hinted at a relationship between predatory dinosaurs and birds. On the other hand, he also suggested a link between birds and the herbivorous bird-hipped ornithischians, because the small dinosaur *Hypsilophodon* he described had a backwardly directed (retroverted) pubis, as did birds.

During the last decades of the 1800s and the opening decades of the 1900s, researchers argued with varying degrees of insistence for a close dinosaur-bird relationship (Haeckel 1875, Marsh 1877, 1880, Williston 1879, Baur 1883, 1887, Dollo 1883, 1884, Wiedersheim, 1885, Osborn 1900, Broom 1906, Hay 1910, Abel 1911; references mentioned in this section and the section titled "The Middle Years" can be found in Witmer [1991]). Within this prodinosaur venue, the particulars varied considerably. Some of the cited researchers suggested that birds and dinosaurs shared a close common ancestor; others favored a direct ancestor-descendent relationship. Both predatory and herbivorous dinosaurs were cited as being closest to birds. The time of divergence between birds and dinosaurs was put as far back as the last period of Paleozoic (more than 250 Myr ago) and as late as the Jurassic (some 150 Myr ago).

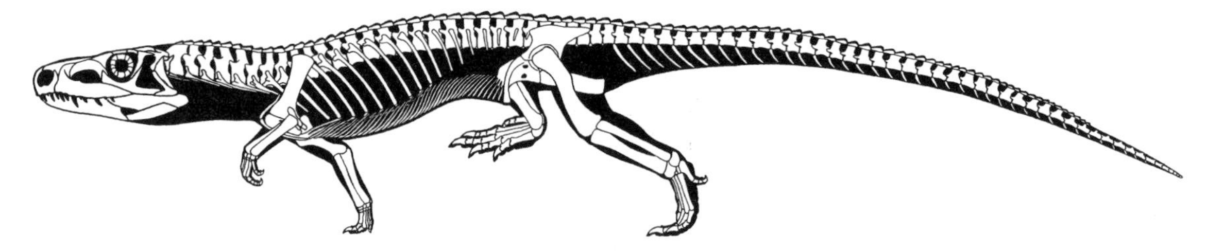

*Figure 1.3.* The classic "thecodont" *Euparkeria.* The discovery of this creature prior to World War I did much to turn the majority of researchers against the notion of dinosaurs as bird ancestors. This lizardlike basal archosaur is certainly close to the ancestors of birds, but only in that it is a suitable ancestral type for all archosaurs.



A special link between dinosaurs and birds was by no means universally accepted. Among the earliest dissenters was one of Huxley's strongest opponents, the arch anti-Darwinian Richard Owen (1875), and he was followed by Mudge (1879), Seeley (1881), Dames (1884), Furbringer (1888), and Broom (1913). The variation within this anti-dinosaur theme was considerable. The proposed relatives or ancestors of birds included pterosaurs (most often), basal archosaurs, crocodilians, and even mammals. An interesting deviation was the notion that birds were not, as generally accepted, members of a single monophyletic group; instead, they were seen as having evolved at least twice from separate ancestors. For example, Vogt (1880) pictured flightless dinosaurs as having given rise to flightless ratites such as ostriches, while flying *Archaeopteryx* was ancestral to birds that were even more volant. Wiedersheim (1885) allied flying pterosaurs, *Archaeopteryx,* and modern birds in one group and put flightless dinosaurs, the Mesozoic diver *Hesperornis,* and ratites in another. Note that in these views, flight was never secondarily lost in the major groups of flightless birds.

An emerging theme of the antidinosaur group was that the similarities between dinosaurs and birds resulted not from shared ancestry but from convergence or parallelism or both. Another argument was that dinosaurs were in various ways too specialized to have been the ancestors of birds. Researchers continue to make such arguments today.

In addition to debates about what birds evolved from, there was a debate over how flight evolved, and this debate became increasingly vigorous over time. On one side were Marsh (1880), Pycraft (1906), Abel (1911), and Beebe (1915), who saw high places—cliffs or trees—as the logical place for avian flight to have begun, probably among protoavian gliders. Others, such as Williston (1979) and the eccentric paleontologist and occasional Balkan spy Baron Von Nopsca (1907, 1923), thought that flight evolved from the ground up. Still others (Hay 1910, Gregory 1916) straddled the fence by suggesting that protobirds were scansorial, that is, adept both on the ground and in high places. Abel (1911) suggested an interesting extension of the arboreal hypothesis: terrestrial dinosaurs had not become arboreal birds; instead, both had descended from tree-climbing ancestors! The discovery in the late 1800s of the seemingly secondarily flightless hesperornithiform diving birds indicated that the story of avian flight in the Mesozoic was not solely one of ever improving aerial performance.

Two serious problems hindered the study of bird origins in those years: the poor understanding of the mechanism of evolution and the lack of a rigorous methodology for restoring the phylogenetic relationships among fossil organisms. The discovery of the computational nature of DNA—the molecule uses four letters or digits to process and transmit information—and the modern synthesis of evolutionary theory, not to mention modern cladistic analysis of relationships, was decades away. However, it is not clear how much difference better scientific methodology would have made in the face of the grossly inadequate fossil record available at the time. Although a second, complete *Archaeopteryx* specimen with spectacular feathers was discovered in 1877, its damaged skull revealed little anatomical detail. The same was true of the skulls of the few later Mesozoic birds known at the time, described in Marsh's classic monograph of 1880. Among predatory dinosaurs known at the time, large ceratosaurs, allosaurs, and tyrannosaurs were becoming well understood, but small examples were rare. Especially scarce were small skulls: the only material available was the smashed crania of *Compsognathus,* the still poorly described head of little Jurassic *Ornitholestes,* and the specialized and often damaged skulls of the Cretaceous ostrich-mimicking ornithomimosaurs.

*Figure 1.4. Top,* the sickle-clawed dromaeosaur *Velociraptor; bottom, Oviraptor,* with its bizarre head and beak. The epic American Museum of Natural History expeditions to Mongolia in the 1920s did not find the hoped-for missing link between humans and apes but did uncover the first partial remains of advanced Late Cretaceous dinosaurs close to the origins of birds. *Oviraptor* did indeed have the wishbone furcula Heilmann was, at that time, saying that dinosaurs did not have. Later discoveries of better remains revealed that both these creatures could fold their arms in the avian manner. Drawn to same scale.

A nondinosaurian discovery that profoundly influenced the issue of bird origins was a complete specimen of the basal, generalized archosaur *Euparkeria* from Triassic sediments in South Africa. Broom (1913) concluded that such primitive lizardlike thecodonts were the ideal ancestors for all the major archosaur groups, simply because they lacked specializations that might hinder their being the ancestors of birds. By that time, it was clear that birds, crocodilians, dinosaurs, pterosaurs, and thecodonts all belong to a major tetrapod group, the archosaurs, characterized in many examples by the presence of a large opening in the side of the skull immediately in front of the orbits.

It is worth noting that Abel's 1911 suggestion that birds and dinosaur shared a common arboreal ancestor incorporates an interesting implication, that is, that all dinosaurs were secondarily terrestrial.

### The Middle Years

The year 1923 saw the first of the American Museum of Natural History's famous China expeditions led by Roy Chapman Andrews. These expeditions did not find the non-African prehumans American paleontologist Henry Osborn back in New York was yearning for, but they did find the incomplete remains of three little predatory dinosaurs, remains that were pregnant with ramifications for the study of bird origins. These remains were briefly described by Osborn the following year: sickle-clawed *Velociraptor,* big-brained *Saurornithoides,* and, perhaps most important among the collection, the partial skull and skeleton of peculiar *Oviraptor.* The underappreciated *Oviraptor* remains have a toothless beak, a very long arm with *Archaeopteryx*-like fingers that is partly folded in the avian manner, a birdlike shoulder blade, and above all else an obvious furcula, or wishbone, which was, alas, mislabeled as an interclavicle! From such little mistakes, big problems arise.

In 1926 came *the* book, the English version of Gerhard Heilmann's *The Origin of Birds.* Interestingly, Heilmann was an artist rather than a paleontologist (Nieuwland 1999), but the force of his detailed and comprehensive work was greater than that of any previous scientist. His seminal study was a combination of astute analysis and missed opportunities. On the positive side, Heilmann favored the arboreal origin of avian flight, he allied birds firmly with reptiles rather than with mammals, and he finally killed off the idea that there was a close link between birds and pterosaurs, explaining that the latter represented a true example of convergence with birds. More questionable was the "proavis" that Heilmann constructed, a speculative arboreal thecodont-bird with initial flight adaptations. This imaginary archosaur stemmed from Heilmann's conclusion that *Euparkeria*-type thecodonts were the direct ancestors of birds, although he thought thecodonts shared a close common ancestor with dinosaurs.

Indeed, Heilmann came within one sentence of more intimately linking predatory dinosaurs and birds to one another, concluding that "it would seem a rather obvious conclusion that it is amongst the Coelurosaurs that we are to look for the bird ancestor." However, Heilmann thought that the clavicles were "wanting in dinosaurs," so he rejected dinosaurs as the ancestors of a group that is so famous for its large wishbones. His conclusion reflected an overly strict adherence to Dollo's law, which deems that once a feature is lost it cannot be regained. What is even more vexing is that had Heilmann looked closely at the oviraptorid wishbone that can be seen in Figure 8 in Osborn's 1924 paper—or better yet, had he gone to New York to see the three little predaceous dinosaurs firsthand—he might have reversed his conclusion. Had he done so, the dinosaurian ancestry of birds would probably be as universally accepted today as is the therapsid ancestry of mammals.

As it was, *The Origin of Birds* had the effect, which Heilmann probably did not intend, of largely suppressing further debate on bird origins, even though the known fossil record continued to remain inadequate to support Heilmann's conclusions. The intellectual entrenchment deepened so quickly that Camp's unambiguous identification of clavicles in an American Jurassic predatory dinosaur in 1936 was blatantly ignored (until recently, when the fact that clavicles are common in predaceous dinosaurs became apparent).

Thus began the barren years in which the how, why, what, and where of bird origins were relatively little discussed. Nopsca (1929) protested Heilmann's arboreal beginning for bird flight, and Boas (1930) continued to argue for a dinosaurian ancestry for birds, albeit with the complication that predatory dinosaurs first evolved from herbivorous ornithischians before birds evolved from predatory dinosaurs. Holmgren's (1955) arguments for a direct ancestor-descendent relationship between predatory dinosaurs and birds were all but ignored. Lowe (1928, 1935, 1944) presented a complex and confusing hypothesis in which the urvogel was a feathered dinosaur unrelated to modern birds. Both Holmgren and Lowe also raised the possibility that some dinosaurs had been feathered. But the Heilmann hypothesis had become accepted as dogma by Gregory (1935), Simpson (1946), De Beer (1954, 1956), Romer (1966), Bock (1969), and Brodkorb (1971). Another dogma accepted at this time was that dinosaurs themselves did not represent a natural group, but were instead two kinds of archosaurs—lizard-hipped saurischians and bird-hipped ornithischians—that evolved independently from basal archosaurs. In this view, generalized thecodonts were the Triassic stem stock from which arose, quite separately, pterosaurs, crocodilians, two dinosaur groups, and birds.

The situation with the fossil record did not help much: dinosaur paleontology had also entered a period of quiescence that began in the Depression–World War II era and lasted for thirty years. Large numbers of complete skulls and skeletons of the small Triassic predatory dinosaur *Coelophysis* were uncovered in the late 1940s in New Mexico, but these remains were not described in detail and therefore had little effect upon the discussion of bird origins. The Solnhofen quarries remained in operation and produced a complete *Archaeopteryx* with a well-preserved skull in the early 1950s, but this invaluable specimen remained unidentified for two decades!

### *Modern Times*

Dinosaur paleontology began to revive in the 1960s. In particular, the discovery and description by John Ostrom (1969) of the well-preserved remains of sickle-clawed Cretaceous *Deinonychus* in Montana began to show how very birdlike certain theropods truly were. Conversely, Ostrom's (1973, 1974a, 1976a) reinvestigation of *Archaeopteryx* showed how dinosaur-like it really was. Wellnhofer's (1974) examination of the until then unappreciated *Archaeopteryx* specimen found in the early 1950s led him also to conclude that it was very dinosaurian. Galton (1970a) had briefly revived a close link between ornithischians and birds, but he quickly abandoned his attempt to do so and joined with Bakker in 1974 to endorse

*Figure 1.5.* The dromaeosaur *Deinonychus*. Dinosaurs as the metabolically energetic ancestors of birds got a big boost with the finding of *Deinonychus* in Montana by a team led by Ostrom in the 1960s. The birdlike form and agility of this Early Cretaceous killing machine are obvious, and dromaeosaurs are now known to have supported such avian features as furculae, large sternal plates, and ossified sternal ribs and uncinate processes. Careful examination of *Deinonychus*'s detailed morphology reveals that dromaeosaurs may be the closest known relatives of urvogels.

a direct ancestor-descendent relationship between dinosaurs and birds, a relationship also favored by R. Thulborn (1975). However, Thulborn retained the now traditional concept of dinosaurs as two distinct groups, whereas Bakker and Galton (1974) broke tradition by uniting all dinosaurs and birds into a single, monophyletic unit.

The view that all dinosaurs and birds form one group above the thecodont level gained support from most of the researchers that used the new cladistic methodology (Padian 1982, Paul 1984a, Gauthier and Padian 1985, Gauthier 1986, Holtz 1994a, Sereno 1997a). That birds descended from predatory dinosaurs has become far and away the majority view expressed in many additional studies. A list of researchers who favor this view is worth citing for the record:

> Gingerich (1973), Osmólska (1976), Kurzanov (1981, 1982, 1983, 1985, 1987), Barsbold (1983), R. Thulborn (1984, 1985), Currie (1985, 1987), Molnar (1985), Ostrom (1985, 1990, 1991, 1994, 1995), Raath (1985), Wellnhofer (1985, 1988, 1993, 1994), Weems (1987), Carroll (1988), Paul (1988a), Chatterjee (1991, 1995, 1997, 1998a,b, 1999a), Novas (1991, 1997), Rayner (1991), Elzanowski and Wellnhofer (1992, 1993, 1996), Sanz and Bonaparte (1992), Sereno and Rao (1992), Currie and Zhao (1993b), Russell and Dong (1993a), Olshevsky (1994), Perle et al. (1994), Chiappe (1995), Chiappe et al. (1996, 1997, 1998), Ji and Ji (1996, 1997a,b), Karhu and Rautian (1996), Horner and Dobb (1997), Larson (1997), Norell and Makovicky (1997, 1999), Novas and Puerta (1997), Britt et al. (1998), Dingus and Rowe (1998), Forster et al. (1998), Griffiths (1998/2000), Gower and Weber (1998), Holtz (1998/2000), Ji et al. (1998), Padian and Chiappe (1998a,c), Clark et al. (1999), Elzanowski (1999), Garner et al. (1999), Schweitzer et al. (1999), Sereno (1999a), Wagner and Gauthier (1999), Barsbold et al. (2000), Brochu and Norell (2000), Burnham et al. (2000), Christiansen and Bonde (2000), Farlow et al. (2000), Gould (2000), Martill et al. (2000), Rensberger and Watabe (2000), Sumida and Brochu (2000), Xu et al. (2000, 2001), Zhou and Wang (2000), Zhou et al. (2000), Hutchinson (2001), Ji et al. (2001), Norell et al. (2001).

As the dinosaur-bird hypothesis has been affirmed and refined, the focus of research has shifted to the determination of what kinds of predatory dinosaurs were closest to birds. Starting with Padian (1982), most of the cladistic studies cited above conclude that sickle-clawed dromaeosaurs were most closely allied with *Archaeopteryx*, a view that has been marvelously reinforced by the discovery of the astonishingly urvogel-like sickle-clawed *Sinornithosaurus* (Xu, Wang, and Wu 1999). Researchers have also noted that troodonts and oviraptorosaurs were exceptionally birdlike. New fossils, such as enigmatic and bizarre *Mononykus* and *Shuvuuia*, both from Asia, have been added to the list of near-birds. Also new from Asia are exquisitely preserved skulls and skeletons of dromaeosaurs, troodonts, oviraptorosaurs, and the recently discovered caudipterygians. The caudipterygians and some oviraptorosaurs even have the bobbed tails so characteristic of birds. Even more dramatic Asian finds are the feathered dinosaurs, plus many hundreds of skeletons of early birds including, finally, a second archaeopterygiform, *Protarchaeopteryx* (Hou et al. 1995, Ackerman 1998, Chen et al. 1998, Ji et al. 1998, Wang 1998). This fossil bonanza, which would have been unimaginable just a few years ago, has been supplemented by important new birdlike dinosaurs and Mesozoic birds from South and North America, Spain, and Madagascar. No one realized that there had been a large-scale Cretaceous radiation of a now entirely extinct group called Enantiornithes, "opposite birds" until Cyril Walker first described them in 1981.

Interestingly, the most birdlike dinosaurs continue to come largely from the Cretaceous. This is one of the factors behind the interesting suggestion made by some researchers that Jurassic *Archaeopteryx* and other birds may not form a clade above the level of nonavian dinosaurs. Instead, *Archaeopteryx,* despite its spectacular feathered wings, is quite dinosaurian in most of its skull and skeleton, whereas later dromaeosaurs, troodonts, caudipterygians, oviraptorosaurs, ornithomimosaurs, avimimids, alvarezsaurs—and even the strange herbivorous therizinosaurs and the great tyrannosaurs—possess various birdlike characters that hint that they may be closer to modern birds than the first bird itself was (Kurzanov 1981, 1982, 1983, 1985, 1987, Paul 1984a, 1988a, R. Thulborn 1984, Chiappe 1995, Elzanowski 1995, 1999, Chiappe et al. 1996, 1998, Novas 1997, Griffiths 1998/2000, Padian and Chiappe 1998a,c). This observation has reopened two questions: Is *Archaeopteryx* more dinosaur than bird, and Are some dinosaurs birds? Also reopened is the issue of whether feathers are limited to birds or were also present in some dinosaurs. An obvious but little-appreciated consequence of the possibility that some dinosaurs were more avian than *Archaeopteryx* is that such dinosaurs may also have been secondarily flightless, a notion supported by the presence of ratitelike features in some of the Cretaceous dinosaurs (Paul 1984a, 1988a, Gould 2000). In this view, the bird clade started to lose the ability to fly almost as soon as that ability evolved in the Jurassic.

Olshevsky (1994) took the hypothesis of early flight loss even farther back in the Mesozoic, reviving Abel's 1911 suggestion that both dinosaurs and birds evolved from arboreal archosaurs in the Triassic. Chatterjee (1991, 1998a, 1999a) argued that the baffling *"Protoavis,"* found during his work in Texas, is closer to birds than *Archaeopteryx,* and he places bird origins in the Triassic. But new evidence challenges these conclusions.

To a certain extent, the modern dinosaur-bird hypothesis has become linked with the ground-up hypothesis of bird flight, the basic idea being that the ground-running predaceous dinosaurs developed flight as an extension of leaping up against gravity after flying insects (Ostrom 1976a,b, 1985, Caple et al. 1983, Balda et al. 1985, Gauthier and Padian 1985, Chiappe 1995, Novas and Puerta 1997, Ji et al. 1998, Padian and Chiappe 1998a,c, Burgers and Chiappe 1999, Easley 1999, Xu, Wang, and Wu 1999). However, some researchers argue that the climbing hypothesis and the dinosaur hypothesis of bird origins are compatible with each other (Paul 1988a, Chatterjee 1998a, 1999a, Garner et al. 1999, Xu et al. 2000). The case for an arboreal origin for avian flight is being strengthened with the discovery of new dromaeosaurs specially adapted for climbing.

A minority body of opinion disagrees, often with great vigor, with both the terrestrial and the dinosaur hypotheses. At about the time Ostrom was redirecting attention toward the possibility that dinosaurs were bird ancestors, A. Walker (1972) revived the idea that basal, gracile crocodilians were the closest relatives of birds. The current minority view prefers modernized versions of Heilmann's arboreal basal archosaur hypothesis, and listing the researchers who favor this opinion is also worthwhile:

> Ellenberger and de Villalta (1974), P. Ellenberger (1977), Whetstone and Martin (1979), Feduccia (1980, 1996, 1999), L. Martin et al. (1980), Tarsitano and Hecht (1980), Hecht and Tarsitano (1982), L. Martin (1983a,b, 1991, 1997), Hecht (1985), Hinchliffe (1985, 1997), Tarsitano (1985,

*Figure 1.6. (opposite)* Newly discovered feathered dinosaurs and birds from the lower Yixian deposits, latest Jurassic or Early Cretaceous lake sediments in northeastern China. *Bottom,* the compsognathid *Sinosauropteryx; center right,* the peculiar dino-bird *Caudipteryx; top,* the near-urvogel dromaeosaur *Sinornithosaurus,* with enough dorsal vertebrae preserved to restore the length of the trunk; *center left,* the archaic bird *Confuciusornis.* In all examples, feathers are preserved. The presence of feathers in the dinosaurs tightly links them to birds and is also evidence that these Mesozoic archosaurs had high metabolism and could conserve heat. Drawn to same scale.

*Figure 1.7.* The newly discovered dromaeosaur *Bambiraptor.* Discoveries related to the origin of birds continue at a fast pace. Among them is *Bambiraptor,* here represented by a nearly complete juvenile skull and skeleton. The shoulder girdle is especially birdlike, in some regards more so than that of *Archaeopteryx.*

1991), Feduccia and Wild (1993), Hecht and Hecht (1994), Welman (1995), Zhou (1995a,b), Hou et al. (1996), Burke and Feduccia (1997), Ruben et al. (1997a, 1999), Geist and Feduccia (2000), Jones et al. (2000a, 2001), Maderson and Alibardi (2000), Ruben and Jones (2000), and Tarsitano et al. (2000).

The postmodern antidinosaur contingent continues to cite convergence as an explanation for what they see as limited and superficial similarities between predatory dinosaurs and birds. They argue that birds must have evolved in the Triassic and note that the most birdlike of dinosaur specimens are found in sediments that postdate the urvogel. Perhaps most importantly, the majority of these researchers have concluded that because flight must have evolved from high places and because predaceous dinosaurs were terrestrial, the latter simply could not have been ancestral to flying birds. Aside from early crocodilians, a variety of small reptiles have been cited as representative of the kind of arboreal creature that birds should have evolved from, such as *Cosesaurus, Megalancosaurus,* and allegedly feathered *Longisquama.* However, as we shall see, the first and second are probably not archosaurs, the status of the third is dubious, *Megalancosaurus* has proven to be a very bizarre chameleon mimic with a prehensile tail, and *Longisquama* is poorly known and not truly feathered. Hence, the nondinosaurian hypotheses of bird origins continue to suffer from the lack of a clear "protoavis" to which they can point as representing the ancestral type for birds.

It is interesting that there is partial correlation between the fossil archosaur group that a particular researcher works on and the archosaur group that that researcher sees as being closest to birds. Few, if any, dinosaur paleontologists dispute that birds are dinosaurs, whereas some, but not all, paleoornithologists refuse to see their feathered flock as a mere side branch of the Dinosauria! Those who oppose the majority include some nondinosaur paleontologists, especially paleoornithologists and a few physiologists. Many of the debates over bird origins are discussed in the invaluable albeit somewhat dated symposium volume *The Beginnings of Birds* (Hecht et al. 1985), which resulted from the 1984 meeting held in Eichstätt, Germany, in the midst of the Solnhofen quarry district from which the protobirds' remains have come. The tentative "Eichstätt consensus" held that avian flight began in the trees and that *Archaeopteryx* could power fly, but these issues remain contentious in the field in general. In 1996, a meeting of the Society of Avian Paleontology and Evolution at the Smithsonian Institution included discussions and papers on bird origins (S. L. Olson 1999). The year 1999 saw the Yale symposium, attended by hundreds, on the origin and early evolution of birds held in honor of John Ostrom. A major conference volume reflecting the state of our knowledge at the end of the twentieth century is in preparation (Gauthier in press). The conference on dinosaur-bird evolution, organized by the Graves Museum in Florida in 2000, will also produce a volume.

These days the study of bird origins is a hot topic in paleontology, and by no means is it a polite debate between dispassionate scholars. (For a popular, and neutral, account of the big bird debate, refer to Shipman [1998]). To a certain extent, the heat is generated by the vocal demonstrations of the antidinosaur minority against the consensus reached by most scientists that *Tyrannosaurus* really is a remarkably close relative of the hummingbird. The rhetorical attacks upon the dinosaur-bird link have sometimes been astonishingly harsh, with some scientists even equating it with cold fusion or the Piltdown hoax and calling it "ideological mumbo jumbo" (Dalton 2000), despite the fact that the dinosaur hypothesis is the dominant paradigm in the paleontological community, a paradigm increasingly supported by the fossil record. Even recent articles in *National Geographic* have generated heated discussions. On a happier note, the debate is also being fueled, and may soon be settled once and for all, by the unprecedented and extraordinary rate of discovery of near-bird and early bird fossils in Mesozoic strata, of which the Yixian and other Jehol beds of northeast China have been the most productive (Stokstad [2001] and accompanying articles give an update on paleontology). From China has come the feathered dinosaurs *Sinosauropteryx*, *Caudipteryx*, *Beipiaosaurus*, *Sinornithosaurus*, and *Microraptor* as well as many hundreds of specimens of the bird *Confuciusornis*. The past few years have even seen the discovery of well-preserved internal organs in dinosaurs, sometimes in three dimensions. Indeed, the fossils are coming so fast that it was hard to keep up with the new data during preparation of this book, and the latest research required the addition of Appendix 6.

CHAPTER 2

# The Science of Bird Origins

Before we can understand a subject like the ancestry and evolution of birds, we must do two things. The first is to set up a rigorous scientific methodology for investigating the problem. The second is to catalog and organize the subjects involved so that we can study them more easily. The best way to accomplish these two tasks is often the subject of vigorous debate, but in the case of bird origins, we must attempt to analyze the evolutionary relationships among the organisms involved and then apply names to the groupings that result.

## Understanding Relationships

### *General Principles and Problems*

Sorting out the genealogical relationships, or phylogeny, of people living today and people who lived in past centuries is frequently difficult, even though extensive written records and DNA comparisons are often available. Determining the evolutionary relationships of nonhuman organisms, living and fossilized, is far more difficult because the methods by which these relationships can be reconstructed are limited: although DNA and other molecules offer powerful tools for restoring the relationships between organisms (Balter 1997, Mindell 1997), such molecules are not always available. The DNA of relatives still living today can be analyzed, but only one potential group of bird relatives, the crocodilians, is still extant. In addition, DNA's ability to survive intact in the remains of long-expired Mesozoic creatures has been questioned (Hedges and Schweitzer 1995, Poinar et al. 1996, Austin et al. 1997, Mindell 1997). Consider the possible discovery of DNA in an ornithischian dinosaur fossil announced at the Florida conference on the evolution of dinosaurs and birds. That the DNA sequence was identical to that of modern turkeys implied a close relationship between the two clades, but it raised suspicions of contamination via a sandwich. Thus, the question of whether direct genetic information can be used to figure out which group of ancient tetrapods birds descended from remains open.

Unless paleogenetics becomes a practical means for restoring relationships, we are stuck with traditional methods of anatomical analysis. Like DNA, soft tissues have rarely been preserved in the Mesozoic tetrapods found to date (the possibility of finding entire dinosaur bodies preserved in perpetually cool polar tar deposits cannot be ruled out). Most of the anatomical work is therefore restricted to comparisons of skeletal anatomy, which suffer from yet another serious limitation in that so much valuable data is missing.

Probably only something like one out of ten of the species of diapsids that lived in the Mesozoic have so far been found. Of those, most are known only from incomplete remains. In some cases, the entire head is missing. In other cases, the skull is complete, but the rest of the skeleton has not been found. In addition, it is sometimes difficult to determine whether a particular head and body actually go together. It helps to know what goes with what when we try to understand what is related to what. Even when the remains are complete, they may be poorly preserved or badly damaged; or critical parts of the inner braincase may be inaccessible (three-dimensional X-ray scanning is starting to ease this problem). There are long stretches of time from which no important fossils have been discovered: witness the dearth of small predatory dinosaur remains from much of the Jurassic.

The paucity of fossils leads to the threshold problem. Paleontologists are fond of ending papers and talks by calling for more fossils to fill in the remaining gaps, and some have wryly observed that paleotaxonomists are perpetually waiting for the next find before coming to a conclusion. However, it is a fact that until enough well-preserved remains are known from enough places and times, we may not be able to discern relationships beyond a certain level.

For example, *Archaeopteryx* skeletons found in the 1800s confirmed that birds arose from flying reptiles rather than from flying fish or flying mammals, but the basic ancestral group remained open to debate because there were not enough skeletons from small dinosaurs and other archosaurs for comparison. By the 1920s, paleontologists arguably had enough remains of Mesozoic predatory dinosaurs and birds to conclude that the latter probably descended from the former, but the specific dinosaur group remained obscure. We now have a good understanding of which Cretaceous predatory dinosaurs are closest to *Archaeopteryx*. However, we still do not have enough transitional forms to determine the exact relationships between the groups closest to the earliest birds, and the Jurassic gap leaves us wandering in the phylogenetic dark when it comes to determining the closest avian ancestors.

In principle, the methods for restoring phylogeny via fossil skulls and skeletons are straightforward. By looking at a large set of fossils from many ages, we can use the changes observed over time to help establish the polarity of a skeletal character (heritable feature); that is, whether a character is "primitive" or "advanced" (Fox et al. [1999] demonstrated the importance of including time in cladistic studies). For example, the earliest reptiles have sprawling legs like those of lizards; erect legs are not seen until dinosaur-like animals show up in sediments some 130 Myr later. Ergo, sprawling limbs are very probably the primitive condition. Note that in this sense, *primitive* and *advanced* are neutral words that indicate the relative time of development rather than describe the relative performance of such legs. To avoid this unintended derogatory implication, paleontologists commonly substitute the terms *basal* and *derived,* respectively.

It is important that only derived characters be used to determine relationships. Using basal characters can be misleading because archaic traits can be retained by two species even after they have diverged strongly from one another. For example, lizards, alligators, humans, cats, and mice all retain five fingers. It is the derived features—such as the complex inner ear in advanced therapsids and mammals or the peculiar system of cranial mobility in birds—that link various creatures into related groups. Once the basal condition is known, a group that possesses the primitive characters is used as the outgroup, against which the more derived species or groups are compared.

Modern phylogeneticists use a special set of terms to describe characters and relationships. An *apomorphy* is a derived character; an *autapomorphy* is a derived character unique to one species or group; and a *synapomorphy* is a derived character shared by two or more species or groups. If the members of a set of organisms evolved from different ancestors, they are *polyphyletic.* A group of organisms that share a single common ancestor is *monophyletic.* These terms are always relative to the level of the phylogenetic tree being examined. For instance, all life is ultimately monophyletic in that it probably descended from a single cell that lived more than 3.5 billion years ago.

For many decades, scientists commonly used both basal and derived characters in determining relationships. Although a lot was accomplished, the results were often questionable. Beginning in the sixties, this system was rejected in favor of cladistics, which uses only derived characters (Hennig 1966, 1981, Wiley 1981, Carroll 1988, Sober 1988). Cladistics has become somewhat controversial, largely because some of its proponents have become rather rigid in their thinking about how the methodology should be applied. Some have even asserted that showing an ancestor-descendent relationship in the fossil record is impossible, and that a fossil's position in time is unimportant. Certainly there is truth to both claims. We now understand that speciation events are often very limited in time and space because they tend to occur in small isolated populations. Because of this "punctuated equilibrium" (Ridley 1996), few trans-species (organisms between two species) were preserved, and most extinct species are not known either. All the gaps in the fossil record make it difficult to sort out exactly who descended from whom. However, ancestral species obviously did exist in large numbers, and some of them must be preserved somewhere in the record. In cases in which the fossil record is exceptionally good, determining ancestor-descendent relationships with confidence may be feasible. As for the time at which a fossilized species lived, we must remember that ancestral species and groups often survive their descendants, much as a grandparent can outlive a grandchild. However, the polarity of a character is determined by its initial appearance in time.

Despite these disputes, the basic cladistic principle of focusing on similarities in derived characters is sound and almost universally employed by modern researchers. As with two people, the more similar two species are, the more closely related they are likely to be. Obviously your lap cat is more closely related to a lion than it is to a horse, and this is apparent even if you examine only the skeletons. The statistical search for the simplest (most parsimonious) pattern that maximizes the number of synapomorphic characters linking species into related groups is the essence of cladistics. The results are expressed in the form of a *cladogram,* a type of family tree that maps out the reconstructed relationships among the species analyzed. However, ancestor-descendent relationships are never shown in classic cladograms. Instead, species and groups are described as "sister" taxa to one another. The points at which sister groups meet on the cladogram are called *nodes.*

As superior as cladistics may be to older methods, it is far from perfect. Two researchers analyzing the same set of fossil taxa can come up with very different results. Why? Because the very procedure feature that is a strength of cladistics leads, ironically enough, to a serious problem with the methodology, one that reveals other problems inherent to restoring phylogeny. A simple but powerful principle in science is that the more data gathered and considered, the better the results. Ergo, the more characters and the more species, the better the resulting cladogram. However, this principle gives rise to two problems. The first is gathering the data. Just observing and recording all the characters is a daunting task. A complete set of remains contains potentially many hundreds of phylogenetically useful characters, and there are many dozens or even hundreds of remains to be examined. Some of the characters can be gleaned from the literature, but others cannot; and the fossils themselves are scattered in institutions around the globe. A secondary problem lies in the errors and contradictory interpretations that inevitably crop up—and sometimes inspire heated disagreements—when data is gathered.

After the data is in hand, the next problem is analyzing it all. Because crunching through so much data is impractical for the human mind, the task is usually left to computers, but even computers are not fully adequate. The critical difficulty is illustrated by the traveling salesperson problem. Say a salesperson has a route covering two dozen cities and wishes to minimize travel time and cost while maximizing efficiency and profit. One might expect that solving this problem would be a simple task for a statistician with some computer time, but it is not (Paul and Cox 1996). It turns out that the permutations proliferate so rapidly that the number of calculations needed to completely solve the problem far exceeds the capacity of all the world's computers put together. Even the most powerful supercomputers would be far from solving the problem when the salesperson died of old age. Large-scale character analyses of large numbers of organisms can fall right into this same trap (although exponential increases in computer power are easing the problem). The situation is only made worse by the gross gaps in the fossil data.

As a result, computer-generated cladograms often fail to produce clear-cut solutions. The researcher is left with a number of plausible alternatives to choose from (Vermeij 1999). I have seen colleagues throwing up their hands in dismay at the ambiguity of the alternative cladograms their computers handed back to them. The fact is that we do not yet have the information-processing systems needed to handle phylogenetic data adequately. On the one hand, digital computers can crunch through lots of data, but they are not able to think problems through and make intelligent judgment calls. On the other hand, the analog-based human brain is very good at absorbing lots of information and then using pattern recognition to derive useful conclusions, but it does so in a sloppy, imprecise, and ill-defined manner that does not fully meet science's requirements for rigorous, repeatable results. What we need are extremely powerful information-processing machines that can analyze enormous databases both with precise digital analysis and with intelligent pattern recognition. Such machines may become available in a few decades (Paul and Cox 1996). Some researchers also believe it may be possible to develop sophisticated procedures that can overcome the computational barrier posed by the traveling salesperson problem.

Until then, and despite the difficulties, working with as much data as possible remains important. One reason is bias in the selection of data. Humans are, well, human, and as they gather characters with which to build their cladograms, they tend to favor characters that they think will support their preconceived notions of what is related to what. Even when researchers do their best to avoid this bias, the selection of data can subtly but significantly alter the results. Such biases can be partly swamped simply by gathering as much information as is practical.

Large character lists are also important for dealing with the three vexations of phylogenetics: convergence, parallelism, and reversal. *Parallelism* occurs when two closely related groups independently evolve the same adaptation. For example, sabre teeth have evolved repeatedly among cats. *Convergence* occurs when two distantly related groups independently evolve similar adaptations. The sabre teeth evolved by certain cat relatives and by some South American marsupial predators constitute one example. Note that convergence and parallelism merge into one another in that sometimes the distance between two groups is neither large nor small, and judging which term best applies is hard. A *reversal* is the loss of a derived character. An example would be the descent of a short-toothed cat from a sabre-toothed ancestor whose own ancestor was short toothed. The confusion these common evolutionary events can cause is obvious. If a cat has short killing teeth, does this mean it retains the basal condition, or is the condition actually derived in that the cat's ancestors had sabre teeth? Sabre teeth are obviously derived relative to the short-toothed ancestral condition but basal relative to the secondary short-toothed condition. What if two fossil cats have sabre teeth? Are they close relatives that inherited their big teeth from a common ancestor with sabre teeth, or are they more-distant relatives that each evolved the teeth on its own?

Before cladistic methodology became the norm, researchers commonly decided for one reason or another that a character shared by the members of two or more groups was the result of convergence, and removed it from the character list before executing the main analysis. The obvious danger of this tactic was that it could bias the conclusions. In well-done cladistics, such prejudgments are avoided, and all derived characters surveyed are analyzed. In this system, it is the results of the analysis that are supposed to sort out what is convergence, parallelism, or reversal.

Paleontologists use three methods to cope with these vexations. One is simply to try to overwhelm them with as many characters as possible. The hope is that the many derived characters shared by close relatives will happen to outnumber those due to convergence, parallelism, and reversal.

Another method is to use critical characters that show that two species belong to distinct groups and, therefore, that any similarities they share must be due to convergence. For example, marsupials and placentals have very different reproductive systems as well as distinctive skeletal features, so a sabre-toothed marsupial is clearly convergent with a sabre-toothed cat. This method, however, assumes that the critical characters that distinguish major groups are already well established, which is not always the case with exotic extinct forms.

The third method deals with reversals and involves analyzing the complexity of a proposed reversal. Complex reversals are less likely than simpler ones. For example, it would be nearly impossible for amniotes to re-evolve gills, which is why even fully marine examples, such as sea turtles and whales, must return to the surface to get oxygen. Nor does it seem possible for tetrapods to re-evolve the lost third "eye," so we humans with our binocular vision are notoriously oblivious to threats from behind. Such implausibilities have led to overly strict application of such rules of phylogeny as Dollo's law. But is this law valid? If, say, the number of teeth declined from twelve in one organism to ten in its descendants, then might not the re-evolution of twelve teeth be easy? The re-evolution of a simple character is especially viable when the coding for that character still exists in an organism's genotype, even though that feature may not be expressed in the phenotype. For instance, "teeth" can be grown in chicks in a process by which the function of quiescent genes is revived (Kollar and Fisher 1980). Heilmann should have recognized that such a disparity between genotype and phenotype was a possibility before he dismissed clavicleless dinosaurs as the ancestors of wishbone-bearing birds.

In any case, it is sometimes difficult to tell whether a proposed reversal is too complex to have been practical. For example, some archosaurs, including crocodilians, have a complicated ankle in which the articular surfaces trace a complex path through the ankle elements, and a very large calcaneal tuber in the heel forms a long backward-directed muscle lever. Dinosaurs and birds, however, have a much simpler, tuberless ankle. Could the complex angle arrangement have been lost in favor of the simpler one, or did the complex ankle type evolve directly from the simple one seen in the most primitive archosaurs? No one knows for sure.

These questions lead us to still another problem: that concerning the weight, or importance, given to characters. It is obvious that the evolution of longer and longer sabre teeth is not as difficult as the evolution of a complex new ankle. If a cladogram fails to account for the difference in importance between simple and complex derived characters, then the results will tend to be skewed in favor of the simple characters. To avoid this difficulty, more weight should be attributed to complex characters. But how much more? There is no known procedure for assigning a value to character complexity.

Yet another problem is how to count characters. The complex ankle, for instance, is itself a set of subcharacters. Because these characters would not exist in the absence of the complex of which they are a part, should we lump them all together under the one major character? Or should we count all these characters individually? On the one hand, counting them individually is a de facto method of increasing their importance. On the other hand, because the characters that eventually make up a structure of greater complexity often evolve in mosaic fashion, they should be counted individually.

In the end, the problem with cladistics is that it is nothing more than a statistical comparison of numbers of derived anatomical characters. It merely shows which species and groups share the most anatomical similarities. This intense focus on anatomy is problematic because anatomy does not always reflect precise relationships. Cladistic parsimony does not, therefore, always exactly recapitulate phylogeny (Huelsenbeck and Crandall 1997, Vermeij 1999, Bower 2000). Who would have guessed that chimpanzees are more closely related to humans than to gorillas (a counterintuitive fact revealed by DNA analysis)? The solution to this problem would be to find complete remains for virtually every species of the group or groups under examination, something that will never happen for Mesozoic fossils. Even then, assessing all the data would be a computational nightmare.

In spite of these drawbacks, skeletons and other aspects of morphology do record relationships. The skeletons of humans and prehumans show that we are descended from apes, not whales. In the absence of a good set of transitional links, morphology-based cladistics is best at determining broad patterns of relationships at a gross level. It is less successful at detailing the intimate relationships between taxa that are part of an intricate branching pattern of mosaic evolution, in which parallelism and reversals are rampant. Parallelism is a particular problem because it is extremely common. It occurs so frequently not only because there are parallel selective pressures but also because parallelism is a necessary side effect of the process by which new forms evolve. Similar regulatory genes possessed by close relatives tend to initiate the development of similar adaptations (Shubin 1998). This fact led Shubin to comment that mosaic parallelism "poses a 'chicken and egg' problem for paleontologists. If independent evolution of key characters is common, how is phylogeny to be reconstructed?" At this time, no method fully addresses this problem. So skeletal cladistics can demonstrate that humans are derived from apes, but it has trouble showing that chimpanzees are closer to humans than to other apes.

Cladistics is a powerful wrench for working on phylogenetics, but it is neither all-powerful nor fully definitive. Good phylogenetics was done before its advent, such as the demonstration of the therapsid ancestry of mammals. Cladistics is one of the more useful tools in the scientific toolbox, but in the end, a good fossil record of transitional types is what provides the most definitive data.

Science has its limits. It is only as good as the evidence it is based upon. Many of the gaps in the fossil record will never be filled. The data threshold needed to discern the detailed interrelationships among some species and groups will prove insurmountable. We can discover a lot about how birds evolved from dinosaurs, but there will always be a lot we will never know.

### *Scenario versus Phylogeny*

Cladistics is not the only method for studying phylogenetics. An alternative method involves deciding how and why one taxon should have evolved from another and then looking for a suitable group that fills the requirements and dismissing groups that do not. For example, we can dismiss whales and bats as having an ancestor-descendent relationship because it is very unlikely that whales evolved from bats, or vice versa. But aside from such glaring relationship faux pas, the problems with this methodology are obvious. Real evolutionary transformations are not always so transparently improbable. Who would have guessed that manatees, rock hyraxes, and elephants are all close relatives? Evolutionary changes that seem strange to human eyes may not actually be strange at all. It is very difficult, sitting here in the Holocene, to determine how evolutionary events transpired. Such problems make most scenario-based hypotheses difficult to test, so much so that understanding how and why one type of organism might have evolved from another requires recourse to the best available phylogenetic methodologies. In other words, phylogeny usually trumps scenario. Only if a scenario-based argument can produce an inescapable problem with the phylogenetic evidence can the former be used to rule out the latter, but this is very improbable.

Scenario-based methodology is most useful when the phylogenetic data are ambiguous. In this case, parsimony can be invoked in favor of the scenario that best explains what happened, without contradicting the phylogenetic data in a fundamental way.

### *Special Principles and Problems*

The general problems outlined so far apply in particular ways to the question of avian descent from dinosaurs. One long-standing and controversial issue is the scenario-based position that birds must have evolved from arboreal ancestors. The proponents of this position hold that if it did not climb, it could not have evolved into a bird. This argument is vulnerable to the problems inherent to scenario-based phylogenetics. In order for the argument to succeed, its proponents must show both that avian flight *had* to begin among climbers and that the group to be dismissed from avian ancestry—in this case, predatory dinosaurs—really were poorly adapted for climbing.

Now we turn to the second major issue, one that is not so controversial, simply because it has been virtually ignored in the debate over bird origins, even though it is at least as important as the arboreal scenario problem. This second issue centers on a special form of reversal, the loss of flight. Researchers have not appreciated the importance of analyzing potential phylogenetic characters in predatory dinosaurs and birds in terms of whether those characters may have been influenced by the development of flight, and by its subsequent loss. For example, we know that the remarkably dinosaur-like shoulder girdles of secondarily flightless birds are not really less avian than those of flying birds (Feduccia 1986, 1996). The former represent a derived condition (furcula often absent, coracoid less reversed or strutlike, supracoracoideus wing-controlling system lost, sternal keel missing) that mimics the preavian condition. This pattern results in a problem with phylogenetically scoring of flight-related characters.

Let us assume, first, that two birds are very similar in form except that one can fly and the other is secondarily flightless and, second, that the latter descended from the former. If we use *Archaeopteryx* as the outgroup and score the characteristics of shoulder girdles in a phylogenetically simplistic manner that does not account for secondary loss of flight, then the flightless bird will be scored as less derived than the other, when actually the opposite is the case.

The academic problem outlined above becomes a practical one when applied to predatory dinosaurs whose overall morphological grade is similar to that of *Archaeopteryx* but whose shoulder girdles have fewer avian attributes. To better establish the phylogenetic weight of characters, we ignore prior presumptions as to the phylogenetic relationships and taxonomic status of the taxa under examination. For example, the traditional position of *Archaeopteryx* as the first bird is given no special status. Nor are rarely preserved characters given extra weight, even if they are traditionally important; in particular, the presence of contour feathers in *Archaeopteryx* cannot be used to show that it is more avian than any given dinosaur until and unless a well-preserved integument has proven that a particular dinosaur definitely lacked contour feathers. Characters strongly influenced by the status of flight are assigned less weight than those that are not. For example, a birdlike braincase is ranked higher than a winged forelimb. The presence of characters known to be associated with flight in some tetrapods is considered evidence for the presence of, or a heritage of, flight capabilities in the taxon under examination.

A computer-generated cladistic analysis of bird origins is premature at this time, in part because some new and important fossils have yet to be described in detail. In fact, the possibility that some dinosaurs were secondarily flightless exacerbates the data threshold problem. Their flying ancestors should have been small, too small to have been readily preserved in the fossil record. In fact, we may be missing the crucial taxa that linked some of the larger-bodied predatory dinosaur groups. In addition, a standard cladistic analysis would probably generate multiple trees with no nodes clearly superior to the alternatives, because insufficient transitional forms have yet been discovered. For these reasons, this study is limited to a non-computer-based set of trees.

## Naming Names

If there is anything more controversial than restoring the phylogenetics of organisms, it is applying names to the groupings that result from such restorations. Scientists have come up with several different naming systems, and all have their own problems. For example, the Linnean classification system in use since its invention in the early 1700s suffers from having been invented long before evolution by natural selection was understood. An attempt to replace the traditional taxonomy with a more modern one has been inspired by cladistic phylogenetics. However, because strict cladistics does not recognize ancestor-descendent relationships, this system too is not fully in accord with evolutionary theory.



### *Naming Systems*

In cladistic taxonomy, only monophyletic groups are named, usually as indicated by the nodes on the cladogram. No attempt is made to rank any of these node-based names in order to generate equivalent levels of families, orders, and so on. *Paraphyletic* groups—that is, groups that are monophyletic but do not include all of the derived descendants of the common ancestor—are never named. For example, in traditional taxonomy, Reptilia is paraphyletic because it includes all the amniotes that share a common ancestor above Amphibia, except for the members of Mammalia and Aves that descended from them. In cladistic taxonomy, Reptilia is limited to those amniotes that share a common ancestor with living reptiles and birds. Therefore, the mammal-like "reptiles" (and mammals) are excluded from Reptilia, whereas birds are included. This kind of cladistic group is called *holophyletic.*

The new system of naming related groups is supposed to be more rigorous and less arbitrary than the old one, and node-based names do have some advantages in these regards. Problems remain, however. For one thing, a system that makes birds into reptiles while excluding the mammal-like reptiles from Reptilia not only overturns long-standing designations but also confuses non-scientists. Such confusion negates the supposed advantages of cladistic taxonomy, its improved stability and the absence of character ambiguity. For example, attempts to define Aves via one or more characters remain a subject of dispute (see Chapter 12). Should the defining character be the presence of feathers (which may be found in dinosaurs) or the reversed hallux (which is not found in kiwis) or a certain kinetic skull (which is not present in *Archaeopteryx*)? A way to avoid this problem is to assign a name to a node that includes the common ancestor of all members of the group. For instance, one could define Aves as the clade that includes *Archaeopteryx,* modern birds, their common ancestor, and all archosaurs that descended from that common ancestor. This group would be stable.

However, cladistics-based stability can be illusive when major uncertainties in phylogeny are involved. An excellent example is the term *Ornithodira,* first coined by Gauthier (1986) to designate the clade that included the seemingly closely related pterosaurs and dinosaurs. In the cladistic system, Ornithodira always includes the common ancestor of pterosaurs and dinosaurs, no matter where that ancestor may lie. This is fine as long as Ornithodira includes only pterosaurs, dinosaurs, their common ancestor, and a few archosaurs within this group (largely protodinosaurs). However, a serious problem has arisen because recent analysis suggests pterosaurs may not be closely related to dinosaurs after all; instead, their ancestry may lie near the base of Archosauria. If so, then Ornithodira includes pretty much all of Archosauria, which involves the myriad "thecodonts" (many cladists do not recognize this group) and crocodilians. This possible arrangement is a radical alteration of the nature of Ornithodira and is the opposite of stability.

Another example of the weak status of cladistic taxonomy is Ceratosauria. Although the term has been widely used in recent years to label a supposedly holophyletic clade of basal tridactyl theropods including coelophysoids, ceratosaurs, and abelisaurs, a number of researchers have recently noted that Ceratosauria is probably not monophyletic. Instead it covers a series of outgroups successively closer to more advanced predatory dinosaurs. In this case, Ceratosauria represents a paraphyletic grade (see Chapter 12).

Why should we not use apomorphic characters to designate groups? This once common technique has fallen out of favor (Sereno 1998, Padian et al. 1999). One obvious problem with this option is that a new character can evolve in more than one group. Another problem is that the onset of a character may be ill defined, or even unknown, especially when it involves soft tissues, such as mammalian fur or avian feathers. In addition, there is the generally insoluble problem of getting everyone to agree on what particular feature should designate an already established major group.

A way to avoid much of the stability problem is to pick a derived character *and* a clade and attach a new name to the group. Although such apomorphy+clade–based names have largely fallen out of use, they were explicitly endorsed by Gauthier—who first laid the foundations for the current system of theropod taxonomy in 1986—at the Yale symposium in 1999. Indeed, apomorphy+clade–based names are more in accord with both evolutionary phylogeny and cladistics than the alternatives. The apomorphy+clade–based names can be used alongside node-based names, the advantages of one complementing those of the other.

The apomorphy+clade method works best when a new grouping is to be named; attempting to revise an old group name with this method is unlikely to be accepted.

When this method is used, the chosen character must be one that is preserved often and well enough to be reliable. It is also best if the character is simple enough so that the question of whether it exists or not can be answered with either a yes or a no. Once chosen, the apomorphic character is tied to a specific clade, so its appearance in a another clade will not result in the name's being applied to both clades. When these criteria are applied, an apomorphy+clade–based name is unambiguous and is less likely to bounce around the phylogenetic chart. Another advantage of single character names is that they do not require the lengthy character lists that accompany node-based names.

The evolution of organisms has been marked by the advent of major innovative adaptations that come to characterize new groups. Cladistic taxonomy does not recognize these developments, but apomorphy+clade–based taxonomy allows all clade members that possess a particular and possibly important character to be given a distinctive title. Another deficiency of node-based taxonomy is that it will almost always leave out at least the basal-most clade members that own a particular feature, because it is very improbable that the very first species to have the character will be found before the name is applied. For example, the tridactyl foot is one of the first birdlike features to show up in predatory dinosaurs. Tridactyl theropods are currently included in the node-based Neotheropoda, but this group does not include the earliest three-toed theropods. Someday, however, a tridactyl theropod that belongs to the same clade but is below the neotheropod node will be found, and yet another name will have to be invented to cover this important group. This will happen repeatedly. Why should we not give these creatures one apomorphy+clade–based name that will forever include all of them?

A related problem is that paleontologists can never know for sure that the very first species with the character has been found. For example, as currently defined, Maniraptora does not and cannot include all predatory dinosaurs in the clade that possessed raptorial hands, and no node-based name ever will.

Cladistic taxonomy is subject to its own form of inherent arbitrariness. Taken to its logical, nonarbitrary extreme, such taxonomy would have to name every single node. The horrifying result would be an explosive proliferation of names to the point that the system would be incomprehensibly complex. Even the most ardent cladists recognize the problem and thus selectively name only what are considered major nodes. The selection is of course arbitrary and subject to dispute, especially when a node is weak in that it is based on just a few minor or incorrect characters.

Purely cladistic taxonomy, in which paraphyletic taxa are forbidden, also results in a taxonomic set of Chinese boxes in which all taxa not on the terminal tips of their phylogenetic branches must be referred to in part by what they are not. For example, dinosaurs outside of Aves are nonavian dinosaurs. Referring to groups this way is sometimes useful, for example, in discussions of the failure of nonavian dinosaurs to survive the terminal Cretaceous extinction. However, not only does excessive use of this practice focus undue attention upon what is only one branch of the diverse Dinosauria, but also naming by exclusion is as logical and necessary as referring to all mammals and insectivores except for bats as nonchiropterian mammals and nonchiropterian insectivores. An extreme example of the problem centers on *Archaeopteryx*. Because it lacked most of the advanced features of birds, some workers occasionally refer to it as a winged dinosaur, meaning a *nonavian* dinosaur in cladistic terminology. But because the urvogel is a bird, in cladistic terms it is inherently an *avian* dinosaur. Obviously, you cannot try to show how dinosaurian *Archaeopteryx* was by calling it a nonavian dinosaur when it is avian. In addition, calling various dinosaurs nonavian is phylogenetically loaded practice in that it presumes that birds descended from the group. As true as this may be, using a naming system that does not make this presumption is often more convenient.

The hypothesis presented in this book that some dinosaurs may be closer than the urvogel to modern birds poses special problems. Current cladistic terminology cannot handle this possibility without resorting to the frequent use of convoluted, multiword taxonomic labels. For example, dromaeosaurs would in this scheme have to be designated non-ornithothoracine avians. Although technically correct, such a designation would confuse most readers. A better way to handle the situation is to devise some working paraphyletic titles that will serve until the phylogenetic situation is better understood.

For all these reasons I have found it very useful to use single words that succinctly describe specific paraphyletic groups. One way to do this

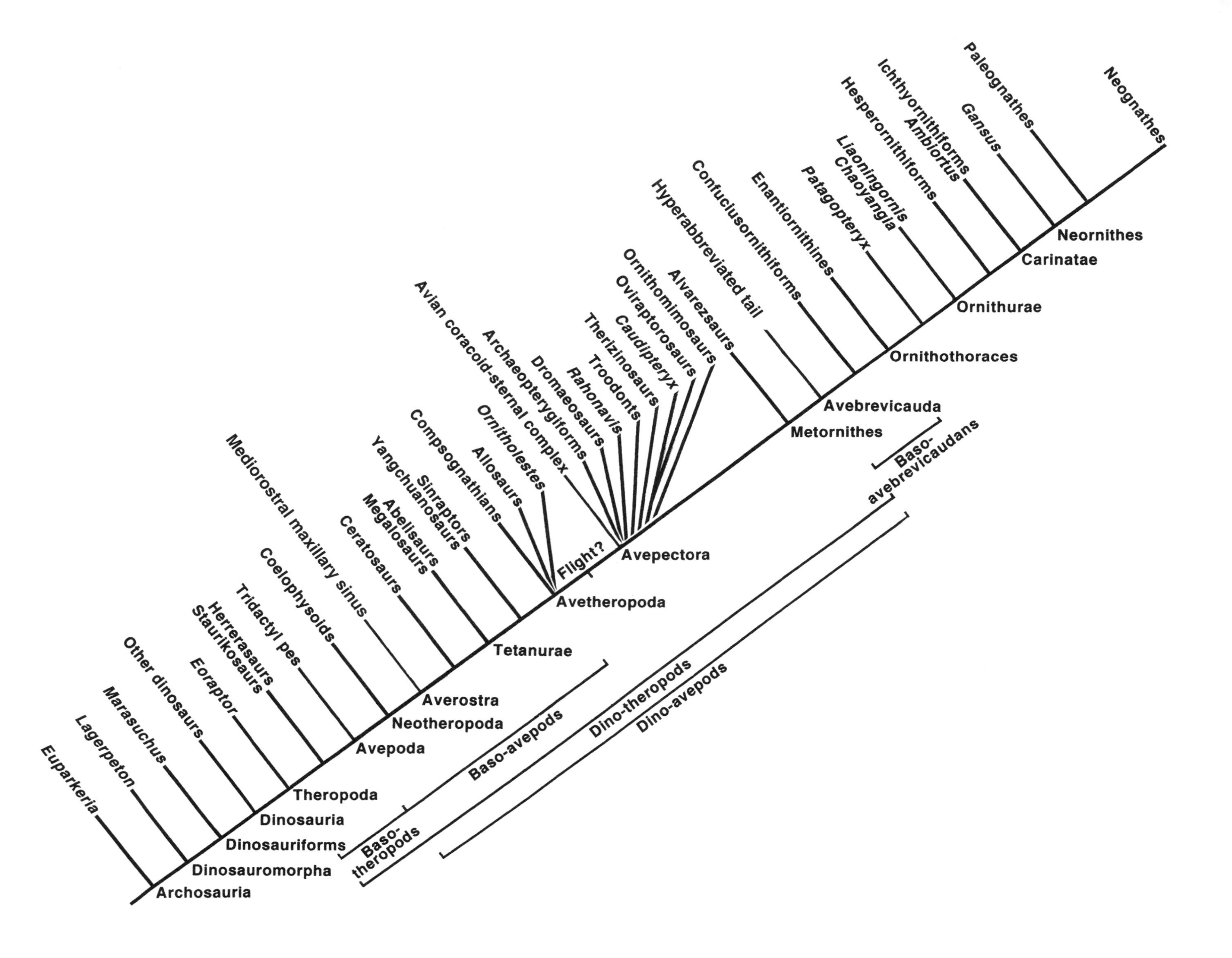

Euparkeria
Lagerpeton
Marasuchus
Other dinosaurs
Eoraptor
Herrerasaurs
Staurikosaurs
Tridactyl pes
Coelophysoids
Mediorostral maxillary sinus
Ceratosaurs
Abelisaurs
Megalosaurs
Sinraptors
Yangchuanosaurs
Compsognathians
Allosaurs
Ornitholestes
Avian coracoid-sternal complex
Archaeopterygiforms
Dromaeosaurs
Rahonavis
Troodonts
Therizinosaurs
Caudipteryx
Oviraptorosaurs
Ornithomimosaurs
Alvarezsaurs
Hyperabbreviated tail
Confuciusornithiforms
Enantiornithines
Patagopteryx
Liaoningornis
Chaoyangia
Hesperornithiforms
Ichthyornithiforms
Ambiortus
Gansus
Paleognathes
Neognathes
Archosauria
Dinosauromorpha
Dinosauriforms
Dinosauria
Theropoda
Avepoda
Neotheropoda
Averostra
Tetanurae
Avetheropoda
Flight?
Avepectora
Metornithes
Avebrevicauda
Ornithothoraces
Ornithurae
Carinatae
Neornithes
Baso-theropods
Baso-avepods
Dino-theropods
Dino-avepods
Baso-avebrevicaudans

is to add a prefix to formal names. For example, the term *baso-theropod* includes the basal theropods that lacked a birdlike, tridactyl foot, and *dino-theropod* includes theropod dinosaurs with the exception of extremely short-tailed birds. I must emphasize that these names are informal in nature. On a more general basis, *dinosaur* can be understood to refer to the classic examples, and birds are birds, even though they are a group of dinosaurs. When it is necessary, one can be more explicit. For example, when one refers to the Cretaceous/Tertiary (K/T) extinction that birds survived, one can note that it was the nonavian dinosaurs that went extinct.

## *A Few Important Names for Predatory Dinosaurs and Early Birds*

Sereno (1998, 1999c) and Padian et al. (1999) have attempted to tighten up the current taxonomy for theropods, and it is not my intention to erect a new, detailed taxonomy for dinosaurs and birds. Even so, the current, primarily node-based, taxonomy is in some regards inadequate, to the point that using it without some apomorphy+clade–based additions profoundly hindered my discussion of the issues in early drafts of this book. I therefore offer, and use, some adjustments and additions (see Fig. 2.1).

Theropoda (coined by Marsh in 1881), meaning beast-footed, has never been an etymologically appropriate title for dinosaurs whose most distinctive feature is a tridactyl, birdlike foot. However, Theropoda has been expanded to include basal, tetradactyl predaceous dinosaurs (Sereno and Novas 1992, Sereno et al. 1993). The latter are very primitive forms that lack many of the avian features present in tridactyl theropods. Neotheropoda (Sereno et al. 1994) is currently the most basal major name for tridactyl theropods, but it cannot include the basal-most bird-footed theropods because it is tied to a specific node above the level of the first appearance of the foot type. A new apomorphy+clade–based title for bird-footed dinosaurs, Avepoda (bird-footed), includes all theropods that either possessed a foot in which metatarsal I did not contact the distal tarsals, or descended from such theropods, and belonged to the clade that includes Neotheropoda. Note that Avepoda is distinct from Avipoda (Novas 1991), which is a junior synonym for Tetanurae (Gauthier 1986). Also note that theropods that developed a bird foot independently from the neotheropod clade are not avepods. Although it is improbable that the first avepod will ever be found, this group is very stable. Avepoda is divided into a paraphyletic group by the addition of the prefix "dino." Dino-avepods (bird-footed dinosaurs) are avepods that are not avebrevicaudans (new names are detailed below). Baso-theropods are tetradactyl basal theropods that are not avepods.

The dino-avepods are divided as follows. Paraphyletic cerato-saurans (includes coelophysoids and ceratosaurs) are dino-avepods that are not averostrans (see discussion below). Baso-avepods (roughly equivalent to Paleotheropoda; Paul 1988a) are dino-avepods that are neither tetanurans (includes procompsognathians, coelophysoids, and ceratosaurs) nor basal Tetanurae (includes megalosaurs and abelisaurs).

Among the most distinctive features to evolve in the Avepoda were accessory maxillary openings that led into complex ossified maxillary sinuses that projected medially toward the center of the snout; even some early birds retained this condition (see Chapter 3 and Appendix 3). A new apomorphy+clade–based title for bird-snouted dinosaurs, Averostra (bird-snouted), includes all avepods that either possessed at least one accessory maxillary opening in the lateral wall of the antorbital fossa that led into a bony mediorostral maxillary sinus, or descended from such avepods, and are members of the clade that includes the Dromaeosauridae. As I explain later, Averostra appears to include ceratosaurs, megalosaurs, and abelisaurs. Dino-averostrans (dinosaurs with bird snouts) exclude avebrevicaudans.

Avetheropoda (Paul 1988a, Holtz 1994a) and Neotetanurae (Sereno et al. 1994) appear to be synonyms, so the latter is junior to the former (contra Sereno 1997a, 1998, 1999c). This conclusion is not negated by my misplacement (Paul 1988a) of some baso-avepods within the Avetheropoda.

Coelurosauria (sensu Gauthier 1986, Holtz 1994a), Maniraptoriformes (Holtz 1996), and Maniraptora (Gauthier 1986) include most small Late Jurassic and Cretaceous avepods, including Jurassic compsognathids and ornitholestids. I show later that the characters used to define these

*Figure 2.1. (opposite)* A simplified phylogenetic diagram illustrating the taxonomic system used in this book (also see Fig. 1.1). Characters listed in Appendix 1 were used to construct this chart, but it is not a strict cladogram; in particular, the interrelationships among avepectorans are obscure.

groups are few and minor in nature and that it is not clear whether these are monophyletic groups. More importantly, the Jurassic examples are comparatively primitive forms whose carpals and hands either do not have well-developed coelurosaurian and maniraptoran features or are poorly known, and these examples may not be closely related to the Cretaceous examples, which are much more birdlike. Sereno (1997a) placed many of the latter in Paraves, but he excluded oviraptorosaurs. However, oviraptorosaurs share with paravians yet another distinctive birdlike adaptation, a shoulder girdle in which the coracoids articulate with a broad sternum in the avian manner. There is a pressing need for a name that includes all theropods with a birdlike shoulder girdle that belong to the clade that includes birds. A node-based name for this group would be too unstable to serve the purposes of this study because naming the node that includes oviraptorosaurs and modern birds would work only as long as the node included archaeopterygiforms. But this study will show that it might not include them, in which case urvogels with birdlike shoulder girdles would fall outside the named group. An apomorphy+clade–based title for the bird-breasted dinosaurs is required. Avepectora (bird-shouldered) includes all averostrans in which the majority of the distal edge of strongly anteriorly facing coracoids articulates with the anterior edge of a broad sternum at an angle of approximately 45–90 degrees from the midline, or that descended from such avepods, and are members of the clade that includes Dromaeosauridae. Avepectora is roughly equivalent to the obsolete Protoavia (Paul 1988a) and is similar to the unnamed node 11 in Figure 4 in Holtz (1994a), but it is not identical to either. Avepectora includes most of the same small Cretaceous avepods as Maniraptora but excludes the Jurassic grade forms. Dino-avepectorans (dinosaurs with bird shoulders) exclude avebrevicaudans. Dino-avepectorans are sometimes informally referred to as protobirds, dino-birds, or near-birds.

Avepoda, Averostra, and Avepectora designate major steps in the evolution of predatory dinosaurs into birds and are the primary theropod taxonomic titles used in this study. This taxonomy is more logical and more precisely defined than the recent taxonomy in use, but in no case is a prior name replaced by any of the new additions.

Another avepod group needing an informal paraphyletic title is the sickle-clawed dromaeosaurs and their close relatives, the archaeopterygiforms and rahonaviforms. These and other avepectorans that share their general body plan are labeled dromaeo-avemorphs. The boundaries of the group are ill defined because the current fossil data is inadequate. Because *Rahonavis* is incomplete and poorly preserved, its membership in this group is tentative. No one has yet demonstrated that incompletely known *Unenlagia* is not a dromaeosaur, and the former is assumed to be a member of the latter group. As I explain in Chapter 11, *Protarchaeopteryx* is considered an archaeopterygiform in this study. Questions surrounding the generic and specific status of various dinosaurs and fossil birds, including *Archaeopteryx,* are not considered in this study.

The high-level classification of Mesozoic birds is also unsettled (Padian and Chiappe 1998c). For example, the classifications offered by Chiappe (1995) and Elzanowski (1995) are very different from each other. Throwing even more confusion on the problem is that there is currently no group name that addresses one of the most critical adaptations marking the majority of Aves, the reduction of the tail into a short appendage tipped by a pygostyle. Several alternatives have been proposed. Until recently, Ornithothoraces included known birds possessing this adaptation, but Chiappe et al. (1999) have taken stub-tailed *Confuciusornis* out of this group. Pygostylia (Chatterjee 1997) also excludes *Confuciusornis*—Chatterjee did not realize it had a pygostyle—and all other short-tailed birds that fall below the designated node. Pygostylia also appears to be a junior synonym of Ornithothoraces (and is not used in Chatterjee [1999a]). Apomorphy+clade–based Avebrevicauda (birds with short tails) includes all Aves in which the free caudals were reduced to ten or fewer, or that descended from such avians, and are members of the same clade as Neornithes. Baso-avebrevicaudans exclude Ornithothoraces.

# SKELETONS, BONES, AND OTHER REMAINS OF THE MESOZOIC

The chapters in Part 2 are the core of this book in that they contain the detailed anatomical descriptions and functional analyses upon which the rest of the study depends. Related aspects of preavian respiratory anatomy and function are detailed in Appendix 3. This is also the most technical section and, for those less familiar with archosaur skeletal anatomy, probably the most difficult. Such individuals may wish to skim through Chapters 3 and 4 and plan to refer back to them when necessary. My reasons for discussing certain anatomical issues may not always be immediately clear, but they should become more so later in the book. For the benefit of nonspecialists, I have attempted to make this discussion more transparent than is normal for a technical work, but only so much can be done to avoid paleontology jargon, some of which really is necessary to explain matters in detail.

Some jargon, though, is not so useful, and there is one area in which it can be minimized. In recent years, a new method for designating the location of body parts has gained favor in anatomical circles. The method uses terms like *rostral* and *cranial* for forward and *caudal* for rear, so that one must use a different term for what is forward depending upon whether one is working with skull or skeletal elements. Even worse is when the terms *cranial* and *caudal* are sometimes used in a manner that makes it difficult to immediately discern whether they are intended to refer to body parts or direction, such as "the caudal vertebrae lack transverse processes." Although there were reasons for the invention of this system—for one thing, it applies equally well to vertical human skeletons and the more horizontal norm of the animal world—it represents an example of technical word inflation that further separates the scientific community from the rest of society. I have therefore opted for the more traditional, and less perplexing, system of describing the location of body parts. In this system, *posterior*, logically enough, refers to something at or toward the rear, and *anterior* to something at or toward the front. *Lateral* refers to things at or toward the side of the animal, and *medial* to things at or near the median vertical plane. *Dorsal* means at or toward the top, and *ventral* means at or toward the underside. *Proximal* means close to the core of the body, and *distal* means at or toward the extremities in all directions. Sometimes these terms are combined. For example, *anterodorsal* means forward and toward the top.

## What Is Available

The first two chapters in Part 2 concentrate on avepectoran dinosaurs and basal birds, with an emphasis on *Archaeopteryx* (see Figs. II.1 and II.2). Recent decades have seen a proliferation of well-preserved remains of dino-birds, and just as important have been the detailed descriptions of these remains. Even *Archaeopteryx* is becoming somewhat less rare and much better known now that seven specimens of varying quality are available. A problem with these remains is that their rarity and their in situ arrangement with feather arrays in place have

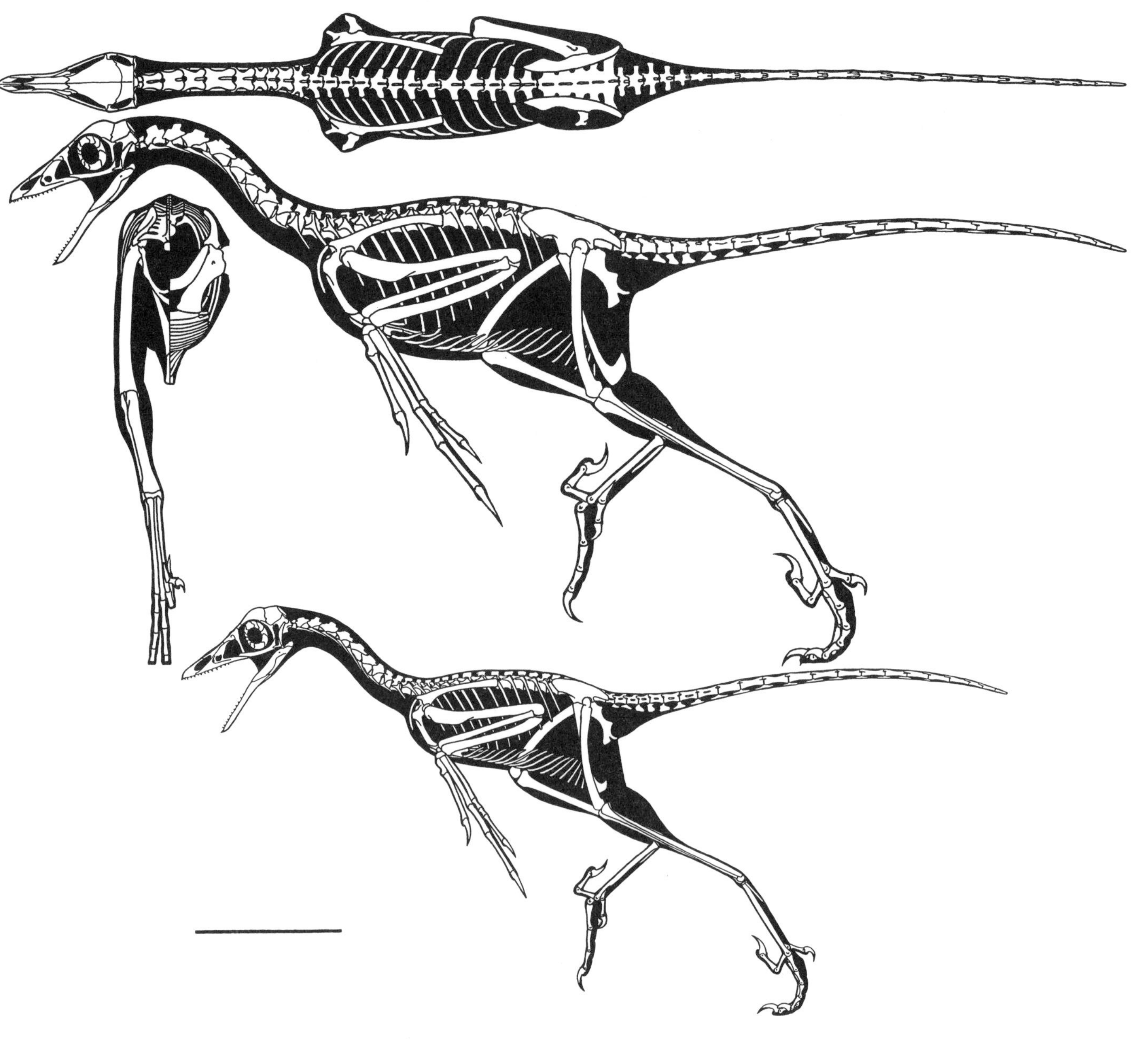

*Figure II.1.* Restorations of *Archaeopteryx* skeletons HMN 1880 and juvenile JM 2257, to same scale. These are the most complete skulls and skeletons yet described. In both cases, poorly preserved or missing parts were scaled in from other specimens when available. The cartilaginous posterior sternum and sternal ribs are restored. Neither the precise orientation of the pubis nor the exact distal joining of the strongly bowed ischia are certain. The second toes are tentatively posed hyperextended. Note the relatively larger skull and shorter arms of the smaller specimen compared with those of the larger specimen. The pectoralis muscles are restored at approximately 10 percent of total mass in 1880, less in 2257. Scale bar equals 50 millimeters.

prevented the removal and full preparation of the bones (Shipman 1998). As a result, many details remain obscure (one can hope that someday a new urvogel skull and skeleton without feather traces will be found and entirely removed from the matrix for a more thorough examination, or that sophisticated three-dimensional X-ray mapping will allow researchers to resolve the details of the bones without uncovering them). It has not been proven that any of the *Archaeopteryx* specimens were those of fully grown, fully ossified adults. However, it would be very unusual for all seven specimens of a land animal found in offshore deposits to be juveniles, so it is assumed that the largest skeletons represent the adult condition.

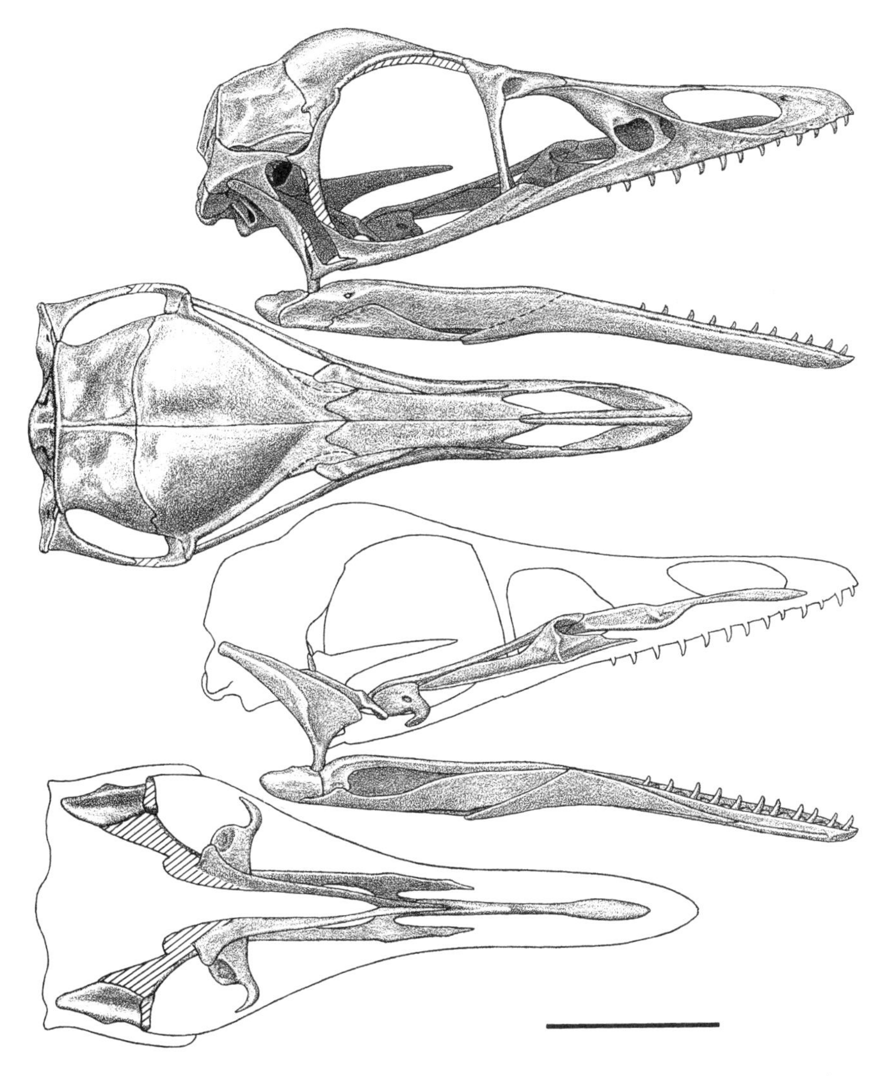

*Figure II.2.* Restoration of *Archaeopteryx*'s skull, proportioned mainly after JM 2257, the most complete skull, with missing parts from other specimens scaled in, including the braincase from BMNH 37001. Unknown parts are indicated with hatching. We do not know whether the postorbital was complete, and the width of the snout is tentative. Scale bar equals 10 millimeters.

Complete and superbly preserved skulls and skeletons have been found for a number of small Cretaceous dino-avepods. Indeed, more complete remains of *Avimimus,* which was first known from incomplete and disarticulated fossils (Kurzanov 1987), have verified that *Avimimus* is indeed a real taxon (Watabe et al. 2000). A contrary situation applies to Triassic remains attributed to "*Protoavis*" (Chatterjee 1991, 1999a). The remains are incomplete, unarticulated, and in many cases badly damaged; and new analysis indicates that this group is probably a chimera. (These fossils are discussed in Chapter 10.)

To ensure clarity in the following discussions, I have used museum specimen numbers to refer to the remains. Institutional abbreviations cited in this study are as follows:

| | |
|---|---|
| AM | Auckland Museum |
| AMNH | American Museum of Natural History |
| BMNH | British Museum of Natural History |
| BSP | Bayerische Staatssammlung fur Palaontologie |
| ChM | Charleston Museum |

| | |
|---|---|
| CM | Canterbury Museum |
| FHSM | Fort Hays State Museum |
| HMN | Humboldt Museum |
| JM | Jura-Museum |
| MCZ | Museum of Comparative Zoology, Harvard |
| MLP | Museo de la Plata |
| MIG | Mongolian Institute of Geology |
| NGMC | National Geological Museum of China |
| NZP | National Zoological Park |
| TM | Teyler Museum |
| TMM | Texas Memorial Museum |
| USNM | United States National Museum |



The *Archaeopteryx* specimens are usually referred to by their present museum locations, but because this practice can be awkward, I instead use the specimen numbers, and simplifications thereof, listed below:

- London, BMNH 37001 (almost complete, partial skull)
- Berlin, HMN MB.1880/81 4598 and 4599 = HMN 1880 (virtually complete)
- Eichstätt, JM SoS 2257 = JM 2257 (virtually complete)
- Haarlem, TM 6428 and 6429 = TM 6428 (incomplete, no skull)
- Munich, BSP 1999 I 50 = BSP 1999 (nearly complete, a few skull parts missing)
- fifth (Maxberg) skeleton = S5 (incomplete, no skull)
- sixth (Solnhofen) skeleton = S6 (virtually complete, except for most of skull)

The last three specimens are in private hands, and of these, S5 is currently missing. I have personally examined the first three specimens and have seen BSP 1999 on display. There are reports of an undescribed, eighth, partial specimen in private hands.

Concerning other, new avepod specimens important to this study, I have directly examined the type specimen (mainslab) and the third *Sinosauropteryx*, the two *Caudipteryx*, and the type specimens of *Protarchaeopteryx, Sinornithosaurus,* and "Archaeoraptor."

Among the fossil birds examined below, Hou and Chen (1999) concluded that the type skeleton of *Liaoxiornis* is not that of a juvenile, despite its diminutive dimensions. However, the relatively large head, which is dominated by a very large orbit; the lack of ossification of almost all but a tiny core of the sternum; the relative shortness of the arms; and the apparent lack of ossification of hand elements indicate that it was not fully grown and was probably not capable of flight. *Eoenantiornis* (Hou et al. 1999) also shows signs of being a larger, more mature juvenile; it exhibits a very large orbit, an incompletely ossified sternum, and a poorly formed hand.

In general, discussions of anatomy start with the front of the skull and continue back through the skull and skeleton, ending with the hind feet and then turning to the integument. I shall follow that tradition here.

CHAPTER 3

# Skulls

When Heilmann (1926) restored the skull of the urvogel, he had only one complete skull, that of HMN 1880, to work with. His restoration (Fig. 3.1A) is now considered unreliable because the surface of the HMN 1880 skull is poorly preserved, and most, but not all, of the details are lost. However, HMN 1880 best preserves the three-dimensional shape of the cranium, except in the posterior-most section (Pl. 6). Although the skull of JM 2257 is crushed and split between two blocks, it nevertheless records numerous details (Pl. 7), as does the less complete and partly articulated BSP 1999 skull (Pl. 5). The best part of the disarticulated and incomplete skull of BMNH 37001 is its nearly complete braincase (Pl. 22A).

Heilmann's restoration reigned unchallenged for years, until JM 2257 inspired a new series of restorations starting in the 1970s. Some included avian details that are not necessarily supported by the preserved remains (L. Martin 1983b, 1991, 1995, Buhler 1985, Chatterjee 1991, 1997, Martin and Zhou 1997, L. Martin et al. 1998), and these restorations were accepted by Feduccia (1996). Wellnhofer (1974) and I (Paul 1988a) restored the skull as more like the skulls of avepod dinosaurs. The competing skull restorations, old and new, are presented in Figure 3.1. Enough of the skull is now known for more than 90 percent to be restored accurately (Fig. II.2 and Pl. 7 show the skull of JM 2257 as preserved). The proportions of the two most articulated skulls differ somewhat: the total skull-length/mandible-length ratio is about 10 percent higher in HMN 1880 than in JM 2257. Because more elements are complete enough to be measured in JM 2257—especially the crucial quadratojugal and jugal, as well as palatal elements and the posterior braincase—I used this skull as the basis of the primary restoration in Figure II.2, despite the skull's being smaller than the more amorphous HMN 1880, for which a much less detailed restoration has been rendered in Figure 3.1G. When bones from JM 2257 were missing or distorted, bones from the other skulls were scaled down to the appropriate size (as per the methods described in Paul and Chase [1989]). Among these bones were the BMNH 37001 braincase and the ventral braincase, squamosal, postorbital process, and palatal elements from BSP 1999. The postorbital may be preserved in HMN 1880. Also important to restoring the skull of the urvogel were newly described skulls of Early Cretaceous birds (see Fig. 10.2O,P), which, via the method of phylogenetic bracketing (Witmer 1995b), allowed the restoration of some of the more obscure aspects of the earlier bird's cranium.

Because the braincase was broad, the width/total-length ratio of the top of the *Archaeopteryx* skull was moderately high at about 2.5 (Figs. II.2, 3.1). The width/depth ratio of the posterior part of the skull was about 1. The L. Martin (1991) and L. Martin and Zhou (1997) restorations were too narrow in top view (Fig. 3.1D) and too deep because these workers misarticulated the quadrate (the skull bone that supports the lower jaw), and therefore their width/length and width/depth ratios of 4.8 and 2 are much too high. The total-length/depth ratio was just under 3. The uncrushed snout of HMN 1880 is slightly upturned, in part because the nasals on top of the rostrum are subtly depressed (as in some dromaeosaurs) (Paul 1988a,b, Fig. 3 in Xu, Wang, and Wu 1999, Fig. 2b in Xu et al. 2000). The snout of *Velociraptor* was transversely very narrow (Fig. 10.3F; Barsbold and Osmólska 1999, Norell and Makovicky 1999), although the narrowness may have been somewhat exaggerated by crushing (compare Figs. 1C and 3B in Barsbold and Osmólska 1999). Associated with the dromaeosaurs' transversely thin muzzle were dentaries that were straighter in dorsoventral view than those in other dino-avepods, in which the anterior ends of the dentaries curved medially toward one another to form a broader lower jaw. The dentaries were similarly straight in all dromaeosaurs (Sues 1977, Currie 1995), indicating that their muzzles were

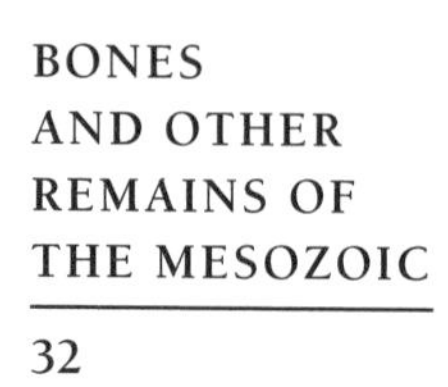


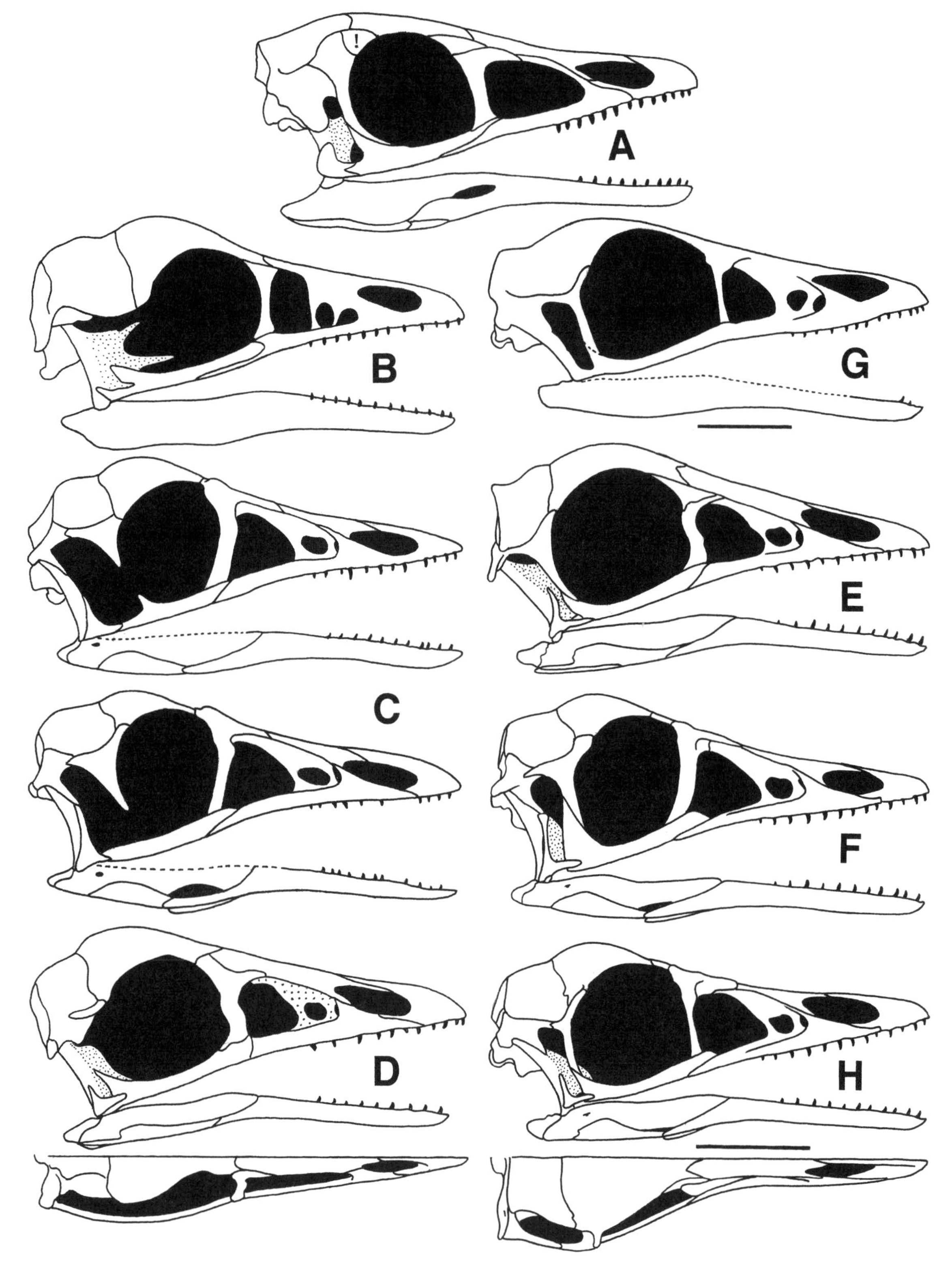
A
B
G
C
E
F
D
H

unusually narrow. The same may or may not have been true of *Archaeopteryx;* that its dentaries are straight may or may not be the result of crushing.

The large opening in front of the eye socket is the antorbital fenestra, the classic feature of an archosaur. The antorbital opening is a window into the interior of the skull in JM 2257, and the opening reveals bones whose nature is sometimes obscure. In birds, the olfactory conchae (thin sheets of bone or cartilage covered by olfactory sensory tissue) are the posterior conchae, which are anchored at their posterior end to the lacrimal (App. Figs. 2D, 5A–F; Bang 1971, Witmer 1995a). The mesethmoid is a vertical, midline sheet of bone that is set at the midlength of the skull in birds (Witmer 1990). A small fragment of bone lies in the dorsoposterior corner of the antorbital fenestra; as preserved, it contacts the lacrimal. This fragment may be part of an ossified olfactory concha, or it may be part of the mesethmoid. In a related matter, Ruben et al. (1996) identified olfactory conchae in the CT scan of a juvenile *Tyrannosaurus* skull (App. Figs. 2Da, 3T). Because the imaged material is set well forward of the lacrimal, it may represent a more anterior concha, but the image is too amorphous for definitive identification of the material. A bone fragment at the anterior end of the JM 2257 jugal sits too low to be within the nasal complex (contra Elzanowski and Wellnhofer 1996), so it is probably a fragment of the maxilla, as originally identified by Wellnhofer (1974).

We now come to the bones ringing the antorbital opening. In HMN 1880 and JM 2257, there is a broad, semibulbous surface of bone anterodorsal to the antorbital fenestra. Whetstone (1983) and L. Martin (1991) identified this bone as the mesethmoid, but it is much too far forward for that to be the case (Witmer 1990). The bone surface in question is clearly a superficial lateral bone that articulates with the nasals. It has the same placement and topography as the antorbital fossa of the maxilla—the depression that surrounds most of the antorbital opening—observed in many dino-avepods; even the "bulbous" transverse curvature of this bone surface is observed in many members of this group (see, for example, the *Velociraptor* skull shown in Fig. 3.3). Therefore, Wellnhofer (1974) and Witmer (1990) are correct that the bone is part of the external maxilla, and their opinion is shared by Ruben and Jones (2000). This conclusion is confirmed via phylogenetic bracketing because a similar antorbital fossa is present in the basal avebrevicaudan *Cathayornis* (Pl. 10; Martin and Zhou 1997).

That the bone surface in question is part of the maxilla means two things. First, the very large fenestra restored by Martin (1991) is too large; the antorbital fenestrae in the two articulated skulls are similar to each other and more normal in size and shape. Second, accessory maxillary fenestrae are preserved in JM 2257 (Wellnhofer 1974) and *Cathayornis* (Pl. 10; Martin and Zhou 1997). The same region of the skull is difficult to restore in the crushed *Confuciusornis* skulls (Chiappe et al. 1999). A restoration (Fig. 10.2O) based on one of the best-preserved specimens (Pl. 9A) suggests that a reduced antorbital fenestra was preceded by an accessory maxillary fenestra. The accessory fenestrae were probably entrances to interior accessory maxillary sinuses (Witmer 1990, 1995a, 1997a,c). There is a subtle kink in the dorsal antorbital rim of the maxilla in JM 2257. It probably marks the bottom edge of the posterior end of the anterior portion of the nasal passage (App. Fig. 2Dc).

*Figure 3.1. (opposite)* A comparison of *Archaeopteryx* skull restorations past and present. *A,* Heilmann (1926) HMN 1880; *B,* Buhler (1985) composite; *C, top,* Chatterjee (1997) JM 2257, and *bottom,* Chatterjee (1999a) BSP 1999; *D,* Martin and Zhou (1997), primarily JM 2257; *E,* Wellnhofer (1974) JM 2257; *F,* Paul (1988a) JM 2257; *G,* new, HMN 1880; *H,* composite proportioned after JM 2257. In *A,* all details are speculative, the postfrontal is absent, the maxillary component of the antorbital depression is missing, and the superior temporal bar is much too thick. *B–D* are birdlike restorations in which either the superior temporal bar or the postorbital bar, or both, is incomplete. In view C *(top),* the postorbital process of the jugal is too far forward; and in the *bottom,* the postorbital process of the jugal is missing. View *D (top)* is overly narrow, and the bone bordering the anterior edge of the antorbital opening (stippled) has been alternatively identified as the internal mesethmoid or the external maxilla. *E–H* are restorations in which the superior temporal and posterior bars are complete. In *F,* the postorbital bar is too thick. Other errors include overly thick lacrimals in lateral view in *A–F;* lacrimals, squamosals, and quadratojugals that are inaccurate to varying degrees in all but *G* and *H;* the absence of the squamosal in *B* and *D;* the overly large postorbital process of the jugal in *F;* the lack of squamosal-quadrate articulation in *B, D,* and *E;* subvertical quadrates in *B–D* and *F;* the posterior pterygoid that is merged with the medial flange of the quadrate in *A, B, D,* and *E.* Deep quadrate surfaces are densely stippled. Scale bars for JM 2257 and HMN 1880 equal 10 millimeters.

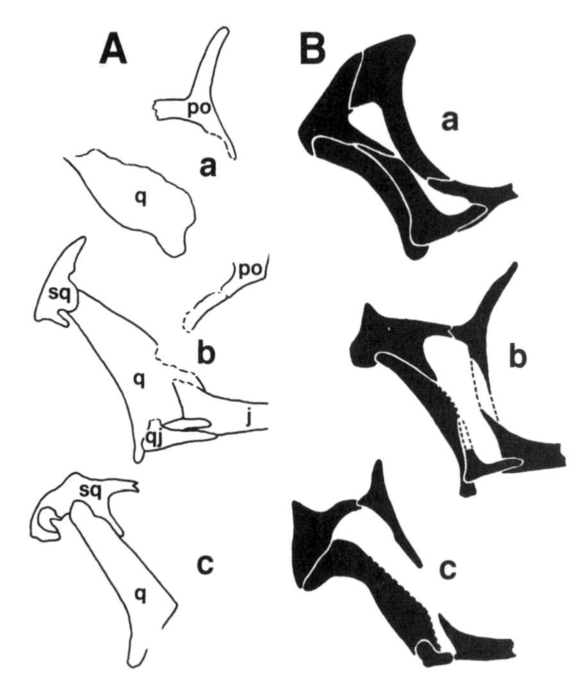


*Figure 3.2.* Right superficial temporal skull elements in side view, drawn to same approximate quadrate height. *A, Archaeopteryx* specimens as preserved with strongly procumbent quadrates: *a,* HMN 1880, in which the full extent of a probably postorbital is not clear (see Pl. 6); *b,* JM 2257, in which the probable fragments of the frontal process of the postorbital process have been shifted ventrally by crushing (see Pl. 7); *c,* BSP 1999 (Fig. 3B in Elzanowski and Wellnhofer 1996); *B,* articulated and restored temporal regions in avepods with a procumbent quadrate and a short postorbital process of the jugal: *a, Ornithomimus* with a complete postorbital bar; *b, Archaeopteryx* in which the bar may have been incomplete and the postorbital is a tentative composite; *c,* juvenile baso-avebrevicaudan bird in which the bar is not complete. Abbreviations are listed in Appendix 7.

The isolated lacrimal of BMNH 37001 is preserved (Fig. 5 in Pl. IX in De Beer 1954). On the skull roof of JM 2257 is a piece of bone that may represent a prefrontal; whether it was originally dorsally exposed is not certain.

The next area of concern is the skull roof over the eye socket, which consists mainly of the large, subtriangular frontals. The placement of the suture between the frontal and the bone immediately behind the eye socket, the postorbital, is obscure in JM 2257. If the suture was placed medially, then there was no well-developed postorbital process on the posterolateral corner of the frontal. If the suture was placed more laterally, then the frontal had a subtriangular postorbital process. A posteriorly facing depression in this area (Pl. 7) may have been similar to the step on the postorbital process of the frontal in dromaeosaurs; alternatively, the depression could have been formed by a medially curving frontal process of the postorbital.

The back of the skull includes the temporal region and the suspensorium that supports the lower jaw. Of the bones in this area, neither the postorbital nor the quadratojugal is yet completely known. The lack of complete information concerning the status of the postorbital has been especially vexing. A major question about *Archaeopteryx* is whether it retained the superior temporal and postorbital bars that border the upper and anterior edges of the lateral temporal fenestra. These bars are present in most archosaurs but not in most birds. Neither of the delicate bars is completely preserved in any of the *Archaeopteryx* skulls, perhaps because they were damaged in ancient times or when the slabs were split. In addition, in both HMN 1880 and JM 2257, the right side of the skull immediately behind the orbit has been dorsoventrally collapsed by crushing, which distorts the true extent, position, and articulation of the elements in this region (the same problem obscures the exact configuration of the postorbital bar in a large number of *Confuciusornis* skulls). The postorbital contributes to both of these bars, and there has been debate about whether this bone existed at all in the protobird.

The issues of the presence of a postorbital bone and of a superior temporal bar can now be considered solved. The frontal process of the postorbital, including its articulation with the frontal, is preserved in JM 2257 (Pl. 7, Fig. 3.2Ab). In HMN 1880 there is a triradiate element on the posterior border of the orbit (Pl. 6, Fig. 3.2Aa; Chiappe et al. 1999). In shape and position, the element has the characteristics expected of a postorbital, and it is identified as such, although its full extent is not clear. Both specimens' postorbitals have a long slender frontal process. The frontal process slopes more posteroventrally in JM 2257 than in HMN 1880, but because both skulls' temporal regions are distorted, it is not clear which is closer to the correct position.

The posterior process of the triradiate HMN 1880 element(s) may include the anterior-most part of the squamosal; if so, then the anterior part of the superior temporal bar is preserved. The posterior half of the superior temporal bar is made by the squamosal, and there is an articulation for the postorbital on the slender anterior process of the squamosal in BSP 1999 (Fig.3.2Ac; Fig. 3B in

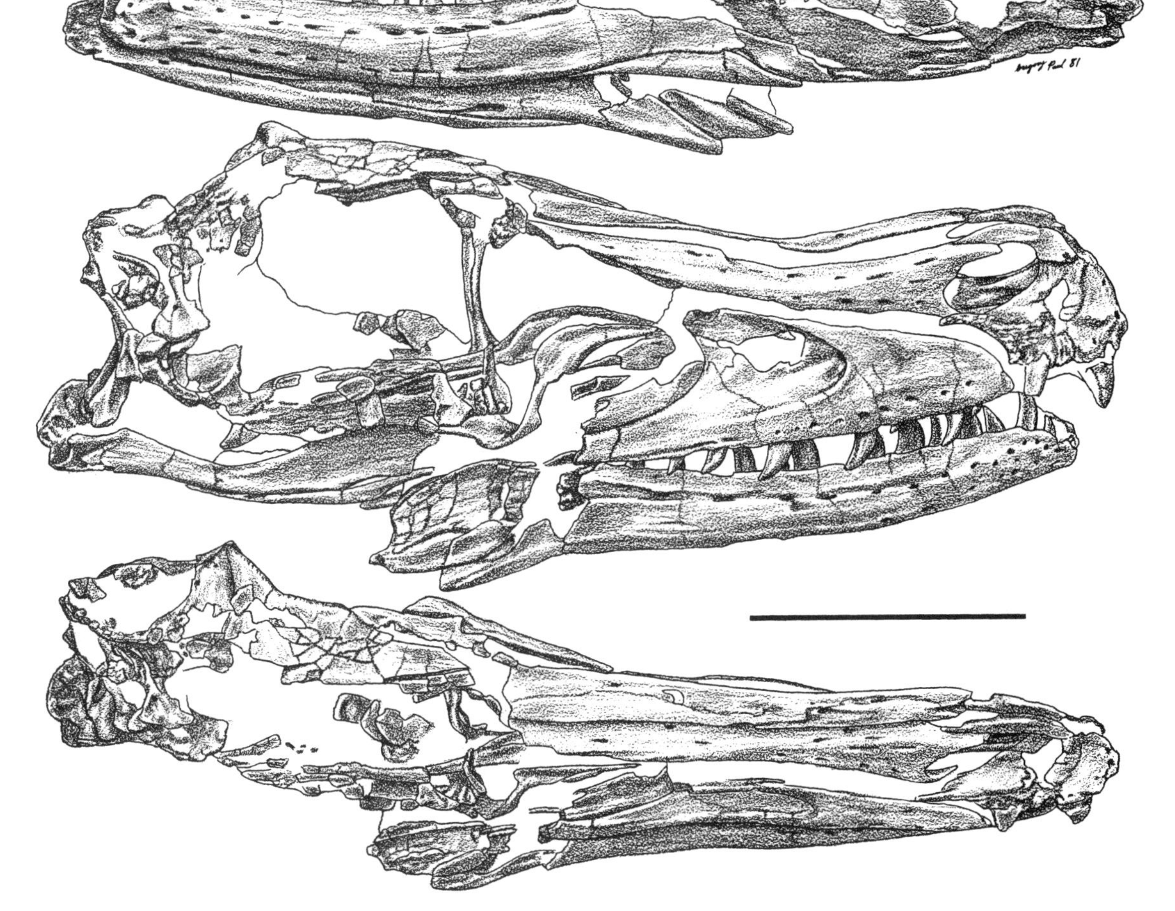

*Figure 3.3.* Skull as preserved of *Velociraptor* AMNH 6515. The arrow indicates the "bulbous" contour of the dorsal process of the maxilla. Scale bar equals 50 millimeters.

Elzanowski and Wellnhofer 1996). The articulation confirms the presence of a postorbital and shows that it helped form a complete superior temporal bar. The presence of a well-developed squamosal and postorbital that unite to form a complete superior temporal bar in various basal birds affirms the presence of the bar in early birds (Fig. 10.2O; Sanz et al. 1997, L. Martin et al. 1998). The failure by Martin and Zhou (1997) and L. Martin et al. (1998) to show a superior temporal bar in *Archaeopteryx* means that their recent restorations of the urvogel skull are obsolete. In *Archaeopteryx,* alvarezsaurs, therizinosaurs, and a juvenile basal avebrevicaudan (Sanz et al. 1997), the bar was slender (unlike to the massive structure restored by Heilmann [1926]), but it was robust in *Confuciusornis* (Fig. 10.2O). It also sloped anteroventrally in these avepods, so the frontal process of the postorbital was tall and nearly vertical. Because the braincase was enlarged dorsally, the bar was set rather low on the skull in basal birds, with an elongated, dorsally oriented and frontal process of the postorbital connecting the bar to the skull roof (Xu, Wang, and Wu [1999] stated that *Sinornithosaurus* lacked the upturned process of the postorbital, but their Fig. 3 shows otherwise). As a result, the superior temporal opening on the top of the skull faced partly sideways. The same combination of an enlarged braincase and short quadrate meant that the superior temporal bar was also set low relative to the braincase in dromaeosaurs, troodonts, therizinosaurs, and avimimids, this placement again tilting the upper temporal fenestra partly laterally.

The ventral process of the postorbital is entirely missing in JM 2257, and its full extent in HMN 1880 is not clear; so its contact if any with the jugal is not preserved. The two available skulls are preservationally and statistically insufficient for determining the status of the postorbital bar. Phylogenetic bracketing is also of limited use because the bar seems to have been barely complete in the dromaeosaur *Bambiraptor* (Burnham et al. 2000), was incomplete in alvarezsaurs (Chiappe et al. 1998), but was complete and even robust in the baso-avebrevicaudan *Confuciusornis* (Pl. 9B, Fig. 10.2O; Feduccia 1996, p. 40, upper right, L. Martin et al. 1998, Chiappe et al. 1999). Martin et al. restored the latter avepod's postorbital bar as having only a minor jugal component, but Chiappe et al. (1999) showed that the jugal was more probably fully triradiate with a well-developed postorbital process. The postorbital is probably complete, albeit slender, in the basal enantiornithine *Protopteryx,* according to Zhang and Zhou (2000). The presence of the complete bar in early birds makes its retention in the urvogel highly plausible. There is no jugal process of the postorbital preserved in JM 2257. The same part of the HMN 1880 postorbital as preserved appears too short to reach the jugal, but it may well be incomplete. However, the elongation of the jugal process in other early birds (Sanz et al. 1997, Chiappe et al. 1998, L. Martin et al. 1998) suggests the same is true in the urvogel. The complete postorbital process of the jugal of BSP 1999 (Elzanowski and Wellnhofer 1996) is small but fully developed (the absence of this process in Chatterjee's [1999a] new skull restoration [Fig. 3.1C, bottom] is perplexing). Was the process of the postorbital long enough to contact the short process of the jugal and form a complete bar bordering the posterior rim of the orbit, as restored by Heilmann (1926), Wellnhofer (1974), A. Walker (1985), and me (Paul 1988a)? Or was it too short to do so, as indicated in the skull restorations completed by L. Martin (1983b, 1991), Buhler (1985), Chatterjee (1991, 1997, 1999a), Martin and Zhou (1997), and L. Martin et al. (1998)?

In order to answer these questions, we must first correctly restore the position of the suspensorium. Many restorations incorrectly show the quadrate—the skull bone that supports the lower jaw—nearly vertical (Fig. 3.1B–D,F). However, the strong downward and forward slope of the descending process of the BSP 1999 squamosal suggests that the quadrate was oriented the same way (Fig. 3.2Ac). In partly articulated BSP 1999 and in the articulated skulls of HMN 1880 and JM 2257, the quadrates slope strongly down and forward, to the point that the jaw joint is beneath the posterior rim of the orbit (Pls. 6 and 7, Figs. II.2, 3.2A; Fig. 3B in Elzanowski and Wellnhofer 1996). The main body of the quadratojugal is very short in JM 2257 and BSP 1999, and the postorbital process of the jugal is close to the quadrate (contra Chatterjee [1991, 1997], who shows the postorbital process too far forward on the jugal). If the quadrate is oriented nearly vertically, then the postorbital process of the jugal almost contacts the lateral surface of the quadrate. The result is that there is no room for a lateral temporal fenestra, the postorbital ramus of the jugal is set too far behind the orbit, and the distance between the jugal and postorbital is too great for them to readily contact each other. Rotating the quadrate so that the jaw joint is set farther forward raises the jaw joint and the posterior end of the jugal. The postorbital process of the jugal now forms the posteroventral edge of a semicircular orbital rim (one can also see that the quadratojugal is so short because the distal end of the quadrate is so far forward), and the postorbital process of the jugal now spans about 40 percent of the distance to the superior temporal bar (Figs. II.2, 3.2Bb). Martin and Zhou (1997) and Chiappe et al. (1999) were therefore incorrect when they stated that the postorbital process of the jugal was too posterior to contact the postorbital in *Archaeopteryx.*

The above factors leave alternative ways to restore the postorbital complex. The procumbent orientation of the quadrate, the reduction of the lateral temporal fenestra, and the shortness of the postorbital process of the jugal are similar in *Archaeopteryx* and ornithomimosaurs (with the latter process spanning 40 percent of the distance to the superior temporal bar; Fig. 3.2Ba,b), and the latter have a complete postorbital bar. A specimen of a juvenile basal avebrevicaudan also has a procumbent quadrate (Fig. 3.2Bc; Fig. 1 in Sanz et al. 1997) and a postorbital process of the jugal that is very similar to that of *Archaeopteryx,* but the process reaches only a third of the way to the superior temporal bar. In the juvenile basal avebrevicaudan, there is a long ventral process of the postorbital, but because it is not long enough to contact the jugal, the postorbital bar is incomplete. In HMN 1880, the jugal process of the postorbital is directed too anteriorly to contact the jugal's postorbital process (Chiappe et al. 1999), but the fact that the postorbital projects somewhat into the orbit suggests it has been rotated anteriorly. In JM 2257, the postorbital appears to have been rotated too posteriorly by the collapse of the side of the skull. The postorbital bar of *Archaeopteryx* may have been complete (as it is in dino-avepods), almost but not quite complete (as per the just described early bird), or intermediate between the two. The first and last options are tentatively considered superior considering the more basal nature of *Archaeopteryx* vis-à-vis the later bird, and the somewhat more robust postorbital and taller dorsal process of the urvogel's jugal compared with those of the juvenile basal avebrevicaudan. The better *Protarchaeopteryx* skulls that can be expected to emerge from the hyperproductive Yixian beds may clarify the status of the postorbital bar in archaeopterygiforms. If the bar was complete, it was not especially robust (contrary to the thicker bar I restored in 1988 [Paul 1988a] and the robust postorbital bar of *Confuciusornis*). A number of other basal birds—such as *Cathayornis,* in which the temporal region is poorly preserved—probably had a procumbent quadrate, a nearly or even entirely complete postorbital bar, and a nearly or entirely complete superior temporal bar.

Also uncertain is the status of yet another bar bordering the lateral temporal fenestra of *Archaeopteryx,* that connecting the squamosal to the quadratojugal. In most archosaurs, the bar is complete; in birds, including most if not all basal examples, it is not (Fig. 10.2P–R). In the dromaeosaur *Bambiraptor,* the two bones appear to barely contact one another (in a manner similar to the postorbital bar, as noted above; Fig. 1 in Burnham et al. 2000). The *Archaeopteryx* skull restorations of Martin and Zhou (1997) and Martin et al. (1998) lack a complete posterior temporal bar, but these restorations are obsolete in that they fail to include the descending process of the squamosal that is well preserved in BSP 1999 (Fig. 3.2Ac). The ascending process of the quadratojugal is incomplete in BSP 1999 where the bone is best preserved. Elzanowski and Wellnhofer (1996) concluded that the slender ascending process was probably not long enough to contact the descending process of the squamosal. They may be correct, but the possibility of contact between the two elements cannot be ruled out. A thin ascending process of the quadratojugal that contacts the squamosal is seen in velociraptorine dromaeosaurs, for example (but the very long ascending process restored by Chatterjee [Fig. 3.1C, bottom] appears to be speculative). If the dorsal process of the quadratojugal of *Sinornithosaurus* is as short as Xu, Wang, and Wu (1999) indicate, then this dromaeosaur's quadratojugal may not have contacted the squamosal. This is a good place to note that the quadratojugal of *Deinonychus* was incorrectly oriented by Ostrom (1969): the articulated skulls of other dromaeosaurs show that the quadratojugal actually had an inverted-T shape (Barsbold 1983, Paul 1988a,b, Currie 1995, Xu, Wang, and Wu 1999). The claim by Ruben and Jones (2000) that the quadratojugal was too short to contact the squamosal in *Caudipteryx* is contradicted by new skulls (Zhou et al. 2000).



Another major question about *Archaeopteryx* is whether it retained a normal archosaurian suspensorium or whether it had the "propulsion joint" between the quadrate and the palate found in most birds. The current absence of a well-preserved and described quadrate and palate of a baso-avebrevicaudan bird hinders our analysis of this question. The possible quadrate of BMNH 37001 (A. Walker 1985) has been identified as a braincase element by Welman (1995), but others have favored its identification as the quadrate (Chatterjee 1991, Elzanowski and Wellnhofer 1996, Gower and Weber 1998). Removal of the remains from the slab may be necessary to clarify the matter. If it is the quadrate, then it has a simple single head. The quadrate of BSP 1999 has been proven to have had a strong articulation with the squamosal at the posterolateral corner of the skull (Fig. 3.2Ac) and to have apparently lacked the strong double head and extensive braincase articulations proposed by a number of workers. If there was a double head that contacted the braincase, the doubling appears to have been a very incipient condition most similar to the condition found in some dino-avepectorans (Russell and Dong 1993a, Maryanska and Osmólska 1997) and some birds (Elzanowski et al. 2000). It is therefore difficult to understand why in 1996 Feduccia continued to present the obsolete data—based in part on a hopelessly ambiguous, low-resolution CT scan—for a strongly double-headed quadrate, and why he ignored Elzanowski and Wellnhofer's (1995, 1996) more definitive data (also see Chatterjee 1991). The medial flange of the quadrate is essentially complete in JM 2257 and BSP 1999. It is a vertical, transversely thin, anteroposteriorly broad plate almost as tall as the quadrate (Pl. 7, Fig. 3.2Ab,c; Fig. 3B in Elzanowski and Wellnhofer 1996). The anterior edge of the plate is a smooth, gentle, anteriorly convex arc, with a sharp anteroventral corner (A. Walker 1985, Paul 1988a). Because a similar structure is present in both dino-avepods and non-ornithurine birds (Chiappe 1995, Sanz et al. 1997), phylogenetic bracketing confirms its retention in the urvogel. There is no evidence for the knife blade–like orbital process of the quadrate that has replaced the medial plate in ornithurines. The supposed orbital process sometimes restored in JM 2257 (by Buhler 1985, L. Martin 1983b, 1991) is actually part of the pterygoid. So is the incorrect subrectangular dorsoanterior corner of the medial flange restored by Martin and Zhou (1997). There is no evidence of an avian propulsion joint on the urvogel's dinosaur-like quadrate.

This brings us the palatal bones, which make up the roof of the mouth (Figs. 3.4, 10.6). In JM 2257 and BSP 1999, the quadrate process of the pterygoid is a large, vertically oriented, transversely thin, deep, anteroposteriorly broad plate connected to the main body of the pterygoid by a thin, flattened, twisted, curved bar (Pl. 7, Fig. 3.4Ac; A. Walker 1985, Paul 1988a, Fig. 3A in Elzanowski and Wellnhofer 1996). In JM 2257, the lateral surface of the vertical platelike process of the pterygoid articulates with the medial surface of the medial plate of the quadrate in the normal archosaurian manner (Pl. 7, Figs. II.2, 3.4Bc); no trace of the classic avian arrangement is present. The long dorsoposterior extension of the quadrate process of the pterygoid in JM 2257 is necessary for full articulation of the pterygoid with the strongly procumbent quadrate, and is additional evidence for the latter orientation. The absence of this extension of the quadrate process of the pterygoid in BSP 1999 is due to breakage (Fig. 3.4Ac; Fig. 11 in Elzanowski and Wellnhofer 1996).

The ectopterygoid is a small bone that connects the palatal bones with the side of the skull below the orbit. Its existence in *Archaeopteryx* was formerly controversial, but its presence in JM 2257 and BSP 1999 is no longer open to debate (Elzanowski and Wellnhofer 1996). It also appears to be present in HMN 1880 (Pl. 6). The ectopterygoid appears to have a small dorsal depression that may be similar to the little excavation found atop the same bone in dromaeosaurs (Ostrom 1969, Sues 1978) and a basal avebrevicaudan bird (Fig. 10.6C). Still debated is the presence of the ectopterygoid process of the pterygoid. The process has the following characteristics: it is on the ventral edge of the main body of the pterygoid immediately anteroventral to the juncture between the main and quadrate sections of the pterygoid; the process is directed ventrally or ventrolaterally; and a lateral ridge progresses from the anterolateral corner of the quadrate process ventrally onto the posterolateral edge of the ectopterygoid process, where it forms the posterior border for the dorsolateral depression into which fits the ectopterygoid (Fig. 3.4Aa,Bb). The pterygoid process tentatively identified as a ventral articulation with the quadrate by Elzanowski and Wellnhofer (1996) lacks the characteristics expected in a propulsion joint, but it has all the listed characteristics of an ectopterygoid process and is identified as such here (Fig. 3.4Ac,Bc). The ectopterygoid process was smaller than in most, but not all, dino-avepods (Fig. 10.6B), but this is to be expected, considering the weak development of the ventral process of the ectopterygoid in *Archaeopteryx*. In addition, the down and forward slope of the posterior border of the process contrasts with the down and backward slope typical of avian quadrate-pterygoid propulsion joints.

The proceeding analysis is confirmed in JM 2257. The ratio between the anteroposterior breadth of the main body of the ectopterygoid and the lateromedial width of the entire ectopterygoid seems to be higher than observed in BSP 1999. The disparity is probably not actually due to differing proportions of the ectopterygoid. Instead, in JM 2257 the ectopterygoid and ectopterygoid process are fully articulated (Pl. 7), and the anteroposterior breadth probably consists of both the main body of the former bone and the posterior ridge of the latter process. In addition, note that the quadrate and the posterior edge of the ectopterygoid process do not contact each other in JM 2257. This is the normal archosaur condition and is further evidence against a propulsion joint (the distance between the quadrate and ectopterygoid process may have been variable; see discussion below on skull kinesis). The dorsal minor components of the *Archaeopteryx* pterygoid observed by Elzanowski and Wellnhofer (1996) are present in some dino-avepods (Fig. 3.4Ab,c,Bb,d). It is concluded that the pterygoid of *Archaeopteryx* is like that of an avetheropod, which is marked by a slender connection between the quadrate and main bodies, with some distinctive but minor autapomorphies.

A small bone articulates with the anterodorsal corner of the quadrate process of the pterygoid in JM 2257 (Pl. 7). This element is in the position and orientation expected for the epipterygoid, is similar in proportions and shape to the small epipterygoid of dromaeosaurs (Currie 1995), and is identified as such (Figs. II.2, 3.4Bc).

At this time we return to the window into the skull afforded by the antorbital fenestra of JM 2257 to view the anterior palate (Pl. 7). As usual, the identification of the bones has been disputed. A slender exposure of bone in the posteroventral corner of the fenestra is either the ventrally displaced anterior end of the right pterygoid, or the right posteromedial process of the palatine in correct position. The posteromedial process of the palatine is so long in BSP 1999 that it either closely approached or contacted the ectopterygoid. Flat bones in the anteroventral and central portion of the antorbital fenestra on the mainslab of JM 2257 are in the correct positions to be the right and left anteromaxillary rami of the palatines.

Of particular interest are two hooklike bones in the antorbital fenestra of the counterslab of JM 2257. These bones were tentatively identified as middle nasal conchae by Elzanowski and Wellnhofer (1996). However, the two bones are too robust in construction, and are set too far posteriorly, to be part of the delicate middle nasal conchae. In addition, it is logical for the strongly built palatines to be absent when delicate nasal conchae are present in the fossil. Indeed, the processes are roughly similar in size and shape to the hooklike anteromedial process of the palatine of BSP 1999 (but may be partly broken off in JM 2257), and they are very similar to each other in shape (suggesting they were originally laterally symmetrical structures rather than being from one side of the skull). Both are in the correct position to be continuations of the rest of the preserved palatines, and they are in the correct position to form the posterior border of the internal nares (which are usually visible in the anteroventral corner of the antorbital fenestra when the skull is viewed directly from the side, as discussed below). In this case, asymmetrical crushing of the skull has displaced the hooklike anteromedial processes of the palatines and the rest of the palatines dorsoventrally relative to one another. Identification of these elements as medial palatine elements contradicts the supposed presence of posterior maxillary sinuses in archaeopterygiforms (Witmer 1990). Because the anterior pterygoid was slender (Fig. 3.4Ac,Bc) and because the medial edge of the posterior palatine was medially concave (Fig. 10.6Be), there was a fenestra between the two elements (Fig. II.2).

In most reptiles and in birds, the roof of the mouth is flat. In many archosaurs, however, the palate was vaulted like the ceiling of a cathedral. Because *Archaeopteryx*'s palate was like that of a dino-avepod—in particular, the anteromedial process was *medially* longer than the posteromedial process—and because both palatines appear to be directed dorsally in JM 2257, the palate of *Archaeopteryx* is restored as vaulted (Fig. II.2). The greater height of the vaulted anteromedial process of the palatine also indicates that the palate was flexed in lateral view, at the juncture of the pterygoids and vomers. However, the anteromedial process was not as tall as it was in most dino-avepods, so the degree of flexion should have been less in the urvogel. The narrow snout probably did not allow room for an extensive palatine shelf on the maxilla; if so, this condition differs from that present in most birds. The articulations between the anterior palatal elements and the medial processes of the maxilla are further described below.

We now turn to some dino-avepods. A new restoration of the coelophysid palate based on articulated American and disarticulated African material (Raath 1977) is presented in Figure 10.6Ac,Bb. The figure shows the beginnings of a fenestra between the palatine and pterygoid (neither palatal restoration in Fig. 43 in Colbert [1989] is entirely accurate). Additional preparation of troodont skull AMNH 6516 (by the author and mainly by Carpenter) revealed details of this avepectoran's palate. The anterior section is basically theropodian in structure, although the size of the anteromedial process of the essentially triradiate palatine is not certain. The posterior half differs greatly from that of most dino-avepods. The posterior pterygoid is a continuous, posteriorly expanding, thin-walled tube, which implies that it was pneumatic (Figs. 3.4Be, 10.6Ag,Bg). The "ectopterygoid" process of the pterygoid is very reduced and no longer in contact with the ectopterygoid. The latter bone is set more posteriorly than is normal in theropods, and the ventral process is absent. The poorly preserved quadrate of AMNH 6516 appears to be thin walled, and it may have been pneumatic. A *Compsognathus* skull element identified as the squamosal and opisthotic by Bidar et al. (1972, Figs. 4 and 5) is actually a complete and well-preserved left quadrate.

One of the most exciting "discoveries" about *Archaeopteryx* in recent times was the realization that most of the braincase was preserved, in old BMNH 37001 no less (Figs. 10.4C, 10.5E). The original description by Whetstone (1983) was inadequate in many regards; A. Walker's (1985) study is superior. Almost the entire braincase of *Archaeopteryx* is known when the elements of BMNH 37001, JM 2257, and BSP 1999 are combined (Fig. 10.4C; Wellnhofer 1993, Elzanowski

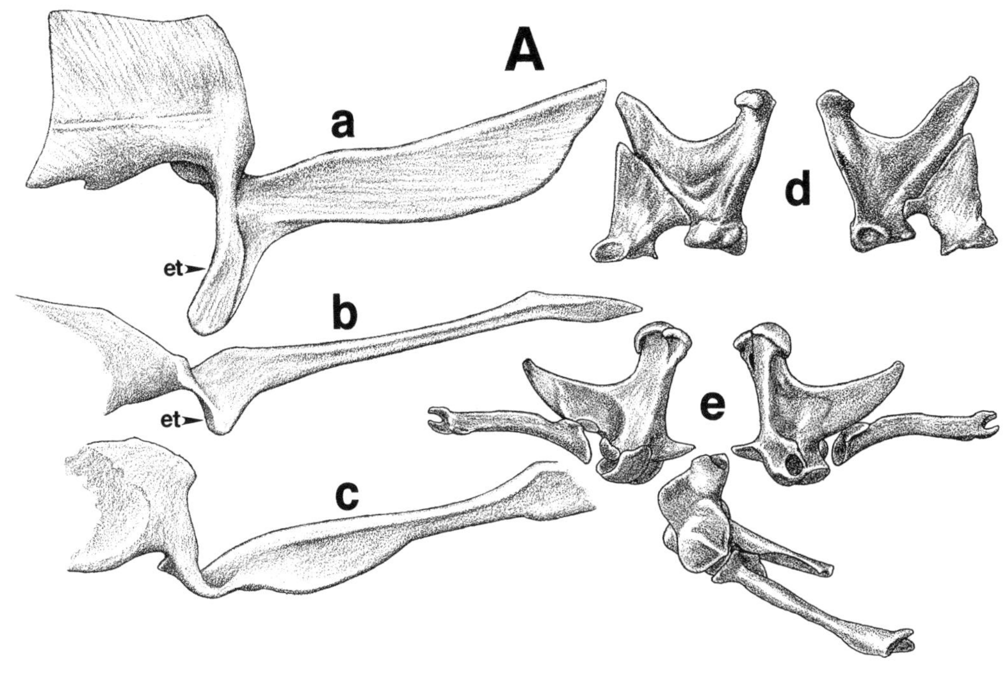


*Figure 3.4.* Archosaur posterior palates and kinetic systems. *A,* avetheropod right quadrates and pterygoids: avepod pterygoids without propulsion joints in side view: *a,* basal *Allosaurus* with large ectopterygoid process (et); *b,* dromaeosaur *Deinonychus* with no ectopterygoid process; *c, Archaeopteryx* BSP 1999 with small ectopterygoid process, posterior quadrate ramus restored after JM 2257. Note the lack of evidence for the propulsion joint seen in following bird quadrates and pterygoids with propulsion joints shown in multiple views: *d,* basal ornithothoracine *Hesperornis; e,* ornithurine loon —in the side and inner views, elements are disarticulated to better expose propulsion joint, and in the bottom view they are articulated. *B,* articulated right quadrates (q), pterygoids (pt), epipterygoids (ep), and ectopterygoids (ec) in side plan view (dashed lines indicate known extent of pterygoid medial to quadrate and epipterygoid): archosaurs with broad, potentially sliding fore-and-aft contact, profile of posterior quadrate ramus indicated by dashed outline: *a,* basal archosaur *Garjainia* (Otschev 1975); dino-avepods: *b, Syntarsus-Coelophysis* composite; dromaeo-avemorphs: *c, Archaeopteryx* BSP 1999 and JM 2257; *d,* composite dromaeosaur; avepectoran dinosaurs with modified palates: *e,* troodont *Saurornithoides* type—the poorly preserved quadrate is not included, and the epipterygoid may be absent; *f,* oviraptor *Ingenia*—the quadrate is fused to the pterygoid and braincase, and the position of the possible epipterygoid is tentative; *g,* loon with orbital process and propulsion joint. *C,* Cranial kinesis in avepods: *a,* opisthostylic dino-avepod in which simple-headed quadrate and quadratojugal can swing backward and retract the mandible, the descending process of squamosal prevents quadrate from rotating farther forward, and the vaulted and flexed palate is ill suited for kinetic action; *b,* streptostylic bird in which double-headed quadrate with orbital process can rotate forward, pushing the mandible and the palate in the same direction; the palate's straight configuration is well suited for lifting the beak, which rotates around a weak point in the roof of the skull.

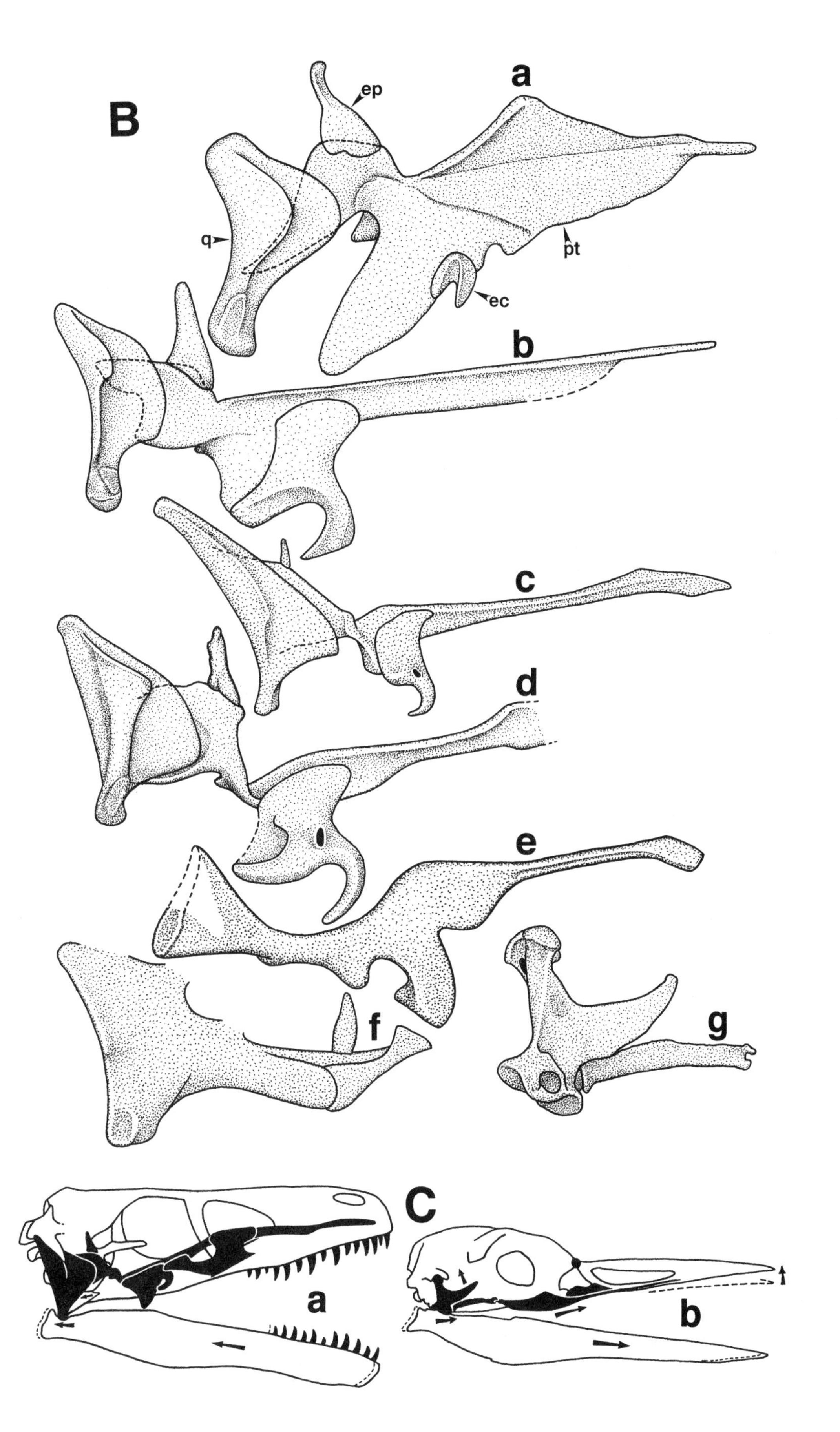
B
a
ep
q
pt
ec
b
c
d
e
f
g
C
a
b

and Wellnhofer 1996). There is no avian lateral otic depression that contains the fenestra ovalis and associated inner ear structures (A. Walker 1985, Currie 1995). The primary uncertainty is the orientation of the partly detached ventral elements relative to the rest of the braincase in BSP 1999. The ventral braincase appears to have been shallow. Because the BMNH specimen appears to have been moderately crushed, we do not know how strongly ventrally or laterally the paraoccipital process was oriented. The basipterygoid processes that connected the braincase to the palate are directed strongly down and forward, an orientation that is in accord with the similarly procumbent quadrate. The occiput that makes up the posterior face of the braincase tilts a little upward in BMNH 37001 instead of facing downward as in most birds, and the occipital condyle is directed straight backward in BSP 1999, as is typical of dinosaurs, rather than downward in the usual avian manner. The neck, therefore, approached the skull from behind rather than from below. The former neck orientation is found in a few birds, such as hesperornithiforms, loons (Fig. 10.4F), moa, and phorusrhacoids; and the latter orientation is present in ornithomimids and especially on troodonts in which the occiput is tilted downward (Fig. 10.4D).

Birds are noted for having cranial kinesis, the ability to flex the beak relative to the rest of the skull (for descriptions of this system, see Buhler [1981], Zusi [1984], and Chatterjee [1991, 1997]). This system is taken to an extreme in parrots: the mobility of parrots' bills relative to the rest of the skull is readily apparent to the casual observer. Did *Archaeopteryx* have such a flexible skull? One of the most important features of the avian arrangement is the propulsion joint (discussed above), by means of which the quadrates can push the pterygoid and the rest of the palate forward, thereby elevating the beak relative to the braincase (Fig. 3.4Cb). Modified articulation of the dorsal head of the quadrate and loss of the postorbital bar permit the quadrate to swing forward freely. In order to form a rod that pushes the beak up, the palate is flat in transverse section and flat from front to back. In the midskull roof, there is a flexible zone or a joint that allows the beak to rotate dorsally. The jaw is also pushed forward along with the quadrate.

As explained above, the only observed articulation between the quadrate and pterygoid in *Archaeopteryx* is a normal archosaurian fore-and-aft arrangement that does not allow the quadrate to propel the palate forward. Nor was the organism's vaulted, flexed palate well suited for pushing the beak forward. Because a descending process of the squamosal was present (contra Chatterjee 1997), the quadrate could not swing far forward anyway. No one has demonstrated that the postorbital bar was incomplete in *Archaeopteryx*, and the apparent presence of the epipterygoid implies that the palate and snout were able to move little, if at all, relative to the braincase. The ventral configuration of the anterior end of the frontals is not known, but the transversely bulbous dorsal contours appear ill suited for flexion. The presence of a lacrimomaxillary bar falsifies Buhler's (1985) restoration of a thin, flexible nasal above the antorbital fenestra. In addition, there is no clear evidence in HMN 1880 or JM 2257 that the elements in the midskull roof were loosely attached, forming a highly flexible bending zone. If any kinesis was present in *Archaeopteryx*, it was probably dino-avepodian, not avian, in function (Fig. 3.4Ca). The down and forward slope of the quadrate is a little less pronounced in JM 2257 than in HMN 1880 (compare Pls. 6 and 7), and the mandible is displaced a little more posteriorly in the former, which implies that a small degree of backward mobility (the opisthosyly described by Chatterjee [1991]) could shift the mandible in the same direction relative to the skull. Quadrate mobility may have been allowed by loose articulations between the quadrate-squamosal, quadrate-pterygoid, and quadratojugal-jugal (all these joints are often disarticulated from one another in JM 2257 and BSP 1999); and the descending process of the squamosal acted as a final stop to forward motion.

The absence of avian cranial kinesis in *Archaeopteryx* is supported by the even greater inflexibility of the skull of *Confuciusornis*, in which the postorbital bar was very robust, and a dorsal bending zone was not present (Chiappe et al. 1999). Oviraptorosaurs (Figs. 3.4Bf, 10.2M) and avimimids (Kurzanov 1987) had a straight palate, and the latter lacked a postorbital bar; but extreme fusion and bracing of palatal elements (to the point that the pterygoid appears to be continuous with the quadrate and has auxiliary articulations with the braincase) prevented kinesis. Interestingly, the quadrate and pterygoid of the basal bird *Patagopteryx* were also fused (Chiappe 1996). At the other extreme, the lack of a complete postorbital bar and the presence of a dorsal bending zone and a straight palate indicate that alvarezsaur skulls (Pl. 12, Fig. 10.2N) were fully kinetic in an avian manner (Chiappe et al. 1998). That alvarezsaurs and non-ornithurine birds lacked an orbital process shows that this anterior extension of the quadrate was not critical to basic avian cranial kinesis. Sanz et al. (1997) stated that the dorsal process of the jugal of a basal avebrevicaudan articulates with the quadrate. Such an unusual contact would prevent the quadratojugal-jugal complex from rotating upon the quadrate during normal kinetic action, but the supposed articulation may be an illusion caused by crushing of the skull.

We wrap up our survey of head parts with the lower jaw. Gingerich (1973) noted that Ostrom (1969) incorrectly restored the mandible of the dromaeosaur *Deinonychus,* and I (Paul 1988a,b) presented a corrected restoration (see Fig. 10.6Ae). Because the posteroventral portion of the splenial of *Archaeopteryx* projects strongly ventrally, this portion may have been visible at the bottom of the lower jaw in lateral view (Fig. II.2). A long slender section of bone in JM 2257 tentatively identified as a hyoid or the quadratojugal by Elzanowski and Wellnhofer (1996) is probably the dorsal rim of the surangular. There appears to be an anteroposteriorly long, dorsoventrally very narrow intramandibular fenestra preserved on the counterslab of JM 2257. Such a shallow profile is compatible with the slender proportions of the mandible.

An intramandibular joint was present in many theropods. This loose connection allowed the front half of the lower jaw, made up of the dentary-splenial unit, to rotate a few degrees relative to the posterior half. Sereno and Novas (1993, Figs. 2B and 3) suggested that an intramandibular joint was present in baso-theropods and baso-avepods, but long overlapping dorsal articulations between the dentary and surangular suggest that any mobility was limited (Fig. 10.2C–E). In dino-avetheropods the surangular-dentary contact was nearly straight and much looser than in more-basal theropods (Fig. 10.2F,J; Pls. 1, 8, and 9 in Madsen 1976); the nature of this articulation in *Archaeopteryx* is not yet known. The splenial and the anterior half of the mandible are free to rotate upon a joint between the angular and splenial in dino-avepods (Fig. 10.6Ae,f; Gingerich 1973). In theropods and *Archaeopteryx* (both mandibles of JM 2257), the long, slender, but strong anterior process of the angular lies in a long, subhorizontal groove of the splenial. The action of this joint has not been analyzed in detail, but vertical rotation of the dentary-splenial unit upon the posterior mandible appears to be favored. The latter possibility is indicated by differential flexion of the articulated mandibles in *Archaeopteryx* (compare HMN 1880 and JM 2257 to BSP 1999) and dromaeosaurs (compare right and left mandibles in Fig. 7 in Currie [1995]). In *Hesperornis,* the angular-splenial contact is shorter and more nearly vertical (Zusi and Warheit 1992), which suggests that rotation was primarily transverse.

The completed *Archaeopteryx* skull restoration is much more like the skull of a dinosaur than that of a modern bird in gross form and detailed osteology (Fig. II.2). This is true of both the superficial views and the palate. The biggest unknown is the status of the postorbital bar. The skulls of other basal birds were also strongly dinosaurian (Fig.10.2O,P).

CHAPTER 4

# Skeletons

Our examination of skeletons starts with the vertebral column. In all the *Archaeopteryx* skeletons, the articulated cervicodorsals at the base of the neck are crushed, degraded, or covered by forelimb bones. Therefore, L. Martin's (1983a,b, 1991) radical restoration of ribless cervicodorsals cannot be either verified or disproven, but such a condition would be extremely unusual for any tetrapod and is therefore implausible. Martin's suggestion that presacral vertebrae 12–14 are fused is more plausible but, again, cannot be verified with the rather poorly preserved vertebrae in the specimens on hand.

Ruben (1991) was mistaken in asserting that the urvogel lacked evidence of pneumatic postcranial elements. The neck vertebrae and ribs may indeed have been pneumatic (Fig. 1B in Wellnhofer 1974, Britt 1995, 1997, Britt et al. 1998, Christiansen and Bonde 2000). The anterior (but not posterior) trunk vertebrae, as well as the ribs, definitely had pneumatic features. It is interesting that De Beer (1954) noted the presence of a laterally excavated vertebra in BMNH 37001, but he dismissed the presence of pneumatic elements because he focused on the limb elements. In contrast to De Beer, Buhler (1992) observed that the question of whether the thin-walled long bones of *Archaeopteryx* were pneumatic remains open, although their pneumaticity is improbable considering their absence in other avepectorans or baso-avebrevicaudans.

So far there is no evidence that the hollow limb elements of any predatory dinosaur were pneumatic. Buhler (1992), Britt (1995, 1997), and Reid (1996, 1997) have emphasized that the excavations common in most avepod dinosaurs' vertebrae and ribs are of the type created by invasion of the bones by pulmonary diverticula that extended from the lungs. Baso-theropods (Sereno et al. 1993, Novas 1994) and alvarezsaurs (Perle et al. 1994, Karhu and Rautian 1996, Novas 1997) lacked pneumatic vertebrae. The extent of postcranial aeration in avepods is outlined by McLelland (1989b), Russell and Dong (1993b), and Britt (1997).

A little-noticed feature of the cervical centra of dromaeosaurs (Fig. 10.7Bc) and ornithomimosaurs (Fig. 3c,d in Pl. LII in Osmólska et al. 1972) is their incipient saddle shape, in which a transversely broad anterior surface articulates with a posterior surface that is taller than it is broad. This condition differs from the more nearly circular, and less flexible, articulations typical of most archosaurs. The condition of the cervical centra in *Archaeopteryx* is not known but was probably nonavian in that *Archaeopteryx*—along with the majority of saurischian dinosaurs—retained the common diapsid condition of having slender, elongated cervical ribs that extensively overlapped one another. The functional implications of such extensive rib overlap are uncertain because this type of overlap is not present in any living animal. Crocodilians have a modest overlap of thick, inflexible ribs that are tightly bound to one another so that they contribute to cervical rigidity (L. D. Martin et al. 1998). But these short, massive-necked forms may not be appropriate models for all diapsids with cervical rib overlap. It is especially implausible that the long, often S-curved necks of theropodian and other saurischian dinosaurs as inflexible as crocodilian necks (contra J. Martin et al. 1998). The cervical ribs of saurischians were so slender that they should have been flexible, and they probably had a sliding articulation with one another instead of being tightly bound together. The large articular surfaces of dino-theropod zygapophyses also favor substantial flexibility. Loss of overlapping ribs in the necks of avepectoran dinosaurs and birds worked together with saddle-shaped centra articulations to further enhance neck flexibility. Whether hypapophyses were present under the cervicodorsals of *Archaeopteryx* (Norell and Makovicky 1999) is unclear; if they were present, they were not well developed.

Zhou et al. (2000) counted 12 neck vertebrae and 9 dorsal vertebrae in *Caudipteryx;* the presence of a fairly long rib on 12 suggests that the correct count should have been 11 and 10, respectively. Barsbold (1983) counted 5 sacrals in dromaeosaurs, but there are actually 6 functional sacrals in *Deinonychus* MCZ 4371. The famous "fighting" *Velociraptor* MIG 100/25 has 8 vertebrae between the ilia, two of which appear to be nonsacrals; and 6 and 5 functional sacrals are reported in adult and juvenile velociraptorines, respectively, by Norell and Makovicky (1997, 1999). Such high sacral counts mean that the functional dorsal count in adult dromaeosaurs should be 12, rather than the 13 typical of most dinoavepods. In *Protarchaeopteryx,* approximately 17 caudals are preserved. A gap at the base of the tail suggests that 3 vertebrae are missing, and distal impressions suggest an additional 3, for a total of approximately 23. In dromaeosaurs, the articular surfaces of the proximal caudal zygapophyses are vertically oriented (Fig. 38 in Ostrom 1969), and the same appears to be true in *Archaeopteryx* (Fig. 7A in Wellnhofer 1974). This adaptation allowed the tail to flex 90 degrees dorsally immediately behind the sacrum; this flexion is observed in some fully articulated tail bases, including those of *Velociraptor* MIG 100/25 (p. 13 in Tomida and Sato 1995) and *Archaeopteryx* S6 (Pl. 4). In the latter, the flexion does not appear to represent strong postmortem dislocation (contra Kemp and Unwin 1997). However, for reasons detailed below, strong caudal dorsoflexion was probably not a habitual tail posture (contra restorations by L. Martin [1995] and Hou et al. [1996]). Too many proximal caudals are missing in *Protarchaeopteryx* for us to determine whether its tail base was hyperflexible. The pelvic, sacral, and proximal caudal elements of *Sinosauropteryx* specimens with strongly dorsoflexed tails appear to be partly disarticulated (Morell 1997, Chen et al. 1998). The same appears to be true of at least some *Caudipteryx* skeletons (Ji et al. 1998), but the situation with other *Caudipteryx* examples is less clear (Zhou et al. 2000). The distal tails of theropod dinosaurs were stiffened by elongation of the prezygapophyses, so that they overlapped the preceding vertebrae; but the tail was usually not a rigid rod, as shown by the naturally gently curled tail of observed in a complete troodont skeleton (Pl. 13A). Ji et al. (1998) and Barsbold et al. (2000) stated that the tail of *Caudipteryx* was distally rigid, but Zhou et al. (2000) disagreed. All of the caudals of *Caudipteryx* are unfused (Ji et al. 1998, Barsbold et al. 2000, Zhou et al. 2000), so it did not have the pygostyle claimed by Feduccia (1999). An oviraptorid pygostyle does not show evidence of being pathological, according to Barsbold et al. (2000). *Confuciusornis* lacked the long bony tail initially attributed to it by some (Chatterjee 1997).

Avepod gastralia articulate along the midline of the belly via a distinctive zigzag interlocking pattern (Fig. 10.7E; Claessens and Perry 1998). In the *Archaeopteryx* specimens, the usual lateral crushing of the gastralial series obscures the nature of the midline articulation, but a few gastralia in BSP 1999 appear to meet one another in the avepod pattern (Fig. 9 in Wellnhofer 1993). The same, albeit reduced, gastralial system is present in confuciusornithids (Fig. 10.7E; L. Martin et al. 1998, Chiappe et al. 1999, Ji, Chiappe, and Ji 1999). Without detailing their reasoning, Martin et al. (1998) questioned the presence of gastralia in *Sinornis,* in spite of the evidence presented by Sereno and Rao (1992) indicating that these slender bones do not articulate with the ribs, that they appear too numerous and posterior to be sternal ribs, and that they are arranged in the manner expected of gastralia.

The anterior and middle dorsal ribs are swept backward in all archosaurs (Paul 1987a, 1988a), and this same arrangement is seen in the *Archaeopteryx* specimens. The preserved anterior (chest) ribs in the articulated urvogel trunks indicate that these ribs were shorter than the mid-dorsal ribs, a condition shared by tetanurans and especially by avetheropods, including birds (Fig. 10.1Be–v, App. Fig. 7D–H; Paul 1988a).

Many years ago, I (Paul 1984a, 1988a,b) noted that the fighting *Velociraptor* has four well-developed ossified uncinate processes, small hook-shaped bones attached to the trunk ribs and common to almost all birds (Fig. 4.1A,C, App. Fig. 7F–H). Similarly curved elements from another dromaeosaur, *Deinonychus,* were identified as sternal ribs by Ostrom (1969). Some of these elements appear to have been transversely flattened, and others are more robust; the flattened elements may represent uncinate processes (Fig. 4.1C top). Ostrom (1969) correctly identified straighter elements as ossified sternal ribs in *Deinonychus.* The doubled sternal ribs tentatively restored by Ostrom in 1969 (Fig. 4.1E left) and copied by Ruben et al. (1997b) were based on disarticulated elements. The doubling of the ribs is falsified by the presence of single sternal ribs immediately posterolateral to the sternal plates in the fighting *Velociraptor* (Fig. 4.1A) and other velociraptorine specimens (Barsbold 1983, Norell and Makovicky 1999), in which there appear to have been three ribs on each side. The sternal ribs of the basal

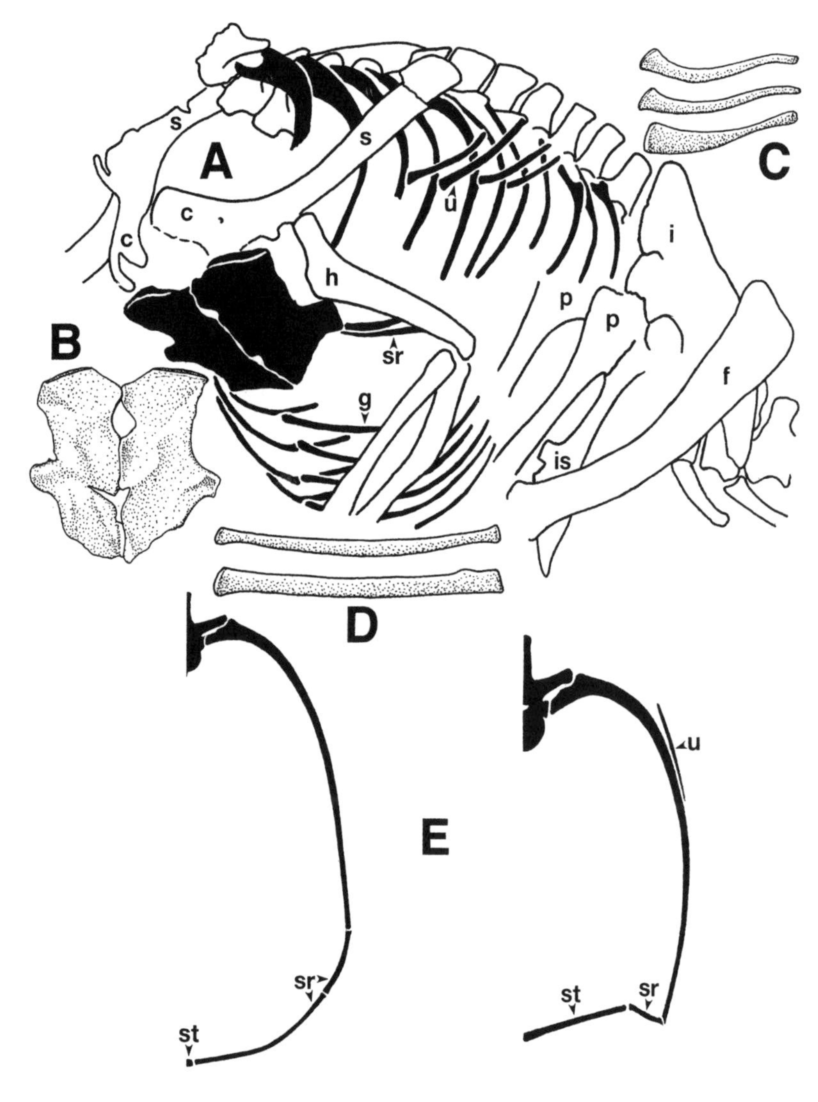


*Figure 4.1. A,* The thorax as preserved of the "fighting" *Velociraptor* MIG 100/25 in oblique anteroventrolateral view (based on Tomida and Sato 1995, 13). Visible are large ossified sternal plates articulated with ossified sternal ribs, as well as a series of ossified uncinate processes. The scapulocoracoids are no longer in articulation with the coracoidial grooves on the anterior edges of the sterna, and the furcula is missing. Ossified thoracic items of dromaeosaurs: *B,* sternum of MIG 100/25. *C,* left uncinate processes: *top, Deinonychus; center,* MIG 100/25; *bottom,* turkey. *D,* sternal ribs: *top, Deinonychus; bottom,* pigeon. *E,* transverse sections of articulated dromaeosaur chests: *left,* obsolete restoration after Ostrom (1969) with double sternal ribs and narrow sternum, at 6th dorsal and rib; *right,* updated version with ossified uncinate, single sternal rib, and broad sternal plate, at fifth segment (see Fig. 10.1Bn). Abbreviations: c = coracoid, f = femur, g = gastralia, h = humerus, i = ilium, is = ischium, p = pubis, s = scapula, sr = sternal rib, st = sternum, and u = uncinate process. Not drawn to same scale.

dromaeosaur *Sinornithosaurus* are single and ossified, and Xu, Wang, and Wu (1999) observed possibly five hinge-type articulations for these elements. Ossified uncinates and sternal ribs seem to be present in *Microraptor* (Xu et al. 2000). Burnham et al. (2000) reported at least four such articulations in the dromaeosaur *Bambiraptor.* The sternal articulations of avepectoran sternal ribs also possess the hinge joint configuration observed in birds, which means that avian-style sternocostal joints were present. Articulated oviraptorid skeletons include ossified uncinate processes and sternal ribs that attach to dorsal ribs 2 through 4 (Pl. 15; Clark et al. 1999). Ossified sternal ribs articulating with a lateral edge of the sternum are definitely present in *Caudipteryx* (Ji et al. 1998). At least one uncinate process appears to be present in the type specimen, and up to six of the processes are present in other skeletons (Zhou and Wang 2000, Zhou et al. 2000). Because the only articulated troodont specimen seems to have been a juvenile (Russell and Dong 1993a), we do not know whether the lack of these ossified elements is due to the age of the individual or whether they are truly absent. Tyrannosaurs may have had partly ossified sternal ribs, according to Lambe (1917), who, along with Currie and Zhao (1993a), observed that only two sternal ribs appear to have been attached to the sternum in nonavepectoran avepods. The absence of ossified uncinates and sternal ribs in any *Archaeopteryx* specimen indicates this was its true condition. The number of costal attachments to the ossified sternum of *Archaeopteryx* is not known, but the shortness of the bone suggests there were only two or, at most, three. Four uncinates were ossified in *Confuciusornis,* and five sternal ribs were ossified in the latter bird and in *Iberomesornis* (Sanz and Bonaparte 1992, Chiappe et al. 1999). It is possible that one or more cartilaginous sternal ribs that did not contact the sternum were present more posteriorly than the ossified examples in these avepods; an ossified example of such an element is present in an oviraptorid (Clark et al. 1999).

The next area for examination is the shoulder—or pectoral—girdle and the forelimb. We will start with the sternal plates (Figs. 4.2A, 10.9). We now know that *Archaeopteryx* had a short but broad sternal plate (Wellnhofer 1993), and *Protarchaeopteryx* possessed a very similar sternum (Fig. 10.9D,E). In none of the *Archaeopteryx* specimens do the anterior-most gastralia closely approach the coracoids, so a cartilaginous extension of the sternum probably doubled the overall length of the plate. A similar space between the coracoids and anterior-most gastralia suggests that a fairly large, partly or entirely cartilaginous sternum was present in many other small dino-avepods (e.g., *Coelophysis:* Fig. 102 in Colbert 1989, Downs pers. comm.; *Compsognathus:* Fig. 7 in Bidar et al. 1972, Fig. 1 in Ostrom 1978; *Sinosauropteryx:* Pls. 13 and 14, App. Fig. 6Aa; *Scipionyx:* App. Fig. 6Ab; *Struthiomimus:* Pl. XXIV in Osborn 1916). A similar coracoid-gastralia gap shows that a broad unossified plate was present in troodonts (Pl. 13A,B); the probable juvenile status of the one complete specimen again prevents us from determining whether the sterna of adult troodonts remained unossified. Anterior ossification of an otherwise cartilaginous sternal complex is observed in ratites and juvenile birds (App. Fig. 6Fa,c,e). Central ossification of what was probably a large sternal plate has been suggested for *Eoalulavis,* a basal bird (Fig. 10.9J; Sanz et al. 1996). There is no gap between the posterior edge of the ossified sternum and the gastralia in dromaeosaurs, caudipterygians, and oviraptors (Figs. 4.1A, 4.2Ae,f). The sterna were well ossified and large in dromaeosaurs, caudipterygians, oviraptorosaurs, basal ornithomimosaurs, and alvarezsaurs (Figs. 4.2Ae,f, 10.9F,G; Barsbold 1983, Perez-Moreno et al. 1994, Perle et al. 1994, Norell et al. 1997, Norell and Makovicky 1997, 1999, Ji et al. 1998, Clark et al. 1999, Xu, Wang, and Wu 1999, Burnham et al. 2000). Dino-avepod sterna are almost always found in the more-complete and more-articulated skeletons, and their absence to date in the dromaeosaur *Deinonychus* is therefore attributable to the disarticulation and incompleteness of the remains. The restoration of a dromaeosaur with a slender sternum in Ruben et al. (1997b) was based on out-of-date data from disarticulated, incomplete remains (Fig. 4.1E). Moderately large, less birdlike ossified sterna are known in *Xuanhanosaurus* (Dong 1984) and *Baryonyx* (Charig and Milner 1997), and smaller ossified sterna have been identified in some other large dino-avepods (*Gorgosaurus,* Lambe 1917; *Carnotaurus,* Bonaparte et al. 1990; *Sinraptor,* Currie and Zhao 1993a). However, Brochu (pers. comm.) has suggested that the gorgosaur's element is actually the central section of the fused anterior gastralia seen in other tyrannosaurs. Less probable is that the fused gastralia are sternal elements with fused ossified sternal ribs. Among these brevisterna avepods, the gastralia start immediately behind the sterna, at least in tyrannosaurs (Lambe 1917). Co-fusion of the paired sternal plates occurred very irregularly in avepod dinosaurs and archaeopterygiforms (Lambe 1917, Barsbold 1983, Bonaparte et al. 1990, Currie and Zhao 1993a, Charig and Milner 1997, Norell and Makovicky 1997, 1999, Burnham et al. 2000) and may have been in part related to ontogeny. Dino-avepod sterna were largely flat plates. There are some exceptions, however, including the keels consistently observed in other fully ossified dino-avepod sterna, the deep keel in alvarezsaurs, and the shallow keels in other sterna. The keels represent ossifications of the central sternal component (described in bird embryos by Fell [1939]). It is therefore possible that a cartilaginous keel was present on incompletely ossified dino-avepod sterna, including those of archaeopterygiforms. These keels were probably shallow, although a deeper one cannot be ruled out in *Archaeopteryx.*

In some articulated avepod specimens, the scapulocoracoids still sit close to their original positions. Among these specimens are many Ghost Ranch *Coelophysis* remains (Figs. 19, 70, and 99 in Colbert 1989), *Compsognathus* (Pl. 19; Bidar et al. 1972), *Struthiomimus* (Pl. XXIV in Osborn 1916), *Sinornithoides* (Pl. 13B), and *Gorgosaurus* (Fig. 7 in Lambe 1917). In such well-articulated specimens, the pectoral girdle is set rather posteriorly on the rib cage, with most of the scapula blade overlapping the chest ribs (Paul 1988a). This posterior position is in accord with and confirmed by neural anatomy (Giffin 1995) and is the usual avian condition (Fig. 10.1Bd–v). In specimens in which the pectoral girdle complex is not properly articulated, the dislocation may have been caused by bloating of the carcass (Paul 1997). Bloating appears to have occurred in all of the complete *Archaeopteryx* specimens (Kemp and Unwin 1997), in which the shoulder girdles appear to be displaced, usually too ventrally. The original position of the scapula on the rib cage is therefore not preserved in the urvogel.

As part of his radical reconstruction of the shoulder region of *Archaeopteryx,* L. Martin (1983a,b, 1991, 1995) showed the acromion process of the scapula articulating with the supposedly ribless cervicodorsal vertebrae, and the furcula articulated solely with the coracoids (whether or not Martin retained the first part of this arrangement in his 1995 study is not entirely clear). The same arrangement appears to be restored in *Confuciusornis* by Hou et al. (1996). No direct articulation between the scapula and vertebrae is observed in any *Archaeopteryx* specimen, or in any tetrapod specimens for that matter, and such articulation must therefore be rejected. The acromion processes were therefore far enough apart to articulate with the proximal ends of the

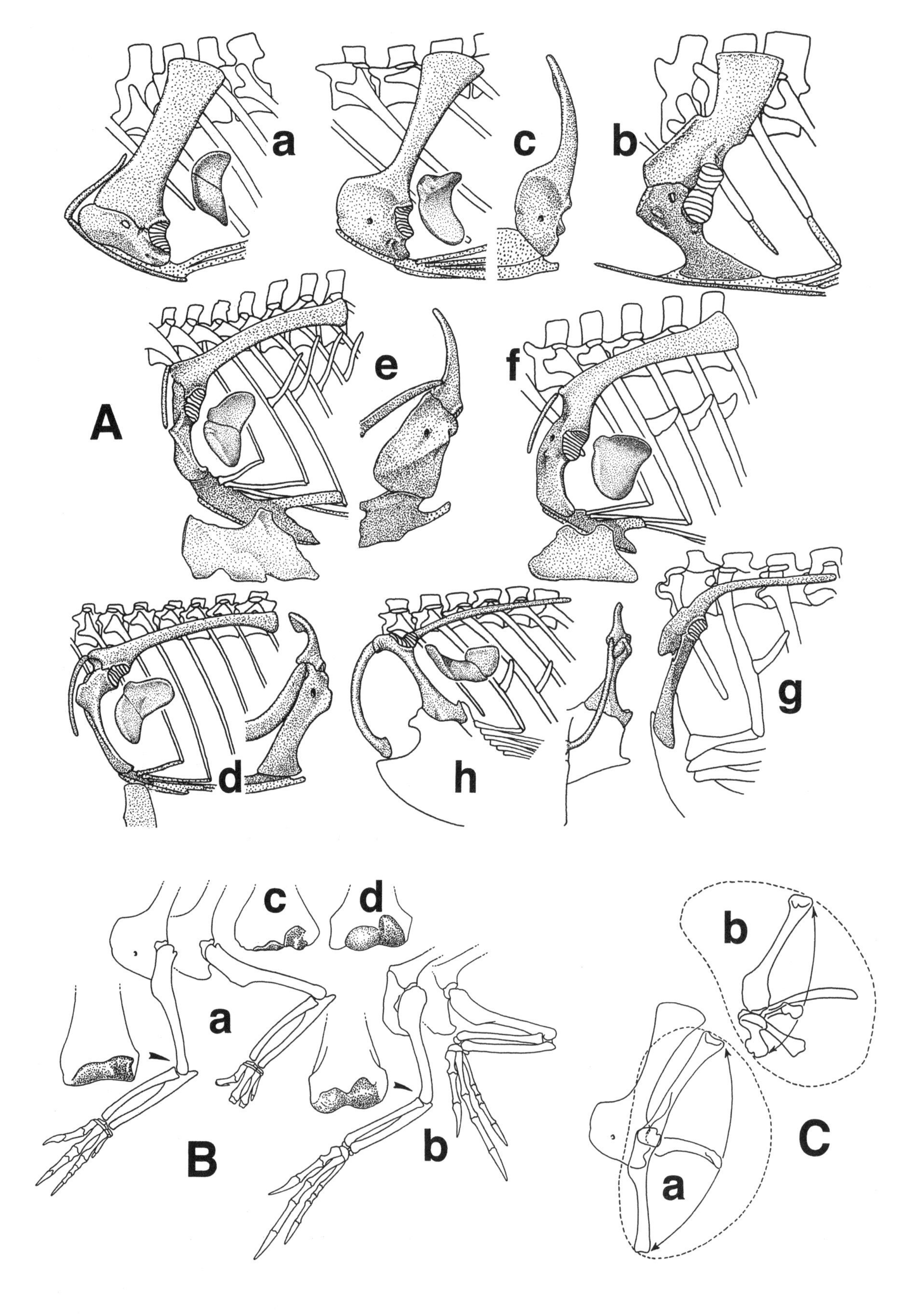
a
c
b
A
e
f
d
h
g
c
d
a
B
b
b
a
C

furcula, as they did in a number of dino-avepods (Pl. 13B; Barsbold 1983, Russell and Dong 1993a, Chure and Madsen 1996, Norell et al. 1997, 1998, Makovicky and Currie 1998, Sasso and Signore 1998, Burnham and Zhou 1999, Norell and Makovicky 1999, Xu, Wang, and Wu 1999) and birds. Feduccia and Martin (1998) incorrectly thought that a dino-avepod furcula articulated with the entire anterior edge of the coracoid (Norell et al. 1998, Norell and Makovicky 1999). In the type specimen of *Protarchaeopteryx,* either a clavicle splint or, more probably, half of a furcula (the majority of which is hidden) articulates with a scapulocoracoid. In the type specimen of *Sinornithoides,* we do not know whether a complete furcula was absent, was not ossified in the juvenile, or was partly lost during preparation.

*Figure 4.2. (opposite)* Shoulder girdles and forelimb action in archosaurs. *A,* articulated pectoral girdles in side and front views with detail of shoulder glenoids *(insets),* and ossified sternal plates in bottom view; heavy stippling indicates cartilage, and *c–f* side views include a restored shallow sternal keel: conventional arrangement in which outer surface of vertical scapulocoracoid faces mainly laterally: *a,* basal archosaur *Euparkeria; b,* modern crocodilian; *c,* baso-avepod *Syntarsus;* avepectorans with birdlike arrangement including horizontal scapula articulating with sharply flexed, vertical coracoid that faces anteriorly and articulates with anterior edge of broad sternal plate: *d, Archaeopteryx; e,* composite dromaeosaur; *f, Oviraptor; g,* flightless rhea; *h,* flying *Apatornis. B,* arm folding in avepods: *a,* baso-avepod *Syntarsus* with normal tetrapod arm action due to little modified, symmetrical humeral articulations *(inset)* and complex wrist; *b,* avepectoran dromaeosaur with avian action due to birdlike, asymmetrical humeral articulations bordered by enlarged distal flanges *(inset)* and simplified, pulley-action wrist; distal humeri in anterior view: *c,* herrerasaur, showing asymmetrical distal humeral condyles and distal flanges that imply that a push-pull folding mechanism may have been partly developed; *d,* tinamou, showing strongly asymmetrical condyles and well-developed distal flanges. *C,* range of possible motion of humerus restored via dry bone manipulation of humerus in shoulder glenoid: *a, Syntarsus* in which laterally facing glenoid *(inset)* does not allow humerus to elevate to vertical but does allow it to describe a flapping arc; *b,* tinamou in which dorsolaterally oriented shoulder glenoid *(inset)* allows humerus to elevate to vertical but prevents humerus from depressing much below horizontal. Not drawn to same scale.

The pectoral girdle of *Archaeopteryx* was placed too low by Yalden (1984). In archaeopterygiforms, the correct position was essentially avian (Ostrom 1976a,b, Paul 1988a, Jenkins 1993, Wellnhofer 1993). The coracoids were sharply flexed relative to the scapula in *Archaeopteryx,* to the point that they were somewhat retroverted (Fig. 10.8L), and the same appears to be true of a poorly preserved scapulocoracoid in the type specimen of *Protarchaeopteryx.* The scapula blade was horizontal and high on the rib cage, with its superficial surface facing dorsolaterally (Figs. II.1, 4.2Ad). As a result the coracoids were vertical, their outer faces faced mainly forward, and their distal ends articulated with the anterior edge of a broad sternal plate via grooves set at a right angle to the body midline. This condition contrasts with that of most archosaurs, in which the scapula was more vertical, the scapula and coracoid were not angled sharply relative to each other, and the superficial surfaces of the scapula and coracoid faced mainly sideways (Figs. 4.2Aa,b,c, 10.1A,Ba–l).

Most avepod dinosaurs also appear to exhibit the latter arrangement, as shown by articulated specimens (Osborn 1916, Lambe 1917, Ostrom 1978, Paul 1988a, Colbert 1989). In particular, restorations that show nonavepectoran avepods with horizontal scapula blades are erroneous (Paul 1987a, 1988a). It is not entirely clear how the coracoids articulated with the small sternal elements sometimes found in these forms. Shallow grooves on the anterior edges of the sterna were identified as the articulations with the posteroventral corner of the coracoid by Makovicky and Currie (1998), and this identification may be correct for tyrannosaurs. However, in *Sinraptor* the grooved edges face slightly medially, at an implausible angle for direct articulation with the coracoids (Fig. 10.9A). It is therefore probable that cartilage filled much of the space between the anterior edge of the sternal plate and the coracoids.

The great majority of researchers have assumed that the general tetrapod pectoral arrangement also applied to the scapulocoracoids of dromaeosaurs, oviraptorosaurs, and troodonts (Ostrom 1974b, Gauthier and Padian 1985, Jenkins 1993, Russell and Dong 1993a, Novas and Puerta 1997, Dingus and Rowe 1998, Forster et al. 1998; the pectoral elements of basal ornithomimosaurs are not yet described well enough to restore). This assumption is a major error—one that has obscured the phylogenetic position and flight heritage of these avepectoran dinosaurs—because they actually possessed fully avian shoulder girdles, as

I was the first to explain and illustrate (Paul 1987a, 1988a). Articulated or co-ossified dromaeosaur, caudipterygian, oviraptor, and troodont scapulocoracoids show that the two elements were strongly flexed and twisted relative to each other (Pls. 13A,B, 23; Figs. 4.2Ae,f, 10.8N; Fig. 13C in Ostrom 1976a, Fig. 16 in Barsbold 1983, Norell and Makovicky 1999, Xu, Wang, and Wu 1999, Burnham et al. 2000, Zhou et al. 2000). The degree of scapulocoracoid flexion in these four groups is usually less than in *Archaeopteryx* and flying birds—the dromaeosaur *Sinornithosaurus* being a notable exception (Fig. 10.8M)—but is similar to that observed in some flightless birds. While we are on this issue, it is interesting to note that the main point of flexion is in the scapula in most avepectoran dinosaurs and some flightless birds; in the scapulocoracoid juncture in the dromaeosaur *Sinornithosaurus* as well as in flying birds and some nonvolant birds; and in the coracoid in *Archaeopteryx* as well as in *Confuciusornis* (Chiappe et al. 1999). The majority of the distal edge of the dino-avepectoran coracoid articulated with a long transverse groove on the anterior edge of the broad sternal plates, so the outer surfaces of the coracoids faced mainly forward (Pls. 13A,B, 23; Figs. 4.1A,B, 4.2Ae,f, 10.9F,G; Fig. 1 in Norell et al. 1997, Fig. 4 in Norell and Makovicky 1997, 1999, Barsbold 1983, Xu, Wang, and Wu 1999, Burnham et al. 2000). Although cartilage probably separated the coracoid from the sternum in avepectorans, as it does in birds (Norell and Makovicky 1997), the cartilage probably was not thick enough to greatly modify the large angle of the articulation between the two bones relative to the body midline. This avian type of articulation is, of course, the defining character of the Avepectora, and it is interesting that the 45–90-degree angle between the scapula-sternum articulation and the body midline seen in dino-avepectorans and some early birds is higher than is usual in modern birds (Fig. 10.9D–O).

In the just described system, the sternum can rise and fall, most especially at its posterior end, rotating upon the transversely broad hinge joint with the coracoids. Because the long coracoid was vertical, the anterior end of the scapula must have been placed high on the chest; therefore, the scapula blade was set high and horizontal in side view on the rib cage (Figs. 4.2Ae,f, 10.1Bn–p; Paul 1987a, 1988a, Norell and Makovicky 1999, Burnham et al. 2000). The nature of the twisting between the scapula and coracoid in these dino-avepectorans is so strong that placing the elements on the rib cage in the conventional dino-theropod manner is impossible, and skeletal restorations should reflect this fact. The birdlike condition appears to apply to the new avepectorans, *Rahonavis* (Forster et al. 1998) and *Unenlagia* (Novas and Puerta 1997). However, Novas and Puerta oriented the transverse plane of the scapula too flat and horizontal, so that the superficial surface of the blade faced straight up. In avepectorans avian and nonavian, the plane of the scapula blade was and is tilted 30 to 50 degrees laterally.

The just described arrangement of avepectoran shoulder girdles has a crucial influence upon the orientation of the shoulder joint, or glenoid. Many workers have assumed that the glenoid faced down and backward in predatory dinosaurs. On the one hand, an avepectoran scapulocoracoid in this conventional position inappropriately directs the glenoid more nearly downward than it really was (*Archaeopteryx,* Bakker and Galton 1974; *Deinonychus,* Jenkins 1993, Novas and Puerta 1997; and *Caudipteryx,* Sereno 1999a). On the other hand, making the blade of the scapula transversely too horizontal rotates the glenoid more upward than it actually was (*Unenlagia,* Novas and Puerta 1997). To put it another way, the orientation of the glenoid relative to the plane of the scapula blade was broadly similar in the various avepectorans, including *Archaeopteryx* and *Unenlagia* (Fig. 10.8J–P). When the scapula of avepectorans is placed correctly and consistently on the rib cage, the glenoid is oriented mainly laterally (Pls. 13A,B, 23, Fig. 4.2Ad–f,h; Fig. 5 in Jenkins 1993, Paul 1988a, Norell and Makovicky 1999, Xu, Wang, and Wu 1999, Burnham et al. 2000). The claim by Burnham et al. that *Bambiraptor* had the most laterally directed dromaeosaur glenoid has not been documented. The lateral orientation of the glenoid in a naturally articulated troodont is so strong that the humeri sprawl strongly to the sides (Pl. 13A,B). This condition contrasts with that in herrerasaurs (Fig. 10.8F), some large dino-avepods (Fig. 10.8H), some flightless birds, and quadrupedal dinosaurs, in which the glenoid faced more downward and backward (Paul 1987a). However, some avepod dinosaurs with conventional shoulder girdles also had a laterally oriented glenoid (Fig. 10.8F,I; Paul 1988a). Therefore, many dino-avepods, including *Archaeopteryx,* could rotate the humerus through an extensive lateral arc (Fig. 4.2C), an arc sufficient to produce the incipient "flapping" motion I described (Paul 1988a), Jenkins (1993), Novas and Puerta (1997), and Xu, Wang, and Wu (1999). However, the humerus could probably not be elevated to vertical, as it can in flying birds; this limitation applies to *Archaeopteryx* (Paul 1988a, Jenkins 1993) and *Unenlagia* (contra Novas and Puerta 1997) and probably to confuciusornithids as well. Alvarezsaurs are interesting in that, although the shoulder glenoid faced only moderately laterally, the humerus head was so strongly angled that the humerus was permanently sprawling (a very large medioproximal projection also prevented the humerus from approaching the side of the body).

In most dino-avepods, the scapulocoracoid was immobile, locked in place by tight articulations with the sternum or a furcula or both (Paul 1987a, 1988a, Makovicky and Currie 1998, Norell and Makovicky 1999). Ornithomimids may have had more-mobile shoulder girdles (Nicholls and Russell 1985, Paul 1988a).

An anteromedial indentation of the coracoid at or near its juncture with the scapula was present in archaeopterygiforms, caudipterygians, troodonts, and basal dromaeosaurs (Pl. 13B, Fig. 10.8J,L,M) and was especially large in *Bambiraptor* (Fig. 1.7; Burnham et al. 2000). This feature was not present in derived dromaeosaurs (Fig. 10.8N; Brinkman et al. 1998), and its status in oviraptorosaurs has not yet been described. The function of this indentation is little studied, but it probably represents the beginning stage of the extreme reduction of the anterior half of the coracoid observed in birds. Ostrom et al. (1999) concluded that basal avebrevicaudans lacked a well-developed supracoracoideus wing-elevator/wing-control mechanism (see Chapter 7). Yet the coracoid of *Iberomesornis* (Sanz and Bonaparte 1992) appears to have had the well-formed, anterodorsally projecting acrocoracoid that is key to the presence of this system. Chiappe et al. (1999) noted that *Confuciusornis* retained a broad, square-tipped scapula blade rather than the pointed tip seen in more-derived birds.

Burnham et al. (2000) claimed that a juvenile *Bambiraptor* had the longest arm among known dinosaurs, but with a forelimb/hindlimb length ratio of about 0.7, it appears to fall below the greater than 0.8 ratio observed in *Sinornithosaurus* (Figs. 1.6, 1.7, App. Fig. 18C; Xu, Wang, and Wu 1999), chiefly because the latter had the longer humerus.

In the wrist, both the radial and ulnare are present in JM 2257, so the latter element is not a crystal, as argued by Tarsitano and Hecht (1980). Avepectorans are generally thought to have two ossified carpals aside from the semilunate block, and this condition appears to exist in the majority of *Archaeopteryx* wrists. L. Martin et al. (1998) seemed to conclude the same, but in the right wrist of JM 2257, Wellnhofer (1974), L. Martin (1991), Feduccia (1996), and Zhou and Martin (1999) figured a total of three elements proximal and lateral to the block. The Martin and Zhou restorations differ in detail from Wellnhofer's map of the specimen, and the two lateral elements may actually be parts of the same element, with the connection being partly obscured by sediment (a demonstration of the need for high-resolution remote scanning of these elements). That there is only one lateral element is probable, considering that only two carpals aside from the semilunate block have been identified in the confuciusornithid specimens (L. Martin et al. 1998, Chiappe et al. 1999) and that adult modern birds share this condition. On the other hand, Xu et al. (2000) tentatively identified two elements lateral to the main block in *Microraptor,* for a possible total of four carpals.

In his *Deinonychus* study, Ostrom (1969) concluded that the articular surface for the medial proximal carpal continued from the semilunate block onto the medioproximal flange of metacarpal I. The proposed wrist joint on metacarpal I is thin, but a smooth and distinct articular surface appears to be present. *If* so, then dromaeosaurs had the ability to rotate the hand strongly anteromedially, an ability not present in modern birds. A smooth-edged medioproximal flange of metacarpal I also appears to have been present in archaeopterygiforms (Fig. 4.3), but the presence of such anatomy and function in dino-avepectorans or basal birds has yet to be confirmed.

Dromaeo-avemorphs (including *Archaeopteryx*), caudipterygians, oviraptorosaurs, troodonts, avimimids, and therizinosaurs all had a large, semilunate carpal block, although the structure is least developed in derived therizinosaurs (Fig. 10.11Cc,d). These dinosaurs also had an enlarged radial condyle on the humerus (Fig. 4.2Bb), with derived therizinosaurs again being the exception because their distal condyles were consistently subequal in size (Perle 1979, Russell and Dong 1993b). The pulley action of the carpal complex worked together with the asymmetrical condyles and other features to form a push-pull wing-folding mechanism (Fig. 4.2Bb; Gauthier and Padian 1985, Paul 1988a). Flattened carpals and symmetrical elbow condyles meant that derived therizinosaurs had the least arm-folding ability among these dinosaurs, but the well-developed semilunate carpal block of basal therizinosaurs (Xu, Tang, and Wang 1999) suggests that their arm-

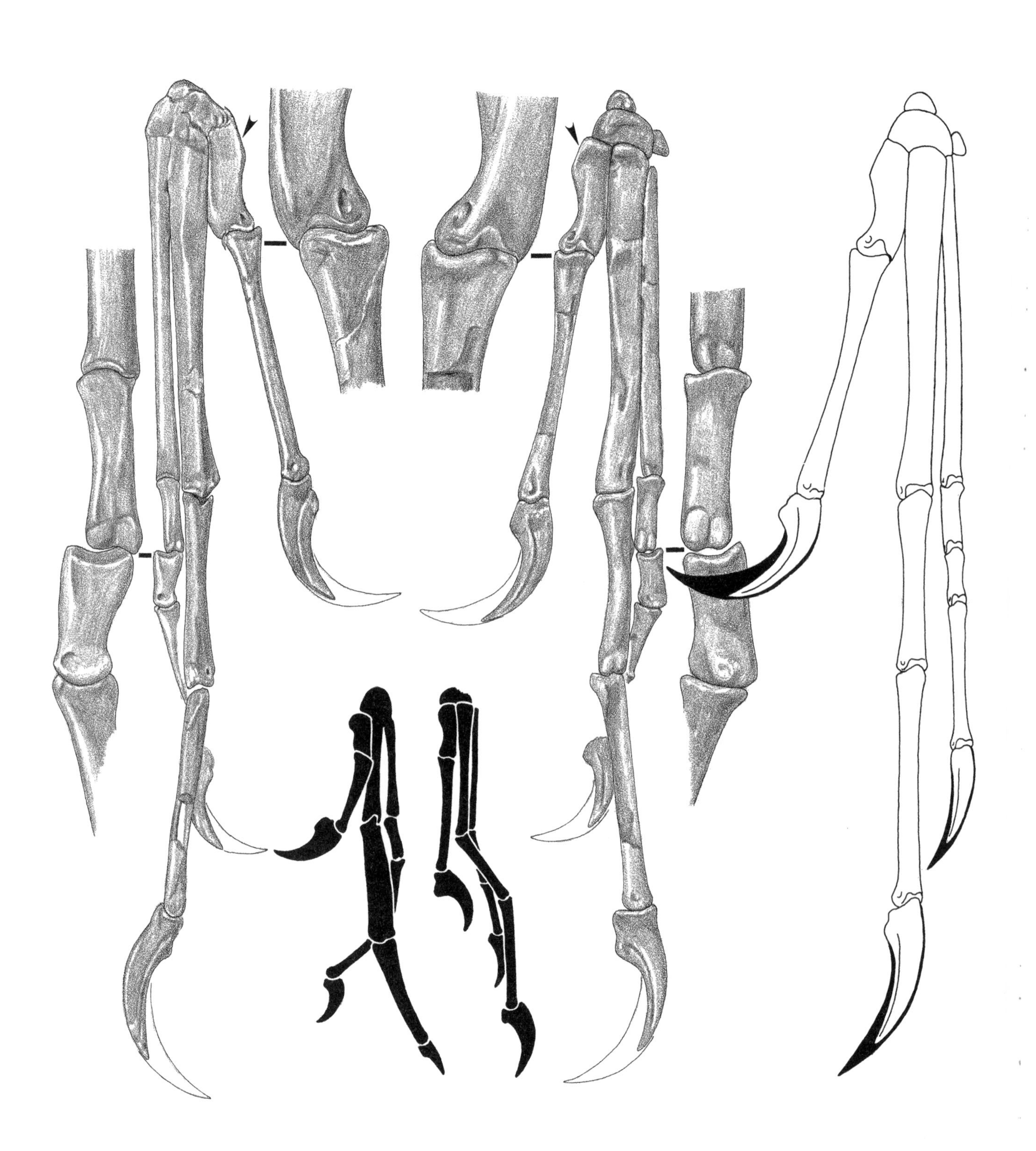

folding gear was as functional as that of other avepectoran dinosaurs. Sereno and Rao (1992) questioned whether any nonavebrevicaudan avepods had a well-developed arm-folding mechanism. The failure of the many articulated *Archaeopteryx* arms to be as tightly folded as they are in other fossil birds, such as *Confuciusornis* (Hou 1995), and the more flexed inclination of the semilunate carpal in avebrevicaudans (Fig. 10.11Bf–j) are evidence that the folding mechanism was not as well developed in *Archaeopteryx*. However, because the wing feathers are in excellent condition in HMN 1880, the arm was foldable enough to have protected them. In *Archaeopteryx* specimens, the wrist is never flexed more than 90 degrees, but because in no example are the completely articulated humerus, radius, and ulna fully tucked up, an additional 10–20 degrees of wrist rotation at maximum distal propulsion of the radius cannot be ruled out. The possible ability of urvogels to rotate the hand anteromedially may also have reduced the tendency of the wrist to be posteriorly flexed. Nor was much more than 90 degrees of wrist flexion needed to protect the wing feathers when the arm was folded (see Fig. 9.2). In some articulated dromaeosaur (Fig. 3b in Xu et al. 2000), oviraptorid (Pl. 15; Fig. 8 in Osborn 1924), and troodont (Pl. 13C) skeletons, the wrist is flexed 90–110 degrees. In the naturally articulated *Sinornithoides* specimen, the right arm (Pl. 13A,B) is not tightly folded at the elbow, and the wrist is not tightly flexed either. The left arm (Pl. 13A,C; Fig. 1b in Russell and Dong 1993a) is tightly folded at the elbow, and the wrist is much more tightly flexed than on the other side. This correlation between elbow and wrist flexion probably records the degree of automatic arm-folding action present in this dino-avepectoran. There is no evidence that dino-avepectorans could collapse the hand all the way onto the radius-ulna, as could avebrevicaudans. It is interesting that the degree of arm and wrist flexion is often greater in troodont and oviraptorosaur specimens than in the *Archaeopteryx* remains, a difference that implies but does not prove that arm folding may have been somewhat better developed in some dinosaurs than in the urvogel.

The presence of asymmetrical humeral condyles (Fig. 4.2Bc; Fig. 4A in Sereno 1993) implies that herrerasaurs had the beginnings of an arm-folding mechanism, but there was no carpal pulley. Many dino-avepods lacked both asymmetrical condyles (Fig. 4.2Ba) and carpal pulleys. Avetheropods such as allosaurs (Fig. 10.11Ca) and coelurids (Fig. 10.11Cb) had distal carpal blocks, but they had a pre-semilunate condition in which the dorsal profile was less rounded than the classic semilunate form, and the humeral condyles were symmetrical. In addition, the articulated hands of nonavepectorans are not preserved highly flexed at the wrist. Therefore, arm folding was weakly developed in nonavepectoran avetheropods and absent in basal avepods.

Zhou and Martin (1999) concluded that avepectoran dinosaurs lacked the distal placement of the proximal end of metacarpal III, as well as the close appression of III to II, seen in *Archaeopteryx* hands. Zhou and Martin's analysis was based largely on dated restorations of the disarticulated hands of *Deinonychus*. However, articulated hands of dromaeosaurs, troodonts, oviraptorosaurs, and other avepectorans show that the two metacarpals did not to diverge distally the way Zhou and Martin asserted that they did (Pl. 13C, Fig. 10.11Ad–i; Fig. 10 in Ostrom 1976a, Fig. 2 in Ostrom et al. 1999, Fig. 4 in Clark et al. 1999, Fig. 8 in Norell and Makovicky 1999). Zhou et al. (2000) showed that the metacarpals are closely appressed in *Caudipteryx* (Fig. 10.11Aj). Whether in life the bones were bound tightly or could spread slightly apart is not known. The fighting *Velociraptor* also appears to have a distally placed proximal end of metacarpal III (Fig. 10 in Ostrom 1976a and unpublished photographs of the prepared specimen), as does an oviraptorid (left hand in

*Figure 4.3. (opposite)* Hands as preserved of *Archaeopteryx* HMN 1880 *(shaded)*, type *Protarchaeopteryx (right black inset)*, and *Confuciusornis (left black inset)*, all in dorsal view, and restored orientation of *Archaeopteryx* claws *(outline figure on right)*. All show crossing of the lateral two fingers, but in feathered *Archaeopteryx* and *Confuciusornis* the central digit is almost always straight, whereas the outer is flexed anteromedially. In both featherless *Protarchaeopteryx* hands, the outer finger is straighter than the central one. In *Archaeopteryx* hands, the claws usually point anteromedially. As emphasized by the enlarged insets, the lateral two fingers repeatedly show a disarticulation of one of the joints, which indicates that the distal digits and their claws have been rotated approximately 90 degrees. Enlarged insets help to show that the thumb remains fully articulated, so claw rotation was minimal. Note that all of the finger joints of *Archaeopteryx* were well-formed, flexible pulleys. Because the small lateral projection on the left central finger is not seen in other urvogel specimens, it is presumed to be pathological. Not to same scale.

Pl. 15A); but this anatomical detail requires further investigation. Interestingly, confuciusornithids lack such a distal placement of the metacarpal (Fig. 10.11Ak) according to L. Martin et al. (1998, Fig. 2C) and Hou (1995), and the same appears to be true of some other basal birds (Sereno and Rao 1992, Zhou 1995a).

In the *Archaeopteryx* specimens, the finger claws usually point anteromedially, but Yalden (1985) concluded that all three finger claws were directed palmward. His conclusion is correct for digits II and III, in which at least one of the joints is disarticulated in the hands of all the specimens because the claws have been rotated either anteromedially by about 90 degrees or posteriorly by the same amount in BSP 1999, owing to sedimentary crushing (Pls. 1–5, Fig. 4.3). However, finger I is fully articulated in HMN 1880, JM 2257, and S6; and in BSP 1999 both thumb-claw points continue to point anteromedially even though most of the other claws face the opposite direction. Therefore, *Archaeopteryx* had an anteromedially divergent thumb. As the hand was flattened onto the plane of the sediment, the forward orientation of the thumb claw tended to guide the other claws forward as well, but as BSP 1999 shows, this did not always occur. The disarticulation in finger III was interpreted as a possible break by Hecht and Tarsitano (1982, 1984). However, this joint is present in the eight articulated hands, as are well-preserved articular surfaces (Howgate 1983, Paul 1984b, Wellnhofer 1985, 1988, 1993). The lateral digits and joints of many *Confuciusornis* hands appear to have undergone similar postmortem anteromedial rotation.

The crossover of fingers II and III present in many but not all of the articulated hands (Pls. 1–5, Fig. 4.3) has been considered to reflect the condition in life (Griffiths 1993, Feduccia 1996), but this crossover is more likely to be a postmortem effect (Kemp and Unwin 1997) caused by the attachment of the outer primary feathers to digit II. The same may be true of the crossover present in many *Confuciusornis* hands (Pl. 8B, Fig. 4.3). The feathers probably kept the central digit straight when they were flattened out on the lagoon bottom. The base of finger III may have also been bound into the postpatagial skin that anchors the wing feathers, but the featherless distal two-thirds of III was free to flex anteromedially. The apparently featherless articulated fingers II and III of *Protarchaeopteryx* are also crossed, but in this case, III is straighter than II (Fig. 4.3; see further discussion in Chapter 5). Martin (1995, 1997), Zhou and Martin (1999), and Martin and Czerkas (2000) have argued that the distal primary feathers and their supporting patigium prevented *Archaeopteryx* from flexing its fingers in life, but they are incorrect. The finger joints were mobile pulleys of the type common to dino-avepods, and there was no fusion or extreme flattening of the joints that would have severely hindered or barred flexibility. The joints of the central finger may have been less well developed than those of the other two digits, in which case the finger that provided the primary support for the distal feathers may have been less flexible than the others were, but we cannot be certain with the current specimens. Rather than bracing the central finger by lying parallel to its long axis, the feathers were instead angled relative to the central finger. Therefore, it is likely that the central finger could flex to some degree toward the palm, with the feathers folding relative to one another in the manner of a circular Amerindian feather headdress. That the largest of the three claws is on the central feather-bearing digit is evidence that it was at least a fairly flexible grasping organ. The inner finger was entirely free of feathers, and the same seems to have been true of some or all of the outer fingers. In *Caudipteryx,* the tight fit between a severely reduced outer finger and the central finger (Fig. 10.11Aj; Zhou et al. 2000) limited the flexibility of the feather-bearing digit, at least at its base (but Martin and Czerkas [2000] are incorrect that the feather array immobilized the central digit, for the same reasons described immediately above). In at least some dromaeosaurs (*Sinornithosaurus, Bambiraptor*), the pulley joints in the central finger appear to be more limited in extent, implying a reduction in flexion. In some other dino-avepectorans, such as oviraptorosaurs, finger II appears to have fully developed pulley joints, but proving that these joints were indeed developed requires additional functional analysis. Transforming the outer wing into a more rigid airfoil required reduction and fusion of the fingers into a truly inflexible unit. Not surprisingly, such reduction and fusion were often associated with the reduction of the central claw in basal birds. For example, the central finger of *Confuciusornis* lacked highly mobile joints, is often preserved straight, and was tipped with a reduced claw (Pl. 8, Fig. 4.3), indicating it was no longer a flexible grasping digit. The other two fingers remained supple and ended with large, hooked claws suitable for grasping, and III was therefore free to cross over II.

Because the hands of all *Compsognathus* specimens are either disarticulated or poorly preserved,

Ostrom's (1978) conclusion that only two fingers were complete cannot be confirmed (Paul 1988a, Gauthier and Gishlick 2000). A truly bidigit condition appears to be contradicted by the presence of three partial metacarpals in one *Compsognathus* specimen (Bidar et al. 1972) and the three complete digits of the compsognathid *Sinosauropteryx* (Chen et al. 1998). The hand of the type specimen of *Ornitholestes* is very incomplete, and the tentative assignment of the separate hand of AMNH 587 (Osborn 1903) is probably incorrect (Miles et al. 1998). Ostrom (1969) restored the disarticulated hand of a dromaeosaur with a straight, robust digit I and a more slender, medially divergent digit II. However, metacarpal I has a medially oriented distal condyle, so it is possible that the proximal phalanx with the twisted shaft articulated with this metacarpal, as it does in other dinoavepods. If so, then the most robust digit in at least some dromaeosaurs is the central one (Paul 1988a); the same is true of *Archaeopteryx*. Finger II is about as robust as I in *Sinornithosaurus* (Fig. 10.11Ai), *Bambiraptor*, and a troodont (Pl. 13C). However, in the articulated hand of *Velociraptor* MIG 100/25 (Fig. 10C in Ostrom 1976a) and another velociraptorine (Fig. 8 in Norell and Makovicky 1999), the thumb is more robust and seems not to be anteromedially divergent, although the difference in robustness in the latter case appears minimal. The unusually great breadth of the proximal bone of right finger II of the type specimen of *Sinornithosaurus* is discussed in Appendix 6.

The last parts of the skeleton to be considered are the pelvic girdle and the hindlimb. Among *Archaeopteryx* ilia, a shallow dorsal concavity is present on BMNH 37001 and S6; in the other specimens the profile is dorsally convex (Fig. 10.12Ca). We do not know whether a pectineal process is present on the anteroproximal corner of the pubis of *Archaeopteryx* (Fig. 10.12Da, Pls. 2–5). That such a process appears to be present in HMN 1880 (Norell and Makovicky 1997) and S6 may be an artifact due to displacement of the pubis on the ilium (Wellnhofer 1988). The process is absent on JM 2257 and BSP 1999 but present in dromaeosaurs (Fig. 10.12Db; Norell and Makovicky 1997).

L. Martin (1991) restored the paired pubic boots of *Archaeopteryx* as flaring strongly laterally, but in all but one specimen, the boots are transversally narrow in the manner typical of avepods (Fig. 10.12B,C). A hypopubic cup at the posterodistal end of the pubis has been reported in *Archaeopteryx* (Martin 1991, Ruben et al. 1997a, 1999, Ruben and Jones 2000), *Rahonavis* (Forster et al. 1998), and *Velociraptor* (Norell and Makovicky 1999). But such a structure is clearly absent in *Sinornithosaurus* and *Deinonychus* (Fig. 10.12Db) and has not been clearly demonstrated in any other dino-avepectoran. In *Archaeopteryx* BMNH 37001, the pubic boot is anteroposteriorly crushed, and the posterior portions of both pubic boots are broken off; so the lateral expansion of the boot, and the supposed cup set in a laterally broad boot, appear to be artifacts of damage and or calcitic crystallization (Norell and Makovicky 1999). In HMN 1880, the indentation on the right side of the pubic boot may be due either to damage or to slippage of the right and left halves. No actual cup is present in any urvogel specimen, and the posteriorly elongated, transversely narrow, and typically dinosaurian form of the boot in the rest of the specimens (Fig. 10.12Da) render the presence of a cup implausible. The pubic boots of basal birds are also too transversely narrow to accommodate hypopubic cups (Fig. 10.12Cc; Pl. 1 in Hou and Zhang 1993, p. 93 in Ackerman 1998, Figs. 2 and 3 in Ji, Chiappe, and Ji 1999, Figs. 40 and 41 in Chiappe et al. 1999). Consequently, the restorations by Martin et al. (1998) and Chiappe et al. (1999) of a *Confuciusornis* pubis that is distally broad or bootless or both are consequently incorrect; specimens preserved on their sides, so that the boot was not distorted (Pl. 8B), show that the boot was moderately developed and narrow in a manner similar to that seen in some other birds (Fig. 10.12Cc,g).

In most archosaurs, the pubic bones that project downward from the hip are procumbent to a greater or lesser extent; in all avebrevicaudan birds, past and present, they have been directed strongly backward (Fig. 10.12A–C). One of the most vexing issues about *Archaeopteryx* has been the orientation of the pubes. Restorations vary from showing the pubes as nearly vertical to as much as 80 degrees posterior to vertical (Ostrom 1976a, Howgate 1985, Paul 1988a, L. Martin 1983a,b, 1991, Ruben et al. 1997a, L. Martin et al. 1998, Norell and Makovicky 1999). Only in BMNH 37001 are the pubes preserved as strongly retroverted as Martin, Ruben, and their co-workers have contended. However, because both pubes are completely disarticulated from the ilia and the pelvis is crushed dorsoventrally (Pl. 1), this specimen is highly misleading in this regard. In the five other pelves, pubic retroversion is only 15–45 degrees, and pubic retroversion has proven to be minimal in the newest specimens (Pls. 2–5). Because the proximal pubis is either articulated with the ilium or only slightly displaced in four of these specimens (the articulation is not preserved in TM 6428), I now conclude (contrary to my earlier restoration [Paul 1998a]) that *Archaeopteryx* was probably not strongly retropubic (Figs. II.1, 10.12Da). Howgate (1985) argued that pubic orientation was different in the various individuals. It is true that the pubis appears fully articulated in BSP 1999, in which it is nearly vertical; and the more highly retroverted pubis of HMN 1880 does not appear to be broken proximally, despite assertions to the contrary. At this time, giving a more firm value for the degree of retroversion in *Archaeopteryx* is impossible; we can say only that retroversion was present in all specimens, that it did not exceed 45 degrees, and that it very possibly was much less. As a result, it is improbable that the distal end of the pubis was posterior to the distal end of the ischium (contra Ruben et al. 1997a). Because the pubes were long as well as oriented subvertically, the body was deep keeled. The deep keel is one reason that all the articulated specimens are preserved on their sides rather than dorsoventrally flattened, as are the skeletons of many flatter-bodied birds, pterosaurs, and bats found in similar fine-grained sediments.

The ischia, behind the pubes, point strongly posteriorly only in BMNH 37001, where they are disarticulated (Pl. 1); in the other four specimens, the ischia are more vertical in accord with the similarly vertical pubes (Pls. 2–5). I conclude that the latter orientation is the original condition (contra L. Martin 1983a,b, 1991). There is no ilioischiatic or ischial fenestra in BMNH 37001, HMN 1880, or BSP 1999 (contra De Beer 1954, L. Martin 1983a,b, 1991, Feduccia 1996); the supposed opening is actually a space between the disarticulated right proximal pubis and ischium in BMNH 37001.

The orientation of the pubis is not preserved in *Protarchaeopteryx*. The lack of a heavy caudal counterweight implies that the pubis should have been retroverted to at least some degree. However, the pubis is vertical in similarly short-tailed *Caudipteryx* (Fig. 10.12Bl), not retroverted as shown in L. Martin and Czerkas (2000). This vertical orientation is seen in NMGC 97-9-A, in which the articulated left pelvis is preserved as bone and impressions, and in other specimens described in 2000 by Zhou and Wang and by Zhou et al. The pubic articulation of *Caudipteryx* also differs from the inverted-V configuration of the often retropubic dromaeo-avemorphs in that the articulation was more complex. Oviraptorosaurs and troodonts were not retropubic (Fig. 10.12Bm,n), contrary to some earlier reports. In dromaeo-avemorphs, pubic retroversion appears to have been markedly variable (Fig. 10.12Br–v). In *Unenlagia* and *Rahonavis*, retroversion was minimal or absent (Novas and Puerta 1997, Forster et al. 1998), and it was modest in *Bambiraptor* (Fig. 1.7; Burnham et al. 2000). Extreme retroversion of the pubis in the velociraptorine *Deinonychus* MCZ 4371, velociraptorine MIG 100/986, and the type *Microraptor* (Xu et al. 2000) resulted partly from dorsoventral crushing, and the proximal end of the pubis is also crushed in the former. However, the retroversion shows that in these dinosaurs the pubis was not procumbent, and the orientation of the pubic facets that articulate with the ilium shows that the pubis was strongly retroverted in the velociraptorines (Fig. 10.12Db; Norell and Makovicky 1999). The articulated, undistorted pubes of velociraptorines MIG 100/25, 100/980, and 100/985 are retroverted by as much as 55 degrees (Fig. 10.12Bt; Perle 1985, Norell and Makovicky 1997, 1999). Although disarticulated from the ilium, the pubes of the type specimen of *Sinornithosaurus* are strongly retroverted relative to the ischia (Fig. 10.12Br; Xu, Tang, and Wang 1999). The impression given by Ruben et al. (1997a) and Tarsitano et al. (2000) that all dromaeosaurs lacked strong pubic retroversion is therefore incorrect (Barsbold 1983, Paul 1987a, 1988a). The degree of pubic retroversion in basal birds is usually strong, but not to the point that the pubis parallels the posterior ilium. Pubic retroversion in *Confuciusornis* has been restored as strong (Hou et al. 1996) or extreme (L. Martin et al. 1998). The tendency of the great majority of the skeletons to be preserved on their bellies or backs favors a pelvis that was shallower than that of *Archaeopteryx*, but the resulting dorsoventral crushing makes the pubis appear even more retroverted than it really was. The last point is confirmed by the few specimens preserved on their sides (Pl. 8B). In these, pubic retroversion is strong rather than extreme, and as in other basal birds, the pubis in these specimens did not parallel the posterior ilium (Fig. 10.12Cc–h). The specimen in Plate 8B also shows that the *Confuciusornis* pelvis restorations by Hou et al. (1996), L. Martin et al. (1998), and Chiappe et al. (1999) are not accurate and that the *Confuciusornis* pelvis was actually more similar to that of *Archaeopteryx* (Fig. 10.12Cc).

Norell and Makovicky (1997, 1999) noted that dromaeosaur ischia are not distally fused to one another, although they are co-joined (Ostrom

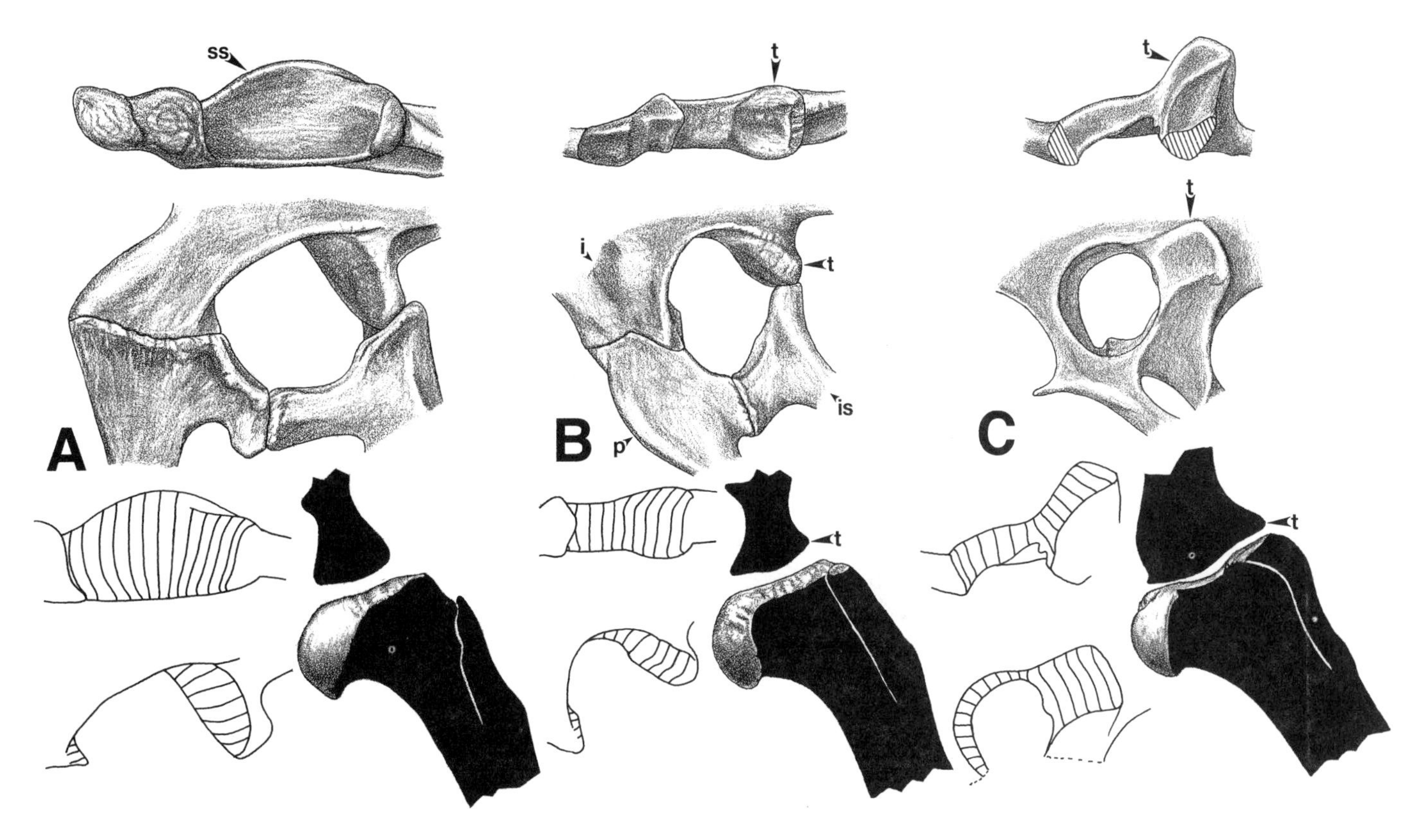


*Figure 4.4.* Avepod hip joint articulations showing loss of supraacetabular shelf and development of antitrochanter; acetabulum in ventral *(top)* and lateral *(center)* view, with contour maps of same *(lower left)*, cross section of posterior acetabular surface/antitrochanter (as per line in lateral view) articulating with protracted femoral head *(lower right)*: *A, Allosaurus* in which the supraacetabular shelf is well developed and antitrochanter is absent; *B, Deinonychus* in which supraacetabular shelf is lost and antitrochanter is developing; *C,* ostrich in which antitrochanter is well developed. Note that femora are partly everted even in erect-legged avepods. Abbreviations: ss = supraacetabular shelf, t = antitrochanter; other abbreviations are given in Figure 4.1. Not drawn to same scale.

1969). The degree of distal attachment of the ischia of *Archaeopteryx* is not certain, but some contact is plausible in view of the weak connections seen in the dromaeosaurs and possibly in some basal birds (Sanz and Bonaparte 1992). Any contact must have been limited to the tips of the distal end, and perhaps to the obturator process, because the ischium remains strongly bowed convex laterally in BMNH 37001 (Fig. II.1).

L. Martin (1991, 1995) restored a strongly laterally everted femur and a semi-erect gait for *Archaeopteryx*. In some cursorial birds, the femur is everted by as much as 20–45 degrees (Fig. 4.4; the latter value is observed in swans for example); yet the feet are held under the body, and leg action is limited to a fairly simple fore-and-aft motion. A truly semi-erect gait in *Archaeopteryx* is contradicted by the consistent pattern of preservation of the body on the side, with both hindlimbs directed ventrally, rather than laterally as is true of the non-erect forelimbs in two specimens. The presence of a partial medial wall in the acetabulum (Fig. 10.12Da) is not indicative of a non-erect gait, because the same is present in protodinosaurs with erect gaits, in various avepod dinosaurs (Norell and Makovicky 1997, 1999, Novas and Puerta 1997, Burnham et al. 2000), and in some ground-running birds and other birds (Fig. 10.12Ba–c,t,u,Ch–j,Db). L. Martin (1991) cited the incomplete articulation of the acetabulum and femoral head in HMN 1880 as evidence against an erect gait, but in this specimen, the femoral head has been displaced ventrally and crushed into the acetabulum. Martin's (1991) articulation of the femur into the acetabulum of BMNH 37001 in a strongly everted posture is unsubstantiated because part of the medial femoral head is missing from the cast he used (the femur remains partly buried in situ on the slab). In addition, the hip joint in *Archaeopteryx* should have been partly cartilaginous—the normal dinoavepodian condition (Paul 1988a)—in which case the original articular surfaces are lost. What is preserved of the femoral head of *Archaeopteryx* is similar in configuration to that of some dino-avepods (Fig. 10.13Be–g), including a strongly inward-turned femoral head; so a similarly erect femoral posture is indicated. The avepod-type, hinge-action hip, knee, and ankle joints of *Archaeopteryx* could work only in a strict fore-and-aft motion, and they precluded a truly semi-erect limb action. Tarsitano (1991) argued that the greater and lesser trochanters of the femora of *Archaeopteryx* and dino-avepods differ so much that a difference in stance is indicated. This is not true, because the structures are actually very similar in the two types (Fig. 10.13Bf,g). L. Martin et al. (1998) stated that the lack of a dorsal articular shelf on the acetabulum of *Confuciusornis* "indicates that the posture found in dinosaurs was not possible." The meaning of this statement is not entirely clear, and no further explanation of the link between dorsal rim design and posture was offered. In any case, the early bird has the normal avepod inturned femoral head that indicates an erect leg posture.

In long-tailed theropods, the fore-and-aft arc through which the femur swings during running and especially walking was probably greater and more vertical than in short-tailed birds, although perhaps not so much as often thought (Galton 1970b, Paul 1988a, Fig. 10 in Gatesy 1991, Jones et al. 2000b). By shifting the hindlimb forward, horizontal femora help a walking bird keep its balance despite its lack of a caudal counterweight (when running the balance requirement is less, and femoral action becomes more extensive so that the propulsive power of the proximal portion of the leg can be fully utilized). A characteristic feature of birds is a large, well-developed antitrochanter set on the posterior rim of the acetabulum; the antitrochanter articulates with the head of the femur only as long as the femur does not retract too far toward the vertical (Fig. 4.4). That ossified antitrochanters were absent in most dino-avepods, including *Archaeopteryx* (Fig. 10.12Da), suggests that femoral action was extensive at all times. However, partly developed antitrochanters were present in protodinosaurs and various dino-avepods both basal (coelophysids) and derived (dromaeosaurs [Fig. 4.4] and troodonts [Russell and Dong 1993a]), and whether or how much these affected femoral posture and action in these long-tailed forms is unclear. That the antitrochanter was well developed in alvarezsaurs (Perle et al. 1994) suggests that femoral action was more limited and subhorizontal. The ilia of *Protarchaeopteryx* and *Caudipteryx* are too poorly preserved for us to tell whether they had antitrochanters, but their very short tails imply that the femora were held more horizontally than is typical of avepod dinosaurs. Considering the development of the urvogel's antitrochanter and tail, femoral posture in *Archaeopteryx* was probably intermediate to that of most theropods and birds, especially at low speeds, although a strong propulsive retraction at high speeds cannot be ruled out.

Contrary to assertions by L. Martin (1983b), there is no "stop" on the lateral femoral condyle of *Archaeopteryx*. The "stop" is the proximal corner of the rectangular lateral femoral condyle common to dinosaurs and birds. The knee of all avepods is fully articulated only when it is flexed (Paul 1987a, 1988a, 1998/2000). Straightening the knee disarticulates the lateral femoral condyle and leaves the knee vulnerable to complete dislocation, so restorations that show avepods with straight knees are inaccurate (see Fig. 10 in Farlow et al. 2000). The hindlimbs of dino-avepods could be tightly folded (Pls. 13A, 15A; Fig. 7 in Lambe 1917). The ossified condyles of most dino-avepod astragalocalcaneums are not developed well enough to allow the ankle to collapse (unlike in mature birds), so cartilage caps probably formed large roller surfaces (as in immature birds).

The fibula is complete distally in JM 2257 and S6, and it contacts the calcaneum in the former. One of the distinguishing features of dinosaurs and birds is found in the proximal half of the ankle. This feature is the ascending, or pretibial, process (see Chapter 10) of the ankle, which is a vertical tongue of bone appressed to the anterior face of the tibia (Figs. 10.14B, 10.18A–C). Tarsitano and Hecht (1980) asserted that *Archaeopteryx* lacked this feature, claiming that the supposed process is actually a deposit of calcite crystals that mimic bone. Ostrom (1976a, 1985, 1991) has shown otherwise, and my observation of the left process of HMN 1880 under direct sunlight showed only a few tiny glittering crystals; the rest of the process is bone. Crushing, damage, and obscuration of the pretibial process conceal its exact form. The bone appears to be moderately tall, and in JM 2257 and S6, it appears to extend strongly medially (Fig. 10.14Bk).

The skeletal evidence speaking to the orientation of the hallux, or first toe, in dino-avepods is ambiguous in some cases and more definitive in others (Ostrom 1976a, 1991, Tarsitano and Hecht 1980, L. Martin 1983a,b, 1991, Paul 1988a, Tarsitano 1991, Norell and Makovicky 1997, 1999, Clark et al. 1999). In articulated feet, the hallux subparallels the other toes instead of being strongly reversed so that it points backward like those of birds. Ostrom (1991) postulated that the first toe of the type specimen of *Compsognathus* may have been partly dereversed postmortem. Interestingly, the degree of reversion of the first toe differs in stored kiwi skeletons, depending upon how firmly the toe is attached to metatarsal II by the remaining ligaments (Fig. 4.5D). If a kiwi foot was fossilized, postmortem alteration might dereverse the hallux as much as is observed in avepod dinosaurs. Therefore, the life position of the dino-avepod hallux may have been more reversed than the fossil bones indicate. Some fossil footprints suggest that the hallux was indeed reversed (Fig 10.19), but Gatesy et al. (1999) concluded that the apparent reversal was an illusion due to the softness of the muds the predatory dinosaurs were walking through. Gatesy's hypothesis will be confirmed if someone shows that seemingly reversed hallux prints do not occur even when the substrate is too firm for this illusion to occur. Some dino-avepod footprints clearly show a half-reversed hallux (Fig 10.19; Gatesy et al. 1999). In the largely articulated right *Sinornithosaurus* foot, metacarpal I is disarticulated and rotated along its long axis, so determining the orientation of this digit is difficult. The hallux of *Archaeopteryx* was clearly strongly reversed, but the full reversal, with digit 1 being 180 degrees opposite toe III, observed in the specimens appears to be due to crushing (Middleton 1999). In the left foot of HMN 1880, the hallux is reversed approximately 120 degrees, and this angle too may have been somewhat exaggerated by crushing. Even less clear is the orientation of the hallux in the articulated but very poorly preserved *Protarchaeopteryx* feet; one appears to be reversed, the other unreversed.

An interesting question surrounds the function and posture of the central three toes of *Archaeopteryx*. Gauthier (1986) and I (Paul 1988a) suggested that the foot had a semididactyl posture in which digit II was hyperextendible (also see Sereno 1999a), an arrangement similar to that seen in the sickle-clawed dromaeosaurs, troodonts, and *Rahonavis* (Forster et al. 1998). Four pieces of evidence support this conclusion. First, in the left foot of JM 2257, and to a lesser extent in the more damaged left feet of BMNH 37001 and HMN 1880, the distal condyles of phalanx 1 are enlarged dorsally, and the ligament fossae on phalanges 1 and 2 are sited dorsally. These adaptations probably allowed the digit to hyperextend (Fig. 4.5). In the other articulated feet, the particular joint is either obscured or too crushed or degraded for its condition to be assessed. The conclusion by Elzanowski and Pasko (1999) that left toe II has been rotated 180 degrees in JM 2257 is incorrect. The entire toe is normally articulated, and the claw is properly oriented. On the counterslab, the shaft of phalanx 2 is correctly dorsally arced, and if the proximal phalanx was dislocated, it somehow returned to a position in seeming articulation with its metatarsal and distal phalanx. The claw on toe II is not enlarged, but the claw on the hyperextendible toe of the dromaeosaur *Adasaurus* is not greatly enlarged either (Fig. 4.5C). Second, among the seven articulated *Archaeopteryx* feet in which digit II is visible, the digit is extended more dorsally at its center joint than are the other two digits in four examples, and is never more flexed than its neighbors (Fig. 4.5Aa). A similar condition is observed in articulated dromaeosaur (*Velociraptor* MIG 100/25; Figs. 5 and 6 in Norell and Makovicky 1997, Fig. 3c,d in Xu et al. 2000) and troodont feet (Pl. 13A; Fig. 5 in Osborn 1924). For the third point, see Figure 4.5Aa,b and Appendix Figure 18A, which show that the foot of *Archaeopteryx* is markedly asymmetrical, with digit II being much shorter than IV (except in S6, in which both digits IV are pathological and missing an element). This condition approaches the asymmetry seen in dromaeosaur and troodont feet (Fig. 4.5Ac, App. Fig. 18A) and means that *Archaeopteryx* bore most or all of its weight on the outer two toes. Fourth and finally, in didactyl dromaeosaurs and troodonts, the primary load-bearing metatarsals III and IV are strengthened by a flange of IV that backs III (Fig. 4.5Ac). A flange of metatarsal IV backs III in the left foot of JM 2257 and in the right foot of S6 (Fig. 4.5a,b). It is therefore possible but not certain that *Archaeopteryx* held digit II clear of the ground either habitually or on occasion (Frontispiece, Figs. II.1, 9.1). Both the second toes of *Protarchaeopteryx* are too poorly preserved for us to determine whether they were hyperextendible, but the symmetry of the foot implies that they were not.

Martin (1991) restored the foot of *Archaeopteryx* as gracile and "birdy," in that metatarsal III is set well anterior to its mates, and toes III and IV gently curve laterally. The latter is an illusion due to the angle at which the accompanying photograph was taken relative to the plane of the slab. The transverse narrowness and anteroposterior depth are due to lateromedial crushing of the metatarsus. When the metatarsus is preserved flat on the sediment plane, as per the right foot of S6, it is more robust and anteromedially flatter

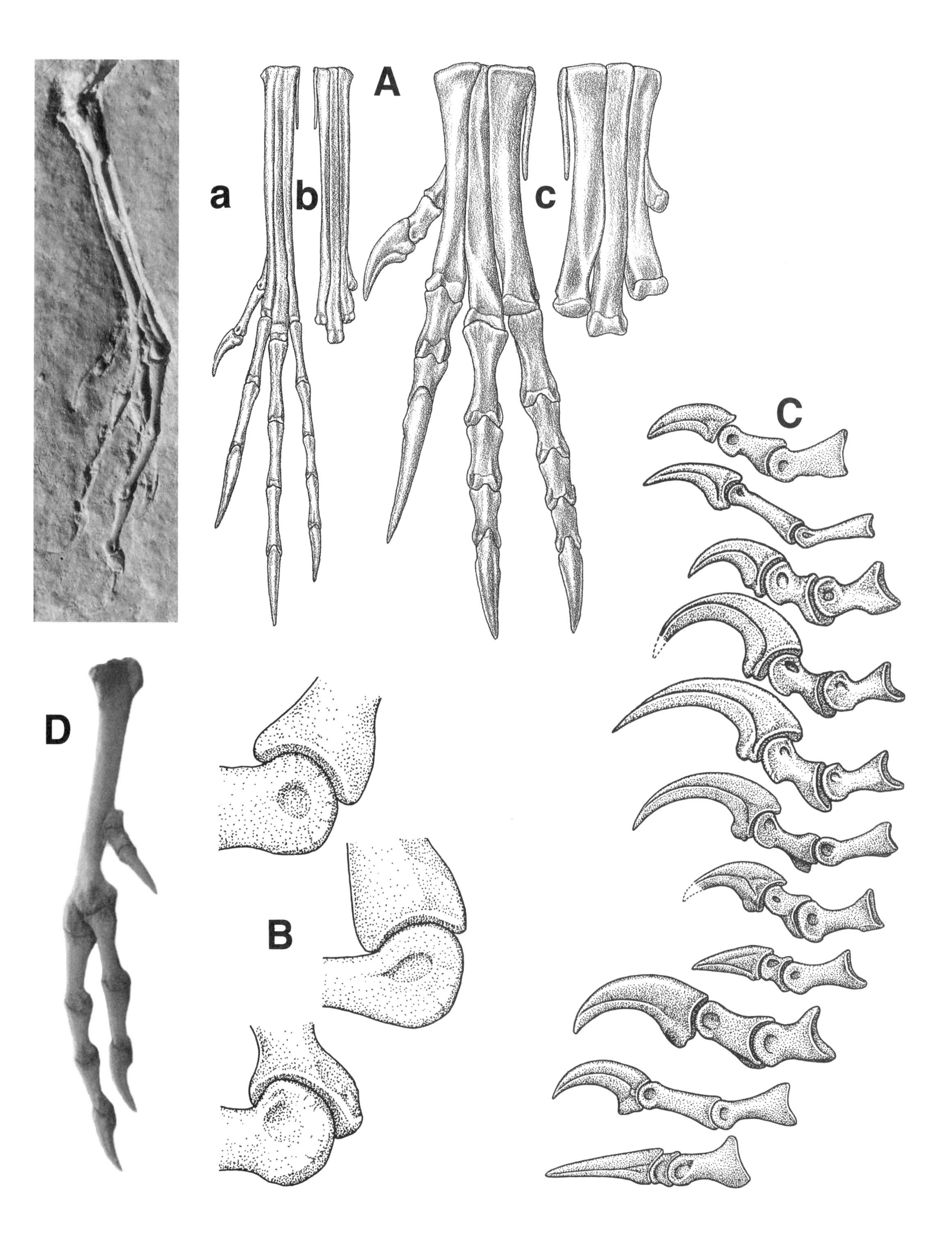

A
a
b
c
C
D
B

(Fig. 4.5Ab) in the manner of dinosaurs and other early birds (Brett-Surman and Paul 1985).

In the Yixian dinosaurs, the bones of the metacarpus, proximal tarsus, and metatarsus are loose and poorly articulated. In Yixian *Protarchaeopteryx,* these bones are tightly bound, but not strongly fused, to one another. This condition is similar to that in *Archaeopteryx.*

We conclude our survey of postcrania by considering some full skeletal restorations. The methodologies for restoring extinct avepod skeletons in multiple views are detailed in Paul (1987a), Paul (1988a), and Paul and Chase (1989). Yalden's (1984) *Archaeopteryx* restoration is too deep chested because he places the pectoral girdle too ventrally. The restorations by Ostrom (1976a) and Elzanowski and Pasko (1999) are a little too deep chested because the anterior dorsal ribs are too vertical. Restorations by L. Martin (1983a,b, 1991, 1995), Hou et al. (1996), and L. Martin et al. (1998) are too flat bodied because they place the pectoral girdle too high and show the pubis as too retroverted. These workers and L. Martin and Czerkas (2000) also restored the body of *Archaeopteryx, Confuciusornis,* and *Caudipteryx* as almost as vertical as the human body, and the tail of the urvogel as habitually flexed dorsally. They did so in part because they restored the dorsal rim of the acetabulum as entirely lacking a supraacetabular shelf and as having been too extremely thin dorsally to bear the weight of the body when the trunk was horizontal (Fig. 30 in L. Martin 1991). Actually, a broad supraacetabular shelf was present in *Archaeopteryx* (Pl. 24), which is not surprising because a hyperthin dorsal acetabulum would have been functionally maladaptive. The shelf was not as well developed as that in most dino-avepods (Fig. 4.4A) but was as broad as or broader than those of most other avepectorans and even tyrannosaurs (Pl. 24, Fig. 4.4B,C). Crushing has obscured the transverse configuration of the acetabulum in *Confuciusornis* and *Caudipteryx* specimens, but the configuration was probably similar to that in other avepods. A vertical body posture would have resulted in malfunction of the ischial-based femoral retractors in archaeopterygiforms and confuciusornithids, because the ischium was not far enough posterior to the proximal hindlimb to allow proper fiber stretching (Paul 1988a). The only dinosaurs with a subvertical body posture were those in which the pelvis was retroverted relative to the dorsal column, such as therizinosaurs (Fig. 10.1Bk,l) and some sauropods that either were very shoulder tall or were adapted for walking while rearing up. Pelvic retroversion allowed the body to be tilted up while the hip remained horizontal, which kept the ischial-based femoral retractors' stretch factors workable. L. Martin and Czerkas (2000) restore *Caudipteryx* with a vertical body because its tail was too short, in their opinion, to have balanced a horizontal body, but horizontal-bodied birds have even shorter tails. Changes in tail length are compensated for by modest adjustments in leg posture. Because a vertical body is atypical for archosaurs and birds in general, and because early birds lacked the pelvic orientation associated with an erect body, restoring them with this stature is illogical (Padian and Chiappe 1998b, Chiappe et al. 1999, Hembree 1999). Less implausible, but perhaps still somewhat too vertical, is the upward pitch of 20–25 degrees restored for *Archaeopteryx* (Elzanowski and Pasko 1999) and confuciusornithids (Hembree 1999). When the body was close to horizontal, the tail base was probably not habitually flexed dorsally in basal birds.

Contrary to the opinion of some researchers, in gross form the skeleton of *Archaeopteryx* was much more like that of an exceptionally long-armed dromaeosaur—most especially *Sinornithosaurus*—than that of a modern bird (Figs. II.1, 10.1Bm,n, 11.1). I base this conclusion upon the analysis and reconstruction of the skeleton of the urvogel; and I did not make my restoration

*Figure 4.5. (opposite)* Hindfoot form and function. *A,* dromaeo-avemorphs: *a,* cast of *Archaeopteryx* JM 2257 left pes; *b, Archaeopteryx* composite left foot in front view and metatarsus in posterior view; *c, Deinonychus. B,* inner view of joint between phalanxes 1 and 2 in toe II: *top, Allosaurus,* in which condyle is normal and digit is not hyperextendible; *center, Archaeopteryx,* in which condyle is expanded dorsally and the digit can hyperextend; *bottom, Deinonychus,* with fully developed extendibility. *C,* avepod second toes hyperextendible and otherwise: *top to bottom, Allosaurus* with unmodified digit; dromaeo-avemorphs: *Archaeopteryx* with minimal modifications; dromaeosaurs and relatives with highly modified phalanxes starting with *Adasaurus* with short claw, *Deinonychus, Velociraptor,* and *Rahonavis* with large claws; *Troodon* with similar adaptations; troodont *Borogovia* with conventional digit; phorusrhacid, seriema, and cassowary with nonhyperextendible digits ending with large claws. *D,* naturally articulated kiwi right foot in inner view showing that the hallux is only half reversed and is less reversed than in *Archaeopteryx;* hallux orientation is the same in skin-covered and living specimens. Not drawn to same scale.

dromaeosaur-like in order to fit phylogenetic or functional arguments (contra Elzanowski and Pasko 1999).

In the above-cited Hou et al., L. Martin et al., and Chiappe et al. (1999) *Confuciusornis* restorations, the furculae are too large, and the hindlegs are too long relative to the arms. Complete specimens show that the humerus was about 15 percent longer than the femur, rather than slightly shorter (Pl. 8, Fig. 10.1Bt).

CHAPTER 5

# Feathers and Other Fluff

Finding and examining superficial integument in fossils is hampered by numerous problems. Feathers not anchored upon bones, and other soft integument, can slough off an animal before burial, even when the skeleton remains well articulated (Davis and Briggs 1998). Even when present at burial, keratin is not nearly as readily preserved as bone. For these reasons, the absence of preserved superficial integument is not definitive evidence for the absence of the same in the live animal, even when its articulated skeleton is found in fine-grained sediments. The presence of feathers or fur can be excluded only when scales or an outermost layer of naked skin can be positively identified, and when scales or skin cover most or all of the body. The presence of bare skin on only part of a fossil is not final proof of lack of superficial integument, because some animals have bare patches as well as superficial body coverings. Ostriches are one example, vultures another. Absence of extensive superficial insulation in some members of a group is not necessarily evidence for the absence of the same in all members of the group. Consider the near absence of fur in elephants, rhinos, some suids, and humans. In a lagerstätten where soft plumages are frequently preserved, consistent absence of the same in a number of examples of one species can be taken as evidence against the presence of plumage. Well-preserved quill nodes on the outer arm bones are evidence for the presence of large feathers. However, the absence of quill nodes is not so definitive; quill nodes are absent in some flying birds, including *Archaeopteryx*.

Feathers differ dramatically in form (see Stettenheim 1974, Spearman and Hardy 1985, Chatterjee 1997). Maderson and Homberger (2000) is a volume on feathers and their origins derived from a 1999 symposium. Most familiar are the intricately constructed contour feathers that adorn flying birds (Fig. 5.1). These pennaceous structures have hooked, interlocking barbules that help maintain the precise form of the feathers. Primaries and other remiges make up the wing, rectrices form the caudal airfoil, and body contour feathers form a streamlined aeroshell around the body. Other feathers include irregular semiplumes and down, both of which lack well-developed central shafts and the hooked barbules. The simplest feathers are bristles (Stettenheim 1974, Spearman and Hardy 1985, Schweitzer et al. 1999). These consist mainly of a single shaft with a tiny barb at the base, for example, the kiwi's large head whiskers, which are similar in gross size and form to cat whiskers. The bristles of some nightjars lack even the barb (Stettenheim 1974). A few birds have display feathers in which the barbules are fused into continuous ribbonlike sheets on both sides of the central shaft. One characteristic that separates even simple feathers from most other dermal structures is that the primary shaft is hollow; this feature is also seen in some mammal fur, such as that in polar bears. The hollow shaft increases both the strength-to-weight ratio and the insulation value of the structure.

Although feathers have been a classic defining feature of Aves, their presence in nonavian diapsids has been asserted a number of times. Indeed, there have been a surprising number of claims that basal archosauromorphs and archosaurs had skin structures that were precursors to feathers, but none of these claims has yet been confirmed (Chiappe 1995). Ellenberger (1977) believed that an aura of "impressions" surrounding a skeleton of the Triassic archosauromorph *Cosesaurus* recorded the presence of feathers. This claim has not been verified or accepted, and the impressions appear to be artifacts created by splitting of the slab.

Triassic *Longisquama*, which may or not belong to the archosaur clade (see Chapter 10), is alleged to have possessed protofeathers or feathers (Sharov 1970, Jones et al. 2000a, 2001). This claim led to considerable controversy (see commentary in Prum 2001, Reisz and Sues 2000, Stokstad 2001, Unwin and Benton 2001). The

*Figure 5.1.* A swan. When we think of feathers, we tend to think of the fine and spectacularly complex feather arrays that adorn flying birds such as a swan. Could it be that these intricate avian marvels first evolved in crude, simple form in the primitive dinosaurs?

actual nature of the preserved structures is unclear, and because only impressions are preserved, biochemical analysis is not possible. Large scales on the neck and along the arms of *Longisquama* are moderately elongated but not featherlike. Much more spectacular is the array of very long leaf- and featherlike asymmetrical structures that appear to emerge from the dorsal midline. These have been restored as either a single or a double row of dermal appendages. Haubold and Buffetaut (1987), for example, restored them as a double row of asymmetrical structures and explained them as gliding airfoils. These workers may be correct, but they also note that there is no actual connection between the structures and the skeleton. The bases of seven of the structures appear above seven dorsal vertebrae, but the most of the bases do not line up closely with the neural spines. It is also disconcerting that additional leaflike structures have been found in the same sediments, with no trace skeletal remains nearby. The structures look suspiciously similar to plant materials, and it is possible that they are really remains of vegetation that, in one case, happened to coincide with a skeleton. Articulation of the structures with another skeleton is necessary for verification of their vertebrate nature. Assuming that they were made of dermal keratin, Jones et al. (2001) showed that the appendages were not typical scales. The central, hollow shaft seemed to be a featherlike feature, but it may be an artifact due to damage of the specimen. The structures appear feathery in terms of superficial morphology and in the details of the central shaft, but otherwise they differ from most feathers in detail. There is no example of either the barb separation or distal fraying inherent to actual contour feathers. The impressions lack the fine barbs that are well recorded in the feather impressions of *Archaeopteryx* and overall differ substantially from the latter (Fig. 5.2). The tissue that makes up the vanes is so thin that some appendages can be seen through overlying examples; in contrast, the thicker feathers of the urvogel completely mask one another. The preceding attributes indicate that the vanes were corrugated medial sheets rimmed by a narrow, smooth, continuous sheet of tissue. The corrugations often arc back toward the appendage base, unlike actual feather barbs, which are consistently oriented distally. Whether or not the structures had a feather-type growth pattern is disputed (compare Jones et al. 2001 with Prum in Stokstad 2000, Prum 2001, Unwin and Benton 2001). It has not been demonstrated that the appendages were true feathers or protofeathers; they may have been scales that only partly mimicked feathers.

The evidence for the presence of furlike pelage on pterosaurs is accumulating (Bakhurina and Unwin 1995a,b, Frey and Martill 1999). Note that these external fibers were not present on the wing membranes, which were naked, as they are among bats (Vaughan and Bateman 1980). Biochemical tests to determine the composition of pterosaur fibers have yet to be conducted, so we do not know whether they were homologous with bird feathers.

Of dubious nature is the Late Jurassic *Praeornis* "feather" (Rautian 1978). Although it has been accepted as a true feather by Davis and Briggs (1995), and as a possible feather by Molnar (1985), other researchers have been much more skeptical (Feduccia 1996). We cannot even be certain whether the object is animal or vegetable in origin, because the necessary biochemical tests have not been done. The structure was rather large (well over 150 mm long), and not enough of it is preserved for us to tell whether it was symmetrical or asymmetrical. The large dimensions reduce the possibility that this Jurassic fossil was avian, but it may be dino-avepodian.

We next considered dinosaurs. Large examples, including avepods, were covered by mosaic scales (Paul 1987a, 1988a, Bonaparte et al. 1990, Czerkas 1997, L. Martin and Czerkas 2000). The last

*Figure 5.2. Left,* impression of probable dorsal integument of *Longisquama. Center,* impressions of primary feathers of *Archaeopteryx,* HMN 1880 counterslab. *Right,* primary feather of a modern bird. All views shown distal end toward the top. The avian feathers are very similar; note how even the distally oriented, fine barbs are well preserved in the urvogel. Quite different are the sheetlike vanes of the longisquamid impressions, whose course ribs arc proximally and are edged by a continuous smooth border region.

two researchers argued that skin made of mosaic scales would have been ill suited for the initial evolution of feathers, but this opinion is speculative since exactly how feathers first developed is not known. Until recently the crucial questions concerning the integument of small dinosaurs went unanswered because neither scales nor feathers had been documented. This problem continues to apply to small ornithischians: the only integument reported in this group has never been described or illustrated in detail and is currently inaccessible (Paul 1988c), so we cannot even be certain that it really is fossil skin. That the supposed skin of the ornithopod was nonscaly casts further doubt upon its true nature. The psittacosaur skeletons found in the Yixian lagerstätten (Barrett 2000) are reported to be scaly, but no specific information has been published.

Fortunately, a burst of new data has suddenly and spectacularly improved the fossil record for the integument of small Mesozoic avepods. Kellner (1996a,b) reported naked, nonscaly skin on part of an incomplete and partly disarticulated Cretaceous example. The skin patches are limited to the hindlimbs, so they may not represent the general body covering. Although the preservation of the fossil tissues is excellent, the preserved skin is very thin, and the outermost integument may have been lost before burial. The nonscaly skin covering Kellner's specimen is very different from the skin of large dinosaurs. This difference—which is difficult to explain because the large dinosaurs evolved from small ones—gives us additional reason to suspect that the preserved integuments do not represent the life condition.

Briggs et al. (1997) reported that the majority of the soft tissue associated with a Cretaceous ornithomimosaur represents muscle tissue; scaly skin covers a throat pouch (contra the impression given by Perez-Moreno et al. [1994] and L. Martin and Czerkas [2000] that scaly skin covered much of the body).

The absence of plumage or skin associated with the *Compsognathus* skeletons preserved in fine-grained sediments (Bidar et al. 1972, Ostrom 1978) does not rule out the presence of body feathers any more than does the absence of the same in complete urvogel specimens from the same lagoonal sediments (Paul 1988a,c). The absence of plumage on the even more exquisitely preserved *Scipionyx* (Sasso and Signore 1998) is equally uninformative because no trace of skin is present.

Various claims concerning trace evidence for feathers have been made. The claim that quill nodes adorn the ulna and metacarpals of "*Protoavis*" (Chatterjee 1991, 1997, 1999a) is not convincing. The irregular "nodes" appear to be the result of postmortem deformation of badly damaged bone surfaces and are not at all like true quill nodes. Also unconvincing is Weems's (1987) inference that the shallowness of dino-avepod trackways resulted from the aerodynamic lift produced by arm feathers. Ellenberger (1974) thought that impressions associated with the footprints of small dinosaurs, which he called carnavians, were made by feathers, but his view has not been accepted (Molnar 1985). Additional trace fossils include possible feather impressions accompanying the belly impressions of resting dinosaurs (Gierlinski 1996, 1997, Kundrat 1998). Although identity of these dinosaurs is not entirely certain, the handprints (which are small, have five digits, and lack a large thumb claw) suggest an ornithopod or a basal theropod, but the footprint (which lacks a fully developed first toe) favors an avepod. In addition, these dinosaurs were fairly substantial at 100 to 150 kilograms. Gierlinski and Kundrat concluded that the impressions record simple feathers rather than contour feathers or down. The possibility that these impressions are pressure-induced distortions of the sediment has yet to be ruled out. Not yet analyzed are possible mid-Cretaceous dinosaur-feather impressions noted by Kranz (1998).

The brooding oviraptorids are interesting in regard to the subject of dinosaur feathers (Norell et al. 1995, Dong and Currie 1996, Clark et al. 1999). Their bodies and arms are arranged in a birdlike manner that suggests they were insulating their eggs, but their slender appendages could not have covered the entire clutch without the aid of well-developed plumage (Fig. 10.16). Although this evidence is highly suggestive, it is not definitive. The same is even true of the presence of a pygostyle at the tip of the abbreviated tail of an oviraptorid (Barsbold et al. 2000). Barsbold and his colleagues proposed that the birdlike structure supported a birdlike fan of contour feathers, a view strengthened by the presence of such an array on the similarly short tail of caudipterygians (see below in this chapter). Likewise, the ridge on the ulna of *Avimimus* implies, but does not firmly establish, the presence of well-developed arm feathers (Kurzanov 1987).

What does establish the presence of avepod feathers is the ulna associated with the remains of sickle-clawed dino-bird *Rahonavis* (Forster et al. 1998). On the bone's posterior surface is a row of exquisitely preserved, classically avian quill nodes, which show that the large arm must have supported well-developed contour feathers.

The first direct evidence for dinosaur feathers came from Cretaceous formations in Asia. One of these sources is the Djadokhta Formation of Mongolia (Schweitzer et al. 1999). Associated with the skeleton of the alvarezsaur *Shuvuuia* found in Mongolia are bristles whose preservation is best described as exquisite (Fig. 13.9). The fibers are still uncrushed, three dimensional, and hollow. On the basis of high-resolution electron microscopy and chemical analysis of the keratin, Schweitzer et al. concluded that the detailed composition of the bristles is most compatible with their having been feathers rather than some other form of epidermal structure. No other Mesozoic integument has undergone such detailed analysis, and yet these results have not received the attention they warrant. These structures establish the presence of external keratin fibers in nonavebrevicaudan avepods and are excellent candidates for bristle feathers.

Another source of evidence for dinosaur feathers is the extraordinary Jehol lagerstätten of China. It has produced three specimens of the compsognathid *Sinosauropteryx*, all covered to some degree by a dense layer of short, fibrous material (Pls. 16–18, Figs. 1.6, 5.3; Ji and Ji 1996, 1997a, Chen et al. 1998, Griffiths 1998/2000). Geist et al. (1997), Feduccia (1999), and Ruben and Jones (2000) have claimed that these fibers represented internal collagen fibers, and the first two references claim that the fibers supported a dorsal fin for swimming. However, what they identified as the body outline outside of the fibers is actually breakage, preparation work, and brushed on sealants. Sealant is extensive on NGMC 2124: it was applied to broad areas surrounding the main skeleton as well as to isolated elements and feather bundles. The sealant stained the sediment to a uniform tone only slightly darker than the untreated, light-toned matrix; actual carbonized body tissues tend to be darker and more variably toned. In most locations, the sealant was thin enough to have dried to a matte finish after having been absorbed into the sediment, but in other places, it was applied so thickly on the matrix that it formed a glossy glaze. Thicker still are the circular, raised-rim, glistening drip marks where drops of sealant beaded after having dropped from the brush onto the sediment. As the brush was stroked across one level of sediment and then over the edge of a shallow (approximately

*Figure 5.3.* Two feathered Yixian dinosaurs. *Top,* the basal avetheropod *Sinosauropteryx,* perhaps bothered by feather-dwelling parasites. *Bottom,* the birdlike avepectoran *Caudipteryx.* Here a male is displaying to a passing female; the banding on the distal tail feathers follows the pattern seemingly preserved in the fossil feathers. The Yixian Formation is part of the Jehol group.

1 millimeter) drop to a lower plane of sediment, the brush often failed to apply the sealant to the space immediately downstroke of the drop-off; the normal color of the sediment in these spaces further falsifies the claim that soft tissue was present. That the same stain continues onto a block of material erroneously attached to the main block (see Appendix 3B) verifies that the stain is sealant applied at the finish of preparation. Biochemical tests that would demonstrate that the stain is fossil organic tissue rather than sealant have not been conducted. L. Martin and Czerkas (2000) claim that a patch of scaly skin on one of the specimens refutes the presence of feathers. This assertion is not documented and is itself refuted by the abundant fibers found on multiple specimens.

Perhaps more amusing but equally unsubstantiated is Feduccia's labeling (in Morell 1997) of feathered *Sinosauropteryx* as the "Pilt-*down* dinosaur." Conducting biochemical tests to determine the composition of the *Sinosauropteryx* fibers has proven impossible because they are too degraded by carbonization. Because the fibers are very fine, tapering, smooth edged, and undoubtedly external, they were almost certainly keratinous in origin (Ji and Ji 1996, 1997a, Chen et al. 1998). The dense packing of the fibers makes it unclear whether they were single bristles or whether they had a more complex branching pattern (see Ackerman [1998, 83], and further discussion later in this chapter). The fibers may have been hollow. They are found on the posterior skull and on the neck, body, tail, arms, and legs. Their widespread distribution and high density indicate that the fibers thoroughly covered the animal, with the possible exception of the snout; the hands and feet were probably bare from the wrist and ankle down, respectively. Bristles seem to adorn the tip of the tail of one specimen (Ji and Ji 1997a, Ackerman 1998, 78–79), but these are on a separate, boneless block (the apparent presence of bone in some photographs is an illusion, and the end of the tail is missing at the edge of the main block), they lack direct attachment to the caudals, and they appear to be drift remains that were placed at the end of the tail by the collector.

Bristles similar to, but longer than, those of *Sinosauropteryx* have also been found on the basal therizinosaur *Beipiaosaurus* (Xu, Tang, and Wang 1999). The bristles are associated with the forearm, thigh, shank, and chest, but the specimen is too incomplete any determination of the full extent and nature of its integument. In particular, no contour feathers are present, even though two hands remain articulated. Although the absence of feathers may reflect a lack of preservation, the pronounced curvature of the central fingers on both hands is also inconsistent with the straightness of the same finger in birds with well-developed digital remiges (see discussion later in this chapter).

Long, ill-defined fibers have been found associated with the skeleton of the little dromaeosaur *Microraptor* (Xu et al. 2000). The dromaeosaur *Sinornithosaurus* also has fairly long bristles distributed over various parts of its body, although exactly which parts is not entirely certain (Figs. 1.6, 6.1; Sloan 1999, 104; Xu, Wang, and Wu 1999; Xu et al. 2001). For additional information on sinornithosaur feathers see Appendix 6 (and Pl. 16). Contour feathers are absent, even though the straight articulation of the central finger in the one complete hand is compatible with the presence of a well-developed set of distal remiges. Although the incomplete articulation in the one specimen, and especially the absence of most of the tail, prevent us from determining whether contour feathers really were absent, there are reasons to think that they were. The absence of remiges in a specimen in which the preservation of the skeleton and the simple fibers is quite good is very significant, and long fibers project from the articulated right forearm and hand with exactly the posterior orientation expected if they were attached to this appendage. The fibers seem less dense astride the hand, perhaps because they were attached only to the slender manus rather than to the thicker arm. In any case, the presence of well-developed middle and distal remiges is therefore excluded. Xu et al. (2001) provide a detailed description of this dinosaurs feathers. They consist of two types: semiplumes without a central shaft and structures with the central shaft and barbs but no barbules. Simple bristles seem to be absent.

Even more spectacular and informative is the Yixian avepectoran *Caudipteryx* (Pl. 14, Figs. 1.6, 5.3; see the discussion of its phylogenetic position in Chapters 10 and 11). Four published skeletons sport both simple feathers on parts of the body, as per *Sinosauropteryx*, and fully developed, symmetrical contour feather fans on the very short hands and on the end of the short tail (Ackerman 1998, Ji et al. 1998, Padian and Chiappe 1998a, Zhou and Wang 2000). The borders of the counter feathers are somewhat irregular rather than crisply defined. Life restorations have shown remiges along the entire arm, but these are absent in NGMC 97-4-A; and fibers posterior to the forearms of NGMC 97-9-A appear to be simple ventral body feathers rather than more-complex brachial examples. Because, as we have seen, the central finger appears to lack a distal expansion of the proximal-most bone, the absence of this feature is not conclusive evidence for the absence of well-developed distal remiges. The presence of simple feathers on the tail base confirms the absence of proximal rectrices. Again, the exact degree of complexity of the simple body feathers is not certain.

The unambiguous contour feathers of *Caudipteryx* effectively settle the question of whether or not at least some avepod dinosaurs were feathered. The apparently hollow nature of other Yixian dinosaurs' simpler fibers indicates that they too were feathers. Indeed, the preservation of these dinosaurs' body plumage is strikingly similar to the irregular, fibrous appearance of body feathers that surround fossil birds (Ackerman 1998, 93; Hoffman 2000, 40; Fig. 6 in Mayr and Mourer-Chauvire 2000). The relatively rigid wing and tail contour feathers are well preserved in the same fossils; the fluffier body-contour feathers collapsed and assumed a fibrous appearance. It is therefore possible that some of the dinosaurs' feathers were more complex than they appear. All of the available evidence, including the evidence provided by the hollow alvarezsaur fibers, supports the conclusion that the fibrous integuments were feathers. Definitive biochemical tests showing that the fibers are not made of feather-type keratin would be required to demonstrate otherwise, and thus for the remainder of this study, it will be assumed that these fibers were indeed feathers. Because the simplest dinosaur feathers appear to have been no simpler than the simplest bird feathers, the former should be considered feathers rather than protofeathers. It is further presumed that the simple feathers formed a nearly complete body plumage. The main question is how far feathers, both simple and complex, extend down the dinosaur phylogenetic tree. Simple fibrous feathers are observed in basal Avetheropoda but may have been present in basal Avepoda and perhaps even lower. So far, pennaceous feathers appear only in Avepectora. Strongly asymmetrical flight feathers should be limited to fliers (see Part 3). It is probable that these feathers belonged to birds, although a flying dinosaur more primitive than the urvogel cannot be ruled out. Since all large avepod dinosaurs descended from smaller examples (Part 5), it is possible that, despite their scaly skins, large dinosaurs retained patches of feathers for display (Fig. 13.10). Mobile tails and arms, including the more-reduced examples, would be primary sites for such arrays. Some workers have speculated that the juveniles of large avepod dinosaurs may have been insulated with feathers (Paul 1988a, Sloan 1999). We can hope that new fossils will eventually answer these questions.



The presence of even the most complex kind of feather on a dinosaur means that we cannot assume that the symmetrical isolated contour and downy feathers found Mesozoic sediments are avian rather than dinosaurian in origin (Paul 1988a, Ji et al. 1998 , although it is possible that such complex feathers did not appear until the advent of some level of flight (Chapter 13).

The last point brings us the to the famous isolated Solnhofen *Archaeopteryx* "ancient" feather, redescribed in detail by Griffiths (1996). He noted that the form of the asymmetrical contour feather is not fully compatible with the form of the feathers adorning the urvogel specimens found in the same sediments, so it is possible that this feather belonged to another flying avepod. Therefore, the common assumption that the anatomy of the very well-preserved isolated feather is directly applicable to the less well-preserved feathers associated with the skeletons is unsubstantiated.

Let us now turn to the feathers attached to skeletons. A drawing of HMN 1880 made before the specimen was prepared seems to affirm that there was a feathery body covering (Feduccia 1996). Traces of this covering around the lower legs may still survive, and we would expect plumage to be present, considering its presence in more-basal avepods and *Protarchaeopteryx* (see below). The drawing does not support the presence of a well-developed feather crest on the head, which is seen on some restorations. We do not know whether the body feathers were simple fibers or more complex, but they probably were similar to those of *Archaeopteryx*'s close relative *Protarchaeopteryx* and the baso-avebrevicaudan confuciusornithids (see discussion later in this chapter). Therefore, the presence or absence of a streamlined aeroshell cannot be confirmed. That there is no trace of body covering present in specimens other than HMN 1880 is not surprising: the wing and tail feathers of the other specimens are more poorly preserved than the spectacular arrays preserved with HMN 1880. The rectrices run along the entire tail instead of being absent from the anterior-most caudals (contra Ji et al. 1998).

Gatesy and Dial (1996) argued that the tail feathers were fixed immobile on their supporting vertebrae. Among the primaries attached to skeletons, none of the non-overlapping examples are well preserved enough for accurate measurements of the breadth of the sections anterior and posterior to the central shaft, but enough is exposed to show that asymmetry was strong (contrary to Speakman and Thomson 1994). In HMN 1880, asymmetry ranges from at least 1.25 to 5, the higher values being seen among distal primaries whose trailing edges are covered.

In my restoration of the wing area of HMN 1880 (Fig. 7.4Aa) and JM 2257, the arms and feathers were reposed in flight position so that the wing feathers would be fully extended, the proximal secondary feathers were assumed to be the same length as the primary feathers, and no patigium was included forward of the inner arm because this typical avian feature is not preserved in any of the *Archaeopteryx* specimens. The results were extrapolated to fit the larger wings of BMNH 37001. The wing area for HMN 1880 (500 square centimeters) is about a third higher than that estimated by Heptonstall (1971a,b) and Yalden (1971a,b), apparently because they underestimated the length of the inner wing feathers. However, Garner et al. (1999) concluded that *Archaeopteryx* lacked proximal wing remiges—in which case the number of primaries is about eighteen in HMN 1880. This interesting observation is compatible with the condition of the specimens but cannot be fully verified because the tucking in of the arms usually leads to the inner wing region being obscured by the trunk, neck, or skull, whose rotting tissues may have destroyed feathers that were originally present. Elimination of the proximal secondary feathers decreases wing area by approximately 25 percent. The total spans of the articulated wings of JM 2257, BSP 1999, and HMN 1880 can be readily measured, and the less articulated wing of BMNH 37001 can be estimated.

In the other archaeopterygiform, *Protarchaeopteryx*, the distribution of feathers in the type specimen (Ji et al. 1998) is perplexing. The distal fan of rectrices is fairly well preserved at the end of the short tail, and simpler feathers adorn other parts of the body, including the base of the tail. The absence of any trace of feathers on the arms is therefore significant and surprising, considering that the arms are so long and that the central finger is the most robust, as in early birds with brachial feathers (whether or not the proximalmost bone of the central finger is distally expanded cannot be determined in this specimen). Even if *Protarchaeopteryx* did not fly, one might it to have had fairly well-developed remiges—especially in view of their presence not only on similar *Archaeopteryx* but also on *Caudipteryx*, which had even shorter arms than *Protarchaeopteryx*. Also significant is the fact that although fingers II and III of *Protarchaeopteryx* are crossed, like the feather-anchoring fingers of *Archaeopteryx*, it is the central digits of both hands that are strongly bent (see Fig. 4.3). These bent digits contrast with the feather-straightened central fingers of the earlier urvogel and suggest that a similar array of feathers was not present in *Protarchaeopteryx* to prevent the central digit from flexing dorsally under the pull of tendons. New remains of this early bird should clear up the matter eventually.

The Yixian early birds *Confuciusornis* and *Changchengornis* (both confuciusornids) and *Protopteryx* (an enantiornithine) had interesting tail feathers (Pl. 8, Figs. 1.6, 6.2; Ackerman 1998, 93; Ji, Chiappe, and Ji 1999, Zhang and Zhou 2000). On one hand, the rectrices were extremely short, to the point that the tail fan was essentially absent (Chiappe et al. 1999), a condition which is confirmed by countless specimens in which the rest of the feathers are present and which contradicts the long rectrices restored on a female in the cover illustration accompanying L. Martin et al. (1998). On the other hand, a small minority of the confuciusornithid specimens sport a pair of hyper-elongated, slender tail feathers that end with a distal expansion (Pl. 8B; Ackerman 1998, 93; L. Martin et al. 1998; Chiappe et al. 1999; Ji, Chiappe, and Ji 1999). These feathers may have been limited to one sex, mimicking the long, paired display feathers that adorn some birds of paradise, drongos, and manakins. The Yixian bird's tail feathers are also unusual in being continuous ribbonlike sheets of the type limited to a few modern birds. The outer primary feathers of *Confuciusornis* are notable in that they are very long and they form an exceptionally large wing (the cover restorations accompanying L. Martin et al. [1998] and especially Padian and Chiappe [1998a] failed to reproduce the full length of these feathers). A patigium is often preserved in confuciusornithids, but alular feathers do not adorn the thumb, as they do in more-derived birds like *Protopteryx* (Chiappe et al. 1999). The body feathers of these birds appear simple and fibrous, but this may be an artifact of the preservation of fluffy but complex body feathers, as discussed above. We, therefore, do not know whether confuciusornithid body feathers formed the streamlined aeroshell seen on modern flying birds. There appears to be a prominent feather crest on the heads of *Changchengornis* (Ji, Chiappe, and Ji 1999) and *Protopteryx* (Zhang and Zhou 2000) and a lesser one on at least some *Confuciusornis* specimens (Pl. 8B, Fig. 16 in Chiappe et al. 1999).

Feathers, both isolated and associated with skeletons, first become numerous at or soon after the Jurassic/Cretaceous (J/K) boundary. Feathered bird remains are actually abundant in the Jehol lagerstätten, which promises to produce many more feathered dinosaurs, dino-birds, and birds.

*Plate 1. Archaeopteryx* BMNH 37001, mainslab and counterslabs. BMNH 37001 was the first skeleton with disarticulated skull parts.

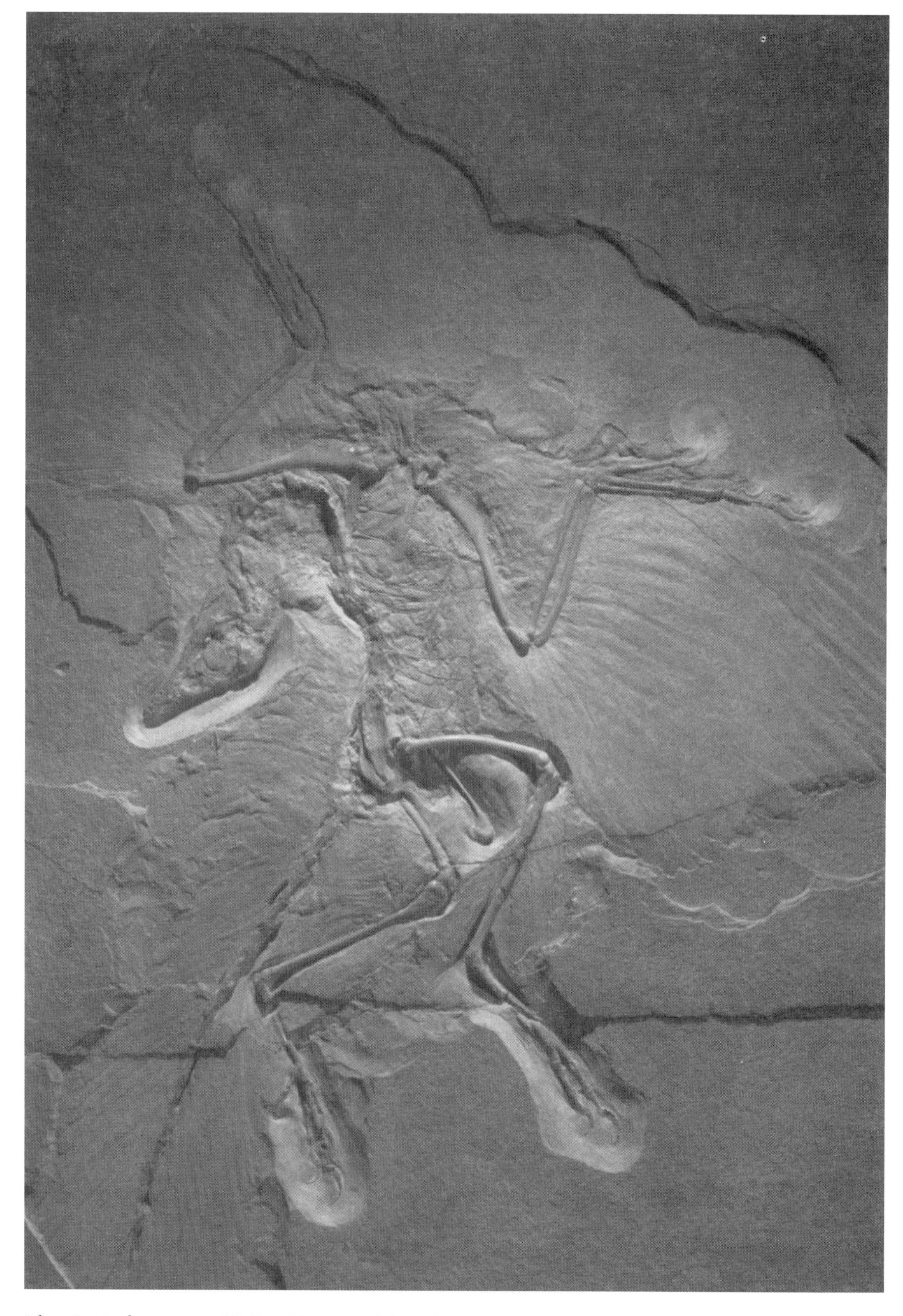

*Plate 2. Archaeopteryx* HMN 1880, mainslab and counterslab. HMN 1880 is the most complete specimen in terms of skeleton and feathers. Photographs courtesy of L. Witmer.

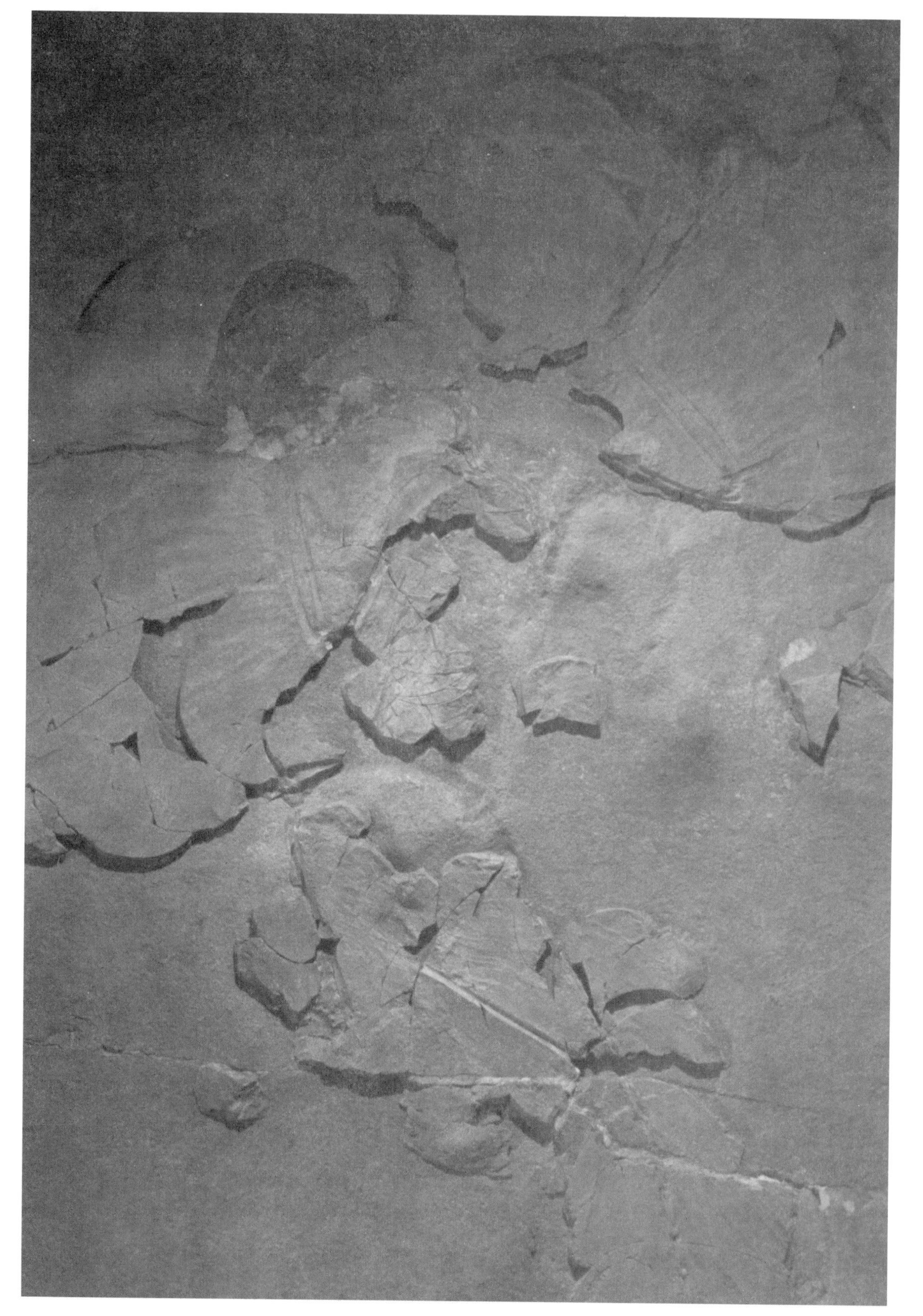

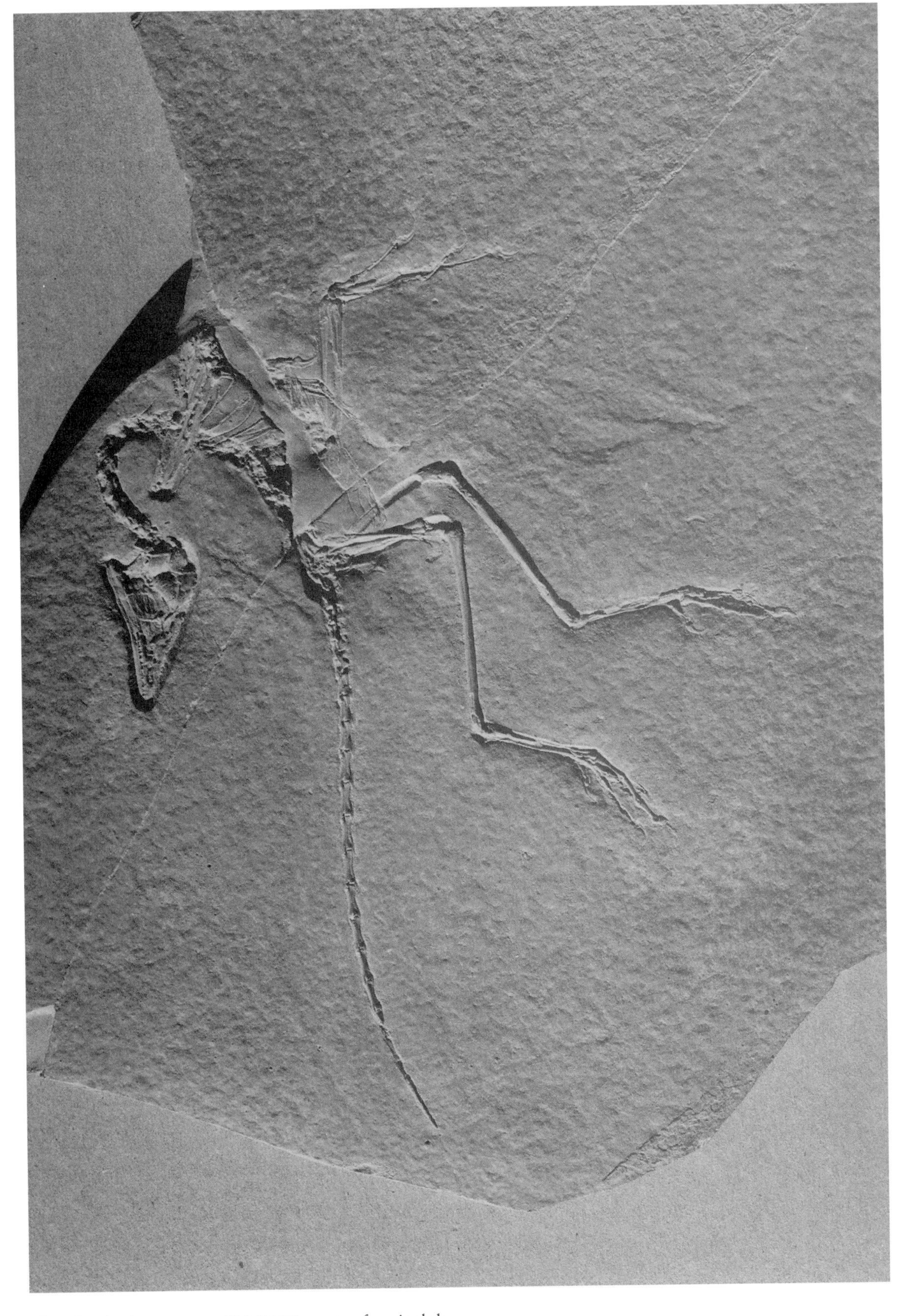

*Plate 3. Archaeopteryx* JM 2257, casts of mainslab and counterslab. The skull and skeleton of JM 2257 are complete.

*Plate 4. Archaeopteryx* sixth skeleton. This largest of the urvogel specimens has a complete skull and skeleton. Photograph courtesy of P. Wellnhofer.

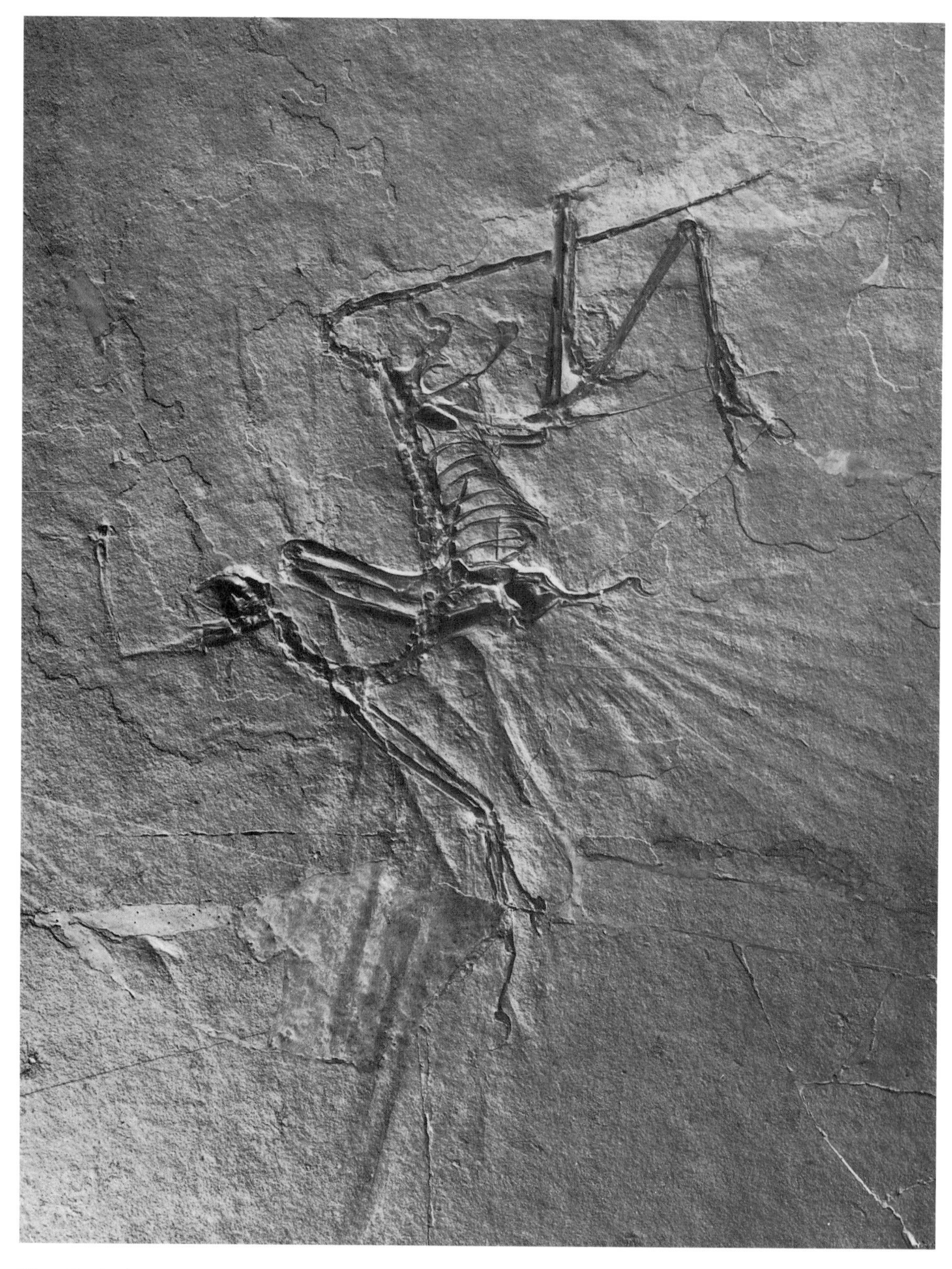

*Plate 5. Archaeopteryx* BSP 1999, nearly complete skeleton with disarticulated skull. Photographs courtesy of P. Wellnhofer.

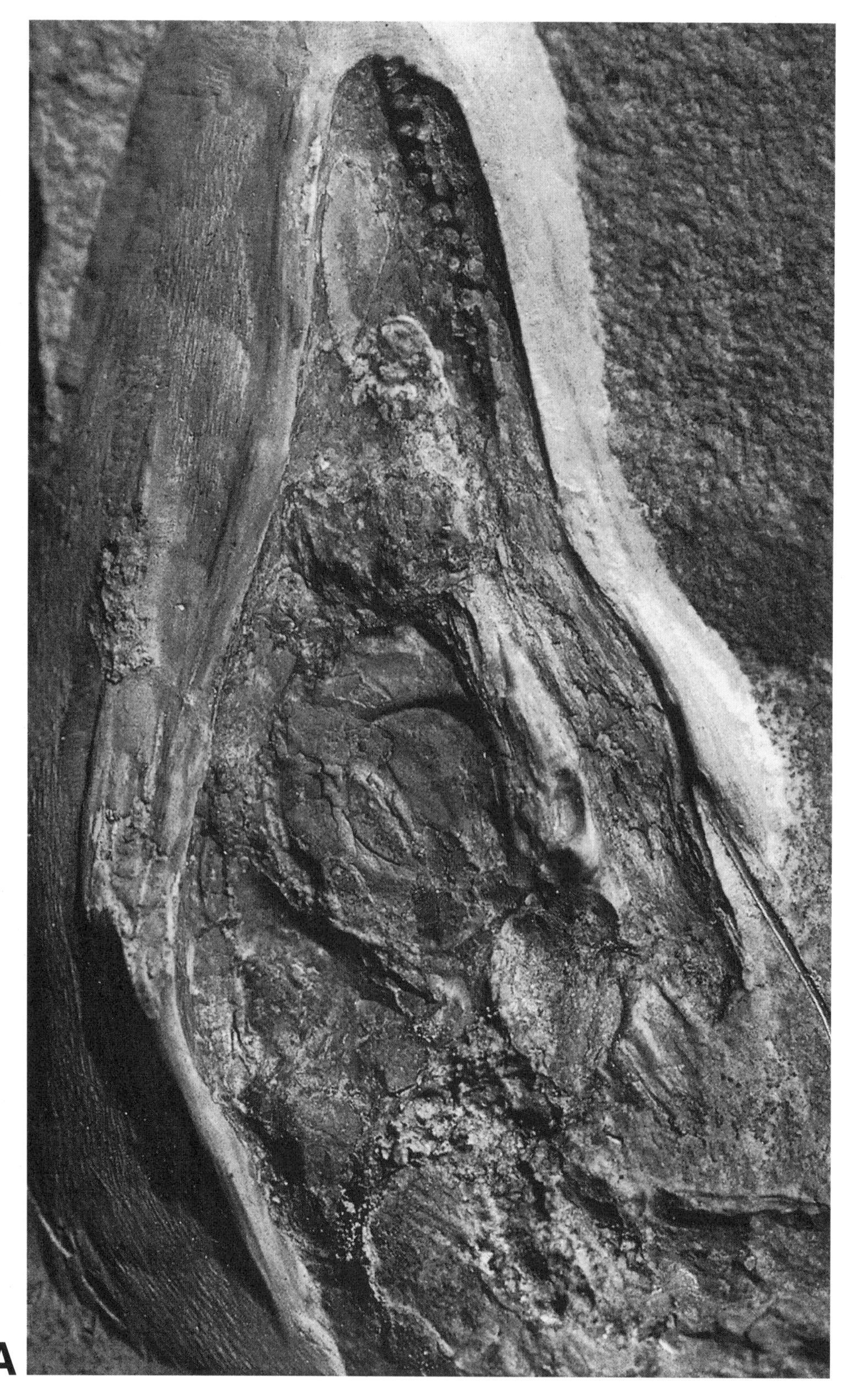


*Plate 6. Archaeopteryx* HMN 1880. *A,* skull (photograph courtesy of L. Witmer); *B,* interpretive sketch. Most but not all details of the skull are lost. Scale bar equals 10 millimeters. Abbreviations are listed in Appendix 7.

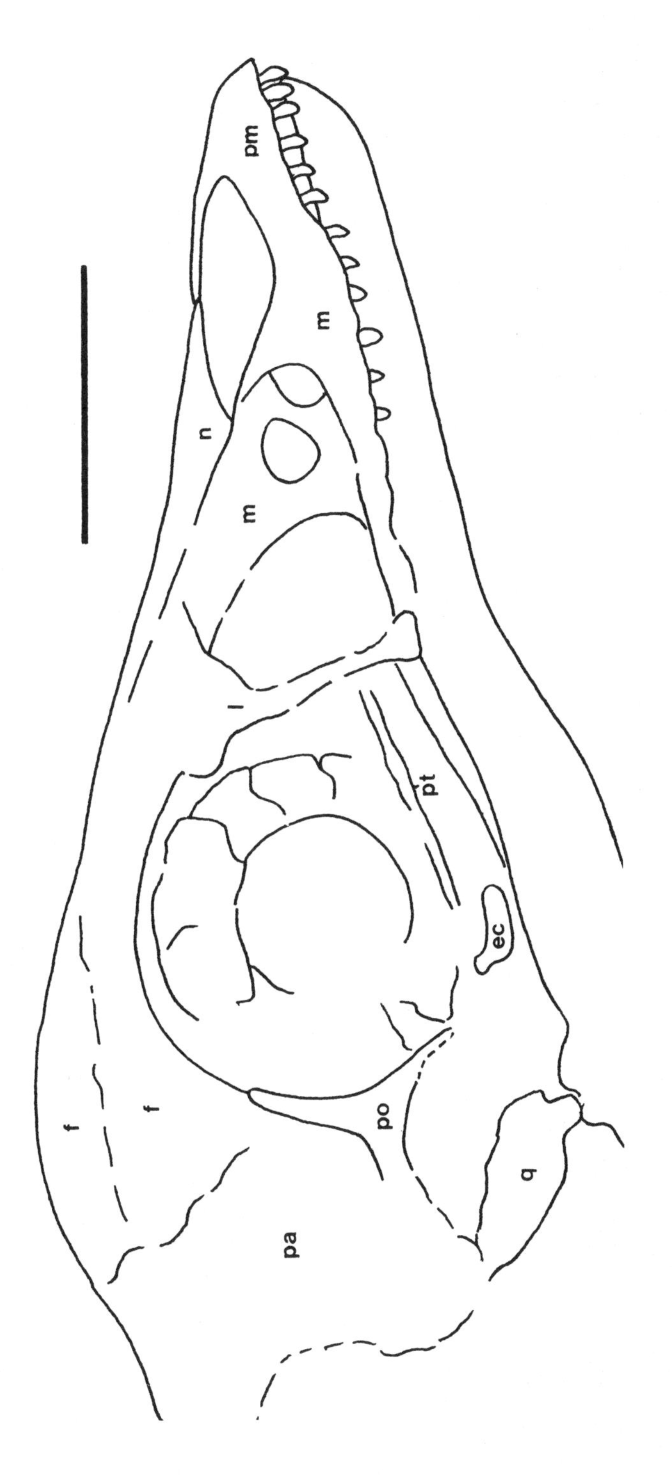
pm
m
n
m
l
pt
ec
f
f
po
q
pa
B

*Plate 7. Archaeopteryx* JM 2257, the best preserved on the urvogel skulls. *A,* skull, mainslab (photograph courtesy of P. Wellnhofer); *B,* skull elements from mainslab *(shaded)* overlain with counterslab elements *(stippled); C,* mainslab right palatal and mandibular elements. *B* and *C* are revised versions of Figure 31 in Chatterjee (1991). Scale bar equals 10 millimeters. Abbreviations: ? = either anterior pterygoid or posteromedial process of right palatine, z = anteromedial process of palatine; the other abbreviations are listed in Appendix 7.

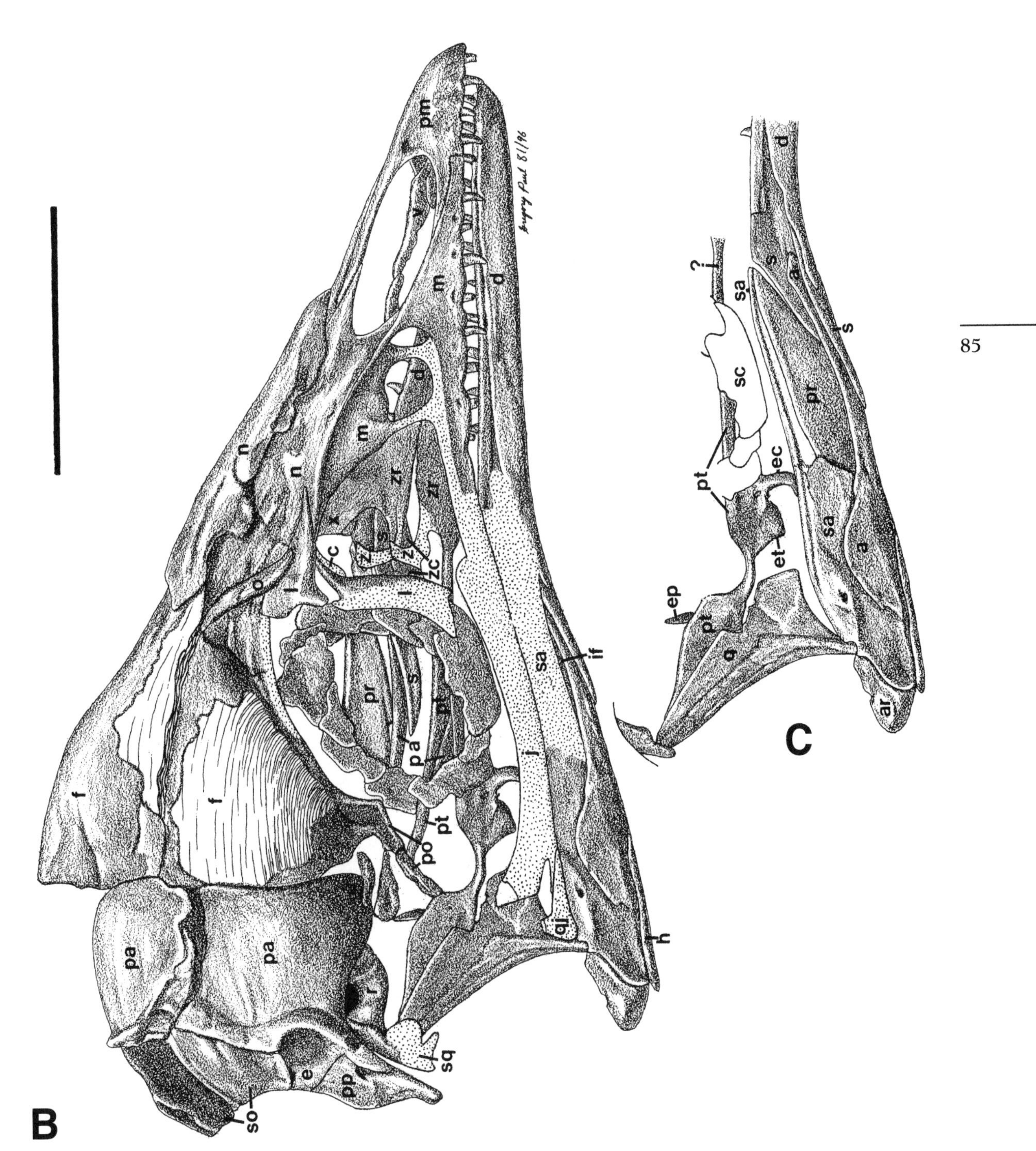
pm
v
m
d
n
o
l
c
s
z
zc
zr
x
pr
pt
pa
f
po
qj
h
if
sa
so
e
pp
sq
r
B
?
sc
ec
et
ep
q
a
ar
C

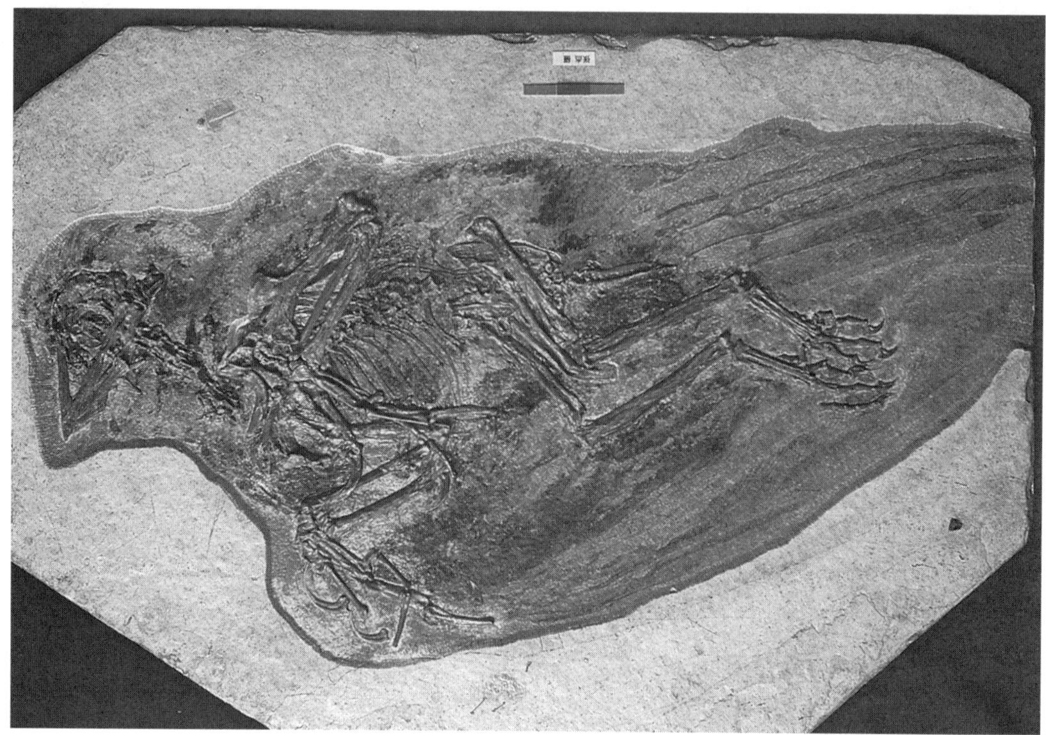


*Plate 8.* Complete *Confuciusornis* skeletons with feathers, two of the hundreds of complete, articulated specimens recently found. *A,* most skeletons are preserved on their backs or bellies (photograph courtesy of B. Mohn). *B,* this skeleton is unusual in that it is preserved on its side, and it records the true shape and depth of the pelvis; feather details are very well preserved, and a crest of head feathers appears to be present (photograph courtesy of L. Hou).

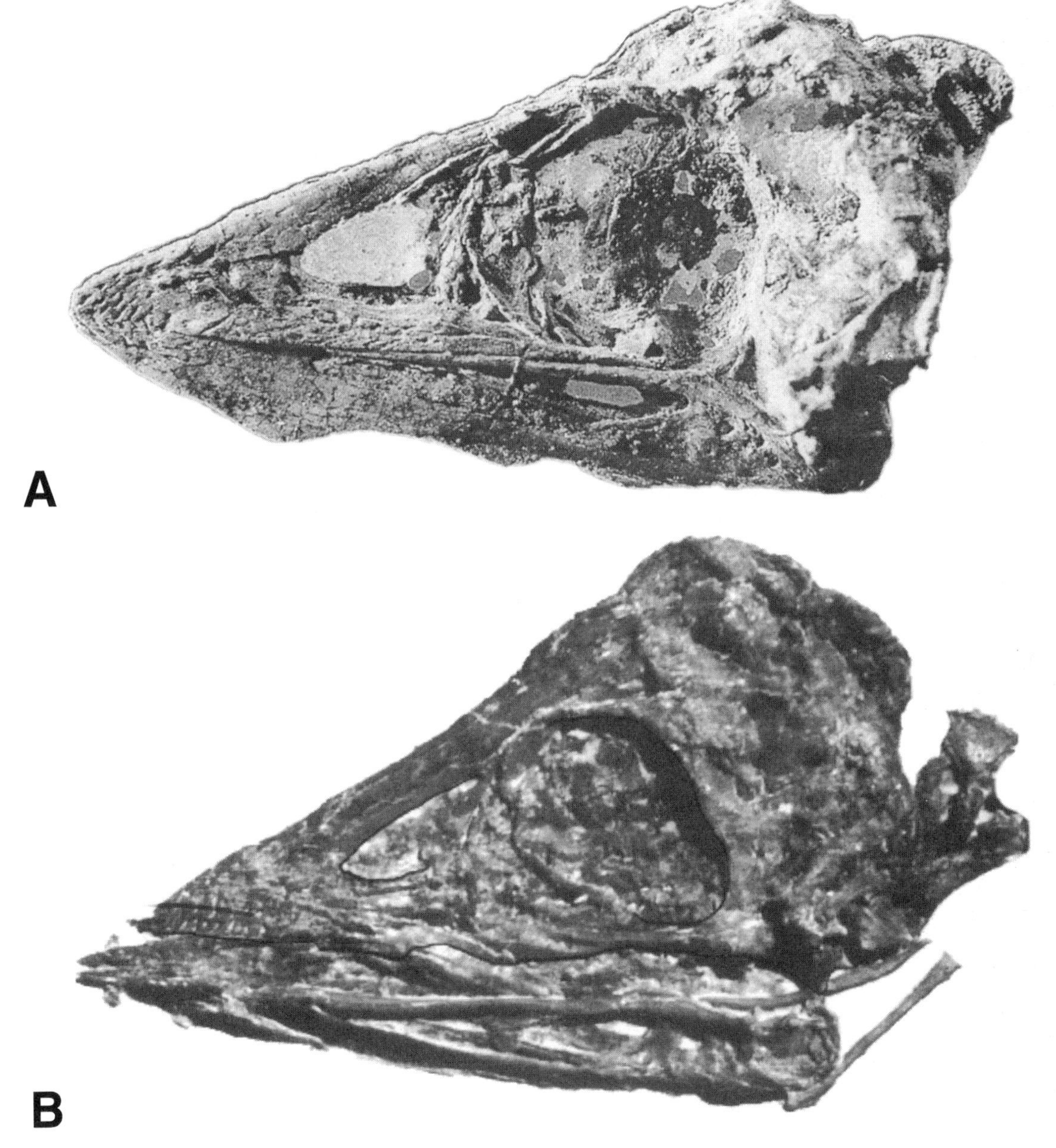


*Plate 9. Confuciusornis* skulls. *A*, the snout and jaws are well preserved in this skull (photograph courtesy of L. Hou). *B*, this example shows a complete and robust postorbital bar immediately behind the orbit.

*Plate 10. Cathayornis* type, nearly complete skull and skeleton. Photographs courtesy of Zhou Z.

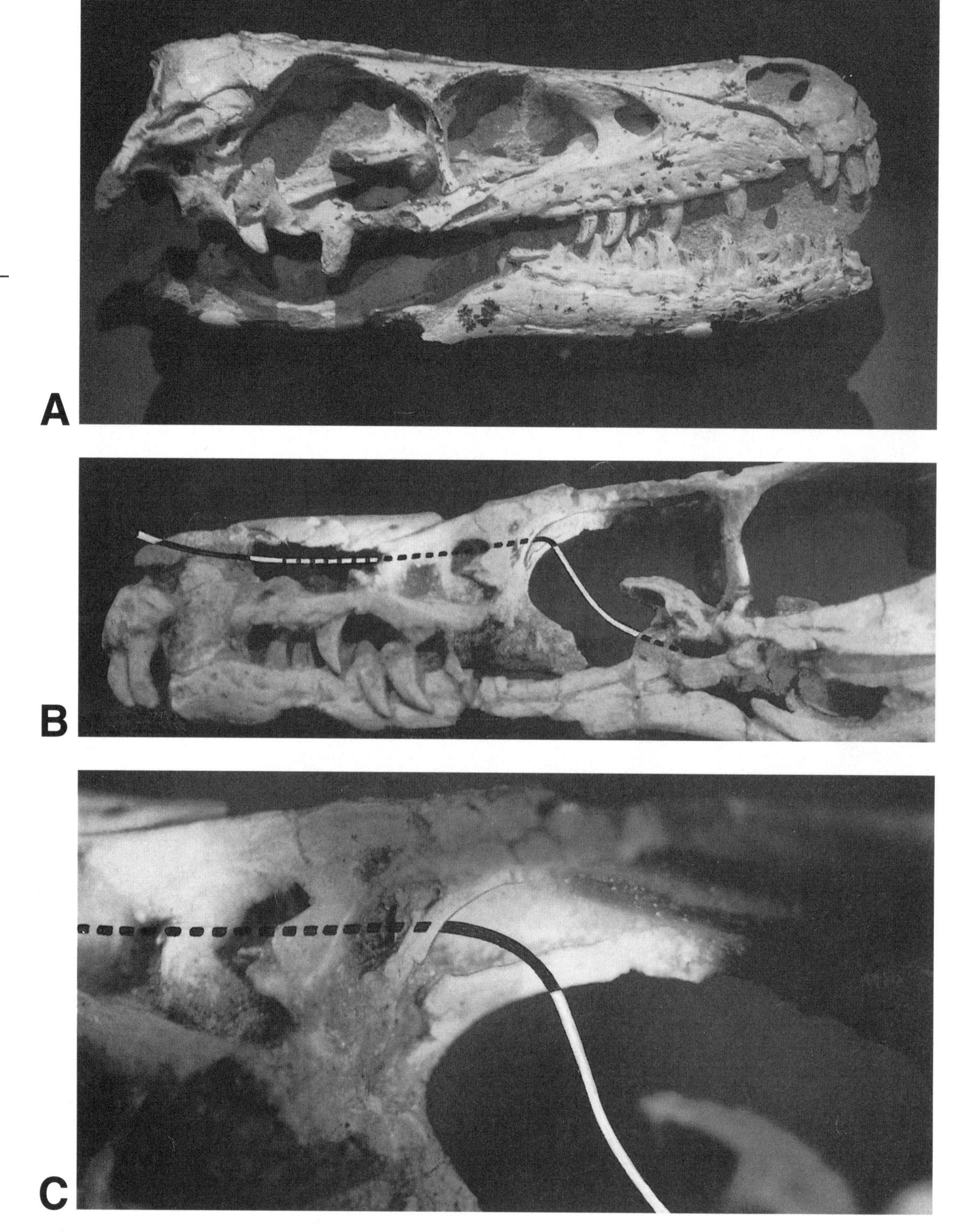


*Plate 11.* Dromaeosaur skulls. *A,* unnamed velociraptorine (photograph courtesy of M. Norell). *B,* snout of *Velociraptor,* showing course of nasal airway, including posterior end of anterior tube shown in enlargement *(C).*

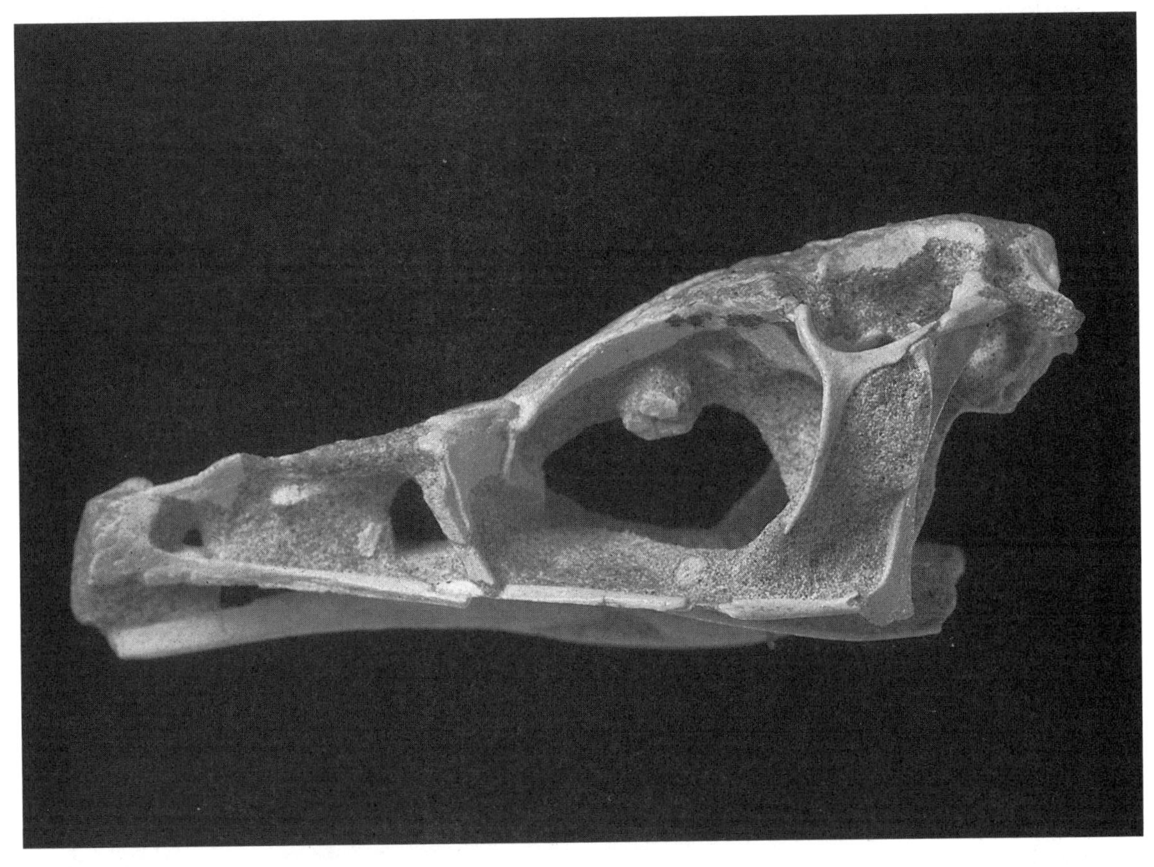

*Plate 12. Shuvuuia* type specimen, skull. This highly avian alvarezsaur lacks the postorbital bar common in dinosaurs. Photograph courtesy of M. Norell.

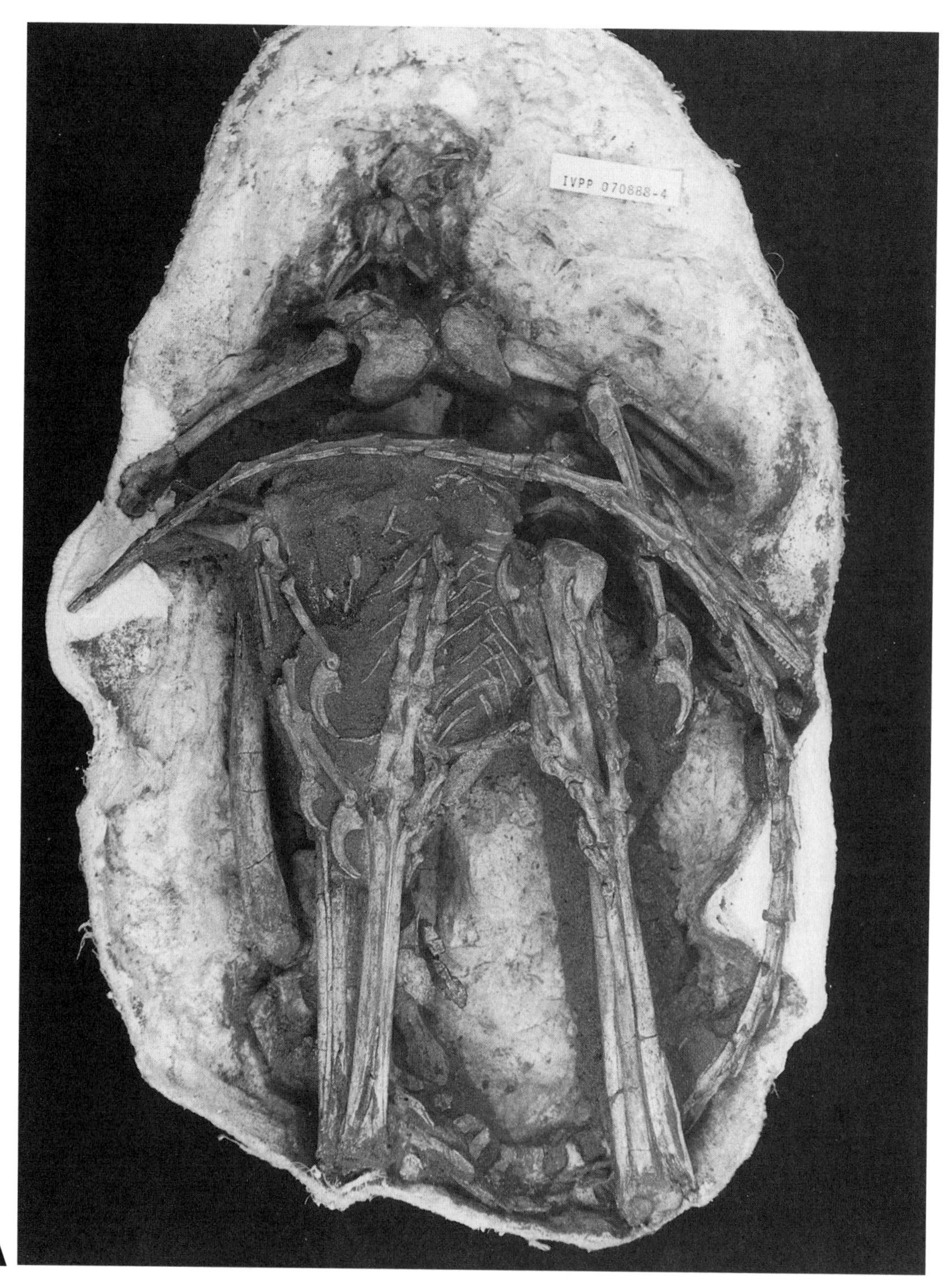


*Plate 13. Sinornithoides* type specimen, complete, naturally articulated skeleton, partly prepared. *A*, entire specimen in bottom view; note that in the shoulder region, the coracoids face strongly forward, the arms and hands are folded, the fingers are not crossed, and the second toe is more extended than the others. *B*, view from front, to the right of and below the shoulder girdle; note that the large coracoids face strongly forward and that their lower edges articulate with the anterior edges of what must have been the large cartilaginous sternum indicated by the gap between the coracoids and gastralia; also note small or partial clavicle in articulation with acromion process of scapula; compare with same view of similar articulated dromaeosaur shoulder girdle in Figure 2 in Norell and Makovicky (1999). *C*, left lower arm and hand, showing tight folding allowed by large semilunate carpal in wrist. Photographs courtesy of R. Day.

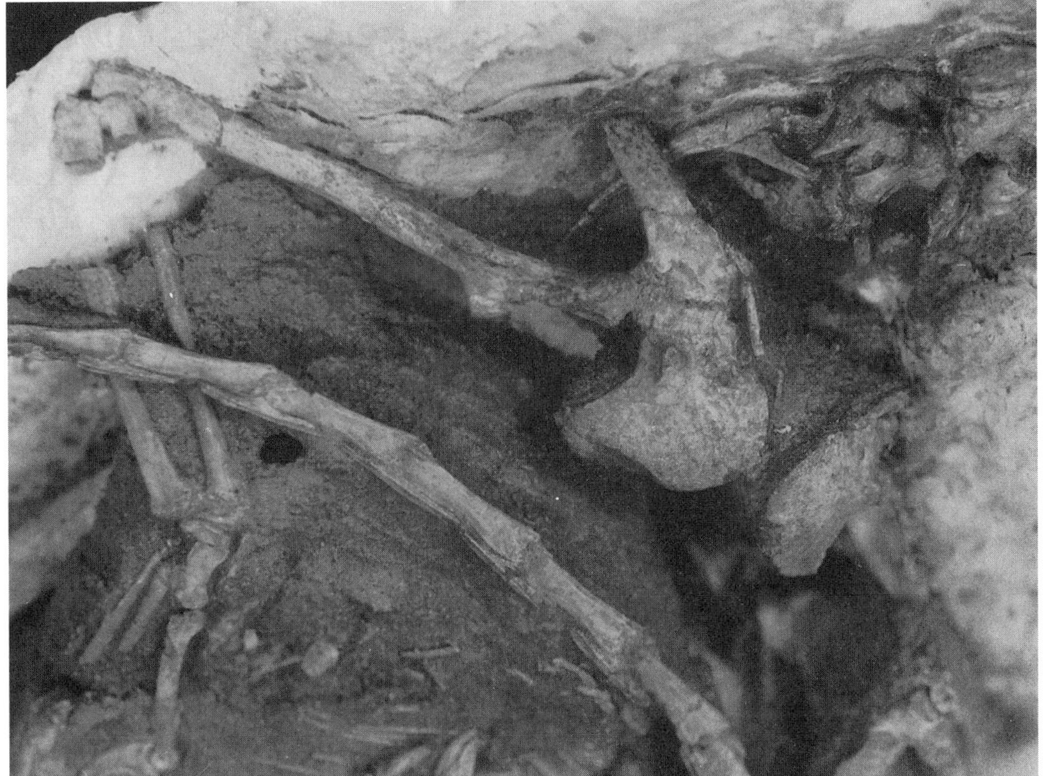
B

C

*Plate 14.* Complete *Caudipteryx* skeletons. *A,* type specimen with well-preserved contour feathers adorning hands and tail. *B,* BPM 0001, with better-preserved bones. Photographs courtesy of Zhou Z.

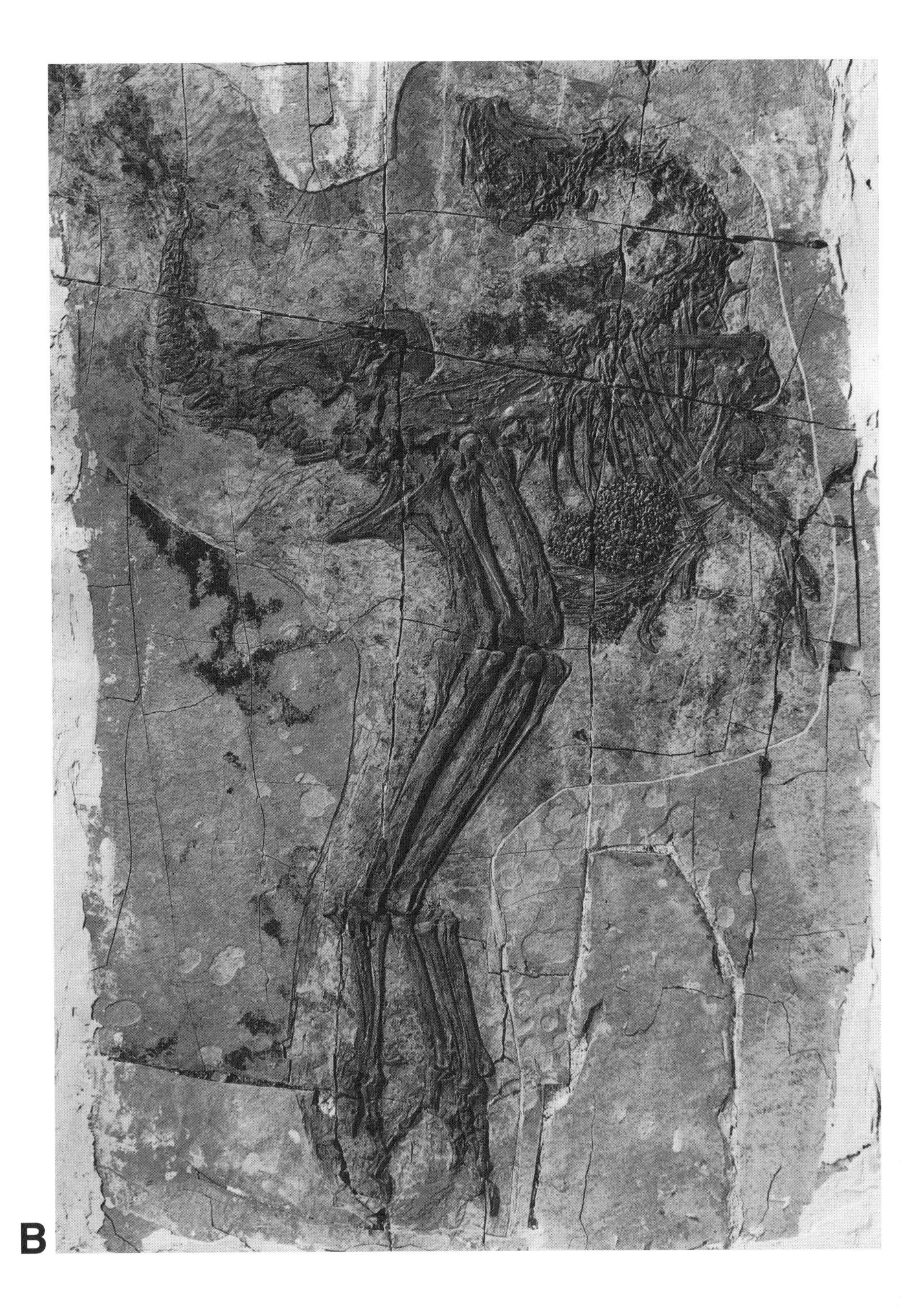
B

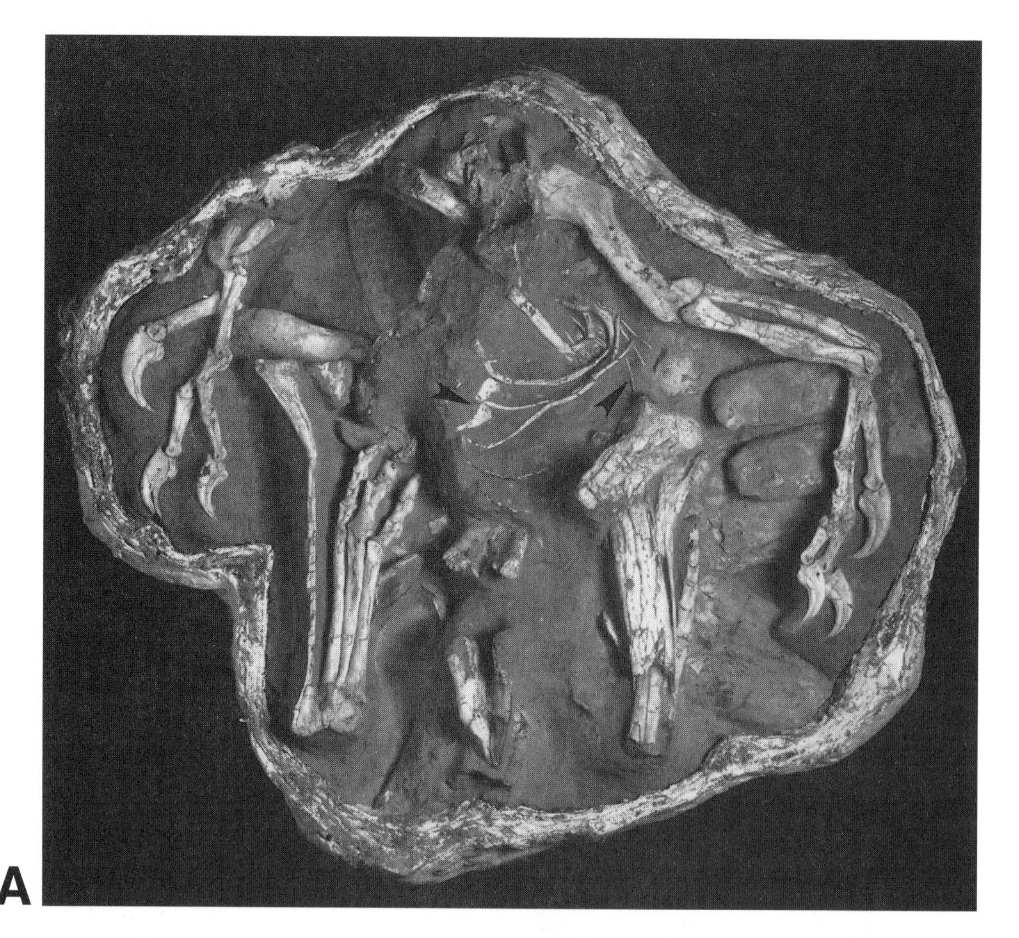


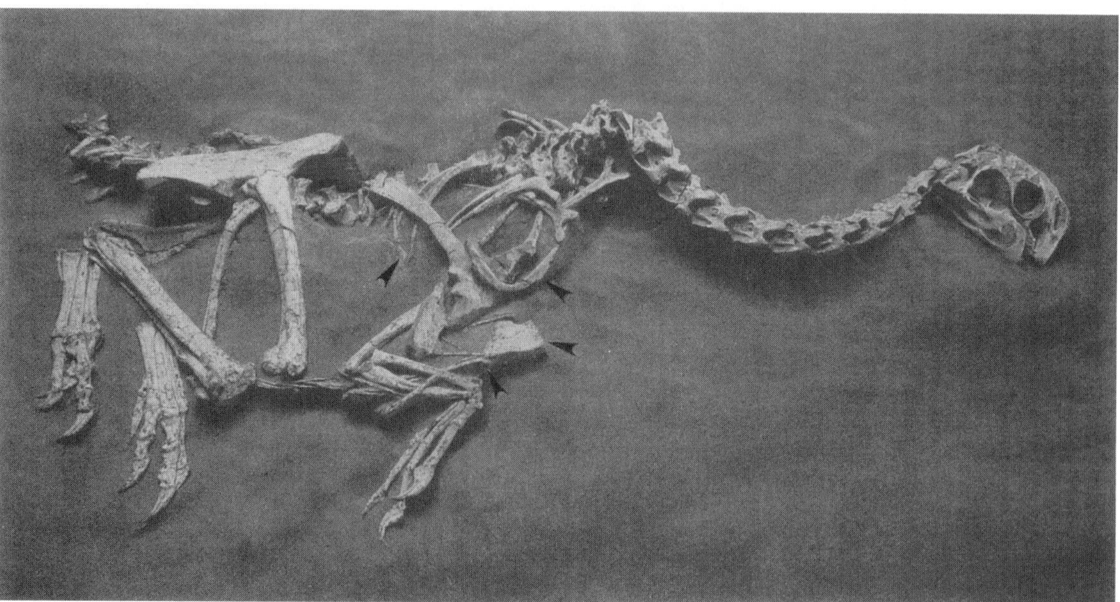


*Plate 15.* Oviraptorids. *A*, partial skeleton in brooding posture; there is direct contact between the sternum and some of the forelimb bones with eggs, which were laid in a two-level ring; note the strong flexion of the wrist, allowed by the large semilunate carpals. *B*, unnamed oviraptorid, almost complete skeleton. Both specimens show some of the remarkably avian attributes of avepectorans, including a large furcula, enlarged sternal plates, and ossified sternal ribs and uncinate processes, as indicated by arrows. Photographs courtesy of M. Norell.

*Plate 16.* Subadult sinornithosaur NGMC 91. Skull and skeleton with feathers in place. Note the unusual breadth of the proximal bone in both central fingers (compare with same finger bone in Pls. 2–5, 8, 10, 14, 15). Photograph courtesy of M. Norell.

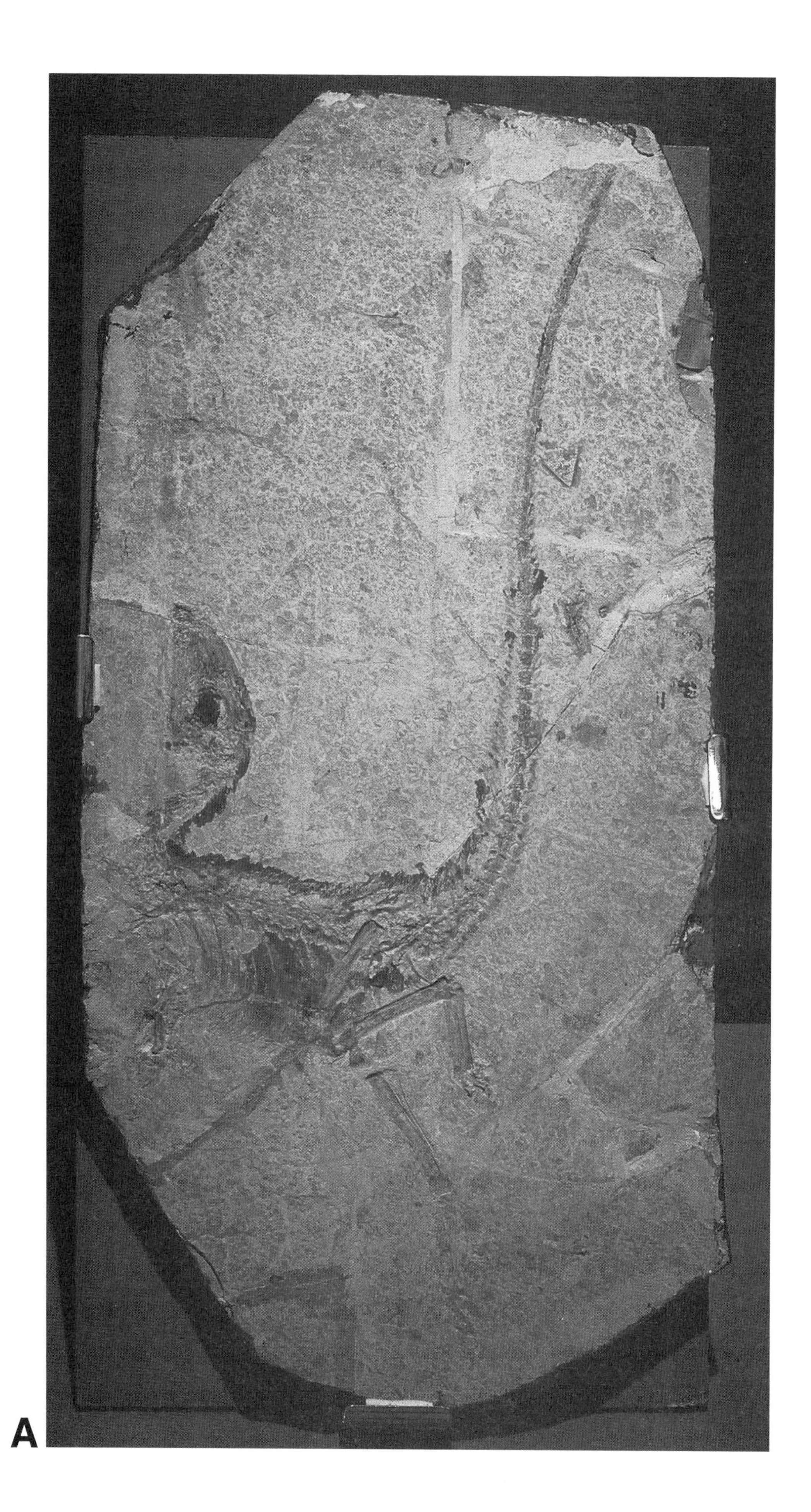
A

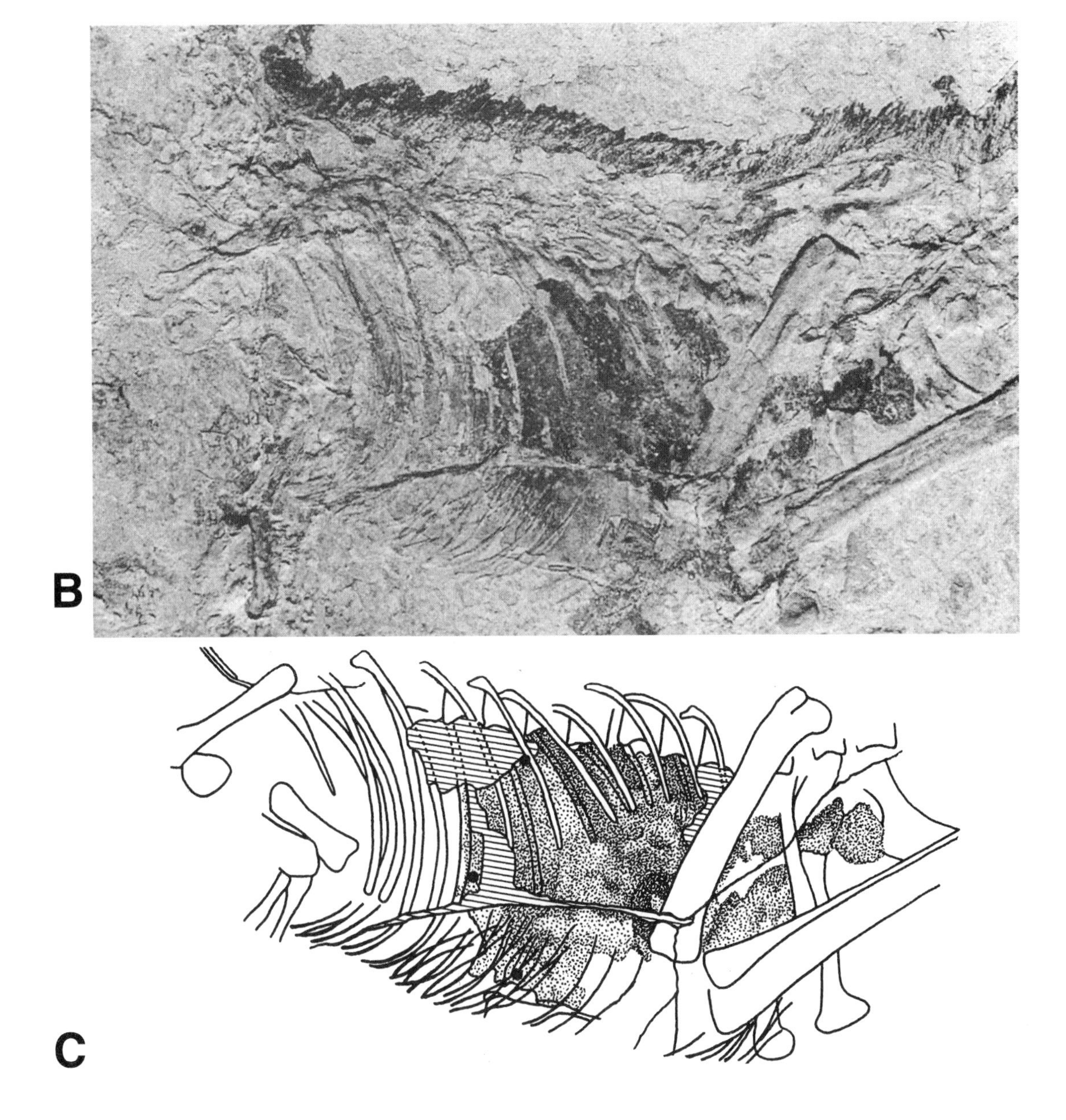


*Plate 17. Sinosauropteryx* type specimen. *A*, complete skeleton with bristle feathers, the first specimen to show that nonavian dinosaurs had these avian structures. Photograph courtesy of B. Mohn. *B*, trunk with soft tissues. *C*, interpretive sketch; shading indicates extent and density of dark material, hatching indicates major cracks and glue infill, solid borders of dark material indicate edges resulting from damage, and dots mark tips of arrows indicating supposed septum in Ruben et al. (1997a). See also App. Fig. 6Aa.

*Plate 18.* Well-preserved, complete *Sinosauropteryx* skeleton with bristle feathers. The subtle darker-toned area around the skeleton and some patches of feathers is preservative; the tuft of feathers at end of tail does not belong to this specimen. Photograph courtesy of B. Mohn.

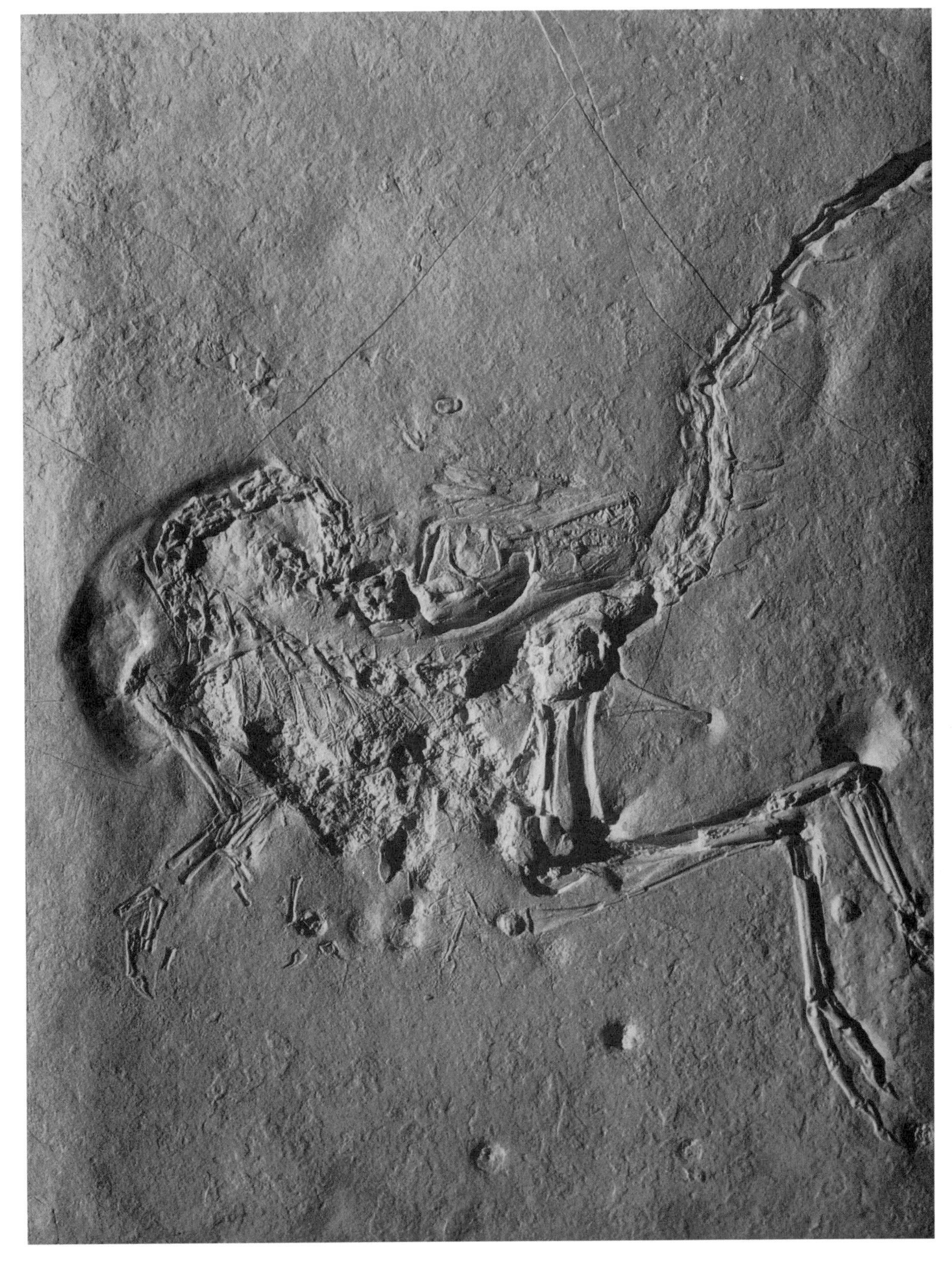

*Plate 19.* Cast of *Compsognathus* type specimen. This specimen is the first discovered and recognized complete skeleton of a predatory dinosaur.

*Plate 20.* Partial skeleton of *Longisquama* type specimen and supposed integumentary structures. Photographs courtesy of H. Haubold.

*Plate 21.* Partial *Megalancosaurus* skeletons. The specimen that consists of only the head to shoulder region is vaguely avian in a few respects, but the other specimen makes it clear that this creature is not a close relative of birds. Photographs courtesy of *Journal of Vertebrate Paleontology.*

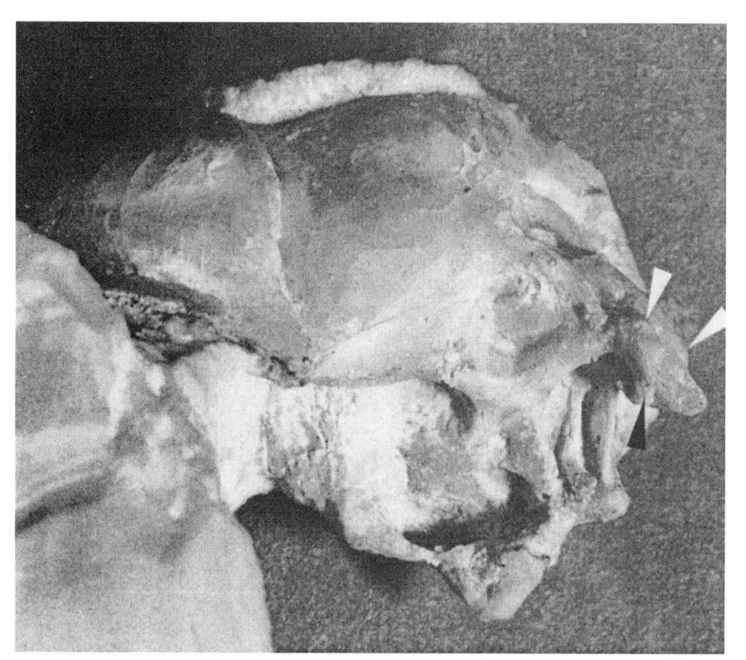

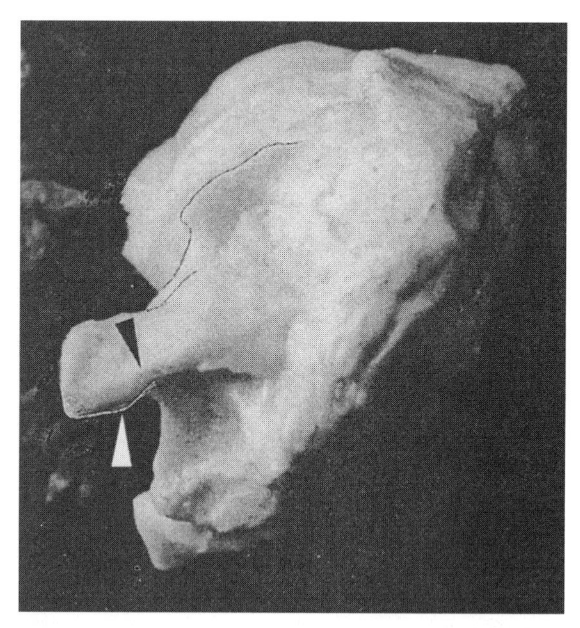

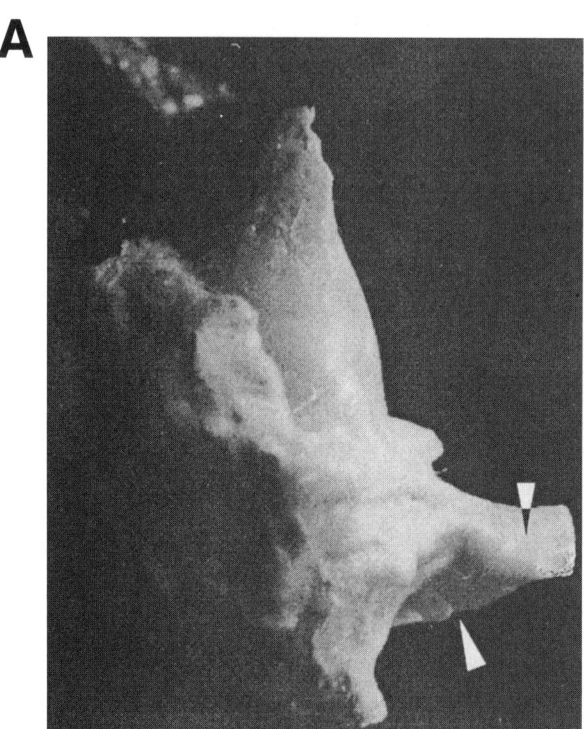


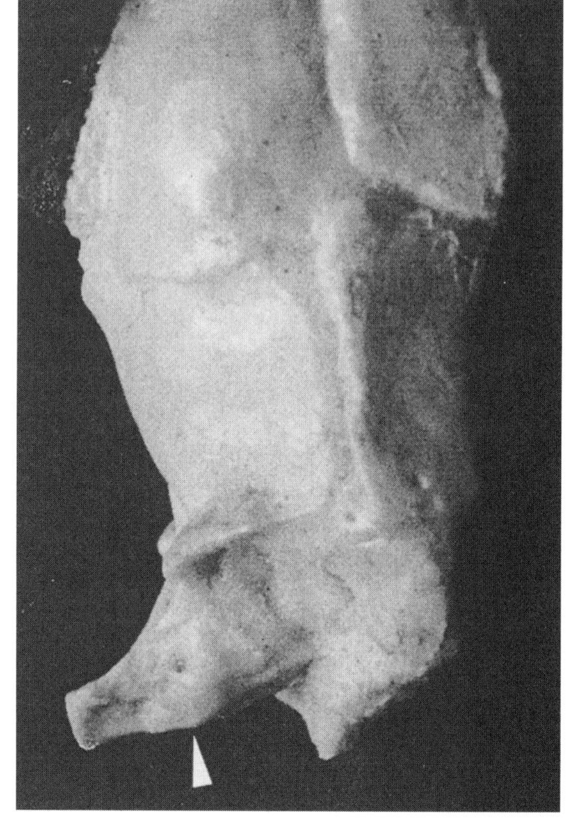

*Plate 22.* Comparison of occipital processes in side, posterior, bottom, and top views. *A, Archaeopteryx* BMNH 37001, side view (photograph courtesy of J. Gauthier; also see stereo views in Fig. 6A,B,C in Whetstone [1983]). *B, Dromaeosaurus* type specimen; arrows point to synapomorphies as explained in text and Appendix 1.

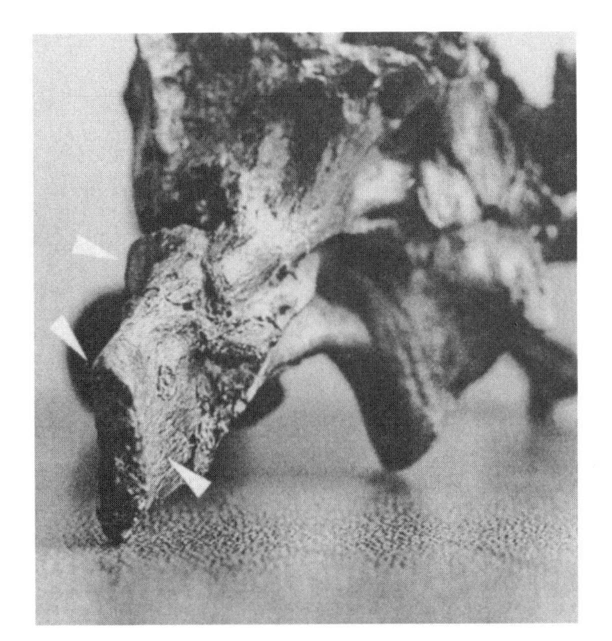

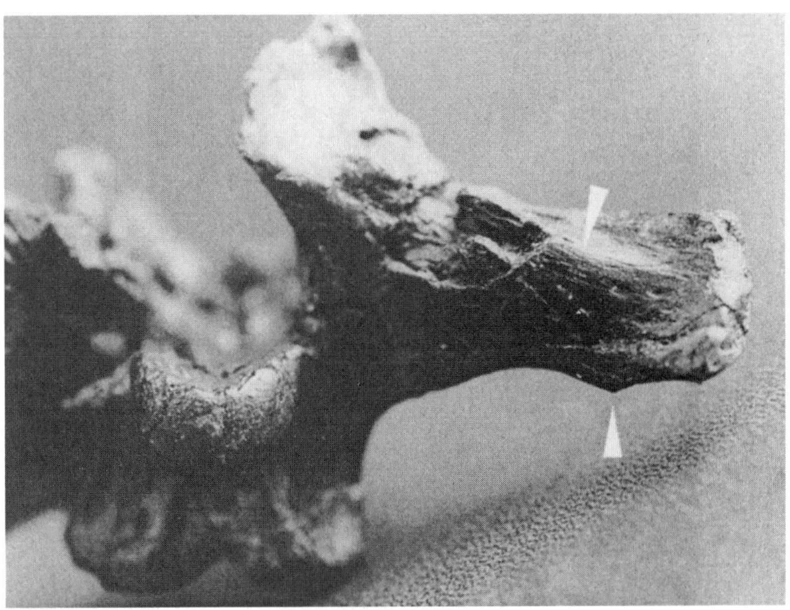

B

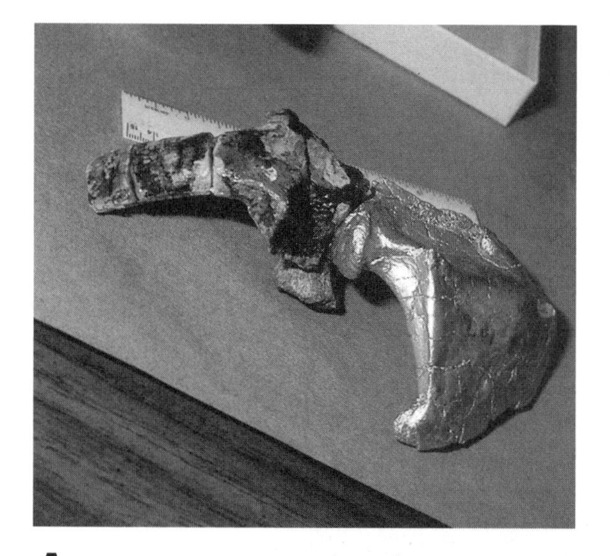

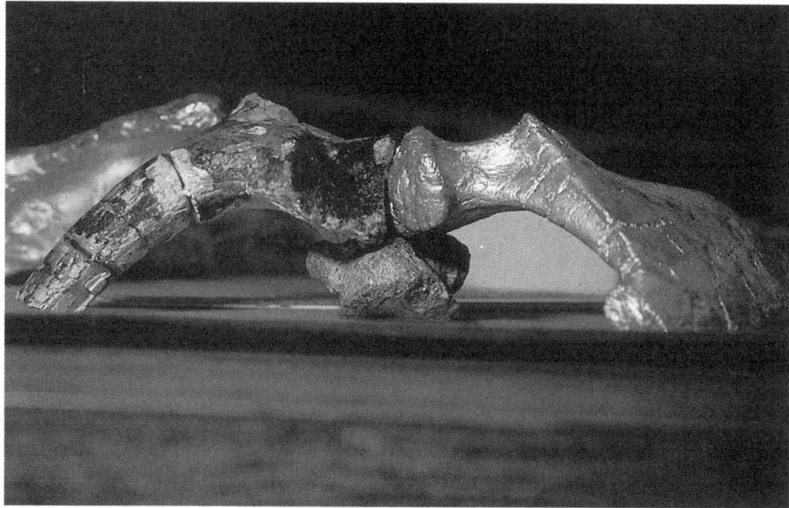

**A**

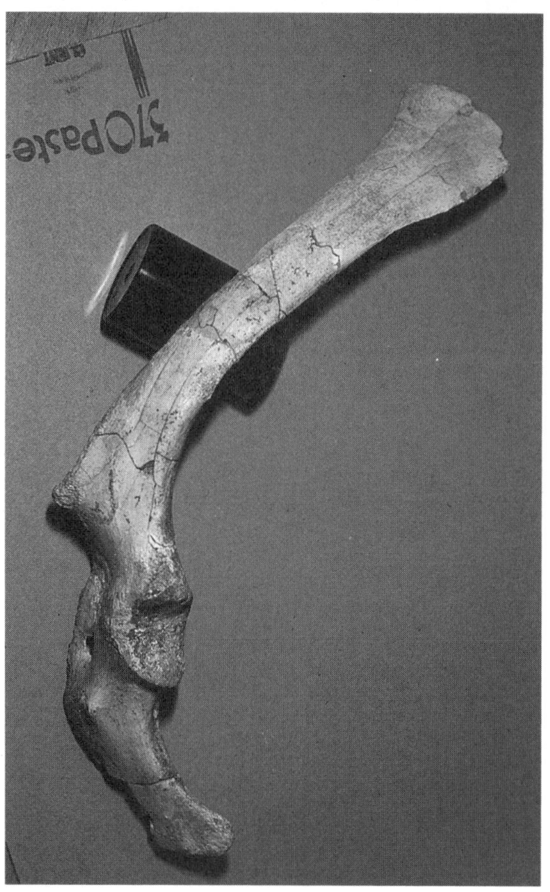

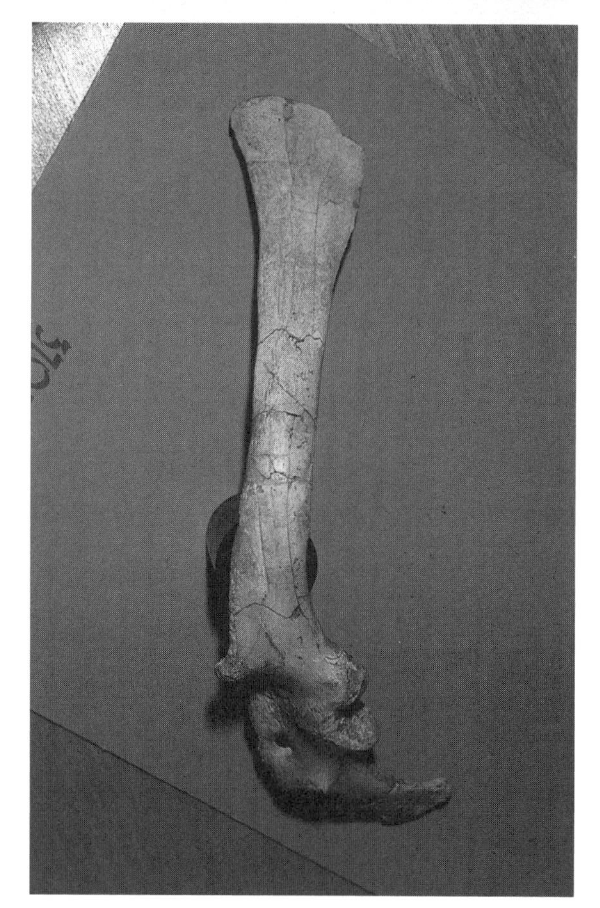

**B**

*Plate 23.* Comparison of strongly flexed scapulocoracoids of avepectorans. *A, Deinonychus* YPM 5236 (cast) and MCZ 4371. *B*, oviraptorid.

**B** a

b

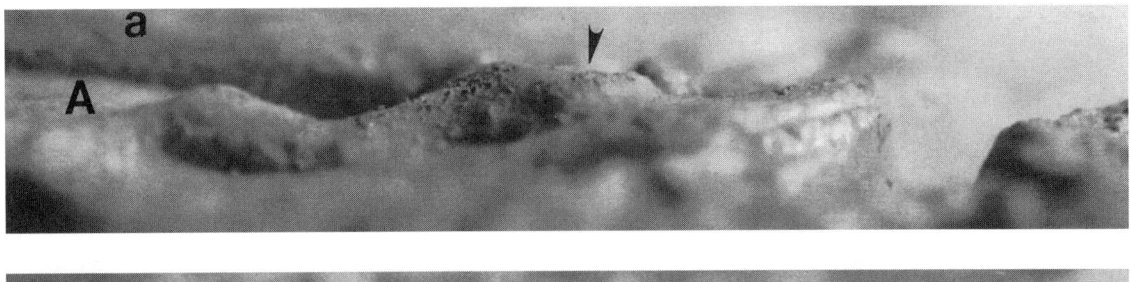


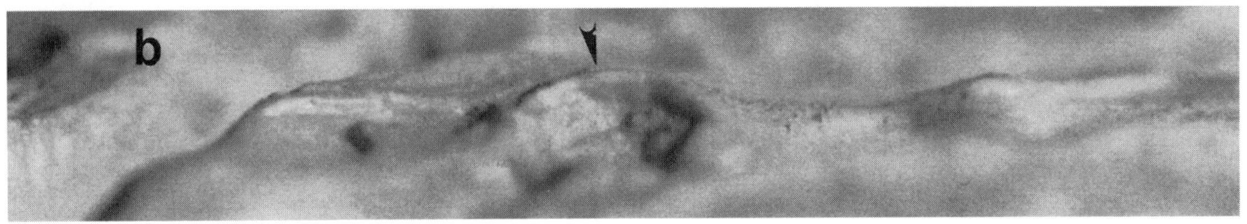


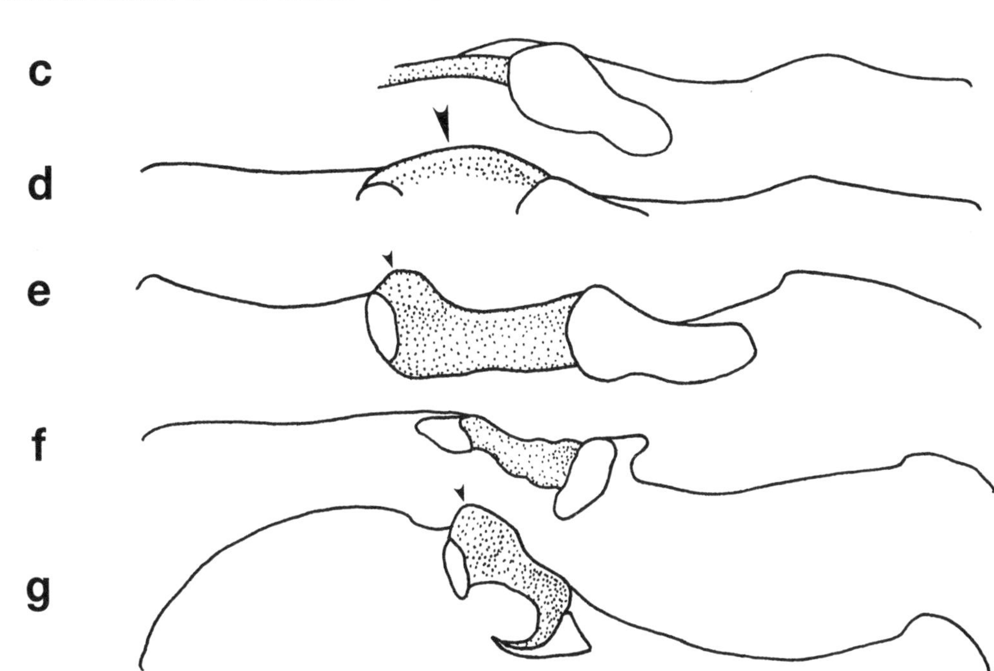


*Plate 24.* Comparison of hip sockets. *A*, right ilium of *Archaeopteryx* BMNH 37001 (cast) in nearly top *(a)* and bottom views *(b)*. *B*, right ilia in direct bottom views, acetabulum stippled: *a*, allosaur; *b*, tyrannosaur; *c*, *Archaeopteryx* 37001, according to Martin (1991); *d*, *Archaeopteryx* after photograph *Ab*, acetabulum partly obscured; *e*, dromaeosaur; *f*, domestic turkey, poorly ossified subadult; *g*, tinamou. Large arrows point to supraacetabular shelves, small arrows to antitrochanters. Not drawn to same scale.

PART 3

# FLIGHT

## *How and Why It Evolves, Why It is Lost, and How to Tell When*

The story of birds is a story of flight, and the story of flight began some 350 Myr ago with insects, which were the first animals to fly. The beginnings of insect flight are obscured by the lack of a good fossil record from that period (Labandeira and Sepkoski 1993, Marden et al. 2000). Among vertebrates, flight evolved at least three times; first among pterosaurs, second among birds, and third among bats (I say "at least three times" because some workers have argued that birds and bats are polyphyletic groups in which flight evolved two or more times). Understanding how these groups developed flight may be important to determining the groups from which they evolved. I say "may be important" because some researchers have argued that understanding how flight developed in these groups is indeed critical to any determination of ancestor-descendent relationships, but others have questioned the importance of this factor.

Part 3 of this volume deals with the how and why of the evolution of flight, especially the avian version. Chapter 6 focuses on how flight may have begun in birds and on the debate about whether or not this issue is crucial for determining their ancestry. Chapter 7 explores the evolution and perfection of avian flight, and Chapter 8 examines the opposite of flight acquisition, that is, flight loss. Failure to recognize instances in which flight may have been lost almost as soon as it appeared may be obscuring our understanding of the relationships among dino-birds.

CHAPTER 6

# The Beginnings of Flight

## *From the Ground Up or from the Trees Down?*

In a field of research that has generated plenty of controversy, the question of whether flight evolved from the ground up or the trees down is one of the most contentious.

### Of the Competing Hypotheses, Which Is the Best?

On one side of the debate are those who argue that flight must have evolved largely or entirely in climbing forms. The large body of contemporary researchers who concur with this climbing hypothesis includes A. Walker (1972), Feduccia (1980, 1993a,b, 1996), Tarsitano and Hecht (1980), L. Martin (1983a,b, 1991), Hecht and Tarsitano (1984), U. Norberg (1985a,b), Rayner (1985a,b), Bock (1986), Pennycuick (1986), Paul (1988a), Tarsitano (1991), Olshevsky (1994), Bock and Buhler (1995), L. Martin and Miao (1995), Chatterjee (1997, 1999a,b), Zhou (1998), Garner et al. (1999), Homberger and de Silva (2000), L. Martin and Czerkas (2000), Ruben and Jones (2000), Tarsitano et al. (2000), and Xu et al. (2000). This hypothesis comes in various flavors. Some hypothesize that climbing took place in trees (the arboreal hypothesis), but others have proposed cliffs as the jumping-off point. In the standard version of the climbing hypothesis, the protofliers are seen as gliders that used rudimentary airfoils, sometimes in the form of wings, to control their descent. Eventually, flapping of the wings allowed protofliers to apply increasing amounts of power in order to lengthen the glides. The wings may already have been large before the gliders began to be fliers, or wings may have started out small and grown in step with the increase in flapping power. In either case, the rate of altitude loss eventually declined to the point that sustained level flight became possible and was eventually followed by climbing flight. Garner et al. (1999) proposed a major variation in the climbing hypothesis by arguing that drag control used during pouncing attacks from above promoted the initial development of flight surfaces.

On the other side of the argument is an equally large body of researchers who propose that avian flight evolved among forms that ran, hopped, or leaped along the ground. A comprehensive list of supporters of this terrestrial (or ground-up) hypothesis includes Ostrom (1974a, 1985, 1986, 1994), Cowen and Lipps (1982), Caple et al. (1983), Balda et al. (1985), Gauthier and Padian (1985), Padian (1985, 1986), Peterson (1985), Rietschel (1985), Gauthier (1986), Weems (1987), Chiappe (1995), Novas and Puerta (1997), Dingus and Rowe (1998), Ji et al. (1998), Padian and Chiappe (1998a,c), Burgers and Chiappe (1999), Easley (1999), Xu, Wang, and Wu (1999), and Earls (2000). Like the climbing hypothesis, the terrestrial hypothesis has several variants. Ostrom's notion that protobirds used protowings as insect snares has been abandoned by all, including Ostrom himself. An even earlier version of the ground-up hypothesis proposed that bipedal runners increased their speed by flapping small airfoils on their arms. This view was largely dismissed because many thought that the early stages of flapping a protowing would have decreased traction without adequately boosting thrust. Therefore, few paid much attention in 1987 when Weems cited shallow-print, long-stride dino-avepod trackways as evidence for the lift generated by brachial airfoils. However, Burgers and Chiappe (1999) and Easley (1999) have revived this hypothesis with new analyses supporting the viability of combining running and wing flapping during take-off.

A leading variation of the terrestrial hypothesis sees protofliers as running insectivores that leaped up after their aerial prey while using small airfoils on the limbs as control surfaces. Eventually, these insectivores increased the height and distance of their leaps by flapping the airfoils, which gradually increased in size and power until sustained flight was achieved.

Another ground-based scenario proposes that wings began as a way to enhance the height of leaps during intraspecific displays, a scenario strengthened by new work showing that most birds take off with a powerful leap rather than with wing power alone (Earls 2000).

Various hypotheses combine aspects of the arboreal and terrestrial scenarios. For example, Pennycuick (1986) and I (Paul 1988a) have suggested that flight evolved among climbers that leapt and then glided through the air. In this scenario, protobirds achieved greater control during leaps between branches or other high places by developing distal control surfaces. The use of airfoils to increase lift extended the range of these leaps. Range was further increased by enlargement of the airfoils and by the application of gradually increasing power via flapping until sustained level flight and, eventually, climbing flight were achieved. This scenario additionally sees the climbers as having evolved from ground runners. Homberger and de Silva (2000) also favor an arboreal-leaping scenario.

Rayner favors gliding as the initial mode of flight, but in 1991 he combined this mode with the terrestrial mode by suggesting that flight evolved in species that ran and glided into strong winds. There is yet another hypothesis that combines flight from high places with flight from the ground: the ridge-gliding hypothesis, in which flight is supposed to have developed among protobirds who glided down steep slopes (Peters 1985, Peters and Gorgner 1992). According to this scenario, protobirds were not climbers, but gravity remained the initial power source for avian flight.

In support of these combination hypotheses, Hopson and Chiappe (1998) found that the foot of *Archaeopteryx* had the proportions that would be expected in a bird adept both at perching and at ground walking. Sereno (1997a) and I (Paul 1988a) also noted that the urvogel possessed both running and climbing adaptations but that the current paucity of protoavian fossils older than the urvogel obscures our understanding of where and how bird flight initially developed, to the point that none of the hypotheses can be considered verified at this time. Despite these limitations, much can be done with the problem of determining the origins of avian flight, as we shall see in the remainder of the chapter.

### *Climbers versus Runners: Using Practical Examples versus Hypothetical Scenarios*

Hypotheses and scenarios can take us only so far without practical examples, which are always superior to even the most well thought out hypothesis. One method of determining whether a scenario could have occurred in a particular case is to find evidence that the scenario actually did occur in some similar case. An alternative strategy is to find conclusive evidence that renders the scenario implausible or impossible.

In an example of the latter strategy, some of the researchers who favor the terrestrial hypothesis have stressed the difficulties that are supposed to have hindered climbers from evolving into power fliers. These researchers start by observing that modern climbing amphibians, reptiles, and nonbat mammals that regularly travel substantial distances through the air do so by parachuting or gliding. Why, they ask, would arboreal gliders that can climb and then travel from tree to tree aerially be under selective pressure to evolve powered flight? In addition, these researchers argue that once passive gliding evolved, switching over to flapping powered flight would be difficult or impossible. These are reasonable opinions. Short-range gliding is itself a sophisticated activity that can indeed result in anatomical adaptations quite different from those seen in active fliers. Compare, for example, the marked difference between the airfoil complexes of flying squirrels and bats. Because of these differences, some workers have a hard time conceiving how the anatomy of a passive glider could be transformed into that of an active flier. They have further argued that the aerodynamics involved in transitioning from gliding flight to powered flight would have been so adverse as to make the transition impossible (this issue is discussed later in this chapter). Ergo, arboreal aerialists should always have become gliders and could never have become flappers.

There are a number of ways to counter such arguments. One is to explain how flapping flight could have evolved from gliding despite the aerodynamic difficulties involved (this explanation is described in more detail later in the chapter). A practical problem with this approach is that little effort has been put into determining how flapping flight developed. Wing flapping—especially when it involves the very flexible, complex appendages of flying vertebrates—is difficult to study because the process is a very complex one involving intricate, turbulent airflow, as well as rapid changes in airfoil shape (see the next chapter). Until recently, there were few practical reasons for scientists to dedicate the time and resources needed to analyze such a difficult subject because mechanical flight is accomplished with comparatively simple fixed and rotating wings. But the situation is changing.

Military requirements for extremely small flying surveillance robots are driving research on the function of similarly small bird and insect wings (Paul and Cox 1996, Wootton 2000 and refs. therein). At the same time, the availability of cheaper and more-capable computers that can simulate complex aerodynamics is making it easier to understand how flapping flight works, and we can therefore expect our understanding of how it got its start to improve dramatically.

Another way to challenge the anti-arboreal argument is to avoid the gliding-flapping transition problem altogether. Instead, we can invoke horizontal leaps as the precursors to powered flight. Of course, any theory involving such leaps must also be speculative in view of the lack of living examples.

At a fundamental level, the basic notion that gliders cannot evolve into power fliers is in danger of denying the transforming power of evolution. Fish have evolved into amphibians; continental ungulates have evolved into marine whales. In comparison to these changes, the shift from gliding to flapping flight seems a modest one, and indeed there is evidence that it did in fact occur.

The best evidence that climbing creatures can learn to power fly is provided by bats. The general view is that most or all bats evolved from insectivores (Yalden and Morris 1975, Carroll 1988, Thewissen and Babcock 1991, Teeling et al. 2000), although some researchers have argued that the bigger fruit-eating megachiropteran bats descended from basal primates (Pettigrew 1995, Simmons 1995). The ground-running abilities of either mice or lemurlike creatures are clearly insufficient for them to leap up to catch flying bugs. This unremarkable ground performance is retained in bats, whose walking and running abilities are modest at best and are often very poorly developed (Yalden and Morris 1975). It is, therefore, widely recognized that bats evolved from climbing ancestors (Jepsen 1970, Yalden and Morris 1975, Caple et al. 1983, Padian 1985, 1986, Rayner 1985a, Pettigrew 1995, Simmons 1995). Although there is no direct evidence for this contention, it is so well founded, and the terrestrial alternative so implausible, that barring a clear demonstration otherwise, bats can serve as an example of a group of climbers that learned to power fly, in spite of the hypothetical barriers that have been proposed. The possibility that climbers evolved powered flight must therefore be ranked as less speculative than ground-based alternatives.

What is less understood about bat origins is exactly how these small climbing mammals became powered fliers. Since no protobats have yet been discovered, we do not know whether they climbed trees or rocks or both. Some researchers suggest that wing membranes were initially used as insect traps that later came to be used for short hovers before level powered flight was achieved. However, hovering flight is itself a very sophisticated, energetically demanding activity that probably appeared only in advanced fliers. Other workers take the horizontal leaping route to avoid the transition between gliding and power flying. Most envision protobats as gliders. In any case, the detailed questions surrounding the origins of bat flight do not undermine the argument that if climbing protobats evolved powered flight, then one or both of the other flying vertebrate groups could have started out from high places as well.

In contrast, there is no unambiguous example of any level of flight having evolved among terrestrial tetrapods. Because the origin of flight in the two potential candidates for this scenario, pterosaurs and birds, is itself the subject of dispute, citing them as examples of this pattern would involve circular logic. Interestingly, there is a viable argument that insect flight evolved from the water up among surface-skimming insects (Marden et al. 2000), but this mode is inapplicable to bigger vertebrates (not that the notion of vertebrates evolving flight from the water is entirely absurd, since some fish glide). At this time, the hypothesis that flight can evolve among ground runners, while not impossible, should be ranked as inferior to the arboreal hypothesis.

### *Why High Places Are Better Than Low*

We can get a better handle on whether powered flight evolved among climbers or among leaping ground runners by looking at how many modern species pursue each lifestyle and by considering the reasons why such species are either common or too impractical to be common, or even exist.

We will start with animals that live in high habitats. That many thousands of species of amphibians, reptiles, mammals, and birds are climbers demonstrates that climbing is a viable mode of life. Within this group, a large proportion leap or glide or both between high places. Tree squirrels are a familiar example of arboreal leapers; flying squirrels are common but rarely seen (being nocturnal) arboreal gliders. Arboreal reptiles, interestingly, are not good leapers. The link between climbing habits and various aerial abilities is obvious. For an animal that normally lived high above the ground, moving horizontally for long distances would require climbing up and down. Doing so is not only arduous but also dangerous for a creature adapted to living in a three-dimensional habitat. Gliding is much faster and less dangerous than walking and climbing between two high places. Gliding is also easier, as many a kitten has learned, in that going down a tree is harder than going up, since gaining purchase with the hindfeet when the body is pointed downward is difficult. Some climbing mammals have specially modified ankles that allow them to direct their feet backward when they are descending: look at the hindfeet of a squirrel the next time you see one upside down on a tree trunk.

Living high also facilitates the use of gravity as a power source for air travel. Indeed, the energy cost of moving between two trees is considerably reduced if gliding rather than ground walking is used for the lateral part of the trip (Rayner 1991, Feduccia 1996, Lamin 2000). The use of some mode of aerial movement, either leaping or gliding, to move between tall trees or rocks is so highly adaptive that this lifestyle has even been adopted by some birds, including two kinds of scansorial New Zealand birds whose island habitat reduces the need for powered flight (see Chapter 8): the kakapo (a parrot) and kokakos (crow relatives) are weak fliers that glide between trees and then climb back up (S. Scott 1974, Feduccia 1980, Livezey 1992, Attenborough 1998). The kokako swiftly regains height with upward hops enhanced by wing fluttering. That some modern avepods are glider-climbers makes it highly plausible that dino-birds were the same.

The situation with terrestrial insectivores that pursue aerial prey is entirely different. Not one living species, avian or otherwise, obtains the majority of its food by leaping up after insects. Land animals may snap at the odd flying insect, and I have seen video of a small jackal repeatedly leaping as it tried to feed on a dense swarm of flying insects. These are, however, opportunistic attempts to take advantage of an occasional food source, rather than sustained lifestyles. All modern animals that regularly pursue airborne insects are themselves adept fliers, either insects (especially dragonflies), birds (including bee-eaters, some falcons and hawks, and swifts and nighthawks), or microchiropteran bats. Indeed, these insect interceptors are all highly adapted to pursue aerial arthropods: all are highly agile or fast or both in the air, some (insectivorous bats and a few birds) have insect-tracking sonar, and many bug-eating birds have oversized mouth traps.

Why do land animals not live by pursuing flying insects, and why must those creatures that do eat flying insects be high performance fliers? Again, the reasons are obvious: hunting flying insects is a form of aerial combat. In such combat, the advantage goes to the craft that possesses greater speed, agility, rate of climb, maximum ceiling, and superior altitude at the initiation of hostilities (Shores 1983). Altitude is especially important because the higher flier is out of easy reach of its opponent and can use gravity as an extra power source to rapidly gain speed. A land animal whose only aerial phase is a ballistic leap is at a severe disadvantage vis-à-vis its winged prey in terms of speed, agility, maximum altitude, and rate of climb because such an animal will begin to fall soon after leaving the earth. It also initiates every engagement from a severe height disadvantage. The problems of catching flying insects are well known to those humans who try to do so. Even though we are reasonably fast and our height, our long arms, and our grasping hands negate some of what we lack in agility, our success rate is so low that we are reduced to employing large nets and other devices to improve the odds. The success rate of an even shorter animal that must make high ballistic leaps in the hopes of snaring in its jaws a fast, evasive insect can be expected to be similarly low. There is no data showing that a terrestrial insectivore can achieve a rate of aerial interceptions high enough for survival. Engaging in ground-to-air insectivory is roughly equivalent to trying to bring down an aircraft with catapult-launched projectiles.

Caple et al. (1983) argued that leaping insectivory is energetically advantageous because leaping insectivores have twice to four times the foraging range of nonleapers. This argument is dubious in terms of physics and energetics because intercepting insects from the ground requires working directly against gravity with every attempt to catch a meal. These antigravity leaps would have to be numerous because most flying insects are small and therefore do not provide much energy. This is an energy expensive way to eat, so much so that the ratio of energy gained to energy lost is probably unfavorable (see Appendix 4).

It is not, therefore, surprising that ground-bound insectivores, such as chameleons and even frogs, that hunt winged insects do not leap after them but rather wait for their prey to land nearby and then ambush them with projectile tongues. Nor is it surprising that those animals that do hunt airborne insects are themselves propelled by fast, agile wings. These hunters are biological Me-262s and F-16s in pursuit of B-17s and MiGs. The hypothesis that ground runners can live by leaping after aerial bugs is therefore entirely speculative and probably impractical. Restorations that show the urvogel or other protobirds chasing swift dragonflies are especially implausible (those that show them chasing butterflies are invalid because butterflies had not yet evolved [Labandeira and Sepkoski 1993]).

In another ground-based scenario, Burgers and Chiappe (1999) tried to revive the running-plus-flapping hypothesis by arguing that the primary function of protowings was to generate thrust rather than lift, because the latter would have reduced ground traction so much that running speed would have declined. In contrast, Easley (1999) has presented data indicating that the "ground effect"—in which air trapped between the ground and wings increases lift—would have made it easier for running protobirds to take off. As theoretically viable as these two hypotheses may be, the absence of living animals that use protowings to increase running speed works against them. The living examples we do have, like birds, tend to fold up rather than flap their wings when they run, and it is unclear why protofliers would have been any different. Moreover, Burgers and Chiappe's focus on big-winged *Archaeopteryx* leaves unanswered the question of how well the combination of flapping and running would have worked in smaller winged pre-urvogels. A problem specific to Easley's analysis is that it seems to require protobird flight muscles to have had the kind of short-term power output seen in reptiles, which is improbable (see Appendix 4). Both of these run-and-flap hypotheses require protobirds to have been specialized runners in the manner of most dino-avepods. Although urvogels were probably able to run fairly swiftly, their somewhat reduced hindlimb musculature suggests they were not as well adapted for running as these scenarios presume (Chapter 9).

Rayner (1991) attempted to get around some practical problems of the insect-intercepting-cursor scenario: he dropped the insectivory and leap-control aspects in favor of the suggestion that avepods learned to fly by making running take-offs into strong winds and then gliding back to terra firma. Rayner postulated the need for head winds because he assumed that small dino-avepods could run at only about 7 kilometers per hour, too slow to glide (Geist and Feduccia [2000] also presume that the dinosaurs were rather slow). However, gracile avepod legs should have been able to propel the little dinosaurs at speeds as high as 30 kilometers per hour for extended periods of time (Paul 1988a, 1998/2000, Christiansen 1999, Farlow et al. 2000, Appendix 4), easily fast enough to allow gliding at take-off. However, the selective forces that would encourage runners to evolve gliding flight are obscure: speed would fall off rapidly after take-off and any energy savings over walking or running the same distance would be at best modest. The modesty of the energy savings may explain why there are no running gliders living today.

All of the above ground-running scenarios have been seriously challenged by Earls (2000), whose observations show that the great majority of birds take off from the ground by pushing off with a leap that produces 80–90 percent of the take-off velocity. (Running take-offs are limited to a few large birds with low-burst-power flight muscles, such as albatross and swans.) It is surprising and somewhat unsettling that researchers took so long to recognize such an obvious and important fact, one that is in line with, and reinforces, the absence of living examples of small bipeds that combine high speed with small wings to achieve a crude level of flight. The legs of small theropods were sufficiently powerful to propel their owners directly into the air, so running was not necessary for the evolution of flight.

The last point makes more interesting the ground-based scenario presented by Stephen (1974) and Cowen and Lipps (1982). They proposed that protobirds initially developed enlarged brachial feather surfaces as intraspecific display structures that were subject to sexual selection. The bigger the display surface was, the more effective it would have been. In addition, the displays included leaps, and higher leaps were favored by sexual selection. In order to boost leap height, these protobirds flapped the airfoils. Increasing the size and power of the airfoil improved both the height of the leaps and the visual impression of the feather displays. This feedback loop continued until sustained powered flight developed. This hypothesis is superior to the ground-running-insectivore model in a number of regards. It eliminates, first, the needlessly speculative running component. Second, we know that sexual selection for impressive displays is a very powerful force—witness the wonderful tails of peacocks, the enormous antlers of Irish elk, and the elaborate mating dances practiced by a variety of creatures—so this scenario logically explains why large surfaces would appear on arms and why the surfaces would be used to gain height in the air. Where the hypothesis is weak is (again) in the lack of living examples, that is, tetrapods that cannot fly or did not descend from fliers but that nevertheless use brachial surfaces for both display and semiflight. However, if flight eventually turns out to have evolved among terrestrial archosaurs, then this little-discussed hypothesis will warrant more serious consideration than the running-insectivore notion.

Yet another hypothesis for the development of large brachial feathers proposes that they initially evolved as thermal coverings for eggs during brooding (Hopp and Orsen 1998) and were only later used as airfoils. Assessing this notion is difficult at this time for a number of reasons. First, we do not know that the dinosaurs Hopp and Orsen focused upon, oviraptorids, really did have long contour feathers on their arms. Second, as discussed later, oviraptorids may have been secondarily flightless. Third, the nesting habits of predatory dinosaurs less derived than avepectorans are not yet known. Even so, this possibility deserves further attention if dino-avepectorans prove not to be neoflightless.

### *How Big Is Too Big to Climb and Glide?*

We can set hypothesis against modern example to address another question about living in and moving among trees: how does size effect climbing and gliding ability? Balda et al. (1985), Pennycuick (1986), Feduccia (1996), Geist and Feduccia (2000), and Tarsitano et al. (2000) have asserted that climbers and gliders must be very small to avoid being injured by falls. Balda et al. calculated that only flat-bodied creatures weighing a mere 8 grams or less can parachute safely with a minimal-area airfoil.

Extant climbing and gliding tetrapods tell another story, however. Pumas, jaguars, leopards, and especially orangutans are excellent climbers, and they reach 90 to 120 kilograms (Nowak 1999). Tree kangaroos weighing 7 to 10 kilograms deliberately leap as far as 9 meters downward between trees and 18 meters down to the ground (Nowak 1999)! At 2 to 7 kilograms, adult Bengal monitors are good climbers, and they regularly survive leaps of 12 to 25 meters with terminal belly flops (Auffenberg 1994). Three-kilogram cats often survive many-story falls onto concrete by slowing their falls with the minimal parachute effect produced by the classic feline spread-eagled posture. Flying squirrels and "lemurs" weigh from 20 to 1,750 grams (Nowak 1999). No squirrel weighs less than 10 grams, and there is no evidence that the ancestors of gliding mammals were as tiny as Balda et al. suggest they should have been. The example of large climbing, falling, and gliding mammals disproves the extreme small size hypothesis.

It is also worth remembering that the Jurassic urvogels were not, by the standards of arboreal or flying creatures, diminutive: they weighed in at half a kilogram (Appendix 2). Truly tiny birds do not appear in the fossil record until the Early Cretaceous. Therefore, bird flight probably began among forms as big as or bigger than pigeons and probably as hefty as crows and gray squirrels. In this regard, avian origins probably differed from those of bats, which seem to have evolved from very small arboreal insectivores.

### *Arboreality in Little Dromaeosaurs*

The terrestrial hypothesis predicts that the dinosaurs closest to the urvogel will be configured for running rather than for climbing. The arboreal alternatives predict that the dinosaurs nearest to the first bird will exhibit more adaptations for climbing than will other dino-avepods. Note that the degree of arboreality does not need to be extreme. The recently described basal dromaeosaurs *Sinornithosaurus* and *Microraptor* have characteristics pertinent to these predictions. As shown in Part 5, these are currently the dinosaurs closest to *Archaeopteryx* not only in form but also in phylogenetic position. As discussed later in this chapter, *Sinornithosaurus* and even more so *Microraptor* were well adapted for climbing, to a degree not seen in any other previously known dinosaurs (Figs. 1.6, 6.1). Yet more compelling is a tiny and nearly complete Cretaceous theropod skeleton (probably a hatchling) described in the year 2000 at the Graves Museum bird origins conference in Florida. The creature's hyperelongated finger and perching, birdlike foot indicate that it was highly specialized for climbing. These new dino-avepods offer fresh and powerful support for the climbing hypothesis, and the fossil record is starting to provide hard evidence favoring the arboreal over the terrestrial view.

### *Arboreality in Early Birds*

The correlations that we have discussed between arboreal versus terrestrial adaptations, on the one hand, and arboreal versus terrestrial origins of flight, on the other, also apply to the earliest birds. A wide array of Mesozoic birds have recently been discovered; not one of the Early Cretaceous examples found so far is a specialized long-legged runner, contrary to what would be expected if their dinosaurian ancestors were specialized runners. Instead, these early birds often show the adaptations for arboreal habits that would be expected if their ancestral stock had spent significant time in the trees (L. Martin and Miao 1995, Feduccia 1996, Hou et al. 1996, Feduccia et al. 1998, Zhou 1998). The most obvious of these features are a hallux that is larger and more reversed—and, therefore, better-suited for perching—than the hallux of running dinosaurs and clawed fingers suitable for grasping branches. That the dino-bird *Archaeopteryx* (see Chapter 9 for further discussion) had such a hallux is particularly important.

*Confuciusornis* (Fig. 6.2) is interesting because it diverged from the usual avian pattern in having a long, narrow body. In this respect, it looked rather like a climbing mammal, which suggests that it converged in body proportions with the latter (Martin and Miao 1995). Hembree (1999) concluded that confuciusornithid claws were better suited for perching than for climbing, but the finger claws would have been useful only for climbing. The retention of the large hooked claws only on the two peripheral and flexible fingers and the sharp reduction of the claw on the aerodynamically stiffened central finger indicate that these hooked claws were specifically retained for climbing (their use for predation is less likely in view of the absence of cutting teeth or a hooked beak). Hopson and Chiappe (1998), Chiappe et al. (1999), and Hembree (1999) concluded that confuciusornithid toes (Fig. 6.3) were pigeonlike in being proportioned in a manner suited both for spending a great deal of time on the ground and for perching (see the discussion of toes later in this chapter and in Chapter 9), and Padian and Chiappe (1998c) restored *Confuciusornis* on the ground. However, the toes of archaic birds were much more distally elongated than those of any running dinosaur or even those of the living roadrunner. In addition, the hyperelongated tail-feather ribbons that adorned some individuals are unlikely to have evolved in creatures that spent much time on the ground, and that there is little abrasion on the delicate structures further suggests that they spent little time on the ground. The very long distal toes of *Sinornis* favor a significant degree of arboreality, as do its large claws (Fig. 6.3). Running birds such as *Yandangornis* and *Patagopteryx* do not show up until later in the Cretaceous. The pattern observed so far favors the arboreal scenario.

### *Distal versus Proximal Airfoils*

Although the exact way that flight evolved in bats is not entirely clear, they may help reveal the anatomical feature that is key to the evolution of powered flight, whether it starts with gliding or leaping. Most gliders develop proximal airfoils, such as the membranes stretched between the wrists and ankles of flying squirrels, or the rib spread surfaces on some lizards. These flight adaptations may be the least suited for developing into flapping airfoils. In bats, the majority of the flight membrane is stretched between the fingers. This suggests that the same configuration was present in the ancestors of bats, which may have resembled flying "lemurs" that also have intrafinger airfoils. That pterosaurs and birds also have distal airfoils borne by fingers is not surprising, since distal airfoils are best placed for the long sweeping movements that generate thrust. Distal airfoils appear, therefore, to be critical to the evolution of powered vertebrate flight.

On a related subject, Chatterjee's (1999b) suggestion that the close-cropped fibrous feathers of dino-avepods such as *Sinosauropteryx* were sufficient to provide a drag-induced parachute effect has to be rejected, since furry arboreal mammals fall rapidly unless they have some form of flattened aerodynamic surface. The longer fiber feathers of *Sinornithosaurus* and *Microraptor* were better suited for slowing falls.

### *Two Legs versus Four*

Some researchers, including Ostrom (1974a), Cowen and Lipps (1982), and Feduccia (1996), have asserted that arboreal protofliers should have been quadrupeds, because arboreal tetrapods are generally, but not always, four legged. Tree kangaroos, for instance, are terrestrial bipeds that have evolved to climb semiquadrupedally to some degree (Grzimek 1990, Nowak 1999). Arboreal birds also show that climbing and bipedalism can mix.

### *Flapping before Flying?*

Cretaceous avepectoran dinosaurs such as *Unenlagia* were too small-armed to fly and were push-

ing the size limits for adept climbers. Yet these creatures had birdlike scapula blades and could flap their arms almost as well as flying birds could (Novas and Puerta 1997; but see Chapter 4). Novas and Puerta concluded that arm flapping evolved before flight among fairly large running dinosaurs, so these workers favor a ground-based origin for avian flight. However, the ability to swing the arm through a wide arc is seen in the earliest avepod dinosaurs (Chapter 4) and is probably an adaptation for arm action during predation. As for the birdlike anatomy and action of dino-avepectoran arms, these features may have been retained from smaller ancestors that climbed and flew. Birdlike arm action in large, nonflying avepectoran dinosaurs does not necessarily favor a terrestrial origin of bird flight over an arboreal one.

### *Pouncing Protobirds: A Theory That Drags Too Much?*

The Garner et al. (1999) hypothesis that avian flight started with ambush leaps from high places includes the premise that the leaps were increasingly better controlled with drag-inducing, rather than lift-producing, distal control surfaces. These workers further explained that their hypothesis requires protobirds to have had symmetrical distal feathers rather than the asymmetrical feathers associated with producing lift. Lacking Jurassic protobird remains, we cannot test this crucial aspect of their hypothesis (as explained later in this chapter and in Section 5, the presence of symmetrical distal feathers in *Protarchaeopteryx* and *Caudipteryx* is not necessarily relevant, because these may not be pre-urvogels).

Theoretically, the pouncing hypothesis, with its gravity assist, is superior to the energetically improbable ground-up hypothesis, and yet there are no living predators that use drag-inducing control surfaces during predaceous leaps. This may be because such systems have very limited lateral steering capacity. Conventional parachutes that rely solely on drag allow a jumper only a limited and imprecise ability to land on a chosen target. Lift-generating parafoil chutes allow jumpers to fly much more laterally and with a much higher degree of control, to the point that formation flying is possible. Therefore, the evolution of drag-inducing control surfaces in leaping predators is questionable. A superior variation of the pounce hypothesis is based on the premise that more effective, lift-generating airfoils immediately resulted in a higher degree of agility. Predatory birds that use lift-producing wings to swoop down upon their prey from above provide living examples of this mode of attack.

### *Putting It All Together: How Powered Flight May Have Evolved, Gliding versus Leaping*

Combining the example of climbing protobats, the absence of ground-to-air insectivores, and the ubiquity of distal airfoils in arboreal fliers, we can use scenarios that blend firm conclusions with alternatives and speculation to generate plausible hypotheses for how powered flight evolved in vertebrates.

The following firm conclusions can be drawn. First, many ground-up hypotheses can be rated as inferior because of the practical problems of capturing small prey while fighting gravity, and their inferiority is confirmed by the lack of living practitioners of ground-to-air insectivory. Unless fossil protavians someday prove otherwise, we can consider it probable that vertebrate flight began in high places. Therefore, we will focus on such scenarios. Second, distal airfoils are probably critical to transforming short-range aerial excursions into sustained level flight and climbing flight. Third, climbing protofliers can be either quadrupedal, bipedal, or a mixture of the two.

To go beyond these conclusions, we must enter the realm of educated speculation because we do not know enough to draw firm conclusions.

---

*Figure 6.1. (opposite)* Dinosaurs in the trees. *Top, Ornitholestes* among the branches. Feduccia (1996) opposed this speculative restoration of a feathered climbing dinosaur, which was first published in 1988. Considering its long, hook-clawed fingers and toes, this small predator may have been more willing and able to go vertical than some paleontologists think. *Bottom,* the basal dromaeosaur *Sinornithosaurus*. Of known dinosaurs, *Sinornithosaurus* was one of the best adapted for life in the trees. In addition to having the grasping features found in other dino-avepods, *Sinornithosaurus* had extremely long arms, toes bones similar in proportion to those of pigeons, and large finger and toe claws, making it at least as good a quadrupedal arborealist as a terrestrial biped. Note that the arms can sprawl to the sides. The long, fibrous feathers preserved on the arms, legs, and elsewhere on the body would have slowed and softened falls, perhaps to the point that *Sinornithosaurus* may have deliberately parachuted on occasion. This dinosaur offers the best evidence to date for the arboreal and neoflightless hypotheses. Drawn to same approximate scale.

*Figure 6.2.* Climbing confuciusornithids. These most primitive of known short-tailed birds had features that suggest they spent considerable time in trees. Although the central finger was specialized for supporting the outer wing feathers, the other two digits retained large, hooked claws suitable for grasping trunks, as shown by a *Confuciusornis (bottom)*. Confuciusornithids, especially *Changchengornis,* which is shown perching *(top),* had a well-developed, fully reversed hallux. The latter bird also exhibits a common confuciusornithid feature more compatible with a life above rather than on the ground: that is, very elongated and delicate tail streamers. Both birds are from the lower Yixian Formation.

We cannot, for example, determine whether cliffs, steep slopes, or trees were the preferred habitat for protofliers—although the much greater diversity of present-day tree dwellers tends to favor tall plants in this regard, especially in the case of bats. It is possible, however, that one flying vertebrate group evolved from each habitat. Also uncertain is whether powered flight evolved with an initial assist from gravity, via gliding, or from largely horizontal leaps that neither directly opposed nor received a boost from gravity.

In the gliding scenario, air surfaces first begin to appear as drag-inducing airfoils that slowed vertical falls, either to make downward jumps easier and safer or to make falls less injurious, or both. The selective pressures that favor this stage of development are obvious. If the drag surfaces were flat, stiffened airfoils rather than soft, irregular fringes, and if the surfaces became enlarged to further improve fall performance, then using the surfaces to start transforming falls into horizontal glides, that is, into a means of travel, became possible.

The assistance of gravity is both a strength and weakness of the gliding hypothesis. On the positive side, gravity supplies a ready power source for initiating the first stage of subhorizontal aerial travel. The usefulness of gravity as a power source

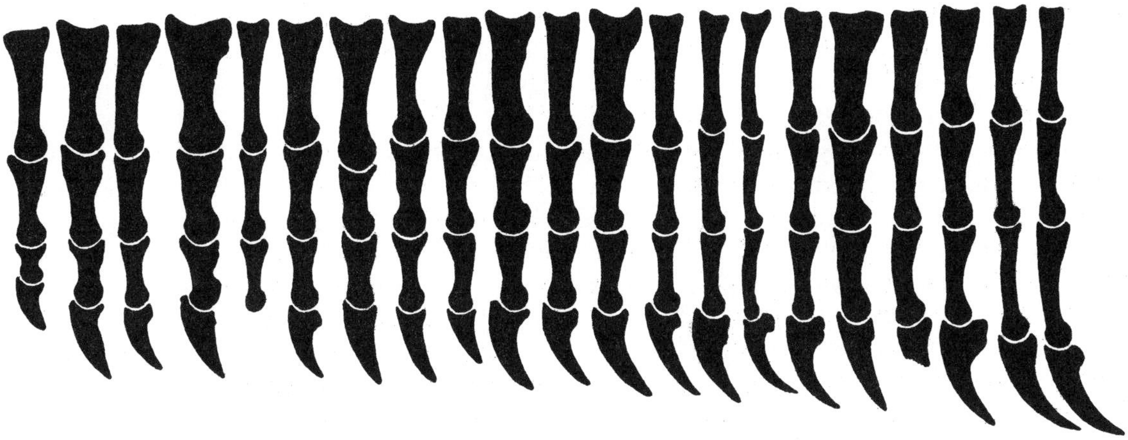

*Figure 6.3.* Comparison of central toe proportions of avepods. The two proximal elements are drawn to the same length in order to emphasize the differing lengths of the distal bones; increasing elongation of the latter progresses from *left to right:* running, never climbing emu with very abbreviated distal elements; *Struthiomimus;* highly terrestrial seriema; *Sinraptor; Rahonavis;* basal troodont *Sinornithoides* juvenile; *Gorgosaurus* juvenile; *Caudipteryx; Compsognathus;* velociraptorine dromaeosaur; basal dromaeosaur *Bambiraptor; Coelophysis;* basal dromaeosaur *Sinornithosaurus; Archaeopteryx;* scansorial pigeon, whose distal elements are elongated in order to better grasp branches; kiwi; basal dromaeosaur *Microraptor; Protarchaeopteryx;* basal avebrevicaudans *Confuciusornis* and *Sinornis;* highly arboreal crow with very elongated distal elements.

is the reason that so many gliders have evolved. On the downside, so to speak, the very ease with which gliders use the planet itself as a power source makes it harder to explain why they would switch to internal power in order to move farther horizontally. Why not improve the glide ratio (that is, the ratio of horizontal travel to vertical descent) simply by evolving a gliding airfoil with more lift and less drag, or even become a soarer by using updrafts as yet another cost-free power source? After all, gliding and soaring are much more energy efficient than power flying the same distance. For a bird, the power output during gliding is only about twice the power output during rest, whereas the energy needed to sustain level flapping flight is ten to twenty times the resting level (Pennycuick 1989, Butler 1991, del Hoyo et al. 1992–99). Even so, there are no known examples of gliders having evolved directly into soarers; the latter have always evolved from powered fliers. For that matter, no gliders have evolved very large airfoils in order to extend glides to very long distances. Why is this so? Probably because high glide ratios require very elongated, narrow wings in small fliers (Ruppell 1975, Pennycuick 1989). This requirement presents a serious problem because long narrow wings are ill suited for woodland fliers that need to maneuver in the tight spaces between branches. Consequently, forest-dwelling birds commonly have shorter, broader wings than those that live in open habitats (Ruppell 1975). It is also important that the updrafts needed for long-range soaring are largely absent within forests. So swifts and gulls, with their narrow, pointed wings, are creatures of the open skies and are not to be found among trees dodging the branches. Because it is difficult for small forest fliers to soar, the selective pressure to increase horizontal range by switching to flapping should be considerable.

However, there are the previously mentioned aerodynamic problems involved in transforming a winged creature optimized for gliding into a flapping flier. When a bird or bat switches from gliding to flapping, it does not do so gradually by starting with small wing beats and then increasing them to full power. Instead, the switch is immediate and total, from static wings to strong flapping strokes. The reason is that small wing flaps do not produce enough thrust to make up for the loss of lift that occurs when wings move from the horizontal plane that produces maximum lift. Consider a glider that has not yet evolved the well-developed arm musculature needed to power strong wing flaps. Trying to boost the length of a glide with the small wing flaps would actually shorten the glide, because a lot of lift would be lost for the sake of only a little thrust. Balda et al. (1985) argued, therefore, that there is a gliding-flapping barrier that prevents gliders from evolving into flappers.

Nevertheless, it is possible to construct a scenario in which gliding turns into flapping flight in forest-dwelling aerialists. This scenario starts with the presumption that selective pressures to increase horizontal range while keeping the wings short can be met only by developing powered flight. Speed of ascent in trees can also be improved by assisting upward leaps with powered flight in the style of kokakos. Aerodynamics expert Ulla Norberg (1985a,b) proposed a model in which a particular pattern of small wing beats might have been used to extend glides. In this model, the wing beats are slow—just two per second rather than the normal six per second expected in small birds—and are asymmetric in that the downbeat is much faster than the upbeat. In Norberg's view, gliders with distal, finger-borne airfoils that form short wings find that incipient flapping motions boost horizontal range. Particularly pertinent in regard to this subject is the kakapo. Its small, weakly muscled wings are incapable of sustaining powered flight, but it occasionally extends the range of its descending glides by flapping its wings (Feduccia 1980, Livezey 1992). How such weak-winged birds make this aerodynamic transition has not been studied, so the Norberg hypothesis has not been tested. In any case, the kakapo proves that a weakly muscled flier can indeed switch from gliding to flapping, and discovering how it does so is important to better understanding how dino-birds may have done so.

Once this knotty transition from gliding to flapping has been made, the next steps are obvious. As the power, sweep, and frequency of the flapping strokes improve, the declining rate of descent both increases the horizontal distance traveled and lessens vertical descent, which in turn decreases the distance that has to be reclimbed at the end of each aerial trip. The speed of upward movement in trees increases as flapping is used to increase the velocity and range of upward leaps. Eventually flapping power becomes great enough to sustain level flight for long distances and then to sustain climbing flight. Despite the possible difficulties associated with transforming gliding into flapping flight, the abundance of living gliders—some with distal airfoils—makes this scenario highly viable.

This scenario avoids the problems that plague the classic gliding and terrestrial scenarios by incorporating the best—and dropping the worst—aspects of both the arboreal and the leaping hypotheses. Imagine a long-limbed climbing archosaur that moves between otherwise inaccessible high places via short leaps, some horizontal, others descending. This archosaur may also have the habit of leaping down upon its unsuspecting prey. Because a primary danger associated with leaping is suffering injury, or worse, after failing to successfully connect with the chosen landing spot, the ability to precisely target prey is advantageous. There is, therefore, substantial selection pressure to improve control of body orientation midleap. Thus, small airfoils gradually develop on the distal arms, where the airfoils' slight aerodynamic effect is magnified. As the archosaur becomes better able to control body orientation during leaps, landing success rises. At this early stage, there is little if any improvement in leaping range, which remains largely or entirely ballistic. As the size of the airfoils continues to increase, control of body orientation and landing performance improve further. Now the airfoils can generate enough lift to begin to significantly increase leaping range, which can be boosted even farther by flapping the growing airfoils to produce thrust. The vertical strokes of the forelimbs are derived from the control actions previously used to orient the airfoils during the ballistic leaps. These transitions are similar to those proposed in the leaping-insectivore hypothesis, with the exception that the problem of working against gravity has been avoided.

The subsequent stages are obvious. Leaping range improves dramatically as the area of the airfoils increases, proximally as well as distally, and there is a corresponding increase in the power of the flapping muscles. When the airfoils become sufficiently large and the muscles powerful enough, the animal is no longer leaping semiballistically but fully flying. Further increases in muscle power both increase horizontal range and make climbing flight possible.

Although there seem to be no practical problems associated with this scenario, the lack of a living vertebrate that uses very small distal airfoils to enhance leaps, but not to glide, may raise objections to this scenario. The way around these objections is the not entirely satisfactory suggestion that aerodynamically controlled leaping is rare and that when it does appear, it is quickly replaced by powered flying. Besides, the same problem afflicts most of the alternative scenarios.

Could a variation of the pounce theory, a variation based on lift generation (rather than drag), be the sole or primary factor driving the beginning of avian flight? Possibly, but why would a small climbing predator that was evolving airfoils to better control prey targeting not simultaneously use the same adaptations to better control interbranch leaps? A combination of leaping and pouncing appears more plausible.

A multifactor scenario combines gliding with leaping. The gliding promotes the development of proximal airfoils stretched between arms and legs. Leaping, which may include predatory pouncing, helps promote the development of distal airfoils. Flapping may develop primarily as a means to extend leaps or glides or both. This multifactor scenario combines the best aspects of the gliding and leaping hypotheses.

Yet another scenario starts with the assumption that long distal arm feathers initially evolved for nonaerodynamic purposes, perhaps display or brooding or both. This scenario eliminates the problems associated with the initial development of distal surfaces. Climbing preavians find that these large, flat surfaces make useful airfoils when moving between high places. The surfaces can be used as parachutes or as lift-generating surfaces to extend leaps and glides. The initially symmetrical feathers become asymmetrical in order to improve their aerodynamic qualities (see Chapter 7 for further details), and the surfaces enlarge. This scenario converges with those described above as the changes to improve range through flapping flight get underway.

The various flying vertebrates may have evolved flight via different modes. Because proximal airfoils are better suited for gliding than for flight control, the well-developed proximal airfoils stretched between the arms and hindlimbs of bats and pterosaurs make them good candidates for the gliding or gliding-leaping scenario. The lesser development of the proximal airfoil in birds makes them the best candidates for the interbranch-leaping scenario explored in more detail in Chapter 13. The persistently bipedal birds are also the best candidates for having initially evolved distal surfaces for nonaerodynamic reasons.

### *Summary and Conclusion*

One of the most vexing problems with figuring out exactly how birds, and for that matter bats and pterosaurs, evolved flight is the absence of living examples of transitional types. That there are no ground-to-air insectivores, few glider-flappers, few leaper-flappers, and few glider-leaper-flappers suggests that the transitional type, whatever it may be, is relatively rare and is quickly displaced by more aerially capable descendants. What we can conclude is that the abundance of arboreal forms with aerial capabilities, the probably arboreal origin of bat flight, and the gross inadequacies of the terrestrial hypothesis strongly suggest that flight started high. But in the end, only direct evidence in the form of fossils can resolve these questions. If protobirds prove to be poorly suited for climbing, then the ground-up hypothesis will be verified. If they turn out to be well adapted for climbing, then the arboreal hypothesis will be supported. The presence of symmetrical distal feathers will confirm the drag variant of the pounce hypothesis. Asymmetrical distal feathers will favor the gliding or leaping hypothesis or both. Evidence for a weak flight musculature will favor the gliding hypothesis (although assessing the strength of protowing muscles is inherently difficult; see Appendix 1).



## Arboreal Scenario versus Terrestrial Phylogeny

The arboreal versus terrestrial dispute has phylogenetic implications. To a great extent, the modern debate over bird origins has gone down a rather narrow path in that the arboreal hypothesis is primarily scenario based, whereas the terrestrial hypothesis is more phylogeny based. In the latter hypothesis, dino-avepods both were the ancestors of birds and were bipedal ground runners par excellence, so flight must have evolved among ground runners. The scenario-based counterargument asserts that because, as detailed above, the climbing hypothesis is superior to the ground-running scenario, protobirds must have been arboreal quadrupeds—the famous "proavis" some still hope to find in early Mesozoic sediments. Since dino-avepods were good runners rather than climbers, they could not be the ancestors of birds. The above outlines are somewhat simplistic; the pro-arboreal group often raises phylogenetic points, for instance. More importantly, not all those who favor the phylogenetic data supporting the dinosaur hypothesis consider the hypothesis incompatible with the arboreal scenario.

### *What about* Caudipteryx?

At first consideration, *Caudipteryx* (Figs. 1.6, 10.1Bo) appears to be a poster dinosaur for the hypothesis that flight and contour feathers evolved in running avepods (Ji et al. 1998, Burgers and Chiappe 1999). Herbivorous *Caudipteryx*—which had short arms, only two short-clawed fingers (of which the largest was not flexible), and very long legs and which lacked both distally elongated toes with large hooked claws (Fig. 6.3) and a large grasping hallux—was clearly a runner rather than a climber (contra Chatterjee 1999b, Garner et al. 1999), yet it sported large contour feathers on its hands and tail (Fig. 5.3). Does this combination favor the ground-up path to flight?

There are reasons to think not. To assess whether or not an avepod was in the process of developing flight on the run, we can list key pre-flight related characters expected to be present in such dinosaurs. In general, protofliers should have flight features that are better developed than those of avepod dinosaurs in general but less well developed than those in *Archaeopteryx*. Protofliers must also meet the following general requirements of the leaping-insectivore or predator hypothesis.



- Tail longer or no shorter than in *Archaeopteryx*.
- Sternum shorter or no longer than in *Archaeopteryx*.
- Arms longer than usual in dino-avepods, because the arms were evolving into wings.
- Long arm contour feathers asymmetrical and streamlined, in order to maximize aerodynamic effectiveness.
- Size small, under 1 kilogram, in order to facilitate ease of development of flight.
- Insectivorous or otherwise predaceous, as per the terrestrial leaping hypotheses.

*Caudipteryx* meets none of these requirements. Instead, its unusually short forelimb and tail, its large sternal plates, its large size, and its herbivorous habits directly contradict these expectations and indicate that *Caudipteryx* was not involved in or representative of the development of flight. Even more telling are its large but symmetrical feathers, which were completely unsuitable for generating either the lift or the thrust integral to the early stages of the development of flight (see the next chapter). The only kind of aerodynamic functions they could have performed for such a large animal were drag-induced air braking and deflective rudder control, either to help the animal maneuver while running or to shorten deceleration at the end of a run. That the feathers had somewhat ragged edges rather than the crisp, clean edges characteristic of avian airfoils also argues against their having had an important aerodynamic function. *Caudipteryx* did not, therefore, have incipient wings any more than does an ostrich (contra Burgers and Chiappe 1999). The argument by Garner et al. (1999) that *Caudipteryx* is an example of a pouncing predator that used its symmetrical distal feathers to control ambush leaps is contradicted by the combination of its atrophied arms and nonpredaceous habits. Without lift, thrust, or strong drag-inducing protowings, *Caudipteryx* cannot in any way be described as a protoflier. One could argue that this avepod descended from running protofliers and that herbivory lead to the shortening of its arms. But this argument neither explains the shortness of the tail nor constitutes positive evidence for the running or pouncing hypothesis.

How then do we explain the flight-related features of *Caudipteryx*? By considering the modern avepods that combine short arms and tails, symmetrical feathers, large size, a small head, and herbivory with long legs. These are ratites and other birds that have lost flight, a point that we pursue in greater depth later in this book. Interestingly, Garner et al. (1999) have noted that loss of flight can explain the characteristics of *Caudipteryx*.

### *Tree-Loving Dinosaurs*

The view that dinosaurs could not climb has been taken to such an extreme that Feduccia (1996) virtually scoffed at the notion of arboreal dinosaurs. Feduccia's skepticism is shared by many others, including Ostrom (1974a, 1986), Cowen and Lipps (1982), Caple et al. (1983), Balda et al. (1985), Gauthier and Padian (1985), Padian (1985, 1997a), Dingus and Rowe (1998), Ji et al. (1998), Padian and Chiappe (1998c), Clark et al. (1999), Gishlick (2000), and Tarsitano et al. (2000). This two-sided dispute, which has dominated the debate on bird origins, has often been dogmatic. Just because some dino-avepods ran on the ground does not necessarily mean that that was all they did. A host of creatures are scansorial, that is, adept at moving both on the ground and in the trees. Among them are many lizards and snakes, a variety of birds, most cats, various weasels, some kangaroos, and a number of primates such as baboons, most probably our small australopithecine ancestors, and children. If you find yourself being stalked by a leopard on some dark African night, there is no point in trying to outrun it. Clambering up a tree will not do you much good either. The big cats climb as well as they run.

Assessing the climbing abilities of predatory dinosaurs is complicated by the mutilfunctionality of their arms and legs. For example, the requirements for grasping prey are not identical to those for grasping plant parts. Therefore, the claws of a predatory climbing dinosaur would probably be a compromise between the two types.

Could bird-footed dinosaurs climb well? Certainly adult tyrannosaurs could not climb, but this fact tells us no more than what the poor climbing abilities of ostriches and lions tells us about the climbing abilities of birds and cats in general. We need to look instead at the many dino-avepods weighing less than 100 kilograms, which, as we have already seen, is the upper size limit for climbers. How small did dino-avepods get? *Sinornithosaurus* weighed just 2.5 kilograms, and *Compsognathus* and *Sinosauropteryx* also weighed in at only a few kilograms. There is tentative evidence (see Chapters 10 and 13) that in Late Jurassic times, there lived advanced, birdlike dinosaurs that weighed only 1 to 2 kilograms. The dromaeo-avemorph *Rahonavis* weighed only a kilogram or so. According to Xu et al. (2000), wee *Microraptor*, the smallest known adult dinosaur, probably did not exceed 200 grams, less than most *Archaeopteryx*.

There were juvenile avepod dinosaurs that began life at only a fraction of a kilogram: baby *Scipionyx*, for example, weighed only 100 grams. Young lizards commonly climb, even when the adults do not (Auffenberg 1981, 1994). The young use trees as a refuge from larger predators, including their adult relatives, and as a source of arboreal prey. The same disparity between the arboreal abilities of the young and those of adults is applicable to many dino-avepods. The babies of even gigantic forms may have climbed (this juvenile behavior may explain the absence of severely abbreviated distal toes in big adults discussed below). A modest parachute effect, plus the ability to reorient the body to best absorb impact, provided by the sort of long, fluffy feathers that adorned the arms and probably the legs of *Sinornithosaurus* and *Microraptor* would have provided an additional safety factor to a falling small avepod dinosaur (Fig. 6.1), so much so that it may have been able to deliberately parachute in the manner of tree kangaroos. Arguments by Balda et al. (1985), Pennycuick (1986), and Feduccia (1996) that all adult dino-avepods were too large to be climbers are false. Size was not a barrier to climbing dinosaurs.

One way to get a better handle on the climbing potential of predatory dinosaurs is to compare the climbing features and abilities of dinosaurs with those of dogs, cats, and monkeys. Dogs lack sharp-tipped, hook-shaped claws, and they do not have long, supple fingers or toes. As a result, they cannot grasp bark or branches and cannot climb well. Cats too have short fingers and toes, so they cannot grasp branches well either; but they do have well-developed, sharp, hooked claws with which to grip bark. As a result, they are good climbers. Note that cats use the claws on their forelimbs and hindlimbs both to climb and to damage prey. Monkeys lack sharp claws, but they have long supple fingers with which to hold onto tree parts; so they too are very good climbers. Climbing lizards have both sharp, hooked claws and fairly long, supple toes and fingers.

Most small dino-avepods had long fingers and toes and sharp, hook-shaped claws, especially on the fingers. They should have been able to grip bark and hook branches with the same claws they used to pierce prey, and they should have been able to grasp branches with the long, supple digits. The presence of large hooked claws in nonclimbing predatory dinosaurs has been cited as evidence that the digits were not used for climbing (see Clark et al. 1999), but this is circular reasoning based on the unsubstantiated assumption that these dinosaurs did not or could not climb. Instead, we can conclude that small predatory dinosaurs had the potential to have climbed better than dogs, possibly better than those arboreal wallabies that are only moderately specialized for climbing (Grzimek 1990), perhaps as well as the most highly arboreal wallabies, cats, and semiarboreal lizards in many cases, but probably not as well as specialized primates.

The reversed hallux present in *Caudipteryx* (according to Zhou et al. 2000) was too small and proximally placed to have been used to grasp objects, but it hints at a climbing heritage in the ancestry of this specialized ground runner. Tarsitano (1991) and Feduccia (1996) cited the usual absence of a reversed hallux as evidence that dino-avepods were poor climbers, but climbing cats lack opposable digits on their hands and feet. In addition, climbing dino-avepods did not need a grasping hallux as much as bipedal arboreal birds did because the dinosaurs were quadrupeds able to use their clawed fingers to help secure a hold. The sprawling action of the dinosaurs' arms would have facilitated climbing (Fig. 6.1).

As touched upon earlier in the chapter, Hopson and Chiappe (1998) noted that toes become increasingly distally elongated as climbing ability improves (Fig. 6.3). Such elongation improves a toe's ability to wrap around and grasp a branch. In contrast, the distal toe bones of running birds are usually strongly abbreviated (the kiwi is an exception, but its toe bones are more robust and correspondingly less well suited for grasping than those of climbing birds). The relative proportions of the toe may underestimate the climbing performance of animals that can also use their arms for climbing. No avepod dinosaur—even the swift ornithomimids—had toes as distally shortened as those of ratites or even those of the predatory phorusrhacoids; and some had distal toes almost as long as those of birds that spend significant time in the trees (Fig. 6.3). No known avepod dinosaur had toes as distally elongated as those of highly arboreal birds, and the dorsal arc of the toe bones common to birds that climb and perch was lacking in dinosaurs (Fig. 6.3), except *Protarchaeopteryx,* which did have arced toe bones. But the toes of many dino-avepods were not proportioned in the manner that favors running over climbing, so it appears that small examples were adapted to spending considerable time moving vertically as well as horizontally.

Feduccia (1996) argued that the stiff trunks of dino-avepods differ from the more supple bodies typical of climbing reptiles and mammals. This is true, but arboreal birds have even stiffer trunks, making dino-avepods the most birdlike of tetrapods in this respect. Among birds, the most dinosaur-like climbers are juvenile hoatzins. These odd baby avians sport long, clawed fingers with which they move quadrupedally among branches. Note that parrots are tripedal climbers, in that they use their hooked beaks as an additional grasping appendage.

What about those avepod dinosaurs, such as *Oviraptor,* that lived in treeless habitats and had hooked claws and long arms? Obviously, they did not climb. Nor do pumas that dwell in dry shrublands; however, pumas that live in forested habitats are frequently arboreal.

The assertion that dino-avepods show few if any adaptations for climbing, as per Feduccia (1996) and Padian and Chiappe (1998c), is not correct. Any small animal with grasping digits ending in recurved claws has significant climbing potential. By the standards of the skeptics, cats and Bengal monitors could be assessed as nonclimbers. This is not to say that most predatory dinosaurs were highly specialized for arboreality. Indeed, the combination of a running heritage and the ability to climb is what best explains how and why some dinosaurs evolved into birds (see the discussion later in the chapter).

Although most small predatory dinosaurs show adaptations suitable for climbing, the level of development varied. Arms tended to be fairly short, and hands and finger claws rather small, in Late Triassic and Early Jurassic coelophysids. Their toes were not strongly distally elongated, although they were more so than those of most other dino-avepods; nor were their toe claws large and hooked (Fig. 6.3), so their climbing abilities can be ranked as modest at best. Judging from *Sinosauropteryx,* compsognathids had somewhat larger hands and arms than coelophysids. Compsognathid arboreal capabilities therefore appear to have been improved over those of coelophysids, but were still modest. Because the nature of the hand and toes of *Ornitholestes* is not yet entirely clear, its climbing ability cannot be fully assessed, but it may have been well developed (Fig. 6.1). As discussed above, unusually short-armed and -toed *Caudipteryx* was ill suited for climbing. Toes that were distally about as short as those of roadrunners suggest *Rahonavis* was not especially well adapted for climbing (Fig. 6.3); the absence of the hands in the specimen prevents an assessment of this aspect of its climbing ability.

Juvenile *Scipionyx,* troodonts, dromaeosaurs, oviraptorosaurs, and *Protarchaeopteryx* should have been capable climbers. However, Gishlick (2000) concluded, to the contrary, that the semilunate carpals of these dinosaurs did not allow the movements needed for climbing. Gishlick seems to have underestimated the degree of action of the arm joints, especially the strong dorsal rotation of the humerus described in Chapter 4. Dromaeosaurs, oviraptorosaurs, and *Protarchaeopteryx* had especially long arms and had fingers ending in large, hooked claws. *Protarchaeopteryx* had exceptionally elongated distal toes, and the claw-bearing bone was arced in a way that enhances branch-grasping ability.

Most interesting were *Sinornithosaurus* (Figs. 1.6, 6.1) and *Microraptor.* Not only were the former's arms 80 percent as long as its legs, but both sets of limbs are markedly longer relative to the rest of the body and the pelvis than are the arms and legs of other dino-avepods. In other avepod dinosaurs the hindlimbs are 2.8 to 5 times as long as the ilium, whereas in *Sinornithosaurus* (Fig. 1.6) the ratio was about 5.25. This ratio exceeds the hindlimb/ilium ratios seen in the swift tyrannosaurs and ostrich mimics. The relatively modest size of the pelvis in *Sinornithosaurus* suggests that the leg musculature was correspondingly modest, an attribute common to climbers (Grand 1977). In addition, the distal leg segments are not especially elongated in *Sinornithosaurus,* and the femur is elongated along with the rest of the leg. The limbs of *Sinornithosaurus* are most reminiscent of the unusually long, proximally slender appendages that spider monkeys, orangutans, and other primates use to increase their reach while climbing among branches. A similarly extreme hindlimb/ilium ratio in the complete *Bambiraptor* skeleton (Fig. 1.7) may have been due to its juvenile condition, and is also the result of distally elongated leg segments common to fast runners, so this appears to have been a running adaptation. *Microraptor* may have had hindlegs elongated in the manner, of *Sinornithosaurus;* incomplete preservation of the ilium makes this uncertain. That the central finger was at least as strong as the thumb in *Sinornithosaurus* and perhaps in some other dromaeosaurs (Chapter 4) further implies that the hand was modified to carry weight during climbing. The retractable, enlarged, sickle claws common to most troodonts and dromaeosaurs may have functioned as climbing hooks and spikes—as do the boot spikes used by workers who climb utility poles—as well as weapons. The large dewclaws of cats have a similar dual function. Although the subequal configuration of dromaeosaur and troodont toes III and IV evolved primarily as an adaptation for the didactyl foot posture associated with a hyperextendible toe II, the resulting symmetry of the two load-bearing toes had the secondary effect of improving their ability to co-wrap around and grasp branches. This symmetrical grasping function seems to have been further enhanced by a classic arboreal feature in *Sinornithosaurus,* whose metatarsals III and IV appear to have been similar in length at the distal end, and even more so in *Microraptor,* in which all three central metatarsals are the same length. In *Sinornithosaurus,* the two toes, as well as the distal toe bones, were both more elongated and more slender than those of other dromaeosaurs or most other dino-avepods, and in this regard they approached those of archaeopterygiforms and pigeons (Fig. 6.3). The distal elongation of the toe bones in *Microraptor* (Xu et al. 2000) appears to be greater than that seen in *Archaeopteryx* or a pigeon, and matches the condition of *Protarchaeopteryx*. The toe claws of *Sinornithosaurus* and *Microraptor* were also exceptionally large—proportionally as big as those of pigeons—and strongly curved. Enlarged flexion tubercles at the base of these claws improved grasping power, especially in *Microraptor.* Lack of better information on hallux reversal inhibits analysis of this feature in *Sinornithosaurus,* but the hallux was not distally placed. The hallux of *Microraptor* was not reversed but was distantly placed to enhance its grasping function. Rejection of a significant degree of arboreality in these lithe basal dromaeosaurs is especially illogical since they exhibit so many clear-cut climbing adaptations (Xu et al. 2000). Indeed, they seem to have been better adapted for quadrupedal climbing than for bipedal running (see Chapter 9 for a comparison to *Archaeopteryx*). If anything, *Microraptor* may have been better adapted for climbing than *Archaeopteryx* because the latter's short distal toes and unequal distal metatarsals were not so well suited for grasping branches. Far and away the most tree-adapted dinosaur yet known is the long-armed Mongolian avepod dinosaur announced at the Florida bird origins conference. With a third finger that was far longer than the second(!) and a foot whose proportions and toe orientation could have been taken from a perching bird, this dinosaur must have been a fully adapted climber that spent little time on the ground.

Chatterjee (1997, 1999a) suggested that dromaeo-avemorphs used their stiff tails as props during trunk climbing, much as woodpeckers do. The manner in which Chatterjee illustrated the tail being used, as a rigid rod with only the tip impinging on bark, is unlikely because the tail was not that stiff or strong. More likely, a long section of the tail was pressed against the trunk; tree kangaroos use their tail this way (Nowak 1999). This function would have been enhanced by the presence of feathers of some sort in order to increase friction between tail and tree. When going up trunks, avepod dinosaurs probably climbed like tree kangaroos and cats, quadrupedally, grasping the trunk with arms out to the sides and the flexed legs held closely together (Fig. 6.1). Among the branches, avepods climbed both bipedally, when moving along the trees (Fig. 6.1), and quadrupedally, when moving between them. When avepods leapt upward between branches, the gait may have been either bipedal, like that of kokakos, or quadrupedal, or both. With its comparatively short arms, *Microraptor* was probably more bipedal in the trees than *Sinornithosaurus.* It is questionable whether dino-avepods could, like lizards (Auffenberg 1994), hang upside down while moving along a branch. With its medially directed palm, the hand could have grasped a branch in this situation, but the fore-and-aft oriented foot with its inflexible ankle would have been ill positioned to do the same. Dino-avepods probably could not crawl down tree trunks head first, because the simple hinge-jointed hindlimbs could not even begin to allow the foot to be reversed. This last point is important. The inability of climbing dinosaurs to descend easily with fingers and toes may have promoted the development of an aerial means of going down.

You are again being stalked by a predator, this time in a Wyoming woodland some 110 Myr ago. You cannot outrun the sickle-clawed dromaeosaur that has decided to taste your strange kind of flesh. If you scramble up a tree, too bad. It will easily outclimb you there as well.

In Chapter 2, I noted that proponents of the scenario-based, arboreal hypothesis must establish that dinosaurs, especially those closest to birds, were poor climbers in order to rule dinosaurs out as bird ancestors. As shown here, the opinion that the largely terrestrial theropods were inherently ill suited for evolving into arboreal forms that in turn evolved into feathered fliers is no more valid than the opinion that early terrestrial ungulates could not have evolved into marine swimmers with flippers. That is what evolution does: it dramatically transforms animals that function in one manner into wholly new forms with entirely different lifestyles. To determine whether or not a particular transformation is possible, we must answer two questions. Is the proposed transformation too extreme to be viable? Does the fossil record support the transformation? As for the first question, I find interesting and perplexing the fact that so many researchers have insisted that creatures with as much potential locomotory flexibility as dino-avepods were limited to traversing only flat terra firma, a view that should be laid to rest as new arboreal examples are found. Let us now turn to the second question. The view that all small, bird-footed dinosaurs were ill suited for climbing proves as mythical as was the long-standing and unsubstantiated opinion that the very same bird-footed dinosaurs refused to swim (Paul 1988a). Quite the opposite is true: after the Triassic extinction of climbing quadrupedal archosaurs with opposable digits, small dino-avepods appear to have been the most adept nonavian archosaur climbers during the rest of the Mesozoic! Of these, we can point to *Sinornithosaurus, Microraptor,* and the new arboreal theropod as the best climbing archosaurs of the Juro-Cretaceous, with the exception of a number of birds; and the climbing dinosaurs indicate that some dino-avepods did make the switch from a predominantly running to an arboreal lifestyle. The absence of medium-sized climbers in mid-Mesozoic trees and bushes may have sparked the evolution of climbing dinosaurs.

As we finally discover in new lagerstätten the specialized climbing dromaeo-avemorphs predicted by the arboreal hypothesis, the majority view that a dinosaurian origin for flight required a primarily running component is being seriously challenged, if not refuted. It is therefore concluded that the probability that the ancestors of birds were climbers does not exclude dino-avepods from the role (Paul 1988a, Olshevsky 1994, Chatterjee 1997, 1999a,b), and the probability that predatory dinosaurs were the ancestors of birds does not require that avian flight evolved from the ground up. If these conclusions are valid, then the phylogenetic implications of the debate over the arboreal scenario versus the terrestrial scenario for the origins of bird flight are not as important as most researchers think they are.

### *What the Proavian* Should *Have Looked Like*

In his classic 1926 book, Heilmann presented and illustrated what he thought the Triassic or Jurassic "proavis," the reptile ancestral to birds, should have looked like. This tradition has been followed by Tarsitano (1991) and Feduccia (1996), the latter with his "protoavis" (not to be confused with "*Protoavis*" [Chatterjee 1991]). All three descriptions are broadly similar. These researchers have assumed that the protoavian was highly arboreal, nonterrestrial, and nondinosaurian. All their protobirds are rather lizardlike forms with necks that are much shorter than their long bodies. These protobirds are fully quadrupedal, the hindleg is not especially long relative to the body, the gait is non-erect, and the hands and feet are rather short. Prefeathery, limb-borne gliding airfoils are present on both the arms and legs in the Heilmann and Tarsitano versions and on the arms only in Feduccia's portrayal. These nondinosaurian proavians also tend to be quite small, in accord with the notion that bird flight got its start in wee tree reptiles. Feduccia's protoavis is based in part upon the drepanosaur *Megalancosaurus* (Fig. 10.1Aa). (Ruben [1998] and Geist and Feduccia [2000] argued that *Megalancosaurus* was a glider whose aerodynamic adaptations might provide insight into the early stages of the development of avian flight. Among its supposed aerodynamic adaptations were limbs supporting flight membranes, including a patigium in front of the upper arm; vertebral shoulder spines fused into a tall shoulder notarium; and a very long and deep vertical tail rudder.) Feduccia (1996) is so confident in his protoavis that he states that "somewhere in pre-Solnhofen sediments a similar animal awaits unearthing."

Well, perhaps not. I have no problem with the arboreal nature of these proavians, but they have numerous problems otherwise. No flight membranes are preserved on *Megalancosaurus,* and its short limbs set low on the body, its deep and narrow body and tail, its prehensile downcurved tail tip, and its tail-heavy configuration appear most unaerodynamic (Fig. 10.1Aa; Renesto 2000). The

*Figure 6.4.* A hypothetical dinosaurian "proavis," the semi-arboreal bird ancestor that may have lived in the Early or Middle Jurassic. This creature may have weighed from one to several kilograms, and it has the form of a protodromaeo-avemorph with moderately elongated arms and an enlarged, partly reversed hallux that would have improved climbing abilities above the level seen in typical dino-avepods. The life restoration shows the protobird leaping between branches, using its incipient feathery airfoils as aerodynamic control surfaces. Will the fossil record verify or deny this speculation? Time will tell.

tall notarium is also unaerodynamic, as well as most unbirdlike, and is better explained as a base for neck muscles and ligaments that helped make the long neck a snap-action organ for catching prey.

Let us now turn to supposedly feathered *Longisquama*. Even if its asymmetrical appendages were dorsal airfoils in the manner suggested by Haubold and Buffetaut (1987), their arrangement was most unavian and appears to have left the appendages without sufficiently powerful control muscles. The appendages' extreme thinness (Chapter 5) makes it unlikely that they could have supported the mass of the body while providing lift. The somewhat elongated scales on the trailing edges of the arms formed a more plausible incipient airfoil. However, because the arms were short and because avian feathers are probably not derived directly from true scales, the rationale for the idea that proavians entirely lost dorsal gliding airfoils while switching to propulsive wings is at best convoluted speculation. The "proavians" are actually not very preavian.

Let us look at the characteristics of birds, especially early ones such as *Archaeopteryx,* and thereby determine the characteristics expected in the real proavians. Mass is substantial, half a kilogram or more. There is no prepatigium or any other wing membrane preserved in any of the specimens. The trunk is so short that it is about the same length as the long neck. The shoulder vertebral spines are shallow and are not fused into a notarium. The tail is reduced and shallow, and it supports a horizontal airfoil. The wings are held horizontally, so the shoulder glenoid faces laterally. In sharp contrast, the hindlimb gait is erect; in particular, the hip joint is cylindrical. Birds are bipeds, so the hindlegs are long relative to the body. The feet and hands are elongated. It is the inner fingers and toes that are divergent and used to oppose the other digits when grasping, rather than the outer digits, as was typical of nondinosaurian archosaurs. The hindleg is completely separate from the brachial airfoil. Feduccia (1996) argues that avian bipedalism evolved because the forelimbs were "released" for flight, but this argument does not explain why birds that evolved from quadrupedal climbers did not continue to climb, and also fly, quadrupedally like bats and pterosaurs. Also pertinent is that the toe proportions of early birds like *Archaeopteryx* and *Confuciusornis* were not fully adapted for either arboreal or terrestrial life (Hopson and Chiappe 1998), because such intermediate toe proportions are exactly what is expected in basal birds whose heritage included both running and climbing. In this view, highly arboreal *Confuciusornis* had pigeonlike toe proportions either because evolution had not had time to completely reconfigure the digits away from the more terrestrial avepod dinosaur arrangement or because *Confuciusornis* could call upon its clawed finger to assist climbing, or perhaps for both reasons.

In the end, the most serious problem with highly arboreal bird ancestors is that they are *too* arboreal, because specialized climbers would probably have produced web-winged, four-legged fliers rather than bipeds that fly with the arms alone.

What, then, should the gliding proavian have looked like? Why, like a long-necked, short-trunked, long-fingered dino-avepod with opposable inner digits! One in which a running heritage has led to the complete decoupling of side-action protowings from long, erect legs, so that wing and tail feathers, rather than wing-leg membranes, are present (Pennycuick 1986, Bock and Buhler 1995). One in which climbing has elongated the forelimbs. One in which a prepatigium is absent. One that has a shallow tail. One that weighs in at the near-kilogram class. Even if it had been a glider, *Megalancosaurus* is no more a suitable model for the early stages of avian flight than is a flying squirrel (see Renesto 2000). If and when the pre-Solnhofen "protoavis" is found, it should look like a long-armed predatory dinosaur rather than like a feathery lizard. It ought not to look like Pennycuick's (1986) "protoavis" either. Although the latter is correctly bipedal, it has no neck, and it glides on wing membranes whose presence in proavians is most implausible.

It is an ironic and underappreciated fact that dino-avepods are such good potential bird ancestors because they fit both the running and arboreal hypotheses. I have executed a speculative skeletal restoration of what I think the dino-proavis may have looked like (Fig. 6.4). This restoration shows a small avetheropod with long arms and an enlarged, partly reversed hallux suitable for climbing, plus the beginnings of an arm-folding mechanism and a somewhat reduced tail. Various details are extrapolations of what would be expected in an ancestor of dromaeo-avemorphs. Of course, no real animal exactly matched my concept, and one change that I might make would be to lengthen its limbs; but I am willing to bet that this depiction will prove to be much closer to reality than other proposed proavians. In this regard, I was pleased to see how closely *Sinornithosaurus* (Figs. 1.6, 6.1) approaches my dino-proavis—which was finished before I was aware of *Sinornithosaurus*—in many of its attributes.

CHAPTER 7

# The Early Evolution of Flight

In evolution, the first organisms to do a particular thing generally do not do it well. The first animals with eyes were barely able to form visual images; the first animals to walk on land were slow and awkward. Likewise, the first flying avepods were far from the skilled aerialists that grace modern skies. There often is intense selective pressure to improve a new capability, and this pressure continues until its ultimate limits are reached. How well did the first birds fly, and how quickly did their descendents upgrade their performance? Answering these questions requires that we take an in-depth look into how flight works, a subject that is frequently misunderstood.

## How Things Fly

### *Flight Is Not as Hard as You Think*

Flight looks magical. The sight of heavier-than-air objects suspended in the atmosphere is so extraordinary that we get the impression that flight must be hard. At first glance, history supports this impression. Vertebrates first power-flew more than 300 Myr after they evolved, and well more than 100 Myr after they first walked on land. All three groups of vertebrates fliers—pterosaurs, birds, and bats—took many millions of years to learn to fly (Fig. 7.1). Humans did not power-fly until December 17, 1903, long after mechanical means of water and land travel had been invented. Because powered flight evolved so long after walking, walking and other means of getting around may seem easier than flying.

In fact, it is harder to walk a mile than to fly one. Radio-controlled aircraft that can take off, climb, do acrobatics, and land are available at the corner hobby store. A child can learn to control them. The first robotic system, the autopilot, was put into widespread use back in the 1930s. Today, the most advanced airliners can take off, cruise at altitude, and land without any intervention from the pilot. One low-budget project produced a fairly realistic radio-controlled flapping flier in the form of a pterosaur (MacCready 1985, Paul 1987b, Paul and Cox 1996). However, despite large research investments, you cannot yet go to a store and purchase a radio-controlled robot that can agilely walk along a flat sidewalk, much less rough terrain. The most advanced robotic labs are only just becoming able to produce such machines (Moran and Van Dam 1996, Paul and Cox 1996, Saunders 2001).

Walking, especially the dynamic walking on erect legs practiced by mammals and birds, is actually very difficult. It requires well-developed senses, including an exquisite sense of balance, coordinated with the operation of complex multi-jointed legs. The amount of information that must be processed in subsecond increments is vast, and there is no break in the task. A single mistake during any step can result in immediate failure and impact with the ground. The ground itself, often highly irregular and strewn with obstacles, is a challenge that increases the overall information-processing and coordination problems many-fold. Running is even more dangerous because higher speed increases the chances of a violent fall or collision with some object.

In comparison, flight, even the flapping version, is a comparatively simple operation. Control of orientation requires the operation of a few tail and wing surfaces whose functions can be easily coded for within the neural network of the brain. Wing flapping is a stereotyped action that also can easily be coded for. Traveling through air is forgiving in that there is a lot of empty space, so the danger of running into something is sharply reduced much of the time, and the time to recover from mistakes is often fairly long. For this reason, children are permitted to fly private aircraft (as long as an adult has access to dual controls), whereas allowing the same youth to drive a car on public roads is strictly illegal (even if the car has dual controls). As children, and small-brained flying insects and pterosaurs, show, flight is not

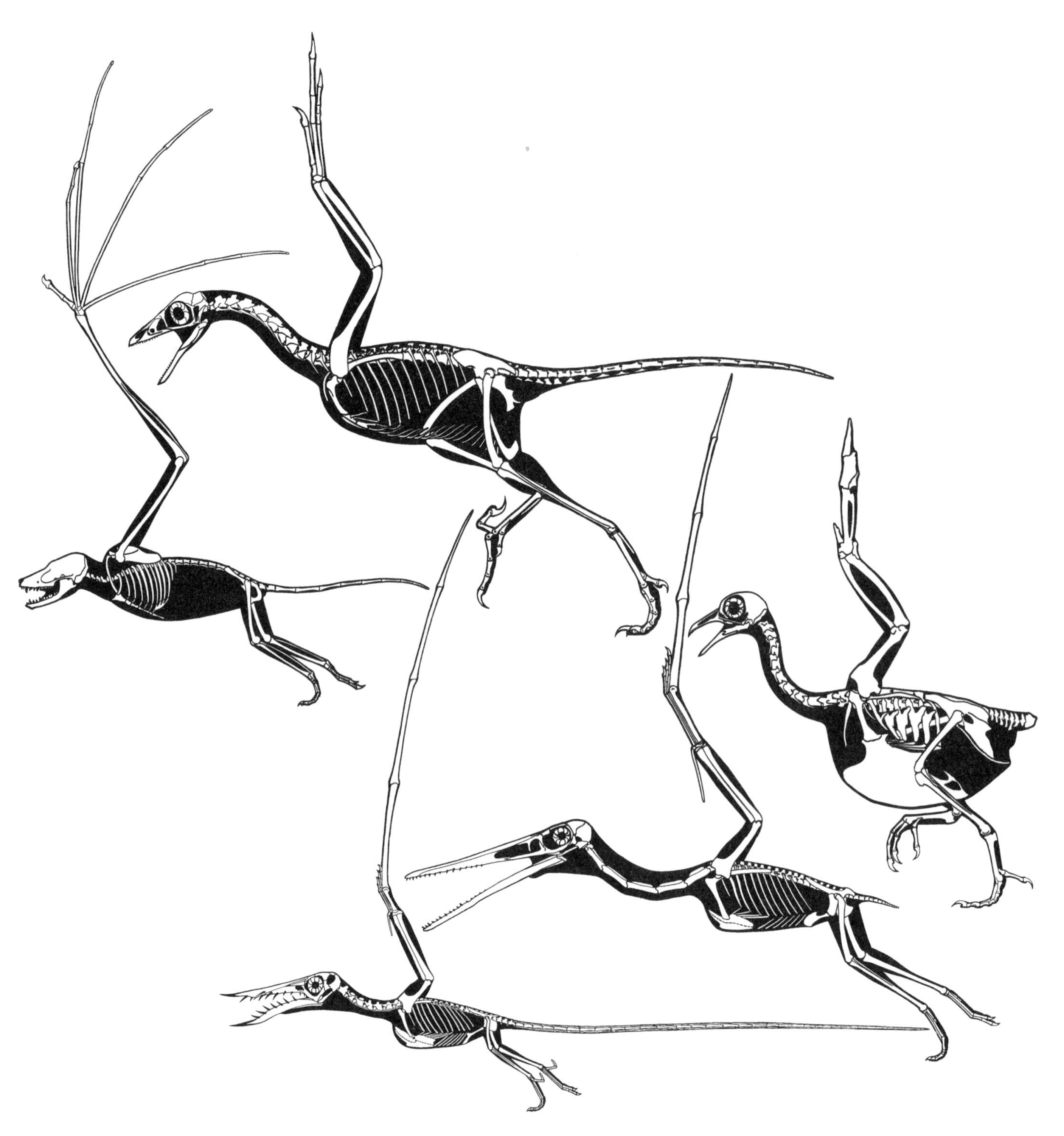

*Figure 7.1.* Representatives of the three groups of power-flying tetrapods. Both of the two archosaur examples included early long-tailed forms, later replaced by more dynamic short-tailed examples. The first archosaurs to power-fly were pterosaurs such as *Rhamphorhynchus,* followed by short-tailed *Pterodactylus.* Birds, including long-tailed *Archaeopteryx* and the pigeon, evolved later. On the left is a representative of the always small-tailed mammalian experiment, the fossil bat *Icaronycteris.* Both pterosaurs and bats stretched membranes over elongated fingers and back to the hindlegs. Bird arms support feathers that are unattached to hindlegs, which are, therefore, entirely free for running. Drawn to same approximate wingspan.

especially challenging from an intellectual standpoint.

Powered flight is, however, more difficult than walking in that it requires a lot of power per unit time; normal walking demands about one-tenth the energy, and swimming even less (Hill and Wyse 1989, Rayner 1991). The power problem is one reason why powered vertebrate flight evolved so long after vertebrates first swam and walked and why humans did not fly until after my still-living grandmother was born. The low aerobic capacity of most fish and all amphibians and reptiles barred them from evolving sustained flight, so "flying" fish, amphibians, and reptiles actually only glide (as did the aeronauts of the 1800s). Not until it became possible to boost the oxygen capacity of the cardiorespiratory system, or to construct internal combustion engines with high power/weight ratios did sustained powered flight become feasible. One reason the Wright brothers were the first to fly was that they were the first to understand the need for a dynamic flight-control system (Culick 1979). However, had their mechanic not built a lightweight engine, and had he not linked it to the first efficient propellers, the Wrights could have built only gliders. Had the classical Greeks had a good motor, they might have gotten off the ground.

### *Lift and Thrust: The Real Story*

The lift created by wings and the thrust generated by propellers are really the same thing; it is just that lift is always up, and thrust can be in any direction. Propellers are just rotating wings that generate lift in order to produce thrust. The Wright brothers' recognition of this fact allowed them to design the first modern propellers and thereby compensate for the rather low power output of their first engines (Culick 1979). A gliding or soaring bird uses its wings to generate only lift. A flapping bird uses its wings to generate both lift and thrust.

Most of us were taught that aircraft and bird wings work in the following way. The bottom surface of the wing is flat, and the top surface is curved (see wing cross sections in Fig. 7.2A). As air flows past the wing, the air flowing over the curved top has to move farther than that moving under the bottom. The air on the top must, therefore, move faster than that on the bottom. The faster a fluid moves past a surface, the lower the pressure on the surface. This phenomenon is called the Bernoulli effect, and it explains why blowing along the top of a piece of paper causes it to rise. The fast-flowing air reduces the pressure on the curved wing top, and the higher pressure on the underside lifts the wing. This nice, neat story, which continues to be repeated in many books (e.g., Ruppell 1975), is misleading.

If this story were the full explanation, the aircraft that fly upside down past air show crowds would plunge to the ground because the curved tops of their wings are directed earthward. In addition, the wings of stunt aircraft are built symmetrical top and bottom so that they will fly well regardless of their orientation. Even more interesting is that the "supercritical" wings of modern airliners are flat on top and curved on the bottom! (This configuration prevents the formation of disruptive shock waves in the higher-speed air that flows over curved wing tops when a plane flies at near the speed of sound.) Pre–World War II aircraft wings were simple cambered sheets with the top and bottom equally curved, and the same is true of bat wings. Bird wings, like some early aircraft wings, are both cambered and more curved on the top than on the bottom. That even a perfectly flat sheet can generate lift is shown by the store-bought balsa wood toy plane.

So how do wings really work? The answer to this remarkably complex and still somewhat controversial question would take up a whole chapter in itself. Lots of mathematics is involved, even true for an idealized "infinite wing" that has no ends, and matters get worse when we turn to real wings, whose tips spill air out to the sides. Even worse are flapping wings that change shape and orientation during each beat. We must consider angle of attack, down thrust of air, and the pressure differentials created by a bound vortex. For example, some airfoils must be angled to generate lift, but others—including the classic shape—can produce lift even with a small negative tilt. In all cases, lift-producing wings do have lower pressure on the upward surface than on the downward surface, but in the end it is the wing's ability to do the work needed to constantly deflect a mass of air downward that keeps aloft an equivalent mass of flying machine; Newtonian physics's reaction for every action makes flight possible. So a plane that weighs 10 tonnes needs to generate 10 tonnes worth of downward thrust (equals lift) with its wings in order to maintain level flight. Those who wish to pursue the physics of flight in more detail are referred to Pennycuick (1989), Wegener (1997), and Anderson and Eberhardt (2001).

Why, then, are most wing tops curved? The answer is that a thin, flat wing set at an angle is neither streamlined nor very strong, so the cross

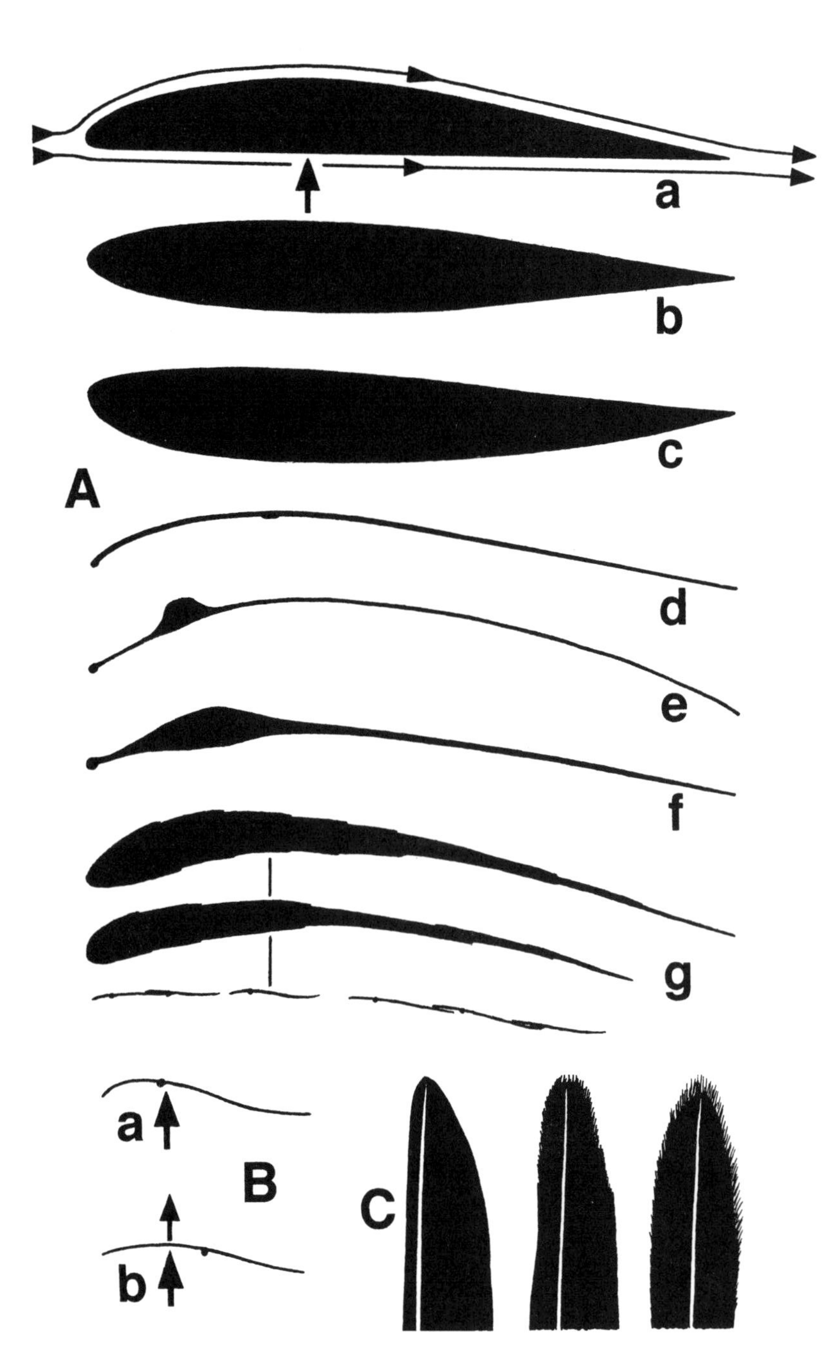


*Figure 7.2.* Cross sections and profiles of lift-producing airfoils. The standard wing form *(Aa)* supposedly generates lift via the speed and pressure differential created by the combination of a curved-top and a flat bottom. However, the symmetrical airfoils *(Ab)* of stunt planes and the flat-topped wings *(Ac)* of airliners such as the Boeing 777, which flies at nearly the speed of sound, are also capable of generating the lift needed to fly. Also capable of generating lift are the thin, cambered wings *(Ad)* of the first airplanes and ultralights, wings that are similar to the membrane wings of bats *(Ae)* and pterosaurs *(Af)*. The majority of a bird's wings *(Ag)* are cambered, but the wings are also thicker in a manner that makes them more streamlined. The outermost part of the wing consists of semiseparate primary feathers that are themselves mini-airfoils. Because the main axis of lift is toward the leading edge *(Ba)*, primary feathers are asymmetrical, with the supporting shaft located at the main axis of lift. Symmetrical feathers will twist along their long axis because of the asymmetrical mismatch between the axis of lift and the main supporting beam *(Bb)*. View C shows wing feathers, leading edges to the left, of the flying crow *(left)*, the semiflightless kakapo *(center)*, and the flightless rail *(right)*.

section needs to be at least somewhat curved in order to satisfy both of these needs. Curving the top of the wing usually works best, because this most favors the creation of a smooth bound vortex while minimizing drag. The classic wing shape is therefore optimal for most purposes, as long as the flyer does not approach the speed of sound or routinely fly upside down. Since birds do neither, and since feathers can form airfoils with enough thickness to make the bottom flatter than the top, this shape has been adopted by flying avepods. Birds, therefore, tend to be fast fliers. Bats are limited to thin, stretched skin, so most must get by with the simple cambered airfoil cross section that was abandoned by mainstream aviation long ago. As a result, bat wings produce too much drag for bats to be generally as speedy as birds. Cambered wings are, however, very good at generating lift at low speeds, which makes them well suited for hard turns—a fact used to advantage by insect-intercepting bats.

The contour feathers that make up the wing surface are the main source of lift and thrust in birds. An airfoil's main axis of lift is located at about one-fourth of the distance from the leading edge to the trailing edge, so the main support spar has to be near the leading edge (Gordon 1978). Flight feathers are therefore strongly asymmetrical in that the main shaft is set well forward (Fig. 7.2B,C).

Although symmetrical feathers can easily generate lots of lift, they tend to twist around the supporting shaft because the axis of lift is well ahead of that shaft (Fig. 7.2B). As the leading edge pitches up, the lift increases, which further increases the twist and so on. During World War I, when this phenomenon was less well understood, the monoplane wings of the Fokker D8 fighter tended to twist off because they were too strong aft. Because feathers are supple, they will not break, but they will continue to twist until the angle of attack becomes so high that the feather stalls out, which leads to the sudden loss of all lift. The feather will then untwist, but the subsequent return of asymmetrical lift begins the cycle again, which continues indefinitely. This twisting presents an additional problem in that when the twisted feather's angle of attack is high, excessive drag is produced. Symmetrical wing feathers, therefore, are worse than useless, and, not surprisingly, no bird uses symmetrical feathers to generate lift or thrust. Strongly asymmetrical feathers smoothly generate a predictable amount of lift with minimal drag, and they help the wings turn subvertical flapping arm action into forward thrust (R. Norberg 1985).

These observations lead us to the emphatic conclusion that if an animal's arm feathers are asymmetrical, then they are used to generate lift and perhaps thrust; if they are symmetrical, then they are not used for either function. Symmetrical feathers are suitable only for a few simple aerodynamic functions: braking and parachuting through drag, and rudder action through deflection of air as it flows symmetrically down the length of a feather. All other forms of aerodynamic control using broad surfaces involve lift. Therefore, symmetrical feathers are not suitable as lift-action aerodynamic devices for controlling yaw, pitch, or roll. In the context of evolution, it is as easy for feathers to be asymmetrical as symmetrical.

These facts lend themselves to the following conclusions. If the first contour feathers evolved to provide the kind of lift or thrust that can be produced when air flows perpendicularly to the long axis of the feather, then they had to be asymmetrical. If they were symmetrical, then they could not have evolved for aerodynamic actions, with the possible exception of the few actions that can be generated when the air flow parallels the long axis of the feathers. In most fully flight-capable birds, primary feather asymmetry ratios—the chord of the feather trailing the central shaft divided by the chord of the feather leading the central shaft—range from two to twelve, with the average being six (Speakman and Thomson 1994). (The symmetry, or lack thereof, of tail feathers is less critical because their long axes tend to lie parallel to the airflow.)

Streamlining improves the function of aerodynamic structures dramatically. Ergo, the less crisp-edged and aerodynamically clean contour feathers are, the less likely it is that they serve an aerodynamic function (Fig. 7.2C).

The shapes that will produce useful lift as long as there is an angle of attack are surprising. Even long, narrow, tubular fuselages generate lift, a fact especially important for level-flight missiles that have only small fins. But flatter surfaces produce more lift than rounder ones, and sustained flapping flight requires large airfoils. We can determine the minimum area and span needed simply by measuring the wing surface–to–body mass ratio (known as the wing loading) in a variety of flying birds (App. Fig. 18G,H). At any given wing size, wing loading varies substantially. At one extreme, forms that soar at moderate airspeeds—such as vultures and condors—have enormous, lightly loaded wings. At the other extreme, birds like ducks and loons that need to minimize drag in order to power cruise at high speeds—and can survive the resulting high-speed landings by splashing down on water—have relatively small, heavily loaded wings. Some barely volant birds have the smallest aerodynamic wings of all (Raikow 1985).

### Power and Control

Flapping flight is actually energy efficient in that for an insect, bird, or mammal of any given body size, flying a given distance requires only about one-third of the energy required to walk or run that same distance (Hill and Wyse 1989, Rayner 1991). Flapping flight is so efficient because the entire wing moves a large volume of air at relatively low velocities to produce thrust (moving a small volume of air at higher velocities to generate the same amount of thrust is much less energy efficient). The efficiency of flight allows butterflies and swans to migrate distances of which landbound creatures can only dream (Appendix 4). However, distances are covered very rapidly in flight, so the cost per unit time of level flying is high, equivalent to that expended by a running

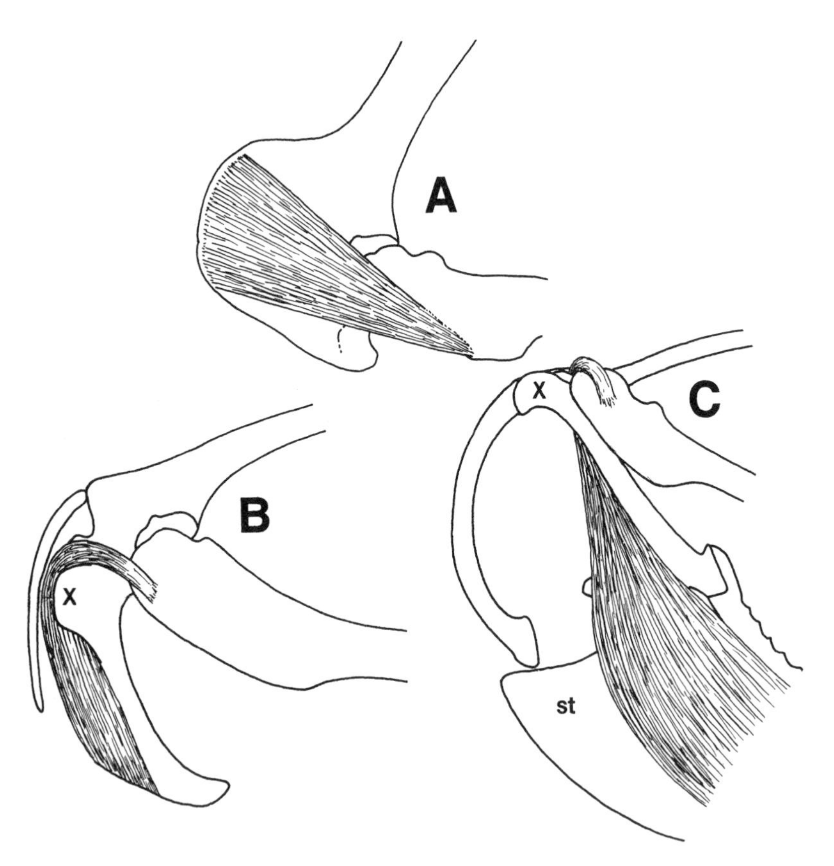


*Figure 7.3.* Supracoracoideus muscle system in *(A)* dino-avepod *Syntarsus,* in which the muscle was conventional; *(B) Archaeopteryx,* in which the supracoracoideus looped partly around the partly developed acrocoracoid; and *(C)* modern bird, in which the greatly enlarged, ventrally placed muscle loops through the triosseal canal before inserting at the humerus. Abbreviations: st = sternum, x = coracoid tuber and acrocoracoid. Not drawn to same scale.

animal. However, no animal can run flat out for more than a fraction of an hour, whereas many birds can fly fast for much of the day. The power requirements also change depending upon the stage of flight. Once cruising altitude is attained, only drag needs to be overcome, so power output is not near the maximum level. Such sustained flight is powered aerobically. In contrast, taking off from the ground requires climbing against gravity as well as overcoming inertia to accelerate; and during this equivalent of an all out sprint, power output is maximal. The intense but brief burst of energy needed for a climbing take-off requires that aerobic power be supplemented with anaerobic power, which is generated in a process that can be used only for short periods.

Generating the power to take off and cruise requires a lot of muscles concentrated in the shoulders, chest, and arms, but perhaps not as much as power as one might expect. In most flying birds and bats, a hefty 15–30 percent of body mass is made up of the flight muscles (Hartman 1961, Greenwalt 1962, Marden 1987). However, in some modern birds with well-developed powered flight and reasonable climb rates (cuckoos, coots, grebes, bitterns, and some flycatchers, wrens, and gulls), the wing-depressing pectorales make up as little as 6 percent of total body mass. The combined pectoralis-supracoracoideus complex makes up only 10 percent or less of the total mass, and the other flight muscles add only another 3–5 percent (Hartman 1961, Greenwalt 1962). Lyrebirds can leap and flutter up to a low branch using slightly shortened wings powered by pectoralis and supracoracoideus muscles that respectively make up only 5.6 percent and 0.6 percent of body mass (Raikow 1985). Raikow further observed that almost flightless scrub-birds that make low flights of only a few meters have shortened wings and lack a furcula, and the pectoralis and supracoracoideus make up 3.7 percent and 0.4 percent of body mass, respectively.

Among flying birds, the coracoid is strongly retroverted relative to the scapula blade. This retroversion is an unusual adaptation; in most tetrapods, the coracoid continues along the same gentle arc as the scapula. The acute angle between the scapula and coracoid in flying birds serves to shorten the fiber length of the proximoventral flight muscles, and this shortening improves the muscles' performance (Feduccia 1996). In flying birds, the furcula has a deep U-shape that enlarges the structure so that it can better support flight muscles (Olson and Feduccia 1979).

The function of the big pectoralis muscle in birds is obvious: it generates the downward power stroke that produces the thrust needed for forward flight. The next largest flight muscle complex is the supracoracoideus complex, which is organized in an unusual way in birds (Fig. 7.3). For many years, the function of these muscles was thought to be understood. Anatomically, the supracoracoideus appeared to help elevate the wings, and this hypothesis seemed to be confirmed by Sy (1936). When he disabled the muscle system in some birds, he observed that they could no longer achieve climbing flight. However, Poore et al. (1997) and Ostrom et al. (1999) have presented new evidence that the primary purpose of the supracoracoideus is to precisely manipulate the action of the forelimb during the complex upstroke. This action is especially important during take-off, low-speed flight, and landing. The implication is that early avians that lacked a well-developed supracoracoideus complex had relatively poor flight capabilities when not flying straight and level. However, bats take off, climb, maneuver, and land very well without a set of modified supracoracoideus muscles. Some birds suffer only modest or minimal disruption of flight performance when the supracoracoideus complex is disabled (Sy 1936, Sokoloff et al. 1994). It is possible that early birds used alternatives to the fully derived supracoracoideus muscle complex, mainly upper shoulder muscles, to achieve similar functions, albeit at a cruder level. The presence of a functioning supracoracoideus wing-elevating and -controlling system is therefore considered a major refinement of avian flight, but this system was not necessary for the initiation of flight or for the early stages of its development.

Another source of power for take-off is the hindlegs, which most birds use to leap straight up into the air before the wings take over (Earls 2000).

The severely abbreviated, pygostyle-tipped tails of birds support a bundle of muscles that can spread the tail feathers into a broad fan for maneuvering, or narrow them into a streamlined bundle (Gatesy and Dial 1996). This ability to dramatically alter the elevator-control surface greatly improves aerial agility and aids landings but, again, was not necessary for initiation and early evolution of flight.

### *Landings*

Landings are a risky part of flight because they require the craft or organism to fly just above its stalling speed. A landing is in essence a controlled crash, and things can quickly go wrong. Most airliner accidents occur during landing. Some researchers have asserted that landing poses such a serious problem for flying creatures that those that did not possess a sophisticated flight apparatus had to take special precautionary measures in order to avoid injury, even during water landings (Chatterjee 1997, 1999a). The need for sophistication is exaggerated: as noted earlier, many arboreal animals land in branches or on the ground after substantial leaps without suffering harm.

### *Strength*

Flapping imposes high bending-loads on the skeleton, so it must be strong enough to resist damage during powered flight (Swartz et al. 1992). In most flying birds, the rigidity of the trunk is assured by fusion of the trunk vertebrae to one another and by ossification of the sternal ribs and the uncinate processes of the ribs. However, bats lack these body-strengthening features; screamers (a type of flying bird) lack ossified uncinate processes (Appendix 3B); and flight-capable juvenile birds such as megapode chicks retain poorly ossified skeletons (including poorly formed arm joints; Sutter and Cornaz 1965). One way to determine the arm strength required to power-fly is to plot the minimum shaft circumference of the humerus as a function of body mass in a variety of birds (App. Fig. 18E).

Another strength factor involves attachment of the primary wing feathers to the trailing edge of the ulna and hand elements via quill nodes. However, quill nodes are absent in some raptors, owls, trogons, and parrots, whose flight abilities are nevertheless excellent.

### *The Kakapo*

The kakapo is particularly useful in this discussion because the unusual flight characteristics of this semiflightless parrot, the largest member of this group, have been studied in detail by Livezey (1992). Although cited as being flightless, the kakapo actually has significant aerial capacity: it can glide a substantial descending distance and is even capable of weak, brief climbing flight. Its flight apparatus has a correspondingly intermediate stage of development compared with that of flightless birds such as ratites, on the one hand, and fully capable fliers, on the other. The kakapo's sternum is fairly large, but the keel is very shallow; and proximal wing muscles are weakly developed compared with those of most flying birds. Unlike in most flightless birds, the coracoids of the kakapo are strongly retroverted. Even more informative are the wings, which are much shorter than those of fully flying birds (App. Fig. 18H). Although the primary feathers are aerodynamically shaped, they are rather stout; the asymmetry ratio of about two is just below the normal range for avian fliers; and the distal edges of the feathers are a little ragged (Fig. 7.2C).

The flight capacity of extinct birds can be assessed by comparing their flight features to those of the kakapo and to those of fully volant birds.

## How Early Birds Flew

### *Archaeopteryx: Flapper or Glider?*

My restoration of *Archaeopteryx* in flying posture reveals a reasonably streamlined, aerodynamic form (Fig. 7.4Aa) and thus contradicts

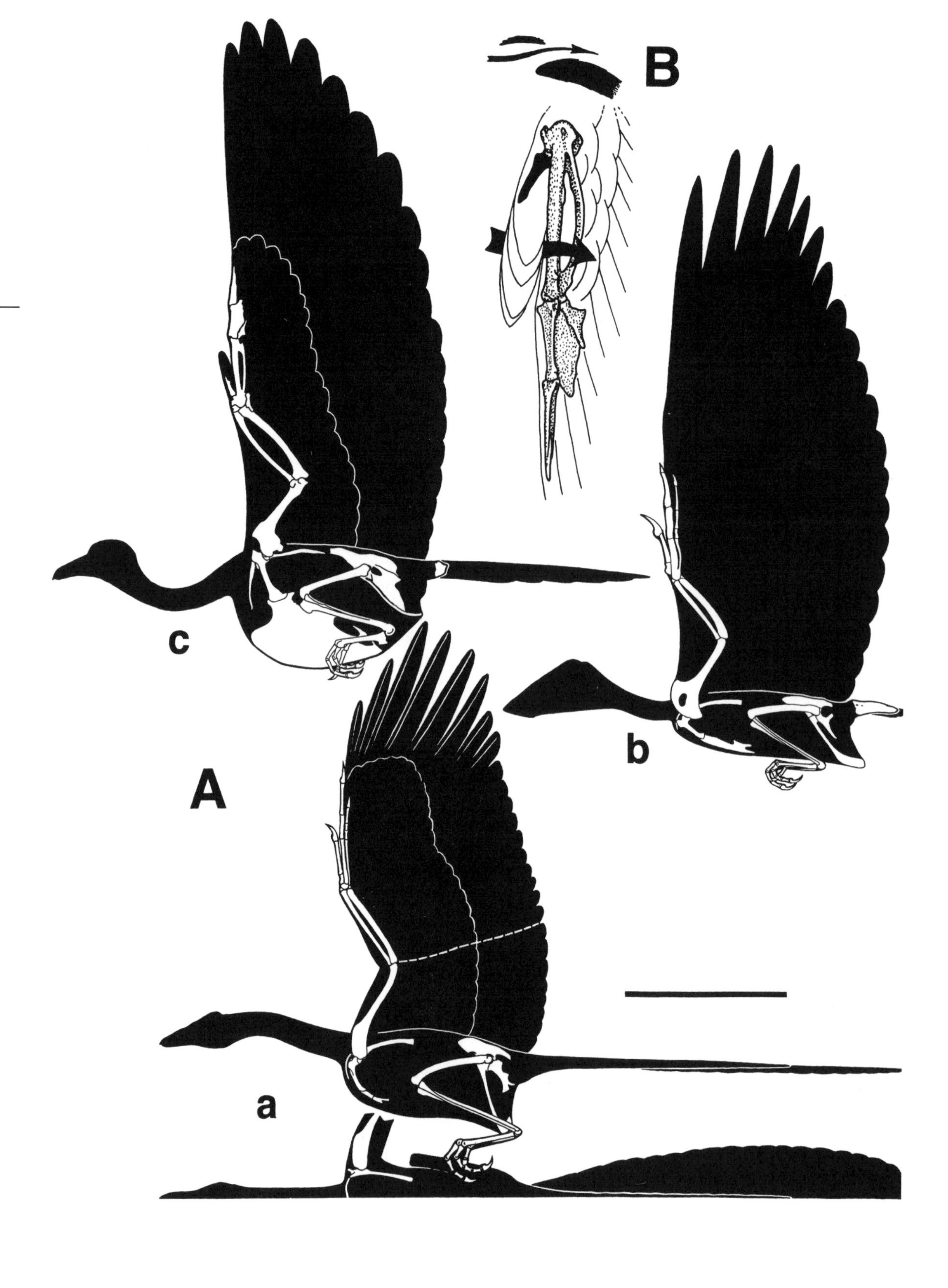
B
c
b
A
a

Thulborn and Hamley's (1985) opinion that the urvogel was nonaerodynamic. Indeed, *Archaeopteryx*'s form appears more akin to that of flapping birds than to that of reptilian and mammalian gliders. Even so, there has been a long and continuing dispute about whether *Archaeopteryx* was largely or entirely restricted to gliding (Heptonstall 1971a,b, Tarsitano 1985, Rayner 1991, Speakman 1993, Speakman and Thomson 1994, Bock and Buhler 1995) or whether it could power-fly to a some extent (Yalden 1971a,b, 1985, Feduccia and Tordoff 1979, Olson and Feduccia 1979, Feduccia 1980, 1993a, 1996, R. Norberg 1985, 1995, Rayner 1985a, Rietschel 1985, Paul 1988a, Sereno and Rao 1992, Vazquez 1992, Ruben 1993, Wellnhofer 1993, Chatterjee 1997, Novas and Puerta 1997, Burgers and Chiappe 1999, Zhou and Martin 1999). On the one hand, its fully developed wing-feather array appears to be aerodynamic overkill for mere gliding. On the other, the absence of many skeletal adaptations associated with powered flight in more modern birds has suggested to many that *Archaeopteryx* could not flap well, if at all (note, however, that most of the above studies date from before Wellnhofer [1993] discovered that the urvogel did have an ossified sternal plate). Especially common have been questions about whether the urvogel could take off from level ground. The "Eichstätt scenario" (Hecht et al. 1985) tended to see the urvogel as a weak flapper that had to take off from high places. Heptonstall (1971a,b) calculated that the wings were too small and that the humerus was too weak for powered flight, but he used a mass estimate that was one-half to twice as high as it should have been, depending upon the specimen being investigated.

*Figure 7.4. (opposite)* Development of flight apparatus in birds from the Jurassic to today. *(A)* flight feathers are known for all subjects (drawn to same scale; scale bar equals 100 millimeters): *(a)* in the Jurassic the still strongly dinosaurian *Archaeopteryx* (based on HMN 1880) retained a long tail, a deep-keeled pelvis, modest flexion of the scapulocoracoid, a poorly developed supracoracoideus wing-control and -elevation system, a small and highly cartilaginous, keelless sternum, and long, slender, supple fingers (the dashed line running across the wing indicates the proximal border of the wing feathers according to Garner et al. [1999]); *(b)* Cretaceous *Confuciusornis* sported an abbreviated tail and a somewhat shallower pelvis indicative of more dynamic flight; areas for flight muscle attachment are expanded via enlargement of the sternum and pectoral crest; the central finger was stronger and more rigid in order to better support the largest flight feathers; a patigium in front of the arm improved the aerodynamics of the wing, which was remarkably large in this early bird; and the extreme brevity of the tail feathers was probably an exceptional specialization; *(c)* today's pigeon is an extremely powerful flier with a hyperexpanded sternum that anchors enormous flight muscles for rapid subvertical climbing flight and high-sustained-speed level flight; the superacoracoideus wing-elevating and -control complex is fully developed; and the hand is completely altered into a streamlined base for the outer half of the wing. Also note the shift from supporting proximal flight muscles primarily on the large pectoral crests seen in the earlier birds to mainly on the sternum. The arm could not actually elevate to vertical in *Archaeopteryx*, and probably not in *Confuciusornis*. In the detail *(B)* of the divergent thumb–borne alula feather complex of a pigeon (left wing in top view and cross section), the arrows show the direction of airflow through the leading-edge slot, which reduces wing stall at high angles of attack.

In quantitative terms, *Archaeopteryx* appears to have had the potential to power-fly. Assuming that the wing feathers were fully developed proximally, the wing loading of the larger specimens falls in the middle of the avian range, indicating that abundant lift was available, even if the tail surface is ignored as a potential source of lift (App. Fig. 18G). If Garner et al. (1999) are correct that the wing feathers were present only distal to the elbow, wing loading is still well within the general avian range, albeit in the lower zone. In either case, *Archaeopteryx* wings were much larger than the wings of some weakly flying birds (such as the noisy scrub-bird; Raikow 1985). The wingspan of *Archaeopteryx* was also in the middle of the avian range (App. Fig. 18H) and was much greater than wingspan of the kakapo. That the smallest *Archaeopteryx* specimen (JM 2257) is near the lower end of the avian range in terms of wingspan and wing area probably reflects its juvenile status rather than limited flight abilities in the protobird.

The large, thick furcula and strong articulations between the coracoids and the broad sternal plate strengthened the *Archaeopteryx* shoulder girdle. The strength of the humerus was as great as in flying birds of similar size (App. Fig. 18D,E). The wrist and hands show adaptations for flapping flight (see Zhou and Martin 1999; some of their conclusions are questionable, as discussed in Chapter 4). Well-preserved wing feathers have the crisp, aerodynamically clean form expected of flight feathers. The asymmetry of the primaries

(Chapter 5) appears to be higher than that observed in the kakapo and in the lower range for flying birds, so the view that the wing feathers were aerodynamic (Feduccia and Tordoff 1979, R. Norberg 1985, 1995) is strongly favored over the opposite view (Speakman and Thomson 1994).

In other respects, the evidence is more ambiguous. The trunk was not as short or as rigidly braced as that of modern flying birds; in particular, ossified uncinate processes and sternal ribs were absent. The angle between the scapula and coracoid was not as acute as that seen in more advanced flying birds, and the elongation of the coracoid was in the low range for flying birds (App. Fig. 18B). The sideways orientation of the shoulder joint encouraged the extensive downward and forward stroke needed to power-fly (Chapter 4), but this orientation prevented the full vertical dorsal reach seen in more-advanced birds. Wrist form and action were not as well developed as they are in modern flying birds (Ostrom 1976b, 1995, Vazquez 1992, Ostrom et al. 1999, Zhou and Martin 1999). There were no quill nodes to firmly anchor the flight feathers to the distal arm bones. The outer wing was not rigid, because the fingers were still flexible. Nor was the hand streamlined; indeed the claws must have produced drag. A thumb-borne, antistall alular feather was not present, but when elevated, the thumb may have acted as a crude leading-edge slat similar to that illustrated in Figure 7.4B. We do not know whether the neck and body were streamlined by a feathery aeroshell.

In terms of power supply, the hinged articulation between the coracoids and the fairly large sternal plate—which was at least partly ossified at least some of the time in a given individual—suggests that a system for ventilating the anteroventral air sacs during flapping flight was developing (Appendices 3 and 4). The same enlargement of the sterna, as well as the large U-shaped furcula (App. Fig. 18F) and pectoral crest, also indicate that the flight musculature was expanded above the tetrapod norm (as described in Appendix 2); the musculature was expanded enough for *Archaeopteryx* to power-fly but was not as large as it is in most carinate birds and bats. The presence of a substantial acrocoracoid is important because it indicates that the supracoracoideus complex was partly developed, more developed than in nonflying birds but not as developed as in modern fliers (Fig. 7.3B). The retroverted coracoid was more that of a flying bird than like that of a nonflying bird and should have improved the stretch factor for the ventral flight muscles, but, as we have seen, the angle between the scapula and coracoid was not as acute as that in more advanced flying birds. Burgers and Chiappe (1999) have calculated that *Archaeopteryx* could have taken off with a running start.

It is difficult to see why *Archaeopteryx* had such large, strong arms and well-developed wings—especially if they were proximally complete—if all it did was glide or pounce upon prey from above. In addition, the characters related to powered flight (propeller-like asymmetric primaries, a strongly flexed scapulocoracoid, and an acrocoracoid that improved the working advantage of the proximal wing muscles) cannot be explained unless these characters were for powered flight. The wings of *Archaeopteryx* appear to have been better adapted for flapping flight than those of the kakapo, even if the wing was limited to area of the arm distal to the elbow (this limitation would have reduced wing area by about one-quarter).

Therefore, the preponderance of evidence suggests that *Archaeopteryx* could power-fly, probably better than the kakapo, but the poor development of the flight complex indicates that its flight was still at a crude level, especially if the wing feathers did not extend proximally. The flapping arc was probably limited, and the power and recovery strokes were less forceful and less well controlled than those of more advanced fliers. If the view just detailed is correct, then, the protobird could probably have taken off from level ground with a leap powered by its long legs, climbed slowly, and sustained level flight. Of course, the more proximally complete the wing was, the better the flight performance would have been.

The large tail-borne airfoil may have provided additional lift if the center of gravity was well behind the wings' axis of lift. Such a body-mass distribution would have made the creature's tail heavier than that of more-advanced birds (Bramwell 1971). This mass distribution is possible, considering that a long, bony, and, therefore, heavy tail was present, but our inability to accurately restore body-mass distribution (Appendix 2) prevents us from knowing for certain. Lack of inner wing feathers would have made any lift provided by the tail more important. L. Martin and Czerkas (2000) suggest that *Archaeopteryx* held its legs stretched backward, which would shift weight aft, causing the tail to produce lift. It is at least as likely that the legs were tucked up, as in modern birds. If the center of gravity was close to the main axis of wing lift, then the tail must have served primarily as a control surface. In any case, the long tail provided stability in level flight in a man-

*Figure 7.5. Archaeopteryx* using its modest flight ability to escape swift running *Compsognathus*. Note that the latter is restored with more feathers than in an earlier version of this illustration (see Fig. 3-8 in Paul 1988a), in accord with the new, feathered fossils of its close relative *Sinosauropteryx*. Also shown is the coniferous, arid-adapted shrub *Brachyphyllum* known from the Solnhofen deposits.

ner similar to the long tail of rhamphorhynchoid pterosaurs.

The deep-keeled body should also have increased stability. The combination of inherent stability and poor flight controls indicates that maneuverability was not high. The relatively small flight musculature probably consisted largely of high-burst-power white fibers (see Appendix 2), so flight range should have been rather short, perhaps more than a few thousand meters. A probable inability to collapse the tail feathers for streamlining (Gatesy and Dial 1996) would have reduced aerial energy efficiency and thereby degraded flying range.

The large tail-borne airfoil also acted as a landing brake when it was erected vertically at the highly mobile base. The extra gripping ability that stemmed from *Archaeopteryx*'s ability to use both the forelimbs and hindlimbs when landing helped make up for landing errors due to mediocre flight controls. Landing speeds may have been so high that slowing down by swooping up to land on a branch was preferred. High-velocity ground landings followed by a bipedal deceleration run should have been as viable as landings at the end of a long leap. Indeed, as we shall see in the next chapter, the habitat of *Archaeopteryx* may have required that it be able to both take off and land from low perches, as well as from the ground. Lack of inner wing feathers would probably further modify our picture of the flight pattern of *Archaeopteryx* away from the pattern seen in later birds with complete wings, but this issue has not been examined well enough for us to consider it further at this time.

In summary, *Archaeopteryx* appears to have been the avian equivalent of one of the Wright Flyers demonstrated in 1908: a workable flying machine that was rather slow and stable, to the point of not being very maneuverable, and had mediocre take-off, climbing, and landing performance and a short range (Fig. 7.5).

## *Other Ancient Birds*

Flight performance dramatically improved in birds that postdate *Archaeopteryx* (Fig. 7.4Ab). Until recently, researchers often assumed that the tail straightforwardly shortened into a proximally flexible, distally stiff pygostyle that bore a fanlike array of elongated feathers. The slender, distally stiffened tails of *Rahonavis* and *Liaoxiornis* hint that some post-urvogels first passed through a rhamphorhynchoid pterosaur–like stage (Fig. 7.1) in which the tail acted as a rigid rudder (also see the discussion on the implications of dromaeosaur tails in Chapter 11). The more severe abbreviation of the tail that accompanied a short, stout pygostyle marked a shift from stable to more dynamic flight. A similar switch occurred in pterosaurs when short-tailed pterodactyloids evolved (Wellnhofer 1991). Long-tailed *Rahonavis* suggests that such basal avepod fliers retained a dromaeo-avemorph–grade tail after the shoulder girdle and wing had experienced considerable upgrades over the urvogel condition.

Wing enhancements seen in post-urvogel birds included improvements of the elevating and control systems. The shoulder glenoid faced more dorsally, allowing the wings to be raised fully vertically (Figs. 4.2Ah, 10.8R,S). The supracoracoideus complex was better developed for better control of wing action. This upgrade included dorsal expansion of the acrocoracoid process until it contacted the furcula. This contact formed the triosseal canal, through which looped the insertion tendon of the supracoracoideus muscle (Fig. 7.3C). The coracoid itself was reduced to a strut and was more retroverted. The latter change further improved the working action of the ventroproximal flight muscles. Ossified sternal ribs better anchored the sternum on the rib cage. The enlargement of the sternal plate and the development of a keel—both at first modest and later extreme—improved the support for the main body of the supracoracoideus. Sternal enlargement also improved support for the pectoralis. The very large size of the pectoral crest of the humerus in many volant Cretaceous birds (Fig. 10.11Ak,m) indicates that the proximal flight musculature was more distally placed than it is in modern birds (in this respect, basal birds paralleled pterosaurs, which also had modest-sized sternal plates and very large pectoral crests). The large pectoral crest means that the modest size of the sternum is somewhat misleading because the pectoralis was not as small as the chest plate alone suggests. The firmer base for the flight muscles, combined with their enlargement, increased the effective power of the downstroke and control of the upstroke. As the sternum elongated posteriorly, the abdominal ribs, or gastralia, were reduced and then lost.

As we look farther out on the wing, we see that the folding mechanism was improved in post-urvogels. A patigium filled in the space in front of the flexed inner wing bones, adding to the wing area while increasing streamlining. Streamlining and rigidity of the outer wing were also enhanced as wrist and hand elements began to fuse; fingers became flatter, shorter, and then reduced and simplified; and claws were reduced and lost. The eventual conversion of the thumb into a support for the alular feather reduced stalling speeds and helped further improve flight control (Fig. 7.4B; Spearman and Hardy 1985, Ostrom 1994, 1995, Sanz et al. 1996). Quill nodes better anchored the flight feathers.

The rest of the body was also upgraded for flight. The trunk became shorter and more rigid. The rigidity stemmed from fusion of the dorsal vertebrae, as well as from ossification of uncinate processes in at least some examples. At first, the pelvis remained remarkably dinosaur-like, but then it broadened; and the pubes and even the ischia split apart and became more retroverted. These changes reduced the ventral body keel that had acted as an aerodynamic stabilizer, and accommodated a further posterior shift of the abdomen. This backward movement of mass was necessary to compensate for the increasingly severe shortening of the bony tail, which was tipped with a pygostyle-borne muscle bundle that operated a deployable fan of feathers.

The results of these modifications were birds that could take off more easily and climb more rapidly than *Archaeopteryx*. Maneuverability and flight control were greatly improved, especially at slow speeds. The enhanced agility and lower stalling speeds improved landing performance, so branch landings gradually shifted from semiquadrupedal to fully bipedal as the hand changed from a grasping appendage into a semirigid airfoil support with an antistalling thumb. Flight range increased in correspondence with the enlargement of the flight muscles. The ability to fold the tail feathers improved streamlining at cruising and dash speeds, thus improving energy efficiency and further boosting range. These post-urvogel birds were the Spads, Camels, and Fokker triplanes of their days, capable craft, yet crude and very limited by more modern standards.

An interesting exception to the general evolutionary trends seen in post-urvogels was *Confuciusornis,* which is the least advanced avebrevicaudan yet known (Fig. 7.4Ab). It had many of the improvements seen in other early avebrevicaudans. But its rectrices—not to be confused with the hyperlong display feathers—were so abbreviated that they could not have had much influence on aerodynamics. Tinamous, grebes, and some other birds with very short tail feathers do fly, but maneuverability could not have been high in *Confuciusornis,* and the lack of an alular leading-edge slat would have further impaired aerial agility. Not surprisingly, then, two of the fingers remained supple, hook-clawed digits fully suitable for quadrupedal landings, as well as for climbing to a degree not seen in any other avebrevicaudans.

*Confuciusornis* was a truly odd early bird because, in contrast to its brief tail, its wings were very large (App. Fig. 18G,H)—not because the arms were especially elongated but largely because the outermost primary feathers were extremely long—and therefore able to produce great lift. No living bird combines such expansive wings with so little in the way of tail. When we add to this the ancient bird's limited (by modern standards flight) musculature; its possible inability to fully elevate the wings; its lack of a highly developed supracoracoideus wing-control system, an alula, and other advanced features; its fairly deep-keeled body; and the male's long drag-inducing tail feathers, the Chinese bird was so unlike anything alive today that attempts to precisely restore its flight performance are confounded. The limited flight-control system and primitive flight apparatus imply that, unless it had some dramatically different feature or skill that allowed it to be a good aircraft, *Confuciusornis* was a rather poor flier. Yet the big lift-generating wings and the numerous remains found in lake deposits imply that it was able to fly at least fairly far over broad expanses of water (see the cover illustration). Considering that its pygostyle probably evolved in association with a well-developed caudal fan intended to increase aerial agility, the stumpy tail and correspondingly poor flight control were probably secondary reversals from the better level of flight control that had already evolved in its ancestors.

By the time advanced enantiornithines, such as *Neuquenornis* and the ichthyornithiforms, appeared in the Late Cretaceous, the avian flight apparatus and associated performance were fully developed. In particular, a very large, deep-keeled sternum suggests that the aerobic capacity of the flight muscles and respiratory tract was as high as that in modern birds. The same feature shows that the bulk of the flight musculature is closer to the body in advanced birds, a change also reflected in the decreased size of the pectoral crest on the humerus (compare Fig. 10.11Ak,l–m). These were the high performance Spitfires and Zero-sens of their distant times. Still later in the Cenozoic would appear fast, diurnal F-104 Starfighter equivalents, such as the peregrine falcon, and owls, the nocturnal avian versions of F-117 stealth attack craft.

CHAPTER 8

# The Loss of Flight

The extraordinary nature of avian flight inspires researchers to fixate on its first appearance and its improvement over time. But there is another side to the evolution of bird flight: the ability to fly was sometimes reduced or even lost entirely during the evolutionary process. In fact, this fairly common pattern is central to a main argument of this book: that the most birdlike dinosaurs may have been so birdlike because they were secondarily flightless, or neoflightless.

## When Flight Becomes a Burden Too Heavy to Bear

### *How Big Is Too Big?*

By necessity, aircraft that carry humans have to be large. The lightest human-powered craft weigh 35 kilograms, without the approximately 70-kilogram human power source on board. The biggest heavy transport aircraft currently weigh 400 to 600 tonnes fully loaded, and much larger machines are under consideration (Fig. 8.1; Wegener 1997). In comparison, nonhuman fliers are small, with the great majority weighing in at less than a few kilograms, and the smallest at less than 2 grams (Matthews and McWhirter 1993). Clearly, self-powered vertebrates cannot grow as large as flying machines built by humans.

Flight among large creatures is often misunderstood. As the size of an animal increases, the energy needed to power fly seems to increase faster than the maximum possible sustained energy output of flight muscles (Pennycuick 1986, 1989, Marden 1994). Therefore, once an animal reaches a certain size, it should no longer be able to power-fly for long distances. Pennycuick (1986, 1989) calculated that animals weighing more than 12 kilograms cannot sustain horizontal powered flight; at best they should be able only to glide and soar. (In terms of power production, there is no maximum mass for energy-efficient gliders and soarers.)

Performance extrapolations like Pennycuick's can be erroneous, though. A better way to determine the maximum mass for powered flight is to abandon hypothesis in favor of specific examples, that is, to find and observe large creatures that actually do fly (see Figs. 8.1, 8.2, App. Fig. 1; Appendix 2). In the 10-kilogram class are condors and albatross. Both must run to take off and must soar to travel long distances. The other living avian giants do not soar. Wild turkeys, also in the 10-kilogram class, have very large, anaerobically capable flight muscles that give them the power to take off without a run, but these muscles can sustain flight only for short distances (Marden 1994). Mute swans, second largest flying extant bird at 15 to 18 kilograms (Matthews and McWhirter 1993), use their modest-sized, aerobically capable muscles to migrate long distances (Alexander 1971, Pennycuick 1989, Marden 1994, Pennycuick et al. 1996). Migrating swans often fly at low altitudes and take frequent rests, especially when they are laden with the heavy fat deposits used to fuel the long trips. On the other hand, whooper swans that can exceed 12 kilograms have been observed flying at 8,000 meters, nearly as high as Mount Everest (Whiteman 2000). Their take-offs require a running start. The same is true of the largest living flying bird, the 15- to 19-kilogram kori bustard. Its flight muscles are small and configured for burst power, so its range is short (Marden 1994).

Extinct fliers were even larger, as detailed in Appendix 2 (also Fig. 8.2, App. Fig. 1). Most researchers view the heaviest flying birds yet known, teratorns weighing more than 70 kilograms (Fig. 8.3), as condorlike soarers (Campbell and Tonni 1983). H. Fisher (1945), however, viewed them as ponderous flappers like herons and pelicans. Yet another possibility is that the largest teratorns were short-range burst fliers. We do not know whether they took off from a standing position or while running. Albatross-like pseudodontorns that approached 50 kilograms had very slender

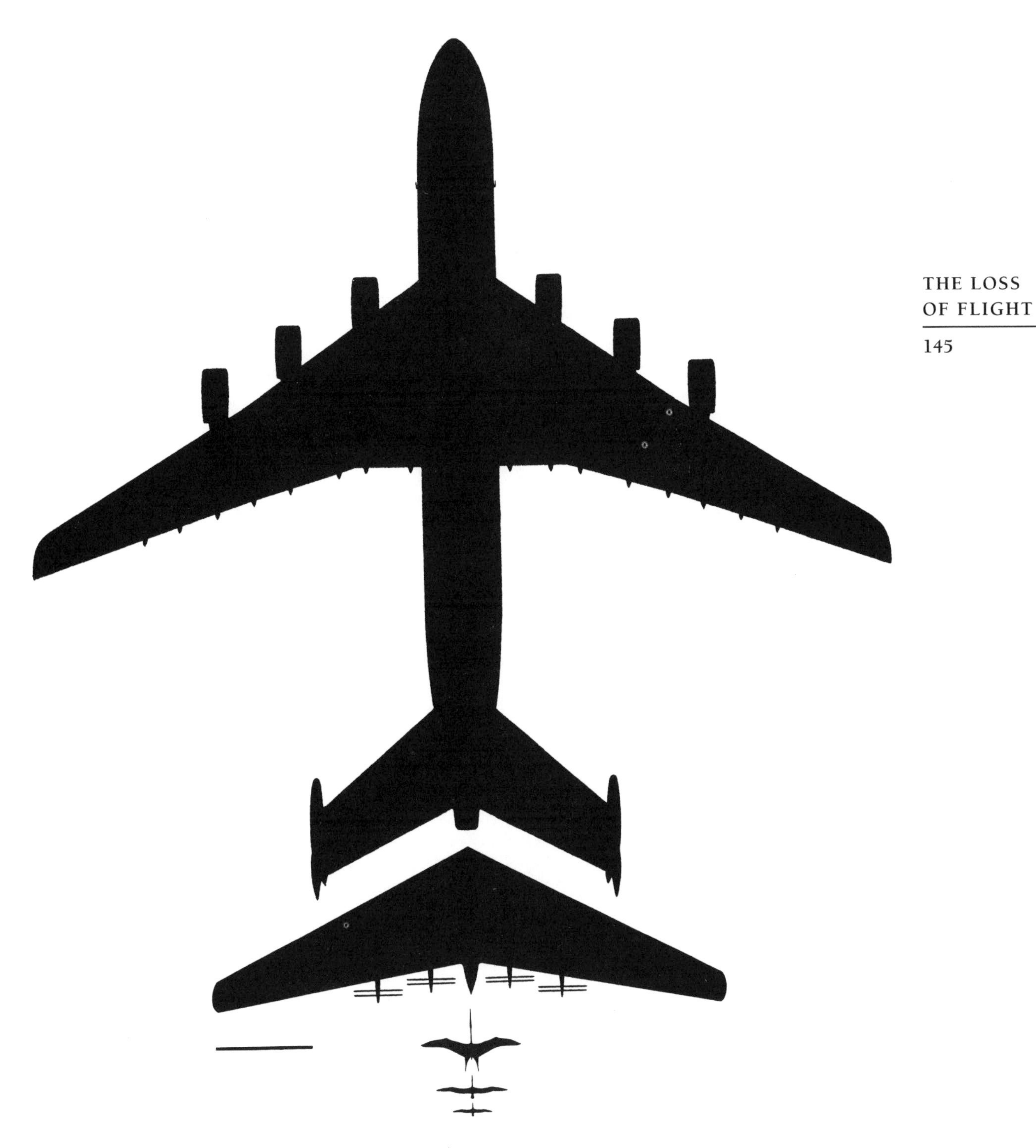

*Figure 8.1.* Giant machines of the air, technological versus biological. *Top to bottom:* largest aircraft, Antonov An-225 Mriya; Northrop YB-35 Flying Wing; largest pterosaur, *Quetzalcoatlus northropi;* extinct bird with the largest wingspan, pseudodontorn; living bird with the largest wingspan, wandering albatross. Scale bar equals 10 meters.

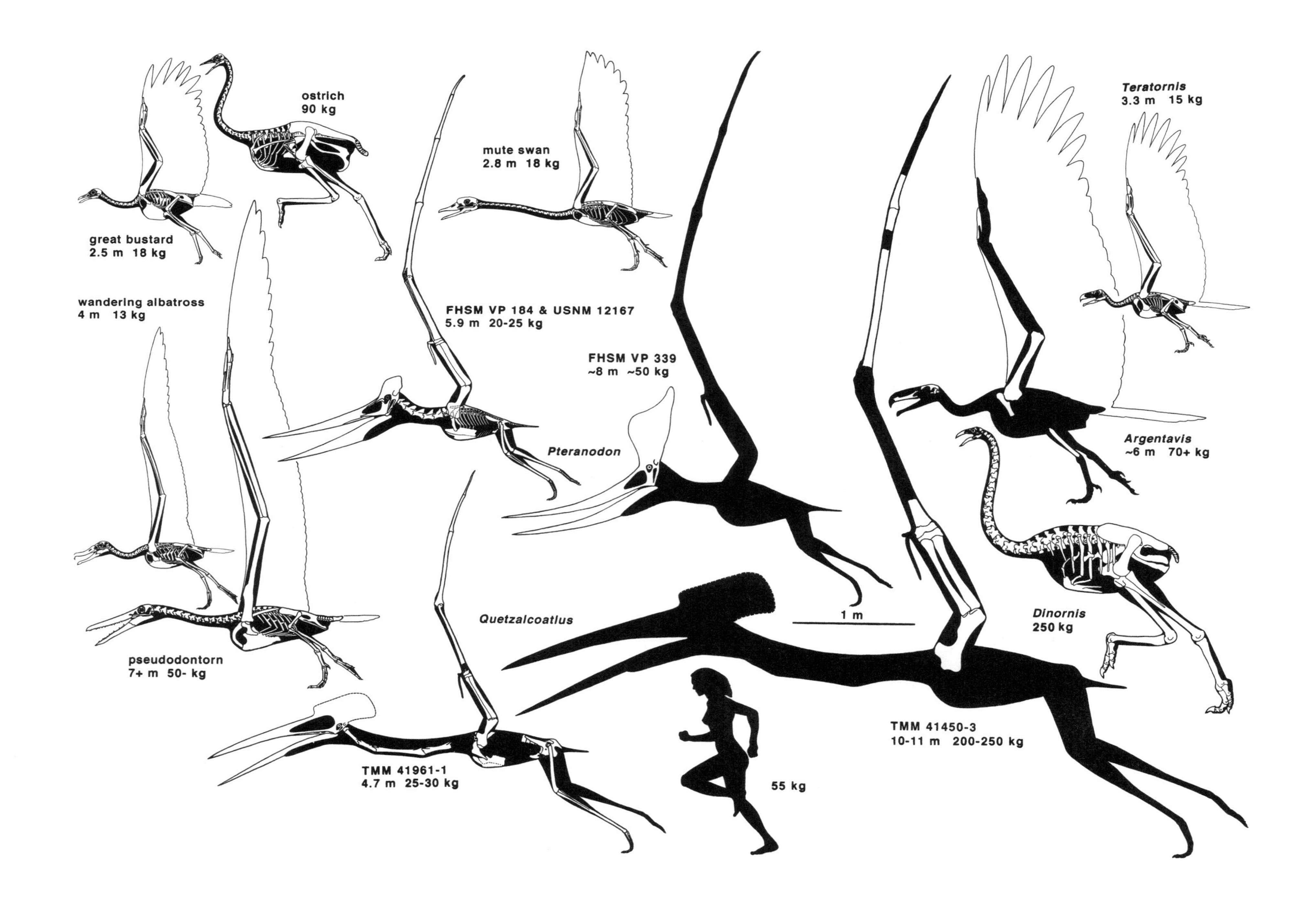
ostrich
90 kg
great bustard
2.5 m 18 kg
wandering albatross
4 m 13 kg
pseudodontorn
7+ m 50- kg
mute swan
2.8 m 18 kg
FHSM VP 184 & USNM 12167
5.9 m 20-25 kg
FHSM VP 339
~8 m ~50 kg
*Pteranodon*
*Quetzalcoatlus*
TMM 41961-1
4.7 m 25-30 kg
55 kg
1 m
TMM 41450-3
10-11 m 200-250 kg
*Teratornis*
3.3 m 15 kg
*Argentavis*
~6 m 70+ kg
*Dinornis*
250 kg

wing bones adapted for wave soaring, to the point that their wing joints were less suited for flapping than those of any modern bird. They must have had to take off with a run.

The largest marine pterosaurs (pteranodonts weighing as much as 50 kilograms) had very long slender wing elements too. However, their well-developed pectoral crests suggest a well-developed flight musculature. The crests and other wing adaptations (Stein 1975) suggest that these pterosaurs were adapted for a gull-like mixture of flapping and soaring flight. Their wings were too long to allow a standing take-off.

The largest pterosaurs (continental azhdarchids, which were perhaps five times as heavy as pteranodonts) had wings that were rather short relative to their bulk, and the extremely large pectoral crests of their stout rather than streamlined inner wing elements probably anchored extremely large pectoralis wing depressors. Although getting a 200-kilogram moa into the air may seem to go beyond what is possible, the giant pterosaur's fully developed wings show that it did indeed fly. In fact, the wingspan/mass ratio of a 200- to 250-kilogram pterosaur is higher than predicted for an average bird with the same wingspan, and is similar to the ratio for slow-flying ultralights and sailplanes; so the pterosaur's wings were able to generate sufficient lift. Marden (1994) showed that a 250-kilogram *Quetzalcoatlus* with 60 kilograms of flight muscles should have had enough power to take off without a run. Azhdarchids appear to have been optimized for power flapping, to the point that their very large proximal wing processes and joints intruded strongly into the airstream, increasing the frontal profile and drag of the wings. The implication is that the streamlining and great wingspan needed to soar were sacrificed in favor of the structural strength and muscle leverage needed for active flight (Paul 1991). The very large flight musculature implied by the great pectoral crest favors a standing-start, short-range flight mode over the running-take-off, long-range alternative.

Many questions remain about the flight of giants. Large swans challenge estimates of the size barrier to powered flight, so much so that we do not entirely understand how they are able to generate sufficient aerobic power to sustain level flight (Marden 1994). There must be a limit to the size of flying vertebrates. A 1-tonne flier would need wings spanning some 18 meters (about 60 feet), and a 10-tonne flier's wings would have to span 45 meters (nearly 150 feet). The practicality of such enormous wings is dubious. We do know, however, that surprisingly large animals—humans, ostriches, and even the moa—have been able to get off the ground. However, that such superfliers are a very rare part of the continental paleofauna suggests that the cost/benefit ratio of being able to fly is marginal. In birds that exceed 10 kilograms, configuring a flight musculature that produces both enough anaerobic power to take off rapidly and enough aerobic power to sustain flapping flight seems impossible. If one is selected for, the other must be dropped. A running take-off is not optimal for escaping predators and probably would not be selected for in large animals living on flatlands. For such animals, flight would be retained only if the combination of rapid take-off and short-range powered flight or soaring conferred advantages that compensated for the costs of the flight apparatus (this trade-off is discussed later in the chapter).

In general, 10 to 20 kilograms seems to be the upper limit for the retention of flight in nonmarine animals; beyond that, selective pressures make the ability to become airborne less likely, but not impossible. The size limit for easy vertebrate flight is important to this study, in that one reason birds can lose the ability to fly is that they become too big.

*Figure 8.2. (opposite)* Same-scale skeletons of giant birds and pterosaurs posed in take-off position with wings extended to fully vertical for comparative purposes. For pseudodontorns, the pelvis, parts of the sternum, and some other elements are not yet known. Moa scaled to cited mass. Scale bar equals 1 meter.



## *The Loss of Flight*

As Raikow (1985) said, "because the basic organization of the avian body is so closely associated with its adaptation for flight, a flightless bird is almost a contradiction in terms." Yet, for all the trouble birds have gone to to develop the marvel and advantages of flight, they are remarkably willing to give it up—unlike bats and pterosaurs, which, as far as we know, have never lost the ability to fly. Why do birds sometimes experience this reversal, a reversal so profound that they become earthbound or water bound?

### Why Flight Is Good

In order to understand why flight capability is sometimes reduced or lost, we need to understand the advantages of flight. The many reasons that birds, as well as bats, pterosaurs and insects,

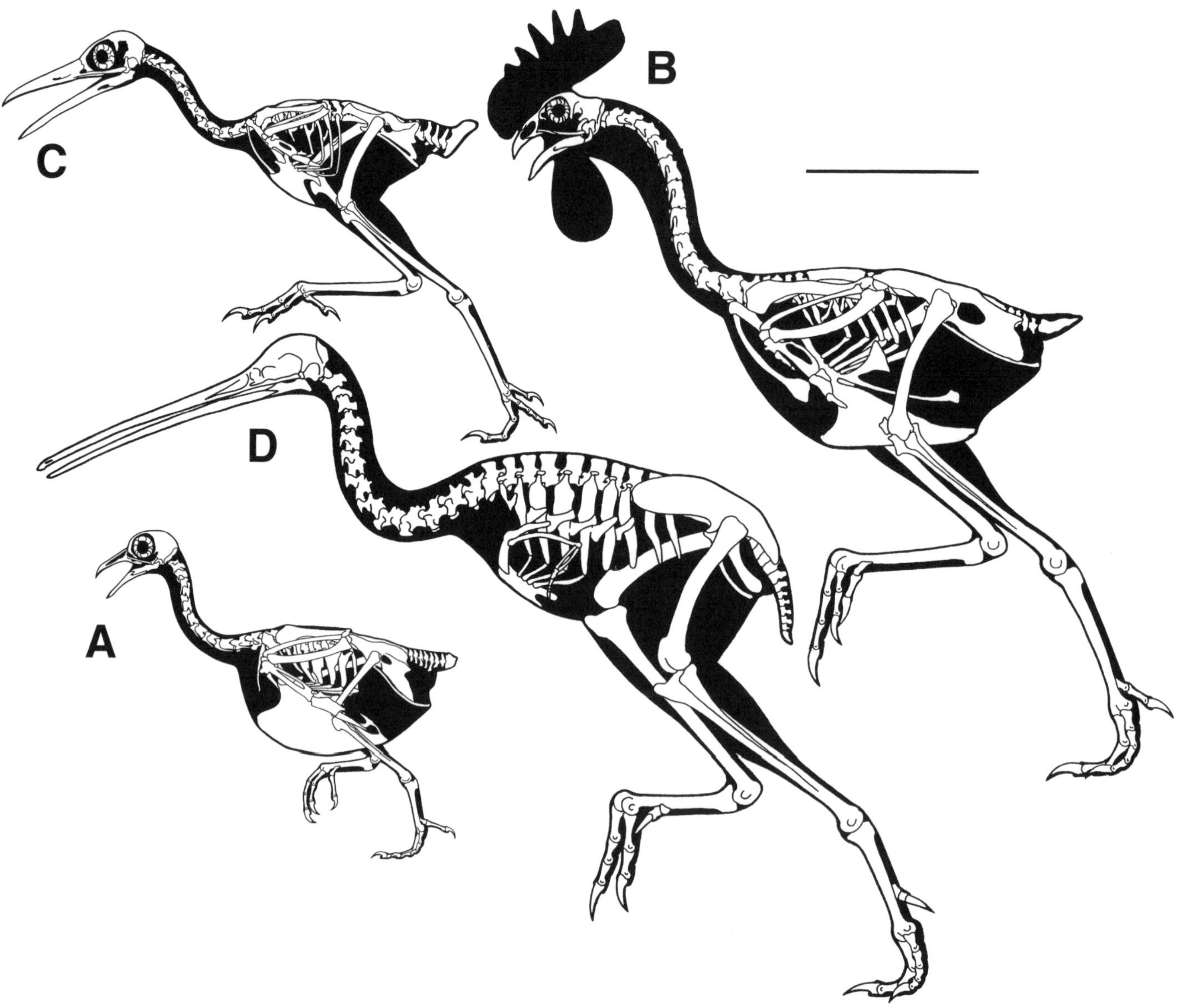


*Figure 8.3.* Various levels of flight in living birds: *(A)* the pigeon, a long-range, high-speed flyer, has large arms, an enormous deep-keeled sternum, a strongly retroverted, narrow coracoid, and a well-developed acrocoracoid that forms a pulley for the supracoracoideus wing-elevator and -control complex; *(B)* the chicken, a powerful but short-range flyer, chicken has relatively shorter arms, and its sternal plate has been reduced to struts; *(C)* the arms of the roadrunner, a weak flyer, are even shorter, the sternum is short and rather weakly keeled, and the number of sternal ribs is reduced; *(D)* the kiwi has lost flight adaptations to a greater degree than any other modern bird; its arms are extremely atrophied; the sternum is very short, keelless, and weakly connected to the ribs; and the coracoid is short, broad, and not retroverted and lacks the acrocoracoid. Drawn to same scale; bar equals 100 millimeters.

evolved and maintain the ability to fly are pretty obvious. For starters, climbing into the air is a superb way to escape attack. Countless earthbound predators have gnashed their teeth in frustration as their avian targets simply flew beyond their reach. This selective advantage is so powerful that some birds—grouse and turkeys, for example—have developed their ability to take off at a steep angle at the expense of long-range flight performance. Other birds, such as pigeons and doves, have managed to achieve both near-vertical take-off and excellent long-range performance (Weis-Fogh 1975, Marden 1994).

The ability to fly aids predators as well as prey. The increased altitude associated with flight is tremendously useful to flesh eaters because it translates into an enormous search range. This increase in range is so valuable that the only full-time scavengers are soaring birds (Houstan 1979). Having spotted still-living prey, aerial predators can use height and gravity to help them ambush or outrace terrestrial or arboreal creatures.

Flying also benefits herbivores in that it allows them to rapidly travel substantial distances in search of plants to eat. Once upon a time, tremendous flocks of passenger pigeons cruised the eastern-central regions of North America in search of the biggest crops of acorns and other mast. Birds and bats are often aerial commuters, flying short and long distances from their homes—caves, nests, tree roosts—to food sources and back; some make the trip once a day; others, many times a day.

As we saw in Chapter 7, the low cost of flying relative to walking or running a given distance also allows fliers to travel almost any distance in search of the most advantageous conditions, whether on a daily or seasonal basis. Many fliers span whole continents and oceans as they follow the sun through its yearly cycle. The most extreme example is the Arctic tern, which flies from the Arctic to the Antarctic and back each year.

### Why Flight Is Bad

For all its advantages, flight also has serious costs. We have already noted that power-flying is very energy expensive on a per-unit-time basis. The flight apparatus is very expensive to grow, maintain, and operate. The large flight muscles and enlarged sternum, for example, are costly to grow, and all those muscles use up considerable oxygen even when at rest (McNab 1994, Feduccia 1996). Because flightless birds have dramatically reduced shoulder and arm muscles (McGowan 1982, Feduccia 1996), their resting metabolic rates are one-quarter lower than those of their flying relatives (McNab 1994). In addition, even when tightly folded, the large flight apparatus can interfere with nonflight activities. Although there are some extra uses for flying wings—they are used as flippers by some divers, as shades for nestlings, as impact weapons, and for display—arms without well-developed fingers have limited utility for nonflight activities. And, as we just discussed, the ability to fly can hinder or prevent creatures from becoming large. Flight gear may not be worth retaining if the advantages of flight do not outweigh the costs.

### What Makes Birds Special?

The reason that birds have repeatedly experienced the reduction or loss of flight capability while bats and pterosaurs have not is the fundamental differences in their locomotory complexes. Birds fly only with their arms, and walk and run only with their legs. They are therefore "two-legged" both in the air and on the ground. Because their legs are entirely decoupled from the flight apparatus, their terrestrial locomotory capability is unhindered by flight needs and therefore high, and birds are able to move about with no help from their forelimbs. The excellent terrestrial performance of bipedal birds allows them to easily jettison their other source of mobility, flight.

In contrast, bats use and very probably pterosaurs used their forelimbs for walking, and the hindlimb is (or was) an integral part of the flight apparatus. Bats and pterosaurs are therefore "four-legged" both in the air and on the ground. Some bats, especially vampires and the New Zealand short-tails, are competent on the ground (Nowak 1999). Most pterosaurs were probably good, albeit flat-footed, walkers (S. Bennett 1997). However, the close coupling of both sets of limbs into both modes of movement hindered these quadrupeds from becoming as adept at ground locomotion as bipedal birds. The close links between the forelimbs and hindlimbs also complicates the modification of either set for new purposes.

There are other, more subtle reasons that pterosaurs and bats have been less prone than birds to evolve flightlessness. Early on, birds evolved large brains, and birds' exceptionally good vision puts them at least on par with land-bound creatures. Pterosaur brains, in contrast, remained reptilian in size, and their eyes do not seem to have become as large as those of birds. Bat brains are not larger than those of other mammals, and although bats are by no means blind, their vision is not exceptional. Their best flight-related sensor system is

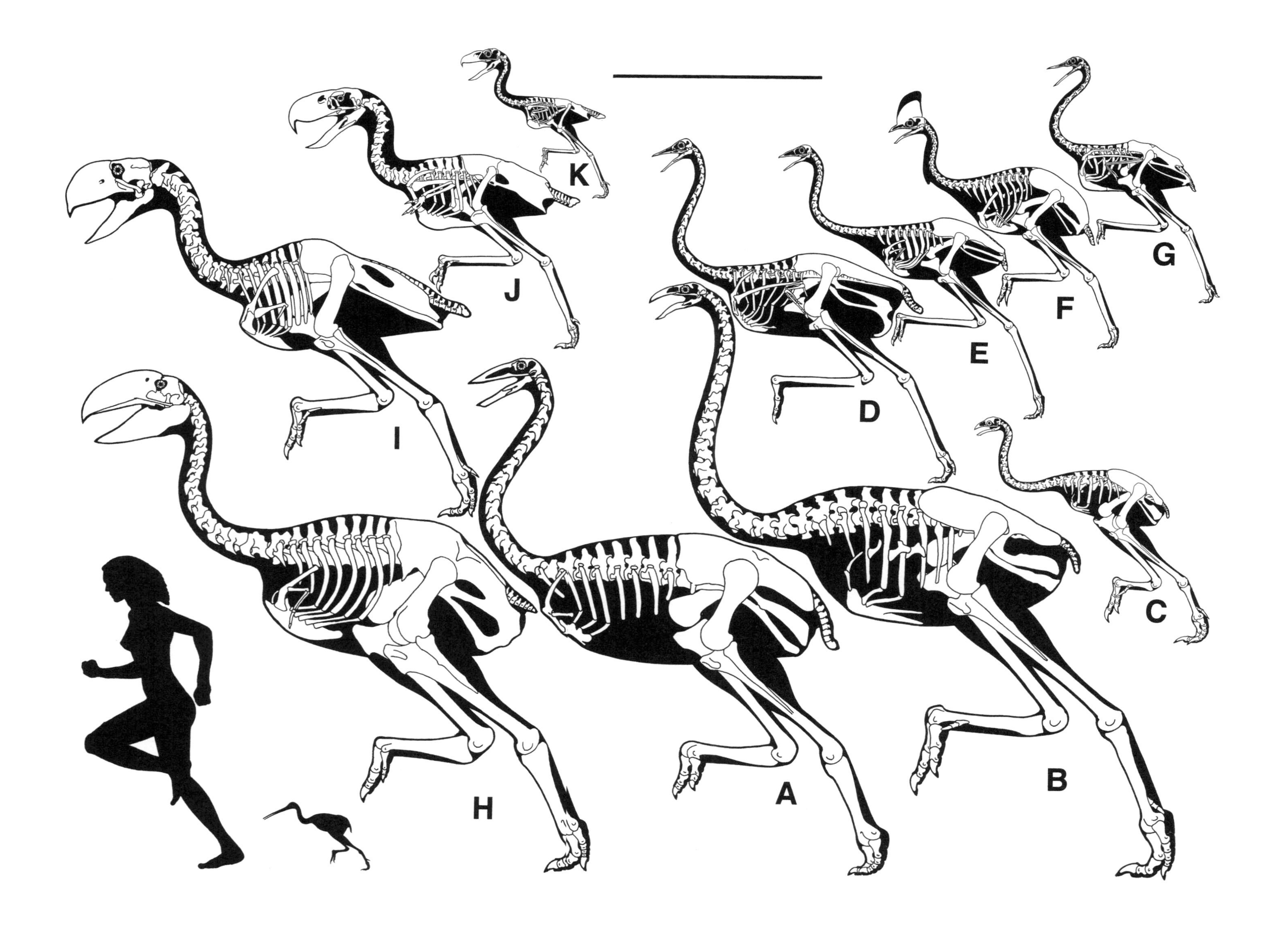
K
J
G
F
E
D
I
C
H
A
B

sonar, which is not of much use to ground dwellers. To the best of our knowledge, island bats have never lost flight.

**Where and Why Avian Flight Is Lost**

Flight has been lost both in marine birds and in terrestrial birds. (Much of the information in this section is taken from S. Olson [1985], Raikow [1985], Chiappe [1995], and Feduccia [1996], and the birds mentioned here are discussed in further detail in Chapter 15.) In the marine venue, the ability to fly has been lost or modified in some diving birds. In foot-propelled swimmers, for example, the flight apparatus is a costly hindrance and has been lost. In other divers, the flight apparatus is retained but modified into propulsive flippers. Flight loss among marine divers started in the Cretaceous, among the toothed hesperornithiforms and perhaps separately in the baptornithids. In the Cenozoic, neoflightless divers included penguins, plotopterids (penguinlike relatives of pelicans), and a few auks, cormorants, and ducks. Some of these divers have been very large, too large to take off from the water. The story of flightless marine divers is of little concern to this study.

In the terrestrial venue, the loss of flight involves two types of bird populations: island birds and continental birds (Fig. 8.4). Loss of flight is very common among island bird populations, for obvious reasons. Since predators are often few or absent, the pressure to retain flight as an escape mechanism is low or nil; and because islands are small, the need to fly hither and yon in search of food is also reduced. Migration is especially superfluous for birds that have no need to leave small lands. In addition, flight loss is a positive advantage in that it eliminates the danger of being blown off the island by strong winds and storms. With the selective advantages of flight about as low as they get, and the advantages of flightlessness high, the costly flight apparatus is readily lost. The size of neoflightless island birds ranges from small—but never as small as common songbirds—to as big as half a tonne. Flightless island birds also tend be rather heavy limbed. This does not mean that they cannot run at all, but they do not need to run fast, because there are no swift predators to flee from. The really big island birds have been enormous ratites (moa, elephant birds), but island ratites can also be small (kiwis, which are the only living examples of this type). Rails are especially prone to losing flight on islands, and a few have become fairly large. Pigeons are fairly common island-bound birds, the big dodo being the premier example. Island ducks, ibises, parrots, megapodes, and owls have also undergone severe reduction of their wings until flight ability was reduced or lost. The tale of flightless island birds is not of primary concern to this study.

*Figure 8.4. (opposite)* Large flightless birds extinct and living. Palaeognathous ratites: *A*, elephant bird *Aepyornis;* moa: *B*, *Dinornis; C, Megalapteryx;* extant examples: *D*, ostrich; *E*, emu; *F*, cassowary; *G*, rhea; neognathes: *H*, dromornithid *Dromornis*, tentative restoration with posterior skull details after *Bullockornis; I, Diatryma;* phorusrhacoid terror birds: *J, Andalgalornis; K, Peleycornis. A, B,* and *H* are scaled to the largest specimens I am aware of. Human (1.63 meters) and kiwi profiles included for scale; scale bar equals 1 meter.

What is most interesting in terms of this study is the loss of flight among the birds living in the rough and tumble world of the great continents, which are filled with dangerous predators and brimming with potential prey ripe for plucking from the air and which offer food sources separated by great distances. The number of bird groups that have lost flight under harsh, dangerous circumstances that seem to favor strong flight performance is remarkable. The first known examples are found all the way back in the Late Cretaceous, when modest-sized patagopterians lost flight. Also known from this time are partial remains of what seem to be nearly ostrich-sized, probably flightless birds (Buffetaut and Loeuff 1998).

Turning to the Cenozoic, which we still live in, we encounter the familiar large ratites (rheas, cassowaries, emus, ostriches), which are fast-running ground birds. Ratites appear to be basal palaeognathous birds related to the small, flying tinamous. The once-common view that some or all ratites are actually advanced neognathous birds that reverted to a more primitive condition is losing favor as the fossil record improves. However, how many times continental ratites have lost flight remains unclear: perhaps once, before they were separated by the dispersal of the southern continents, or more probably, multiple times on various continents after the break-up. The Cenozoic also saw the appearance of ratitelike and largely flightless fleet-footed neognathous ergilornithids, which lived during much of the era. The Neogene dromornithids or mihirungs of Australia were thought to be robust ratites, but new remains show that they were oversized duck relatives.

All of the continental birds mentioned so far were or are herbivores or omnivores. More ambiguous from a dietary perspective was another group of duck relatives, the Paleogene diatrymids, which appear to have been closely related to, if not the same as, the large gastornithids that lived at about the same time. There is little doubt that the hook-beaked "terror birds" were arch-predators. Seriemas—which are the living relatives of phorusrhacoids and belonged to a group that also produced the semipredaceous Paleogene bathornithids, idiornithids, and other phorusrhacid relatives—are medium-sized birds with little or no ability to fly.

You may have noted the common theme of continental flightless birds: they range from fairly large to gigantic. Although some small land birds fly poorly and some birds prefer to run rather than fly away from danger, all of these birds retain some ability to become airborne. It seems that the danger of not being able to escape into the air is too great for small continental birds.

However, unless they give up flight, continental birds cannot enjoy the advantages that accrue to large, fully terrestrial animals. Larger animals can walk farther than smaller ones; large size is itself a source of protection from predators. These are neutral trade-offs; the real advantages of being big center on feeding.

For herbivores, large size allows them have large fermentation chambers to digest low-quality plant materials that smaller animals cannot handle (Morton 1978, Demment and Soest 1985, Farlow 1987). However, herbivory becomes a problem for flying birds because they are hard pressed to loft into the air with the fermentation chambers necessary for increased digestive efficiency. Therefore, leaf-eating flying avian herbivores are rare; they are either weak fliers with fermentation chambers (hoatzins; Morton 1978), or they have very inefficient digestion (geese; M. Owen 1972). Therefore, being an efficient herbivore generally requires being larger than is usual for fliers.

For predators, being large is a necessary adaptation for attacking big prey. The burden flightless avian predators bear is the difficulty they have in converting their wings into grasping weapons comparable to those of their dinosaurian ancestors or to the front paws of felids. Yet, new evidence suggests that the arms of at least some of the great phorusrhacoids were modified into weapons bearing one or two claws, one on a mobile thumb (Chandler 1994).

What it all comes down to is that the feeding advantages associated with being large have overcome the advantages of flight among avian herbivores and predators at least half a dozen, and possibly a dozen or more, times.

### Losing Flight Is So Very Easy to Do

Whereas flight takes millions of years to evolve, the reversion to the flightless condition can occur much more rapidly. Some island birds seem to have lost flight and dramatically reduced their flight apparatus in just generations rather than millennia, and complete alteration to ground-bound forms that differ significantly from the flying ancestors may require only a million years or less (Feduccia 1996). This is not surprising, since most birds are so well adapted for ground locomotion that they do not have to go through the long process of evolving land legs. Once flight is abandoned as a major mode of movement, the selective pressures to reduce and eliminate the costs of growing and maintaining the flight apparatus are very strong; and the reduction of an appendage is easy in genetic and developmental terms. This trend has been taken to various extremes. In some flightless birds, the wings are reduced in size but retain all of the features normally associated with the avian wing. Ostriches and rheas still have fairly long arms; those of cassowaries, emus, and kiwis are much more degenerate. Elephant birds had only a small humerus. The moa carried arm reduction the furthest: the arm was entirely missing, and even the scapulocoracoid was reduced to a small splint.

Another reason losing flight is easy is that, according to some researchers, all a bird has to do is fail to mature completely. Juvenile birds lack such features as big arms, large, deep-keeled sterna, strongly retroverted coracoids, and well-formed flight feathers. If, as evolution proceeds, these juvenile features are retained in adulthood—such retention is known as neoteny—the result is a bird with a shoulder girdle, arms, and feathers that are too poorly developed for flight (Dawson et al. 1994, Feduccia 1996). The genetic changes needed to promote this phenomenon may be minimal. Indeed, the neotenic loss of certain flight adaptations in birds may have been induced by thyroid deficiency. (Elzanowski [1988] disagreed with simplistic views of flightless adults as birds that never quite grew up.)

### Poor Fliers Make the Best Nonfliers

A number of birds that have lost flight descended from such strong fliers as hawks, owls, ducks, and pigeons. However, the flying palaeognathous birds that may be ancestral to ratites, the tinamous, are short-range fliers. The rails, which have produced so many neoflightless island dwellers, tend to run rather than fly away from danger.

Weak fliers should be especially prone to losing flight because they have less to lose in terms of the advantages associated with air travel.

## Losing Flight When It Has Only Just Begun

### *Why It Should Happen*

Until the 1950s, the hypothesis that flightless birds evolved directly from flightless ancestors was still very common, but this hypothesis has been contradicted to the point that very few current researchers consider it viable. In fact, all the generally accepted cases of flight loss to date have centered around birds whose ancestors clearly possessed sophisticated levels of flight, such as *Hesperornis* and *Patagoapteryx* in the latter half of the Cretaceous. We know that flight was lost fairly often among sophisticated flying birds in the Mesozoic and Cenozoic.

Surprisingly little consideration has been given not only to the possibility that avian flight might have been abandoned in its initial stages but also to the possibility that such reversals may be *especially likely in basal fliers, that is, when aerial performance is still poor and flight adaptations are only weakly developed.* As noted above, low-performance fliers should have been more susceptible to flight loss than more sophisticated fliers. Because the arms of basal fliers retained long, supple fingers with large hooked claws, they could easily be used as raptorial appendages for hunting prey or for other purposes. Because a long, bony tail was still present, its traditional use as a counterbalance and dynamic stabilizer when running and leaping could have been reemphasized. Because the skeletal modifications needed to return to a flightless condition were few, the time required for the transformation should have been even briefer than that observed for modern birds. In evolutionary terms, reconfiguring an unsophisticated flier into a nonflier should have been easy, considerably easier than revamping a sophisticated flier into a nonflier (Paul 1988a, Gould 2000).

Also little appreciated is the notion that nonfliers that evolved from primitive fliers may have evolved flight-related adaptations that accidentally gave them a competitive edge over nonfliers whose ancestors had never flown. We have noted that avepod fliers, more than pterosaurs and bats, upgraded their brain and other neural functions, their vision, and their skeletomuscular systems in order to better meet the needs of moving at high velocities in a three-dimensional world. Upon returning to a terrestrial life, neoflightless forms may have found themselves superior in these regards to competitors whose ancestors had never flown.

We know that flight is fairly often lost among sophisticated flying birds in the Mesozoic and Cenozoic. Even more easily able to lose flight, early avepod fliers should have been more prone than advanced birds to becoming secondarily flightless.



This probability should not prompt us to mistakenly assume that certain nonavebrevicaudan avepods *were* neoflightless. However, because the hypothesis that some avepod dinosaurs were secondarily flightless is so plausible, we should not assume either that the burden of proof falls upon the hypothesis. We do not yet know, and we must see which way the evidence leans, and how far.

### *How to Detect Early Loss of Flight*

It is not difficult to tell whether an advanced avebrevicaudan bird is secondarily flightless. For one thing, such birds have all the nonflight skull and skeletal characters that place them among advanced birds, so it is obvious that their inability to fly represents a reversal. In addition, they possess characters that strongly indicate that their ancestors could fly. For example, the avian pygostyle is an obvious flight adaptation, so any avepod that has one almost certainly had flying ancestors. The following is a list of changes that occur when flight is lost in birds (changes marked with an asterisk always occur; changes not so marked are less consistent):

- Number of free caudals increases.
- Pygostyle is reduced.
- Furcula is absent.
- Tip of scapula blade is squared off.
- Scapulocoracoids are reduced* or lost.
- Coracoids are less flexed relative to the scapula blade.
- Coracoids are broad.
- Acrocoracoid is reduced.
- Sternum is reduced to a greater or lesser extent.*
- Sternal keel is very reduced or lost.
- Arm is reduced or lost.*
- Pectoral crest is reduced.*
- Wing-folding mechanism is lost.
- Size and complexity of carpals, especially smaller elements, are reduced.
- Fingers are better developed.

- Finger claw(s) are well developed.
- Feathers are simplified.
- Wing feathers are symmetrical.*
- Hallux is lost or no longer fully reversed.
- Flight-related characters are progressively lost within the group.

Detecting the neoflightless condition becomes more difficult as one gets closer to the origins of avian flight. For example, *Archaeopteryx* had neither a pygostyle nor a highly modified hand. Conversely, modern birds already lack a long tail, so it cannot be further shortened. Dinosaurs had long tails, and secondarily flightless examples might have inherited a tail that had previously been shortened for dynamic flight. It is also possible that loss of flight in the initial stages of its evolution may involve less in the way of neoteny than its loss in more modern birds. Because early fliers lacked a deep-keeled sternum and their flightless descendants may have retained large arms, the strong degeneration of the arms typical of flightless birds should not be expected in neoflightless dinosaurs. Because preservation of the arm feathers is unlikely, assessing their condition may not be possible. Ergo, the retention of juvenile characters into adulthood may not be as good an indicator of neoflightlessness as it is in more advanced birds.

How, then, do we identify a possible flightless descendent of such an early protobird? The basic method is as follows. The first stop is to list flight-related characters that did appear or would be expected to have appeared in the early stages of the evolution of avian flight and that might have been retained in flightless descendants. These are what I call potentially neoflightless characters. Note that the changes associated with early flight loss may sometimes differ from those associated with the loss of more advanced flight. Consider arm length. The arms of nonavepectoran avepods tended to be moderate in length. In birds, arms tend to be either exceptionally long (in fliers) or unusually short (in nonfliers). The shoulder girdle, especially the furcula, and the arms are often severely reduced or lost in flightless birds because the wing is not co-opted for any important alternative use. In raptorial-handed protobirds, the arm was more likely to have remained long, because it could have been used as a powerful weapon; and the shoulder elements should consequently have remained well developed in order to provide a suitable base for the arms. Ergo, if a dino-avepod's arms were either unusually long or unusually short, they qualify as a potentially neoflightless character.

*Archaeopteryx* is an obvious source of information about the kind of characters expected to be associated with flight loss in the early stages, but some predictions about what we should expect in slightly less advanced or somewhat more advanced fliers are necessary. The most obvious potentially neoflightless characters center on the adaptations that made the shoulder girdle suitable as a base for wings; note that many such adaptations are found in the flying pterosaurs as well. The potentially neoflightless characters associated with the arm are more limited because this appendage is remarkably little modified from the predatory dinosaur condition in *Archaeopteryx*. There are some characters in the axial skeleton that may also have been linked to flight. For example, ossified uncinate processes may have strengthened the rib cage or aided respiration or both. Various tail characteristics were involved in making the appendage into an aerodynamic stabilizer and are found only in fliers such as *Archaeopteryx* and long-tailed pterosaurs, as well as in some protobirds. Abbreviation of the tail beyond the degree seen in the urvogel or the presence of a pygostyle is especially important. Some neural and sensory upgrades may be associated with flight control and guidance. Finally, nonavepectoran predatory dinosaurs were just that, predators equipped to attack and kill large prey. The first small fliers were too small to have been such arch-predators. Therefore, deviations from the standard pattern of dino-avepod predation qualify as a potentially neoflightless character.

The importance of the characters varies. The possession of a birdlike shoulder girdle—which implies the presence of an arm musculature that was optimized at one time for flapping flight—and folding arms is much more compelling evidence for loss of flight than is the presence of forward-facing eyes and a big brain. Symmetrical contour feathers on the arms are markedly more important than reduction of olfaction. Details can be important too. For example, neoflightless birds often retain a posterolateral bowing of the third metacarpal and a posterolateral flange at the base of the central finger, even though the large flight feathers these adaptations were designed to support are no longer present.

The case for secondary flightlessness becomes strongest only when a number of potentially neoflightless characters are combined in a potentially flightless subject. The combination of a near avian trunk (including all the shoulder girdle adaptations listed below), as well as ossified sternal ribs and uncinate processes, is telling because this combination indicates that the forequarters and rib cage were strengthened and modified to allow high-level respiration during strong flapping flight. The presence of most of the following characters—which will be discussed in further detail in Chapter 11 and Appendix 6—in a taxon shifts the preponderance of evidence in favor of neoflightlessness.

- Furcula is U-shaped.
- Shoulder girdle is large and strongly built.
- Scapula blades are horizontal.
- Scapula tip is pointed.
- Acromion process is developed as well as or better than in *Archaeopteryx*.
- Coracoids are proximally as narrow as or narrower than in *Archaeopteryx*.
- Coracoids are subvertical and more flexed relative to scapula than is normal in tetrapods.
- Coracoids are retroverted.
- Coracoids are long (or large in general).
- Outer surfaces of coracoids face strongly anteriorly because they articulate via transversely long articulations with anterior edge of sternum.
- Most of sternum is ossified.
- Sternum is a broad plate.
- Sternum is a large plate, especially it is long.
- Sternal ribs are ossified.
- Sternocostal articulations are more numerous than in *Archaeopteryx*.
- Pectoral crest of the humerus is large and hatchet shaped.
- Arm is longer than in nonavepectorans.
- Arm is shorter than in nonavepectorans.
- Arm-folding mechanism is present.
- Carpometatarsus is fused.
- Metacarpal III is bowed posterolaterally.
- Proximal phalanx of finger II has a posterolateral flange.
- Finger II is at least as robust as I.
- Flexibility of finger II is reduced.
- Finger III is tightly appressed to finger II.
- Finger bones are less numerous than in *Archaeopteryx*.
- Arm feathers are pennaceous and symmetrical.
- Ossified uncinate processes are present.
- Tail base is hyperflexible, especially dorsally.
- Distal tail is stiffened by one or both of the following adaptations: centra are greatly elongated; distal tail is ensheathed with long, slender ossified rods.
- Tail is reduced in length or caudal count or both relative to nonavepectoran dinosaurs.
- Tail length or caudal count is as reduced as or more reduced than in *Archaeopteryx*.
- Majority of tail is slender.
- Pygostyle is present.
- Pubes are strongly retroverted (retroversion indicates that ventral body keel was reduced to increase aerial maneuverability).
- Hindlimb/dorsosacral length ratio is higher than in *Archaeopteryx*.
- Progressive loss of potentially neoflightless characters within group is exhibited.
- Volume of the braincase is enlarged above the reptilian maximum.
- Eyes have overlapping fields of vision—often indicated via triangular frontals on skull roof.
- Olfaction is reduced—indicated by reduction of olfactory nasal space or olfactory lobes of brain or both (the sense of smell is weakly developed in most birds because they are unable to utilize it in the air).

To the flight characters that can be expected to be retained in neoflightless protobirds, we can add the flight characters we might *not* expect to see, either because they are not present so early in avepod fliers or because they are normally lost along with flight or perhaps for both reasons:

- Coracoids may not be strongly retroverted.
- Coracoids may not be narrow.
- Acrocoracoid may not be dorsally elongated and thereby help form a triosseal canal.

The next step is to determine where the nonflight-related phylogenetic characteristics of the subject place it within the avepod clade, a task that is discussed in detail in Chapter 11. Then we must combine all the cladistic and flight character data. At the same time, we must consider the possibility that the potentially neoflightless characters actually evolved to serve nonflight functions and therefore merely mimic and converge with the flight condition. In general, the greater the number of potentially neoflightless characters that suggest a neoflightless condition, and the more phylogenetic characters that suggest the subject was closely related to birds, the more likely it is that the avepod was secondarily flightless. A neoflightless condition is especially probable if there are potential post-urvogel characters that suggest the avepod was closer to modern birds than to flying protobirds such as *Archaeopteryx*. For the purpose of the following discussion, I have assumed that the subjects always lack arms large enough to support flight.

Consider an avepod whose cladistic placement is below the level of avepectorans. Such an avepod has no potentially neoflightless characters. The possibility that this dinosaur was secondarily flightless is extremely low. However, if a non-avepectoran avepod has a few potentially neoflightless characters, then the possibility that its ancestors flew is higher but still too low to be considered probable. In this case the potentially neoflightless characters probably represent only convergence. However, if a non-dino-avepectoran has numerous potentially neoflightless characters, then the possibility that its ancestors flew is increased greatly. However, the possibility that the characters only mimic the flight condition cannot be ruled out.

Now consider a dino-avepectoran that has a modest set of potentially neoflightless characters. In addition, its ossified sternum is no longer than that in *Archaeopteryx,* and its sternal ribs and uncinate processes are not ossified. For this dino-avepectoran, the possibility that its ancestors flew is higher than for the nonavepectoran avepod with a similar set of adaptations because the former is closer to the origins of flight than the latter. Even so, convergence cannot be dismissed. If the arm is very long, but not as long relative to the hindlimb as in *Archaeopteryx,* then this dino-avepectoran may either have been involved in the development of flight or have been neoflightless to some degree. In either case, the level of flight was at most no higher than that seen in *Archaeopteryx.* If the number of post-urvogel characters is zero or low, then this dino-avepectoran can be assessed as both phylogenetically and aerodynamically less derived than the urvogel. If the number of post-urvogel characters is high, then the situation becomes more complicated and ambiguous. Then the dino-avepectoran is a candidate for having evolved in parallel with birds. Alternately, it is a neoflightless post-urvogel that has undergone severe reduction of its flight-related features.

Next we consider an avepectoran that has a greater number of potentially neoflightless characters. Its ossified sternum is longer than that in *Archaeopteryx,* and the sternal ribs and uncinate processes are ossified. Yet the arms are too short to have sustained flight, and contour feathers are either absent from the arm or they are symmetrical. The possibilities that this avepectoran was neoflightless, and that the stage of flight that had been achieved by its ancestors was more advanced than that seen in *Archaeopteryx,* must be ranked as high. These possibilities are even more likely if the avepectoran's tail is shorter than that of *Archaeopteryx* and if a pygostyle is present. If the number of other post-urvogel characters is also high, then the most logical conclusion is that the avepectoran lies closer to modern birds than *Archaeopteryx* does. If the number of other post-urvogel characters is low, the situation is less clear. The lack of post-urvogel features may be the result of a general reversal to a pre-urvogel condition associated with a return to a land-bound life, in which case a post-urvogel status remains likely. Alternately, the avepectoran descended from a clade of dino-avepectorans that evolved an advanced stage of flight independently of the main avian clade. Or the apparently flight-related characters actually resulted from lifestyle requirements that had nothing to do with flight.

If a cladistic analysis reveals that a dino-avepectoran possesses numerous avian characters absent in *Archaeopteryx,* then it may have been secondarily flightless even if it possesses no potentially neoflightless characters. However, it is also possible that the avepod merely converged with birds above the level of archaeopterygiforms. Or it is possible that *Archaeopteryx* was a member of a clade that developed flight independently of the clade that includes modern birds, in which case avepods closer to birds may have not yet evolved flight.

We need to consider an additional factor when we assess a potential neoflightless condition. If the increasingly derived members of clade are in the process of developing flight, then we should expect a corresponding increase in the adaptations for flight. If instead we find that flight-related adaptations appear to decrease in the increasingly derived members of a clade, then it is more likely that they descended from a volant ancestor.

It is apparent that assessing whether an avepod dinosaur was actually neoflightless or just appears to have been is a complicated task full of phylogenetic pitfalls. But making the effort is worthwhile because the results, whether positive, negative, or inconclusive, promise to be informative. We will apply these methods to predatory dinosaurs in Chapter 11, but before we do so, we need to look at the habitat and lifestyle of what remains the ultimate fossil dino-bird, *Archaeopteryx.*

PART 4

# THE *ARCHAEOPTERYX* PROBLEM

Some fourteen decades after the discovery of *Archaeopteryx,* the debate about bird origins still centers on this famous early bird. *Archaeopteryx*'s place in this debate is to a certain degree an accident. No other fossil avepod yet found is such an obviously transitional form and also in such a good state of preservation. In *Archaeopteryx,* we have a toothed, long-fingered, long-tailed reptile-bird known from seven largely articulated skeletons, of which five are complete or nearly so; three have useful skull remains, and all are adorned with feathers. Although in recent years many dozens of feathery skeletons of later Mesozoic birds have been found, they are more fully avian and therefore less informative in regard to bird origins. A few researchers assert that earlier "*Protoavis*" gives us more information about the origin of birds than *Archaeopteryx* does, but "*Protoavis*" is known only from disarticulated, damaged remains that probably represent more than one kind of creature, none of them closer to birds than Triassic dinosaurs (Chapter 10). Eventually, somewhere, somehow, some other creature will displace *Archaeopteryx* as the premier ancestral bird type. Someday.

In the meantime, there are many problems with *Archaeopteryx.* Despite the fact that this archosaur and its habitat have been studied for so long, its biology remains poorly understood and controversial in almost all regards. Was it a glider or a powered flier (Chapter 8)? Was it an arboreal climber, a terrestrial runner, or both? Was it an insectivore, a small predator, a shorebird that either waded or swam in pursuit of fish, or a combination of these? All of these possibilities have been proposed by one researcher or another. No one agrees on what it ate, how it moved, whether it power-flew or just glided, or even exactly where it lived. These questions complicate our attempts to understand how birds and their flight evolved.

CHAPTER 9

# The Lifestyle of the Urvogel

## The Solnhofen Habitat

The problems with *Archaeopteryx* begin with the location of its remains. They lie in fine-grained sediments deposited at the bottom of a marine lagoon in what is now northern Bavaria (Viohl 1985, Barthel et al. 1990, Kemp and Unwin 1997). In the northern end of the lagoon were small islands, and farther to the northwest and northeast were larger islands that made up what was then the European Archipelago. To the south lay the great Tethys Ocean, of which the Mediterranean is but a remnant. The nature of this location leaves open a number of possibilities for habitat and lifestyle. Did the protobird live in the lagoons? Did it live on the small islands? Did it live on the more distant lands? Or were these habitats combined in some fashion?

The excellent condition of the best *Archaeopteryx* skeletons—HMN 1880, JM 2257, S6, BSP 1999—as well as the *Compsognathus* skeleton found in the same lagoon leaves little doubt that they did not drift a long distance after death (Kemp and Unwin 1997, Davis and Briggs 1998). They either died in the water—while swimming or after having falling in while flying between islands—and sank immediately, or they drifted from the nearest, small islands. The number of *Archaeopteryx* skeletons is some one-thirtieth to one-fortieth the number of pterosaur skeletons found in the same beds (Wellnhofer 1991, Davis 1996). Most of the Solnhofen pterosaurs—mainly *Rhamphorhynchus* and *Pterodactylus*—are considered to have been rather like gulls and shorebirds in their habits, and they may have formed large flocks. The comparative scarcity of *Archaeopteryx* means either that it was much less common and less social than the pterosaurs or that the bulk of the *Archaeopteryx* population lived far from the lagoon. However, because there are a fair number of *Archaeopteryx* skeletons and because the protobird's flight performance was mediocre, it seems unlikely that they represent "accidentals" that had wandered far from inland homes on the larger islands 50 kilometers or more to the north. Therefore, for the rest of the discussion, I will assume that *Archaeopteryx* lived in the coastal zone of the larger lagoonal islands, on the small lagoonal islands, or in some combination of these habitats.

Classic *Archaeopteryx* restorations show it climbing about in tall trees (Burian in Spinar and Burian 1972, Fruend in Ostrom 1985, Knight and O'Neill in Feduccia 1996). These restorations are problematic because there may not have been any trees for the protobird to climb in. Although the small Solnhofen islands themselves have been destroyed by erosion, their structure, climate, and flora are fairly well understood (Buisonje 1985, Viohl 1985, Barthel et al. 1990). The islands were low lying, but we do not know whether they were edged with short cliffs or wide beaches or some combination of the two. The climate was subtropical and semi-arid, with a long dry season and a short rainy season, which may in fact have not been very rainy. Indeed, the coastal strip along the northwest border of the lagoon may have been virtually rainless, like the coastal deserts along the western edges of South America and Africa. Freshwater was present at least seasonally farther inland, but because it was not abundant, the flora was adapted for dry conditions. The flora consisted largely of trunkless shrubs—mainly coniferous—less than 3 meters tall. The absence of tree trunks in the lagoons suggests that trees were at best rare if not altogether absent. Particularly telling is the absence of wind-borne tree pollen in the Solnhofen sediments. If any trees were present on the islands surrounding the ancient lagoons, they were probably limited to the inland portions of the larger islands, and dense woodlands were almost certainly absent. Although a few workers, such as Feduccia (1996), have continued to argue for the presence of Solnhofen woodlands, there is no positive evidence for this view. The terrestrial habitat of *Archaeopteryx* appears to have been more akin to the islands in the Middle East than

*Figure 9.1.* The habitat of *Archaeopteryx*. The urvogel lived not in woodlands with tall trees but on arid islands covered with dry-adapted shrubs. In this scene, two individuals display to each other, flashing underwing color patterns (speculative) and hyperextending their second toes. Flocks of the little pterosaur *Pterodactylus* pass overhead.

to a lush tropical paradise. The paleoecological data has inspired newer restorations showing *Archaeopteryx* on a more barren island habitat (Figs. 7.5, 9.1; Hallett in Schlein 1996).

Complicating the urvogel habitat problem is the discovery of urvogel-like teeth in the somewhat earlier Guimarota assemblage of Portugal. Although Weigert (1995) assigned these teeth to *Archaeopteryx,* significant differences suggest they would be better placed among archaeopterygiforms in general (Elzanowski pers. comm.). Schudack (1993) proposed that the Guimarota habitat was swampy. If it was, then some archaeopterygiforms lived in wet places; we do not know whether they preferred tall trees or more-open areas. The Cretaceous archaeopterygiform *Protarchaeopteryx* lived in a lush, heavily wooded habitat (Wang 1998).

## Climber or Runner?

The seeming barrenness of the Solnhofen islands poses a paradox for those trying to restore the lifestyle of its most famous inhabitant, since there are indeed reasons to think *Archaeopteryx* was adapted for climbing. In the previous chapter, we discussed whether or not long-armed dino-avepods were adapted for climbing as well as running, and this discussion also applies to *Archaeopteryx*. Many researchers have insisted that it was primarily adapted for life on the ground (Ostrom 1976a,b, 1985, Cowen and Lipps 1982, Caple et al. 1983, Balda et al. 1985, Gauthier and Padian 1985, Padian 1985, 1986, Peters 1985,

Peterson 1985, Rayner 1985a, 1991, Peters and Gorgner 1992, Chiappe 1995, Padian and Chiappe 1998a,b). A similar number of other workers have argued that the protobird was a specialized climber (Tarsitano and Hecht 1980, L. Martin 1983a,b, 1991, R. Norberg 1985, 1995, Rietschel 1985, Tarsitano 1985, 1991, Yalden 1985, Feduccia 1993a, 1996, Chatterjee 1997, 1999a, Zhou 1998, Zhou and Martin 1999). Still others favor the climbing hypothesis but also think that *Archaeopteryx* may have been a dual-purpose creature adept on the ground as well as in high places (Paul 1988a, Holtz 1994b, Sereno 1997a, Hopson and Chiappe 1998).

In part, the debate over whether *Archaeopteryx* was a climber or a runner is driven by the debate over whether avian flight started from high places or low places (see Chapter 6). The anatomy of the protobird also has a lot to tell us about where it preferred to travel.

The most obvious running adaptation of *Archaeopteryx* was its long legs, which were fairly gracile appendages similar in proportions to those of terrestrial avepods. However, many climbing creatures, including various primates, also have very long limbs. It is therefore important that *Archaeopteryx*'s pelvis and knee were not especially well adapted for running. In running dino-avepods and most living birds, the ilial plate was quite long (Figs. 10.1B, App. Fig. 18I), but *Archaeopteryx* was distinctive in having an ilium that was markedly shorter relative to body mass. Its hindlimb/ilium ratio of about 5.5 exceeded even that of *Sinornithosaurus* (Chapter 6). The short ilium and cnemial knee crest indicate that the thigh muscles probably were narrower and therefore smaller than those in running avepods. The small cnemial crest also means that the shank musculature was only modestly developed. The combination of gracile legs with a modest hindlimb musculature suggests that the protobird was competent on the ground, but not as well adapted for running as terrestrial dino-avepods and ground birds. As noted in Chapter 6, reduction of leg musculature vis-à-vis that of terrestrial counterparts is characteristic of many arboreal animals (Grand 1977), and some of these animals (especially various primates) have long, gracile legs.

When we turn to the feet of *Archaeopteryx,* we find that the tarsometatarsus was not especially elongated. This condition resembles that of birds with some arboreal capabilities: avepods—especially birds—specialized for running tend to have more elongated upper feet (Zhou 1998). This point should not be taken too far, however: the distal segments of *Archaeopteryx* were longer and more gracile relative to the femur than those of the swift coelophysids. Hopson and Chiappe (1998) concluded that the degree of distal elongation of the toes indicates that the urvogel was scansorial in the manner of pigeons; certainly the toes were more distally elongated than those of any known running dinosaur (Fig. 6.3). The toe claws were also large and well curved, which improved their grasping ability, but the urvogel's toe bones were not strongly dorsally arced, as is common in birds that frequently climb and, for that matter, in *Protarchaeopteryx* (Figs. 6.3, 9.2). Nor were the *Archaeopteryx* metatarsals subequally long; subequal metatarsals are a strong arboreal feature likewise lacking in pigeons. The urvogel's central three toes show that it very probably did climb on a regular basis, but just how well and how often are obscured by two things: first, by the possibility that, to a certain degree, its toes retained proportions and a lack of arcing that originally evolved in running dinosaurs and, second, by how much less heavily it relied on its hindfeet for climbing than do birds that lack clawed, supple fingers.

Several other hindlimb features are profoundly important in assessing running versus climbing ability. The hallux, or first toe, of *Archaeopteryx* was larger, more distally placed, and more reversed than that in the running dino-avepods. This condition should not have hindered running performance any more than do the reversed toes of roadrunners, fowl, and many other land birds. What the reversal did do was render the hallux fully capable of cooperating with the central three toes to grasp a branch. The result was an improved ability to grasp branches with the feet, something no other avepod dinosaur could do as well (with the possible exception of *Microraptor,* as discussed in Chapter 6). Modern running birds probably inherited their reversed halluxes. No one has yet explained why urvogels would have evolved a reversed hallux if they retained their ancestors' running habits; the structure makes sense only if a shift toward climbing was underway. On the other hand, the hallux was not as distally placed or as long as it is in modern climbing and perching birds (Fig. 9.2). The hyperextendible second toe may also have been useful in climbing vertical surfaces, with the claw acting as a spike.

One of the important facts about *Archaeopteryx* is that its arms were its primary tools for locomotion, as indicated by several pieces of evidence. The forelimb (as well as its girdle) was stronger and larger than the hindlimb (Fig. II.1,

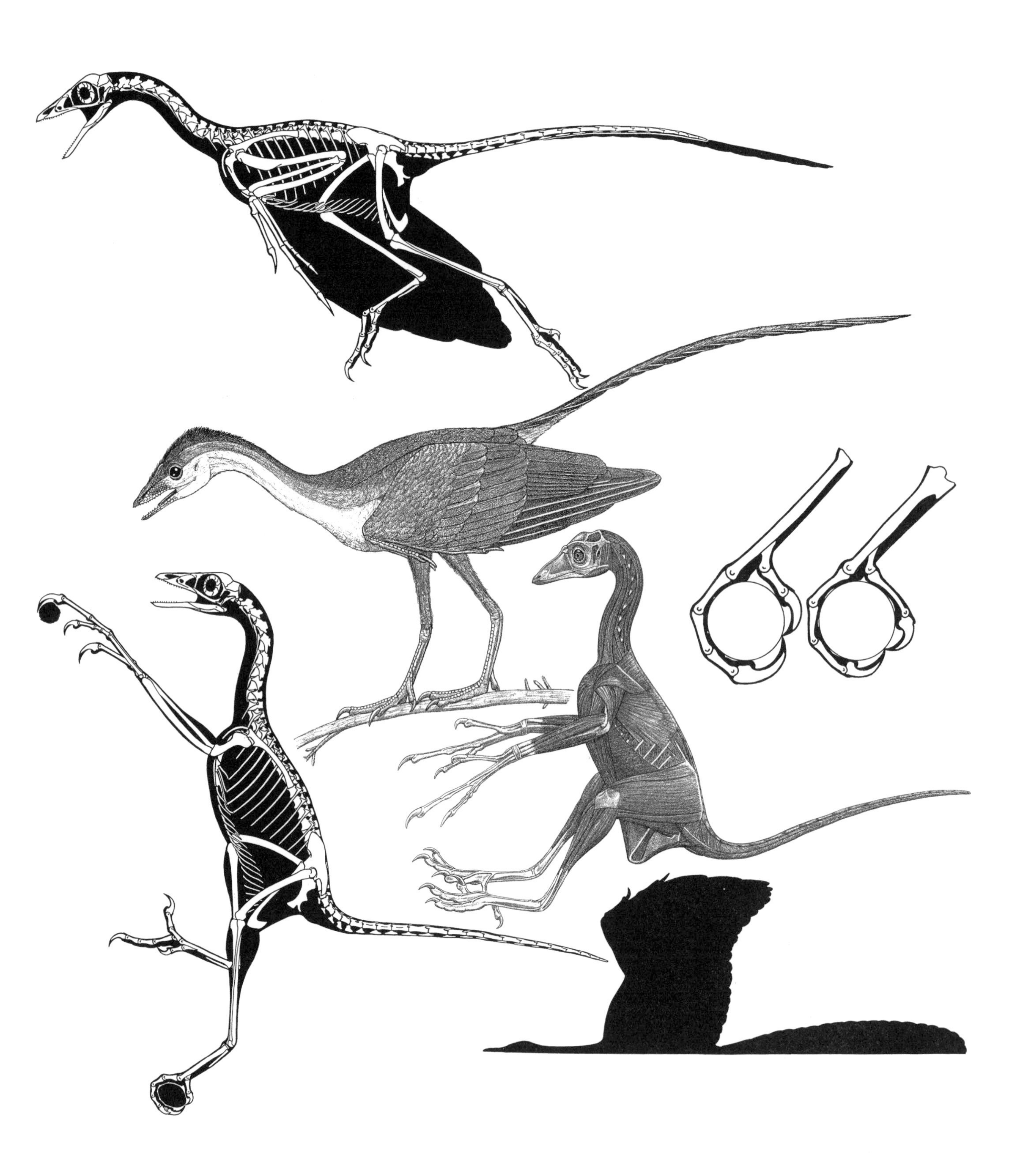

App. Fig. 18C,D). The humerus was thicker than the femur. The strongest finger—which was the central digit rather than the thumb weapon, as in most avetheropods—was thicker than the central and strongest toe. The arm was slightly longer than the leg in the adults. Unlike the reduced cnemial crest of the tibia, the pectoral crest of the humerus was very large. Opposite to the reduction of the pelvis, the enlarged furcula, coracoid, and sternal plate were larger than those of terrestrial dino-avepods. As discussed earlier, these enlarged elements supported an enlarged musculature. Of course, the enlargement and strengthening of the arm bones and muscles were associated with the modification of the anterior appendages into feather-supporting, flapping wings. The strengthening of the central digit, in particular, was associated with its being the primary anchor for the outer feathers. However, because the hands had long, supple fingers, the combination of muscle power, bone strength, and digital dexterity indicates that the arms should also have made very good climbing organs. If so, *Archaeopteryx* would have been a quadrupedal climber whose arms did more work than its legs. This mimics the situation in bats, which retain a clawed thumb that is used for climbing. However, the urvogel's antero-medially divergent thumb claw was more weapon than climbing hook because it was not as well suited for grasping as the other two fingers, whose claws pointed toward the palm.

*Figure 9.2. (opposite)* Locomotory adaptations of *Archaeopteryx,* whose form was suited for both running and climbing. The hindlegs were long, but a relatively small pelvis implies *Archaeopteryx* was slower than other dino-avepods. That the arms were larger and stronger than the legs indicates that the former were the primary organs of locomotion and could have been used for quadrupedal climbing. Longer and stronger than the other fingers, the central finger may often have been the only one used to hold onto a branch. The foot of *Archaeopteryx (left),* with its well-developed grasping abilities and its large reversed hallux (based on HMN 1880), is compared with that of a pigeon *(right),* which has an even larger hallux that is better adapted for perching. Note that when *Archaeopteryx* was climbing, the arms were sprawling, the legs erect. The life restoration shows the urvogel striding along a branch with folded wings, and the muscle study shows the arms and legs ready for a branch landing. A speculative swimming configuration shows the wings partly folded.

We next turn to the claws of the urvogel, which have been the center of debate. Rietschel (1985) concluded that the finger claws were used for grooming the feathers. This is very possible, but one claw per hand, rather than three, would have been adequate if grooming was their primary function. Indeed, it is difficult not only to understand why six well-developed claws were needed for grooming but also to visualize how so many could be used for the purpose.

Rayner (1991), Peters and Gorgner (1992), and Padian and Chiappe (1998c) argued that the claws of *Archaeopteryx* were not suitable for climbing and, for evidence, pointed out that the claw tips were not worn in the expected manner. However, the lack of wear is not a valid argument, because continuous growth of the keratin sheath keeps claw tips sharp (but Feduccia [1996] is incorrect in believing that cats sharpen their claws; they actually shed them). Besides, one might also use the lack of wear on the toe claw tips to argue that *Archaeopteryx* did not run on the ground. In any case, there are other explanations for the lack of wear. For example, roadrunners' sharp claws do not become worn, because their claws are persistently elevated off the ground (Fig. 8.3). This explanation does not apply to *Archaeopteryx* claws, because they are not so strongly extended in most of the articulated toes.

Ostrom (1976b) and Peters and Gorgner (1992) concluded that the protobird's claws were most similar to those of ground-running birds, but Yalden (1985) and Feduccia (1993a) concluded that they were most similar to those of climbing birds. What is true is that the claws were strongly recurved, sharp-tipped hooks that could easily grasp branches and grip bark. This was especially true of the finger claws, which were very strongly recurved. That the central finger claw was the strongest strengthens the case that it was a strong grasping organ. Because the central finger was the longest, it would often be the only one to hook onto a branch (Fig. 9.2); therefore, the digit should have been strengthened to carry much or most of the creature's mass.

It was not logically consistent for L. Martin (1995, 1997), Zhou and Martin (1999), and Martin and Czerkas (2000) to argue that *Archaeopteryx* was arboreal while stating that it could not flex its feather-bearing fingers to the degree that would be needed to grasp branches. If the claws were limited to acting as spines, the effectiveness of the fingers would have been severely limited. As explained in Chapter 4, the unfused fingers were very probably free to curve toward the palm.

If predation were the main purpose of the claws, one would expect the thumb claw to be the largest, as in other dino-avepods. The toe claws were less strongly arced than the finger claws, but they were more curved than those of other small dino-avepods. In the end, the urvogel's claws appear to have been general-purpose tools suited for a range of tasks, from climbing to predation to grooming.

The condition of *Archaeopteryx*'s feathers has also been part of the debate. Feduccia (1993a, 1996) observed the lack of fraying of the tail feathers and suggested *Archaeopteryx* did not drag its feathers along the ground as some long-tailed ground birds do. This lack of fraying is compatible with arboreal habits, but *Archaeopteryx* could have carried its strong tail clear of the ground. Chatterjee (1997, 1999a) noted that the stiff tail could have been used as a prop during climbing, and because a broad feather splay was present, this is certainly possible. However, the lack of tail feather fraying may work against this notion. All of these discussions about feather wear may be moot, however. Only in HMN 1880 are the feather tips really well preserved, and this individual may have undergone a feather-rejuvenating molt just before it expired. Chatterjee also suggested that the retroverted pubis of *Archaeopteryx* improved the ability of the flattened body to press against tree trunks, but as detailed in Chapter 4, the pubis was not strongly retroverted, and the body was deep keeled. In addition, we must remember that there may not have been any tree trunks for the urvogel to scamper up.

On the one hand, statements such as those by Feduccia (1993a) that *Archaeopteryx* was a "perching bird" clearly go too far, in that the urvogel's hallux, toe proportions, and toe-bone arcing were not as well developed as in perching birds; its medially directed thumb claw was not strongly adapted for climbing; and its quadrupedal form of climbing was not like that of most modern birds. Nor was *Archaeopteryx* as adapted for climbing as the squirrel-like, semiquadrupedally climbing birds of the Cretaceous, such as *Confuciusornis*. On the other hand, assertions such as Rayner's (1991) that the protobird "shares the cursorial morphology of other theropods, and its skeleton shows no features associated with climbing or arboreality," as well as Padian and Chiappe's (1998b) statement that it "shows no arboreal characters," neglect its reduced hindlimb musculature and pigeonlike toe proportions and do not take into account the very well-developed grasping abilities of its arms and feet. Indeed, the greatest failing of hypotheses that do not apply a strong climbing component to the urvogel's lifestyle is that they have been unable to explain why it did have such a large reversed hallux. Lacking a positive explanation, they have been forced to try to explain away or ignore the reversed hallux.

As we saw in Chapter 6, long-fingered dino-avepods were well adapted for climbing. Little *Archaeopteryx* was less adapted for running and even better adapted for climbing than the same dino-avepods (except that *Sinornithosaurus* may have equaled the urvogel's abilities, *Microraptor* may have exceeded them, and the new arboreal dino-avepod certainly did). The toes may not have been better proportioned for climbing than those of the pigeon—which tends to fly from branch to branch—but the modern bird does not have grasping fingers to help it scamper among branches. *Archaeopteryx* probably compensated for the absence of a better-developed hallux with the grasping abilities of its fingers. Rather than being a perching bird or a running dinosaur, *Archaeopteryx* probably was a climbing and running avepod. It was scansorial.

That an archaic near-dinosaur such as *Archaeopteryx* was not as well adapted for perching as parrots, crows, songbirds, and so forth is not surprising. The best modern avian analog for *Archaeopteryx* is the famous juvenile hoatzin, which uses the claws on its inner two fingers to clamber about branches quadrupedally (del Hoyo et al. 1992–99). The claws of this peculiar bird are lost as it matures, and perching reverts to the normal bipedal mode. *Archaeopteryx* probably climbed in the bipedal-quadrupedal manner described for similar avepod dinosaurs and early birds, with the exception that its larger arms suggest that the quadrupedal component was stronger than that in tree-loving dinosaurs. Like these dinosaurs, *Archaeopteryx* could not travel head first down a trunk, but it is unlikely to have been the climber of tall tree trunks that Yalden and Feduccia envision anyway. However, in treeless deserts and islands, tall shrubs often contain climbing birds, so the lack of tall trees may not have been a barrier to an arboreal *Archaeopteryx*.

## Swimmer or Wader?

That *Archaeopteryx* remains have been found in the same parts of the lagoon as all of the terrestrial creatures (Barthel et al. 1990, Davis and Briggs 1998) implies that the urvogel was terrestrial rather than marine. However, because well-preserved *Archaeopteryx* skeletons are found in the bottom of lagoons, the possibility that it lived, or at least fed, in the lagoons needs to be considered. A number of otherwise terrestrial animals with unwebbed feet are good swimmers that feed underwater, including some shrews and weasels among mammals (Nowak 1999) and water dippers among birds (Grzimek 1973). Juvenile hoatzins also swim, after dropping into the water to escape danger (del Hoyo et al. 1992–99). Hoatzins and dippers use strokes of their short-span, broad-chord, partly folded wings to propel themselves underwater. The short, broad, strongly muscled wings of *Archaeopteryx* may have formed similar hydrofoils (Fig. 9.2). Its long, feathered tail is similar in proportions to the tails of diving anhingas and may have been an effective rudder.

Thulborn and Hamley (1985) argued that the urvogel was a wader that used its wings to create shade under which fish congregated before they were snapped up, but *Archaeopteryx* lacked the very long legs typical of such avian waders. Nor did it have the elongated toes and the distally placed, elongated hallux that improves traction on soft mud in waders. The wading hypothesis is therefore weak.

## What Did It Eat?

What did *Archaeopteryx* eat? As it happens, none of the fossils found to date has provided any direct evidence in answer to this question: no preserved stomach contents have yet been found (we can hope that this situation will change). There is wide agreement that *Archaeopteryx* was not herbivorous or strongly omnivorous, which leaves invertebrates and vertebrates for it to dine upon. We have already seen (Chapter 6) that *Archaeopteryx* probably did not leap after flying insects, and it was rather large to have been a terrestrial insectivore because it would not have been able to find enough noncolonial insects and spiders to meet its energy needs (Appendix 4).

The large hooked claws on the hands provide some clues regarding *Archaeopteryx*'s diet. The evolution of such claws was not necessary for snatching insects. Could the claws have been used to hold larger prey? We have already discussed why (contrary to Zhou and Martin [1999]) the feathered raptorial fingers could grasp well enough to help grab items such as prey. Holtz (1994b) observed that the creature had many of the adaptations of a miniature dromaeosaur, including the large, laterally flattened hooked claws on the hands and feet, as well as a hyperextendible second toe and a tail with a hyperflexible base. These features suggest that, like dromaeosaurs, *Archaeopteryx* was a leap-and-pounce predator, able to capture large vertebrate prey.

Although *Archaeopteryx* had claws suitable for capturing prey, it lacked the array of serrated, blade-shaped teeth predators typically use to slice their victims' flesh. Its teeth are small, short, and widely spaced. The unserrated crowns are subconical, with recurved tips that are somewhat compressed mediolaterally, a sharp anterior cutting edge, a rounded posterior edge, and a base that is rounded in cross section (Howgate 1984a,b, Wellnhofer 1988, L. Martin 1991, Elzanowski and Wellnhofer 1996). There is a slight or modest constriction between the crown and root of each tooth, and the roots are slightly bulbous. Such teeth are suitable for grabbing and swallowing prey whole or for chopping up large items. In contrast, small insectivorous lizards tend to have larger, more closely spaced teeth with roots that continue straight into the crowns (Hotton 1955), teeth that are suitable for piercing the hard but thin exoskeletons of arthropod prey.

The nature of the urvogel's teeth hints that there may have been an aquatic component to its diet (Thulborn and Hamley 1985, Paul 1988a). With their sharp cutting edge, *Archaeopteryx* teeth are similar to the teeth of marine ichthyornithiform birds. In addition, the constricted waists and large roots of the protobird's teeth are features common to the teeth of numerous aquatic predators, such as crocodilians, some monitors, mososaurs, plesiosaurs, ichthyosaurs, many whales, and most strikingly the diving hesperornithiforms. Small, widely spaced, conical teeth seem to suffice for aquatic predators, which need to hold their prey only momentarily before swallowing it whole. (The claws of *Archaeopteryx* are also compatible with fish catching in that they are similar in shape to those of fish-eating bats [Fig. 9.2].) Bulbous roots may allow teeth to be loose at the base and partly mobile in their sockets, which would improve the ability of the teeth to hold onto struggling, slippery fish and cephalopods while remaining securely attached to the jaws via large amounts of connective tissue. When the prey was too large to swallow intact, the cutting edges could have

*Figure 9.3.* With its large eyes, *Archaeopteryx* probably had excellent, full-color vision.

helped cut it into smaller pieces. However, some seemingly terrestrial birds also had teeth with thick crowns and constricted waists (Hou et al. 1996, L. Martin and Zhou 1997, Zhou and Martin 1999). So did troodonts and a dromaeosaur, except that their teeth were at least partly serrated and much more densely packed.

In summary, the dentition of *Archaeopteryx* was somewhat more generalized than that of most marine predators and could have been used for multiple purposes. The ability to cut up items suggests that *Archaeopteryx* often fed on creatures too large to swallow in one piece, which also tends to favor vertebrates and larger invertebrates over insects, although the latter cannot be completely stricken from the menu.

Hunting larger creatures rather than insects is energetically advantageous since the hunter gains a large amount of energy with each catch. Only ten or so 5-gram prey items, or two 25-gram victims, would satisfy the daily energy needs of an animal with the same size and high metabolic rate of *Archaeopteryx*. Nonvolant creatures—especially disabled fish—are also easier targets than flying insects.

## Sensory and Information-Processing Equipment

*Archaeopteryx* probably had excellent vision. Its subcircular orbits were very large and are nearly filled by sclerotic rings, so the eyeballs were so big that they filled the entire eye socket (Figs. II.2, 9.3). Because the frontal bones above the orbits were triangular and the snout was narrow, the eyes should have faced partly forward, and therefore a substantial field of overlapping vision must have been present.

How well could *Archaeopteryx* smell and hear? There was sufficient room in the rostrum to contain well-developed olfactory conchae, and the possibility that the conchae were ossified (see Chapter 3) implies that *Archaeopteryx*'s sense of smell was at least fairly sensitive. The presence of a well-developed, deep, external auditory meatus suggests hearing was good by archosaur standards (the lack of the external and inner ear complexes enjoyed by mammals limits reptilian and avian hearing in general).

The brain size of *Archaeopteryx* plots well above the reptilian and pterosaur maximums and falls in the lower avian range (App. Fig. 16). The urvogel should have been on the bright side as far as thinking power, which probably approached or equaled that of ratites but was inferior to that of most modern birds.

## Putting It All Together: A Very Tentative Restoration of the Lifestyle of *Archaeopteryx*

In *Archaeopteryx* we have a fairly intelligent, sharp-eyed dino-bird found in lagoons that lapped upon the shores of shrubby desert isles. This dino-bird had teeth somewhat like those of aquatic predators and could run, climb, fly, and perhaps swim fairly but not especially well and was a miniature dromaeosaur in the form of its body, limbs, and tail. What should we make of this disparate set of characteristics?

The similarities between *Archaeopteryx* and hoatzins are instructive in putting together a picture of *Archaeopteryx* lifestyle. Especially pertinent is the fact that hoatzins live in the trees and bushes that sometimes border saline lagoons and that they swim (when young) in those waters (del Hoyo et al. 1992–99). A very suggestive outline emerges when we consider that *Archaeopteryx* was hoatzin-like in form, that it had teeth like those of marine bird and claws like those of fishing bats, and that it is preserved in lagoons. Water-loving habits do not necessarily contradict the arboreal and land-hunting adaptations seen in the urvogel's claws and limbs. In addition to climbing and swimming, baby hoatzins, fishing cats, and martins swim and climb well enough to hunt both in the water and out (Nowak 1999).

*Archaeopteryx* can be envisioned as a semi-arboreal, semiterrestrial, semi-aquatic Jack-of-all-trades, patrolling stream banks and the shoreline, climbing among the branches of dense shrubbery bordering watercourses, hunting for whatever terrestrial and aquatic prey came along. Its mode of arboreality was a combination of quadrupedal scrambling and leaping, assisted by short flights and vertical flutters. On the land, *Archaeopteryx* hunted like a little dromaeosaur, first leaping upon its prey, then holding and killing larger victims with its hand claws and its hyperextendible second toe. It could snatch smaller creatures with its hands or jaws or both. It moved about by means of bipedal striding and flying. It may also have made forays into the dry open spaces. It could have dived into water after aquatic prey, which might have included dead and disabled fish and invertebrates, as well as small terrestrial animals that had dropped into the water. *Archaeopteryx* may even have flown out over the lagoon to dive from low altitude, ambushing and swimming after aquatic prey or picking up the drifting bodies of the unhealthy or expired. *Archaeopteryx* probably captured small aquatic prey in its jaws and large prey with its hand and toe claws. The scarcity of *Archaeopteryx* remains suggests that the urvogel did not move in large flocks but rather was solitary or traveled in small groups. The presence of a well-preserved winged juvenile in the Solnhofen lagoon suggests that flight was achieved before growth was completed. However, the relatively small wings of juveniles (Chapter 7) imply that flight was either limited or not yet present, so it remains possible that the body of the youngster merely drifted into the lagoon.

In the above scenario, scansorial *Archaeopteryx* is a part-time shorebird, an occasional climbing bird, and an archosaurian "cat" at home on the ground, in xeric shrubs, and in the water, with an aerial component added to the mix. However, this scenario is speculative and difficult to test, as are all reconstructions of the lifestyle of the urvogel. *Archaeopteryx* may have spent little or no time in the water. Or it may have avoided traveling on the ground as much as possible. At the other extreme, the climbing adaptations of *Archaeopteryx* may merely represent a heritage from more-arboreal mainland ancestors, a heritage retained but little used in what was a recently evolved desert island isolate. We are not certain whether the habits and habitat of *Archaeopteryx* were typical for protobirds or represent marginal conditions at the edge of their habitat range. Maybe most *Archaeopteryx* did, after all, live in woodlands well to the northwest of the lagoons.

Most importantly, it is very unlikely that *Archaeopteryx* is *the* ancestor of later birds. Much more likely is that *Archaeopteryx* was on one of a number of dead-end branches of avian evolution present in the Late Jurassic. The urvogel's habits may not fully represent the habits of the early birds that were in the mainstream of evolution that produced the majority of Aves.

The result of all these uncertainties is that the lifestyle and habitat of *Archaeopteryx* can tell us only so much about the origins of birds and avian flight. The dinosaurian nature of *Archaeopteryx*'s limbs supports the argument that a running theropodian was among its ancestors, and at the same time, its climbing adaptations bolster arboreal scenarios for the origin of flight. It is unlikely that dino-birds living on arid, treeless islands would be under selective pressure to evolve arboreal abilities; such abilities are more likely to have been inherited. The evidence, at any rate, is not definitive.

PART 5

# WHO IS RELATED TO WHOM, AND WHY?

Building on the data and conclusions presented in Parts 2–3, Part 5 explores the questions surrounding the relationships between early birds and their ancestors. Part 2 detailed aspects of the anatomy of various avepods. In Part 3, we saw that although avian flight probably evolved among climbers, this factor is not necessarily critical to determining the ancestry of birds. In addition, we analyzed the features expected to accompany the loss of flight in the earliest stages of its evolution. In Part 4, we learned that information about the habits and habitat of *Archaeopteryx* is not as useful for understanding bird origins as one might hope. Where, then, can we find the information we need? We cannot turn to genetics, because we lack DNA samples from nonavian dinosaurs. The only source of the abundant data needed to determine which tetrapod group spawned birds is the skeletons of Mesozoic archosaurs, supplemented by other fossil remains and traces, including rare but critical evidence from softer tissues, as well as information on reproductive patterns.

The problems surrounding the phylogeny of bird origins are many and complex, and the idea that flight may have been lost very early in the evolution of birds may have profound implications for how we work out the detailed phylogenetic relationships at the transition between dinosaurs and birds. Because the subjects involved in restoring relationships are tightly interlinked, we must discuss them together. As a result, Part 5 is extensive, and the chapters within it cover a number of topics. Chapter 10 contains a group-by-group examination of potential bird ancestors, which is followed by an analysis of the antidinosaur argument and a defense of the dinosaur-bird link. With that basic question settled, Chapter 11 addresses the major hypothesis of this book, the possibility that some avepod dinosaurs were secondarily flightless and closer to modern birds than *Archaeopteryx*. The chapter also includes a group-by-group analysis of these exceptionally birdlike dinosaurs. Chapter 12 brings together the analyses in the previous chapters to survey and summarize the relationships among predatory dinosaurs and Mesozoic birds. Chapter 12 also includes a brief tabulation of what we currently know and do not know about bird origins. Finally, the issue of what is and what is not a bird is given a quick look.

CHAPTER 10

# Looking for the True Bird Ancestor

So many groups have been proposed as the closest relatives to birds that analyzing the various alternatives is a large task. (The characters used to restore the relationships between birds and other archosaurs are listed in Appendix 1A.) Cladistics can be tricky. For example, some researchers have assumed that a derived character present in any three or more groups is not a valid synapomorphy (Tarsitano 1991), but this is true only for potentially related groups. The presence of a particular derived character in group A and in groups B and C has no phylogenetic relevance if group A is clearly unrelated to group B or C. For example, the possession of erect, bipedal limbs both in kangaroos and in dinosaurs and birds is irrelevant. The methodology I use here presumes that the presence of a birdlike character in a group that is clearly unrelated to birds is not of phylogenetic importance, and I use derived characters only from those taxa that are currently considered relatives of birds. For example, I ignore the presence of birdlike pneumatic vertebrae in pterosaurs and sauropods because neither the flying creatures nor the giant herbivores are plausible bird relatives. The character analysis is therefore limited to longisquamians, drepanosaurs, basal archosaurs, crocodilians, and predatory dinosaurs.

## Fish and Mammals

Birds and mammals share four-chambered hearts, high metabolic rates, and soft insulation. These superficial similarities inspired some early researchers to a favor a close link between these two groups, a link that Richard Owen formalized as the group Haemothermia. However, both the fossil record and the fundamental differences in the detailed anatomy and physiology of birds and mammals have falsified this hypothesis. Indeed, one cannot find two more distantly related amniotes than birds and mammals; all their derived similarities are classic examples of superficial convergence. Nevertheless, a few workers still cling to the belief that haemotherms represent a true clade (Gardiner 1982, Lovtrup 1985). Janvier (1983) even illustrated a projected common ancestor of the warm-blooded tetrapods, a creature that combined mammalian whiskers and external ears with a birdlike, tridactyl foot bearing a reversed hallux! Gauthier (1986) and Gauthier et al. (1988) carefully refuted the mammal-bird link, and *Dinosaurs of the Air* does the same.

The only thing that might be said in favor of the supposed mammal-bird link is that it is superior to Hai's (1993) startling notion that birds evolved from flying fish. How gilled vertebrates could have evolved avian lungs was not explained (Hassenpflug and Kopp 1997).

## Enigmatic Diapsids

It is among the diapsids that the ancestors of birds are to be found, and in recent years, some small Triassic examples have garnered enough support as potential bird relatives for Tarsitano (1991) and Feduccia (1996) to label them avimorphs. All of these supposed bird relatives fail to meet the requirements for that role, in part because not a single synapomorphy uniquely links *Cosesaurus, Longisquama,* and drepanosaurs to birds. Nor do these fossil diapsids show evidence of having been neoflightless. These reptiles are therefore better labeled pseudoavimorphs, although even this term does not sufficiently emphasize how nonavian they were.

### *Cosesaurus*

*Cosesaurus* created a stir when Ellenberger and de Villalta (1974) and Ellenberger (1977) claimed that the one poorly preserved Spanish skeleton—which consists of only an impression and may be a juvenile—was surrounded by feather impressions and had a birdlike skull, a furcula, and a pelvis with a retroverted pubis. Because no one knew whether this lizardlike form was an archosaur or simply related to archosaurs, a nonarchosaur origin for birds was implied. Ellenberger (1977) went so far as to suggest that *Cosesaurus* and birds formed a clade distinct from diapsids and that *Archaeopteryx* was a diapsid that converged with birds! Further examination has shown that the feathers and furcula of *Cosesaurus* are illusory, the supposedly enlarged braincase and pelvis are too crushed to be accurately restored, and the jaws are not really beaklike. The supposedly avian features of the skull—large size, big eyes, and short, tapering snout—are not unique to birds and are commonly observed in many small reptiles and in some dinoavepods, especially juveniles, as well as in birds. There is no evidence that the gait was nonsprawling. Sanz and Lopez-Martinez (1984) concluded that *Cosesaurus* was one of the prolacertiforms, which were archosauromorphs related to the more-derived archosaurs. A bird relationship is no longer considered viable by researchers. It is not even clear whether *Cosesaurus* was an arboreal, a terrestrial, or a semi-aquatic form.

#### *Longisquama*

*Longisquama* is known from Kyrgyzstan via a single, rather poorly preserved skeleton that consists of the front half of the body, including the skull (Pl. 20; Sharov 1970, Haubold and Buffetaut 1987, T. Jones et al. 2000a, 2001; the skeleton is not completely known, as claimed by Feduccia in 1996). Poor preservation of the bones has so far hindered a detailed osteological analysis. *Longisquama* is often considered an archosaur because an antorbital fenestra—the hallmark of the group—may be present, but damage to the skull makes its presence difficult to confirm (see Sues in Stokstad 2000, Unwin and (Benton 2001), despite a photograph presented in Jones et al. (2000). The short neck and conventional arms are not avian. One potentially birdlike feature is the triangular skull with its large orbit, but as we have seen, this inconclusive character continuously attracts the attention of the unwary. Much more interesting is the U-shaped set of clavicles (Fig. 10.10A), but examination of a high-resolution cast of the specimen failed to reveal whether these are a pair of separate clavicles (Sharov 1970, Unwin and Benton 2001) or a true but damaged furcula (Feduccia 1996, Jones et al. 2000a, 2001). Feduccia (1996) cites this quasi-furcula as evidence for some sort of link between *Longisquama* and birds, while at the same time denying that the even more birdlike unified furculae common to predatory dinosaurs are evidence for avian relationships. As discussed in Chapter 5, the possibility that this diapsid possessed feathers, as claimed by its describers, is open to serious question. Equally dubious is the proposition that the dorsal frill is related in some way to the avian mode of flight (Chapter 6). Nothing like this frill of supposed feathers has ever been seen in any bird, even archaic *Archaeopteryx* (see Fig. 5.2). The frill may well be a unique display structure with no flight function. Although *Longisquama* is usually and plausibly assumed to be arboreal, a crucial point to those who see it as a proavian, its arboreality is not certain, because the hands, hindlimbs, and tail are not preserved. Hindlimb posture is correspondingly unknown. *Longisquama* is far from being the ideal bird ancestor its promoters claim it to be (in Stokstad 2000); what is known about it makes it such a poor candidate for a bird relative that it must be rejected as one, unless new and better remains show compelling reasons to think otherwise. At most, it converged with birds in developing a furcula and vaguely featherlike structures with possibly hollow central shafts and asymmetrical vanes.

#### Drepanosaurs

The strange drepanosaurs are archosauromorphs according to Dilkes (1998) and Renesto (2000). (They also see *Megalancosaurus* as a drepanosaur, so the opinions of Calzavara et al. [1980] and Feduccia [1996] that *Megalancosaurus* is a thecodont are obsolete.) Although completely known from a series of skeletons (Pl. 21, Fig. 10.1Aa; Calzavara et al. 1980, Renesto 1994, 2000), the skull is not entirely well preserved. Renesto (2000) noted numerous skull similarities between *Megalancosaurus* and *"Protoavis,"* and what is known of the quadrate of *Megalancosaurus* also appears similar to that of *"Protoavis."* Even more telling are their distinctive and closely matched cervical vertebrae (Fig. 10.7Ba; Downs pers. comm. 1996; Fig. 14 in Renesto 2000). Because it is implausible that such extreme similarity is due to convergence, discussion of the posterior skull and neck of *"Protoavis"* will improve our knowledge of drepanosaur crania. Their grasping hands and feet and prehensile tales indicate that drepanosaurs were highly arboreal (Calzavara et al. 1980, Renesto 1994, 2000, contrary to Padian and Chiappe 1998c, who suggested they were primarily aquatic). Their arboreality has led Tarsitano (1991), Feduccia and Wild (1993), and Feduccia (1996) to view megalancosaurs as bird relatives par excellence. This hypothesis was very weak but not

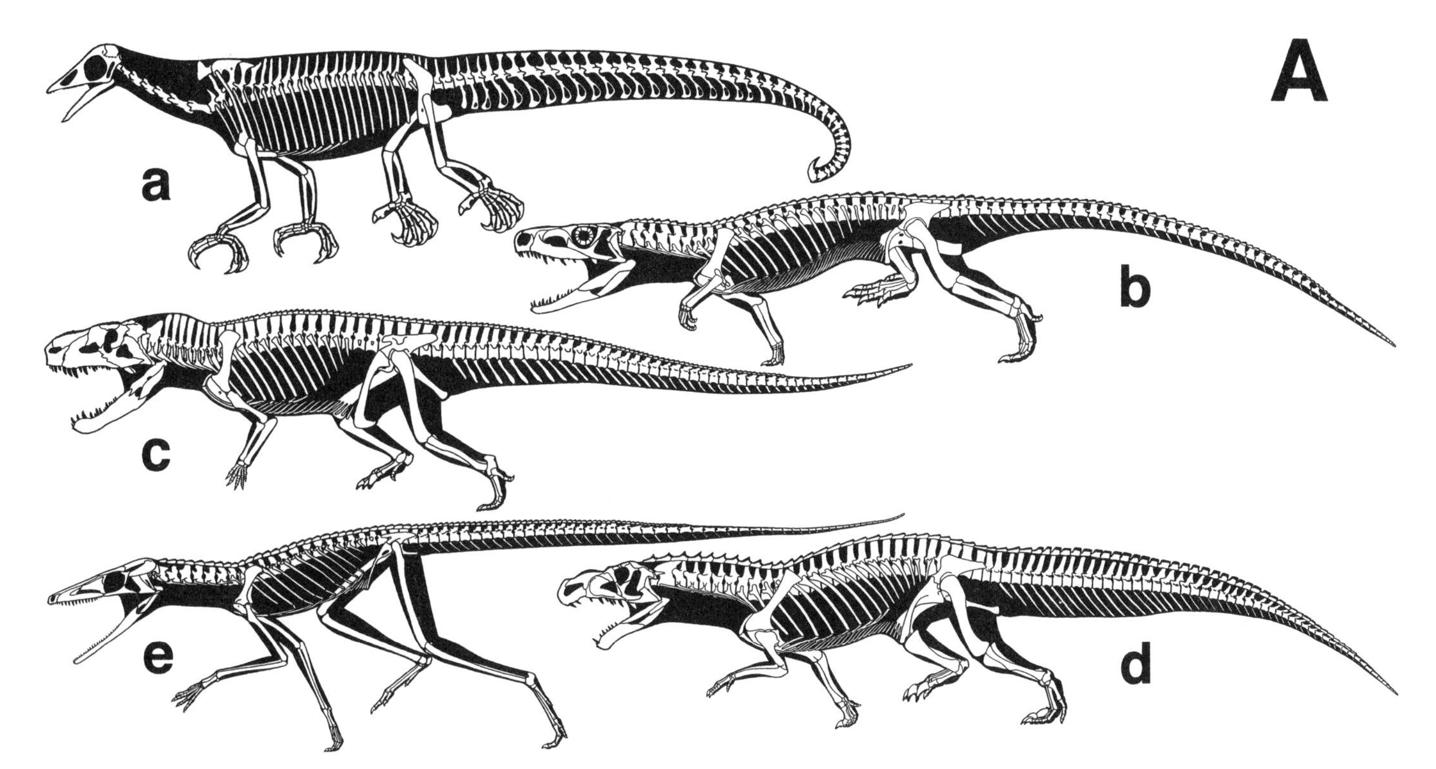


*Figure 10.1.* Skeletons. *A,* nondinosaurs: *a,* drepanosaur *Megalancosaurus; b,* basal archosaur *Euparkeria; c,* poposaur *Postosuchus; d,* ornithosuchian *Riojasuchus; e,* sphenosuchian crocodilian *Terrestrisuchus; B (pages 174–175),* dinosaurs: *a,* protodinosaur *Marasuchus;* baso-theropods: *b, Eoraptor; c, Herrerasaurus; d,* cerato-sauran *Syntarsus;* avetheropods: *e, Allosaurus; f,* tyrannosaur *Gorosaurus,* juvenile; *g, Compsognathus; h, Ornitholestes,* restored with nonavepectoran shoulder girdle; *i, Scipionyx,* hatchling; *j,* ornithomimid *Gallimimus,* juvenile; avepectorans: therizinosaurs: *k,* moderately derived *Alxasaurus; l,* derived, composite; *m, Archaeopteryx; n,* dromaeosaur *Deinonychus; o,* oviraptoromorph *Caudipteryx; p,* oviraptorid (mainly after Pl. 15B); *q,* troodont *Sinornithoides,* juvenile?; *r, Mononykus* with skull of *Shuvuuia; s, Avimimus,* tentative restoration; avebrevicaudans: *t, Confuciusornis; u,* enantiornithine *Cathayornis; v,* neognathe *Messelornis,* soft crest as preserved. Views *m* and *n* include details of occiputs in posterior view (also see Fig. 10.5E,F); left paraoccipital process *(shaded)* in left, side, center, posterior, bottom, and ventral views; pubic peduncles of the ilium (also see Fig. 10.12D); and cross sections of the pubes. Small arrows point to avepod synapomorphies as explained in the text and in Appendix 1, and larger arrows point to synapomorphies of dromaeoavemorphs and in some cases troodonts. C *(page 175), Allosaurus* left paraoccipital process for comparison with same views of this process in *Bm* and *Bn.* Not drawn to same scale.

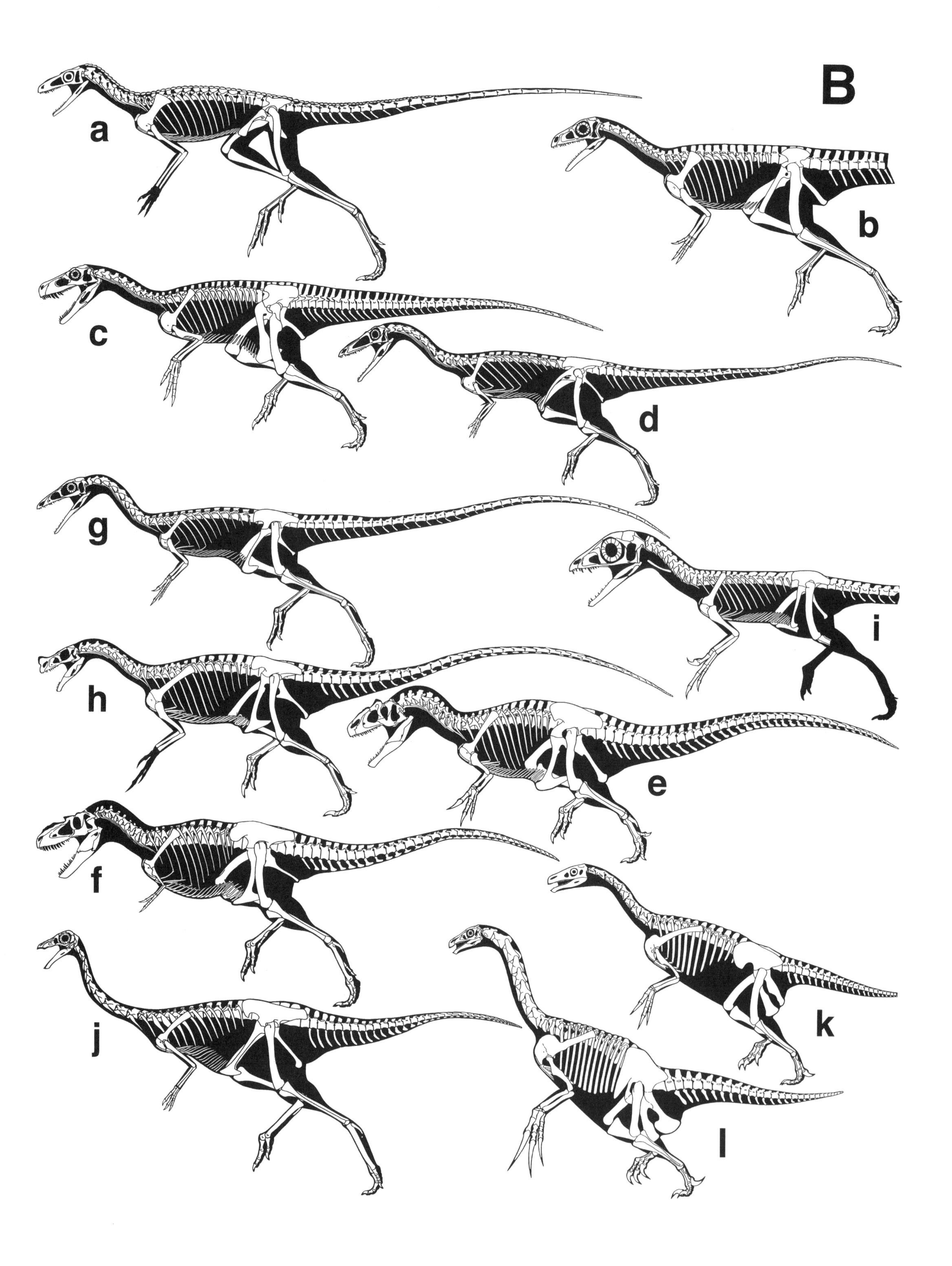
B
a
b
c
d
g
i
h
e
f
k
j
l

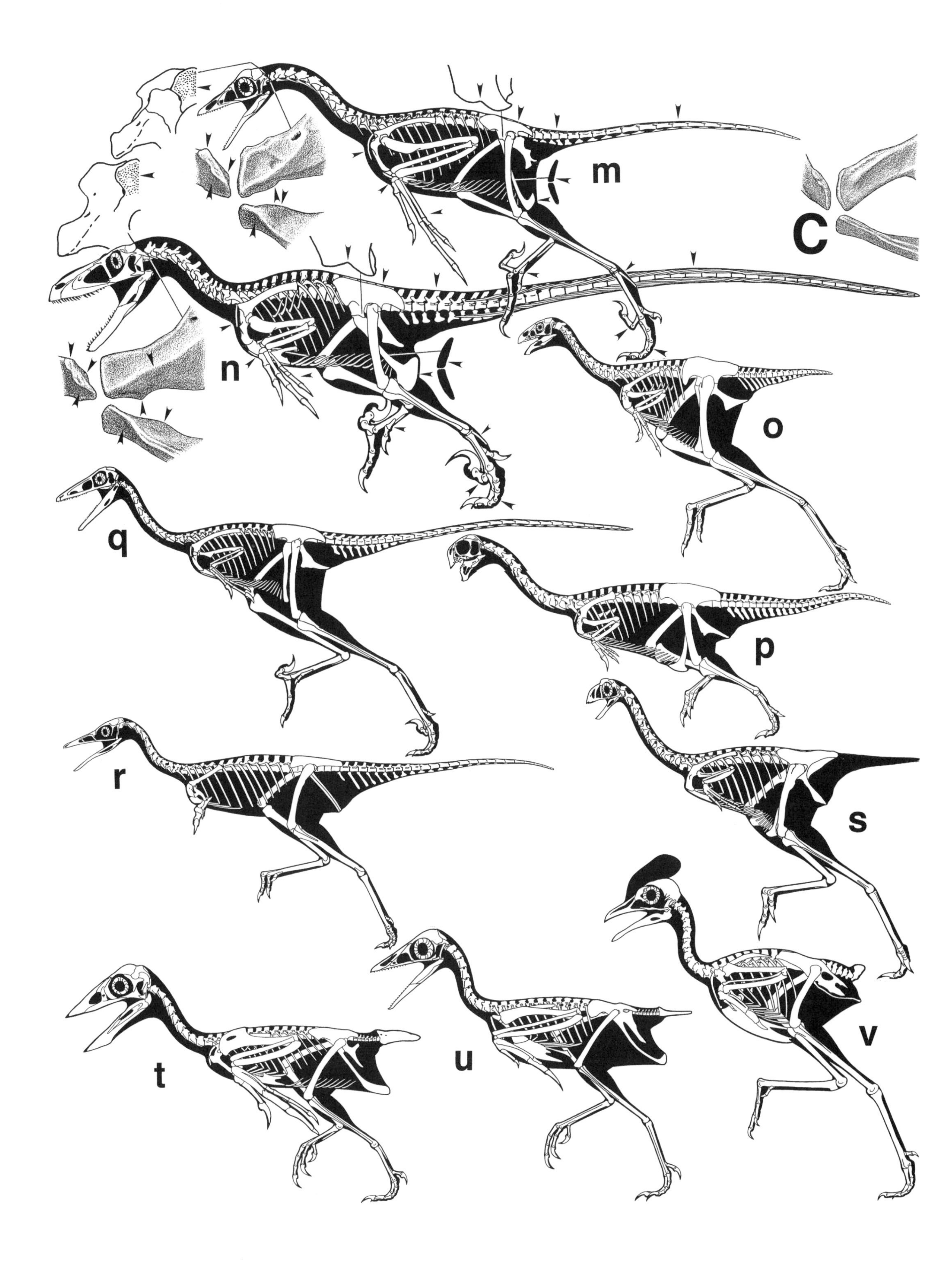
m
C
n
o
q
p
r
s
v
t
u

entirely unreasonable when only the front half of the creature was known. However, the hypothesis has become much less tenable since Renesto (1994, 2000) revealed the animal to have been a very unusual, chameleon mimic.

The drepanosaurs represent another example in which large eyes and short triangular skull have been mistaken as avian. The posterior elongation of the external nares is birdlike, but such elongation is also seen in some avepod dinosaurs. There is no evidence that *Megalancosaurus* had the true "beak" Feduccia claimed it had, and the severe reduction of its conical teeth is matched in some dino-avepods but not in *Archaeopteryx*. *Megalancosaurus* is also unlike the urvogel in that the frontal in the roof of the skull is anteriorly very narrow and posteriorly winglike.

The two supposedly avian quadrates with orbital processes are poorly preserved in *"Protoavis"* and differ significantly from one another. Chatterjee (1999a) restored the jugal of *"Protoavis"* as shallow and the quadratojugal as very short. However, the jugal appears to be significantly deeper in photographs, and the quadratojugal much longer. The *"Protoavis"* occiput retained a posttemporal fenestra (Fig. 10.5B), a feature lost in all avepods. Similarities of the paraoccipital processes between the two drepanosaurs and *Archaeopteryx* are superficial, and drepanosaurs lacked the distinctive paraoccipital twist seen in the urvogel and in dromaeosaurs. The braincase and middle ear of *"Protoavis"* were similar to those of avepods, and in some ways, they seem close to those of much later forms. For example, Currie and Zhao (1993b) noted similarities to troodonts. Chatterjee (1997, 1999a) stated that the braincase was like that of *Archaeopteryx*. The presence of an opening to the pneumatic sinus on the paraoccipital, and a diamond-shaped dorsal supraoccipital, were distinctive features otherwise limited to the urvogel and dromaeosaurs (Currie 1995). Currie (1995) concluded that the *"Protoavis"* braincase shows few avian characters that are not also found in various theropods.

Drepanosaur cervicals had shallow neural spines and hypasphenes like those of birds, but the drepanosaurs' long, narrow-waisted, laterally L-shaped anterior cervicals were rather unbirdlike. Chatterjee pointed to the neck vertebrae articulations as one of the most birdlike features of *"Protoavis"* and, by implication, of other drepanosaurs as well (Fig. 10.7Ba). The anterior centra facets were very broad and shallow; the posterior facets much narrower and deeper. This condition mimics the saddle-shaped condition found in birds. However, rather than enhancing overall neck flexibility, the drepanosaur arrangement may have evolved as part of a fore-and-aft snap-action head-neck-shoulder system that, like a chameleon's tongue, was used for capturing prey. The drepanosaur system included enlarged, distally expanded shoulder spines that formed a tall withers and that were adorned with an extra supraneural element; none of these features is seen in any bird. Initial claims that a furcula was present in drepanosaurs have not borne out. The slenderness of the drepanosaur scapula was matched in some dino-avepods. Numerous skeletons show that the scapula blade was directed up and a little forward; only the specimen with the least complete shoulder region shows the scapula blade horizontal in the manner of dino-avepectorans and birds (Fig. 5A–C in Renesto 1994; Renesto pers. comm.).

Tarsitano's claim that *Megalancosaurus* had a semilunate carpal block was not confirmed by Renesto. Drepanosaur hands and feet do not resemble those of birds in any way. Quite the opposite is true: the drepanosaurs' elongated proximal carpals (which converge with those of crocodilians) and opposable digits could hardly have been more unbirdlike. In the hand, the outer two digits were divergent from the rest; chameleons have a similar arrangement, but in birds the thumb is divergent (Fig. 10.11Aa). In drepanosaur feet, the opposable digit is I, but the club-tipped appendage is otherwise entirely unlike the avian hallux (Fig. 10.14Fa). Metatarsal I was not reduced and was as divergent as the rest of the digit. The severe reduction of elements in the central fingers and toes III and IV was strongly divergent from the initial avian pattern. Drepanosaur limb posture is not certain. The tail vertebra form a spike at the end of a distally long, supple, and distally downcurved prehensile tail that in no way constitutes a suitable ancestral type for the stiffened, reduced, and straight distal tail of birds.

*Cosesaurus* and *Longisquama* are so poorly preserved that they are somewhat difficult to analyze. Not so drepanosaurs, and in most respects, their well-preserved skeletons could hardly diverge more from the avian pattern. The similarities shared by drepanosaurs and birds—which are largely limited to the head and neck—are not close enough to overcome the massive dissimilarities; and these similarities are often absent in the most basal birds. Drepanosaurs and birds share only 15 derived characters, and all the characters are also found in dino-theropods. The similarities are best attributed to a moderate degree of convergence, moderate because of the disparate functional requirements (snap-action predation versus pecking action; Renesto 2000). Because drepanosaurs make such implausible bird relations, it is disturbing that in 1996 Feduccia not only continued to press claims that *Megalancosaurus* was a birdlike avimorph but also failed to reproduce the new skeletal remains and skeletal restoration that two years earlier had shown how extremely unbirdlike the animal really was. If this is the best that those who advocate nondinosaurian origins of birds can produce, then their hypothesis is truly flawed.

## "Protoavis"

It is difficult to know where to discuss the set of bones assigned to "*Protoavis*"—which has been offered as evidence that birds originated in the Triassic (Chatterjee 1991, 1995, 1997, 1998a, 1999a, Kurochkin 1995)—since they probably belonged to more than one potential bird ancestor. Because part of the skull and the neck appear to be those of a drepanosaur, I have analyzed these features in the section on drepanosaurs above. The rest of the remains consists of the largely disarticulated elements of two or more small individuals. Preservation of the bones ranges from good to poor. Chatterjee insists that the bones belong to just two specimens from the same taxon and cites nonduplication of parts in support of this view, but the lack of articulation leaves open the possibility that the bones represent an assortment of animals. The scapula and coracoid are so small relative to the vertebrae (Fig. 7 in Chatterjee 1991, Pl. 2 in Chatterjee 1999a) that their association with the rest of the skeleton must be challenged, even if we assume that these bones belong to a juvenile. The identity of some bones remains in dispute; the rather simple depictions of some bones do not closely match the photographs; and older and newer depictions of the same bones sometimes differ from one another. The strong similarity of some "*Protoavis*" elements to dino-avepod elements, on one hand, and drepanosaur elements, on the other, affirms the presence of at least two taxa. At this time, the identity of the "*Protoavis*" bones is so unclear that unified skull and skeletal restorations are unjustified and probably misleading. In addition, the taxonomic status of the name "*Protoavis*" is open to question because the type specimen may include parts of different taxa. Parts of the referred specimen may or not belong to the same taxon as the type.

I saw the "*Protoavis*" fossils not long after their discovery, and although an exhaustive consideration of "*Protoavis*" is beyond the scope of this study, some observations on the nondrepanosaur material are in order. (Keep in mind that some of the following bones may in fact be drepanosaurian.) The maxilla appears to have an antorbital fenestra but lacks the complex auxiliary sinuses characteristic of averostran dinosaurs and early birds. Although the presacral vertebrae are thin walled, it is not clear whether this represents the nonpneumatic condition observed in basotheropods or marks the presence of extensive pulmonary diverticula. The scapula blade is much broader than that of birds. There is a strutlike element that does look rather like the coracoid of a flying bird; but this element is poorly preserved, and its identity has not been confirmed by others. The supposed hypocledium of the furcula is too fragmentary to be identified as such with certainty. The same is true of the bones identified as sternal elements. Although the distal carpals are very large, the wrist and hand are otherwise entirely nonavian. The hand is short and broad, all four metacarpals are long, and the fingers are short and stubby. The supposed quill nodes on the ulna and metacarpals are large, ill-formed bumps—not at all like the subtle feather anchors of birds, including the dino-bird *Rahonavis* (Chapter 5)—that appear to be the result of irregular crushing of the bones. The thumb was not medially divergent in the dinosaurian-avian manner. The pubis appears to have been retroverted (Fig. 10.12Bd). The astragalus is that of a basal theropod or avepod; in particular, the short ascending process contrasts with the tall process and other features observed in more birdlike dinosaurs and basal birds. Although "*Protoavis*" is restored with a fully reversed hallux, whether this restoration is confirmed by fully articulated elements is not clear.

What, then, is to be made of these Triassic bones? Chatterjee's conclusion that "*Protoavis*" was a flier whose forelimb functioned as a wing is implausible. His conclusion cannot be verified because the pectoral elements are either uncertain in identity or too fragmentary. In the smaller "*Protoavis*" specimen, the supposed scapula blade and coracoid are each a mere two vertebrae long. In all flying birds, even the early ones, the scapula and the coracoid are at least four dorsal vertebrae in length. The supposed fragment of a scapula associated with the bigger specimen appears to be relatively larger, but this bone is too fragmentary for unambiguous identification or restoration of its dimensions. Therefore, the very large shoulder girdle restored by Chatterjee (Fig. 22 in 1999a) cannot be verified. Quite to the contrary, the girdle looks too small to have anchored even a minimal set of flight muscles (see Appendices 2 and 4). In addition, the arm is not very long, and the short, stout, heavy hand is the opposite of aerodynamic. Thus there is no evidence of a light, streamlined wing. The lack of any aerodynamic shaping of the hand also argues against the possibility that *"Protoavis"* was a neoflightless archosaur with a secondarily reduced forelimb.

The dinosaurian nature of some *"Protoavis"* elements and the birdlike form of others have caused Chatterjee to conclude that it was close to the link between dinosaurs and birds and that it was closer to modern birds than *Archaeopteryx* was. These conclusions cannot be verified, because the supposedly birdlike characters are not well preserved and are often found on the drepanosaur elements. In addition, *Archaeopteryx* and other early birds shared major features not seen in the Triassic animal, including auxiliary maxillary sinuses, a semilunate carpal block, slender tridigitate hands with long fingers, and the urvogel-type pelvis typical of Late Jurassic–Early Cretaceous dromaeo-avemorphs and birds. Some of these features exclude *"Protoavis"* from Averostra and Avetheropoda as well as from Aves. Chatterjee's placement (1995, 1999a) of *"Protoavis,"* with its long tail and its four palm bones, well into Pygostylia and even Ornithothoraces—above the likes of *Confuciusornis, Sinornis,* and *Iberomesornis*—was extreme and untenable (Chatterjee himself rejected this placement in 1997). The many problems with *"Protoavis"* have led most researchers to express varying degrees of skepticism as to its avian or near-avian status (Paul 1988a, Witmer 1991, Chiappe 1995, Currie 1995, Feduccia 1996, Ostrom 1996, Padian and Chiappe 1998c, Sanz et al. 1998). These researchers consider it likely that the remains represent one or more creatures that independently evolved some avepectoran features. In 1988 (Paul 1988a), I tentatively suggested that *"Protoavis"* was a semi-arboreal baso-theropod that mimicked later avepectorans in various regards, in an extreme expression of the previously described tendency of baso-theropods, such as herrerasaurs, to parallel avians in development. My hypothesis is weakened by the fact that the most birdlike parts of *"Protoavis"* have been assigned to drepanosaurs. Similarly tenable is Feduccia's (1996) idea that at least some parts of the *"Protoavis"* specimens are those of a thecodont that converged with birds. In any case, *"Protoavis"* does not pose a serious challenge to the hypothesis that birds are the product of Jurassic rather than Triassic dinosaurs.

## *Scleromochlus* and Pterosaurs

Pterosaurs possess an antorbital fenestra and are therefore generally considered archosaurs. Their birdlike shoulder girdles and ankles encouraged some early researchers to conclude that these aerial diapsids were close relatives of birds. It is now universally agreed that the great dissimilarities in the wings and other parts of the skeleton show that there is no close link between these two groups; the flight adaptation similarities that do exist are merely examples of parallelism among diapsids. Pterosaurs were traditionally viewed as one of the many archosaur groups that independently evolved from basal archosaurs. In recent years, a number of workers have concluded that the pterosaurs are close relatives of dinosaurs, with which they form Ornithodira (Padian 1984, Paul 1984a, Gauthier 1986, Sereno 1991, Novas 1996). Most of the similarities that link the two groups are in the hindlegs. Larson (1997) argued that the shared presence of pneumatic skeletal elements supports the monophyly of pterosaurs, theropods, and sauropods, but the absence of skeletal pneumaticity in protodinosaurs and basal dinosaurs negates this link. At the other extreme, Wild (1984) argued that pterosaurs were not even archosaurs but rather an independent diapsid group that paralleled archosaurs in developing the preorbital opening. Wellnhofer (1991) tentatively supported this view. In the most extensive study of the issue to date, S. Bennett (1996) concluded that pterosaurs were a sister group to the rest of the archosaurs, in which case the similarities between the dinosaur and pterosaur legs resulted from convergence.

Like the various hypotheses concerning bird origins, hypotheses about pterosaur origins often rely on supposed links between lifestyle and ancestry. A close link between pterosaurs and dinosaurs is tied to a terrestrial origin for archosaur flight, and nondinosaurian origins are tied to arboreal habits in both pterosaurs and birds. These connections may be equally invalid for pterosaurs and for birds, and an arboreal beginning to pterosaur flight is not incompatible with the flexible locomotory adaptations of protodinosaurs.

Pterosaur origins remain obscure because there are few suitable ancestral types in the fossil record. In this regard, pterosaurs are more like bats, whose early flying ancestors are as yet unknown, than like birds. The exception to this gap is thought by some to be Triassic *Scleromochlus,* a small Scottish diapsid known from a small set of skeletal impressions (Woodword 1907, Benton 1999). Unfortunately, the impressions are often vague and amorphous, so much so that interpretations based upon them are questionable. A large antorbital fenestra seems to be present. Whether *Scleromochlus* was arboreal or terrestrial, erect limbed or sprawling, is not clear. Some derived features shared with pterosaurs have led Padian (1984) and Sereno (1991) to conclude that *Scleromochlus* is the closest known relative of pterosaurs. A problem with this view is that *Scleromochlus*'s arms, and especially the hands, are quite short (App. Fig. 18C), whereas the arm and the wing finger are extremely elongated in pterosaurs (Wellnhofer 1991). It remains possible that *Scleromochlus* is a member of a short-armed group closely related to the longer-armed ancestors of pterosaurs. However, S. Bennett (1996) placed scleromochlids as a sister group to dinosaurs, only distantly related to pterosaurs, whereas Benton (1999) considered scleromochlids to be basal archosaurs, a sister taxon to both pterosaurs and dinosaurs. In any case, there is no evidence that this weak-armed creature was closely related to birds, as was suggested by L. Martin (1983a).

## *Euparkeria*

Aside from *Archaeopteryx,* no other extinct creature played a greater role in the debate on bird origins for much of the twentieth century than *Euparkeria,* a rather small, lizardlike predator from the Early Triassic of South Africa. A set of specimens makes our knowledge of the skull and skeleton almost complete. Unfortunately, the mediocre illustrations in the primary modern description by Ewer (1965) have left many of the details of this important fossil still inadequately documented after all these years. *Euparkeria* is the classic, generalized, basal archosaur (some cladistic classifications have placed among the Archosauriforms [Sereno 1991], but I follow the traditional arrangement, as do Gower and Weber [1998]). Until the sixties, this basal archosaur was the central character in the "pseudosuchian" hypothesis of bird origins.

*Euparkeria* is a very primitive archosaur that was not as gracile as it has sometimes been presented (Figs. 10.1Ab, 10.2A). Because the forelimbs were rather short and the legs semi-erect (Fig. 10.13Da; Parrish 1986), it is often shown in a bipedal pose. However, its very large head made it so front heavy that the normal gait was probably quadrupedal and similar to the gait of "high walking" and running juvenile crocodilians, which also have short forelimbs (Paul 1987a). Although *Euparkeria* was probably primarily terrestrial, opposable digits on its hands and feet suggest a well-developed ability to climb.

*Euparkeria* is so generalized and primitive that it does not share any derived skeletal characters with birds, and the skull is largely if not entirely nonavian as well (Figs. 10.2A, 10.3). Welman (1995) cited similarities in the braincase of *Euparkeria* and *Archaeopteryx* as evidence for a close relationship between the two. Gower and Weber (1998) refuted the bulk of Welman's argument by showing that the characters Welman cited are generally primitive for Archosauria and that the braincases of the two archosaurs are, in fact, not especially similar (an *Archaeopteryx* element that Welman identified as part of the braincase may not actually belong there; see Chapter 3). Instead, the urvogel's braincase is generally more similar to that of dino-avepods than to the much more primitive complex of *Euparkeria* (Fig. 10.5). For instance, no one has shown that *Euparkeria* had the lateral depression in the prootic that characterizes avepods avian and nonavian. In a few respects, *Euparkeria* is not so generalized: its massive palate, its reduced ectopterygoid, and some other related features are not typical of the more open palates of most other archosaurs (Paul 1984a). *Euparkeria* was also like other thecodonts, and unlike birds, in that its outer fingers and toes, rather than the inner ones, were divergent and opposable (Figs. 10.11Ab, 10.14Fb).

Few dispute that *Euparkeria* is a suitable ancestral type for birds. But it is suitable only in the same sense that an early Cenozoic, primitive, generalized primate is a suitable ancestral type for humans. In addition, there is no evidence of flight-related adaptations. *Euparkeria* is important to bird origins only in that it represents the earliest Mesozoic and most primitive archosaur stem stock from which the entire dinosaur-bird clade later evolved. For that matter, *Euparkeria* is a good ancestral type for virtually all archosaurs.

## Gracile Protocrocodilians

At first, the notion that massive, armored, semi-erect, water-loving gators and crocs may be close relatives of lightly built birds seems counter-intuitive. However, in the Triassic and Jurassic,

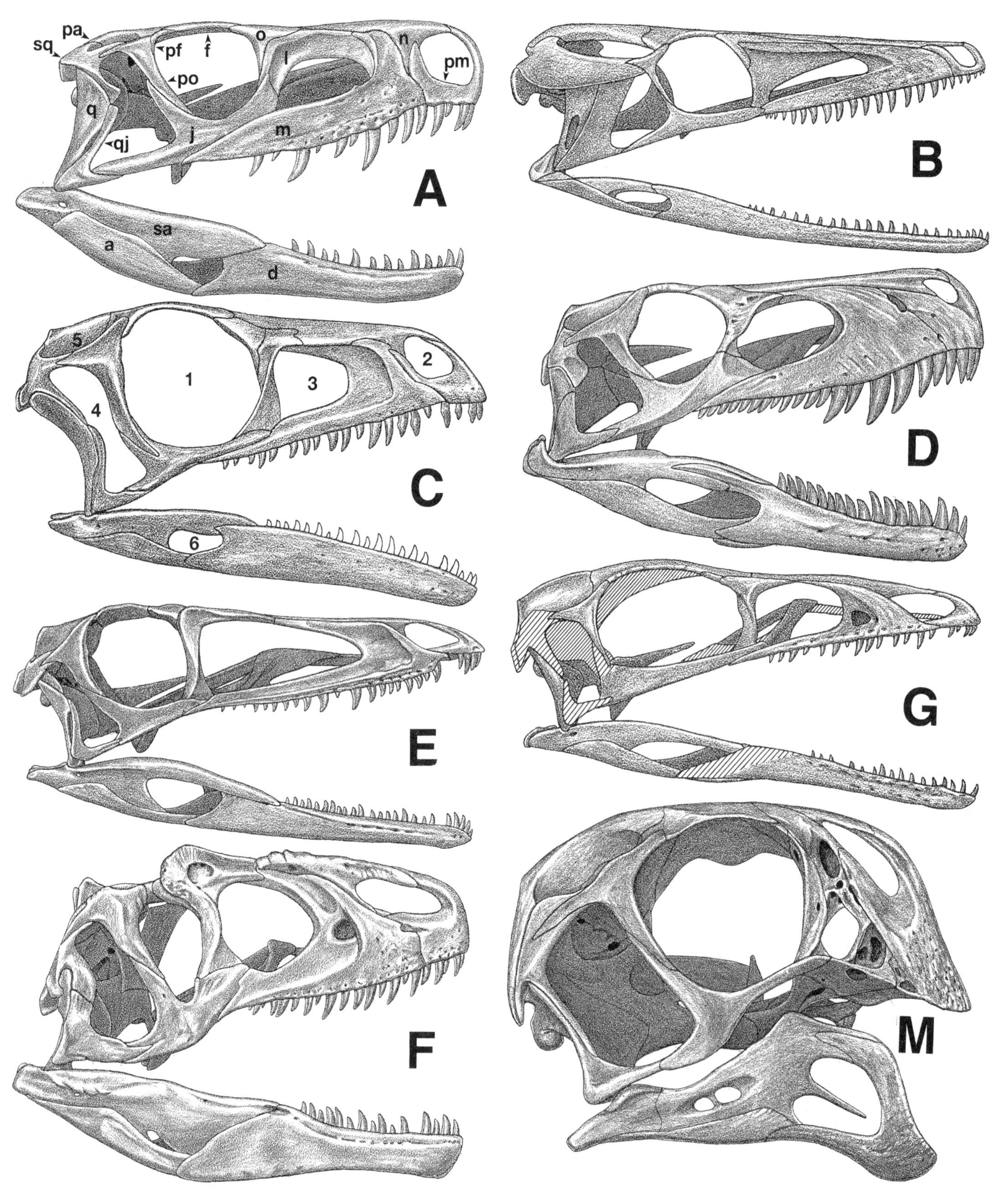


*Figure 10.2.* Skulls. *A,* basal archosaur *Euparkeria; B,* sphenosuchian protocrocodilian *Terrestrisuchus;* theropods: baso-theropods: *C, Eoraptor; D, Herrerasaurus;* avepods: *E,* cerato-sauran *Syntarsus;* averostrans: *F, Allosaurus; G, Compsognathus; H,* ornithomimid *Gallimimus;* avepectorans: *I,* therizinosaur *Erlikosaurus; J,* dromaeosaur *Velociraptor; K, Archaeopteryx; L,* troodont *Saurornithoides* with posterior mandible after *Sinornithoides; M,* oviraptorid *Ingenia; N,* alvarezsaur *Shuvuuia;* avebrevicaudans:

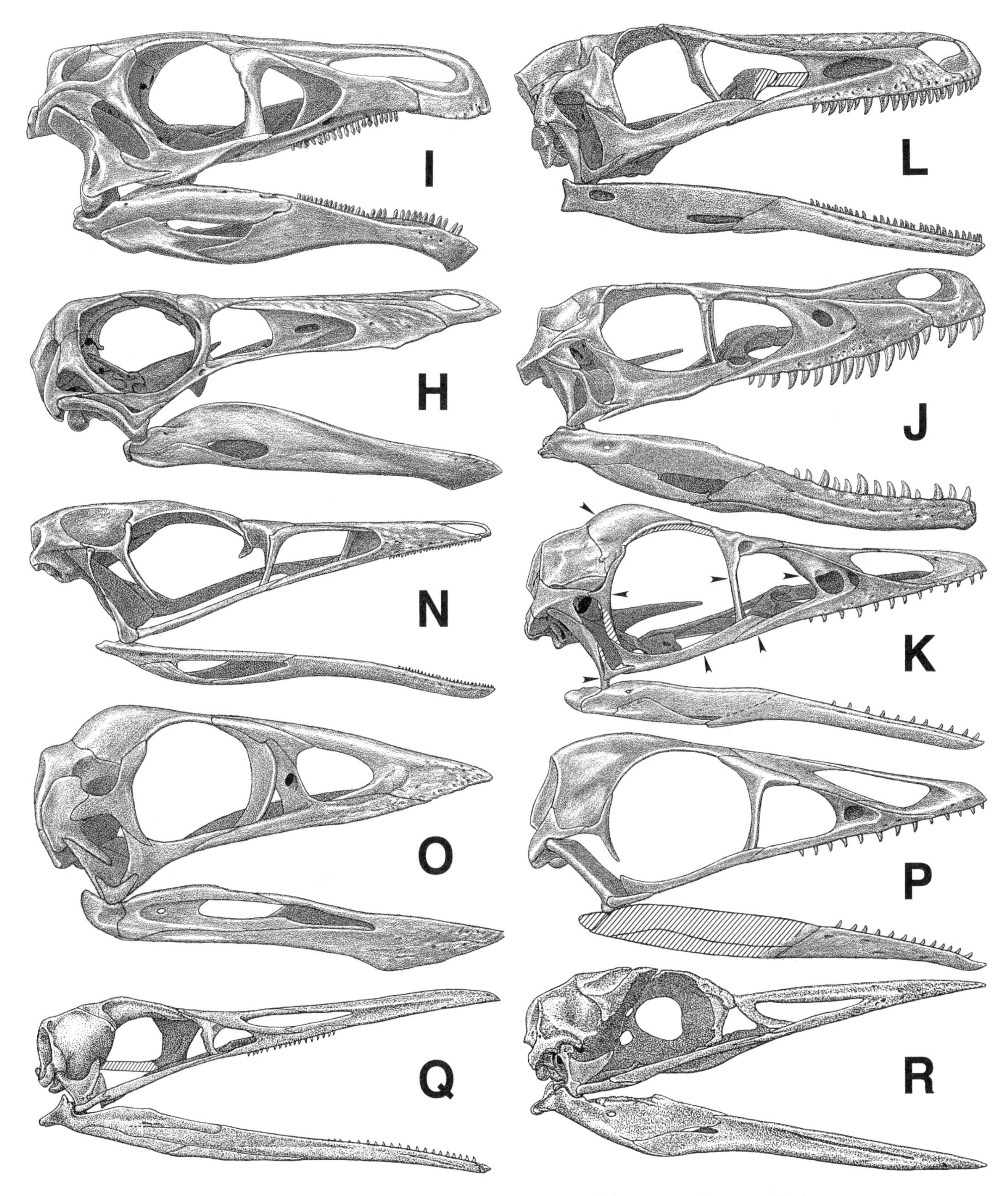


*O, Confuciusornis; P, Cathayornis; Q, Hesperornis; R,* loon. Abbreviations are listed in Appendix 6. *1,* orbit, *2,* external nares, *3,* antorbital fenestra set within antorbital fossa, *4,* lateral temporal fenestra, *5,* superior temporal fenestra, *6,* intramandibular fenestra. Arrows in *K* point to avepod synapomorphies, as explained in the text and in Appendix 1. Drawn to same length.

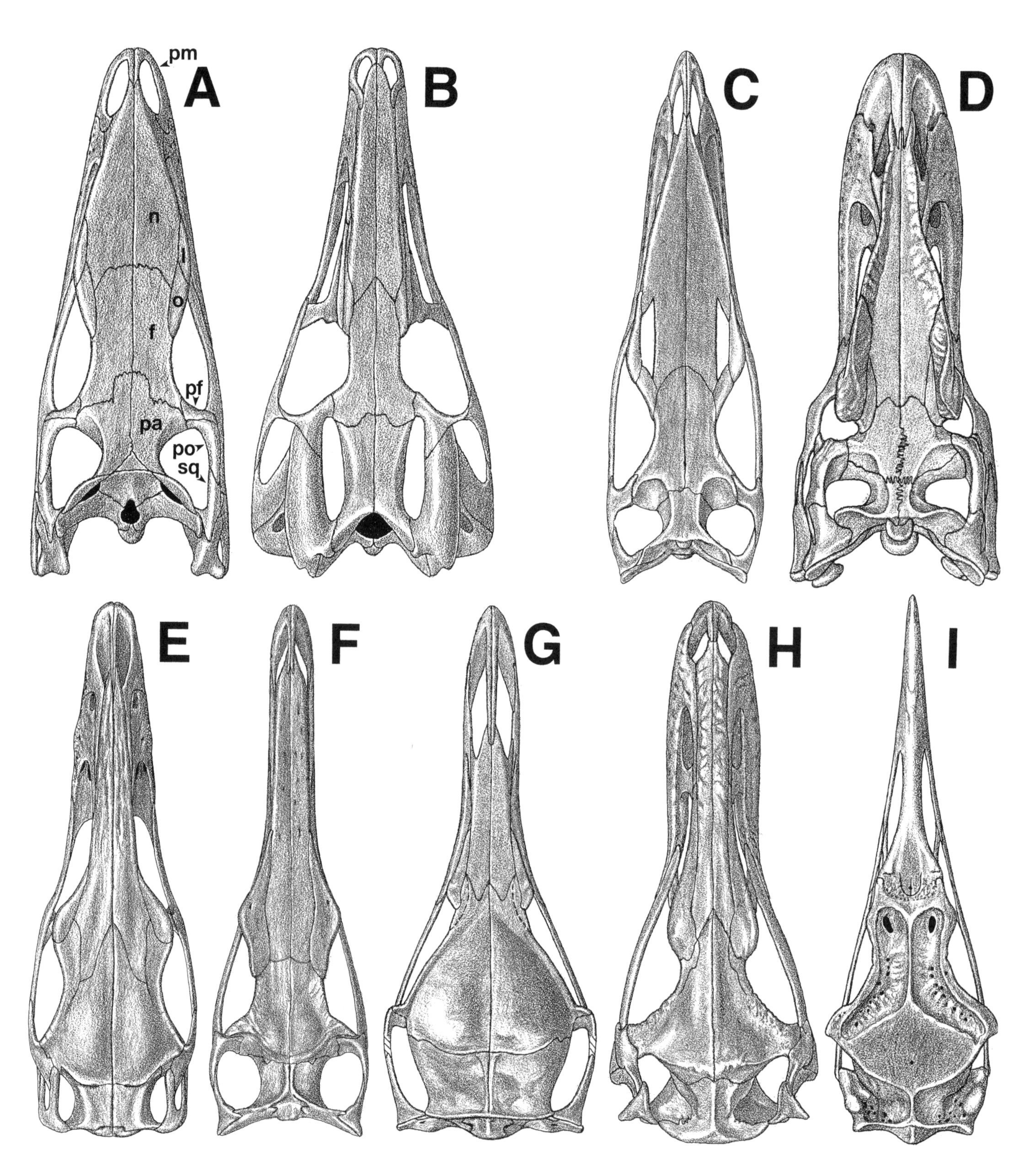


*Figure 10.3.* Skulls in top view. *A,* basal archosaur *Euparkeria; B,* sphenosuchian protocrocodilian *Terrestrisuchus;* theropods: avepods: *C,* cerato-sauran *Syntarsus;* averostrans: *D, Allosaurus; E, Ornithomimus;* avepectorans: *F,* dromaeosaur *Velociraptor; G, Archaeopteryx; H,* troodont *Saurornithoides; I,* loon. Abbreviations are listed in Appendix 7. Drawn to same length.

basal sphenosuchian "protocrocodilians" were small, gracile forms with long, erect, digitigrade limbs and medially open hip sockets (Figs. 10.1Ae, 10.2B). A. Walker (1972), Whetstone and Martin (1979), L. Martin et al. (1980), L. Martin (1983a,b, 1991), and Zhou and Martin (1999) have concluded that birds probably share a common ancestor with crocodilians that is distinct from the ancestor of other archosaurs. In 1984 (Paul 1984a), I proposed that crocodilians were close relatives of dinosaurs.

A. Walker (1972) suggested that protocrocodilians were well adapted for climbing and that they were therefore suitable ancestors for arboreal protobirds. Juvenile modern crocodilians do show some climbing abilities (Cott 1961). If anything, the gracile sphenosuchians should have been better climbers than are their living aquatic brethren. On the other hand, protocrocodilians lacked the opposable toe seen in basal archosaurs, and their digits and claws were much shorter and less suited for grasping than those of theropods, both nonavian and avian. Therefore, protocrocodilians can be ranked as inferior climbers. The gracile and erect-limbed sphenosuchians must have been good runners. However, the shoulder glenoid faced semilaterally, the hip socket was only partly medially open and was not deep, and the spherical femoral head was only partly turned inward. These features indicate that protocrocodilians easily adopted a semisprawling posture, something no avepod can do. These and other features—in combination with the reduction of arboreal adaptations; the beginnings of nonvaulted, crocodilian-type secondary palates set within skulls that tended to be flatter than those of other predaceous terrestrial archosaurs; and the persistent tendency of crocodilians to adopt aquatic habits—imply that at least some sphenosuchians were frequently aquatic, albeit less so than more derived crocodilians. Although archaic birds may have taken to the water too (Chapter 9), their avian style of swimming followed the stiff, avepodian body pattern, a pattern quite different from sculling pattern characteristic of crocodilians with their flexible bodies and tails.

Why, then, have crocodilians been proposed as bird relatives? The answer has to do with quadrate similarities. In most archosaurs, the quadrate, which supports the lower jaw at its vertical end, has at its dorsal end a single head that articulates solely with the squamosal in the posterodorsal corner of the skull (Chapter 3). However, birds share with the crocodilian clade a double-headed quadrate that articulates with braincase elements. This similarity in the suspensorium helped inspire the crocodile-bird hypothesis, which was encouraged by unwarranted speculation that *Archaeopteryx* had the same condition. Ironically, it was A. Walker (1985) who falsified this link by positively identifying an *Archaeopteryx* element as a single-headed quadrate (Chapter 3) and by showing that the croc-bird similarity represents convergence. In fact, the convergence is actually divergent in that the similarities developed for different purposes. In crocodilians, the suspensorium is a massive structure, and the double-headed and immobile quadrate strengthens the support of the heavy jaw. The dorsal end of the long quadrate extends into a typically small reptilian braincase. The quadratojugal is massive and often reaches dorsally to the superior temporal bar. In birds, however, the suspensorium is delicate, and the quadratojugal is very small. In this case, the quadrate is double headed in order to increase fore-and-aft mobility, and the enlarging braincase overrides the head of the short and vertical or procumbent quadrate. The systems could hardly be more different.

A number of additional skull characters have been offered as demonstrating a link between crocodilians and birds. These include highly pneumatic braincase elements and quadrates, a fenestra pseudorotunda in the middle ear, and teeth that have constricted waists and whose replacements are set within pits in the roots. However, because *Archaeopteryx* lacks a pneumatic quadrate, these skull similarities are again convergent. In addition, Ostrom (1991), Elzanowski and Wellnhofer (1996), and Gower and Weber (1998) have shown that some of the dental similarities have been exaggerated or are probably the result of convergence. As we saw in Chapter 9, teeth of this form are common in aquatic predators, and short roots may force the placement of replacement teeth in pits. Besides, all of the skull characters discussed above are also found in dino-avepods (see later in this chapter). Crocodilians, especially derived examples, share with birds an elongated coracoid that is strongly flexed relative to the blade of the scapula (Fig. 10.8C,D, App. Fig. 18A). But the long crocodilian coracoid is part of a clavicle-free arrangement that enhances shoulder mobility (Fig. 4.2Ac), an arrangement that is quite different from the avepectoran system, in which the large, flexed coracoid is part of a furcula-dominated system that increases the rigidity of the shoulder girdle.

A. Walker's (1972) argument that crocodilians evolved a birdlike arm-folding mechanism has never been accepted, and crocodilians lack the

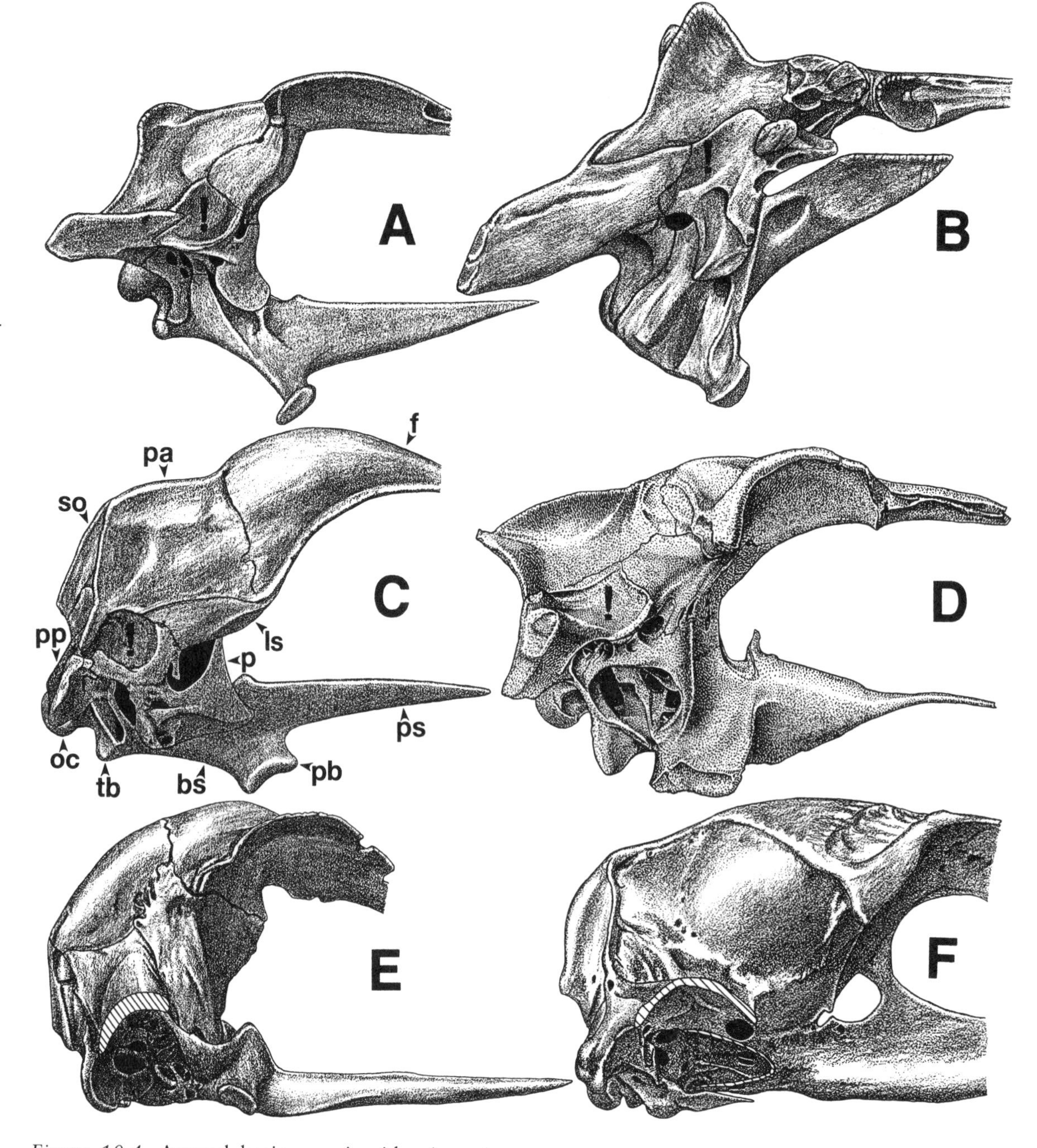


*Figure 10.4.* Avepod braincases in side view. *A,* cerato-sauran *Syntarsus; B,* large averostran *Allosaurus; C, Archaeopteryx,* mainly BMNH 37001 with ventral elements from BSP 1999, occipital condyle restored; *D, Troodon* (modified from Currie and Zhao 1993b); *E,* juvenile ostrich (hatched area indicates bone that has been cut away for improved view); *F,* loon. Abbreviations are listed in Appendix 7; ! = prootic depression. Not drawn to same scale.

pulleylike wrist bones so characteristic of birds. His suggestion that the Jurassic sphenosuchian *Hallopus* had a retroverted pubis cannot be confirmed because the puboischial base for the pubis is not preserved, and the pubis was probably procumbent, as in all other crocodilians.

In addition, the suggestion that an early sphenosuchian (Bonaparte 1984, Paul 1984a) possessed a dinosaur-like ankle has been shown to be incorrect (Parrish 1987). Instead, crocs belong to the Crurotarsi, a large clade of archosaurs that had complex crurotarsal ankles (Fig. 10.14Bc; Sereno 1991).

All of these points refute a close relationship between dinosaurs and crocodilians. Protocrocodilians share only eighteen derived characters with birds, and all are found in dinotheropods as well. In addition, protocrocodilians show no flight-related adaptations. What sphenosuchians do have are derived crocodilian characters quite different from those observed in birds, including unusual elongated proximal carpals that converge with those of megalancosaurs (Fig. 10.11Aa,c). There is no compelling reason to conclude that gracile crocodilians were close bird relatives, and there are many reasons to think they were not. In 1985, even A. Walker (1985) criticized the crocodilian-bird link.

## Predatory Dinosaurs

We now turn to dinosaurs, specifically the predatory examples, as potential bird relations. For overviews of predatory dinosaurs, see my *Predatory Dinosaurs of the World* (Paul 1988a), as well as Weishampel et al. (1990), Currie and Padian (1997), Glut (1997, 1999), and Perez-Moreno et al. (1998/2000). In the following sections, I consider the flesh-eating dinosaurs in a series of increasingly avian groups, and then I discuss their interrelationships and consider the possibility that some or even all theropods may have been secondarily flightless.

### *Protodinosaurs and Baso-Theropods*

The "protodinosaurs" *Marasuchus, Lagerpeton,* and *Pseudolagosuchus* reveal the nature of the small, gracile, erect-limbed archosaurs that form the base of the dinosaur clade (Bonaparte 1975, Arcucci 1986, 1987, Sereno and Arcucci 1993, 1994). None of these protodinosaurs is completely known, and the absence of good skull material is especially vexing (Fig. 10.1Ba). The most primitive members of Dinosauria proper are the medium-sized *Staurikosaurus, Eoraptor,* and *Herrerasaurus,* which constitute the baso-theropods (Galton 1977, Sereno and Novas 1992, 1993, Sereno 1993, Sereno et al. 1993, 1998, Novas 1994). Among these baso-theropods, the skulls and skeletons of *Eoraptor* and *Herrerasaurus* are now well known, although *Eoraptor* has not been thoroughly described (Figs. 10.1Bb,c, 10.2C,D). One of the features that mark these archaic forms is the retention of a rather short, reptilelike ilium in the pelvis (Fig. 10.12Ba–c), a feature also retained in the herbivorous prosauropod dinosaurs that lived at the same time.



What is most important about all these early predators is that they were the first archosaurs to show an overall form that was beginning to become recognizably preavian: they had a light skull at the end of an S-curved and therefore more flexible neck (Fig. 10.7C); they were at least partly bipedal; their legs were erect in the dinosaur-avian manner (Fig. 10.13D); they had a simple "mesotarsal" cylindrical ankle joint with the beginnings of a pretibial process (Fig. 10.14e); and they had a digitigrade, laterally compressed foot in which the outer toe was reduced to a splint (Fig. 10.14Fd,e). To these features can be added the cylindrical hip joint with a medial opening that first appears in the baso-theropods (Fig. 10.13D), and a long raptorial hand in herrerasaurs. Also well underway was the shift from a condition in which the outer digits were well developed and divergent to a more birdlike condition in which the outer digits were reduced and the thumb and first toe were divergent. This combination of characters leads us to the point that the dinosaur clade includes birds. According to the main character list (Appendix 1A), protodinosaurs and nonavian theropods share with birds 347 derived characters not seen in the nondinosaurian alternatives; and dino-theropods and birds share 305. Tetradactyl predatory dinosaurs are to birds what the earliest apes are to humans.

Before we finish with basal dinosaurs, it is worth noting that *Herrerasaurus* had a few extra avian attributes: the scapula blade was narrow (Fig. 10.8F), asymmetrical humeral condyles hint that an arm-folding mechanism was incipient (Chapter 4), and the pubis was booted and retroverted about as much as in *Archaeopteryx* (Fig. 10.12Be,Ca). These features were largely absent in the otherwise even more birdlike basoavepods, so they probably represent a case of mild parallelism.

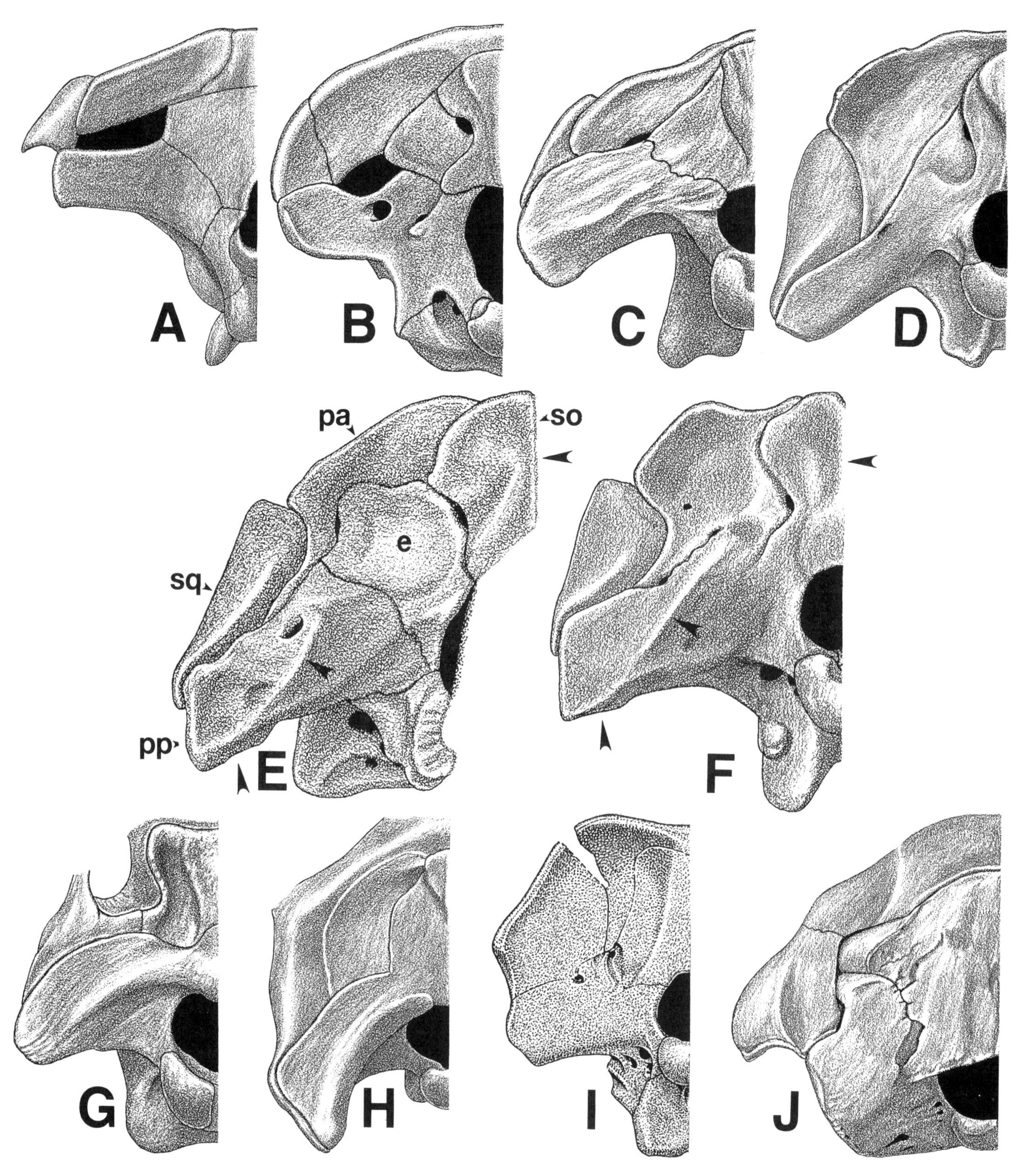


*Figure 10.5.* The left halves of occiputs of archosaur braincases with squamosal, in posterior view. *A,* basal archosaur *Euparkeria; B,* drepanosaur *"Protoavis"; C,* baso-theropod *Herrerasaurus;* avepods: *D,* ceratosauran *Syntarsus;* avepectorans: dromaeo-avemorphs: *E, Archaeopteryx; F, Velociraptor* (largely after Barsbold and Osmólska 1999, Fig. 2A); *G,* ornithomimid *Gallimimus; H,* oviraptor *Ingenia* (modified from Currie and Zhao 1993b); *I, Troodon; J,* juvenile ostrich. See Figure 10.1Bm,n for additional details. Abbreviations are listed in Appendix 7. Arrows in *E* and *F* point to synapomorphies, as explained in text and Appendix 1. Not drawn to same scale.

## *Avepods*

### Coelophysoids

Coelophysoids are well known from a very large number of complete skulls and skeletons. Gracile Late Triassic *Coelophysis* and nearly identical Early Jurassic *Syntarsus* are the preeminent small-bodied examples (Figs. 10.1Bd, 10.2E; Raath 1977, Paul 1988a, 1993, Downs 2000; the figures in the Colbert [1989] monograph are often inaccurate). *Dilophosaurus* was a big coelophysoid (Welles 1984). It was in these earliest of the avepods that the avian body form became predominant. Most especially notable is the tridactyl foot in which the metatarsal of the inner toe no longer contacted the ankle (Fig. 10.14Fg–m). In addition, metatarsal V was just a simple splint. With such changes, the shift from the outer to the inner toe as the divergent digit further approached the avian condition. Interestingly, the hand too became trisymmetrical, although in this case via an asymmetrical reduction of finger III so that it was shorter than II. Other externally visible avian attributes included a tapering skull; a short, compact, rigid body; an increase in the leverage of forelimb and hindlimb muscles as long bone crests expanded in size; a long sacrum and a large ilial plate supporting broad thigh muscles; and a large cnemial crest anchoring a drumstick-shaped set of shank muscles.

Important, newly developed avian internal features of the coelophysoids were pneumatic neck vertebrae and mobile posterior ribs suitable for housing ventilating air sacs (Appendix 3B). There are also skull details—including a jugal reduced to just three prongs instead of the usual four, a hook-shaped ectopterygoid process, a fenestra between the pterygoid and palatine, and a shallow, circular depression on the side of the braincase (a pneumatic feature, according to Witmer 1997c)—that later show up in *Archaeopteryx*. The preavian furcula may have first developed in basal avepods (see later in this chapter), although an earlier appearance is possible. An important but little-appreciated preavian attribute of early dino-avepods was lateral rotation of the shoulder joint so that the arm could swing laterally, a feature that aided the development of horizontal wings. Also underappreciated is that dino-avepods shared with birds an unusual interlocking zigzag articulation of gastralia along the midline of the belly, a feature that may have been in the process of evolving in coelophysoids and that differed markedly from what is found in other archosaurs (Fig. 10.7E).

The small coelophysoids are so birdy that Raath (1985) proposed them as the closest relatives to birds, and the fusion of the pelvic elements common in these dinosaurs is especially birdlike. However, more advanced dino-avepods are even more avian in most regards, and because fusion of the pelvis does not occur in some early birds, this fusion is an early example of intra-avepod parallelism. In fact, in one respect coelophysoids are nonavian: they tend to have short arms and hands, although the latter are raptorial. Coelophysids and dilophosaurids are also divergent from the avian clade in having a subnarial gap at the front end of the upper jaw. Welman (1995) asserted that the coelophysid braincase was too specialized to be ancestral to that of birds. If he is correct, his assertion only confirms that coelophysoids diverged from the dino-avepod–bird clade. Coelophysoids were not the closest relatives of birds, but they represent an early ancestral stock. The number of synapomorphies shared by avepod dinosaurs and birds is 292. These basal avepods are to birds what advanced apes are to humans.

### Averostrans and Avetheropods

The similarities between dinosaurs and birds become increasingly extensive when we consider the averostran avetheropods of the Jurassic and Cretaceous. Although basal examples, such as ceratosaurs, were still like coelophysoids in most regards, the level of organization in the more derived examples approached the avian condition. Classic and new basal averostrans known from good skulls and skeletons include big *Ceratosaurus, Yangchuanosaurus, Sinraptor, Allosaurus,* and *Monolophosaurus,* as well as small *Compsognathus, Sinosauropteryx, Scipionyx,* and *Ornitholestes* (Figs. 10.1Bg–e, 10.2G,F; Madsen 1976, Ostrom 1978, Currie and Zhao 1993a, Zhao and Currie 1993, Chen et al. 1998, Sasso and Signore 1998, Madsen and Welles 2000). In fact, *Archaeopteryx* was itself an avetheropod in that, as far as we know, it had all the appropriate cranial (including palatal) and postcranial characters—a fact that has been obscured by the all-too-common tendency to restore the skeleton and especially the skull as more birdlike than it really was. The averostran head was marked by the development of maxillary sinuses that altered the main nasal passage into an L-shape, a feature now known to have been retained in some Cretaceous birds. In the roof of the mouth, the fenestra between the pterygoid and palatine became larger, and the bar that connected the anterior and posterior halves of the pterygoid became increasingly slender (Fig.

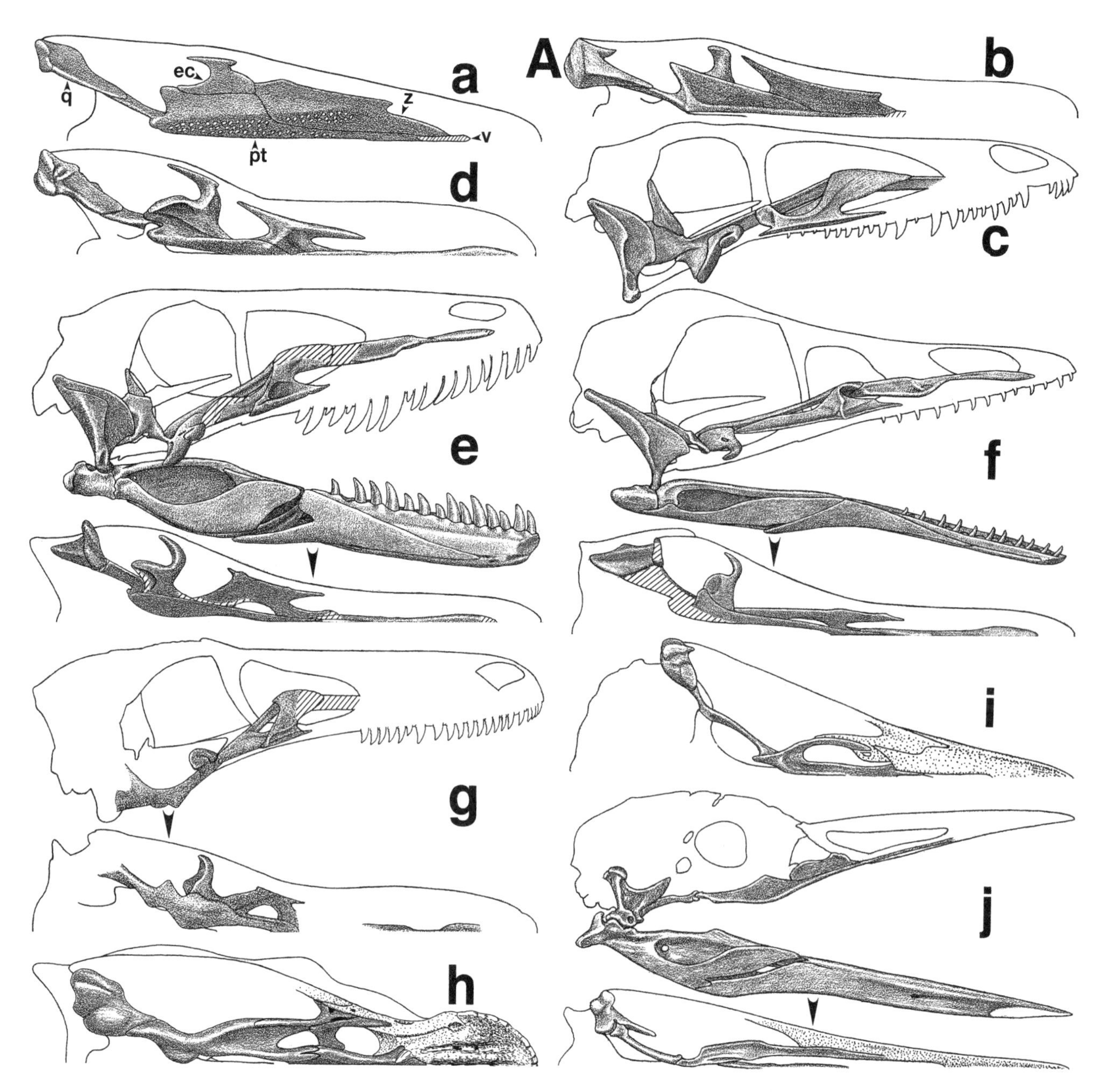


*Figure 10.6.* Interior structures of archosaur skulls. *A,* articulated right palates in side view, bottom view, or both, and inner views of left mandibles: *a,* basal archosaur *Euparkeria; b,* baso-theropod *Herrerasaurus;* avepods with interpterygoid palatine fenestra: *c,* cerato-sauran *Syntarsus-Coelophysis* composite; *d,* basal avetheropod *Allosaurus;* avepectorans: *e,* dromaeosaur composite; *f, Archaeopteryx* composite; *g,* troodont *Saurornithoides* type; *h,* oviraptorid *Ingenia; i,* palaeognathous tinamou; *j,* neognathous loon; premaxillary and maxillary shelves are coarsely stippled, restored elements are hatched. Arrows show which side and bottom views belong together. *B,* articulated right pterygoid, epipterygoid, ectopterygoid complex and/or palatine in side plan view: *a,* basal archosaur *Garjainia* with massive pterygoid, epipterygoid, and palatine; avepods with gracile pterygoids and hook-shaped anteromedial palatine processes: *b, Syntarsus-Coelophysis* composite with broad connection between anterior and posterior pterygoid; tetanurans with increasingly slender connection between pterygoid sections: *c,* basal tetanuran *Sinraptor; d, Allosaurus;* dromaeo-avemorphs with reduced epipterygoids: *e, Archaeopteryx; f, Deinonychus; g, Saurornithoides; h, Ingenia; i,* therizinosaur *Erlikosaurus;* birds with highly modified kinetic pterygoids and lost epipterygoids: *j,* early ornithurine *Hesperornis; k,* palaeognathous tinamou; *l,* loon. *C,* right avepectoran ectopterygoids with dorsal pits in top view: *a, Deinonychus; b, Archaeopteryx; c,* enantiornithine *Gobipteryx* (after Fig. 2 in Elzanowski 1995). Abbreviations are listed in Appendix 7. Arrows in *Be* point to theropod synapomorphies, as explained in text and Appendix 1. Not drawn to same scale.

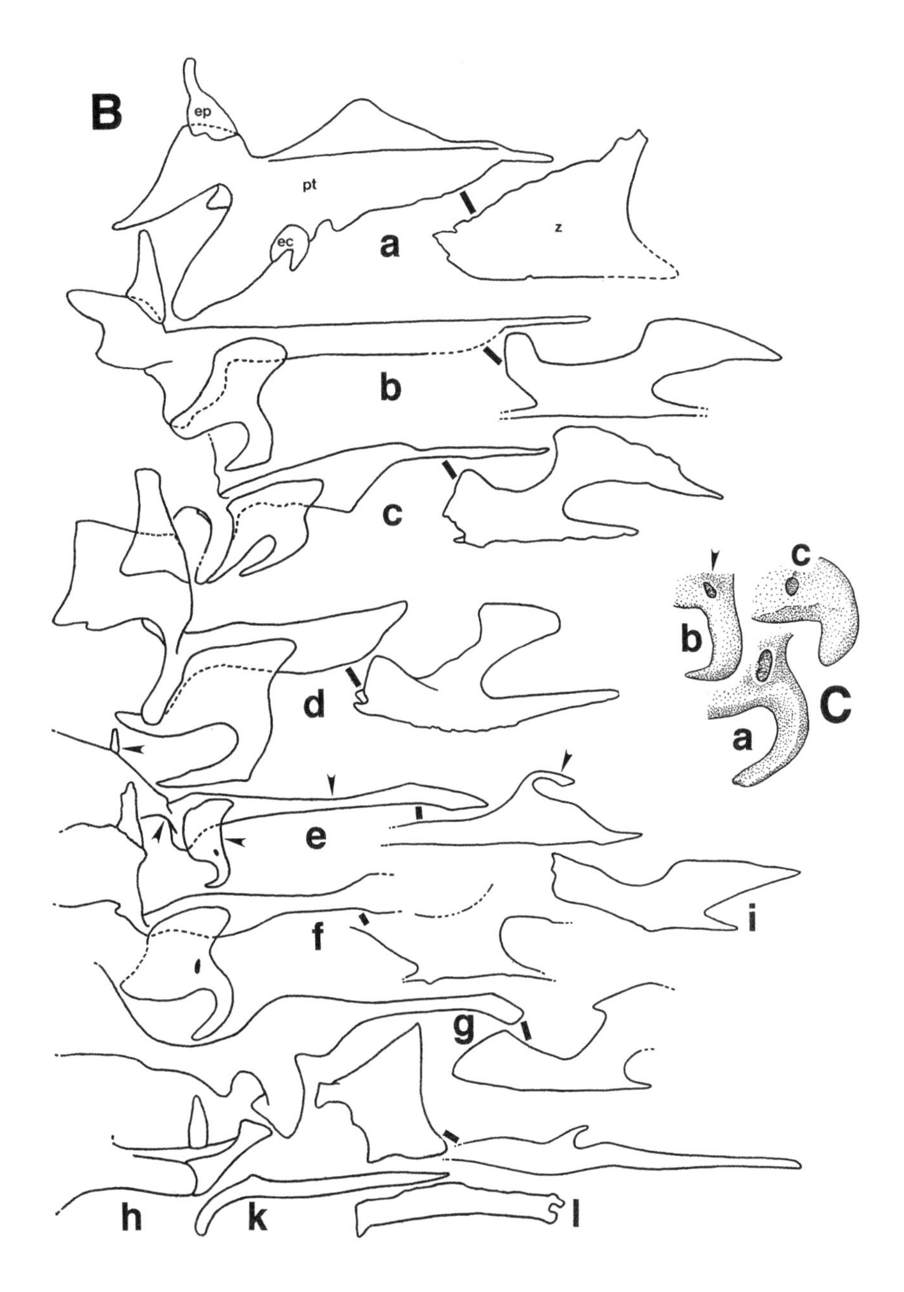




10.6Bc–f). The air sac system probably expanded dramatically as dorsal vertebrae started to become pneumatic, and anterior ribs shortened and posterior ribs lengthened in tetanurans and even more so in avetheropods (Appendix 3B). The most visible avetheropod characters were the three-fingered hand (Fig. 10.11Ae–j) and the distal carpal blocks. Also notable were the boot commonly found at the end of the pubis (Fig. 10.12Bg–o, q–v,Ca–e), the elongation of the ascending bone of the astragalus (Fig. 10.14Bh–p), and the proximal narrowing of the central metatarsal of the foot (Fig. 10.14Fb–m). *Compsognathus* has sometimes garnered particular attention as a potential bird relative, largely because it is found in the same lagoon deposits as the urvogel. But compsognathids were no more avian than other basal avetheropods and were less so than derived examples. Averostran and avetheropod dinosaurs share with birds 263 and 233 derived characters, respectively, and the former are to birds as australopithecines are to humans.

### Dino-Avepectorans

Avepectoran dinosaurs are now well known from a wide variety of Cretaceous forms, although some species remain poorly described. The group certainly includes dromaeosaurs, troodonts and other sickle-clawed forms, therizinosaurs, caudipterygians, oviraptorosaurs, alvarezsaurs, and avimimids (Figs. 10.1Bk–s, 10.2M–N), all of which are further detailed immediately below and in the next chapter. We will see later that some other groups may also belong to this group. To say that these sophisticated dinosaurs were nearly avian is an understatement: they were in most regards nearly or fully as birdlike as, and in some respects more birdlike than, the avepectoran *Archaeopteryx!* Common dino-avepectoran features of the skull included two accessory maxillary fenestrae leading into the mediorostral maxillary sinuses, and subtle and not so subtle avian features in the palate. Expansion of the braincase not only reflected improved brainpower but also tilted the upper temporal fenestra so that it faced partly sideways. Reduction of the olfactory lobes of the brain was another avian attribute (Dorst 1974). The most obvious avian skeletal features were the little-appreciated birdlike shoulder girdle and sternum and the folding arms (Figs. 4.2Ad–h,Bb,d, 10.8K–U, 10.9D–O). The distinctive anteromedial indentation at the coracoid-scapula juncture is limited to archaeopterygiforms and other avepectoran dinosaurs. In most dino-avepectorans, loss of overlapping ribs and incipient saddle articulations made these creatures' necks almost as flexible as those of avebrevicaudan birds. In the ankle, the ascending process of the astragalus was quite tall (Fig. 10.14Bi–o). The skeletal characters that imply that these dinosaurs were closer to birds than the urvogel are addressed in Chapter 11.

### Dromaeosaurs and Other Dromaeo-Avemorphs

How very close birdlike dinosaurs were to birds becomes most apparent when we compare dromaeosaurs and other sickle-clawed forms with retroverted pubes to the urvogel (Figs. 1.6, 1.7, 10.1Bm,n). The dromaeosaurs, which ranged from quite small to somewhat large, are fairly well documented (Paul 1988a,b). Basal *Sinornithosaurus* is known from a fairly complete and partly articulated adult skull and skeleton (Sloan 1999, Xu, Wang, and Wu 1999) and a completely articulated subadult (Appendix 6). Tiny *Microraptor* is know from the majority of a skull and skeleton (Xu et al. 2000). *Deinonychus* is known from fairly complete, largely disarticulated remains (Ostrom 1969), and some new partial material has been published (Brinkman et al. 1998). The very similar but smaller *Velociraptor* is known from excellent skulls and skeletons (Sues 1977, Barsbold 1983, Barsbold and Osmólska 1999), although the skeletons have yet to be fully described. Norell and Makovicky (1997, 1999) described the partial, well-preserved skeletons of Asian velociraptorines. Nearly complete remains document *Bambiraptor* (Burnham et al. 2000). *Dromaeosaurus* is represented mainly by a partial skull (Colbert and Russell 1969, Currie 1995). *Achillobator* (Perle et al. 1999) so far consists of parts of the skull and skeleton, although the assignment of some of the bones has been challenged. The increasing mix of basal dinosaur features and derived birdlike features in the above avepods has caused some workers to question whether all of them are true dromaeosaurs, and so has the presence of some troodont-like features in *Sinornithosaurus* and *Bambiraptor* discussed at the Florida bird origins conference. We do not know enough of *Unenlagia* (Novas and Puerta 1997) to tell for sure whether it is a dromaeosaur stricto sensu, but if it is not, it appears to be a very close relation. Little *Rahonavis* (Sampson et al. 1997, Forster et al. 1998) is very incomplete, and the two studies' restorations of the connection between the arm elements and the rest of the skeleton are probable but not absolutely proven. The placement of rahonaviforms in this grouping is provisional until more complete remains become available.

A number of workers have noted similarities between sickle-clawed dromaeosaurs and *Archaeopteryx* (Paul 1984a, 1988a,b, Gauthier and Padian 1985, Gauthier 1986, Ostrom 1990, Holtz 1994a, Currie 1995, Elzanowski and Wellnhofer 1996, Chatterjee 1997, Sereno 1999a, Xu, Wang, and Wu 1999, Xu et al. 2001). These archosaurs share a similar dino-avepectoran body form

*Figure 10.7. (opposite)* Avepod axial elements. *A,* presacral vertebrae, *left to right* (not all vertebrae are shown in each view): axis, midcervical, dorsocervicals (with hypapophysis stippled), anterior dorsal, middorsal, posterior dorsal: *a,* basal archosaur *Euparkeria; b,* protocrocodilian *Pseudohesperosuchus* and other taxa; avepods: *c, Coelophysis-Syntarsus; d, Archaeopteryx; e, Deinonychus,* including front views of cervicodorsals; *f, Ichthyornis. B,* midcervicals in multiple views: *a,* drepanosaurs *"Protoavis" (left), Megalancosaurus (right); b, Archaeopteryx; c, Deinonychus; d, Ichthyornis. C,* neck curvature, cervicals articulated in neutral posture with centra facets flat on to one another and zygapophysis fully articulated: *a, Euparkeria; b, Deinonychus; c,* loon. *D,* left cervical ribs: *a, Euparkeria; b, Coelophysis-Syntarsus; c, Archaeopteryx; d, Deinonychus; e,* loon. *E,* articulated gastralia of archosaurs, in ventral view, anterior is toward the top, *top to bottom: Euparkeria;* crocodile; avepods: *Syntarsus;* dromaeosaur; basal bird *Changchengornis,* after Mook 1921, Ewer 1965, Raath 1977, Norell and Makovicky 1997, and Ji, Chiappe, and Ji 1999. Not drawn to same scale.

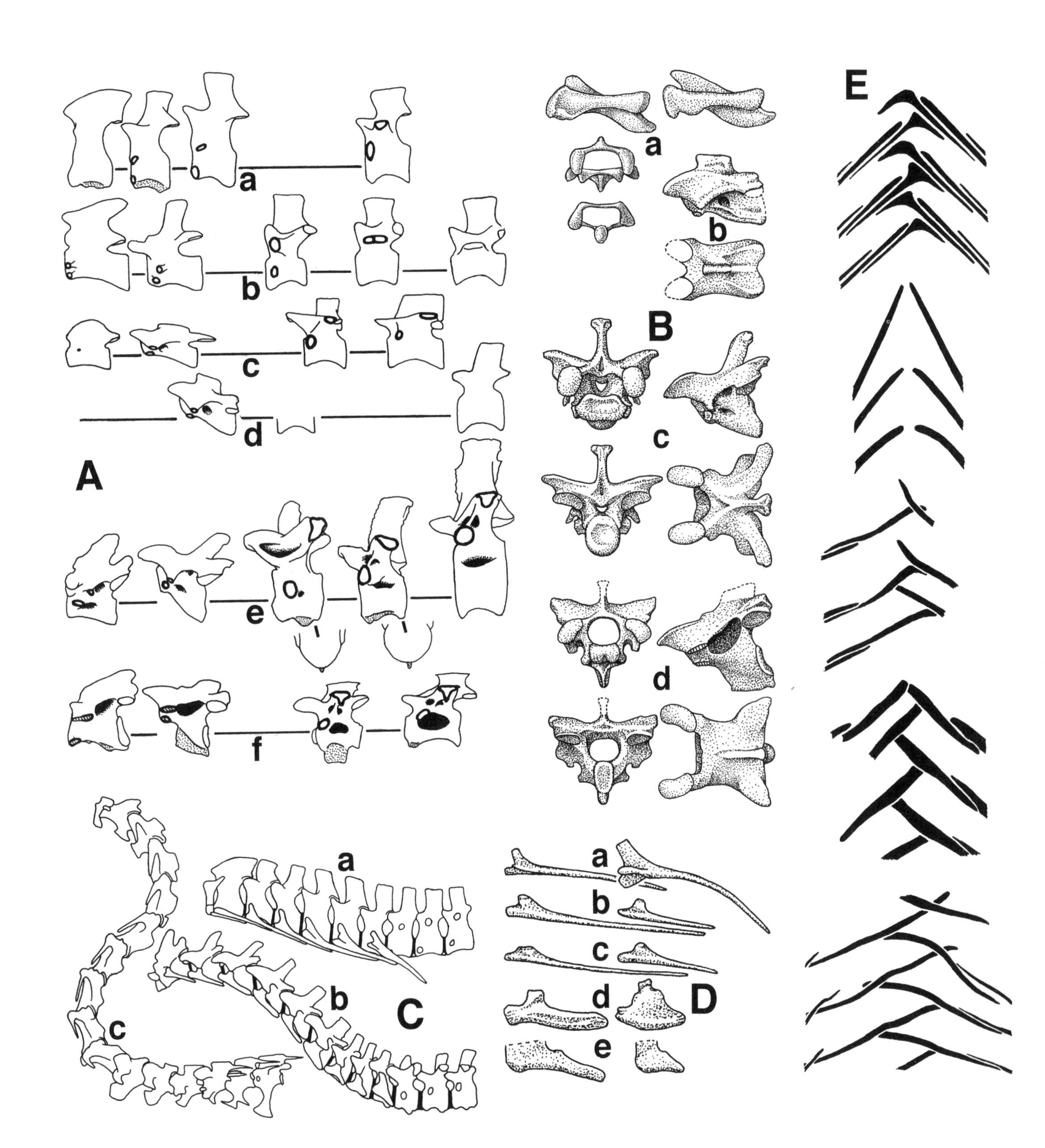

a
b
c
d
A
e
f
E
a
b
B
c
d
a
b
c
C
a
b
c
d
e
D

with long arms, birdlike hips, and slender, proximally flexible tails.

In the skull of dromaeosaurs and archaeopterygiforms, the occiput on the back of the braincase is an area of exceptional and distinctive similarity (Figs. 10.1Bm,n, 10.5E,F, Pl. 22). In addition to an overall dino-avepod profile, these two forms share a distinctive diamond-shaped dorsal section of the supraoccipital bone above the spinal opening (the only other archosauromorphs that are close to having this condition are drepanosaurs, but their non-avepod-like occiputs are otherwise quite different from those of dromaeosaurs and archaeopterygiforms; see Fig. 10.5B). Even more striking is how the winglike paraoccipital processes are similarly twisted and distally expanded in order to form a deep, external auditory meatus. (This similarity was briefly noted by Currie in 1995 but has since been largely neglected.) This twist is so strong that the posterior face of the process has a bulge. The rectangular expansion at the end of the process also helps form the auditory space. I have examined a wide array of archosaurs, including other avepods, and so far have not come across any with this exact pattern. Elzanowski's (1999) dismissal of similarity between the palates of archaeopterygiforms and dromaeosaurs is based in part upon a misinterpretation of parts of the palate, as discussed in Chapter 3.

As striking as the cranial similarities are, they are matched by the unique (outside of some basal avebrevicaudan birds) postcranial details shared by dromaeo-avemorphs, including archaeopterygiforms, details that center on the urvogel-type pelvis and hindlimb (Figs. 4.5A–C, 10.12Br–v, Ca,b,D). These include not only a parallelogram-shaped ilium, an often massive and retroverted pubic peduncle, and an ischium only 50 percent as long as the pubis, but also such small details as the inverted-V articulation between the ilium and pubis and the shallow V-shaped cross section of the flattened pubic shafts. These features are not observed in other archosaurs. Most like *Archaeopteryx* is *Sinornithosaurus,* whose avian scapulocoracoid, furcula, and extra-long arms present extraordinary similarities (Figs. 10.8L,M, 10.10Be,j, 10.11Ah,i, 11.1). Indeed, *Sinornithosaurus* is an ideal transitional form that directly links archaeopterygiforms and dromaeosaurs. The reduced tail and waisted teeth of *Microraptor* are exceptionally urvogel-like. *Rahonavis* had pubes and ischia little different from those of *Archaeopteryx*. The urvogel even shows partial development of an asymmetrical, semididactyl foot with a hyperextendible second toe, a feature that is very well developed in even more advanced *Rahonavis*. Asymmetry of the toes is also characteristic of baso-avebrevicaudans (App. Fig. 18A), which suggests that this is a dinosaur-avian synapomorphy.

As incomplete as they are, newly found *Rahonavis* and *Unenlagia* are proving especially valuable in filling in some of the character gaps that until now separated dinosaurs and birds. They share with urvogels a pubic boot without an anterior projection. Both also have an ischium with dorsal and distal projections similar to those of *Archaeopteryx, Rahonavis* is most like the urvogel in this respect, and both *Sinornithosaurus* and *Bambiraptor* also had a dorsal ischial projection.

Although Currie (1995) suggested that dromaeosaurs are not as birdlike as others have supposed, they are actually so similar in form and detail to *Archaeopteryx* that we can describe the latter as a miniature, flying dromaeosaur and the former as flightless, big game–hunting archaeopterygiforms. Both types had autapomorphies (see Currie 1995) and other distinctive characters not found in the other, but the evidence clearly shows that they and other dromaeo-avemorphs were much more intimately related to one another than to any other dinosaurs or birds.

Nearly all dino-avepectorans and birds (at least *Archaeopteryx*) share 206 synapomorphies. These most birdlike of dinosaurs are to birds what early members of the genus *Homo* are to modern humans.

### Feathered Dinosaurs

The discovery of the feathered dinosaurs *Sinosauropteryx, Caudipteryx, Beipiaosaurus, Sinornithosaurus,* and *Microraptor* is the icing on the phylogenetic cake. Certainly these are just the first among many—although the possibility that some small predatory dinosaurs, especially early ones, were not feathery remains open. The presence of both simple and complex feathers on dino-avepods extends the evidence for the dinosaur-bird link beyond the bones to soft tissues—indeed, to the very soft tissues that have long been considered the most distinctly avian.

### Bone Fine Structure

Rensberger and Watabe (2000) observed that at the microscopic level, ornithomimids share with birds a bone organization that is distinct from the more mammal-like arrangement in ornithischians. Further work demonstrating that additional ave-

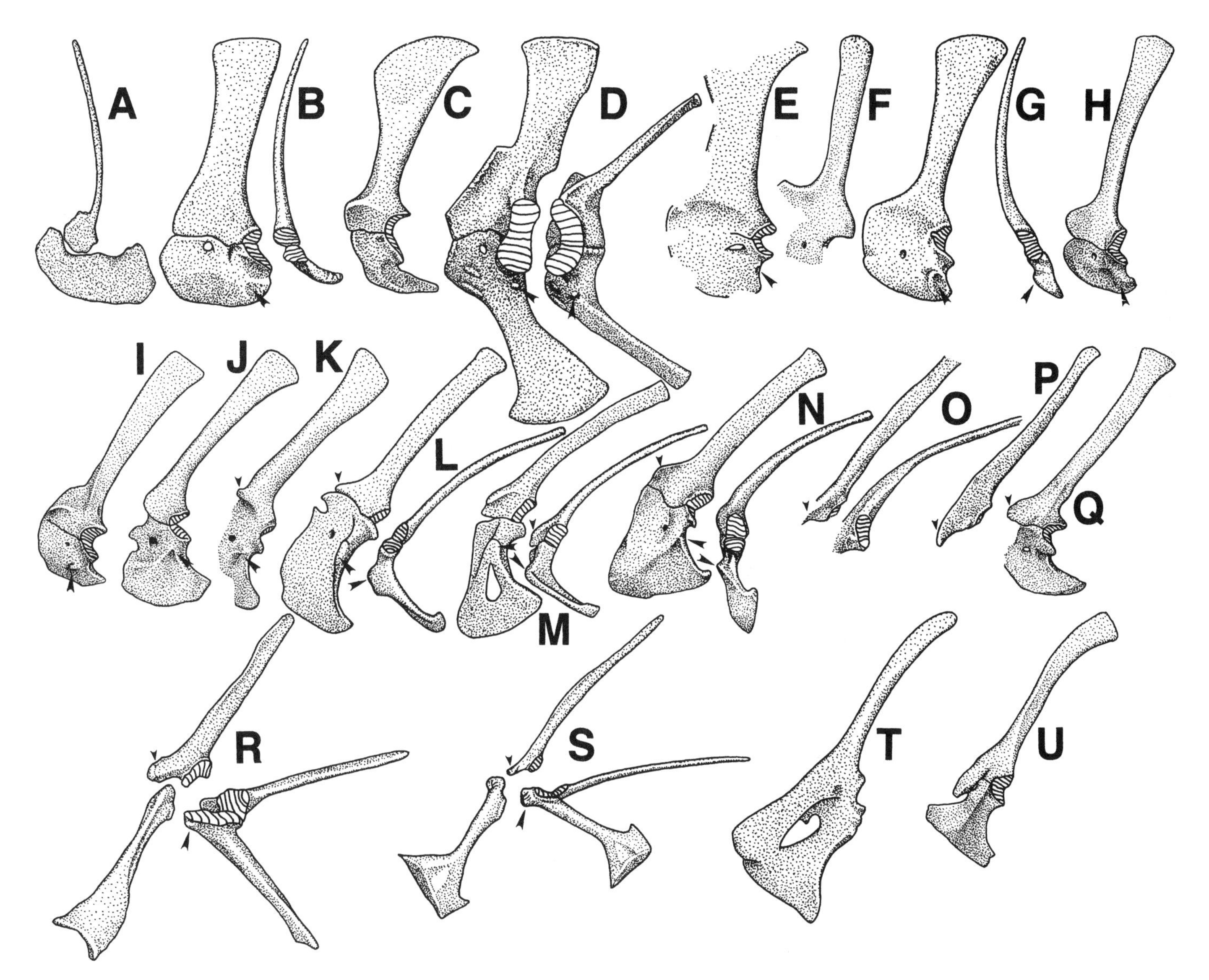


*Figure 10.8.* Scapulocoracoids, drawn to same approximate scapula length so that the size of the coracoids can be compared. Coracoids and scapulae are shown in consistent comparative views showing orientation of glenoid relative to the plane of the scapula, a side view flat on to the main plane with the two elements flattened onto a single plane, and an edge on to main plane of the scapula posterior view of naturally articulated elements, which shows degree of flexion between the two elements. *A,* drepanosaur *Megalancosaurus; B,* basal archosaur *Euparkeria; C,* protocrocodilian *Terrestrisuchus; D,* alligator; *E,* protodinosaur *Marasuchus; F,* basotheropod *Herrerasaurus; G,* cerato-sauran *Syntarsus; H,* basal avetheropod *Allosaurus; I,* ornithomimid *Gallimimus;* avepectorans: *J, Caudipteryx; K, Oviraptor;* dromaeo-avemorphs: *L, Archaeopteryx* mainly BMNH 37001; *M, Sinornithosaurus,* profile of poorly preserved elements; *N, Deinonychus* with additional information from other dromaeosaurs; *O, Unenlagia,* showing that the glenoid is not more dorsally oriented than in dromaeosaurs; *P, Rahonavis; Q, Mononykus;* flying birds: *R,* enantiornithine *Enantiornis; S,* ornithurine *Apatornis;* flightless birds: *T,* ostrich; *U, Diatryma.* Small arrows point to the acromion process; large arrows to the coracoid tuber/acrocoracoid; hatching indicates contours of glenoid. Note the presence of two planes of coracoid in view *N.*

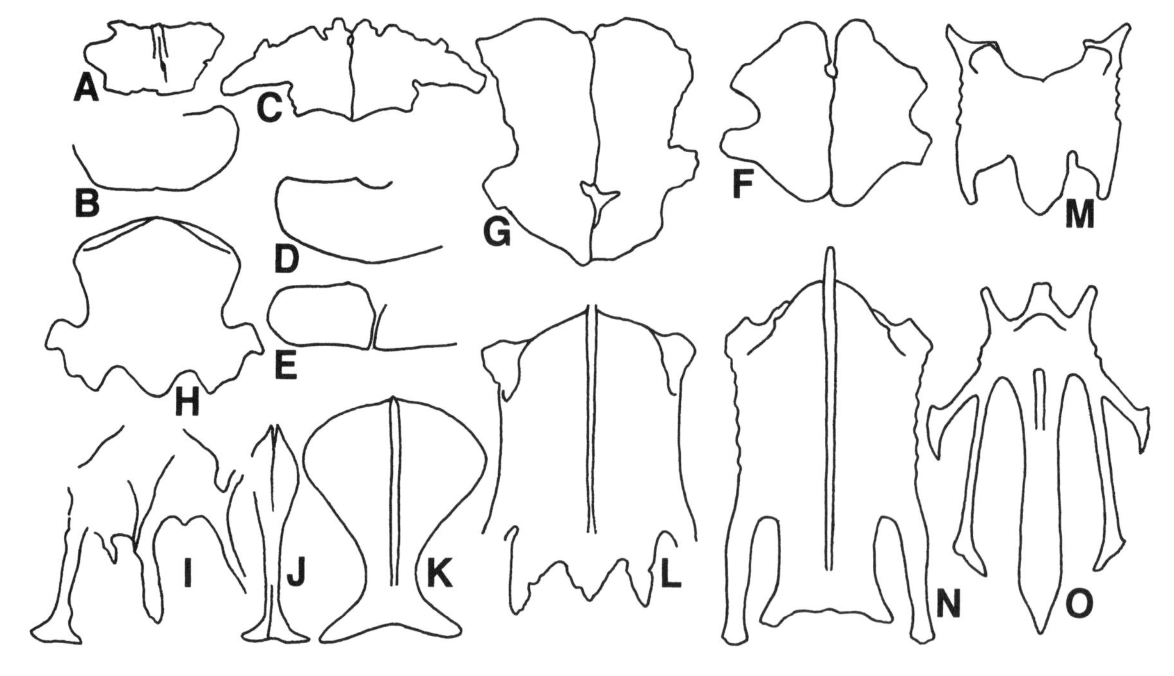


*Figure 10.9.* Avepod sterna in bottom view. Large basal tetanurans: *A, Sinraptor; B, Xuanhanosaurus;* C, tyrannosaur *Gorgosaurus;* avepectorans: *D, Archaeopteryx; E, Protarchaeopteryx; F, Oviraptor;* G, dromaeosaur *Velociraptor;* avebrevicaudans: *H, Confuciusornis;* enantiornithines: *I, Concornis; J, Eoalulavis,* in which only the central section appears to be ossified; *K,* basal ornithurine *Liaoningornis; L, Apatornis;* modern birds: *M,* flightless kiwi; *N,* burst-flying chicken; *O,* long-range flying duck. Not drawn to same scale.

pod dinosaurs possess birdlike bone fine structures not seen in other archosaurs would provide additional support to the dinosaur hypothesis of bird origins.

### *Preavian Reproduction*

It is becoming apparent that the similarities between dino-avepectorans and birds extend even to their modes of reproduction. To begin, avepod dinosaurs and birds share distinctive eggshell microstructure characteristics not observed in other amniotes (Fig. 10.15; Norell et al. 1994, Chatterjee 1997; Zelenitsky and Hirsch 1997; Xu, Tang, and Wang 1999, Horner 2000). This similarity extends the derived morphological similarities between the two groups beyond their skeletons and integument, a remarkable fact that remains underappreciated.

Before we continue, we must define brooding. Brooding is not merely sitting on the nest, which, if the eggs are deeply buried, is only nest guarding (contra Geist and Jones 1996). Brooding refers to the practice of protecting or incubating eggs that are exposed or only shallowly buried and thus in close or direct contact with the body (Bellairs 1970).

Reptile reproduction has been described by Cott (1961), Bustard (1967), Bellairs (1970), Shrine (1988), Coombs (1989), Varricchio et al. (1997), and Pough et al. (1998). Reptiles have two functioning oviducts. Their eggs range in size from small to large relative to the size of the female, and no living reptile egg exceeds 200 grams. The number of eggs laid ranges from a few to very many, and in the latter case, the eggs are deposited en masse (Bustard 1967, Bellairs 1970). Eggs are laid in a random pattern in the nest. They may be deposited in sheltered positions (crevices, debris piles, burrows), soil nests, or vegetation mounds. Eggs are not turned. A few lizards and a number of snakes brood eggs by curling around or among them. This occurs in sheltered nests and apparently provides protection and, in some cases, humidity control. Only pythons use muscular contractions of their sinuous bodies to incubate their exposed eggs (A. Bennett and Dawson 1976). The living reptiles closest to birds, crocodilians, deposit large numbers of relatively small eggs in soil nests and vegetation mounds. They often guard their eggs by lying partly atop the nest, but they never brood them.

Avian reproduction has been described by Calder (1978), Seymour and Ackerman (1980), Coombs (1989), del Hoyo et al. (1992–99), and Varricchio et al. (1997). Birds have only one functioning ovary-oviduct, the exception being kiwis, whose two working units are probably a secondary adaptation. The eggs are large relative to the size of the female, with the largest eggs reaching 9 to 12 kilograms. Some birds lay only a few eggs, while others lay many, but all birds lay only one egg at a time. Megapode fowl bury and incubate their eggs either in soil or in fermentation mounds; this behavior is probably a secondary adaptation. All other birds incubate their eggs with body heat. In most cases the eggs are exposed, but in a few cases (such as the Egyptian plover) they are shallowly buried at least part of the time. Brooded eggs are positioned into flattened arrays, upon which the incubating parent rests in a stereotypical posture upon folded legs, carefully placed to prevent damage to the shells. Feathers completely cover the brood in order to both trap body heat (much of which is transmitted via a naked brood patch) and protect the eggs from exposure to the elements. Eggs are frequently turned. Some birds construct elaborate, multiunit nests, and others make no nest at all.

Knowledge of the egg-laying and brooding habits of dino-avepods is growing rapidly and has been described by Norell et al. (1995), Dong and Currie (1996), Horner and Dobb (1997), Varricchio et al. (1997), Chen et al. (1998), Hopp and Orsen (1998), Carpenter (1999), Clark et al. (1999), and Horner (2000). The fact that the eggs preserved in the body cavity of *Sinosauropteryx* were paired, and that deposited eggs were paired, indicates that the egg-laying pattern in dino-avepods followed the reptilian pattern. At the same time, the egg pairing suggests that dino-avepods laid each egg pair separately over a period of time, which would indicate that the egg-laying pattern was closer to that of birds. Also avian were the large size of the eggs in both relative (approximately 40 percent tibia length in *Sinosauropteryx*) and absolute (4.5 kilograms in therizinosaurs) terms and the modest numbers of eggs found in predatory dinosaur nests. The eggs of oviraptors (Pl. 15A; Fig. 10.16) and troodonts appear to have been only partly buried, a pattern intermediate to reptilian and avian norms. There is no evidence for the presence of fermenting vegetation in these two dinosaurs' nests (contra Martin and Simmons 1998). The multiple layers of deposited eggs, retention of the original paired deposition, and partial burial suggest that their eggs were not turned. The brooding adult sat in the center of the nest, legs tightly folded, with parts of its body in direct contact with the inner ends of the eggs and with its arms symmetrically draped over part of the egg ring in a birdlike manner. Multiple examples establish that the described posture was habitual in oviraptors and troodonts.

Horner (2000) and others have correctly observed that by laying a clutch of eggs over a period of time, arranging them in flat arrays, and then brooding the eggs on folded legs with the arms arranged symmetrically over the egg ring, oviraptorids and troodonts exhibited preavian nesting behaviors not observed in any reptile. The limited number of eggs and their large size also followed avian rather than crocodilian patterns. The stereotyped dinosaur brooding posture was very different from that of pythons, and it was not at all like crocodilian nest-guarding, in which a female drapes part of her body in irregular poses atop a nest within which her eggs are deeply buried. Martin and Simmons (1998) and Carpenter's (1999) suggestion that oviraptorids sat in the middle of their semi-exposed egg rings in order to protect them from predators is not the most logical explanation, since the presence of a fairly large animal in a brooding posture atop a nest would have tended to alert predators to the presence of the nest. Although the flattened posture of brooding dino-avepods would have reduced the chances of detection, the sprawled-arm pose would have made these creatures more vulnerable to attack and less able to spring up and rebuff predators; keeping the arms folded next to the body would have been a better method if quick action to defend against predators had been the primary goal. Eggs are much better protected when they are buried deeply and when an adult either remains close enough to the nest to spring to their protection or leaves the area to reduce the possibility of detection. Nest humidity levels are also better controlled when eggs are buried. A nest site that leaves the eggs exposed to the elements is advantageous only if the eggs are to be incubated via body heat produced by an adult that covers the nest as completely as possible (Appendix 4). The partial burial of dino-avepectoran eggs was an adaptation that allowed heat from both soil and body to be used for incubation. It is improbable that the switch from reptilian soil incubation to avian body incubation occurred without an intermediate phase in which both heating modes were exploited at the same time. Some avepod dinosaurs appear to have practiced exactly the dual incubation predicted in preavians, and their reptilio-avian form of reproduction links them with birds. To falsify this link, one would have to demonstrate that predatory dinosaur reproduction was not preavian in character or that other potential bird ancestors exhibited birdlike reproduction or both.

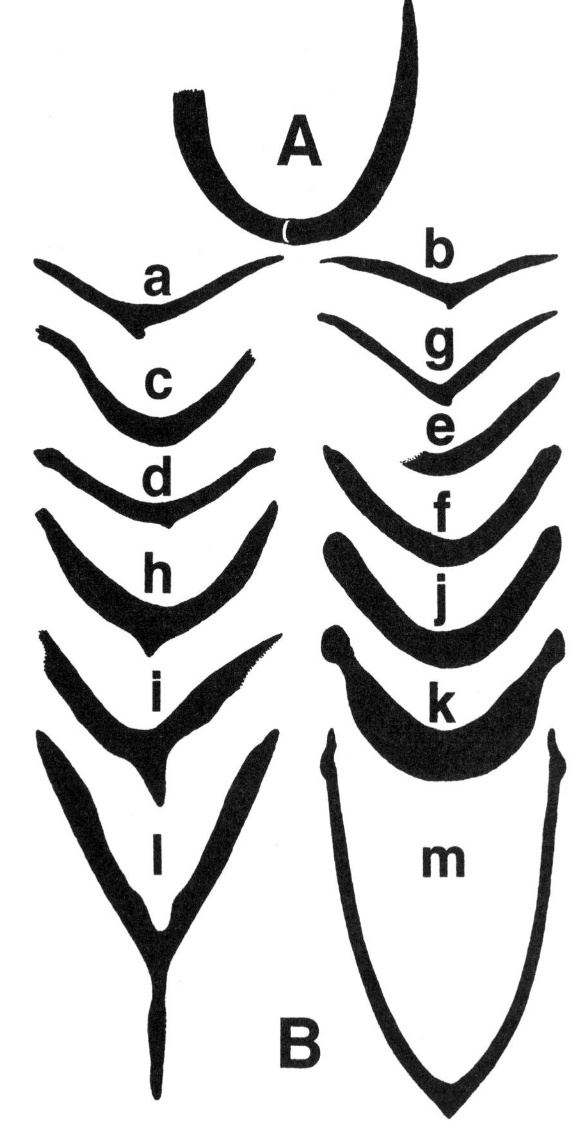


*Figure 10.10.* Diapsid wishbones. *A, Longisquama* (it is not certain whether the elements were fused); *B,* basal avetheropods: *a, Scipionyx; b, Allosaurus;* tyrannosaurids: *c, Gorgosaurus; d, Daspletosaurus;* dino-avepectorans: *e, Sinornithosaurus; f,* velociraptorine dromaeosaur; *g, Velociraptor; h, Ingenia; i,* the *Oviraptor* furcula that could have helped settle the bird origins question in the 1920s; *j, Archaeopteryx;* avebrevicaudans: *k, Confuciusornis; l, Cathayornis; m,* pigeon. Drawn to same width.

## What Best Explains Why Predatory Dinosaurs Were So Birdlike, Convergence or Ancestry?

The total number of derived characters shared by predatory dinosaurs and birds is very high. Of the 382 avian characters also listed outside of the avebrevicaudans, 100 percent are also present in protodinosaurs and dinosaurs, 91 percent are limited to the latter, and 76 percent are limited to avepod dinosaurs. The number of avian characters found in the most birdlike dinosaurs is so high that many "avian" characteristics, whether they be gross form and function or anatomical details, first appeared in the predatory dinosaurs. Indeed, a remarkably small number of derived characters separate primitive Cretaceous ornithurine birds such as *Confuciusornis, Cathayornis,* and *Sinornis* from the most birdlike "nonavian dinosaurs" living at about the same time, and only half a dozen are not directly involved in flight improvements. In addition to all the new avepod dinosaurs sporting characters previously known only among birds, we have the remarkable fact that every well-preserved anatomical detail of *Archaeopteryx* either is found in dinosaurs or is similar to and readily derived from them. Sereno (1999a) emphasized this fact by observing that the number of characters currently separating the basal-most birds from the most birdlike dinosaurs is less than the number separating some of the major known theropod dinosaur clades! In this study, *Archaeopteryx* has only seven unambiguous avian characters not also found in known predatory dinosaurs, and only two are not flight related. Birds are as much avepods as avepods are theropods.

Can we expect this situation to change with further research? The answer is yes, in that the already great number of derived features shared by dinosaurs and birds will probably increase substantially as the morphology of fossil and modern forms is further detailed. For example, Mark Norell (in Morell 1997) mentioned new dromaeosaur braincase characters that closely match those of birds that have yet to be published. Also crucial is the fact that there is a series of increasingly birdlike dino-avepods; such a step-by-step pattern is exactly what one would expect in the group ancestral to birds. This evolutionary pattern of relationships is a Darwin's delight; one could hardly ask for anything better. Dinosaurian and avian avepods are so much alike that researchers sometimes confuse their bones (Sternberg 1940, Cracraft 1971, Harrison and Walker 1973, 1975, Brett-Surman and Paul 1985, Paul 1988a, Jensen and Padian 1989, Feduccia 1996), and whether some of the elements are dinosaurian or avian remains uncertain. One of the urvogel specimens was even first thought to be a dinosaur. No other tetrapod group comes close to being so similar to birds, and none shows a progressive development of increasingly avian features. Indeed, as explained above, no one has yet shown that any

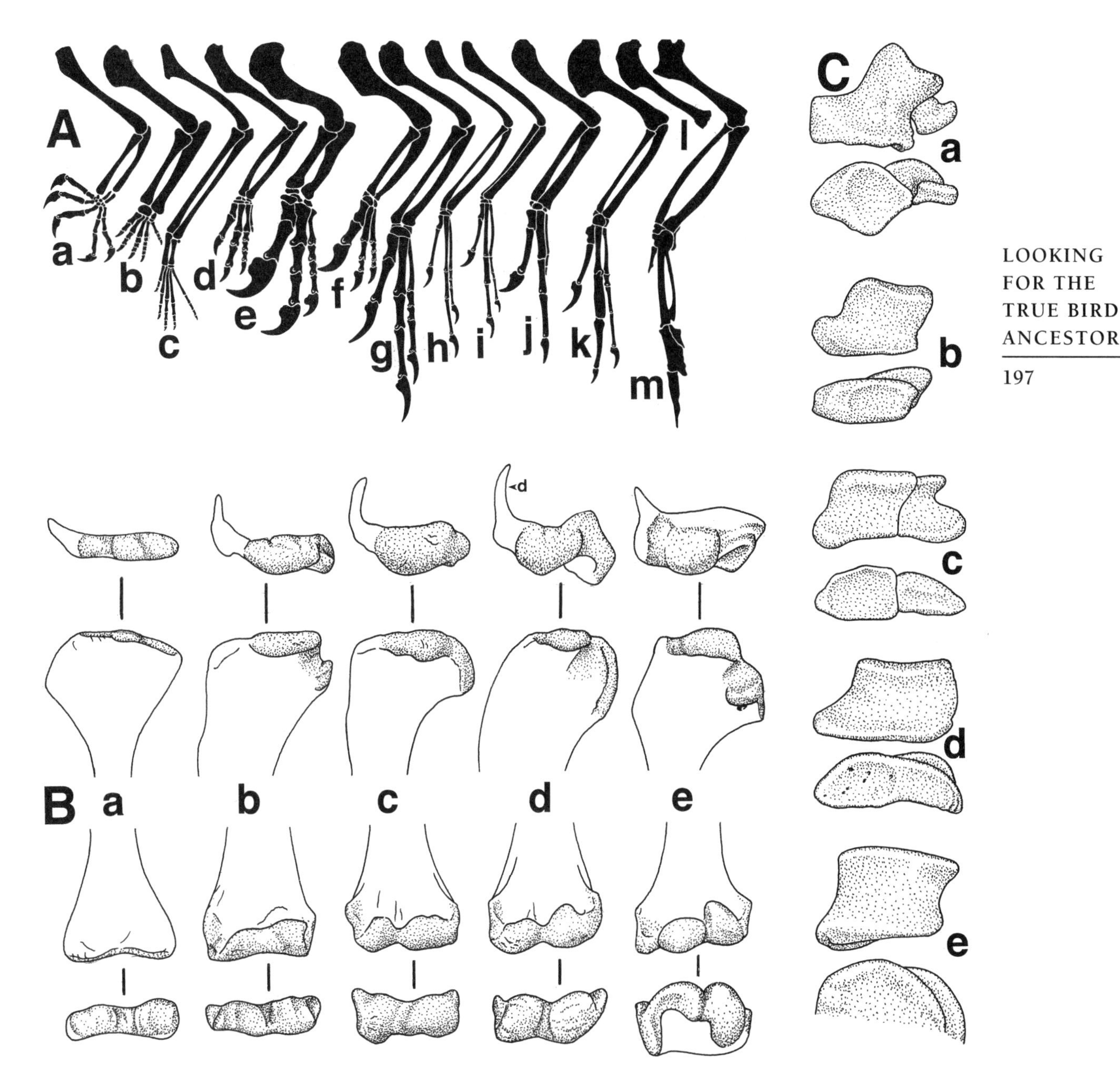


*Figure 10.11.* Arms and wrists, all lefts. *A,* forelimbs drawn to same humerus length in order to contrast relative length of forearm and hands; distal fingers and claws drawn flat on to viewer to facilitate comparison: *a,* drepanosaur *Megalancosaurus; b,* poposaur *Postosuchus; c,* protocrocodilian, composite; avepods: *d,* cerato-sauran *Syntarsus; e,* basal avetheropod *Allosaurus; f,* oviraptorid *Ingenia; g,* derived dromaeosaur *Deinonychus; h, Archaeopteryx; i,* basal dromaeosaur *Sinornithosaurus; j, Caudipteryx; k,* basal bird *Confuciusornis; l, Ichthyornis* (humerus only); *m,* pigeon. *B,* increasingly avian shoulder (top two rows) and elbow joints (bottom two rows); joints are either in end or posterior views: *a, Euparkeria;* avepods: *b, Syntarsus; c, Allosaurus; d, Deinonychus; e,* tinamou; articular surfaces, medioproximal flange and distal flanges stippled. *C,* enlarged and co-joined avepod distal carpals in proximal and dorsal view: basal avetheropods with partly developed pulley-action complex: *a, Allosaurus; b, Coelurus* (medial carpal only); avepectorans: *c,* basal therizinosaur *Alxasaurus;* with fully developed semilunate structure: *d,* dromaeosaur *Deinonychus; e,* pigeon. Not drawn to same scale.

nondinosaurian archosaur shares a single synapomorphy with birds that is not also seen in predatory dinosaurs.

Assertions that dinosaurs and birds share few synapomorphies (Gardiner 1982, L. Martin 1983a,b, 1991, 1997, Tarsitano and Hecht 1980, Tarsitano 1991, Feduccia 1993b, 1996, L. Martin in Stokstad 2000) are therefore in gross error, and such assertions appear to result from a lack of a detailed knowledge of the anatomy of dinoavepods and even of *Archaeopteryx*. The lack of contact between a hallux-bearing metatarsal I and the ankle in avepods, both dinosaurian and avian, alone falsifies Gardiner's 1982 allegation that "the dinosaurs share not a single unique feature with the birds." Also misleading is L. Martin's 1997 cladistic study disputing the dinosaur-avian link. The number of characters he includes is limited compared with the number included in recent analyses that support the link, and the number of errors in Martin's study is high. Currently, the opponents of the dinosaur-bird link cannot point to any character or characters in *Archaeopteryx* that clearly separate it from dinosaurs. Because so many characters link the two groups, the problem faced by those who oppose a dinosaur-bird link is serious. To disprove the link, they must discredit the phylogenetic validity of the many dinosaur-bird synapomorphies, either by disavowing their existence or by showing that they are nonhomologous and convergent. In lieu of finding a set of fossil alternatives to dinosaurs, their best hope is to show that something in the nature of predatory dinosaurs actually bars them from being ancestral to birds.

The researchers cited in the previous paragraph have tended to focus on a limited number of characters supposedly shared by dino-avepods and birds and have either disputed their validity or claimed that certain attributes of birdlike dinosaurs prevented them from being ancestral to birds. The issues are as follows.

### Cranial Potpourri

In 1991, L. Martin continued to claim that *Archaeopteryx* lacked extra accessory maxillary fenestrae leading into mediorostral sinuses. However, not only were they present in the urvogel (Fig. 10.2K, Chapter 3), but also Martin and Zhou (1997) have found them in the early bird *Cathayornis* (Fig. 10.2P). That such complex structures would evolve twice is improbable. Therefore, these birds are exemplary members of the Averostra.

L. Martin (1991), Tarsitano (1991), Feduccia (1996), and L. Martin and Stewart (1999) have continued to assert that theropods lacked supposedly crucial avian skull characters: pneumatic braincase elements and quadrate, strong contact between the quadrate head and the braincase, a fenestra rotunda in the middle ear, subconical teeth crowns separated from roots by a constricted waist, replacement teeth set within pit in root, and so on. However, there are two problems with these assertions. The first is that the characters are not as crucial as these researchers insist. Since these are mostly derived characters, their absence in the ancestral group would not necessarily have any phylogenetic importance. We have already seen that the presence or absence of a quadrate with a

*Figure 10.12. (opposite)* Pelves in side view, sometimes with pubis in front or back view. *A,* nondinosaurs: *a,* drepanosaur *Megalancosaurus; b,* basal archosaur *Euparkeria;* crurotarsi: *c, Ornithosuchus; d,* poposaur *Postosuchus;* basal crocodilians: *e, Terrestrisuchus; f, Nothochampsa. B,* dinosaur clade: mesoschian examples: *a,* protodinosaur *Marasuchus;* basal theropods: *b, Staurikosaurus;* and *c, Herrerasaurus; d, "Protoavis,"* provisional; longoschian avepods starting with cerato-saurans: *e, Syntarsus; f, Elaphrosaurus;* averostrans: *g, Yangchuanosaurus; h, Allosaurus; i,* tyrannosaur *Tarbosaurus; j,* ornithomimid *Gallimimus; k,* composite therizinosaur; avepectorans: *l, Caudipteryx,* with position and orientation of ventral elements from articulated left elements and impressions of NGMC 97–9-A; *m,* oviraptor *Ingenia; n,* troodont *Sinornithoides;* alvarezsaurs: *o,* basal *Patagonykus;* and *p,* derived *Mononykus; q, Avimimus;* urvogel-type pelves, starting with dromaeo-avemorphs: *r, Sinornithosaurus; s, Deinonychus; t,* unnamed dromaeosaur MIG 100/985; *u, Unenlagia; v, Rahonavis. C,* birds: *a, Archaeopteryx* composite and ilium of HMN 1880; *b, Protarchaeopteryx;* avebrevicaudans: *c, Confuciusornis;* continuing with enantiornithines: *d, Sinornis; e, Cathayornis; f, Enantiornis;* a bird of uncertain position: *g, Chaoyangia;* Cretaceous ornithurines: *h,* flightless *Patagopteryx,* and *i,* flying *Apatornis;* modern ornithurines: *j,* flightless palaeognathous emu; *k,* flying neognathous domestic turkey, subadult in which elements remain unossified. *D,* detailed studies of *a, Archaeopteryx* composite, detail of iliopubic articulation after BMNH 37001; *b, Deinonychus* composite, detail of iliopubic articulation after AMNH 3015; both include a cross section of pubes *(solid black),* and detail of pubic peduncle, arrows indicate characters not observed in other dino-avepods. Abbreviations: o = obturator process; other abbreviations are listed in Fig. 4.1. Not drawn to same scale.

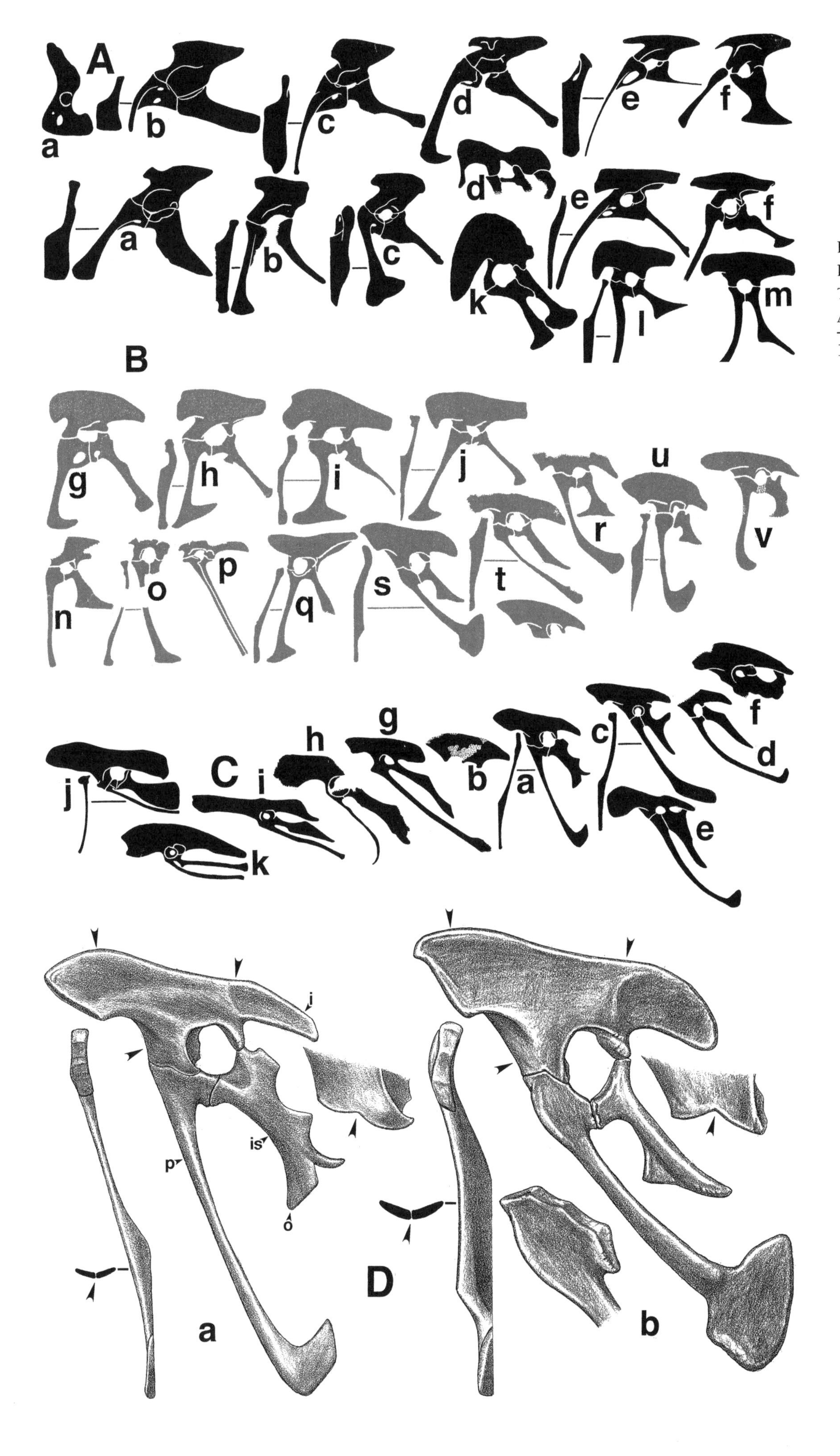
A
a
b
c
d
e
f
a
b
c
d
e
f
k
l
m
B
g
h
i
j
u
r
v
n
o
p
q
s
t
C
j
i
h
g
b
a
c
f
d
e
k
D
a
b
i
is
p
o

well-developed double head is especially irrelevant, since such a quadrate was not present in *Archaeopteryx* itself. The second problem is that Tarsitano, Martin, and Feduccia have failed to understand that various small dino-avepods actually do have many of the above features (as demonstrated in Barsbold 1983, Currie 1985, 1987, Molnar 1985, Barsbold et al. 1987, Witmer 1990, Elzanowski and Wellnhofer 1992, 1993, Currie and Zhao 1993b, Perle et al. 1993, Russell and Dong 1993a, Norell et al. 1994, Maryanska and Osmólska 1997, Ji et al. 1998, Makovicky and Norell 1998, Elzanowski 1999, Xu et al. 2000), Zhou et al. 2000. When Whetstone (1983) first described the braincase of *Archaeopteryx*, he unfortunately did not know enough about dino-avepod braincases to realize that they often do have *Archaeopteryx* braincase characters he thought they did not have: a posttemporal fenestra reduced to a small foramen, pronounced occipital vein grooves, and a lateral prootic depression. Had Whetstone examined the dromaeosaur braincases that were then available, the dromaeosaur-like nature of the urvogel braincase would have been made apparent at that time.

Some of these avian skull characters are indeed important, most especially the maxillary sinus complex. The significance of others is more ambiguous. Pneumatic braincases and quadrates are known in nondinosaurian archosaurs, and a fenestra rotunda is a widespread amniote character. The issue of contact between the quadrate head and the braincase in troodonts and oviraptorosaurs is interesting. On the one hand, the nature of the contact in these two groups—in which the quadrate is vertical and the braincase is expanded onto the top of quadrate—was more birdlike than the crocodilian arrangement. On the other hand, the quadrate contacted different braincase elements in the dinosaurs than in birds, leaving open the possibility of parallelism. Feduccia's statement (in Morell 1997) that "there's no way you can derive a pegged tooth from one that's shaped like a steak knife" represents an astounding failure to understand the most elementary evolutionary principles. Because diapsid teeth are simple in form, they are actually exceptionally plastic. Within the single genus *Varnus*, teeth range from serrated and bladelike to blunt and conical. The same is true in crocodilians. Within dromaeosaurs, teeth vary from large, serrated blades in large, big-game-hunting examples to less serrated, more conical teeth better suited for grabbing small prey in little *Microraptor* (Xu et al. 2000), an excellent example of dental plasticity due to differences in size. That the teeth of *Protarchaeopteryx* are both serrated and peg shaped demonstrates a potentially transitional stage between the two types. Equally interesting is that the teeth of juvenile troodonts (not dromaeosaurs) were conical and unserrated (Norell et al. 1994, Norell and Makovicky 1999; also see Elzanowski 1999). Retention of this tooth form into adulthood would result in the avian condition. For that matter, a recently discovered adult troodont had unserrated teeth similar in shape to those of *Archaeopteryx* (Norell et al. 2000).

### Thoraxes and Pulmonary Complexes

Ruben et al. (1997a, 1999) have contended that the presence of a crocodile-like pelvovisceral pump in avepod dinosaurs should have precluded derived examples of these dinosaurs from having been the ancestors of birds, in part because the pressure differential between the fore and aft body cavities associated with a pelvovisceral piston should have prevented breaching of the airtight septum by posteriorly extending air sacs. It is possible to disagree with this view. Considering the radical alterations in form and function that evolution has often generated, the possibility that a

---

*Figure 10.13. (opposite)* Legs and their joints, all lefts. *A*, left hindlimbs, some with tibias in inner view, drawn to same femur length: *a*, *Euparkeria; b*, protocrocodilian, composite; avepods: *c*, *Coelophysis-Syntarsus; d*, *Velociraptor; e*, *Archaeopteryx*. *B*, left femoral heads in proximal, anterolateral, and, in some cases, side views: *a*, basal archosaur *Euparkeria; b*, protodinosaur *Marasuchus;* theropods: *c*, *Staurikosaurus;* avepods: *d*, cerato-sauran *Syntarsus; e*, basal avetheropod *Allosaurus; f*, *Archaeopteryx;* coalescence of greater trochanter and femoral head: *g*, dromaeosaur *Deinonychus; h*, alvarezsaur *Mononykus; i*, ostrich. *C*, profiles of distal femur and proximal tibia-fibula in end views, anterior at top: *a*, *Euparkeria; b*, modern crocodilian; theropods with increasingly large cnemial crests: *c*, *Staurikosaurus;* avepods with well-developed femoral condyles: *d*, *Syntarsus; e*, *Allosaurus; f*, dromaeosaur *Deinonychus* MCZ 4371; *g*, ostrich. *D*, archosaurian leg posture and hip socket design in *a*, basal *Euparkeria*, semi-erect, closed acetabulum; *b*, derived thecodont *Saurosuchus*, erect, acetabulum closed but facing strongly downward; *c*, basal theropod *Herrerasaurus*, erect, acetabulum faces laterally but is open and cylindrical with right-angle femoral head. Abbreviations: cc = cnemial crest, lt = lesser trochanter, gt = greater trochanter, pt = posterior trochanter. Not drawn to same scale.

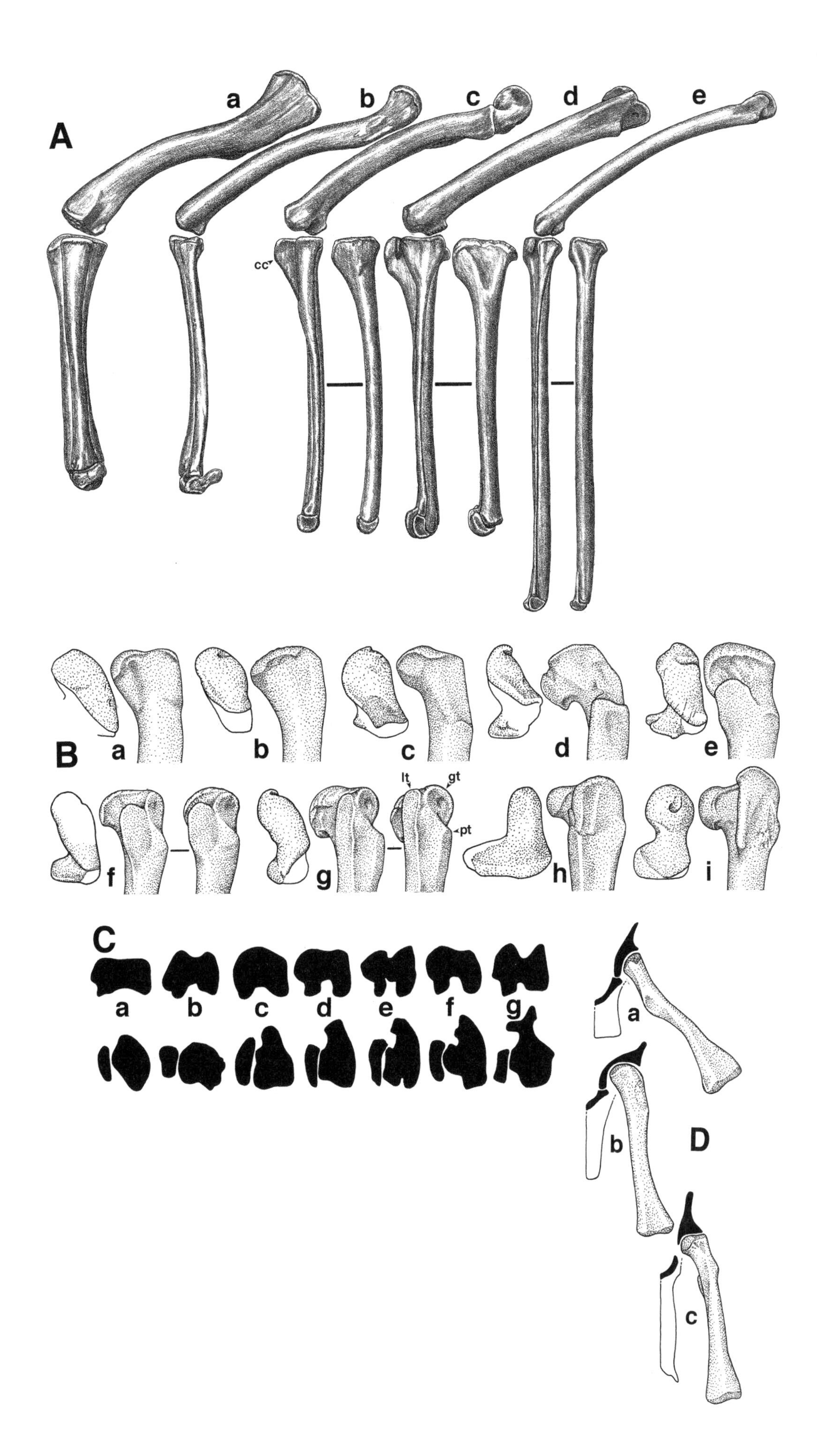
A
a
b
c
d
e
cc
B
a
b
c
d
e
f
g
lt
gt
pt
h
i
C
a
b
c
d
e
f
g
D
a
b
c

crocodilian-like respiratory complex evolved into the avian system cannot be ruled out. However, such a transformation can be ranked as improbable, in part because there is no apparent reason for such a major shift. After all, bats power their flight with a lung system whose function is more like that of crocodilians than that of birds. Such speculations are rendered academic because, as explained in Appendix 3B, Ruben and co-workers' analysis is without sufficient basis and must be rejected. In addition, Ruben and his co-workers have not placed the appearance of a dino-avepod pelvovisceral pump in a plausible evolutionary context, one in which the various features of the system progressively appear in noncrocodilian archosaurs. For instance, Ruben and company have failed to explain why, if dino-avepods had a crocodilian-like respiratory system, they shared with basal birds a distinctive zigzag interlocking of the gastralia that is very different from the arrangement found in crocodilians. The question of whether the avian air sac complex evolved among avepods that began with a pelvovisceral pump is irrelevant because there is no convincing evidence that any saurischian dinosaur had the latter system in the first place. Instead, the progressive development of a pneumatic skeleton and many other osteological characters associated with air-sac-driven lung ventilation in dino-avepods—to the point that sternocostal complexes of avepectorans dinosaurs were more avian than in *Archaeopteryx*—is powerful and compelling evidence that the avian deep breathing began to evolve in dinosaurs.

### Furcula and Sternum

Ossified furculae appear to have been widespread in dino-avepods from coelophysoids to dino-avepectorans (Fig. 10.10B; Barsbold 1983, Chure and Madsen 1996, Norell et al. 1997, 1998, Ji 1998, Makovicky and Currie 1998, Sasso and Signore 1998, Burnham and Zhou 1999, Norell and Makovicky 1999, Xu, Tang, and Wang 1999, Xu, Wang, and Wu 1999, Burnham et al. 2000, Downs 2000).

Feduccia (1996) and Feduccia and Martin (1998) concluded that the furculae of predatory dinosaurs and birds were probably not homologous. They reached this conclusion because they think dinosaur shoulder girdles are very different from those of birds, which is not true, because the dinosaur furculae are not identical to those of flying birds, and because the furculae are better developed in the nonvolant dinosaurs than in flightless birds. Of course, dinosaurs that used their arms for vigorous activity may have more use for a furcula brace and muscle attachment than grounded birds with degenerate wings, and dinosaurs' fused clavicles should not be expected to be exactly the same as those of fliers. The close match of the detailed morphology of the "boomerang" furculae of some dromaeosaurs with the furcula of *Archaeopteryx* and other basal birds is especially telling (Fig. 10.10Be,f,j,k; Burnham and Zhou 1999, Xu, Wang, and Wu 1999).

According to Feduccia and Martin's logic, the presence of the character whose supposed absence once falsely barred dinosaurs from bird ancestry is not only dismissed out of hand but also used as evidence against a relationship between dinosaurs

---

*Figure 10.14. (opposite)* Ankles and feet, all lefts. *A*, tibias in distal view, anterior at top: *a, Euparkeria; b, Staurikosaurus;* avepods: *c*, cerato-sauran *Liliensternus; d, Allosaurus; e, Deinonychus; f*, moa. *B*, articulated tibia, fibula, astragalus, and calcaneum in front view: *a*, drepanosaur *Megalancosaurus; b*, basal archosaur *Euparkeria; c*, crurotarsal modern crocodilian; *d*, protodinosaur *Marasuchus;* theropods with increasingly tall pretibial processes: *e*, basal theropod *Herrerasaurus;* avepods: cerato-saurans: *f, Syntarsus; g, Ceratosaurus;* avetheropods: *h, Allosaurus; i*, ornithomimid *Ornithomimus; j*, dromaeosaur *Deinonychus; k, Archaeopteryx*, tentative; *l, Caudipteryx; m, Troodon; n*, moa; *o*, emu; *p*, owl. *C*, articulated astragalus and calcaneum in distal view, anterior at top: *a*, modern crocodilian; *b, Marasuchus; c*, basal tetanuran *Sinraptor; d, Deinonychus; e, Troodon; f, Patagonykus; g*, juvenile ostrich. *D*, fibula plus calcaneum and/or tibia plus astragalus in side view: *a*, dromaeosaur *Deinonychus; b, Troodon; c*, moa. *E*, profiles of metatarsals I-IV or II-IV in proximal view, anterior toward bottom: *a, Marasuchus;* theropods: *b, Herrerasaurus;* avepods: cerato-saurans: *c, Syntarsus; d, Ceratosaurus;* averostrans: *e, Sinraptor; f, Ornitholestes; g, Allosaurus; h*, tyrannosaur *Tarbosaurus; i*, ornithomimid *Ornithomimus, j*, dromaeosaur *Deinonychus;* birds: *k*, enantiornithine *Avisaurus; l*, juvenile ostrich. *F*, metatarsus in front view, I and V solid black: *a, Megalancosaurus; b, Euparkeria; c*, protocrocodilian *Terrestrisuchus; d, Marasuchus; e, Herrerasaurus; f*, therizinosaur *Segnosaurus;* avepods in which I does not contact ankle and V is reduced to at most a splint: *g, Syntarsus; h, Allosaurus; i*, ornithomimid *Ornithomimus, j*, dromaeosaur *Deinonychus; k, Archaeopteryx; l*, enantiornithine *Avisaurus; m*, juvenile ostrich. Abbreviations: a = astragalus, ca = calcaneum, pp = pretibial process. Not drawn to same scale.

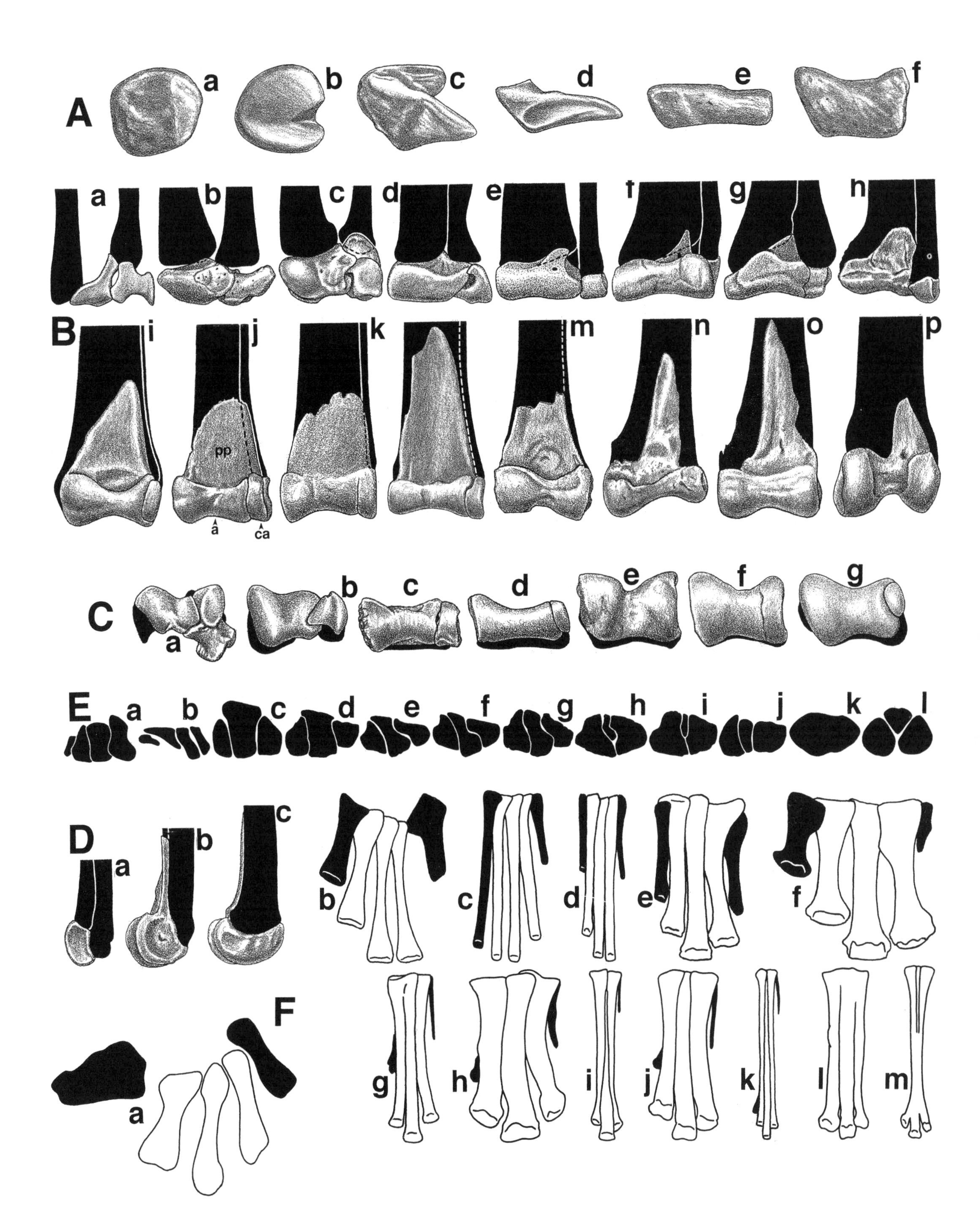
A
a
b
c
d
e
f
B
a
b
c
d
e
f
g
h
i
j
k
m
n
o
p
pp
a
ca
C
a
b
c
d
e
f
g
E
a
b
c
d
e
f
g
h
i
j
k
l
D
a
b
c
b
c
d
e
f
F
a
g
h
i
j
k
l
m

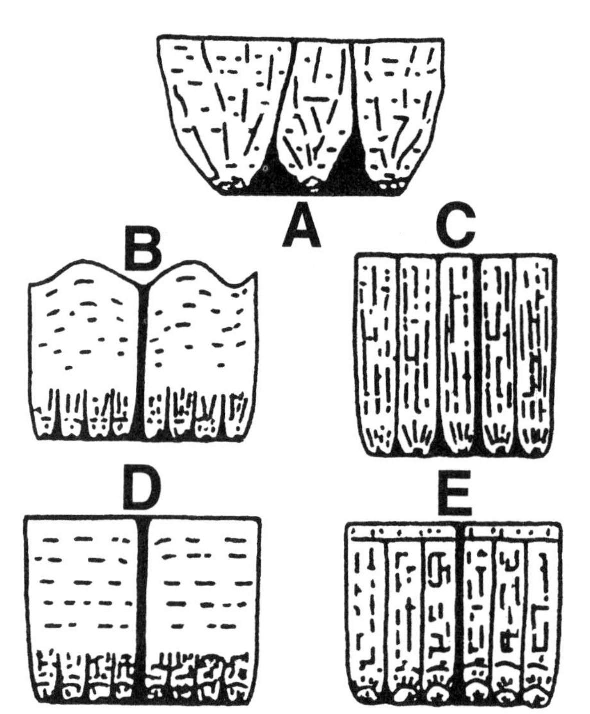


*Figure 10.15.* Eggshell microstructure. Crocodilian eggshells *(A)* are very different from those of the dino-avepods *(B) Oviraptor* and *(C) Troodon* and from those of *(D)* ratites and *(E)* neognathous birds. Not drawn to same scale.

and birds. In other words, when dino-avepods were thought to lack clavicles, Martin and Feduccia argued that dino-avepods could not be bird ancestors, but now that we know that they actually do have birdlike furculae, Martin and Feduccia suggest they must have merely converged with birds! This amounts to stacking the phylogenetic deck. Feduccia and Martin also violate the rules of objective analysis when they cite the clavicles of *Longisquama*—whose possible furcula is no more birdlike than the dinosaurs' and whose shoulder girdle is truly unbirdlike—as evidence that it was an "avimorph." Arguments of this sort leave one wondering whether some opponents of a dinosaur-bird link would reject any synapomorphic dinosaur-bird character, no matter how obvious the connection, and accept any character that seems to link birds with the non-dinosaurs they happen to prefer as bird ancestors.

The sound conclusion is that although the dinosaur furcula shows that dinosaurs were suitable ancestors for birds, it is a wash as a general synapomorphy because of its possible presence in another potential bird relative. However, that the most birdlike furculae are found in dino-avepectorans is additional evidence that they are the closest of the dinosaurs to birds.

The presence of a large sternum with a strong keel in alvarezsaurs has been cited as evidence of avian affinities by Perle et al. (1993) and others, and dismissed as convergence due to functional divergence (digging versus flying) by Feduccia (1996) and others. Resolution of this issue is difficult. Further obscuring the problem is the presence of a keel, albeit a small one, on the little sternum of the basal avetheropodian avepod *Sinraptor* (Fig. 10.9A; Currie and Zhao 1993a).

### Scapula and Coracoid

In a confusing and internally contradictory exposition, L. Martin (1997) asserted that dinosaur and bird shoulder girdles are too different for one to have evolved from the other. However, as shown in Chapter 4, entirely the opposite is true, the scapulocoracoids of *Sinornithosaurus* and *Archaeopteryx* being especially similar (Xu, Wang, and Wu 1999). Martin also argued that the backward-facing shoulder glenoid of dinosaurs was ill suited for ancestry of the laterally facing avian shoulder joint, which is more similar to that of sprawling reptiles. This argument is more logical, but the fossil record suggests that the sprawling avian shoulder joint did indeed evolve from a more erect basal dinosaur glenoid and that the reversal took place within avepod dinosaurs. Tarsitano (1991) and Tarsitano et al. (2000) asserted that dino-avepod coracoids are incompatible with those of birds because the former were too short and because they lacked the two distinct planes seen on the coracoids of basal archosaurs and *Archaeopteryx*. Both of Tarsitano's assertions are false. Dino-avepectoran coracoids were often unusually long and dromaeosaur coracoids did have two distinct planes (Fig. 10.8M,N). In 1977, Walker argued that the coracoid or "biceps" tuber on the coracoid of dromaeosaurs could not have anchored the biceps muscle and could therefore not be homologous with the avian acrocoracoid. The presence of an *Archaeopteryx*-type proacrocoracoid in *Sinornithosaurus* invalidates Walker's reasoning. Because avepectoran coracoids shared the anteromedial indentation that would eventually produce the strutlike bird coracoid, they were actually uniquely similar to the bird coracoid.

It is interesting that Feduccia (1986) has acknowledged that the close similarity between the scapulocoracoids of dromaeosaurs and ratites (Fig. 10.8N,T) may constitute evidence for a close relationship between the two groups! However, he has before and since preferred convergence (to the point that he did not cite his own

*Figure 10.16.* A brooding oviraptorid. This figure is based on the remains shown in Plate 15A; the positions of the skeleton and the eggs have been reproduced exactly. Note that all the eggs were in contact with the body. In order to leave the eggs visible, the feathers have been kept short; in real life they should have covered the entire nest.

1986 paper in his 1996 book), although he has not explained why the shoulder girdles of nonflying dinosaurs and neoflightless birds would converge so closely.

**Arm Length**

We now come to the common yet erroneous argument that never dies (see Feduccia 1999, Geist and Feduccia 2000, Tarsitano et al. 2000), no matter how often it is killed. The thesis is that the arms of predatory dinosaurs were too short for dinosaurs to have been ancestral to the long-armed birds and that a progressive shortening of dino-avepod arms over the Mesozoic was the opposite of the avian trend. Ergo, the argument goes, the ancestors of birds must lie among early quadrupedal archosaurs with long arms, and the birdlike arms of avetheropods were merely convergent with birds. None of this is true.

Since the beginning of this century, we have known that ornithomimosaurs were long armed: the arms of the possible giant ornithomimid *Deinocheirus* exceeded 2 meters in length! Dromaeosaurs and oviraptorosaurs are long armed as well, a fact ignored by Feduccia (1996). We now know that the arm length of dromaeosaurs is as much as 80 percent of hindlimb length (Xu, Tang, and Wang 1999, Burnham et al. 2000). If the arms currently placed with *Rahonavis* do indeed belong to the rest of the skeleton, then it is a very long-armed dino-bird. Appendix Figure 18C shows that the total-forelimb/ total-hindlimb length ratio in dromaeosaurs and oviraptorosaurs was as high (about two-thirds or more) as in quadrupedal basal archosaurs and basal crocodilians.

The assertion by Tarsitano et al. (2000) that the dinosaur hypothesis requires avepod dinosaurs with arms as long as the legs can be countered by demanding that they point to other nonavian archosaurs with arms so long. Besides, protoflyers should not be expected to have arms as long as the legs. Although it is true that the shortest-armed dino-avepods—especially the tyrannosaurs—are

Cretaceous, the long-armed dino-avepods also come from the late Mesozoic, whereas Triassic coelophysids were short armed. Assertions that dino-avepods lack sufficiently long arms to be bird ancestors are particularly illogical. Many dino-avepods are actually the most birdlike of archosaurs in that they are the only other long-armed bipeds with long, gracile hands in Archosauria (Fig. 10.11Ae–g,i. Poposaurs were also obligatory bipeds, but their hands were short and broad; see Fig. 10.11Ab).

### Wrist and Hand

L. Martin (1997) has argued that the arm- and wrist-folding mechanisms in dino-avepods and birds are dramatically different. His use of my older skeletal restorations (Paul 1988a) as the basis of his analysis was a mistake in that my skeletal studies were osteological blueprints intended to standardize limb posture and proportional comparisons for artistic purposes; they were never intended to imply that *Archaeopteryx* dragged its wing feathers on the ground! (Elzanowski and Pasko [1999] similarly mistook this artistic convention for a deliberately theropod-like grasping pose of the hand.) Martin ignored the much more important fact that I and most other researchers have shown that avepectoran dinosaurs and birds share the same arm-folding mechanism, one not seen outside Avetheropoda.

The lunate carpals that allow wrist folding in avepectoran dinosaurs and birds are unique and very similar (Fig. 10.11Cd,e). Opponents of dinosaur ancestry for birds inaccurately label this similarity as superficial (Feduccia 1996) and attempt to demonstrate that the elements are not really homologous. Feduccia (1996) and Hinchliffe (1997) go so far as to assert that the rarity of semilunate carpals in dinosaurs makes them less plausible homologies for the avian carpal. This assertion is both odd and incorrect: odd because it takes only one example of a feature in the potential ancestral group to make it a valid synapomorphy and incorrect because such carpals are found in coelurids, dromaeosaurs, troodonts, caudipterygians, oviraptorosaurs, basal therizinosaurs, and avimimids. In fact, most small theropods of the late Mesozoic probably had this feature. Feduccia and Hinchliffe's assertion is also misleading, because there is no example of a potential nondinosaurian bird relative with a birdlike wrist. This is another example of how a derived similarity between dinosaurs and birds is arbitrarily turned into evidence against a close relationship. Poposaurs had an enlarged distal carpal (Chatterjee 1985), but they were not viable bird relatives; the poposaur carpal was not as birdlike as the carpals of dinosaurs (Fig. 10.11Ab), and poposaurs lacked the asymmetrical humeral condyles integral to an arm-folding mechanism.

At a deeper level, some workers have questioned whether the half-moon carpal is made of the same elements in dinosaurs and birds. In birds, the carpal is a mediodistal element. Ostrom (1969) misidentified the dinosaurs' semilunate carpal as being made of proximal carpal elements, but it is actually made of one or two mediodistal elements (Fig. 10.11C; Gauthier 1986, Padian 1997a, Chatterjee 1998b). Feduccia (1996) seriously misunderstood the configuration of dino-avepod carpals, especially in his belief that dino-avepectoran wrists were missing an entire row of carpals. There is no firm evidence that dinosaur and bird semilunate wrist bones are not the same structure, although the block no longer originates from more than one element in birds. Feduccia (1996) and Zhou and Martin (1999) also seem confused when they cite the presence of three carpals aside from the semilunate block in *Archaeopteryx* as an avian feature not seen in avetheropod dinosaurs, which may have had only two such carpals (but see Chapter 4). Yet L. Martin, Feduccia, and co-workers (1998) noted the presence of only two of these bones in *Archaeopteryx* and *Confuciusornis,* and the same configuration is present in adult modern birds. Such inconsistent observations are hardly convincing. The carpals and hand elements of *Archaeopteryx* were somewhat more birdlike than those of avepectoran dinosaurs (Zhou and Martin 1999), but one would expect the wrists of flightless dinosaurs to be less avian than those of a flying relative. The distal placement of metacarpal III that Zhou and Martin cite as an avian, non-dinosaurian feature appears to be inconsistently absent and present in dino-avepectorans and birds, respectively (Chapter 4). Zhou and Martin also failed to note that the hands of some dino-avepectoran also had avian flight features, some not present in the urvogel (close distal contact between the outer two metacarpals, a strongly bowed outer metacarpal, and a robust and stiffened central finger).

No other question of avetheropod-bird homology has received more heated attention than the origin of the two groups' three fingers. The focus on this question has only increased as character after character previously thought to have been limited to *Archaeopteryx* and other birds has been found in new dinosaurs, leaving the finger homology issue a last line of defense for the pro-

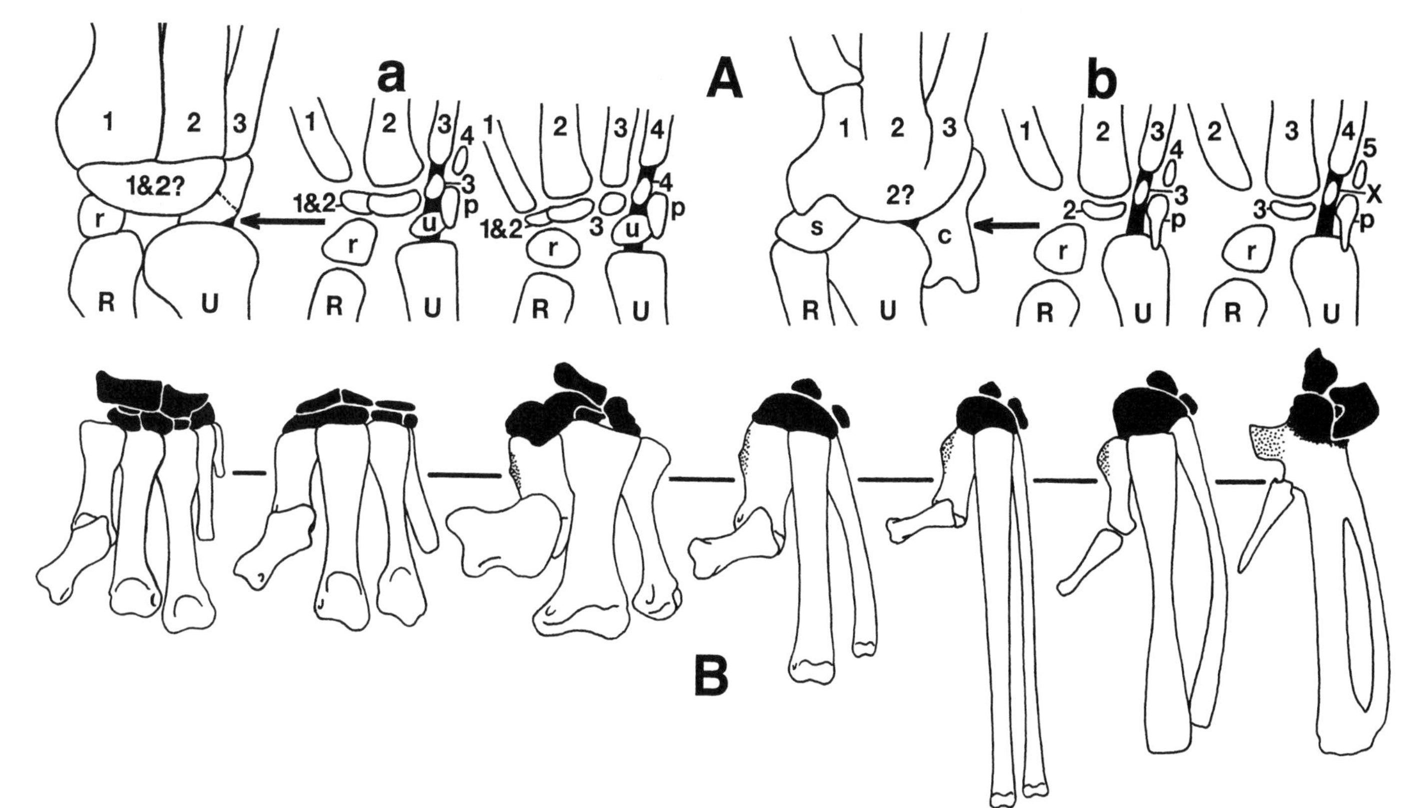


*Figure 10.17.* Evolution and embryology of theropod hands. *A,* semischematic renderings of right hands in dorsal view, with primary axis of growth indicated by solid black zone: *a,* generalized avetheropod/basal bird: *left,* adult with only digits 1–3 and primary axis through III, number of lateroproximal carpals uncertain; *center,* probable embryonic condition, with only I–III large, IV severely reduced, and primary axis through III; *right,* implausible embryonic condition with primary axis through a large digit IV; *b,* generalized ornithurine bird: *left,* adult; *center,* embryo with the probable retention of digits I–III and distal carpals 2–3 indicated; *right,* embryo showing common assumption that primary axis runs through IV, digit I is lost, and element X is unidentified. *B,* left hands in dorsal view, showing loss of lateral two digits and appearance of semilunate carpal (carpals *black*) and extensor process of metacarpal I *(stippled)* in theropods with medially divergent thumbs; *left to right:* basal theropod *Herrerasaurus;* cerato-sauran *Syntarsus;* tridigitate avetheropods: basal *Allosaurus;* dromaeosaur *Deinonychus; Archaeopteryx;* a juvenile ornithurine; and an adult ornithurine. Abbreviations: R = radius, U = ulna, c = cuneiform, p = pisiform, r = radiale, s = scapholunar, u = ulnare. Not drawn to same scale.

ponents of the antidinosaur hypothesis. Obviously, the tridigit hands of avetheropod dinosaurs, avepectorans especially, and birds are remarkably alike. If someone could demonstrate that the fingers of the two groups are actually fundamentally different in origin, then their similarity would stand as a shining example of convergence, one that would refute a close dinosaur-bird link.

The fossil record clearly shows that the reduction of the fingers in predatory dinosaurs was strongly asymmetrical (Fig. 10.17B). Digit I was always a strong, big-clawed thumb. Digit V was just a metacarpal splint in baso-theropods and was lost in cerato-saurans. Digit IV was just a small metacarpal tipped by at most a single abbreviated finger element in baso-theropods and baso-avepods, and was lost in avetheropods. Digit IV could not have evolved into a well-developed finger with five bones including a large claw, while the strong thumb was lost. Digit III was shortened relative to II in avepod dinosaurs. This skeletal data refutes arguments (Thulborn and Hamley 1984, R. Thulborn 1993) that avetheropod fingers are fundamentally II–IV. Thulborn and Hamley identified as digit I the small medial impressions observed in the handprints found in the trackways of some tridactyl-footed dinosaurs. However, these impressions are too indistinct for confirmation that they record a small digit I. For that matter, the handprints are so short fingered that it is improbable that a theropod made them. It is more likely that the trackways were made by basal semiquadrupedal ornithischians whose toe IV was too short to make an impression (as per heterodontosaurs). The asymmetry of theropod digit reduction is unusual and appears to reflect the need for retention of the thumb as a large weapon in bipeds whose hands were no longer important for locomotion.

There are three alternatives regarding the origins of avetheropod and bird fingers. The first is that bird fingers are the same inner three as in dinosaurs (Van Tyne and Berger 1976, Howgate 1983, Paul 1984c, 1988a, Gauthier 1986, Proctor and Lynch 1993, Shubin 1994, Burke et al. 1998, Chatterjee 1998b, Garner and Thomas 1998, Padian and Chiappe 1998a,c, Zhou et al. 2000). The second alternative is that bird fingers are II–IV, in which case their hands are only convergent with those of avetheropods (Hecht and Tarsitano 1982, 1984, Hinchliffe 1985, 1997, Tarsitano 1991, Hecht and Hecht 1994, Feduccia 1996, 1999, Burke and Feduccia 1997, Burke et al. 1998, Feduccia et al. 1998). The third alternative is that there was a shift in digit count as the asymmetrical trifingered hand evolved in avepods (Wagner and Gauthier 1999). Proponents of the second alternative often argue that bird fingers *should* be II–IV, because this symmetrical digit reduction is the norm in tetrapods. But it is not the norm for bipedal archosaurs, as shown by the example of theropods.

The primary evidence for avian digits being II–IV is embryological, but there are many problems with this evidence. The first problem is that although we can observe the growth pattern of avian digits in bird embryos, we have no avetheropod embryos for comparisons. Second, there are no bird embryos that have five digits, from which the true digit numbering could be determined. In addition, positive identification of avian digits via molecular coding is not yet possible, and according to Dahn and Fallon (2000), it may never be possible. Nikbakht and McLachlan (1999) have suggested that manipulation of growth factors in avian digits may eventually solve this problem. Until then, the count of the four digits, of which the outer is lost with growth, must be inferred on the basis of assumptions that remain untestable with current biotechnology. Despite attempts to change the situation with additional work (Hinchliffe 1985, Burke and Feduccia 1997, Burke et al. 1998), the barriers to reliably counting bird embryo fingers remain as strong as they were when Hinchliffe (1977) cautioned that "the embryological convention that digit I is missing [in bird embryos] is not based on firm evidence." Nikbakht and McLachlan (1999) summarized the situation well when they stated that "studies of the existing elements will probably never resolve this argument, since the information is too scanty."

In tetrapods the lateral-most digit next to the pisiform in the side of the wrist is V, so Hinchliffe (1985) assumed that the same was true in bird embryos (Fig. 10.17Ab). However, Burke and Feduccia (1997) noted that in tetrapods, digit IV normally grows in line with the ulna and ulnare and forms the "primary axis" (Fig. 10.17Ab) from which the rest of the inner digits develop via a medial arc. Therefore, these workers identified the digit at the end of the ulna-ulnare axis—which is immediately medial to the lateral-most digit—in bird embryos as IV. According to this view, the remaining two digits are II and III, and I is lost during development. Because Hinchliffe identified the digit second from the medial side as III, he was forced to identify the distal carpal that caps the digit, and grows into a semilunate, as 3. This left a laterodistal element as mysterious "X."

In adult avetheropods, finger III, not IV, was in line with the ulna and ulnare (Fig. 10.17Aa). If finger IV started out in line with the lateral wrist elements at the beginning of embryonic development, then III would have had to shift laterally in order to take the place of IV during ontogeny. There would have been no need for such a convoluted growth pattern if III initiated growth in line with the lateral arm elements. In this case, there are two possible scenarios. Either the primary axis was a centered on a reduced, laterally placed IV that was later lost into the ulnare (Garner and Thomas 1998); or the primary axis shifted to III (Shubin 1994, Chatterjee 1998b, Garner and Thomas 1998), and IV was reduced to a stub lateral to the primary axis before it was lost (Fig. 10.17Aa). In the latter scenario, IV mimics V in being small and laterally placed; and laterally shifted III mimics IV by being the primary axis throughout growth. Shifting of the axis of finger growth is observed in other tetrapods (Chatterjee 1998b).

Let us look at the matter from another aspect. The atypical asymmetry of the avetheropod hand may have forced a correspondingly atypical change in its ontogeny, perhaps via a fundamental shift in the primary axis. This change appeared in avepod dinosaurs and may be operating in birds (Fig. 10.17A). If this view is correct, then the identity of the distal carpals in birds is now more obvious: the mediodistal carpal is 1 or 2 or both, and element "X" is distal carpal 3.

A third hypothesis, proposed by Wagner and Gauthier (1999), held that the early avepod thumb was I, that the avian thumb is II, but that the latter descended from the former. In this interesting view, the developmental difficulties of growing an asymmetrical set of digits forced the embryological condensation that led to the thumb shifting from I to II, and the rest of the fingers had to shift laterally as well. In this case, the thumb was digit II in tridactyl theropods. Wagner and Gauthier supported their reasoning with the important observation that initial condensation and digit identity are inconsistent in the two-fingered hands of kiwis. The possibility of such shifts in finger identity is supported by new work by Dahn and Fallon (2000) showing that digit identity is not fixed. The Wagner and Gauthier hypothesis may be correct, but it currently suffers from the same problems of verification that afflict the alternatives.

The final problem with the antidinosaurian version of the II–IV hypothesis is that it does not account for the divergent nature of the innermost digit in both dino-theropod and bird hands, nor does it explain how the finger-element count in birds happened to exactly mimic that in avetheropods: 2 bones in the thumb, 3 in finger II, and 4 in finger III. Burke and Feduccia's (1997) attempt to explain away this shared condition in dinosaurs and birds proposes chance events rather than positive selective forces. The Burke and Feduccia hypothesis is improbable, but it is not strange. The same cannot be said of Hecht and Tarsitano's (1982, 1984) scenario: they suggested that what is usually considered to be a joint in the lateral finger of *Archaeopteryx* might actually be a break, one that resulted from the final impact of the urvogels with the water. (Howgate in 1983 made some appropriately caustic comments on this notion!) In this case, the finger-element count of *Archaeopteryx* would be 2-3-3, and its hand would not be so similar to that of avetheropods. However, as discussed in Chapter 4, because the joint is present in all specimens, it is real and, thus, the finger had four bones, as in dinosaurs. Finally, Feduccia (1996) stated that the distinctive crossing of the outer digit under the central digit in the various *Archaeopteryx* hands is a nondinosaurian character, but as discussed in Chapter 4 this is more likely to be a minor postmortem distortion.

If researchers working on hand homology had noted that the general tetrapod pattern of digit development poses a problem in identifying bird fingers as I–III, they could have made an interesting point. If the dinosaur-bird hypothesis was not much stronger than the alternatives, then the embryological data could be an intriguing but not definitive argument against the hypothesis. It would help the antidinosaur case if all embryologists agreed that that bird fingers are II–IV. The problem is that some workers have insisted that this issue poses a major challenge to—and perhaps even conclusively falsifies—the dinosaur-bird hypothesis. This is not at all convincing when embryologists themselves disagree on the developmental origin of avian fingers! Overall, the evidence favoring the dinosaur-bird link is too overwhelming to be seriously challenged by such nondefinitive evidence. Instead, the wrists and hands of avepectoran dinosaurs and birds are so similar in digit count, morphology, and function—to the point that *Sinornithosaurus* and *Archaeopteryx* share remarkably similar appendages (Fig. 10.11Ah,i)—that those attempting to demonstrate that they are superficially convergent rather than homologous must supply definitive evidence that the avian digits are II–IV and could not have been derived from dinosaur digits I–III. Zhou et al. (2000) have reinforced this point by emphasizing that the hand of *Caudipteryx* so closely matches the avian condition—outer finger severely reduced and appressed tightly to the central finger (Fig.10.11Aj,m)—that it supports the homology of dinosaur and avian fingers. The presence on the central finger of a dromaeosaur of a distinctive flight feature otherwise limited to the central finger of birds further supports the homology of dinosaur and bird hands at the expense of convergence (Appendix 6). At this time, the embryological data is absent in the case of dinosaurs, the means to reliably number avian digits is lacking, and no one has demonstrated that major condensation shifts cannot occur. Therefore, researchers cannot form a consensus on the issue. Such ambiguous information cannot be used to overturn the probable homology of the tridactyl avepod dinosaur-bird hand. The statement by Feduccia et al. (1998) that "dinosaurs retain digits one, two and three, and birds have digits two, three and four" is a plausible but unverified opinion rather than a scientific fact. Feduccia also ignores the fact that dino-avepods had made the shift from having a divergent outer digit to having a divergent inner digit, a condition that is retained in birds. The remarkable likeness of the distinctive wrists and tridactyl hands of birdlike dino-avepods and birds, especially *Archaeopteryx* and other avepectorans, is yet another reason to ally the two groups.

### Pelvis

Because the pelves of the *Archaeopteryx* specimens are so similar to those of theropods, minor differences have been cited as evidence that the similarity is a phylogenetic illusion due to parallelism. Some researchers allege that the pubis of *Archaeopteryx* is more retroverted than that of any dinosaur, but the pubes of some dromaeosaurs and alvarezsaurs are if anything directed more backward than the pubis of the urvogel (Chapter 4). In most theropods, the main shaft of each pubis is a rod, and sometimes a medial sheet connects the two rods. In *Archaeopteryx* the pubic shafts are flattened, subtriangular plates that meet at a shallow angle. This condition could constitute a major difference with dino-avepods, except that the cross sections of dromaeo-avemorph pubic shafts have the same features (Fig. 10.12D), so much so that the evidence firmly supports the probability that the retroverted pubes are homologous rather than being yet another example of convergent retroversion, which was common among archosaurs.

L. Martin (1991), Feduccia (1996), Ruben et al. (1997), and Ruben and Jones (2000) have pointed to the lack of an anterior projection of the pubic boot, and the presence of a hypopubic cup, as evidence the urvogel was no avepod. But a number of some dromaeo-avemorphs also lacked an anterior process of the pubic boot; for that matter a number of theropods with large pubic boots lack this process (Fig. 10.12Bb,g,r,t–v; Burnham et al. 2000). Therefore, the phylogenetic significance of this character is at best weak. In addition, as discussed in Chapter 4, there is solid evidence that *Archaeopteryx* or any other avepod had a hypopubic cup. Feduccia's (1996) assertion that a long, booted pubis represents the primitive archosaur condition is by no means correct. Early archosaurs had short, broad pubes; and the boots independently evolved only among theropods and poposaurs (Fig. 10.12Ac); and the poposaur example is unimportant because the forms with crurotarsal ankles are not considered potential bird relatives.

L. Martin (1991a), Tarsitano (1991), and Feduccia (1996) have stressed that the auxiliary processes present on the dorsal edge of the *Archaeopteryx* ischium are absent in avepod dinosaurs. Actually, ischial dorsal processes can be found in a wide variety of dino-avepods (Fig. 10.12Bf,g, r–v), ranging from rather primitive *Elaphrosaurus* (Janensch 1925), giant *Yangchuanosaurus* (Dong et al. 1983), the urvogel-like dromaeosaur *Sinornithosaurus* (Xu, Wang, and Wu 1999), and birdlike *Unenlagia* (Novas and Puerta 1997), *Bambiraptor* (Burnham et al. 2000), and *Microraptor* (Xu et al. 2000) to the even more birdy *Rahonavis,* in which the processes are most similar to those of the urvogel (Sampson et al. 1997, Forster et al. 1998). Hutchinson (2001) carefully detailed the evolution of the avian pelvis and its musculature from that of dinosaurs. He emphasized that there is no sharp demarcation between the hips of dromaeosaurs and basal birds, and he found no evidence that this extreme similarity resulted from convergence. When the evidence is viewed objectively, it is apparent that no archosaur had a more birdlike pelvis than the retropubic sickle-clawed dinosaurs (Fig. 10.12Br–v,Ca–g,D).

### Acetabulum, Femoral Head, and Leg Posture

In 1997, L. Martin continued to contend that *Archaeopteryx* had a semi-erect leg posture and a partly closed off hip socket. If his contention is true, then more-advanced birds converged with dinosaurs in developing an erect bipedal gait. However, his notion is false because most dromaeo-avemorphs tended to have partly closed off acetabulum, as do some birds, and medial closure of the hip socket is not associated with a semi-erect gait in either dinosaurs or birds (Chapter 4). Tarsitano (1991) asserted that the trochanters at the femoral head of *Archaeopteryx* are dissimilar to those of dino-avepods. Actually, those of the urvogel are very similar to those of dromaeosaurs (Fig. 10.13Bf,g).

### Ankle

Only the members of the dinosaur-bird clade have a proximal tarsal process that is exposed on the anterior surface of the tibia (Fig. 10.14B, 10.18A–C). This synapomorphy has been challenged on two grounds. First, some have claimed that because the ascending process is a feature that evolved repeatedly (Martin 1991, Tarsitano 1991, Feduccia 1996), it is not a homologous synapomorphy limited to dino-avepods and birds. However, the ascending process first appears in incipient form in protodinosaurs, and it is a general dinosaur feature that from its modest beginnings developed into a taller structure in various dino-avepod clades (Paul 1988a, Sereno 1991). Therefore, only the height of the process increased independently, and there is no evidence that the dinosaur ascending process is not a shared homologous synapomorphy.

The second challenge to this synapomorphy is based on the claim that in dinosaurs the process is a centrally placed extension of the astragalus in dinosaurs, but in birds it is a more laterally placed and separate ossification—the pretibial bone—that mainly contacts the calcaneum rather than the astragalus. Welles (1984) showed that the ascending process is a separate ossification in dinosaurs too (Fig. 10.18C), but his vital study was not cited by L. Martin (1991), Tarsitano (1991), or Feduccia (1996). However, there is no consistent difference between the placement of this process in dinosaurs and birds. In some dino-avepods, the apex of the process is laterally placed, and contact of the process with the calcaneum ranges from nonexistent to extensive; the process is therefore both centrally and laterally placed within the alvarezsaur group, and the process is centrally placed and contacts only the astragalus in palaeognathous birds and in a few other birds (Figs. 10.14B, 10.18A–C; Ostrom 1976a, 1991, McGowan 1984, 1985, L. Martin and Stewart 1985, Paul 1988a). The obvious conclusion is that the ascending process and pretibial bone are the same structure, the pretibial process, which simply shifts position laterally or medially

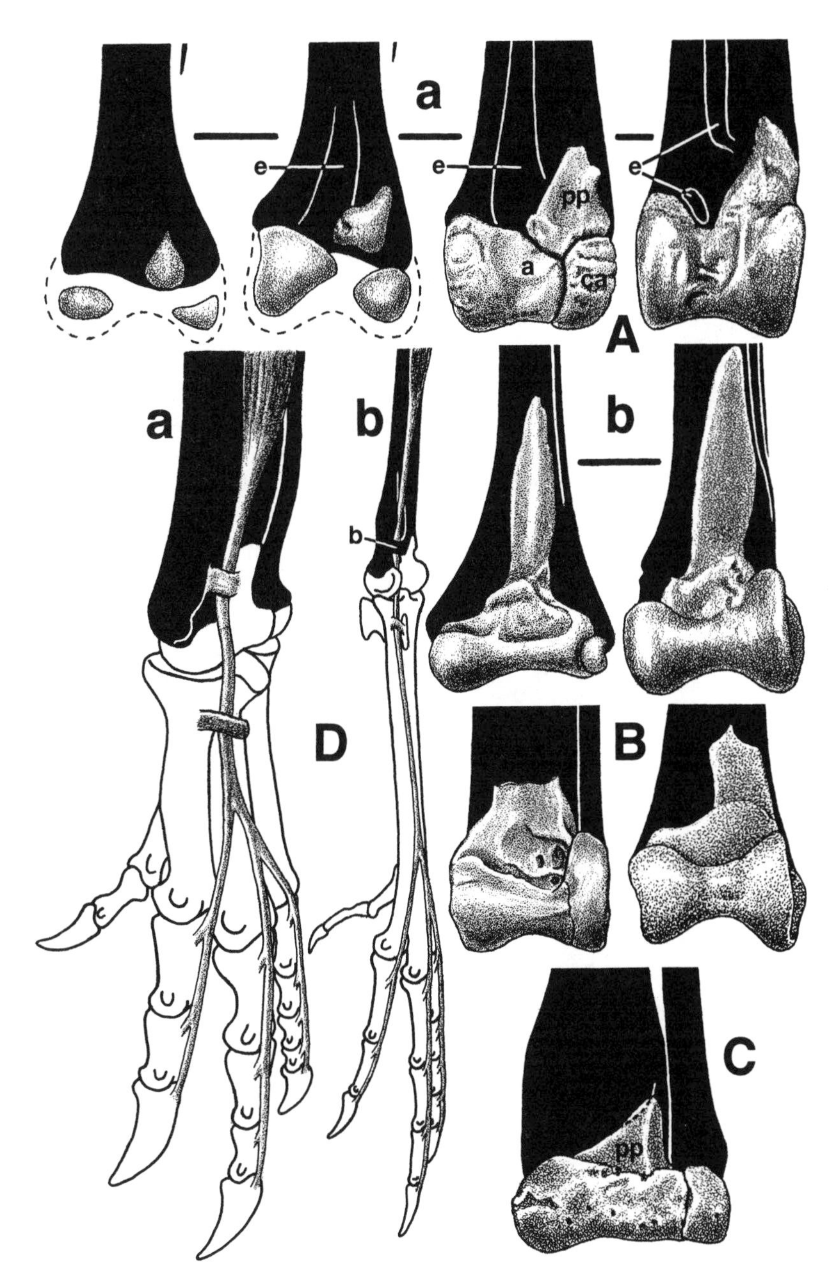


*Figure 10.18.* Variation and ontogeny of avepod pretibial processes, all lefts. *A,* development of the ankle elements: *a,* chicken (in part after Jollie 1977): *left to right:* 1.5 days, 1.5 weeks, 6 weeks, fully mature; *b,* ostrich juvenile and adult. *B,* alvarezsaurs: *left, Patagonykus; right, Mononykus. C, Dilophosaurus,* showing division between pretibial process and astragalus. *D,* lower legs and feet, showing how the development of a extensor canal (e) and supratendinal bridge (b) to accommodate the extensor digitorum longus tendon influences the position of the pretibial process: *a,* dino-avepod *Allosaurus,* in which the canal is shallow, the bridge is absent, and the process extends medially (see Fig. 10.14Bh); *b,* chicken, in which the canal is deep, the bridge is present, and the process is therefore laterally placed. Additional abbreviations listed in Figure 10.14. Not drawn to same scale.

*Figure 10.19.* Avepod footprints drawn to same scale, all presented as rights. In the set of supposed bird tracks from the Late Triassic and Early Jurassic at the *upper left,* the slender, widely splayed toes and seemingly reversed halluxes are birdlike, but some are suspiciously large compared with the print of an average-size male paleontologist *(upper right).* In the *lower center* are probable bird prints that indicate that birds large and small were roaming shorelines in large numbers by the Cretaceous. The two prints on the *lower right* are those of an emu and a goliath heron on the *right.* On the *left* is an Early Jurassic impression of a small dino-avepod that includes a possible trace of a reversed hallux. The giant print that underlies the other prints was probably made by *Tyrannosaurus,* and it appears to show a partly reversed hallux. Scale bar is 200 millimeters.

in response to functional needs. In particular, a well-developed supratendinal bridge on the anterior surface of the distal tibia has pushed the ascending process to the side in most birds (Fig. 10.18A,D). Because there is no consistent difference in the origin and placement of the proximal extension between the two groups and because the process is not found in nondinosaurs, it should be considered a valid dinosaur-bird synapomorphy. The burden of proof falls upon those who wish to show otherwise.

### Hindfoot and Hallux

The absence of a fully reversed hallux in a dino-avepods has long been cited as a crucial failing of the dinosaur-bird hypothesis. This is a mistake. Even though this feature is lacking in the dinosaurs, its absence means only that the dinosaurs were not yet fully avian in this regard; ancestors cannot be expected to have all the characteristics of their descendants (Witmer 1997b)! Although the reported reversal of the hallux in *Caudipteryx* (Zhou et al. 2000) supports the dinosaur-bird connection, the importance of a fully reversed hallux as an avian marker should not be overemphasized. Consider, for example, that the kiwi hallux is no more reversed that that of *Tyrannosaurus* (compare Figs. 4.5D and 10.19). There are three anatomical details that are more phylogenetically important than hallux reversal: dino-avepods and birds are the only potential bird ancestors in which metatarsal I fails to contact the ankle; dino-avepods were the one nonavian archosaur group that, like birds, made such a profound shift from having a divergent outer digit to having a divergent inner digit; and dino-avepod and basal bird tarsometatarsi are so similar that they have been confused for one another (Brett-Surman and Paul 1985).

### Body Posture

Feduccia (1999) cited the transformation from the dino-theropods' "balanced seesaw body plan to the avian model" as a serious problem. Quite the opposite is true. It was the already gracile-tailed and highly bipedal theropod dinosaurs, especially the dino-avepectorans with their further reduced tails, that were the archosaurs best suited for evolving into stub-tailed bipeds.

### Feathers

Until now, the evolution of feathers was one of the least understood topics in the study of bird origins. The problem was a simple gap in the fossil data, a gap that is only now being filled by the long-awaited discovery of nonavian archosaurs with feathers. The factors involved in the origin of feathers are discussed and debated in Maderson and Homberger (2000), as well as in Brush (1993, 1996, 2000) and Feduccia (1985, 1986).

Feathers are not, as often thought, simply elongated reptile scales with frayed edges gone complicated. At the molecular level feathers are unique structures that differ from scales, so they are novel in origin. Brush suggests that feathers were derived from tubercles and that the first feathers should thus have been simple, hollow bristles. Feduccia and others argue that feathers first evolved as aerodynamic flight surfaces and only later were used for insulation. In the latter view, complex body contour feathers are overkill as thermal shields because furlike structures will do the job and because in birds the body feathers originate from well-defined tracts instead of being spread evenly over the body. Ergo, according to his argument, feathers neither could have evolved in dinosaurs nor have been simple fibers spread densely over the body. Feduccia's 1996 version of the "proavis" therefore has a scaly body, with feathers limited to the wing and tail airfoils. This is an example of the phenomenon whereby researchers decide on what they believe should have happened and then derive subsidiary conclusions from that belief. Fortunately, conclusions derived in this way can be falsified by fossil discoveries; for example, the simple hollow fibers observed on dino-avepods are proving to have been homologous with feathers (Chapter 5).

That a basal avetheropod such as *Sinosauropteryx* had feathers means two things. First, a feathery body covering first evolved as simple, hollow, fiber insulation spread over the body of small predatory dinosaurs, rather than as contour feathers for avian flight. Second, feathers have gone beyond being a speculative possibility in predatory dinosaurs to constituting an intimate link between them and birds. Where Feduccia may be correct is in his assertion that complex *contour* feathers may not have evolved until the advent of flight, perhaps in the Avepectora.

The argument by Jones et al. (2000a, 2001) that *Longisquama* had true feathers is inconsistent and ironic on multiple counts. Jones and his co-workers deny that the keratin fibers found on dinosaurs can be identified as feathers because biochemical examination has yet to conclusively verify the hypothesis, even though the data so far favors it. Yet they concluded that the longisquamid structures were true feathers even though it is not possible to use biochemical analyses to test whether the impressions were even animal rather than vegetable in origin. Jones et al. further asserted that that feathers are very unlikely to have evolved twice, but these workers have no trouble hypothesizing that many similarities shared by dinosaurs and birds are due to convergence. The reverse is the case. The few similarities shared by longisquamid dorsal appendages and feathers—central shaft and associated details, asymmetrical vanes—are statistically much more likely to have resulted from convergence than the massive skeletal and integumentary similarities in predatory dinosaurs and birds. The claim that dorsal appendages that may have evolved for display were true feathers is also in danger of contradicting the argument that avian-type flight was the driving force behind the initial development of feathers.

Even if the dorsal structures associated with *Longisquama* prove to be homologous with feathers, this homology is not in and of itself evidence against the dinosaur-bird hypothesis. To provide evidence against the hypothesis, someone must demonstrate that the dinosaur's fibers were not homologous with feathers. We do not understand the integument of basal archosaurs well enough to assess whether protofeathers evolved in basal archosaurs—and were then inherited by dinosaurs as well as pterosaurs—or did not appear until after the development of dinosaurs. It is possible that future fossils will reveal that various extinct diapsids and archosaurs bore an array of unusual integumentary structures.

The most telling fact is that the longisquamid impressions are different from all known feathers, whereas the three-dimensional alvarezsaur fibers are indistinguishable from modern bristle feathers.



### Reproduction

The assertion by Feduccia et al. (1996) and Martin and Simmons (1998) that dino-avepod and bird eggshells are divergent has been falsified by the discovery that the microstructure of dino-avepectoran eggshells follows the avian pattern. This similarity means that both the skeletons and the eggs of avepod dinosaurs and birds shared derived morphologies, which further reduces the possibility of convergence. Also contradicted by the evidence are the claims by Feduccia et al. (1996), Geist and Jones (1996), and Martin and Simmons (1998) that the reproduction of avepectoran dinosaurs was within the reptilian norm. Instead, the mixture of reptilian and avian egg-deposition and nesting habits is exactly what would be expected in archosaurs that are not yet birds but are very close to them.

### Predatory Dinosaurs Dragged Too Much

Tarsitano et al. (2000) claimed that all dino-avepods had vertical or procumbent pubes and were therefore were so deep keeled that their bodies were not streamlined enough to be protofliers. These workers ignored the abundant evidence that some dromaeosaurs were at least as strongly retropubic as *Archaeopteryx* and, therefore, no more deep keeled than the flying dino-bird. The demand by Tarsitano et al. that the aerodynamic qualities of small predatory dinosaurs must be verified in wind tunnel tests is incorrect; instead it is up to them to convincingly demonstrate that dromaeosaurs were markedly less streamlined than the urvogel they so closely resembled.

### Lifestyle: Arboreal or Terrestrial?

We briefly return to the debate about arboreal and terrestrial origins of avian flight, addressed in depth in Chapters 6 and 9, to repeat that the arboreal hypothesis does not contradict the dinosaur-bird hypothesis, for three reasons. First, many avepod dinosaurs could climb well; *Sinornithosaurus, Microraptor,* and the recently discovered arboreal dino-avepod in particular refute arguments that they could not. Second, dino-avepods were well suited for both the climbing and the running hypotheses. Third, hard anatomical data takes precedence over inherently speculative scenarios about how flight should or should not have evolved. In the end, phylogeny will tell us more about how flight evolved than vice versa.

### Size

Tarsitano et al. (2000) restated the old argument the dino-avepods were too big to be the climbing, gliding ancestors of birds. Always a weak contention, it has been demolished by the finding of adult dino-avepectorans such as *Microraptor* that were at least as small as *Archaeopteryx*.

Besides, the antidinosaur group has reversed the logic of the issue yet again. They have proposed nondinosaurian proavians weighing from one to a few dozen grams. However, the earliest known flying bird may have weighed as much as half a kilogram, and we know that smaller birds appeared only later. Thus, in this antidinosaur scenario there is a temporary and inexplicable size bump in early avian evolution, an awkward inconsistency that has yet to be addressed by those who propose this scenario.

Why then was *Archaeopteryx* so big, when the evolution of flight should have been even easier at a smaller scale? Perhaps because birds descended from dino-proavians that weighed a kilogram or two? It is interesting that there is evidence, discussed in the next subsection, for a radiation in the later Jurassic of a class dino-avepectorans that weighed a kilogram or so and were small enough to glide.

### Time, Trackways, and Bits and Pieces

A constant theme among the proponents of the antidinosaur thesis is that the dinosaurs evolved too late to have been the ancestors of birds (Martin 1991, Tarsitano 1991, Feduccia 1996, 1999, Geist and Feduccia 2000). Obviously, birds had evolved by the Late Jurassic, at the latest. If dinosaurs were ancestral to birds, then the most birdlike dinosaurs should be found in the Jurassic, if not earlier. Instead, the birdlike avepectoran dinosaurs are found in the Cretaceous and, therefore, the argument goes, cannot not be ancestral to the earlier *Archaeopteryx*. Ergo, the dinosaurs only converged with birds.

This argument ignores several crucial facts. First, it is well known in phylogenetic circles that using time as a basis for argument can be tricky because ancestral types commonly survive after some of their descendants have become extinct. For example, chimpanzees are not much different from the long-gone apes that gave rise to both humans and chimps. Ratites and tinamous are survivors of an early radiation of birds that gave rise to neognathous birds. No one argues that Cretaceous dino-avepods were ancestral to birds. However, the late Mesozoic dino-avepectorans do appear to be descendants of earlier species that *were* ancestral to birds (Brochu and Norell 2000).

Second, no convincing *nondinosaurian* bird predecessors have been found in the Jurassic prior to *Archaeopteryx* either. Aside from the highly dubious "*Protoavis*" remains, the only potentially avian fossils from earlier than the Late Jurassic are birdlike footprints from the Early and Middle Jurassic (Lockley et al. 1992, Chatterjee 1997, Gierlinski and Sabath 1998). The footprints are interesting. Some are oddly asymmetrical and not especially birdlike. More tantalizing are the prints that do indeed look like those of birds, having very slender toes, widely divergent digits II and III, and a strongly or fully reversed hallux (Fig. 10.19). However, are those who suggest that these prints are avian risk repeating pioneer paleoichnologist Edward Hitchcock's classic mistake? In the early 1800s, Hitchcock identified all the tridactyl prints he described as having been made by birds of the early Mesozoic, when in fact they were actually made by dinosaurs (Mikulic 1997). In some of the recently described birdlike prints, the hallux is half reversed, no more reversed than it is in some avepod dinosaurs. The footprints in which the hallux seems to be fully reversed may represent the same artifact postulated for some dino-avepod prints (Chapter 4). Most importantly, some of the prints are as big as human footprints (Fig. 10.19), which means that the mysterious avepods were as large as herons and emus! There are three possible explanations for these prints. The first explanation, that big birds were wandering about as early as the Early Jurassic, is not impossible, but definitely improbable. The second, that the small prints were made by birds and the large ones by a group or groups of avepod dinosaurs whose hindfeet happened to exactly mimic those of the little birds, is possible, but still improbable. The third, that all of the prints were made by small (or juvenile) to medium-sized dino-avepods whose feet paralleled or preceded the avian condition at least in terms of slenderness and divarication and perhaps in terms of hallux reversal, is not only possible but also probable.

This last point brings us to the fact that the most avian of earlier Jurassic archosaurs were predatory dinosaurs. Feduccia's (1999) statement that "Triassic theropods are devoid of birdlike features" is of course false, since it is in these creatures that birdlike features first appeared. It is also important that theropods with long-fingered hands suitable for climbing (coelurids and basal therizinosaurs) are found as early as the Late Triassic (in herrerasaurs), as well as near the J/K boundary. For these reasons, the Mesozoic fossil record favors rather than disfavors the dinosaur hypothesis over the alternatives.

We must remember a basic fact of paleontology: that the fossil record is spotty. One pertinent example is the lack of skeletons of the makers of the Jurassic birdlike footprints. For that matter, there are no reasonably complete skeletal remains of any kind of small predaceous dinosaurs for most of the Jurassic. They were there, but we just do not know much about them. Late Jurassic sediments have produced only three good skulls and skeletons representing just two kinds of dino-avepods, compsognathids and ornitholestids. That these happened to be avetheropods less birdy than Cretaceous dino-avepectorans may have been just the luck of the draw. But the situation is changing. Ornithomimosaur-, dromaeosaur-, and troodont-type teeth are known from the same time, as are a few isolated bones that may belong to dromaeosaurs or their predecessors (Jensen and Padian 1989, Chure 1994, 1995, Zinke 1998). Dromaeosaur- and troodont-type teeth have been found as early as the Middle Jurassic (Evans and Milner 1994, Metcalf and Walker 1994). It is interesting that most of the remains of these dinosaurs are quite small, suggesting that these creatures weighed only a kilogram or two. Similarly diminutive *Microraptor* appears to be a Cretaceous representative of this size class. Small size explains why we are not finding more such small dinosaurs; being tiny dramatically reduces the chances of preservation (that they were highly energetic predators near the top of the food chain also made them scarce and even less likely to appear in the fossil record). Someday, better remains of the small dinosaurian "proavians" of the Jurassic will probably be found. Those who oppose the idea that birds were dinosaurs appear unfamiliar with this tantalizing data.

More ambiguous is the therizinosaur lower jaw that supposedly pushes avepectorans back to the Early Jurassic (see Chapter 11). Also ambiguous at this time are the caudipterygians and therizinosaurs of the Yixian Formation, which are Early Cretaceous in age (see Chapter 13). The presence of such sophisticated and divergent avepectorans at these times adds to the evidence that the clade first appeared well back in the Jurassic. As discussed in Chapter 13, the Late Jurassic appearance of birds actually favors a dinosaurian ancestry of birds.

### "*Protoavis*"

"*Protoavis*" is involved in a variation of the contention that dinosaurs evolved too late to have been the ancestors of birds: some researchers have concluded that Triassic "*Protoavis*" was a protobird or basal bird that, because it was too early in time, contradicts a dinosaurian ancestry of birds. However, dinosaurs lived at the same time as "*Protoavis*," so they are not temporally excluded on the basis of "*Protoavis*." Besides, as we have seen, some of the bones attributed to "*Protoavis*" have dinosaurian attributes. This fact prompted Chatterjee (1991, 1999a) to argue that "*Protoavis*" is a dinosaur-like early bird that links the two groups. Although this argument is flawed, the very fact that Chatterjee makes it means that "*Protoavis*" by no means offers telling anatomical evidence against the dinosaur hypothesis. The drepanosaur parts of "*Protoavis*" do demonstrate convergence, that is, convergence between birds and nondinosaurs. The removal of "*Protoavis*" as a Triassic avian weakens the empirical case for a Triassic origin of birds.

### *Archaeopteryx* and *Protarchaeopteryx* Were Birds, Not Dinosaurs

Some researchers have argued that the original birds were just that, birds, rather than dinosaurs. This view was especially favored by Feduccia (1996). More recently, some have argued that *Protarchaeopteryx* was a neoflightless bird. Of course, even if this contention were true, it would not necessarily disqualify dinosaurs from being the ancestors of archaeopterygiforms. Besides, the urvogel lacked many the avian features that have been attributed to it. Feduccia (1996) was particularly out of step with the latest information on *Archaeopteryx* when he cited older studies (in particular Whetstone 1983 and Haubitz et al. 1988) that had concluded without justification that it had a nondinosaurian braincase and suspensorium. Feduccia failed to consider the more recent work that uses better data from the newest specimens to disprove the earlier work (especially Elzanowski and Wellnhofer 1995). For the same reason, failure to consider recent work, L. Martin et al. (1998) overemphasized how close the urvogel was to the much more birdy *Confuciusornis*.

The avian kinetic system was not present in the urvogel. Except for the short tail, the vertebral column and rib cage were no more derived than those of avetheropod dinosaurs in general. The pectoral girdle was that of an avepectoran dinosaur, not an advanced bird; and the supple-fingered hands were much more dinosaur-like than birdlike. The pelvis and hindlegs were closest to those of other dromaeo-avemorphs, except for the birdy reversed hallux. Today, even the feathers of *Archaeopteryx* fail to clearly separate it from dinosaurs. In anatomical grade, *Archaeopteryx* and its larger relative *Protarchaeopteryx* were dino-birds, rather than early birds convergent with dinosaurs.

The fact Ostrom pointed out in 1973 remains true today: if not for their feather impressions, the *Archaeopteryx* specimens would have been classified as long-armed dinosaurs, specifically, as close relatives of dromaeosaurs. They probably would have been classified as arboreal (in view of the long, grasping hands and reversed hallux) and possibly capable of flight (considering the birdlike shoulder girdle and big arms). Perhaps someone might even have proposed that they were feathered. Likewise, a featherless *Protarchaeopteryx* would have been labeled a short-tailed dinosaur, again, possibly feathered. The meaning of their few avian features would have been hotly disputed, and many would have cited them as evidence for a close affinity between dinosaurs and birds. The verification of Ostrom's thesis is that JM 2257—in which the feather impressions are especially faint—was classified as a theropod dinosaur for twenty years!

### *Caudipteryx* Was a Bird, Not a Dinosaur

The discovery of this feathered avepod *Caudipteryx* led to immediate claims that it was really a secondarily flightless bird closer to the likes of *Confuciusornis* than to dinosaurs (Martin in Gibbons 1998b, Feduccia 1999, Feduccia in Fischman 1999, L. Martin and Czerkas 2000, Ruben and Jones 2000). Specifically, some have argued that that because *Caudipteryx* was an herbivore, it must, therefore, have been a bird rather than a predatory dino-theropod. This argument does not follow: consider that oviraptorosaurs, ornithomimosaurs, and therizinosaurs were herbivorous to some degree. Nor were the abbreviation of the tail, and the ossified sternal ribs and large sternal plates, distinctly avian feature: some dino-avepectorans had these features.

One reason that those who argue that *Caudipteryx* was a bird feel justified in doing so is because this is a new taxon rather than a previously well-known dinosaur now found with unambiguous feathers. To challenge these claims, we must show that the skeleton is indeed that of a dinosaur. We can do so, again, by considering what might have happened had the feathers not been preserved. We would have been left with only the skeletons for analysis, and in this case, claims that these feather-free dinosaurs confirmed a dinosaur-bird relationship would certainly have been met by counterclaims that they were the skeletons of dinosaurs that merely paralleled birds! The many dinosaurian features would have been pointed out, and the absence of avian attributes noted. Among the nonavian attributes is the probable contact between the quadratojugal and the squamosal in the skull. The short, downturned, posteriorly bifurcate mandibles are otherwise seen only in oviraptorosaurs. The unusual configuration of the palate (see Chapter 11) is otherwise limited to oviraptorosaurs. In the hindquarters, the deep, rectangular ilial plates contrast with the urvogel-type pelvis (Fig. 10.12Bl). So do the nonretroverted pubes, which are most similar to those of oviraptorosaurs. The ischia lack the dorsal processes characteristic of birds, but they do have the large triangular obturator process like those of dromaeosaurs, troodonts, and especially oviraptorosaurs. The shallow pelvic canal (Fig. 10.12Bl) is unlike the deep pelvic canal of birds. The slenderness of the distal fibula cited as avian by Feduccia (1999) is also seen in some dino-avepods. The contact between the fibula and calcaneum, the very tall yet broad ascending process of the astragalus, and the small oviraptorosaur-like dorsal projection of the astragalus over the calcaneum are all dinosaurian (Fig. 10.14Bl). So is the lack of tight articulation between the metacarpals, tarsals, and metatarsals. There is nothing in the skeletons that separates *Caudipteryx* from the dinosaur clade; instead *Caudipteryx* appears to be either a sister taxon to known oviraptorosaurs (Zhou and Wang 2000, Zhou et al. 2000) or a basal oviraptorosaur (Sereno 1999a, Barsbold et al. 2000).

Because *Caudipteryx* specimens definitely do have contour feathers, and because it is considered improbable for such complex feathers to have evolved twice, those who oppose the dinosaur-bird hypothesis are forced to assert that *Caudipteryx* was a bird. Otherwise their entire position collapses. At the same time, the circular logic of this group seems to be that if it had feathers, it must have been a bird and could not have been a dinosaur; thus, *Caudipteryx* was a bird. This logic is apparent in Feduccia's (1999) statement that one reason *Caudipteryx* was a bird is that it had "wing feathers attached as in archaic birds." However, mere assertions that these extraordinary remains represent birds will not suffice. Remarkably avian features are present in *Caudipteryx,* some not seen even in *Archaeopteryx,* but these features automatically do not exclude it from the Dinosauria. Not only must the avian features be demonstrated, but also the many dinosaurian attributes have to be plausibly dismissed or explained as parallelism. The skeletons are so dinosaurian in most respects—to the point that *Caudipteryx* can be allied with a specific group of avepectoran dinosaurs—the convergence required to make *Caudipteryx* a dinosaur-like bird would take convergence to extremes never observed in the history of evolution.

Ironically, those who label *Caudipteryx* a bird are surrendering to the proponents of the dinosaur-bird hypothesis without realizing it. As Zhou et al. (2000) has explained, if it was a bird, it was also a dinosaur that retained dinosaur characters not seen in other birds, a dino-bird that strongly supports a close relationship between the two groups. Perhaps it was a secondarily flightless "bird" closer to modern avians than *Archaeopteryx,* just as a number of other avepods we call dinosaurs may have been.

### *Rahonavis* Was a Bird and a Dinosaur

If *Rahonavis* was what its describers' claim it was—a winged dino-bird with a sickle-clawed second toe—then opponents of the dinosaur-bird link are faced with an obvious transitional type between the birds and dinosaurs. The skeptics have tried to dismiss *Rahonavis* as a chimera by claiming that a bird's wings were incorrectly attached to a dinosaur's hindquarters (Ruben and Martin in Gibbons 1998a). This argument is plausible in that the association of the arm bones with the rest of the skeleton has yet to be proven via an articulated specimen. It is also largely irrelevant not only in that the hindquarters themselves combine dinosaurian features with avian features, but also in that the avian features are those long cited as being absent in dinosaurs. These features include a pubic boot that does not have an anterior process; an ischium with dorsal auxiliary processes and a distally placed, hook-shaped obturator process; and even a fully reversed hallux! Indeed, it is ironic to hear those who have asserted that a reversed hallux and climbing ability are strictly nondinosaurian features state the *Rahonavis* foot is dinosaurian because they do not want it to be associated with the wing bones. What is important is that this dino-bird combines a dinosaurian second toe with an avian first toe. Also important is that *Rahonavis* ends up in much the same phylogenetic position whether the arms are included or excluded (Forster et al. 1998, Padian 1998b).

### "Archaeoraptor" Was Not a Dino-Bird

At first cited in a popular publication as an example of a dino-bird combining a dromaeosaur tail with an otherwise fairly derived bird anatomy (Sloan 1999), "Archaeoraptor" subsequently proved to be a chimera produced by an amateur: the tail actually belongs to the skeleton of *Microraptor* (Simons 2000, Xu et al. 2001). The mistake was quickly recognized and announced, and the false data were never incorporated into the dinosaur-bird hypothesis in any technical study. Therefore, this "find" no more falsifies the proposed dinosaur-bird link than the Piltdown hoax falsified the ape-human link. The errant data resulting from the *"Protoavis"* chimera and from the *Sinosauropteryx* misanalysis (Appendix 3B) have had much more serious and lasting influences upon the antidinosaur hypothesis.

### Avepectoran Dinosaurs Were Birds

That argument that avepectoran dinosaurs were birds has not yet been presented but can nevertheless be anticipated. Let us assume that feathered *Caudipteryx* is an oviraptorosaur relative and, further, that unambiguous contour feathers are found on the likes of dromaeosaurs, troodonts, oviraptorosaurs, or therizinosaurs. According to the hypothesis that feathered archosaurs are birds with fingers II–IV rather than dinosaurs with digits I–III, the avepectorans were members of the bird clade, a clade that evolved independently from theropods that lacked large ossified sternal plates! The problem with this notion is that avepectorans share far too many derived characteristics with other predatory dinosaurs for us to accept the extreme degree of convergence required to produce this version of dinosaur polyphyly. The closest known sister taxa to avepectorans clearly are the less-derived averostran avetheropods such as *Allosaurus, Ornitholestes,* and *Coelurus.* Consider that opponents of the dinosaur-bird link have long thought that the teeth of dromaeosaurs and troodonts are nonavian, so making these avepectorans post-urvogels only serves to reinforce the link. The presence of feathers in nonavepectoran dinosaurs also works against the possibility that dinosaurs were not monophyletic.

**Pterosaurs Converged with Birds**

The argument that pterosaurs converged with birds observes that because some skeletal features of pterosaurs were birdlike even though birds and pterosaurs were only convergent, the same might also be true of predatory dinosaurs (see Feduccia 1999). It is true that the isolated remains of pterosaurs have sometimes been mistaken for those of birds (Jensen and Padian 1989), but in the end, the convergence argument works against those who promote it. The only pterosaur feature that is very birdlike is the ankle. In their totality, pterosaurs were very different from birds. The skulls, vertebral columns, rib cages, pelves, most of the hindlimbs, and even shoulder girdles and wings of these two groups are easily distinguishable. This fact shows that even flying archosaurs do not converge enough to resemble one another closely. Ergo, if nonvolant predatory dinosaurs were merely convergent with flying birds, then they should be less birdlike than flying pterosaurs. This expectation brings us to the conclusions presented below.

### *The Problems with Convergence*

The basic idea that avepod dinosaurs look so much like avepod birds because the two groups converged with one another is not inherently absurd. Convergence and parallelism are indeed normal in evolution and are rampant among the Avepoda. However, for convergence to become the preferred explanation for the similarities between birds and dinosaurs, two requirements must be met. First, the convergent similarities must not be so extensive and so thoroughly pervasive in the skull and skeleton as to become unreasonable. Second, there must be an evolutionary mechanism that explains the observed level of convergence. The convergence hypotheses advocated by L. Martin (1991), Tarsitano (1991), and Feduccia (1996) have failed to meet either of these requirements. These workers have proposed massive convergence on an unprecedented scale, yet they have failed to propose an evolutionary mechanism that explains how *terrestrial* dinosaurs converged with *arboreal* protobirds.

Let us consider some examples in which convergence makes sense. The doglike Tasmanian thylacines and canids look so alike because they evolved to do much the same thing. Likewise, sabre-toothed marsupial thylacosmilids, nimravids, and felids are similar in form because they evolved under similar selective pressures. Because dino-avepods were land runners (small forms have a secondary arboreal capability), protobirds could converge with them only if protobirds too had a strong terrestrial component to their lifestyle. Feduccia (in Morell 1997) tacitly acknowledged this problem when he stated that "you expect to see similarities" between avepod dinosaurs and birds "because both groups developed adaptations for bipedal running." Yet Feduccia explicitly (1996) rejected a bipedal, running ancestry for birds. This gross contradiction negates Feduccia's running convergence hypothesis. If we assume that protobirds were specialized for arboreality, why would quadrupedal climbers evolve three-fingered hands and folding arms indistinguishable from those of predatory bipeds? The possibility that bird fingers would happen to exactly match the 2-3-4 finger bone count of avetheropods is one of many unlikely coincidences required by the convergence hypothesis. If protobirds were climbing nondinosaurs, then they should have evolved fingers that were more symmetrical, both from left to right and along their length, to better their grasping ability. While it makes sense that running bipeds evolved long hindlegs with cylindrical joints limited to a vertical, fore-and-aft action; a pretibial process that braces the ankle; and a compressed tridactyl foot, why would quadrupedal climbers have evolved such inflexible legs? Conversely, why would running bipeds have undergone a shift whereby the divergent digits on the hands and feet went from being on the outside to the inside, just like quadrupedally climbing preavians? Even more importantly, why would climbers and runners have evolved the same distinctive locomotory apparatus in *both* the forelimbs and the hindlimbs? Having one set of convergent limbs is bad enough; having two sets is extraordinary. Why, then, would arboreal, flying *Archaeopteryx* have converged so closely with running, nonvolant dromaeosaurs if they were doing such different things?

The convergence hypothesis also fails to explain why flightless dinosaurs developed the same flight adaptations seen in urvogels, horizontal scapula blades and folding arms among them. Even more telling, why would *Archaeopteryx* and a ground-bound (in the convergence scenario) basal dromaeosaur like *Sinornithosaurus* share similarly configured furculae, the same retroversion of the coracoid, the prominent pro-acrocoracoid, and even the distal expansion of the proximal-most central finger bone that have been considered classic flight adaptations?

Equally telling is that the similarities between advanced theropods and birds are not limited to gross body form but also include an extensive series of small details that are difficult to explain by even the most extreme examples of convergence. How does convergence between running dinosaurs and climbing protobirds explain the diamond-shaped dorsal supraoccipital, the similarities in the outer auditory meatus, or the distinctive iliopubic articulations shared by dromaeosaurs and *Archaeopteryx?* Why did *Archaeopteryx* evolve maxillary sinuses and a palate so similar to those of tetanuran avetheropods? How does convergence between climbers and runners explain why the urvogel and dino-avepods both developed a gracile palate with a palatine hook, a fenestra between the (sometimes triradiate) palatine and pterygoid, and a slender juncture between the anterior and posterior sections of the pterygoid? Why would archaeopterygiforms and avepectoran dinosaurs separately evolve a subcircular indentation on the anteromedial edge of the coracoid? Why would predatory dinosaurs with (in the convergence scenario) crocodilian-like lungs evolve hinge-action sternal plates and sternal ribs so like those of flightless birds, as well as the same gastralial midline articulation that is seen in basal birds but is so unlike that present in crocodilians and other archosaurs? Convincing answers to these questions have not been offered. Feduccia (1999) cited examples of remarkable convergence between different bird groups as analogies for the similarities between dino-avepods and birds. However, in reality, the birds merely paralleled one another because they all began with a body plan that was already fully avian. The degree of convergence required to make dino-avepods and birds that evolved from lizardlike basal archosaur ancestors so extremely similar is many times greater.

As important as the similarities between avepod dinosaurs and birds is the absence of major differences between the two groups. Although a thylacine looks very much like a wolf, an anatomist can easily tell that they represent two different groups that have converged with one another: numerous skeletal details clearly distinguish them from one another. Most obvious is the pelvic "marsupial bone" that helps support the pouch, and more-subtle dental characters are also important. Likewise, details of nimravid osteology allow them to be separated from cats. However, no unambiguous features that clearly segregate derived avepod dinosaurs from basal birds have yet been identified. For instance, the attempts to demonstrate nonhomology between predatory dinosaur and bird teeth, respiratory tracts, and fingers have not borne definitive fruit.

Sabre-toothed nimravids and felids are so similar not only because they evolved to do the same thing but because their common basal aeluroid ancestor was already a fairly derived carnivore not drastically dissimilar from its later descendents. In terms of evolutionary changes, the amount of convergence is modest. If birds formed a feathered clade with forms like *Longisquama* separate from the avepod dinosaur clade, then the common ancestor of the two clades was extremely different from either its avian or dinosaurian descendents; yet the latter two groups ended up nearly identical in form. The degree of evolutionary convergence required by this hypothesis, both in amount of change from the common ancestor and in ultimate degree of similarity, exceeds that exhibited by any other known example.

The similar reproductive patterns of dinosaurs and birds stretch convergence even farther beyond the plausible. How can convergence explain the similar eggshell microstructures or the near-avian nesting habits of dino-avepectorans?

The totality of the convergence required by the nondinosaur hypotheses is, therefore, so extreme as to be absurd. Consider *Oviraptor.* According to the convergence hypothesis, *Oviraptor* mimicked birds in the palate and other features of the skull and in brain expansion, the shoulder girdle (including the large sternal plate), the folding arms, the wrist and tridigitate hands, the avepod hindlimb, bone microstructure, eggshell microstructure, and even egg deposition patterns and brooding posture! At the same time, dromaeosaurs and *Archaeopteryx* somehow managed to converge in detailed occiput, outer ear, scapulocoracoid, furcula, pelvic, and tail morphology. *Sinornithosaurus* and *Archaeopteryx* particularly stretch convergence well beyond the breaking point. Convergence requires alvarezsaurs to have independently evolved a skull more like that of a bird than that of *Archaeopteryx* and to have featherlike dermal structures that are not really feathers. The amount of convergence required to explain the similarities between birdlike dinosaurs and birds is unprecedented among tetrapods, to a point beyond what is reasonable and plausible. To put it another way, critics of the dinosaur-bird relationships have failed to show that birds are to dinosaurs what Tasmanian "wolves" are to canids.

The critics have the logic of the problem reversed, in terms of both time and phylogeny. Their view that Cretaceous avepod dinosaurs converged with birds that had first evolved in the Triassic so closely that no unambiguous differences falsify an intimate relationship is incorrect. Instead, the convergence with the late Mesozoic birds occurred in the Triassic, when drepanosaurs, longisquamids, and some other diapsids were rather birdlike in a few ways but were so nonavian in other regards that a close bird link can be falsified.

### *The "Dogma" Is Right: Birds Are Dinosaurs*

Feduccia (in McDonald 1996) has decried the dinosaur-bird hypothesis as "dogma"—of the unsubstantiated type—adhered to by dinosaur paleontologists who so desire to believe that the great dinosaurs did not go extinct that they fail to see the evidence to the contrary. That the hypothesis has become a dogma is true, but this is not necessarily a bad thing. Feduccia's use of the term *dogma* is rhetorical propaganda. The dinosaur-bird hypothesis is scientific dogma in the same way that the geocentric theory that Earth is a sphere orbiting a nuclear fusion reactor is scientific dogma. The theory of evolution is also dogma. Dogma is appropriate in science when it is so well substantiated by the data that it must be the established consensus, unless something crucial comes along to show otherwise.

The antidinosaur argument has failed to do what it needs to in order to falsify the legitimate dogma of the dinosaur-bird relationship or present a viable alternative. It has neither shown that dino-avepods were merely convergent with birds nor proposed a plausible mechanism that explains the massive convergence required by the hypothesis. Nor has it demonstrated that the origin of avian flight is incompatible with the habits and function of small predatory dinosaurs. A viable set of nondinosaurian protobirds that bridge the gap between nondinosaurian archosaurs and birds—the classic "proavis"—has not been found, and after so many years of fossil hunting, the situation is not likely to change. The major work advocating a nondinosaurian descent of birds, Feduccia's 1996 book, contains too many errors and omissions—and demonstrates Feduccia's lack of understanding of dinosaur anatomy—to be considered a serious challenge to the dinosaur hypothesis (Norell and Chiappe 1996, Padian 1997a, Sereno 1997b). Similar problems afflict the effort to show that avepod dinosaurs lacked preavian lungs and air sacs.

The antidinosaur argument is largely a description of how evolution should have been occurred: bird ancestors *should* have been arboreal quadrupeds, the primary axis of finger development *should* have run through digit IV, the initial radiation of avians *should* not have included odd creatures like alvarezsaurs. The activities of the should-have-been school of bird origins have largely been propagandistic, in that the proponents have often vocally protested the linkage but have not been able to present the definitive evidence needed to refute avian descent from dinosaurs. In 1999, Feduccia summarized the antidinosaur argument as follows:

- The similarities between dinosaurs and birds are due to convergence between running versus climbing bipeds [highly implausible].
- The body and tail postures and proportions of theropods and bird are too divergent from one another [the opposite is true].
- Theropod and bird tooth replacement are too divergent from one another [the opposite is true].
- Theropod and bird hands are not homologous [the data are too scanty to be conclusive].
- Theropods had a crocodile-like, not birdlike, respiratory complex [incorrect].
- Dino-avepod arms were too short [incorrect].
- Running dino-avepods could not evolve flight [but many small theropods could climb well].
- The most birdlike dino-avepods first evolved well after birds first appeared [probably not true, and does not necessarily contradict a close relationship anyway].

The arguments the antidinosaur hypothesis is based upon are at best superficial; not one has been substantiated, and most have been discredited. The hypothesis also lacks an evolutionary context founded on a set of increasingly avian nondinosaurian fossil protobirds. The case against birds being dinosaurs is, therefore, very weak, so weak that one is forced to wonder—in a reversal of the Feduccia scenario—whether in some quarters the rejection of the hypothesis is based entirely on science, or more upon a desire to keep the glorious birds separate from the archaic monsters. Feduccia's 1996 statement that the hand of *Archaeopteryx* "does not closely resemble that of a theropod dinosaur" appears to demonstrate that he is biased and in a state of denial, in view of the fact that no other known archosaur's hand was as similar to the bird's than that of avepectoran dinosaurs even prior to the discovery of the near-urvogel *Sinornithosaurus*. Such statements bring us to the fact that the arboreal, nondinosaurian hypotheses are so strongly theory-driven that it is not clear what fossil evidence, if any, their advocates would accept as definitively falsifying their hypotheses.

After all, only automatic dismissals have accompanied recent discoveries such as urvogel-like *Rahonavis*, *Sinornithosaurus*, and feathered dinosaurs that not only support the dinosaur-bird connection but also fill in specific gaps in the dinosaur-bird hypothesis that its opponents previously pointed to. The antidinosaur researchers have proven disturbingly adept at dropping arguments that they once considered crucial (for example, the dorsal ischial process and the reversed hallux, which were formerly seen as strictly avian features, suddenly became irrelevant when they showed up in the dinosaurian hindquarters of *Rahonavis*). Nor is the antidinosaur argument convincing when profoundly unbirdlike drepanosaurs are labeled "avimorphs" while dromaeosaurs that closely match the morphology of urvogels are dismissed as merely convergent. The same problem arises when impressions associated with not very birdlike *Longisquama* are proclaimed to be true feathers even though they are only partly featherlike, they form a most unavian dorsal frill, and they cannot be biochemically examined, whereas the more truly birdlike integuments of small birdlike dinosaurs are not accepted as feathers until absolutely definitive biochemical tests are done. Equally disconcerting is how convergence between the longisquamid integument impressions and avian feathers is being rejected in favor of far greater convergence between dinosaur and bird skeletons. It seems that an unduly extreme burden of proof is being placed on the dinosaur-bird hypothesis, whereas the alternative hypotheses are accepted on the basis of much less rigorous criteria. Such bias has caused some researchers to accuse some members of the antidinosaur minority of resorting to methodologies that are disturbingly close to those used by creationists (Dalton 2000). An undue emphasis placed by some researchers upon the "Archaeoraptor" problem as a critical flaw in the dinosaur-bird hypothesis only serves to reinforce this concern.

In comparison, the hypothesis that birds evolved from dinosaurs has been built upon the much more meticulous work of researchers such as John Ostrom, Peter Wellnhofer, Rinchen Barsbold, Philip Currie, Larry Witmer, Andrzej Elzanowski, Xing Xu, and the American Museum of Natural History team, just to mention a few. It should also be understood that the debate is not an equally matched between cladists on one side and those who favor arboreal preavians on the other. Cladistic methodology is not crucial; even if it had not yet been invented, the dinosaur-bird link would still be obvious because the fossil osteological, trace, and even behavioral evidence that dinosaurs are birds overwhelms the alternatives even when cladistics is left out of the scientific bag of tools. (Note that the therapsid-mammal link was well established on scientific grounds before the advent of cladistics.) Specifically, the strength of the dinosaur hypothesis of bird origins rests upon the following points of evidence:

- All of the few derived characters shared by pseudoavimorphs are also present in dino-avepods.
- Unlike the alternatives, small-bodied dino-avepods with long arms and legs fit into both the arboreal and terrestrial hypotheses of bird flight.
- Dino-avepods are by far the most birdlike of all tetrapods.
- Basal birds share an extremely large number of distinctive derived characteristics with birdlike dinosaurs; and there are very few avian characters that are not also found in the latter, to the point that it is hard to tell where dinosaurs end and birds begin. A specific group of predatory dinosaurs, dromaeosaurs, especially *Sinornithosaurus*, can be shown to be the closest known relatives of *Archaeopteryx*, demonstrating that the knowledge base supporting a dinosaur-bird link has reached a high level of sophistication and resolution.
- *Archaeopteryx* is a diminutive dromaeo-avemorph dinosaur in gross form, and in

detailed osteology. All of its well-preserved characters either are found in at least some dinosaurs or are very similar to and readily derived from the dinosaur condition; and no character clearly distinguishes the urvogel from dinosaurs.

- If *Archaeopteryx, Protarchaeopteryx,* and *Caudipteryx* had been found without feathers, they would have been assigned to Dinosauria, and their avian attributes would have been dismissed as convergent with those of birds by opponents of the dinosaur origin hypothesis.
- The characters linking dinosaurs and birds cover not only skull and skeletal anatomy, including bone microstructure, but also functional aspects, egg size, eggshell morphology, reproductive behavior, and even integument.
- Dino-avepod integument is more featherlike than that of longisquamids, and the degree of convergence required to explain the similarity of longisquamid and avian integument is far lower than that needed to explain the similarity of avepod dinosaur and bird skeletons.
- Some avepod dinosaurs possess the same flight-related adaptations seen in urvogels.
- Convergence is inadequate to explain such extreme similarity, especially when the proposed common basal archosaur ancestors were so unlike either birds or predatory dinosaurs.
- No evolutionary mechanism has been proposed that explains how predatory dinosaurs and birds became so extremely similar to another after independently evolving from a grossly dissimilar common ancestor.
- None of the major avepod synapomorphies has been demonstrated to be nonhomologous.
- There is nothing in the characteristics of the birdlike dinosaurs that bars their being the ancestral type for birds.
- That there is a progressive development of avian characters in dino-avepods places the descent of birds from dinosaurs in the required evolutionary context.
- The dinosaur-bird hypothesis's prediction that an increasing number of transitional forms linking the two groups will be found is proving to be correct, the latest and perhaps most telling examples being *Sinornithosaurus* and *Microraptor.*
- There is evidence for a radiation of very small, very birdlike dinosaurs in the later Jurassic before *Archaeopteryx.*

The last three points are important because they mean that an ancestor-descendent rather than a common ancestor relationship between the two groups is strongly favored. Note that there is no need to apply the principle of parsimony to the question of bird origins; the evidence that birds are dinosaurs is so overwhelming that the alternatives are buried under its weight. The question of avian origins is no longer an open one, unlike the issue of the origins of pterosaurs and turtles. The current case for an avepectoran dinosaur ancestry for birds is nearly as strong as that for a cynodont therapsid ancestry for mammals, so much so that the burden of falsification weighs heavily on the opponents of a dinosaur-bird link. After all, unlike in the case with pterosaurs and bats, we have many of the transitional forms that link birds to terrestrial ancestors, and they are dinosaurs. Although there are still gaps to be filled—in particular the lack of adequate dino-bird remains prior to the Late Jurassic—falsification would require the finding of a series of nondinosaurs at least as birdlike as dino-avepods or firm evidence that the characters shared by dinosaurs and birds are not homologous, or both.

The dino-avepod hypothesis of bird origins also explains the birds' distinctive combination of bipedalism and leg-free wings by combining the best aspects of the terrestrial and arboreal hypotheses of the origin of flight. If birds had evolved from highly arboreal creatures, then birds should be fully quadrupedal forms in which the airfoil is attached to the hindlimbs (and, as a result of this attachment, loss of flight would have been inhibited). Instead, the strongly terrestrial habits of early avepods resulted in cylindrical hinge joints in the hindlimbs that inhibited the later adoption of a less erect, fully quadrupedal climbing body form. In the semi-arboreal leaping dino-avepods, the erect hindlimbs were not suitable for bearing horizontal airfoils, which were therefore restricted to the forelimbs. The dino-avepodian nature of a rooster can still be seen as it struts on two strong legs across the barnyard.

We are living at a time when sickle-clawed dinosaurs are showing more and more characters once thought to have been restricted to urvogels; the *Archaeopteryx*-like nature of *Sinornithosaurus* and the *Rahonavis* hindquarters are especially telling. We now have Jehol fossils that show that feathers simple and complex developed in predatory dinosaurs. The smoking gun has been found. Birds are dinosaurs. We can even show that dromaeosaurs are the dinosaurs closest to the urvogel.

Had the complete skulls and articulated skeletons of dromaeosaurs, troodonts, oviraptorosaurs, and alvarezsaurs known today been available earlier in the 1900s, the dinosaur-bird link would long ago have become the established dogma in the paleontological and ornithological communities. As we enter the 2000s, the dinosaur hypothesis of bird origins is not the equivalent of cold fusion, notwithstanding claims to the contrary. Instead, it is the nondinosaur hypotheses that are equivalent to the ether hypothesis of the 1800s, which was overturned by the relativity hypothesis of the 1900s.

Therefore, the question is no longer whether birds evolved from predatory dinosaurs, but how, when, and why. For instance, the finger homology problem no longer effectively challenges the dinosaur-bird link; the question now is how the bird hand developed from that of dinosaurs. An arboreal origin of flight does not contradict the dinosaur-bird link; instead we ask whether avian flight originated among climbing or running dinosaurs, or both. The dinosaur origin hypothesis is so advanced that it is exploring and answering the question of exactly which avepod dinosaurs are closest to birds. Less clear is exactly how the various avepectoran dino-birds are related to other dino-avepods and to birds, because each type has avian characters not found in the other. This problem raises more questions, questions that ask not whether birds descended from dinosaurs but what happened as they did so.

CHAPTER 11

# Were Some Dinosaurs Also Neoflightless Birds?

The standard dinosaur-bird hypothesis presumes that *Archaeopteryx* was closer to birds than any dinosaur was and that the birdlike dinosaurs were a persistently earthbound lot. The proponents of this view hold that all of the flight-related adaptations found in theropods are exaptations: that is, preadaptations that originally evolved for some other purpose and subsequently proved useful for flight. This view runs so deep that its proponents have applied birdlike attributes to *Archaeopteryx* that it did not really have and have denied the avian features that dinosaurs did have. For instance, in photographs of the fighting *Velociraptor* taken in the quarry where it was found, I could see what were almost certainly bony uncinate processes on the ribs. But I was on occasion chided for making such dromaeosaurs too avian—after all, *Archaeopteryx* lacked ossified uncinates—and was sometimes required to remove the processes from my skeletal restorations. Dromaeosaur uncinates have since been verified (Chapter 4).

Was the story more complicated than the conventional wisdom allows? Are some flightless dinosaurs really post-urvogels and neoflightless dino-birds? This possibility has not been ignored. Aside from myself (Paul 1984a, 1988a), a number of researchers—starting with Osmólska (1976) and continuing with Kurzanov (1981, 1982, 1985, 1987), Molnar (1985), Perle et al. (1993, 1994), Olshevsky (1994), Elzanowski (1995, 1999), Sereno (1997a), and Chatterjee (1999a)—have noted the presence in certain dinosaurs, mainly avepectorans, of flight-related or avian characters, or both, not found in *Archaeopteryx*. Griffiths (1998/2000) and Padian and Chiappe (1998c) explicitly noted the possibility that alvarezsaurs were neoflightless, and T. Jones et al. (2000b) and Zhou et al. (2000) described the evidence for the neoflightlessness in *Caudipteryx*. *Protarchaeopteryx* is also a candidate for being secondarily flightless and closer to birds than its close relative *Archaeopteryx* is. Note that Feduccia (1996) erred in implying that I have argued that basal coelurosaurs were neoflightless; my hypothesis of neoflightlessness is limited to avepectorans. Conversely, a number of researchers have implied that in 1988 I limited the hypothesis to dromaeosaurs (Paul 1988a), when in fact I actually included all the avepectoran dinosaurs (Fig. 12.1I).

In Chapter 8, the characters expected in secondarily flightless dino-avepods were detailed (also see additional features detailed in Appendix 6). How often did these characters appear in predatory dinosaurs? Which avepod dinosaurs had characters that were closer to those of birds more advanced than the urvogel? Did some of the birdlike dinosaurs combine flight-related features *and* avian attributes not seen in *Archaeopteryx*? The short answer is yes.

## Signs of Flight Lost

Did certain of the most birdlike attributes of avepectoran dinosaurs evolve before the advent of flight or after? In principle, this question is best answered by phylogeny. If a particular avian character that is potentially flight aided is present in a taxon that evolved before flight then that character is at best only an exaptation for flight. If the character is present only in fliers or their descendants then it evolved for the purposes of flight. The problem is that we do not understand the phylogeny of the small number of dino-bird fossils well enough to rely on the above phylogenetic methodology at this time. Indeed, the possible misidentification of pre-urvogel exaptations as post-urvogel characters may be leading us to incorrectly restoring the phylogeny of avepectorans.

Until we have enough skeletons to more reliably restore dino-bird phylogeny, we will have to make do with an analysis of the available fossils in terms of the presence or absence of potentially neoflightless characters that may indicate descent from flying ancestors. In Table 11.1, these characters are listed for various theropods. The analy-

*Table 11.1. Presence or Absence of Potential Neoflightless and Related Characters in Theropod Dinosaurs*

| | *Herrerasaurus* | Compsognathids | *Allosaurus* | *Ornitholestes* | *Archaeopteryx* | *Protarchaeopteryx* | Dromaeosaurs | Troodonts | *Caudipteryx* | Oviraptorosaurs | Therizinosaurs | *Avimimus* | Alvarezsaurs | Ornithomimids | Tyrannosaurs |
|---|---|---|---|---|---|---|---|---|---|---|---|---|---|---|---|
| Furcula U-shaped | – | – | – | | + | p | s | | | p | – | | | – | – |
| Overall shoulder girdle large and strongly built | – | – | – | – | + | / | + | + | + | + | + | | + | – | – |
| Scapula blades horizontal | – | – | – | | + | + | + | + | p | + | | | – | – | – |
| Scapula tip pointed | – | – | – | | – | – | s | – | – | – | – | | – | – | – |
| Acromion process as well developed as in *Archaeopteryx* | – | – | – | | \ | + | + | + | – | + | – | | – | – | – |
| Acromion process better developed than in *Archaeopteryx* | – | – | – | | \ | – | s | – | – | – | – | | – | – | – |
| Coracoid proximally as narrow as in *Archaeopteryx* | – | – | – | | \ | + | s | – | – | – | – | | – | – | – |
| Coracoid proximally narrower than in *Archaeopteryx* | – | – | – | | \ | – | s | – | – | – | – | | – | – | – |
| Coracoids strongly flexed relative to scapula and subvertical | – | – | – | | + | + | + | + | + | + | | | – | – | – |
| Coracoids retroverted | – | – | – | | + | + | s | – | – | – | | | – | – | – |
| Coracoids long | – | – | – | | + | + | s | / | / | s | + | / | – | – | – |
| Outer coracoid surfaces face strongly anteriorly owing to long transverse coracoid-sternum arcticulation | – | – | – | | + | + | + | + | + | + | p | | – | – | – |
| Most of sternum ossified | | – | – | | + | + | + | | + | + | | | + | s | |
| Sternal plates broad | – | – | – | | + | + | + | + | + | + | | | – | u | – |
| Ossified sternal plates longer than in *Archaeopteryx* | – | – | – | | – | – | + | + | + | + | | | + | s | – |
| Sternal ribs ossified | – | – | – | | – | | + | | + | + | | | | – | |
| Pectoral crest of humerus large and hatchet shaped | – | – | – | + | + | | + | – | – | + | | – | – | – | – |
| Arm/leg length ratio between 0.5 and 1.0 | – | – | – | n | + | + | + | – | – | + | + | – | – | + | – |
| Arm-folding mechanism present | / | – | – | | + | + | + | + | + | + | + | + | – | – | – |
| Arm folding as well developed as in *Archaeopteryx* | – | – | – | | + | + | + | + | + | + | n | + | – | – | – |
| Carpometacarpus fused | – | – | – | | – | – | – | – | – | – | – | + | + | – | – |
| Metacarpal III bowed posterolaterally | – | – | – | | + | + | + | – | – | – | – | | – | – | – |

*(continued)*

*Table 11.1. (continued)*

| | *Herrerasaurus* | Compsognathids | *Allosaurus* | *Ornitholestes* | *Archaeopteryx* | *Protarchaeopteryx* | Dromaeosaurs | Troodonts | *Caudipteryx* | Oviraptorosaurs | Therizinosaurs | *Avimimus* | Alvarezsaurs | Ornithomimids | Tyrannosaurs |
|---|---|---|---|---|---|---|---|---|---|---|---|---|---|---|---|
| Metacarpal III bowed more posterolaterally than in *Archaeopteryx* and large distal contour feathers are absent | – | – | – | | \ | p | s | – | – | – | – | | – | – | – |
| Posterolateral flange on proximal phalanx of finger II is better developed than in *Archaeopteryx* and large distal contour feathers are absent | – | – | – | | \ | u | s | – | – | – | – | | – | – | – |
| Finger II at least as robust as I | – | – | – | | + | + | s | + | + | – | + | + | – | – | – |
| Flexibility of finger III reduced | – | – | – | | | | s | n | + | n | n | | + | – | – |
| Finger III tightly appressed to finger II | – | – | – | | \ | – | – | – | + | – | – | | | – | – |
| Number of finger bones fewer than in *Archaeopteryx* | – | – | – | | \ | – | – | – | + | – | – | | + | – | – |
| Well-developed symmetrical arm contour feathers present | | – | | | | n | / | | + | | | | | | |
| Ossified uncinate processes present | – | – | – | | – | – | + | | + | + | | | | – | – |
| Tail base hyperflexible, especially dorsally | – | – | – | – | + | p | + | + | u | – | – | | – | – | – |
| Caudal count below 40 | – | – | – | | + | + | + | + | + | + | + | | + | + | + |
| Caudal count below 24 | – | – | – | – | \ | + | – | – | + | – | – | | – | – | – |
| Caudal/dorsosacral length ratio below 1.7 | – | – | – | – | \ | + | – | – | + | s | p | | – | – | – |
| Caudal/dorsosacral length ratio below 1.0 | – | – | – | – | \ | + | – | – | + | – | n | | – | – | – |
| Distal tail slender owing to shallow chevrons beyond caudal ~15 | – | – | – | + | + | + | + | + | + | – | | | – | / | / |
| Distal tail stiffened | – | – | – | – | + | | + | + | u | s | | | – | – | – |
| Pygostyle present | – | – | – | – | – | – | – | – | – | s | – | – | – | – | – |
| Pubes strongly retroverted | / | – | – | – | u | | s | – | – | – | / | – | – | – | – |
| Hindlimb/dorsosacral length ratio higher than 1.9 in nonclimber | – | – | – | – | / | | p | | + | – | – | p | – | – | – |
| Progressive loss of potentially neoflightless characters within group | – | – | – | – | p | p | + | | / | / | + | | | p | |
| Brain enlarged above reptilian maximum | – | – | – | – | + | p | + | + | p | + | + | + | + | + | – |

| | *Kerrerasaurus* | Compsognathids | *Allosaurus* | *Ornitholestes* | *Archaeopteryx* | *Protarchaeopteryx* | Dromaeosaurs | Troodonts | *Caudipteryx* | Oviraptorosaurs | Therizinosaurs | *Avimimus* | Alvarezsaurs | Ornithomimids | Tyrannosaurs |
|---|---|---|---|---|---|---|---|---|---|---|---|---|---|---|---|
| Overlapping vision fields | – | – | – | – | + | | + | + | – | – | | + | / | + | + |
| Olfaction reduced | – | – | – | – | + | | / | + | / | – | / | | u | + | – |
| Standard dino-theropod predation modified or absent | – | – | – | + | + | + | + | + | + | + | + | + | + | + | + |
| Significant post-urvogel characters present | – | – | – | – | | + | + | + | + | + | + | + | + | + | – |
| Potentially neoflightless character | | | | | | | | | | | | | | | |
| Score | 1 | 0 | 0 | 2.5 | | 23.75 | 38.25 | 17.5 | 24 | 22 | 10.75 | 7.5 | 9.25 | 8 | 2.5 |
| Grade | 2.3 | 0 | 0 | 16 | | 66 | 83 | 45 | 53 | 49 | 32 | 60 | 23 | 18 | 6 |

NOTE: Characters were scored from the basal-most known member(s) of a group whenever possible. The scoring procedure is outlined in the text.

Progressive loss of potentially neoflightless characters is scored as partly developed for caudipterygians and oviraptorosaurs on the basis of the fact that these forms represent their own clade, the first group being less derived than the second.

The following symbols are used in this table: + = character or attribute is present, p = probably present, – = not present, n = probably not present, s = sometimes present, / = partly developed, \ = based on this taxon, u = preserved but uncertain. If a character is unknown at this time, the cell was left empty.

sis covers the early flier *Archaeopteryx* as well as a broad range of types, from those that are not expected to show potentially neoflightless characters to those that might do so. The inclusion of all these types allows us to test the method, ensures that we do not miss potentially neoflightless characters where we do not expect them, and makes it possible for us to compare the variation in potentially neoflightless characters among theropods.

In brief, the analysis shows the following results. There is considerable variation in the number of potentially neoflightless characters present in theropods, ranging from low in baso-theropods and the less birdlike dino-avepods to very high in avepectoran dinosaurs. There is also a positive correlation between the presence of potentially neoflightless characters and the presence of other avian characters not observed in *Archaeopteryx*. The two results suggest that a real phenomenon is present. The results also strongly suggest that those nonavepectoran theropods that lack many potentially neoflightless characters did not descend from flying ancestors.

The last point reveals that a few characters that may appear to be flight related are not directly so. For example, straplike scapula blades are present in herrerasaurs and allosaurs, furculae are present in baso-avepods and allosaurs, and laterally oriented shoulder joints are seen in early avepod dinosaurs. Posteriorly bowed ulnas distinguish flying birds from the most dino-avepods, which typically have straight forearm shafts, but bowed ulnas are also found in some coelophysids and allosaurs. The shortening and stiffening, although not the fusion, of the dorsal vertebral column that is associated with avian flight began to develop in early theropods. Because none of these theropods shows other significant evidence that their ancestors flew, it is probable that these birdlike features evolved for reasons unrelated to flight; that is, they were exaptations for flight.

The analysis also shows that the shoulder girdles of dromaeosaurs, troodonts, oviraptorosaurs, and caudipterygians have many or all the characters expected in neoflightless forms. Or, to put it another way, these dino-avepectorans' shoulders are of the type otherwise observed only in flying and secondarily flightless birds and pterosaurs (Figs. 4.2A, 7.1, 10.8). To recap, the narrow scapula blade is horizontal. The coracoid is often large and flexed relative to the scapula so that the former is vertical or even retroverted. The superficial surface faces strongly forward, and it articulates with the anterior edge of a broad sternal plate via a long hinge. The shoulder girdles of these dinosaurs are especially similar to those of flightless birds, in which the scapulocoracoid is often less flexed, and the coracoid shorter and broader, than in more aerial forms (Feduccia 1996). In these respects, the pectoral girdles of these theropods are as avian as those of *Archaeopteryx*, and in some regards more so.

Another potentially neoflightless character is the strongly foldable forelimbs of dromaeosaurs, troodonts, caudipterygians, oviraptorosaurs, basal therizinosaurs, and avimimids, which were at least as well developed as those of archaeopterygiforms (Figs. 4.2B, 10.11A). It is interesting that the arms of potential avepectoran dinosaurs tended to follow the avian pattern in being either unusually long—to an extreme degree in *Bambiraptor* and especially *Sinornithosaurus*—or short. The avepectoran dinosaurs also exhibited most of the more peripheral characters associated with flight, including the neural and visual ones. It is also intriguing that the dino-avepectoran tails always had potentially neoflightless characters—if nothing else they are usually shorter than is typical—not seen in other avepod dinosaurs. The highly modified tails of dromaeosaurs, troodonts, caudipterygians, and *Protarchaeopteryx* are especially notable because similar structures were otherwise limited to *Archaeopteryx* and rhamphorhynchoid pterosaurs.

Perhaps the most spectacular potentially neoflightless characters are the large, symmetrical contour feathers that sprouted from the forelimbs or tails or both of *Protarchaeopteryx* and especially *Caudipteryx*. If distal contour feathers so well developed yet so ill suited for flight were found in a bird, they would be taken as sure evidence for the loss of flight. The absence to date of similar feathers in other avepectoran dinosaurs may or may not be an artifact of preservation and thus does not count against their being neoflightless (unless their absence is convincingly demonstrated with fossils according to the criteria cited in Chapter 5).

### *Neoflightlessness versus Exaptation*

Can selective advantages related to nonflight or preflight functions explain the potentially neoflightless characters present in avepectoran dinosaurs? The answer is yes concerning the few adaptations mentioned above that are present in the less advanced avepods. For example, the lateral orientation of the shoulder joint that allowed wing flapping probably evolved in order to free up the arm so that it could be swung out to the side for grabbing prey. The alternative explanation for the sideways orientation—that is, that it was a climbing adaptation—is inferior because such shoulder joints are present in short-armed coelophysids. The nonflight selective forces that may have promoted the evolution of the theropod furcula and straplike scapula blade are explored in this chapter.

Explaining the majority of the brachial and caudal adaptations present in avepectoran dinosaurs as exaptations for flight is another matter. The characters and other attributes associated with the development of flight, and the method for distinguishing them from those expected in neoflightless avepods, were discussed in Chapter 8. We will start with tails.

The proximally flexible, distally stiff tails of troodonts and especially dromaeosaurs are usually thought to have evolved as dynamic stabilizers in leaping predators. On the one hand, these may have also evolved for stabilization during interbranch leaps (tree kangaroos use their tails for a similar purpose when climbing bipedally; Grzimek 1990) or for running leaps in semi-aerial protofliers. On the other hand, most predatory dinosaurs hunted well without dynamic caudal stabilizers, and no other nonfliers, predatory or otherwise, have developed tails of this particular kind. The only other group of archosaurs that did evolve slender, dynamically stabilizing tails was the rhamphorhynchoid pterosaurs. We therefore know that such tails can evolve for purposes of flight control, so a pterosaur-like tail made very slender by shallow neural spines and chevrons along much of its length is a leading potentially neoflightless character.

Even more difficult to explain as exaptations for flight are the short tails of various dino-avepectorans, most especially the much-abbreviated tails of *Protarchaeopteryx*, caudipterygians, and some oviraptorosaurs. Short tails did not evolve in any nonavepectoran avepods, and there is no obvious reason why a bipedal nonflier would lose its posterior mass balance by evolving a tail even shorter than that of flying *Archaeopteryx*. Doing so would degrade any display function. Short tails are especially inexplicable in herbivores, since herbivores do not need a tail to enhance agility while attacking prey; and the absence of a long tail to counterbalance the large gut would be a particular disadvantage. Improving dynamic flight agility in the manner of pterodactyloids and avebrevicaudans is a powerful selective agent for the evolution of a short tail. In nonavepectoran avepods the tail had 40 or more vertebrae, and a count below this number constitutes a potentially neoflightless character. The fewer the number of caudals there are, the more this point is true, especially if they are as few as or fewer than the 20–23 seen in *Archaeopteryx.* A tail that is exceptionally short relative to the rest of the body is also a potentially neoflightless character, especially if it is as short as or shorter than the urvogel tail, which is about one and two-thirds as long as the dorsosacral series.

Secondarily flightless running birds often have very long legs relative to the length of the trunk. In particular, the distal segments below the knee are elongated. As discussed in Chapter 4, this elongation is related to changes in femoral posture and action that result from severe tail reduction, which is in turn a result of advanced level flight. No nonavepectoran avepod had legs that were longer relative to the combined trunk and hip vertebrae than those of *Archaeopteryx,* but some avepectorans do. An adult leg/dorsosacral series length ratio higher than 1.9 in a ground runner with a short femur qualifies as a strong potentially neoflightless character (T. Jones et al. 2000b).

Although a furcula is not in and of itself a flight-related feature, the configuration of this element appears to be. In flying birds, including *Archaeopteryx,* the furcula forms a deep U or V so that it can better support flight muscles. In nonavepectoran dinosaurs and many avepectoran dinosaurs, the shallow V-shaped furcula was adequate for whatever purpose or purposes it served. A deep, U-shaped furcula is considered a good potentially neoflightless character.

Among avepods, a sharp-tipped scapula blade is limited to flying birds and their descendents, and there is no apparent reason for this feature to evolve in nonfliers. Therefore, its presence qualifies as a potentially neoflightless character. In addition, there is no apparent reason for climbers or predators to evolve horizontal scapulae that were attached at a sharp angle to large vertical or even retroverted coracoids and whose outer surface faced strongly forward. This is probably why no arboreal nonavians have such unusual shoulders. Indeed, the slender but vertical scapulocoracoids of highly arboreal drepanosaurs and especially chameleons mimic those of conventional dinosaurs (Paul 1987a). In addition, the great majority of predatory dinosaurs retained conventionally oriented and sized scapulocoracoids. Ostrom's (1974b) suggestion that some dromaeosaur coracoids were distally enlarged in order to better facilitate predation is plausible but out of line with the example of other predatory dinosaurs. Expansion of the coracoid does occur as flight muscles enlarge, and is often retained in neoflightless birds, especially newly neoflightless examples. Some anterior rotation of the coracoid's superficial surface should be expected in a protoflier, but the rotation should not be as extreme as that seen in *Archaeopteryx.* A better explanation for a strongly flexed and anteriorly rotated scapulocoracoid is that it improves the stretch factor of flight muscles.

Improvement of arm action during predation is a plausible but speculative explanation for development of a birdlike arm-elevation and -control system that includes elongation of the acromion process and narrowing of the proximal coracoid. Flight is a significantly better explanation because birds do in fact use this arrangement for flying. This explanation is even truer if the elevation and control system is better developed than that of *Archaeopteryx,* in which case predation as the selective agent for such an advanced complex becomes implausible, and a post-urvogel level of flight becomes highly probable.

Chatterjee (1997, 1999b) proposed that the large avepectoran sternal plate initially developed in order to support enlarged pectoralis muscles for climbing. A problem with this seemingly appealing notion is that limb muscles of climbers are not especially large (Grand 1977), which may explain why no climbers have evolved such large, subrectangular, birdlike sternal plates. In addition, such plates are unknown in any other predators that use their arms to hold onto prey. Large sternal plates are even more difficult to explain as exaptations when they are present in dinosaurs with very short arms, all the more so if the dinosaurs were herbivorous. Protofliers should have ossified sterna no longer or shorter than the sternum of *Archaeopteryx.* The large ossified sterna and their hinged articulation with the transversely broad coracoids were probably involved in the air sac ventilation system (Appendix 3B). However, it is doubtful that these features evolved expressly for this purpose, since they seem to be unnecessary in flightless avepods with well-developed air sacs. The same applies to the presence of ossified uncinate processes and sternal ribs. Instead, the presence of birdlike bellows-action sterna, plus well-braced rib cages with intracostal joints, is best explained as an adaptation for high-capacity respiration during flapping flight. After all, other than dino-avepectorans, only neoflightless birds have birdlike, keelless sternal plates larger than those of *Archaeopteryx.* The extra large sterna of

*Sinornithosaurus* and *Bambiraptor* are especially difficult to explain as having evolved for nonflight purposes. As part of this flight complex, an increase in the number of sternocostal attachments above the number observed in *Archaeopteryx* is one of the potentially neoflightless characters listed in Chapter 8. Unfortunately, the uncertain number of places at which the sternal ribs attached to the sternum in the urvogel (Chapter 4) hinders examination of this feature, but a sternocostal articulation count of four or higher is considered a potentially neoflightless character.

The forelimbs of nonavepectoran avepods tend to be little more than half the length of the legs or less, with the highest values observed in allosaurs (App. Fig. 18C). Although it is possible that dino-avepectorans evolved longer arms for predation, it was not vital for them to elongate the arms. A heritage of flight or climbing or both better explains the evolution of long arms, which would tend to be retained in neoflightless dinosaurs that could find a use for them. At the opposite extreme, severe arm reduction is fully compatible with loss of flight. However, there are viable hypotheses for arm reduction that do not involve flight, and some basal dino-avepods had very short arms (ceratosaurs, abelisaurs). As explained in Chapter 7, protofliers should have arms longer than the norm for dino-avepods. Middleton and Gatesy (2000) observed that flightless ornithurine birds have shorter lower arm segments than do fliers, but this is probably because hands of the former are so specialized for flight that they atrophy after flight is lost. In addition, some non-avepectoran dinosaurs also have shortened lower arms.

Protofliers probably needed to fold their arms to protect their elongating arm feathers. However, the feathers of protofliers should have been shorter than those in *Archaeopteryx*, and the folding mechanism less well developed. The avian-type wing-folding mechanism may have evolved as a way to retract and protect the sharp claws of predaceous dinosaurs, a crude version of the retractile claws of cats. This possibility applies most obviously to the derived therizinosaurs, whose sabrelike finger claws were exceptionally long. However, as will be shown below, therizinosaurs underwent a reduction in their ability to fold their arms at the same time that their finger claws were lengthening. In addition, it is not obvious why such a complex system evolved in order to protect claws that were held well clear of the ground anyway. This system would not have been necessary in most theropods with either large hand claws or long arms like those of ornithomimids. Gauthier and Padian (1985) and Ostrom (1996) envisioned predatory avepectoran dinosaurs using a snap-action arm-unfolding mechanism to reach out and grab prey. But snap-action devices (projectile tongues, hinged forelimbs) evolve in sit-and-wait (frogs) or slow-stalking (praying mantis, chameleons) predators that need a fast-moving body part to suddenly reach out and catch prey before it can flee. They are neither needed by, nor of apparent benefit to, predators that use swift legs as their prey-capture mechanism. Why would complex arm action have been an advantage for fast running predatory dinosaurs that, like the many examples with conventional arms, could simply dash alongside and apply their clawed fingers to their hapless victims? In addition, the herbivorous caudipterygians and therizinosaurs, both basal and derived, did not need their folding arms to grab prey.

Chatterjee (1997, 1999a,b) suggested that the arm actions involved in climbing tree trunks promoted the evolution of the avepectoran folding forelimb. But the actions associated with climbing do not include the ability to tightly fold the wrist, and no other climbers have evolved such folding arms.

Another non-flight-related explanation for folding arms is that they evolved to protect and deploy long arm feathers that developed for egg brooding (Hopp and Orsen 1998). Another possibility is that the folding arms evolved to protect arm feathers that had initially become lengthened for sexual display. These interesting hypotheses may prove especially important if avepectoran dinosaurs prove not to be neoflightless, but in competition with the neoflightless hypothesis, they suffer from some deficiencies. Low-power uses of brachial feathers do not adequately explain the development of the avian-type scapulocoracoid, or the ossification of the large sternal plates and their ribs. Even large feathers do not weigh much, and a more conventional, less powerfully muscled shoulder girdle would have provided an adequate base for arms that merely draped the feathers over a nest or held them out for a potential mate to see. As for the folding arms, brooding and sexual display will apply the most evolutionary pressure for arm modification if the feathers were present for much or all of the year and if they adorned both sexes. If long arm feathers were seasonal display or brooding structures or were limited to one sex, then the selective forces for protecting them via arm folding would be lower. Also consider that brooding and display feathers do not need to be kept in perfect condition. Male peacocks drag their display feathers on the ground. Ostrich arm feathers, used for both brooding and display, are irregular in form. These observations bring the question of why the arm feathers would need to evolve such complexity, and be supported by such large folding arms, merely to cover eggs. Big living ratites have lost well-developed pennaceous wing feathers, and their arms are reduced to a greater or lesser extent. Especially telling are the extinct flightless birds that lost not only the folding mechanism but also essentially the arms themselves. If selective forces have not been strong enough lead to the retention of folding arms bearing large arrays of feathers for brooding in flightless birds, then the possibility that such complexes initially evolved primarily for brooding must be ranked as low. After all, birds can use an overall body covering of long feathers to cover even large nests like a blanket (del Hoyo et al. 1992–99).

The most logical explanation for the presence of collapsible arms operated from large, birdlike shoulder girdles in flightless dino-avepods both predatory and herbivorous is that they evolved in flying forms that needed to fold up delicate aerodynamic dermal structures in order to keep them in prime condition. In this view, these major adaptations first developed in order to meet the rigorous requirements of flight and were then secondarily exploited to recruit mates and brood eggs in a new manner. In other words, flight, not reproduction, best explains why birds have collapsible wings. Finally, the brooding hypothesis does not explain the evolution of dinosaurian tail fan feathers. Even the shorter tails seen on some avepectoran dinosaurs, such as oviraptorosaurs, extended beyond the outer edge of preserved nests (Fig. 10.16).

What about hand and finger adaptations? In flying birds, metacarpal III is bowed posterolaterally. This feature, which appears to expand the hand area needed to support the arced spread of the largest distal flight feathers so they can better resist the bending forces experienced during flight, is often but not always retained in flightless birds. An alternative explanation for bowing of the outer metacarpal has not been offered, and conceiving of such an explanation, especially considering that caudipterygians with large distal remiges lack a bowed metacarpal III, is difficult. In addition, this feature is not found in any nonavepectoran and is, therefore, considered an excellent potentially neoflightless character, even more so if the bowing of the metacarpal is greater than that seen in the urvogel, in which case the bowing qualifies as both a post-urvogel character and as evidence of a more advanced level of flight. If the strongly bowed metacarpal is accompanied by an absence of well-developed flight feathers that the structure could help support, then the neoflightless potential of this combination is doubled.

Flying birds have a stiff central finger in order to better support the distal most flight feathers that attach to this appendage, and this feature is retained in neoflightless birds. It is difficult to develop a hypothesis explaining why stiffening of the central finger would have evolved in predators that use the hand to grasp prey, and the flexion potential of the finger appears to have been high in oviraptorosaurs that may have had well-developed distal remiges. Reduced flexion in finger II is therefore rated as an excellent potentially neoflightless character; the lack of better data concerning *Archaeopteryx* hinders assessment of its potential as a post-urvogel character. What qualifies as both a neoflightless and post-urvogel character of the highest order is stiffening of the central finger by close appression of a severely reduced outer finger, and the loss of bones in the outer finger associated with this arrangement is an additional avian attribute. This advanced outer wing feature is very difficult to explain outside the context of flight, all the more so since it is absent even in *Archaeopteryx* and other basal birds with well-developed outer wings.

Another feature of the hand, verified too recently to be detailed in the main text, may be the best single indicator of a post-urvogel level of flight. The feature is an expansion of the base of the central finger that improves the support for flight feathers, even when large wing feathers are absent. (See Appendix 6 for further discussion.)

This brings us to the complex yet symmetrical hand feathers of tail feathers, or both, seen in the Yixian dino-birds. Unlike the simple feathers covering entire bodies, these distally placed structures did not evolve for insulation. Display for purposes of enhanced reproductive success offers a more plausible selective cause for the development of such intricate dermal adornments, but similarly complex integumentary display structures have not evolved for display in other amniotes. The apparent absence of large display feathers in *Sinosauropteryx*—another cursorial, feathered dinosaur that lacks evidence of a flight heritage—works against the display hypothesis. Their development as airfoils during the initial stages of the evolution of flight from the ground up suffers from the problems inherent to the scenario (as detailed in Chapter 6). The symmetrical hand feathers of *Caudipteryx* were especially useless for generating lift or control in even a preflier. Aerodynamic needs do, however, offer the best explanation for the development of large, intricately constructed contour feathers. In addition, display offers the best explanation for the retention of large, symmetrical feathers at the ends of arms and tails in the neoflightless, with a possible secondary function as maneuvering surfaces during high-speed running. The best explanation for these dinosaurs' contour feathers is that they first evolved for flight and then were retained and modified for display—and possibly for aero maneuvering and braking—after flight was lost (also see Griffiths 1998/2000). Because the status of feathers in other avepectoran dinosaurs is unknown, they cannot be assessed in this regard.

Although there are reasonable non-flight-related explanations for many of the features seen in dino-avepectorans, and such explanations may in the end prove correct, they tend to be ad hoc and piecemeal and in many cases do not come together to form a single, compelling hypothesis. Brooding, display, and prey capture are, therefore, relatively weak explanations for the avian form of birdlike dinosaurs. Flight and its loss offer a single, eminently logical explanation for the evolution of all of the birdlike pectoral, brachial, caudal, and dermal adaptations seen in avepectoran dinosaurs. It follows that the reason the only creatures with avepectoran type scapulocoracoids are avepectorans and pterosaurs is that they all either flew or descended from fliers. The broad sternal plates of dino-avepectorans first evolved to anchor enlarged supracoracoideus muscles, with a secondary respiratory function for the plates, as part of a flight complex that may have been more advanced than that of *Archaeopteryx*. Folding arms first evolved as folding wings to protect the all-important, large, and complicated yet delicate flight feathers. Reduced, dynamic tails evolved under the same selective pressures that directed the development of similar tails in basal birds and pterosaurs.

Ultimately, it is the totality of the evidence that is most telling. Explaining an arm-folding system *or* a dynamic tail as unrelated to flight is one thing. However, explaining flexed scapulocoracoids *and* enlarged proacrocoracoids *and* hinged coracoid-sternal articulations *and* folding arms *and* reduced dynamic tails in dinosaurs as unrelated to flight is a much more challenging proposition. In dromaeosaurs, troodonts, caudipterygians, and oviraptorosaurs, it is not just the sternum but also the wing-folding mechanism and the scapulocoracoid that are avian to varying degrees; and all these dinosaurs have tails that are similar to those of early birds and pterosaurs. If one were to design a neoflightless dinosaur, it would look very much like a dromaeosaur or like *Caudipteryx*. Further burdening the nonflight argument are the avian characters that are found in these dino-birds but not in *Archaeopteryx*. Explaining any one of these adaptations as an exaptation or as a result of parallelism is plausible, if sometimes difficult. Explaining all of them in these terms stretches plausibility to near its limits, because no nonvolant tetrapod, arboreal or predatory, has combined the pectoral, arm, and tail features observed in avepectoran dinosaurs. Conversely, it is difficult to explain why dinosaurs that never used their forelimbs and tails for flight activities would develop shoulder girdles, folding arms, and dynamic caudal stabilizers that are so similar to those of pterosaurs and birds. To do so, one must invoke parsimony: the reason that some dinosaurs' forequarters and tails look like those of flying and, especially, neoflightless birds is simply because dinosaurs too were neoflightless.

It is possible that the potentially neoflightless characters, both skeletal and dermal, seen in avepectoran dinosaurs were inherited from preavian dinosaurs whose flight performance was substantially less developed than that of known urvogels. But it is only a possibility, not a probability, because there is evidence that the ancestors of avepectoran dinosaurs flew better than *Archaeopteryx* did. That the tails of *Protarchaeopteryx* and *Caudipteryx* were even more abbreviated than that of the urvogel suggests that they descended from fliers that had achieved a more dynamic flight performance. Further evidence centers on the ossified sternal plates of *Protarchaeopteryx*, dromaeosaurs, caudipterygians and oviraptorosaurs, which were larger than those of *Archaeopteryx*. The implication is that the flying ancestors of the dinosaurs had larger sterna than the urvogel and were thus more powerfully muscled and therefore more advanced fliers. If the dorsal reach of the arm of *Unenlagia* really were more vertical than in *Archaeopteryx*, as Novas and Puerta (1997) suggested, then this would be evidence of descent from advanced fliers, but neither the upswing of the arm of *Unenlagia* (Chapter 4) nor the flight performance of its ancestors appears to have been greater than in other avepectoran dinosaurs.

There are other characters potentially associated with flight, and although they are less directly related to flight than the characters we have examined so far, they are nonetheless well worth considering. Interestingly, avepectoran dinosaurs persistently deviated from the normal dino-theropod pattern in terms of brain size, sensory equipment, and feeding habits. Among dinosaurs as whole, large brains and overlapping fields of vision were largely limited to avepectoran dinosaurs that had other potentially neoflightless characters, which suggests a possible link. The reduced sense of smell seen among some dino-avepectorans is especially interesting, since most dino-theropods appear to have had a well-developed olfactory apparatus useful for hunting, whereas birds—which have little use for detecting odors on the wing—are noted for their usually limited olfactory powers (Dorst 1974). Olfactory performance can be estimated via the relative size of the olfactory lobes of the braincase and the volume of the posterior antorbital cavity. Almost all nonavepectoran theropods used a combination of lateral vision, nonsaltorial running, a moderate number of large, blade-shaped teeth with fine symmetrical serrations, and modest-sized arms to catch, kill, and cut up their large prey (Paul 1988a,d). Considering that this classic dino-theropod hunting mode worked well during the entire span of the dinosaur reign, selective pressures to veer away from this pattern should not have been strong as long as size remained at least modest. The miniaturization associated with flight, however, should have forced an abandonment of this type of hunting. Any neoflightless descendants of small fliers either should have hunted in a different manner—most likely by exploiting a subaerial leaping mode—or should have become omnivores and herbivores. It is therefore pertinent that this is exactly what one sees in all dino-avepectorans. None hunted in the classic dinosaurian manner.

The case for neoflightlessness is less direct in some other advanced dino-avepods. Ornithomimosaurs, for example, have some avian attributes, both flight related and not, but these attributes were not as strongly developed as in most avepectorans. Even more interesting in this respect are alvarezsaurs, which are so strongly avian in the skull and parts of the skeleton that the possibility that they were closer to birds than the urvogel is substantial. Yet alvarezsaurs lack many of the neoflightless features observed in other dino-avepectorans. This lack leads to the observation that if alvarezsaurs with their few potentially neoflightless characters were neoflightless, then the possibility that other avepectoran dinosaurs, with their numerous potentially neoflightless characters, were neoflightless must be considered as even higher. The possibility of a very strong reversal from the flying ancestral condition to the nonvolant condition needs to be considered in the case of ornithomimosaurs and alvarezsaurs. This brings us to the subject of reversal.

### *Going into Reverse: A Problem Overblown?*



The most serious difficulty with the hypothesis that some predatory dinosaurs were neoflightless is that the same creatures lack various avian attributes seen in *Archaeopteryx*. If the avepectoran dinosaurs were secondarily flightless post-urvogels (and if we assume the urvogel did not parallel more-derived birds in these particular attributes), then either the dinosaurs represent parallel efforts at flight or they experienced reversals back to the more general dinosaurian condition. The latter is certainly possible; after all, the less sharply flexed, broad coracoids of flightless birds recapitulate the dino-avepod condition. The loss of any sternal keel and shortening of the arms are also normal reversals for flightless birds. Remember that the absence of the semilunate carpal block and the arm-folding mechanism does not necessarily disprove a neoflightless condition, because these features are sometimes lost in flightless birds.

In addition to lacking some avian attributes, these dinosaurs also have some nonavian features. Most obvious among these are the procumbent pubes of troodonts, oviraptorosaurs, avimimids, and perhaps some dromaeosaurs and the conventional tails of oviraptorids and alvarezsaurs—although the required degree of pubic deretroversion is not major, considering the subvertical pubes of *Archaeopteryx*. As it happens, possible reversals of these pubic and caudal features are probably linked to one another. The dino-avepod combination of long tail, nonretroverted pubis, and vertical-action femur was one that worked very well and was lost largely or entirely under the selective pressures associated with the development of avian flight. Indeed, the old dinosaur arrangement might have been superior in terms of natural selection for flightless avepods, in which case there would have been selective pressure to re-evolve it when flight was lost. This kind of reversal never occurs in neoflightless birds: their tails always remain short, their pubes retroverted, and their femurs subhorizontal, probably because birds are too far down the avian path to return to the ancestral configuration. The tail in neoflightless birds is such a truncated pygostyle that relengthening it is not feasible; therefore, the pubes remain retroverted in order to keep the abdominal mass as far aft as possible. In addition, because avian pubes are split from one another and surround the belly, deretroverting them is difficult. These constraints would not have applied to protobirds. In flying protobirds, the pubes were still joined along the midline and supported the abdomen from behind, so they should have been able to swing in whatever direction best enhanced balance, or any other function related to their orientation. In addition, the tail was still fairly long, so it could re-elongate, redeepen (by means of an increased number or span, or both, of chevrons), and rebroaden (by means of an increased number or span, or both, of transverse processes) upon the loss of flight and again fulfill the functions normal to terrestrial bipeds. It is pertinent that the number of free caudals tends to increase and the pygostyle tends to become reduced in flightless birds, to the point that moa almost lacked a pygostyle. It is also interesting that the mammals most similar to bipedal dinosaurs, kangaroos, have reevolved tails that are longer and heavier than typical for mammals. The re-enlargement of the avepod tail would force anterior rotation of the pubes, which would shift the abdominal mass forward again and cause the femur to become more nearly vertical again. The combination of a vertical pubis with a retroverted pubic peduncle in *Rahonavis,* an avepod otherwise more derived than the more retropubic *Archaeopteryx,* supports the possibility that *Rahonavis* underwent pubic reversal to a more basal condition. Contradicting this hypothesis, at least to some degree, is the absence of pubic retroversion in caudipterygians and in those oviraptorosaurs that had only an abbreviated tail to balance their bellies.

Although pubic deretroversion was not remarkable, the same cannot be said of the recontact of metatarsal I with the ankle in therizinosaurs; in no other tetrapod group has such a severely reduced digit regained its original, fully developed form (Paul 1984b, Xu, Tang, and Wang 1999).

Reversal issues also apply to the skull. Until recently, the complete bars bordering the posterior rim of the orbits of potentially neoflightless dinosaurs were seen as a major challenge to their also being closer to avebrevicaudans than to *Archaeopteryx.* Not any more. The postorbital bar missing in most birds may have been complete in *Archaeopteryx* and was complete and quite broad in the more derived *Confuciusornis.* Such a stout bar would not be considered out of place on a Triassic diapsid, so its presence in *Confuciusornis* is as remarkable as it is important. If a fully developed bird could have a postorbital broader than that of a number of dino-avepods and still be well above the level of the urvogel, then so too could avepod dinosaurs. In the post-urvogel–neoflightless scenario, we can postulate that the complete and sometimes robust postorbital bars common to avepectoran dinosaurs were a retained character. Alternately, the complete postorbital of *Confuciusornis* may be the result of a reversal, and a remarkably strong one at that. One reason for the redevelopment or restrengthening of the bar would be increased stress loads due to changed feeding habits, such as large-game predation in dromaeosaurs and troodonts or herbivory in ornithomimosaurs, caudipterygians, therizinosaurs, and *Confuciusornis.* In any case, the presence of a well-developed postorbital can no longer be cited as compelling evidence against a post-urvogel status. (For that matter, *Archaeopteryx* cannot be scored as having either a complete or incomplete postorbital bar, since we do not know at this time.) By the same token, the absence of avian cranial kinesis in *Confuciusornis* shows that this condition does not bar an avepod from being a post-urvogel. Because the completeness of the squamosal-quadratojugal bar is not certain in *Archaeopteryx,* we cannot determine whether the complete bars characteristic of most avepectoran dinosaurs represent potential reversals. If the complete bars were the result of reversals, the most likely explanation, again, is that the bars strengthened the suspensorium for feeding.

Returning to the skeleton, if a dino-bird's hand changed from being an anchor for the outer wing back to being a weapon, bowing of the outer metacarpal and stiffening and robustness of the central finger may be reduced or lost. Avepectoran dinosaurs lacked the avian fully reversed hallux, and this condition could represent a reversal from the reversed inner toe of the urvogel. Dereversal of the hallux is as compatible with a neoflightless, post-urvogel condition as is the unreversed first toe of some birds, such as the kiwi (Fig. 4.5D). This ratite shows that flightless birds that do not need a fully reversed hallux can lose it.

It is also important to consider the possibility that at least some potential reversals may not have occurred even if the avepectoran dinosaurs are closer to modern birds than was *Archaeopteryx.* It is quite possible that at least some of the avian attributes of the urvogel evolved in parallel with those of avebrevicaudans (this possibility is further addressed later in the chapter). In addition, some of the birdlike characters of *Archaeopteryx,* such as the highly circular orbit and slender posterior cranial bars, may have resulted from its small size.

### Going into Reverse in the Australian Thunder Birds: Extensive Pseudophylogenetic Reversal Associated with Loss of Flight in an Avian

First discovered in the early 1800s, the big dromornithid "thunder birds" were for a long time known almost exclusively from skeletal material (Fig. 8.4); skull parts were fragmentary and badly preserved. The dromornithid postcrania were similar enough to those of ratites to cause researchers to conclude that dromornithids were yet another example of this group (Rich 1979). But S. Olson (1985) dissented, and new and better skull material has since demonstrated that dromornithids were definitely neognathes, probably closely related to anseriforms (Fig. 15.5; Murray and Megirian 1998). Here we have an example of neognathes that, after having lost flight, continued reversing their form until it mimicked that of palaeognathes. If such extensive reversals of even non-flight-related features can occur in neoflightless modern birds, then it is all the more plausible to propose that neoflightless avepectoran dinosaurs could have experienced reversals extensive enough to return them to a pre-urvogel, avetheropod-like condition.

### Going into Reverse: Positive Evidence for Flight Loss

As noted in Chapter 8, a decrease in potential flight-related adaptations in the increasingly derived members of a clade is the pattern expected if the ultimate ancestor of the clade was a flier. Why would dinosaurs develop such flight-type characters for nonflight needs such as predation and then lose them even though they remained predators? The fossil record of avepectorans is too incomplete for us to assess this pattern in full, but, as will be shown below, this form of reversal normally associated with loss of flight is observed in ornithomimids, therizinosaurs, and dromaeosaurs.

### *Placement in Time*

An analysis of the time frame for the appearance of various features favors the hypothesis that some dinosaurs were neoflightless. Predatory dinosaurs failed to develop birdlike shoulder girdles, foldable arms, and specialized tails during the Triassic and much of the Jurassic. These features appear in the fossil record only when flying avepods were already present. This temporal pattern suggests that rather than being exaptations for flight, the characters under discussion instead evolved as a result of flight. If we carry this suggestion one step farther, we can explain why the most birdlike dinosaurs date from the Cretaceous: because they were offshoots, rather than predecessors, of the early fliers in the Jurassic. Even potentially neoflightless *Protarchaeopteryx* postdated its close, flying relative *Archaeopteryx*. Also significant is that within avepectoran clades such as ornithomimids, therizinosaurs, and dromaeosaurs, there appears to have been a decrease in the number of potential flight-related adaptations with time—such a decrease is to be expected among the neoflightless.



## Group Analysis

Having reviewed the general evidence for neoflightlessness and major reversals in predatory dinosaurs, we will now consider the evidence specific to each group of avepectoran dinosaurs and analyze the characters that support a post-urvogel status. To recap, the greater the number of potentially neoflightless characters that are present in a particular group, the more likely it is that the group was neoflightless. Numerous potentially neoflightless characters may also suggest that a relatively advanced stage of flight was lost. Few potentially neoflightless characters may suggest that only a crude, early stage of flight was lost, or may be the result of extensive reversals from an advanced flight condition. Potentially neoflightless characters can also be post-urvogel characters if they represent a condition more advanced than seen in the urvogel. The greater the number of post-urvogel characters that are present in a group, the more likely it is that the group is closer to modern birds than the urvogel is. If both numerous potentially neoflightless characters and post-urvogel characters are present, then the case for neoflightlessness is strengthened. However, if numerous potentially neoflightless characters are combined with only a few post-urvogel characters, flight still may have been lost, but it may have been lost among fliers less capable than and/or more distant from modern birds than archaeopterygiforms. Because numerous post-urvogel characters imply descent from more-basal protobirds, they always favor flight loss, even if potentially neoflightless characters are few. By assessing the balance between potentially neoflightless characters and post-urvogel characters within and between groups, we can get a better idea of where each group may fall on both the spectrum of neoflightlessness and of their phylogenetic placement relative to *Archaeopteryx* and avebrevicaudan birds. After looking at the various groups, we will use the accumulated data to delve further into the question of whether exaptation or loss of flight combined with a post-urvogel status best explains the birdlike form of avepectoran dinosaurs.

In order to quantitatively assess the problem, forty-six characters or attributes are scored and then graded for each taxon in Table 11.1. The combination of either a metacarpal III that is more posterolaterally bowed than that of *Archaeopteryx,* or a finger flange, with the absence of large distal contour feathers is scored at 2. Of these forty-six attributes, four involving neural or sensory organs and feeding habits are scored at half the value of features more directly related to flight. When a potentially neoflightless character is partly developed or probably, as opposed to definitely, present, it is counted at half its full value. Characters that are uncertain or probably absent are not scored. After scoring, each taxon is graded in order to minimize disparities due to varying amounts of available information. The highest potential grade is 100 out of whichever of the attributes is available for scoring in a given taxon. *Archaeopteryx* is included in the table for comparative purposes.

### Nonavepectoran Theropods

As Table 11.1 shows, nonavepectoran theropods have either no potential neoflightless and post-urvogel characters or far too few of them for these theropods to be considered neoflightless. It is notable that a grade of 0 was awarded to the compsognathids, including *Sinosauropteryx,* whose lack of any potentially neoflightless characters in its uniquely long (up to 64 caudals!) tail is positive evidence that its ancestors never flew. This grade indicates that simple feathers of *Sinosauropteryx* cannot be considered evidence that it was neoflightless.

### Coelurids and Ornitholestids

*Coelurus* was once poorly known (Ostrom 1980a), but new remains promise to fill in our knowledge when they are fully described (Miles et al. 1998). The coracoid is not especially large or flexed relative to the scapula, and the lozenge-shaped distal carpal block is not as well formed as the semilunate structure of avepectorans. The ulna is bowed, as in avepectorans, but this is at best a poor indicator of a flight heritage. The weak potentially neoflightless characters are compatible with a flightless ancestry, or perhaps descent from incipient fliers much less aerially capable than *Archaeopteryx.*

Assessment of the neoflightless condition of *Ornitholestes* is hindered by the lack of critical parts of the skeleton. The slender hand long assigned, but not attached, to *Ornitholestes* actually probably belongs to *Coelurus,* and the rest of the arm and shoulder girdle are also unknown in the former avepod (Fig. 10.1Bh). This lack of material also prevents us from determining whether it should be placed inside or outside of the Avepectora. So too does the fact that the dinosaur's nearly complete skull and partial skeleton have yet to be carefully described, despite their discovery a century ago. *Ornitholestes* did have shallow, anteroposteriorly elongated chevrons under the distal section of the tail. Their presence is an attribute of a dynamic stabilizer, albeit a weakly developed one. However, the tail was quite long. The arm was fairly long. The head was quite small, and the teeth have reduced serrations. With only eighteen characters available to assess, the grade of 16 is tantalizing but ambiguous. The potentially neoflightless characters may or may not be a signs of loss of some level of flight, perhaps at a very early gliding/leaping stage.

Considering the overall grade of the available remains for *Coelurus* and *Ornitholestes,* neither appears to be strongly birdlike. The pubes are completely conventional in design in both groups. Neither the neoflightless status nor the phylogenetic placement of coelurids and ornitholestids is clear at this time.

### Tyrannosaurs

At first consideration the big tyrannosaurs of the Late Cretaceous would not seem to belong in this discussion (Fig. 10.1Bf). They do not have a large number of either potentially neoflightless characters or post-urvogel characters. Tyrannosaurs shared with allosaurs some distinctive characters, detailed later in this chapter. However, Holtz (1994a, 1996) argued that tyrannosaurs were members of the Arctometatarsalia, which includes troodonts, elmisaurs, ornithomimosaurs, tyrannosaurs, and avimimids. If correct, then tyrannosaurs were avepectorans. This possibility is supported by the rather small size of early tyrannosaur fossils (Manabe 1999). The "tyrant lizards" did have characters observed in other members of the clade, including a pneumatic quadrate and a rather short tail. Shallow, anteroposteriorly elongated chevrons under the distal-most section of the tail qualify as a minor potentially neoflightless character. Severe abbreviation of the arms and hands is compatible with flight loss but can also be explained in terms of weight reduction to allow for increased speed and to compensate for increased head size. The absence of an arm-folding mechanism could have been due to the severe forelimb reduction. In any case, abandoning the arms as important weapons in favor of using vision to direct bites made with a scooplike array formed by premaxillary teeth with a D-shaped cross section diverged from the classic dino-theropod hunting style (Paul 1988a,d). The overlapping fields of vision, especially, hint at flying ancestors. But the tyrannosaur hunting mode lacked a leaping component and in this regard is a logical extension of the standard dino-theropod system. Aside from being exceptionally robust, most of the teeth were conventional in form. So were the long, overlapping cervical ribs, which strengthened the neck of these big-headed monsters at the expense of some flexibility; then again, *Archaeopteryx* and birdlike *Caudipteryx* had overlapping cervical ribs too. The number of reversals required in a neoflightless scenario for tyrannosaurs is very high. The necessary reversals include the redevelopment of a conventional dino-avepod ilium and an ischium with a long distal rod. Although the potentially neoflightless characters grade of 6 is low, it is higher than typical of dino-avepods, and the extraordinarily avian and probably neoflightless alvarezsaurs do not have much in the way of potentially neoflightless characters either. In any case, the possibility that tyrannosaurs descended from fliers must be ranked as lower than observed in the following groups.

### Ornithomimosaurs

The classic Late Cretaceous ostrich-mimic ornithomimids are well known via complete skulls and skeletons (Figs. 10.1Bj, 10.2H; Osmólska et al. 1972, Russell 1972). New remains are beginning to reveal the nature of the earlier types (Barsbold and Perle 1984, Barsbold and Osmólska 1990, Perez-Moreno et al. 1994), although the status of *Pelecanimimus*—known from only part of the skeleton—as a basal ornithomimosaur has been challenged (Taquet and Russell 1998). The spectacular arms of *Deinocheirus* may belong to a giant ornithomimosaur (Ostrom 1972). Chatterjee (1993) suggested that the nearly complete skull of *Shuvosaurus* represents a Triassic ornithomimosaur, but this form may instead have been an ostrich-mimic that belonged to one of the so-far-headless gracile and semibipedal crurotarsaled archosaurs of the same time and place; until the issue is settled this fossil cannot be properly analyzed. Ornithomimosaurs had a number of the attributes observed in avepectorans. In particular, the modified palate of ornithomimosaurs had some birdlike attributes, and avian features not observed in *Archaeopteryx* are found in the ventral braincase, which contained a brain as large as the brain of a ratite. Overlapping fields of vision and a rather short tail might be remnants from flying ancestors. The tail is on the short side at 36 to 39 vertebrae. The distal tail had the shallow, anteroposteriorly elongated chevrons that suggest it may have acted as a dynamic stabilizer. The more conventional nature of the rest of the tail may represent a reversal among these nonpredators that no longer needed a well-developed caudal stabilizer. Especially interesting is the presence of large sternal plates in basal ornithomimosaurs (assuming that *Pelecanimimus* is one) and the apparent absence of such plates in the derived ornithomimids. The former is a potentially neoflightless character. The more-conventional shoulder girdles of more-advanced forms (Fig. 10.8I) hint that a reversal from an ancestral flying condition to a neoflightless condition was underway. Specifically, the realignment of the sternocoracoid articulation from strongly transverse to more fore-and-aft may have been integral to the development of shoulder girdle mobility associated with the feeding actions of the arms in these herbivores (Nicholls and Russell 1985, Paul 1988a). As discussed later in the chapter, the same shift seems to have occurred in the similarly herbivorous therizinosaurs. The reduction of arm folding observed in therizinosaurs (below) may also apply to the nonfolding arms of ornithomimids. The same kind of reversal may be true of the flexible central finger of ornithomimids (Nicholls and Russell 1985). The conventional design of ornithomimid pelves may contradict the neoflightless scenario, but secondary reversion is also plausible. Having gastrolithic mills and blunt-beaked, tubular snouts with either very reduced teeth or none at all, ornithomimosaurs clearly had broken from the classic dino-theropod hunting techniques, so much so that some form of omnivory or herbivory is probable (Paul 1988a, Kobayashi et al. 1999). This shift in feeding fits with the avepectoran pattern of dietary modification that may have been initiated in small fliers. The potentially neoflightless characters grade of 18 is not impressive but may understate the importance of the large sternal plates in basal examples. Indeed, the absence of more-basal ornithomimids may be hiding a neoflightless heritage.

Short cervical ribs are a post-urvogel character that indicates preavian enhancement of neck flexibility. Although ornithomimosaurs are the least likely of the gracile dino-avepods considered in this section to have evolved from flying ancestors, there is significant evidence for the neoflightless hypothesis.

### Archaeopterygiforms

The one *Protarchaeopteryx* skeleton so far described is not well preserved, but the diagnostic characters that are present suggest that it is a very close relative of *Archaeopteryx* (Ji and Ji 1997b, Elzanowski pers. comm.). The archaeopterygiform characters of *Protarchaeopteryx* include small, conical teeth; a strongly flexed coracoid seen among classic dinosaurs only in *Sinornithosaurus;* a rather small, short yet broad subrectangular sternum (breadth/length ratio 3.5 to 4) with a narrow anteromedial indentation (Fig. 10.9D,E); slender fingers, II and III being crossed and II the most robust (Fig. 4.3); a small, low, parallelogram-shaped ilium with a slender, downward-projecting, sharp-pointed posterior process (Fig. 10.12Ca,b); an apparently deepened pelvic canal; and nonfused but tightly articulated metacarpals, tarsals, and metatarsals. Nonarchaeopterygiform features include symmetrical feet and deep proximal chevrons. Until better remains become available, *Protarchaeopteryx* is best interpreted as an archaeopterygiform.

Remarkably little is distinctly avian—that is, not also seen in avepod dinosaurs—in these two urvogels. A fully triradiate palatine—due to loss of the usually well-developed posterior maxillary process (Fig. 10.6Bb,d)—in the roof of the mouth is one of the more-avian features of *Archaeopteryx* (Fig. 10.6Be,j), but this condition is approached or matched in dromaeosaurs, troodonts, therizinosaurs, oviraptorids, and perhaps ornithomimids (Fig. 10.6Bf–i; Fig. 2A in Osmólska et al. 1972). Avian features of the wrist and hand (Zhou and Martin 1999) are largely flight-related features that can be expected to have been lost in secondarily flightless avepectoran dinosaurs. Avian cranial kinesis was not developed, the absence of a postorbital bar is open to question and observed in other dino-avepectorans, and the middle ear is no more avian than in most dino-avepods. The skull, occiput, and skeleton of archaeopterygiforms are basically those of dromaeosaur avepectorans, modified with a shortened tail, as well as large wings and, in *Archaeopteryx,* a larger, fully reversed hallux. The skeleton of *Archaeopteryx* is so primitive that it is a suitable ancestral type for all birds or for a dinosaur that developed flight independently from true birds.

It is tempting to assume that the partial development of toe asymmetry and hyperextendibility represents an incipient condition. This may well be true, but a reversal from a more derived condition cannot be ruled out. Such a reversal occurred in troodonts (see the section below on troodonts in this chapter), and, indeed, a much bigger reversal is possible. In 1997, Chatterjee made the unusual suggestion that the poor flight performance of may represent a reversal from a higher level! This is not an outlandish notion. Indeed, it is compatible with *Archaeopteryx*'s island habitat (see Chapter 8) and explains the creature's weakly ossified sternum. However, wing reduction is typical in birds with reduced flight performance (Raikow 1985, Feduccia 1996). The very large arms and wings of the protobird suggest that that it was trying to be as aerodynamic as its level of organization would allow. Therefore, the urvogel probably did not evolve from better, earlier fliers.

As for *Protarchaeopteryx,* it shows strong signs of having been neoflightless and of having undergone reversals related to this condition. Flight-related features include long, folding arms, and sternals about as well developed and as broad relative to the dorsal vertebrae as those of *Archaeopteryx.* Although the tail has as many vertebrae as that of *Archaeopteryx,* it was 60 percent shorter relative to the length of the dorsal vertebrae. Such severe abbreviation is expected in a flier that flew more dynamically than the earlier urvogel. However, *Protarchaeopteryx* was about three times heavier than its relative and perhaps even heavier in view of the fact that the nonfusion of the sternals implies that the *Protarchaeopteryx* specimen was a juvenile. *Protarchaeopteryx*'s arms were more than a third shorter relative to both its trunk and hindlimbs than those of *Archaeopteryx* (in which the arms were slightly longer than the legs), and are no longer relative to the legs than are the arms of some dromaeosaurs and oviraptors (in which arm length is about two-thirds leg length) (App. Fig. 18C). The *Protarchaeopteryx* humerus was much more slender than the femur, and relative to body mass, such slender arms are another feature of nonvolant creatures (App. Fig. 18D,E). The reduction of the arms despite the increased mass indicates that flight abilities were markedly inferior to the capabilities of the smaller, bigger-armed bird. Indeed, if as it seems the arms did not bear feathers, then *Protarchaeopteryx* was flightless. The outer metacarpal was bowed more strongly than that of *Archaeopteryx,* but because there probably were no large flight feathers for this arrangement to support, this neoflightless-bird-like condition qualifies as both a strong potentially neoflightless character and a post-urvogel character. Combining the very high potentially neoflightless characters score of nearly 24 and the grade of 66, along with the seemingly paradoxical combination of the dramatically shortened tail expected in a dynamic flier with limited or zero flight abilities, strongly suggests that *Protarchaeopteryx* descended from an archaeopterygiform that flew better than *Archaeopteryx* did. The flexible tail and its feather array may have acted as a rudder and air brake when *Protarchaeopteryx* was running. In any case, *Protarchaeopteryx* affirms that the tendency to lose flight may have been operative very early in avian evolution. According to the neoflightlessness scenario, the serrations on the teeth may have re-evolved to allow *Protarchaeopteryx* to feed on prey that was larger than that pursued by little *Archaeopteryx.* The apparent redeepening of the proximal chevrons at the base of the tail relative to the same in the earlier urvogel suggests that this kind of redeepening is yet another reversible neoflightless feature.

### Rahonaviforms

As we have seen, the remains of *Rahonavis* are incomplete, but if we assume that the ulna, which bears quill nodes, and the scapula, which is like that of a flying bird, do belong to the sickle-clawed hindquarters, then this big-armed creature certainly was not neoflightless. Indeed, the tapering scapula blade and quill nodes indicate that *Rahonavis* was a more advanced flier than *Archaeopteryx* (*Rahonavis* was more derived than *Confuciusornis* in having a tapering scapula). Quill nodes and a tapering scapula are also post-urvogel characters that suggest that rahonaviforms were closer to birds than the urvogel was (as per Forster et al. 1998). Alternatively, they may have evolved an advanced level of flight independently from the main bird clade. The pubis of *Rahonavis* was more vertical than that of *Archaeopteryx* (Fig. 10.12BV), and this may represent a reversal.

### Dromaeosaurs

With a character score exceeding 38 and a grade of 83, the long-armed, sickle-clawed dromaeosaurs (Figs. 1.6, 1.7, 10.1Bn, 10.2J) possess more in the way of potentially neoflightless characters than any other dinosaurs. One of the few features they definitely did not have was the fused carpometacarpus. Only dromaeosaurs had a nearly avian flight-related package that included all the pertinent rib, shoulder girdle, arm, and tail characteristics, including pterosaur-like ossified caudal rods. The number of vertebrae was below the dino-avepod norm at 36–40; in *Microraptor* it was only 24–25. The brain was enlarged, and the rather narrow olfactory lobes and the compressed snout imply that dromaeosaurs' sense of smell was inferior to that of most avepod dinosaurs. The presence of subtriangular frontals or very narrow snouts or both indicates that fields of vision overlapped. The combination of this type of vision with a dynamic tail, long, grasping arms, and sickle-shaped claws in dromaeosaurs indicates that they may have employed a leaping mode of hunting (Paul 1988a,d) indicative of major shift from the classic dino-theropod pattern, a shift of the sort expected in neoflightless hunters.

With its small body, very long arms, exceptionally elongated sternum with as many as five costal attachments, *Archaeopteryx*-type furcula, retroverted coracoid with a prominent proacrocoracoid, and the uniquely avian expansion of the bone at the central finger base (see Appendix 6), *Sinornithosaurus* must have been extraordinarily close to having the ability to fly (Figs. 10.8L,M, 10.11Ah,i, 11.1). Had the arms been just a little longer and had long, asymmetrical flight feathers been present, it would have been a true flier. As it is, the absence of a wing shows that *Sinornithosaurus* was quite flightless, despite all of its well-developed flight features. In addition, dromaeosaurs are the only dino-avepectorans known to have had the strong potentially neoflightless attribute of a posterolaterally bowed outer metacarpal (Fig. 10.11Ag,i). In *Sinornithosaurus,* and perhaps *Bambiraptor,* the bowing of the metacarpal is greater than observed in *Archaeopteryx* (Fig. 10.11Ah,i). Added to the neoflightless potential of this feature is the fact that *Sinornithosaurus* lacked the large flight remiges the bowed hand bone and expanded finger base were so suited to support. It will be interesting to see if a plausible alternative hypothesis can explain these classic attributes of secondarily flightless birds; until an alternative is presented, the dromaeosaur hand and other parts of the skeleton must be seen as recording an advanced flight heritage.

It is interesting that while *Sinornithosaurus* had the most urvogel-like scapulocoracoid among known dromaeosaurs, *Bambiraptor*'s scapula tip was more pointed, its scapular acromion process

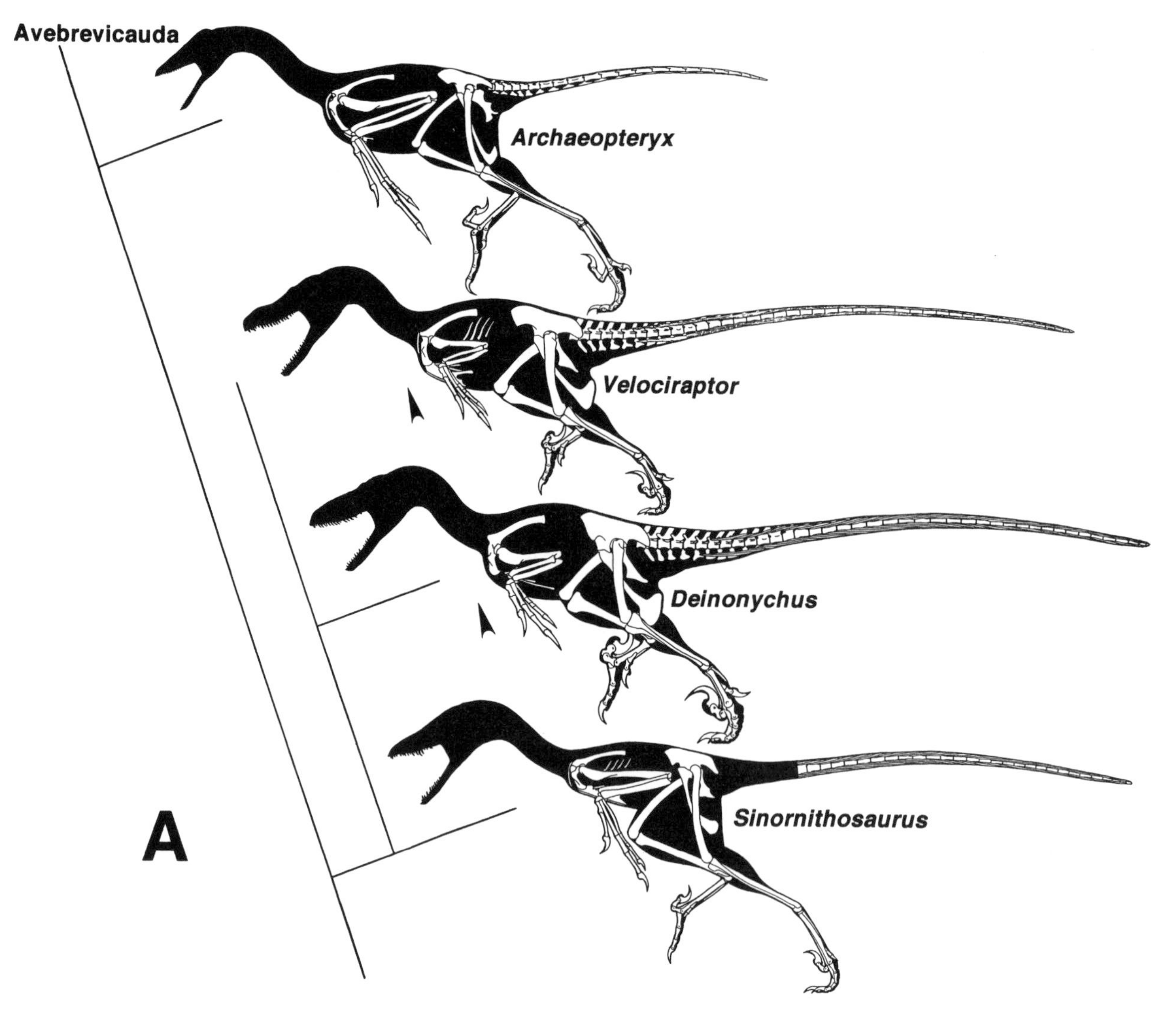


*Figure 11.1.* A comparison of the conventional versus the neoflightless hypotheses, focusing on dromaeosaurs and *Archaeopteryx*. The skeletal restorations include only those flight-related adaptations that are preserved in the fossils. Arrows indicate general direction of reduction of flight-related features. *A,* conventional hypothesis: flying *Archaeopteryx* is closer to modern birds than flightless dromaeosaurs are, despite *Sinornithosaurus*'s having had flight-related adaptations more advanced than those of the urvogel (very large ossified sternum, ossified sternal ribs and uncinate processes, well-developed posterolateral flange on the proximal bone of the central finger). *B,* neoflightless hypothesis: with its primitive flight features, *Archaeopteryx* is placed basal to *Sinornithosaurus,* which inherited its more advanced flight-related features from a flying post-urvogel; reduction of flight-related features then occurred in more-derived dromaeosaurs (furcula shallower, sternum shorter, posterolateral flange of the central finger lost). The thin dashed line indicates an alternative scenario in which archaeopterygiforms and dromaeosaurs form a monophyletic clade, and basal flying dromaeosaurs developed advanced flying adaptations independently of the avian mainstream.

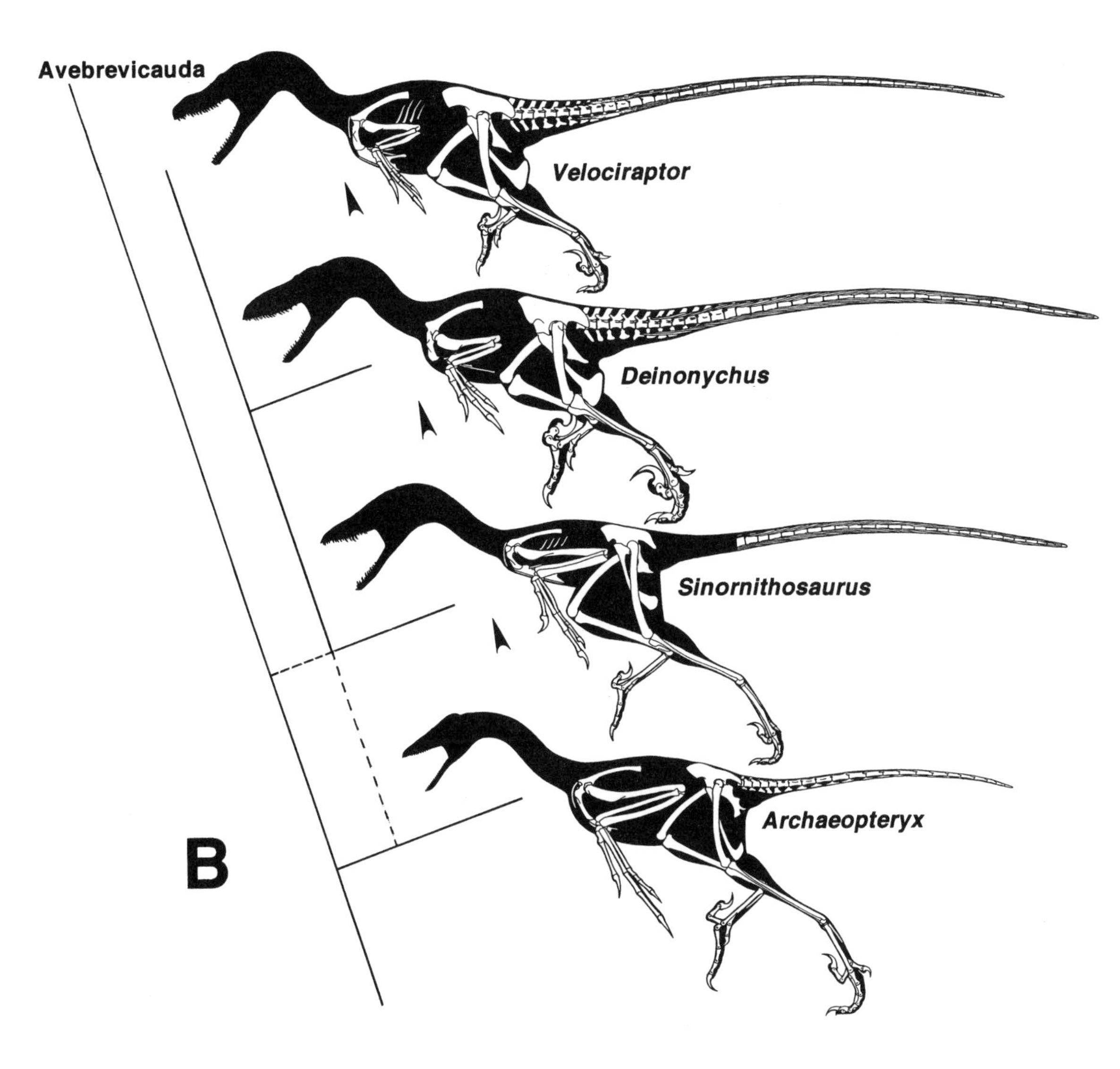


was better developed, and its coracoid was more proximally narrowed, more so than in *Archaeopteryx,* which indicates that the *Bambiraptor* had a better-developed arm-elevation and -control system than the flying bird. Exactly what this complicated pattern means is unclear, but it is difficult to explain how shoulder girdles so like those of today's neoflightless birds—including very long sterna (App. Fig. 10B)—evolved unless dromaeosaurs' ancestors flew better than *Archaeopteryx.*

Xu et al. (2000) observed that *Microraptor* was more birdlike than later dromaeosaurs. That *Microraptor, Sinornithosaurus,* and *Bambiraptor* were the earliest and/or most basal dromaeosaurs provides a clear indication that as dromaeosaurs evolved, they reduced their potential flight-related adaptations by deretroverting and shortening the coracoids (the coracoid of *Deinonychus* was so large that it was misidentified as the pubis [Ostrom 1969, 1974b], and later *Velociraptor* has shorter coracoids [Norell and Makovicky 1999]) and the arms, by making the furcula more slender and less curved, by squaring off the tip of the scapula blade, by reducing the scapular acromion process, by broadening the proximal coracoid, by reducing the posterolateral bowing of metacarpal III, by losing the flange in the central finger, by increasing the robustness of the thumb relative to that of finger II, by increasing the length of the tail (very reduced in *Microraptor,* apparently short in *Bambiraptor* [Fig. 1.7]), and possibly by increasing the flexibility of finger II and reducing the number of sternal ribs. A better pattern of neoflightless reversals could hardly be imagined, and this pattern directly contradicts exaptation as the primary evolutionary force behind the initial development of these features. Reversals such as re-elongation of the tail and deretroversion of the coracoids are no longer speculative; they are demonstrated by the fossils. Why would dromaeosaurs or their ancestors develop for nonflight purposes the same features flying birds further adapted for flight, only to reduce or lose these features as these predators continued to evolve? A much more satisfactory hypothesis argues that the features evolved in association with flight in the ancestors of dromaeosaurs and were selected against after flight was lost. This pattern explains the size increase that occurred as dromaeosaurs evolved in the Cretaceous.

The possibility that dromaeosaur ancestors once flew is ranked as high. The presence of very large sternal plates, ossified sternal ribs, and strongly retroverted pubes strongly suggests that dromaeosaurs evolved from ancestors whose respiratory and flight muscle power systems were superior to that of *Archaeopteryx*. The retention of very long arms in *Sinornithosaurus* and, to a lesser extent, in *Bambiraptor* can be seen as a climbing adaptation, the arm feathers being only long enough to make falls safer. Although the legs of *Sinornithosaurus* appear to have been relatively longer than those of *Archaeopteryx*, the probability that *Sinornithosaurus* was a long-reaching interbranch climber, and the great length of the femur, negate the neoflightless potential of this feature. The very long, distally elongated arms and legs of less arboreal *Bambiraptor* do qualify as a potentially neoflightless character.

Turning to other characters, we see that non-flight-related post-urvogel features of dromaeosaurs include an essentially avian-grade respiratory system featuring the large sterna, ossified uncinate processes, and sternal ribs, none of which are seen in urvogels. Dromaeosaurs also exhibit some additional, nonflight-related post-urvogel characters, including the absence of an ectopterygoid process of the pterygoid (Ostrom 1969, Currie 1995, Xu, Wang, and Wu 1999), short cervical ribs that indicate enhanced neck flexibility, strongly retroverted pubes, and a birdlike lateral edge of the proximal end of the femur. As noted in Chapter 3, the postorbital and posttemporal bars were less complete and the jugals were more slender in *Sinornithosaurus* and *Bambiraptor* than in more-derived dromaeosaurs (Xu, Wang, and Wu 1999). The possible redevelopment of more robust or complete posterior skull bars and the loss of the dorsal process on the ischium in more derived dromaeosaurs appear to represent reversals from urvogel-like conditions back to basal dino-avepod conditions. We do not know whether *Sinornithosaurus* had a more birdlike palatine than its more-derived relatives did. It is interesting that a large dromaeosaur-type sickle claw for hunting is absent in *Archaeopteryx* but present in the more advanced flier *Rahonavis*. The implication is that the dromaeosaur sickle claw is a post-urvogel character. However, dromaeosaurs have fewer post-urvogel characters overall than other avepectorans. The reduced second toe claw of *Adasaurus* (Fig. 4.5C) may represent a reversal. Dromaeosaurs may have re-evolved their distinctive, asymmetrically serrated, blade-shaped teeth as they re-adapted to big-game hunting. The subvertical pubes of *Unenlagia* and *Achillobator* may or may not represent a reversal from the more strongly retroverted condition seen in basal *Sinornithosaurus*. Consider the fact that the pubes of *Rahonavis* and *Archaeopteryx* are no more retroverted than in *Unenlagia* and *Achillobator*. Also, possible reversals, or nonavian characters, are the shallow pelvic canal, conventional pubic boot, and rod-like distal ischium that appear to characterize *Achillobator*.

### Troodonts

The sickle-clawed troodonts are fairly well known. Late Cretaceous *Saurornithoides* is known from some fairly complete skulls (Fig. 10.2L) and parts of the skeleton. The very similar *Troodon* is less well known (Barsbold 1974, Russell 1969, Currie 1985, Osmólska and Barsbold 1990, Currie and Zhao 1993b). Late Cretaceous *Byronosaurus* is known from most of the skull and undescribed pieces of the skeleton (Norell et al. 2000). Older, smaller *Sinornithoides* is almost entirely known from a spectacular juvenile skull and skeleton (Pl. 13, Fig. 10.1Bq), which have been partly described by Russell and Dong (1993a). Although the troodont tail lacks ossified rods, it was shorter than the dino-avepod norm, with a caudal count in the low thirties. In addition, the distal tail of the early troodont *Sinornithoides* is very similar to that of *Archaeopteryx* (Russell and Dong 1993a). We do not know whether troodonts lacked birdlike ossified sterna, uncinate processes, sternal ribs, and a furcula, because the absence of these features in the only *Sinornithosaurus* skeleton maybe due to its juvenile status or to damage. It is difficult to tell whether metacarpal III was slightly bowed (Pl. 13C), but it certainly was no more bowed than in the urvogel. The flexion potential of finger II has not been assessed either. Arm length was moderate, so this tells us little. The brain was large, and very narrow olfactory lobes and a slender snout suggest that the sense of smell was sharply reduced. Triangular frontals mean strongly overlapping fields of vision. The unusually small, coarsely serrated teeth and the sickle claws, dynamic tails, and overlapping fields of vision indicate that the troodont hunting mode (Paul 1988a,d) had shifted from the conventional mode to one more compatible with a flight heritage. In juvenile *Sinornithoides,* the legs were relatively longer than in the urvogel, but this character cannot be scored because the legs of adults were probably proportionally shorter. With a strong—especially in view of the absence of some crucial data—potentially neoflightless characters score of 17.5 and a grade of 45, the possibility that these sickle-clawed forms were secondarily flightless is ranked as high.

The skull was a blend of the dino-avepodian and the strongly avian. The braincase and inner ear were strikingly birdlike (Barsbold 1983, Currie and Zhao 1993b), much more so than those of other theropods and *Archaeopteryx* (Figs. 10.4DC, 10.5E,I). The post-urvogel braincase features included a semilunate occiput face into which the squamosal was subsumed, subhorizontal basipterygoid processes, a tubular basisphenoid process, and a distinct lateral otic depression. The quadrate contact with the braincase was a probable post-urvogel character. There was an incipient antitrochanter in the acetabulum, and the laterally compressed, roller-surfaced proximal tarsals were so birdlike (Fig. 10.14Bm,Ce,Db) that some possible troodont ankle bones have been mistaken for those of birds (Paul 1988a)—in this case giant owls (Harrison and Walker 1975)! Loss of cervical-rib overlap and the corresponding improvement in neck flexibility qualify as post-urvogel characters. The posterior palate was highly modified, but the anterior section was conventional (Fig. 10.6Ag,Bg). That tooth serrations were smaller (or even absent) in basal troodonts than in more derived examples is interesting because this is the pattern that would be expected if a reversal from the unserrated avian condition had occurred. The very robust postorbital bar in *Saurornithoides* and *Troodon* is opposite the usual avian condition but similar to the condition in *Confuciusornis;* and as in *Archaeopteryx,* the jugal lacked a long quadratojugal process. The loss of the pubic boot was a birdlike condition (Fig. 10.12Bn). The nonretroversion of the pubes was nonavian, but the presence of this condition in such a birdlike, seemingly neoflightless form is tentative evidence that this was a reversal (the ilium, one of the most diagnostic postcranial elements, is poorly known in troodonts). It is possible that troodonts re-evolved their very distinctive blade-shaped teeth as they readapted to big-game hunting. In some troodonts, the second toe appears to have secondarily reverted to a more normal dino-avepod condition (Fig. 4.5C; Osmólska 1987), opening up the possibility that the same occurred in other Cretaceous avepods.

### Therizinosaurs

As of yet there are no complete specimens of therizinosaurs, also know as segnosaurs, but most parts of the Late Cretaceous examples of these herbivorous dinosaurs are known from various remains, including a complete skull (Figs. 10.1Bk,1, 10.2I; Barsbold 1976, 1983, Perle et al. 1981, Barsbold and Maryanska 1990, Russell and Dong 1993b, Clark et al. 1994, Maryanska 1997). Also in poor condition are the remains of the earlier *Beipiaosaurus* (Xu, Tang, and Wang 1999). *Beipiaosaurus* is especially informative because it was the most basal member of the group; *Alxasaurus* is less basal, and other therizinosaurs were advanced. Some of the latter were gigantic: the finger claws of *Therizinosaurus* were up to three-quarters of a meter long in life. Arguably the strangest of dinosaurs, derived therizinosaurs appear to have been designed by a committee, and a badly coordinated one at that. Various parts, and the entire animal, have been assigned at various times to disparate tetrapod and dinosaur groups, as discussed in detail below. Reanalysis of the complete skull of *Erlikosaurus* and of the basal members of the group—*Alxasaurus* and especially *Beipiaosaurus*—demonstrates that these are avepod dinosaurs with avepectoran attributes. Therizinosaurs even have an unusually long outer toe, quite possibly inherited from sickle-clawed ancestors. More open to question is the claim by Zhao and Xu (1998) that a partial mandible from the Early Jurassic is that of a therizinosaur. They may be correct, but a single fragmentary specimen is not adequate to demonstrate that such advanced avepods appeared so early.

*Beipiaosaurus* is important because its basal condition shows that therizinosaurs were avepectorans that started out with many potentially neoflightless characters and that the less birdlike form of later and more-derived therizinosaurs exhibits secondary adaptations, probably for herbivory. The shoulder girdle of derived therizinosaurs was not simply conventional but actually primitive by dino-theropod standards (Paul 1984c). But the coracoid of *Beipiaosaurus* was subrectangular, like the coracoid of avepectorans, and therefore more birdlike. This evolution from birdy to primitive appears to have occurred in order to enhance shoulder mobility among plant-eating theropods increasingly distant from their volant ancestors (see above discussion on ornithomimosaurs). In a like manner, the semilunate carpal was well developed, well ossified, and fully birdlike in *Beipiaosaurus,* less so in more derived *Alxasaurus,* and was strongly flattened and divided into two elements in *Therizinosaurus* (Fig. 10.11Cc; Fig. 6 in Barsbold 1976). The fact that the decrease in arm folding occurred concurrently with elongation of the claws—to two-thirds of a meter in *Therizinosaurus*—directly contradicts the possibility that therizinosaurs evolved arm folding in order to protect their claws. One cannot ask for a better example of a reversal away from the avian flight pattern, in this case probably driven by a lessened need for folding arms. This scenario fully solves the problem with therizinosaur carpals that has started to vex researchers trying to sort out the evolution of the semilunate block, while assuming therizinosaurs were pre-urvogels with no flight heritage (for example, Chure 1999; note that in the scenario presented in my study, secondarily reduced carpals blocks of derived therizinosaurs are not very useful for restoring the development of the preavian wrist). The elbow is not well documented in *Beipiaosaurus,* and in derived therizinosaurs, the distal humeral condyles were not strongly asymmetrical as they are in other avepectorans. Another potentially neoflightless character was the three-fingered hand, which was most gracile in the classic dino-avepectoran manner in *Beipiaosaurus.* The hand of the latter is also birdlike in that the central digit was the most robust. The fact that this robustness is gone in more-derived therizinosaurs is another example of a reversal away from the flight condition. The great length of therizinosaur arms may reflect their descent from wings. The tail is not fully known, especially in *Beipiaosaurus,* but it appears to be markedly shorter than the tails of most avepod dinosaurs and even *Archaeopteryx* (Russell and Dong 1993b), and it lacks a highly flexible base. Explaining why herbivores with such heavy bellies would have evolved such stunted tails is difficult unless they inherited the latter from fliers. The retention of retroverted pubes presumably indicates a need to keep the bulk of the big belly as far aft as possible in these light-tailed forms. Herbivorous habits offer a reversal explanation for the absence of forward vision. The extreme modifications for herbivory mean that predation pattern of therizinosaurs, like that of avepectorans, diverges from classic dino-theropod predation pattern. Loss of flight explains why basal therizinosaurs were smaller than later examples. When the potentially neoflightless characters were originally scored without *Beipiaosaurus,* the resulting grade was rather low but not insignificant. When the basal therizinosaur was factored in, the grade nearly doubled to 32, although this value is tentative owing to an absence of knowledge about many aspects of basal therizinosaur anatomy.

The lack of sufficient data for basal therizinosaurs also hinders the assessment of their post-urvogel characters, but a lateral otic depression that contained the inner ear is a strong post-urvogel character. Elzanowski (1999) presented some other avian cranial features not found in archaeopterygiforms. The absence of bowing of the outer metacarpal and the flexibility of finger II are potential reversals. As we learn more about the nature of early therizinosaurs, they are coming to stand as outstanding examples of massive reversal in dino-avepods, and the evidence is mounting that these dinosaurs were secondarily flightless members of the post-urvogel clade.

### Caudipterygians

*Caudipteryx* is known from a rapidly growing number of largely articulated and complete specimens (Pl. 14; Fig. 10.1Bo; Ji et al. 1998, Zhou and Wang 2000, Zhou et al. 2000).

The caudipterygians' many potentially neoflightless characters include rather large ossified sternal plates, which were connected to the rib cage via ossified sternal ribs. Sereno's (1999a) observation that the lack of a laterally facing shoulder glenoid contradicts a flight ancestry is incorrect because *Caudipteryx* probably had a sideways-directed shoulder joint, and some neoflightless birds do not (Chapters 4 and 8). Zhou and Wang's (2000) scoring of the seemingly nonretroverted coracoid as a nonavian character represents a classic example of how flight-related features are confused for phylogenetic ones; and Zhou and his co-workers (2000) went on to observe that less crushed caudipterygian scapulocoracoids do show strong retroversion. *Caudipteryx*'s arms were quite short, yet the hands bore an array of large, symmetrical, poorly streamlined contour feathers whose aerodynamic function was at best limited to air braking and steering by air deflection. The extreme shortness of the arms further argues against their having been protowings, because the latter should have been relatively longer than was typical for avepod dinosaurs. Instead, the combination of very short arms with long but symmetrical contour feathers is what one would expect in an herbivorous avepod (see Chapter 3) that had lost flight and did not need to use its arms to catch prey. Nonpredaceous habits, a nonarboreal form, and the shortness of the arms leave flight as the best explanation for the large sternal plates. The combination of short arms with hands bearing only two fingers—the central finger was at least partly stiffened and unusually strong relative to the thumb—with very long hindlimbs appears to make *Caudipteryx* ill suited for the climbing habits postulated by Chatterjee (1999b) and Garner et al. (1999), but this combination is commonly seen in the herbivorous, neoflightless, and cursorial ratites. The sharply reduced outer finger that braces the feather-bearing central finger is a remarkable neoflightless character not yet found in any other avepectoran including *Archaeopteryx* (Fig. 10.11Aj). Alternative explanations for this extraordinarily avian configuration are hard to devise, especially in view of the modest dimensions of the feather fan. The tail was distally stiffened and very short—both in having a low caudal count of twenty-two (Ji et al. 1998, Barsbold et al. 2000, Zhou et al. 2000) and in being 60 percent shorter than the urvogel's; and it bore a distal fan of large contour feathers that may have served as rudder and brake. Such a lightweight appendage makes little sense in a herbivorous dinosaur, especially one made front heavy by a big belly filled with a large bundle of gizzard stones, unless the appendage first evolved in dynamic fliers that were close to evolving a pygostyle. At the other dimensional extreme, the legs of this runner were exceptionally long, especially distally, being about twice the length of the combined dorsosacral series. As T. Jones et al. (2000b) observed, this ratitelike condition is difficult to explain outside the context of flight loss. Indeed, the combination of big sternal plates anchored by ossified sternal ribs; very short arms with a strongly modified hand that could have supported a wing much better developed than the one actually present; a dramatically reduced tail; and long, symmetrical distal feathers is fully compatible with the loss of flight; and, furthermore, imagining a condition more strongly indicative of neoflightlessness is difficult. Small heads, much reduced teeth, a very deep and therefore capacious gut, and gizzard stone bundles indicate that caudipterygians represent an omnivorous or herbivorous break with prevolant predatory dinosaurs (Ji et al. 1998, Zhou et al. 2000). We are not yet certain whether a furcula was present, so this feature is not scored in Table 11.1. The potentially neoflightless character score of 24 and grade of 53 are matched or exceeded only by *Protarchaeopteryx,* dromaeosaurs, and avimimids, so the possibility that the ancestors of these odd dinosaurs with ratitelike shoulder girdles and stub tails is ranked as high. The short tail and the extreme modification of the outer finger indicate that ancestral flight had reached a level of power and dynamic control well above that seen in the urvogel and other avepectorans with longer tails. Sereno's (1999a) conclusion that "there is no evidence of flight function in the ancestry of *Caudipteryx*" is becoming increasingly untenable.

In addition to the characters already detailed, *Caudipteryx*'s other post-urvogel characters include a small femoral head, an essentially avian-grade respiratory complex, and the dramatic reduction of the maxilla and its fossa and fenestrae. On the basis of examination of the two published specimens, Longrich (pers. comm.) has restored the anterior palate as sitting more forward than in most other dino-avepods and as projecting strongly below the margin of the upper jaws. This arrangement is similar to, albeit less well developed than, that seen in oviraptorosaurs; as noted below, such a palate contains post-urvogel characters. The reversed hallux reported by Zhou et al. (2000) is a potential post-urvogel character, as is the increase in the number of neck vertebrae at the expense of the dorsals. The same research team cited the complete posttemporal bar as a pre-urvogel feature, but the latter point is open to question because of the uncertain status of the bar in *Archaeopteryx*. The long, overlapping cervical ribs are not a post-urvogel character. The tall, rectangular ilium, shallow pelvic canal, and lack of pubic retroversion are potential reversals; the absence of a retroverted pubis is in any case perplexing in an herbivorous biped with only a short tail to counterbalance the heavy belly.

### Oviraptorosaurs

Cretaceous oviraptorosaurs include the caenagnathids, which in turn include the elmisaurs (Sues 1997) and microvenators (Makovicky and Sues 1998). Of these, only the Late Cretaceous, deep-jawed oviraptorids such as *Oviraptor, Conchoraptor,* and *Ingenia* are known from complete skulls and skeletons (Pl. 15; Figs. 10.1Bp, 10.2M). The skulls are well described (Osmólska 1976, Barsbold 1981, 1983, Elzanowski 1999), the skeletons only partly so (Barsbold et al. 1990, D. Smith 1992, Clark et al. 1999). Barsbold et al. (2000) described parts of an incomplete but interesting new oviraptorid, whose complete tail and hindquarters are shown in Sloan (1999).

The short, deep skulls of oviraptorosaurs are unusual, to say the least. When illustrations of the first well-preserved examples were published, at first glance it was hard to tell which end was the front and which was the back. The rostral crest of *Oviraptor* is as odd as it is spectacular. Yet the unusual nature of these forms should not be allowed to obscure their near-avian nature. The shoulder girdles, ossified uncinate processes and sternal ribs, and long, folding arms all had potentially neoflightless features. Longrich (1999) noted that the semilunate carpal of oviraptors was laterally extended onto metacarpal III in a modern avian manner, a potentially neoflightless character. The brain was large. The tails of most oviraptorosaurs have not yet been described in detail. They were conventional in form, but with only 27–32 vertebrae in at least some examples (Barsbold et al. 2000), they were shorter than those of most avepod dinosaurs (which hints at a flight heritage) but longer than the tail of *Archaeopteryx*, Much more interesting is the new short-tailed oviraptorid (Barsbold et al. 2000). Its tail consists of just 24 short vertebrae, the last 5 of which are fused into a pygostyle of the avian type, with chevrons included in the fusion (Sloan 1999, Barsbold et al. 2000). The lack of a complete dorsal series precludes an exact calculation of the body/tail length ratio, but it appears to have been markedly lower than in *Archaeopteryx* but higher than in *Caudipteryx*. The number of transverse processes was also higher than in *Caudipteryx*. Such an abbreviated, birdlike tail is an obvious potentially neoflightless character. Yet the caudal series was not as heavily modified as that of avebrevicaudans, so the short tail retained the potential to revert to a more basal condition. The greater length of other oviraptorosaur tails may have resulted from a partial reversal among neoflightless dinosaurs that did not engage in leaping predation; indeed, in this scenario the twenty-four-count tail would represent a reversal from the urvogel condition. This scenario appears to be supported by the greater abbreviation of the apparently more basal sister taxon to the oviraptorosaurs, *Caudipteryx*, but cannot be further assessed until we know whether basal oviraptorosaurs had such short tails and whether their tails had as few bones as the tail of *Archaeopteryx*. Blunt beaks and unusual jaws again demonstrate that oviraptorosaurs represent a break in the classic predation pattern. With a potentially neoflightless character score of 22 and a grade of 49, these dino-avepectorans are nearly as likely as *Caudipteryx* to have been secondarily flightless. The presumption by Barsbold et al. (2000) that oviraptorosaurs are nonavian theropods, despite their sometimes avian tails, is not affirmed by this analysis. Elzanowski's (1999) suggestion that oviraptorosaurs may have been neoflightless post-urvogels is supported.

Consider that the conjoined mandibles of oviraptorosaurs differed greatly from those of other most other avepod dinosaurs, and the mandibles of caenagnathids were so birdlike that isolated examples have been misidentified as bird jaws (Sternberg 1940, Cracraft 1971). The oviraptorosaur quadrate and especially the palate also differed from the dino-avepod standard to the point that the roof of the mouth was much more birdy than that of *Archaeopteryx* (Osmólska 1976, Maryanska and Osmólska 1997, Elzanowski 1999), which results in a large number of post-urvogel characters. The enlarged, beaked premaxilla, the markedly reduced maxilla, and the associated fossa and fenestrae are post-urvogel characters, and the extreme development of these features in the derived oviraptorids may represent partial parallelism with birds. The quadratojugal and jugal were more slender than in *Archaeopteryx*—although the jugal was not the true rod described by Elzanowski (1999)—and the postorbital bar was slender. However, avian kinesis was not present: in fact, the quadrate and pterygoid were often fused. A lateral otic depression in basal oviraptorosaurs is a post-urvogel character. Improved neck flexibility due to cervical ribs that were too short to overlap was a post-urvogel character. So was the avian-grade respiratory apparatus. The pubes were vertical. However, the pubes were proximally directed vertically or even a little posteriorly, and they curve anteriorly progressing distally down the shaft (Fig. 10.12Bm). This leaves open the possibility that oviraptorosaur pubes evolved from a more retroverted condition. The shallow pelvic canal, straight outer metacarpal, and possibly flexible central finger are potential reversals. The lack of forward vision may have also been a reversal, one associated with less predatory habits.

### Alvarezsaurs

The manner in which alvarezsaurs were first described (Perle et al. 1993) was rather unfortunate, because the analysis of these extraordinary Late Cretaceous dino-birds was so preliminary and limited that few saw much that was clearly avian. That the skull was barely known did not help matters. As a result, there has been a series of similarly inadequate attempts to dismiss alvarezsaurs as close bird relatives, each attempt considering only a few characters (Ostrom 1994, Wellnhofer 1994, Kurochkin 1995, Zhou 1995b, Feduccia 1996). L. Martin (1997) considered more characters, but numerous errors tainted his results. Wellnhofer (1994), Karhu and Rautian (1996), and Sereno (1997a) gave more balanced views of the uncertainties surrounding the phylogenetic status of these archosaurs; however, contrary to Sereno's contention, the absence of a fully reversed hallux does not exclude alvarezsaurs from having been more avian than *Archaeopteryx* (the same is true for the hallux of kiwis). More recent work has greatly expanded our knowledge of the complete skulls and fairly complete skeletons of *Mononykus* and *Shuvuuia* (Pl. 12; Figs. 10.1Br, 10.2N; Perle et al. 1994, Chiappe et al. 1996, 1998), *Parvicursor* and less advanced *Patagonykus* are not as well known (Karhu and Rautian 1996, Novas 1997). What makes all these forms peculiar is the very short, hyperstout arms and the hyper-asymmetrical near elimination of all fingers but the first, which is unique among tetrapods. Indeed, extreme and rapidly evolved adaptations for digging into insect nests and feeding upon the inhabitants may obscure the relationships within the group (Longrich 2000), much as adaptations for herbivory have hindered understanding the position of therizinosaurs.

The body plan of alvarezsaurs looks a little like that of ornithomimosaurs—gracile, tubular-snouted head; long, slender neck; moderately long tail; very long gracile hindlegs—but this resemblance is superficial. In most respects, the detailed morphology of alvarezsaurs is dramatically different from that of ornithomimosaurs (compare Figs. 10.2Bj and r, 10.12 Bj and o,p, 10.14Bi and 10.18B) and of all other avepods for that matter (Perle et al. 1994, Chiappe et al. 1996–98, Padian and Chiappe 1998c). L. Martin (1995, 1997) cited the superficial similarities between ornithomimosaurs and alvarezsaurs as reasons to ally them, and Sereno (1999a,b) has noted specific characters shared by the two groups, mainly in the skull and hand. However, a close and comprehensive examination reveals that alvarezsaur skeletons, and even more so their skulls, have numerous and well-developed post-urvogel characters not observed in other dinosaurs, or in *Archaeopteryx*. As noted by Padian and Chiappe (1998c), the fact that alvarezsaur postcrania appear strange and, at first glance, not very birdlike does not mean alvarezsaurs were not close to birds.

Most spectacularly avian is the skull, which shows evidence of a well-developed avian-type cranial kinetic system, one much better developed than that of the urvogel. The adaptations include an incomplete postorbital bar, a rodlike jugoquadratojugal, a mobile quadratojugal-quadrate joint, a double-headed and mobile quadrate head, the absence of a descending process of the squamosal, the absence of an ectopterygoid that contacts the jugal, and a flexible zone in the midskull roof. The nature of the propulsion joint, if there is one, has yet to be described; as in basal birds, the quadrate lacks an orbital process. The elongated, posteriorly directed posteromedial process of the triradiate palatine rivals that of *Archaeopteryx* in the degree to which its configuration is avian. A tubular snout, severe tooth reduction, and the hypermodified arms combine to indicate a complete divergence from the classic dino-theropod predation mode. The small premaxillae and external nares and, most especially, the extremely ventrally elongated basipterygoid processes are nonavian.

The rest of the skeleton exhibits mixed attributes. Well-developed hypapophyses under the cervicodorsal vertebrae are birdlike, but the lack of saddle-shaped articulations between the vertebrae and their nonpneumatic centra are not. In the hindquarters, a well-developed antitrochanter, the confluence of the lesser and greater trochanters (Fig. 10.13Bh), a reduced femoral head and pubes that meet only at their distal ends, and anteroposteriorly well-developed roller-shaped proximal tarsals (Fig. 10.14Cf) are post-urvogel characters; the presence of a overhanging supra-acetabular shelf is not. The tail is conventional, albeit rather short with a caudal count somewhere in the twenties.

The primary source of confusion and controversy is the shoulder girdle and the arms. The scapulocoracoid is actually fairly conventional in organization (Fig. 10.8Q), and does not articulate with the sternum in the avian manner. The sternum itself is large and has a keel (two avian attributes), but it is not the broad, light, shallow plate seen in neoflightless birds. Instead, it is massively constructed, narrow, and deep keeled. The carpometacarpus is fused, and the extreme shortness of the arms fits into the neoflightless pattern; but otherwise the nonfolding arms are not birdlike. It is not possible to tell how and why the forequarters had a keeled sternum and a fused wrist. Did they evolve entirely for digging, or were they retained from flying ancestors and modified for this purpose? Or did the keel disappear after flight loss and then re-evolve for digging?

What is dubious is the argument that the arms of alvarezsaurs could not have evolved from those of basal birds. The proponents of this almost antievolutionist view forget not only that radical transformations are frequent in evolution but also that alvarezsaur arms had to have evolved from gracile tridactyl hands whether those hands belonged to dinosaurs or early birds. Conversely, adaptations for eating insects may have made alvarezsaurs more birdlike than they really were. In particular, consider the light construction of the skull: the ingestion of tiny invertebrates does not require a strong skull (Longrich 2000). The conventional tail may or may not represent a reversal in these nonpredators.

Although the alvarezsaurs' potentially neoflightless character grade of 23 is modest, their very strong set of post-urvogel characters suggests that they were very close to, if not within, Aves and therefore potentially secondarily flightless (Padian and Chiappe 1998c). Ostrom's (1994) denial that the ancestors of alvarezsaurs ever flew cannot be confirmed at this time. Note that a close relationship between alvarezsaurs and ornithomimosaurs does not necessarily refute a neoflightless post-urvogel status for the former, since the ostrich mimics may have possessed these two conditions. It is worth repeating that accepting alvarezsaurs, with their limited potentially neoflightless characters, as neoflightless, while rejecting other dino-avepectorans with more extensive potentially neoflightless characters as neoflightless, is illogical.

### Avimimids

The remains of *Avimimus* are fairly complete (Kurzanov 1981, 1982, 1983, 1985, 1987, Norman 1990, Watabe et al. 2000), although the midsection of the skull, most of the hand, and various other parts are missing, and some new material has not yet been described. This dino-bird is extremely birdlike in many features of both skull and skeleton (Fig. 10.1Bs). Although the ulna ridge may have born feathers, the small arms were ill suited for flight, contrary to the opinion of Kurzanov (1987). The fingers are not yet available for comparison to those of *Caudipteryx* and birds. The legs were proportionally very long, probably more so than in *Archaeopteryx*. The eyes faced strongly forward, and the brain was large. The incompleteness of the described remains hinders potentially neoflightless character analysis. The high grade of 60 would probably be further boosted if the nature of the shoulder girdle, tail, and other key structures were better known. Most probably, avimimids were secondarily flightless avepods.

What is known of the skull is virtually avian, most especially the very slender quadratojugal and the complete absence of a postorbital bar. Such features normally indicate avian kinesis, but fusion of the quadrate with surrounding elements prevented cranial movement. Perhaps this condition represents a reversal from a more mobile arrangement. The pelvis (Fig. 10.12Bq) was a mix of the conventional dinosaurian and the avian. The avian attributes included pubes that were joined only toward their distal ends, the broad flaring of the posterior ilia, and a well-developed antitrochanter. That a pelvis so avian could have procumbent pubes suggests a reversal from the avian arrangement back to a more conventional dinosaurian one. A small femoral head was another post-urvogel character, and the rest of the legs and feet had strongly avian attributes.

### *Yandangornis*

*Yandangornis,* a recently described avepod from the early part of Late Cretaceous (Cai and Zhao 1999), is of interest despite the poor preservation of the one, nearly complete articulated skull and skeleton. The only missing parts are the wrist elements and most of the hand, but all the dorsal elements are obscured because the skeleton is preserved on its back. It seems that no furcula is preserved, but what this means is not clear. *Yandangornis* appears to be some form of toothless dino-bird with a tail proportionally about as long as that of *Archaeopteryx,* but according to Cai and Zhao's description, the tail of the former has only 20 tail vertebrae, which is a potential post-urvogel character. So is the long, ossified sternum with ossified sternal ribs, as well as the small finger claws and a number of other skull and skeletal features (see Chapter 12). The small-clawed fingers and the long, gracile legs with short toes also bearing small claws (including the hallux) indicate that *Yandangornis* was a runner with little climbing ability despite its small size (about 2 kilograms).

Judging from the complete humeri, we can tentatively say that the arms of *Yandangornis* were not quite as large relative to the body as in *Archaeopteryx,* and they were markedly shorter than the legs. The sternum was keelless. Because no feathers are preserved, we cannot reliably restore the flight performance of this avepod, but the smaller arms hint that it may have been less aerially capable than the urvogel. When we combine *Yandangornis*'s possibly reduced flight capabilities, its advanced sternal complex, and its almost certain post-urvogel status, we can speculate that *Yandangornis* may have been a long-tailed—and therefore nonavebrevicaudan—ground bird whose flight capacity was secondarily reduced or lost.

## A Synthesis

### *Do Not Let a False Sense of Progressive Evolution Hide Early Losses of Flight*

The second to last of Stephen Gould's (2000) famous series of essays in *Natural History* magazine was a commentary upon the bird origins debate, including the possibility that *Caudipteryx* was neoflightless. He suggested that the almost automatic rejection of the neoflightless hypothesis by a majority of researchers stems from the persistent tendency of evolutionary scientists to think of evolution as inherently progressive, even though the same people intellectually understand that it is not so. If evolution were persistently progressive, then flight abilities would only improve. In reality, there was no inescapable evolutionary pressure for the first avian fliers to retain flight. As Gould explains, they were free to lose flight if selective pressures directed them to do so, and we should be surprised if they did *not* do so. The question is did they do so?

### *A Tale of Wee Tails and a Search for Parsimony*

We now know that in addition to avebrevicaudan birds, two kinds of dinosaurs and a probable archaeopterygiform had strongly abbreviated tails that either ended in a pygostyle or were at least distally stiffened. To explain these, the conventional scenario requires at least two, and perhaps three or even four, independent developments of stub tails (Barsbold et al. 2000). It is also hypothesized by various researchers that the bobbed tails evolved for different purposes, either to support a major flight-control system in flying birds or to support a display structure in the land-bound dinosaurs. The neoflightless scenario holds that the short, distally rigid tail may have evolved only once or twice (in derived archaeopterygiforms and in a caudipterygian-oviraptorosaur-avebrevicaudan clade), always as a flight adaptation with any display function being secondary. The neoflightless hypothesis is the most parsimonious explanation for avepod caudal abbreviation and distal rigidity. But the tail is only one part of the story of the beginning of birds. What about the larger picture?

### *Parallelism versus Phylogeny Redux: Why Neoflightlessness Edges Out Exaptation*

We now know that in at least two groups of dinoavepectorans, the basal members were more birdlike than the more-derived examples. The question is whether this pattern was the logical result of loss of flight, or due to other factors. *Sinornithosaurus,* being a basal dromaeosaur, provides a test of the conventional versus the neoflightless hypotheses (Fig. 11.1). The conventional scenario predicts that such a basal dromaeosaur should have fewer potential neoflightless characters than the urvogel. However, the opposite is true. So the conventional view results in a convoluted situation in which dromaeosaurs are less derived than *Archaeopteryx:* the basal-most dromaeosaur *Sinornithosaurus* has flight-related features better developed than those of *Archaeopteryx,* but more derived dromaeosaurs exhibit reduction or loss of some of these features. The situation becomes much more straightforward when dromaeosaurs are placed above *Archaeopteryx.* The neoflightless hypothesis predicts the existence of basal dromaeosaurs like *Sinornithosaurus;* specifically, the hypothesis predicts that basal dromaeosaurs should have more potentially neoflightless characters than *Archaeopteryx* on the one hand and more-derived dromaeosaurs on the other. In this scenario, the advanced flight-related features of basal dromaeosaurs are inherited from a more sophisticated flier than *Archaeopteryx,* and the subsequent reduction of these features in more-derived dromaeosaurs is exactly the pattern of reversal that would be expected in a clade that had lost flight. *Sinornithosaurus* has all the signs of being a secondarily flightless climber. One can barely ask for better fossil confirmation of the neoflightless hypothesis: it fulfills the detailed predictions of the hypothesis and possesses no features that contradict it.

Those who oppose this hypothesis must explain why the dromaeosaur had better-developed flight-related features than either the urvogel or its more derived dromaeosaur relatives. They must also explain why the clade evolved flight-type adaptations for such non-flight-related purposes as predation and then proceeded to undergo reduction of the same features even as these predators continued to evolve.

Let us consider *Caudipteryx* next. As we first noted in Chapter 6, exaptation and development of flight do a very poor job of explaining why this large, heavy-bellied herbivore had such long yet symmetrical and nonaerodynamic feathers on its arms and a tail shorter than the urvogel's, as well as larger sternal plates and ossified sternal ribs. These same features are exactly matched in many neoflightless birds.

To make matters worse, explaining how a group of animals that never flew became so much like flying birds requires parallelism on a massive scale. In contrast, loss of flight easily explains why the complex feathers had evolved in the first place (as airfoils); why the arm feathers were short, limited to the hands, and symmetrical (because they no longer served an aerodynamic function); why the arms were so reduced (because they no longer served an important function in a running herbivore); why the sternum and ribs were so well developed (again, because they had no important function in a running herbivore); why *Caudipteryx* was on the heavy side for a flier (because flight had been lost); why the legs were so long (the length was a ratitelike adaptation for running with a short tail); and why the tail was so short (because *Caudipteryx*'s ancestors flew better than *Archaeopteryx*). The importance of the last feature is reinforced by the similar tail abbreviation seen in some oviraptorosaur relatives of *Caudipteryx:* tail abbreviation may represent a basal feature of the caudipterygian-oviraptorosaur clade. Explaining the astonishing modification of *Caudipteryx*'s hand into a reduced, stiffened wing-support without invoking flight verges on the hopeless.

Neoflightlessness easily and fully explains why *Caudipteryx* had the features expected in a flightless avepod; exaptation and flight development provide a grossly inferior explanation

Nevertheless, *Caudipteryx* is a confusing feathered avepod because it possesses a mixture of characters that suggest, on one hand, that it was more avian than and descended from fliers more advanced than archaeopterygiforms and, on the other hand, that it was farther from other birds than was the urvogel. If instead caudipterygians were consistently more avian than *Archaeopteryx,* then their phylogenetic position and neoflightless status would be clear. By the same token, if dromaeosaurs were consistently less avian and had potentially neoflightless characters that suggested they descended from fliers less well developed than archaeopterygiforms, then the former's position vis-à-vis the latter would also be clear. What should we make of the mix of post-urvogel characters and anti-post-urvogel characters in the skeletal features of alvarezsaurs, especially when we consider these characters in combination with the highly kinetic skulls? If all avepectoran dinosaurs were consistently less avian than *Archaeopteryx,* then their birdlike shoulder girdles, arms, and tails could be explained as exaptations for flight. If all avepectoran dinosaurs were consistently more avian than *Archaeopteryx,* then their birdlike shoulder girdles, arms, and tails could be explained as resulting from the loss of flight. As it is, all the avepectoran dinosaurs with potentially neoflightless characters show an ambiguous mix of post-urvogel characters and non-post-urvogel characters. This mixture indicates that the urvogel is not consistently more avian than any particular dino-avepectoran. This inconsistency means that the issue of whether or not some dinosaurs were post-urvogels is not clear-cut, and it leads us again to the problem of parallel evolution versus inheritance.

There is no doubt that a great deal of mosaic parallelism occurred among the Avepectora. Consider some examples. *Mononykus* and *Parvicursor* have some avian characters in the pelvis and hindlimb that are absent in their advanced relative *Patagonykus* and in basal birds: these characters include slenderer, more retroverted pubes and ischia that are distally split and seem to lack a boot, fusion of femoral head trochanters, a fibula that does not reach the ankle, and a laterally placed pretibial process. These probably represent parallel developments. Whereas *Sinornithosaurus* had a more flexed, *Archaeopteryx*-like scapulocoracoid than *Bambiraptor,* with its more modern birdlike coracoid, the latter had a more *Archaeopteryx*-like distal pubis than *Sinornithosaurus,* which sported a pubic boot more like that in confuciusornithids and *Chaoyangia.* Even the possible closure of the intramandibular fenestra in *Archaeopteryx* probably parallels the same in many birds because this space remains open in some baso-avebrevicaudans. The outer finger was much more reduced in *Caudipteryx* than in confuciusornithiforms (Fig. 10.11Aj,k). This surprising reversal of the expected pattern may be the result of reduction of this finger in the fliers immediately ancestral to caudipterygians: because these ancestors were becoming cursorial herbivores, they no longer needed the finger to facilitate climbing. But a return in the basal, climbing avebrevicaudans to the old condition cannot be entirely ruled out. The presence of complete postorbital in *Confuciusornis* and its absence in alvarezsaurs represent an especially glaring contradiction that leads to three viable alternatives. Alvarezsaurs retained the condition from urvogels if the former too lacked the bar. Alvarezsaurs paralleled the majority of birds in losing both the connection between the postorbital and jugal and the accompanying kinetic system. The avebrevicaudan bird experienced a reversal to the dinosaur condition.

The problems created by mosaic parallelism are well illustrated by *Archaeopteryx* and its closest relatives, the dromaeosaurs. For sake of this discussion, we will assume that the latter were neoflightless. *Archaeopteryx* and dromaeosaurs each have a significant number of avian characters that the other lacks (Appendix 1B). There are a number of possibilities. One possibility is that *Archaeopteryx* is closer to birds than dromaeosaurs are and that the latter paralleled birds in a number of regards. If dromaeosaurs paralleled birds, then the former may have even descended from fliers that were more advanced than *Archaeopteryx* but were outside the mainstream of avian evolution. A second possibility is that dromaeosaurs were closer to birds than the "original bird." In this case, either *Archaeopteryx* paralleled birds or dromaeosaurs experienced a number of reversals from the avian condition or both. How can we discriminate between these alternatives? By the numbers? The analysis presented in Appendix 1B indicates that *Archaeopteryx* has about two-thirds more avian attributes than known dromaeosaurs. When characters that may have not been present in baso-avebrevicaudans are dropped from the analysis, the number of avian attributes for *Archaeopteryx* and dromaeosaurs becomes 31 and 18, respectively. But 10 of the *Archaeopteryx* characters are potentially neoflightless characters subject to reversal with flight loss, so the final numbers are 21 and 18 (these figures do not take into account the lack of certain data in the basal dromaeosaur *Sinornithosaurus,* such as information about the form of its palatine). In a cladogram, this numerical superiority scores the *Archaeopteryx* as closer to birds than the sickle-clawed dinosaurs, but not at all by an overwhelming margin.

However, as discussed in Chapter 2, parsimonious character counts do not necessarily equate with true phylogenetic closeness (Zhou et al. [2000] made this observation when considering the possible loss of flight in *Caudipteryx*). To put it another way, the degrees of parallelism postulated by various scenarios of which dino-avepectorans were closer to birds are in no case so extreme that any of the scenarios have to be rejected outright; they are well within the realm of the plausible. The degree of reversal proposed for neoflightless avepectoran dinosaurs appears comparable to that associated with the reconstruction of the neognathous thunder birds into palaeognathe mimics. The cladistic numerical advantage enjoyed by the urvogel over the dromaeosaurs, as we have seen, is modest. And some of the characters at issue may not be as important as first seems. Some of the avian features seen in the urvogel but not in dromaeosaurs, such as the enlarged premaxilla and external nares, are not seen in the seemingly more-derived alvarezsaurs. This lends support the possibility that at least some of the avian features of *Archaeopteryx* developed independently.

Then there is the issue of the sophistication of flight. The dromaeosaurs' large sternal plates, which imply that these creatures descended from fliers more advanced than *Archaeopteryx,* suggest but do not prove that dromaeosaurs are closer to birds than the crude flying urvogel, regardless of the other characters. In addition, as discussed in detail below, dromaeosaurs share a number of distinctive characters with other avepectoran dinosaurs that suggest that dromaeosaurs form a clade with birds above the level of archaeopterygiforms. In this case, reversals required for readaptation to a terrestrial lifestyle may have driven down the number of avian features in dromaeosaurs, and this decrease creates the illusion that dromaeosaurs are less birdy than archaeopterygiforms. Or the latter may have fooled us by developing avian characters merely in parallel with birds. Or perhaps both factors were at work.

If either archaeopterygiforms or dromaeosaurs were far more avian in their features than the other, then the problem of their position relative to birds would disappear. As it is, all the alternatives are fairly similar in terms of plausibility. Truth be told, we do not yet have the phylogenetic tools or, more importantly, the fossils to reliably choose between these alternatives.

When we consider all of the avepectoran dinosaurs together, the problem becomes somewhat clearer because all of the dino-avepectorans with birdlike shoulders, arms, and tails have additional avian features that potentially place them closer to modern birds than the Solnhofen bird. In addition, these near-birds possess many characters that one way or another suggest that the avepectoran dinosaurs form a clade above the level of *Archaeopteryx*. These characters are those that suggest a neoflightless condition, those that link the dinosaurs with birds, and those that link various avepectorans together. For example, dromaeosaurs, troodonts, oviraptorosaurs, and avimimids share subtriangular obturator processes under their ischia. Especially notable is that all protoavians with folding arms possess avian characters not observed in archaeopterygiforms; the number of such characters totals 89. A certain degree of parallelism can be expected between birdlike dinosaurs and birds (Kurzanov 1987, Paul 1988a), but the large number of avian characters in the dinosaurs seriously challenges the status of archaeopterygiforms as the sister taxon to birds above the level of dinosaurs. That at least some avepectoran dinosaurs were closer to birds than *Archaeopteryx* is very possible if not probable, although by no means certain.

Neoflightlessness even explains why all avepectoran dinosaurs diverged from classic theropods in terms of brain development, olfaction, vision, and feeding habits: that is, because their ancestors were small, smart fliers that lived in a three-dimensional world and did not hunt big game.

In the neoflightless scenario, parsimony and logic favor flight as the causal agent of the birdlike pectoral girdles, arms, and tails of dino-avepectorans. The potentially neoflightless characters of avepectorans are so extensive that they stretch exaptation and parallelism to or beyond their limits. Extensive exaptation and parallelism between fliers and nonaerial forms are therefore inferior hypotheses. Blanket assertions that the flight-related features of predatory dinosaurs were exaptations for flight are consequently open to challenge. So the possibility that the Cretaceous avepectoran dinosaurs descended from flying Jurassic urvogels must be ranked as considerable. Assertions that the pre-urvogel status of avepectorans shows that their contour feathers, folding arms, long sternal plates, and other birdlike features evolved prior to the advent of flight are at the least premature.

The hypotheses that avepectoran dinosaurs were post-urvogel or neoflightless or both are still only working hypotheses; they are relatively new and incomplete, and have not yet been extensively analyzed by researchers. They do not, therefore, have the verified status of the general theory of the dinosaurian ancestry of birds. However, with the discovery of new fossils such as bird-handed *Sinornithosaurus* and *Caudipteryx*, stub-tailed oviraptorosaurs, and birdlike basal therizinosaurs, crucial predictions of the neoflightless hypothesis are being fulfilled. As a result the hypothesis is rapidly gaining strength, so much so that it is markedly stronger now than it was when I began this book—and is on its way to becoming superior to the alternatives. (See Appendix 6 for additional comments.)

Although the inconsistent presence of post-urvogel characters and potentially neoflightless characters in dino-avepectorans complicates our understanding of their phylogenetic position relative to archaeopterygiforms and other birds and our determination of their flight status, this character mix is exactly what one would expect to see according to the neoflightless hypothesis. The dino-avepectoran show signs of being a classic product of mosaic evolution combining neoflightless reversals with intense parallelism, the latter induced by a combination of close genetic and anatomical similarities and similar selective pressures.

## *Parallel Losses of Flight*

While massive parallelism due to exaptation may not be a viable option, more modest parallelism within the neoflightless scenario is another matter. The possibility that flight was lost early in the development of avian flight leads—perhaps unfortunately—to a proliferation of alternative scenarios. Flight may have been lost only once, either before or after the stage represented by *Archaeopteryx*, and that loss led to the entire radiation of neoflightless avepectorans from *Protarchaeopteryx* to alvarezsaurs. Or flight may have been lost more than once, a possibility that leads to another series of alternatives. All of the neoflightless may have descended from fliers less sophisticated than the urvogel or from fliers more sophisticated, or some of the neoflightless from one level, some from the other.

A single loss of flight in dino-avepectorans is possible but improbable. We can begin to understand why by considering the reduction or loss of flight within archaeopterygiforms. This event probably occurred independently of the loss of flight in other dino-avepectorans. Why? Because it is improbable that the rest of the dino-avepectorans descended from the likes of *Protarchaeopteryx*. If they had, the degree of parallelism between birds and dino-avepectorans would have to have been high, to a degree that is rather difficult to explain. For example, the large sternal plates of many dino-avepectorans would have to parallel the avian condition, and do so without the selective pressures imposed by flight.

The evidence suggests instead that there were multiple losses of flight, occurring at various stages of avian evolution. *Protarchaeopteryx* seems to represent the most basal known example to have experienced reduction of flight, from a stage where the ossified sternal plates were still very short. Competing for status of earliest are dromaeosaurs, which are so primitive and urvogel-like compared with other avepectorans that they seem to represent another group that returned early to the ground. Indeed, dromaeosaurs are among the least likely of avepectorans to be closer to the rest of Aves than archaeopterygiforms. Yet the sickle claws appear to have descended from fliers that had developed a flight-muscle power level above that exhibited by the urvogel either independently from the avian mainstream (and evolved pterosaur-like tails) or as part of the mainstream (in which case the tail re-elongated). The more-avian braincases of troodonts suggest that they split from the avian mainstream at a higher phylogenetic level higher than dromaeosaurs. Caudipterygians appear to have descended from shorter-tailed, and presumably even more advanced, fliers than troodonts. Skull similarities imply that caudipterygians and oviraptorosaurs form a single neoflightless group. Alternatively, they descended from a group of fliers with oviraptorosaur-like heads and lost flight independently from each other. At the other extreme, dromaeosaurs, troodonts, caudipterygians, and oviraptorosaurs all share distinctive postcranial features that suggest they are closely related (see Chapter 12); perhaps they all descended from a common ancestor whose flight capability was reduced or lost. Alvarezsaurs and especially avimimids are so highly avian that they seem to have split off from much more derived fliers.

### More Complications

We can add even more scenarios to the alternatives presented so far. It is possible that the potentially neoflightless condition of the avepectorans is misleading. So far in this study, it has been assumed that flight evolved only once in dinosaur-birds. What if flight evolved two or more times? What if a group of dinosaurs very close to the *Archaeopteryx*–bird clade independently evolved flight? Also assume that some or all of the avepectoran dinosaurs evolved from this second group. On the basis of these assumptions, we can postulate that the birdlike dinosaurs' ancestors could have flown better than *Archaeopteryx*. This possibility would explain the advanced flight-related features, especially the large sterna. In this scenario, the avian attributes of the dinosaurs represent strong parallelism.

A related scenario has the very first stages of flight evolving only once, but the subsequent development occurs along two lines (this is a viable option, considering that the last advancement of flight may have happened at least twice, once in enantiornithines and once in ornithurines). Again, this scenario has some or all of the avepectorans descending from relatively advanced fliers while paralleling birds (Fig. 11.1B).

It is also possible that flight or its subsequent improvement or both evolved more than twice, with the various avepectorans evolving and losing advanced flight independently. It is even possible that *Archaeopteryx* evolved flight independently from a clade consisting of avepectoran dinosaurs and avebrevicaudan birds. Parallelism, however, is an inferior explanation for the fully birdlike complex of wing feathers that adorn *Archaeopteryx*.

Other questions arise. Let us assume that *Sinornithosaurus* was neoflightless. Were its flying ancestors living back in the Jurassic when the first dromaeosaur-type teeth show up in the fossil record, or did it descend from Cretaceous fliers similar to *Rahonavis*? Or did dromaeosaurs develop a high level of flight independently from archaeopterygiforms or birds or both?

If the preceding complications were not enough, there is yet another. What if in some cases flight appeared, was lost, and then reappeared again, perhaps to be lost a second time or maybe even more times? Since neoflightless dinosaurs would have often been similar in form to protofliers, repeat losses of flight are a real possibility. This pattern has not been observed in birds, but when they lose flight, they often reduce the flight apparatus too far to re-evolve it. Although parsimony tends to favor the single origin of flight and minimal losses of flight, nature is not always so simple. At this time, we cannot reliably choose among the many, speculative alternatives, so many alternatives that they threaten to rapidly grow into a set that is as lengthy as it is speculative. speculative because they cannot be rigorously analyzed at this time owing to our inadequate database. As a result of the lack of data, the post-urvogel and neoflightless hypotheses are difficult to assess at the detailed level. Finally, it remains possible that no dinosaur evolved from a flying ancestor and that all the post-urvogel features of all dinosaurs are the result of repeated and extensive parallelism and convergence with birds.

### *Post-Urvogel Avepods May Not Look Like Post-Urvogel Avepods*

It is worth reemphasizing that—in view of the remarkably nonavian cranial features found in *Confuciusornis*—the absence of what have been presumed to be critical avian features in a dinoavepod does not necessarily exclude it from being a post-urvogel. The most unbirdlike derived therizinosaurs—with their prosauropod-ornithischian-like heads, basal theropod–like scapulocoracoid, and sauropod-like bellies and ilia—are the prime example of this phenomenon because we now know they descended from basal therizinosaurs whose form was much more like the urvogel and other dino-avepectorans. The molelike arms of alvarezsaurs may be another extreme example of evolution away from the avian pattern.

### *Cladistics, Flight Loss, and the Lack of Little Fossils*

Cladistics will be a useful but limited tool for further investigating the neoflightless dinosaur hypothesis. At the same time, the hypothesis has important implications for cladistic studies of bird origins, as discussed in Chapter 2.

In addition to mosaic evolution, another problem vexes our efforts to understand early loss of flight, a problem that promises to be difficult to overcome. If we assume that various avepectoran dinosaurs are secondarily flightless, then they may have evolved from an extended series of small, pygostyleless fliers whose flight development and closeness to modern birds varied. In other words, the transitional links between dromaeosaurs, troodonts, caudipterygian-oviraptorosaurs, alvarezsaurs, and even tyrannosaurs may lie among an array of flying dino-birds that weighed less than 1 kilogram. To date we have found only two of these little long-tailed early fliers, *Archaeopteryx* and *Rahonavis*. This is not surprising; such small creatures are relatively unlikely to be preserved, especially if they cannot fly well enough to form large flocks whose remains end up in the middle of fine-grained sediments of lake and lagoon bottoms. The result is that we are sampling the larger and therefore more preservable flightless end branches of this evolutionary bush; the smaller-bodied transitional fliers as the base of the branches are so far largely absent.

One way that we can use cladistics to help overcome the problems posed by mosaic evolution and the potential absence of transitional fliers, as well as to tentatively test the neoflightless hypothesis, is to execute a series of comparative cladistic character analyses using homogenous data sets. In one run, we include all characters, including those often lost with neoflightlessness. In another run, we exclude the potentially neoflightless characters subject to reversal with the loss of flight (as we did in the comparison between *Archaeopteryx* and dromaeosaurs). In another run, we also exclude those non-flight-related characters subject to reversal with flight loss (tooth form, pubis orientation, etc.). The differences, if any, between the runs may expose biases introduced by the inclusion of characters commonly reversed with the loss of flight.

Let us consider an avepectoran dinosaur group that in conventional cladistic analyses plots on a node below *Archaeopteryx*. If when we exclude reversible flight characters, the dinosaur group plots above the urvogel, then this is good evidence that the dinosaur is secondarily flightless and closer to birds than *Archaeopteryx*. However, such cladistic evidence is not definitive. The possibility remains that the highly birdlike dinosaur group merely parallels birds. If the group falls below the level of *Archaeopteryx* in every run, then parallelism with birds in the urvogel or reversals not related to flight loss (or both) may hide the true placement of the dinosaurs, which are nearer to birds. In order to successfully challenge the neoflightless hypothesis, an extensive cladistic analysis must show that the exaptation hypothesis dramatically reduces parallelism and reversals in comparison with the neoflightless alternative.

We must remember the caution that although such a series of cladistic tests will be informative and may favor one hypothesis over another, they will not definitively verify or falsify any given hypothesis, again because we lack the definitive data.

For the sake of discussion, let us assume that a set of dinosaurs evolved from flying post-urvogels. Let us assume further that while losing flight, the dinosaurs lost most of the avian characters that had so far appeared—both those directly related to flight and those that were unrelated—as they re-evolved dinosaur characters better adapted for an earthbound existence. Cladistic methodology would be unable to reconstruct this course of events and would instead place the neoflightless post-urvogels basal to *Archaeopteryx*.

Consider the increasingly complex situation surrounding dromaeosaurs, some of which seem to have had pubes and ischia little different from those of standard avepod dinosaurs, and others of which have strongly birdlike ventral pelvic bones. In standard cladistics, this pattern would indicate either that dromaeosaurs evolved birdlike pelves independently of birds or that dromaeosaurs are paraphyletic. However, if dromaeosaurs were secondarily flightless post-urvogels, then we may be seeing a reversal from an initially avian condition back to a seemingly primitive state. The latter scenario is made more likely by the fact that the basal dromaeosaurs *Sinornithosaurus* and *Bambiraptor* had the most urvogel-like pubis and ischium. Again, cladistics has difficulty coping with this kind of situation.

## *We Need More Transitional Forms*

The limits of cladistic methodology mean that no matter what the results of contemporary cladistic analyses or investigations of flight heritage are, in the end, only an improved set of transitional fossils will reveal the actual situation. Therizinosaurs show us why this is true. If we had only the specialized derived examples to examine for phylogenetic and flight-related characters, the evidence that they were secondarily flightless avepectorans would be at best modest. The better-preserved basal dromaeosaurs *Sinornithosaurus* and *Microraptor* are even better examples of transitional forms whose level of flight adaptations appears ideally intermediate between an urvogel-like condition and the entirely grounded condition seen in later, larger dromaeosaurs. The evidence for a neoflightless condition and for avepectoran relationships has been significantly reinforced by the discovery of basal examples that retain a more birdlike form. More broadly, the specialized nature of other avepectoran groups for which we currently have good remains only for derived examples may be misleading us as to their true placement vis-à-vis *Archaeopteryx* and other Aves. We can hope that there are lagerstätten out there that preserve more of the small fliers that interlink the dino-avepectorans, if they exist. One hopes so. The uncovering of dromaeosaurs even more basal than *Sinornithosaurus* and *Microraptor* may reduce or eliminate the superiority the urvogel currently enjoys in terms of avian characters. For example, *Microraptor* has only one or two more tail vertebrae than *Archaeopteryx*, just enough to place the former farther from birds than the latter. But what if they were basal dromaeosaurs with tails as short as *Archaeopteryx*'s? Only the discovery of more skulls and skeletons will provide the information we need. Even so, we may never be able to reach the necessary fossil threshold. If not, then our understanding of the detailed interrelationships between Mesozoic avepectorans will always be limited.

## *Implications for the Phylogeny of Functional Anatomy*

The neoflightless hypothesis has profound implications for our understanding of the evolution of functional anatomy during the dinosaur-bird transition. For example, Gatesy (1990) postulated that reduction of the caudofemoralis complex began before the development of flight in dromaeosaurs and ornithomimids, but if these forms are neoflightless post-urvogels, then derived avepod caudofemoralis reduction was a consequence of flight. The manner in which the oviraptorosaur pygostyle evolved strongly depends on whether it did so before or after the advent of flight. When Chure (1999) examined the detailed evolution of the avepod dinosaur semilunate carpal, he assumed that the rather weakly developed and poorly fused carpal blocks of therizinosaurs represented a basal stage in the development of this complex. If therizinosaurs are neoflightless post-urvogels, then their weakly developed wrists represent an atrophied condition, and using them to reconstruct the basic evolution of the avepod wrist is potentially misleading. Rather than evolving for a snap-action predatory stroke (Gauthier and Padian 1985, Dingus and Rowe 1998, Gishlick 2000), the semilunate carpal block may have evolved for flight and only later have been applied to prey capture. Chiappe et al. (1999) and Clarke et al. (1999) suggested that *Archaeopteryx* may have had ossified uncinates and sternal ribs because they were present in avepods these workers considered less avian than the urvogel, this in face of the absence of these bones in all the articulated urvogel specimens. In the neoflightless scenario, the lack of these bones in *Archaeopteryx* accurately reflects its position below the rest of the Avepectora. How we view the evolution of all the potentially neoflightless characters depends upon the validity of the neoflightless hypothesis versus the conventional alternative.

## *Loss of Climbing*

Related to the neoflightless problem is the question of whether a number of theropods descended from climbers. Those theropods with broad scapula blades and short hands—staurikosaurs, baso-avepods, compsognathids—lack adaptations for climbing, but there are hints of arboreality in other theropods. Climbing is a possible explanation for the avepod furcula if a similar structure was present in the possibly arboreal longisquamids (but see the discussion in this chapter on longisquamids). The furcula may have initially adapted in such a way as to strengthen the shoulder girdle so it could better absorb the stresses imposed by quadrupedal climbing. Climbing may also have been the selective force behind the evolution of straplike scapula blades in herrerasaurs and most avetheropods; the same feature evolved in highly arboreal drepanosaurs and chameleons

(Renesto 1994, Paul 1987a). The long arms and hands of herrerasaurs and most avetheropods may also have developed in climbing ancestors, although the possibility that they developed to improve prey capture is an equally viable explanation. The odd combination of distally elongated toes and short claws in *Compsognathus* hints that it descended from a bigger-clawed ancestor that used long toes to better grasp and move among branches. Avepectoran forward vision may have evolved in order to improve the ability to navigate in the complex three-dimensional arboreal world, and avepectoran brains may have enlarged in order to process the large volume of incoming sensory information.

*Protarchaeopteryx* is interesting with regard to potential reduction of arboreality. With its larger size, its shorter arms and fingers, and its lack of a large, distally placed hallux, *Protarchaeopteryx* can be judged as less arboreal than its earlier relative *Archaeopteryx*. The very long legs of *Protarchaeopteryx* also favored running over climbing. The Cretaceous archaeopterygiform appears to have been able to climb, but no better than dromaeosaurs and oviraptorosaurs. Yet *Protarchaeopteryx* retained short metatarsals like those of climbing *Archaeopteryx* and other tree-loving birds. The short foot may have been an arboreal feature retained from its arboreal ancestors.

*Sinornithosaurus* and *Microraptor* are even more informative in terms of the potential reduction of arboreality. As detailed in Chapters 6 and 9, these basal dromaeosaurs approached or exceeded *Archaeopteryx* in climbing abilities and were much more arboreal than derived dromaeosaurs. This is exactly the pattern to be expected in a clade that saw a reduction of the arboreal component of its locomotory repertoire.

### *Loss of Flight Does Not Contradict the Dinosaur-Bird Link*

One reason some may resist the concept of neoflightless birds is that the antidinosaur group has claimed that *Protarchaeopteryx* and *Caudipteryx* were secondarily flightless birds. But as we have seen, these are dinosaurs, whether they were neoflightless or not, whether they were post-urvogels or not. *Caudipteryx* was no more or less a neoflightless bird than were its close oviraptorosaur relatives. By the same token, an ostrich is both a neoflightless bird and a dinosaur. In the end, the validity of the neoflightless dinosaur hypothesis is irrelevant to the validity of the dinosaur-bird hypothesis.

### *Are Dinosaurs Neoflightless Birds?*

George Olshevsky (1994) has taken the hypothesis that flight was lost at an early stage to an extreme. He argues that *most or all* dinosaurs descended from feathered Triassic ancestors that were volant or arboreal or both. This is an extension of Abel's (1911) argument that dinosaurs were secondarily terrestrial archosaurs that shared with birds a common arboreal ancestry. In Olshevsky's view, birds did not descend from dinosaurs. Instead, dinosaurs descended from early birds! As outlandish as it may sound, this interesting concept is not impossible, and it requires consideration. Verification would require the finding of a set of unambiguous, volant dinosaur ancestors in the Triassic, or strong evidence of neoflightlessness in nonvolant dinosaurs of the Triassic and Jurassic. Concerning the needed Triassic protodinosaurs, Olshevsky cites longisquamids and *"Protoavis"* as examples, but we have seen that the flight abilities and phylogenetic relevance of these creatures to birds are dubious, all the more so with the elimination of *"Protoavis"* as an early dino-bird. In addition, the narrow-footed protodinosaurs show no strong signs of having been arboreal; running seems to have been their forte. As will be discussed in more depth in Chapter 12, the dearth of well-documented flying dinosaurs in the Triassic and most of the Jurassic argues against the presence of flying dino-birds during those times. We have also seen that there is not much in the way of neoflightless characters in the theropods of the Triassic and earlier Jurassic. Most Triassic-Jurassic predatory dinosaurs lacked big brains or forward-facing eyes, and they lacked flight-modified pectoral girdles, arms, and tails. Baso-theropods even lacked laterally facing shoulder joints, and dinosaur furculae are not yet known outside the Avetheropoda. Olshevsky suggested that the pneumatic vertebrae of predatory dinosaurs were developed in early fliers, but they are absent or poorly developed in Triassic and Early Jurassic theropods. That potentially neoflightless characters do not become numerous in avepods until the later Jurassic, and especially the Cretaceous, tells us that flight probably did not develop until well after dinosaurs had evolved from nonvolant ancestors in the Triassic. Ergo, birds are probably dinosaurs, rather than the reverse.

### *A Challenge*

Those who oppose the neoflightless hypothesis must explain how, if it did not have ancestors that passed through a stage of flight more advanced than that seen in the urvogel, wingless *Sinornithosaurus* came to possess a very large sternum, numerous ossified sternal ribs, and a remarkably broad, streamlined proximal central finger bone, all of which are more avian and flight adapted than those of wing-bearing *Archaeopteryx*. They must then explain why these features were reduced or lost in more-derived dromaeosaurs.

## A Quick Wrap-Up

At this time *Protarchaeopteryx*, dromaeosaurs, troodonts, caudipterygians, oviraptorosaurs, and therizinosaurs provide telling evidence of having descended from fliers more advanced than *Archaeopteryx*. All also possess post-urvogel characters. Of these dino-birds, dromaeosaurs were the least likely to have been closer to avebrevicaudans than the urvogel, but the possibility that they were closer remains viable if not superior to their conventional placement. Ornithomimosaurs and alvarezsaurs show fewer signs of flight loss, but some potentially neoflightless characters are present, and the lack of other such attributes may be due to strong reversion to life on terra firma. This is especially true of alvarezsaurs, whose potentially neoflightless characters are exceptionally well developed. Tyrannosaurs are the least likely to have been neoflightless and postarchaeopterygiform, but the possibility cannot be ruled out. This possibility is especially true for therizinosaurs. Had we not found a basal example that retained so much of its avepectoran character, we would never have known how a group could undergo a reversal that modified its anatomy so far from the avian pattern, to the degree that therizinosaurs have often been taken for non-theropods!

CHAPTER 12

# A Look at the Phylogenetics of Predatory Dinosaurs

Having considered the increasingly birdlike anatomical organization of theropod and avepod dinosaurs and the possibility that some of the latter were secondarily flightless and closer to birds than *Archaeopteryx,* we need to try to sort out the relationships among the bird ancestors (Figs. 1.1, 2.1). This section of the chapter continues to focus on how these dinosaurs are related to birds, but it is not intended to be an exhaustive and definitive analysis of all the taxa nor to arrive at definitive conclusions. We do not yet have enough data in the form of transitional species to tell exactly what is related to what, as revealed by the contradictory cladograms presented by various researchers (Fig. 12.1) This section presents an overview of the question, an overview that points out probabilities and problems, raises issues that have not been adequately addressed by the paleocommunity, and considers some of the difficulties posed by cladistic techniques. To review the fossil taxa discussed below, refer to Chapter 10; many of the characters considered below are listed in Appendix 1A.

## Relationships from Protodinosaurs to Birds

### *Protodinosaurs*

Because what is known of *Marasuchus* is so primitive and generalized, this creature is an ideal ancestral type for all dinosaurs. *Lagerpeton* was an early offshoot of the protodinosaur clade, because of its oddly asymmetrical foot that shows it had diverged from the dinosaurian norm of pedal symmetry (Sereno and Arcucci 1993).

### *Dinosaur Monophyly*

Until the early seventies, dinosaurs were considered an unnatural group made of the "lizard-hipped" Saurischia and the "bird-hipped" Ornithischia, which evolved independently from basal archosaurs. This view began to give way when the first new remains of the earliest generations of dinosaurs—the marasuchian protodinosaurs, the baso-theropod staurikosaurs, and the herrerasaurs, as well as early herbivorous ornithischians—began to show how interrelated these groups were. At first, a few researchers opposed the dinosaur monophyly proposed in 1974 by Bakker and Galton (R. Thulborn 1975, Charig 1976), but attempts to refute the evidence that dinosaurs form a distinct clade failed. So did the attempts by A. Walker (1964) and Chatterjee (1985) to argue that the ancestors of various giant dino-avepods were ornithosuchians or poposaurs with crurotarsal ankles. One of the major achievements of cladistics has been to verify the monophyly of dinosaurs, to the point that it is almost universally accepted (Padian 1982, Paul 1984a, 1988a, Benton 1985, Gauthier and Padian 1985, Gauthier 1986, Carroll 1988, Novas 1991, 1996, Sereno 1991, 1997a, Olshevsky 1994, Sumida and Brochu 2000). Even Chatterjee (1997) now appears to accept dinosaur monophyly. The continued

*Figure 12.1. (opposite)* Major cladograms analyzing the relationships among predatory dinosaurs and birds. *A,* Gauthier's (1986) cladogram, which initiated the modern cladistic method of dinosaur classification. *B,* a major effort produced by Holtz (1994a). *C,* Chatterjee (1997); note the proposed position of *"Protoavis." D,* Makovicky and Sues (1998). *E,* Sereno (1998), part of the most extensive analysis of dinosaur relationships to date. *F,* attempt by Padian et al. (1999) to more precisely define the taxonomy of major theropod clades. *G,* a second major analysis by Holtz (1998/2000). *H,* analysis of early birds by Chiappe et al. (1996). *I,* theropod phylogenetics from Paul (1988a) (for current scheme see Fig. 2.1). Note that there is consistent agreement that dromaeosaurs were very close relatives of *Archaeopteryx.*

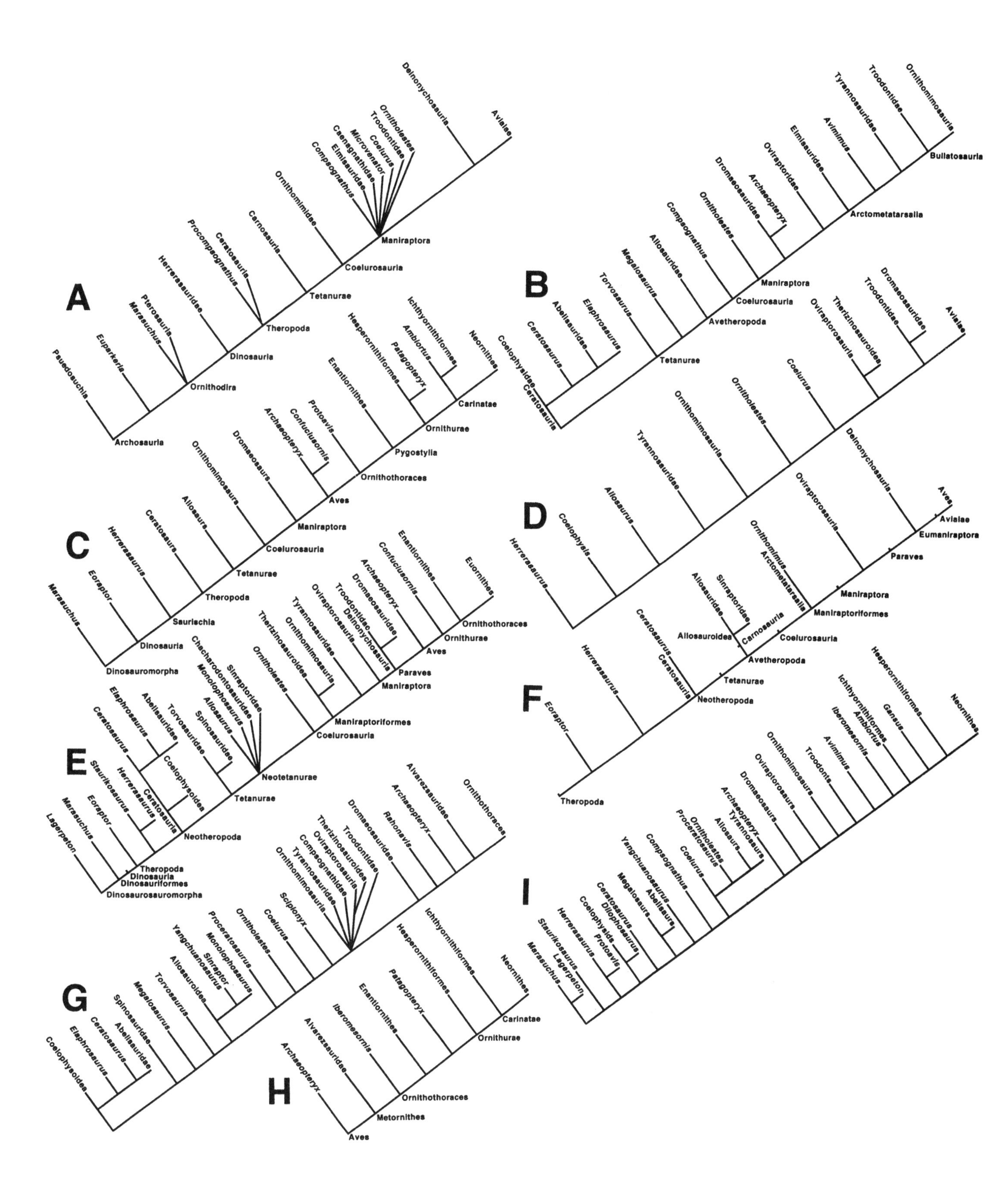

acceptance of multiple origins for dinosaurs by Tarsitano (1991) and Feduccia (1996) is therefore anachronistic and unsubstantiated. At this time, the postcranial characters that unite dinosaurs center on the erect, mesotarsal-ankled, digitigrade hindlimb (Novas 1996), but the S-curved neck and other details are involved as well.

Although the polyphyletic split of dinosaurs into Saurischia and Ornithischia has ceased, the two groups continue to exist as the major division of the Dinosauria in most studies. On the one hand, Saurischia is a rather weak clade because so few features are involved in the delineation of the clade. On the other hand, the stiffening of the dorsal column and the beginnings of skeletal pneumaticity seen in basal saurischians are attributes of birds, which belong to this group.

### Baso-theropods

Eoraptors, staurikosaurs, and herrerasaurs were in most regards so primitive that they were suitable ancestral types for the rest of the Dinosauria. However, with complete skulls and skeletons on hand by the early 1990s, we learned that they shared some characters with avepods. Hence, they have since been generally considered basal theropods (Sereno and Novas 1992, 1993, Sereno et al. 1993, Novas 1994, Currie 1997, Sereno 1997a). However, the interrelationships between early dinosaurs remains poorly understood owing to a lack of sufficient fossils, and the nature of the first herbivorous dinosaurs is particularly obscure. *Eoraptor* is interesting in that its differentiated dentition had adaptations found in both predatory and herbivorous dinosaurs. It remains possible that the archaic baso-theropods, especially the generalized eoraptors and staurikosaurs, were not closer to theropods than to other dinosaurs.

### Dino-avepods

The dino-avepod group is so large, has lasted so long, is so diverse, and yet, in some ways, so poorly known that sorting out the interrelationships between the many members has long been a challenge. It continues to be so. Cladistic studies have produced both consensus concerning the broad view of how the tridactyl dinosaurs were related to one another and to birds and disagreements on the details.

Most researchers generally agree that avepod dinosaurs form a monophyletic clade. Actually, this view is quite old, and cladistics has served to reinforce it to the point that no one currently challenges dino-avepod monophyly with original data. Proponents of this view hold that a tridactyl foot in which the first metatarsal does not contact the ankle evolved only once among dinosaurs. In particular, scenarios that might try to split the most birdlike avepod dinosaurs and birds from nonavepectoran avepods are not viable.

There is also general agreement about where most of the major dino-avepod groups stand in basic relationship to one another. The small coelophysids and larger dilophosaurids are recognized as a distinct clade, marked by the presence of a subnarial gap in the upper jaw. The presence of this gap suggests that this Coelophysoidea is a side group off the mainstream of avepod evolution. Clearly these lie near the base of the Avepoda, as do the stocky-horned ceratosaurs and gracile elaphrosaurs of the Late Jurassic, even though at least the former may have been an averostran. Despite being specialized and having lived in the Cretaceous, abelisaurs too were basal avepods.

As we climb up the avepod family tree, we find that researchers generally agree that a number of large Jurassic Cretaceous avepods exhibit a series of moderately advanced characters. This "middle" grade of avepods includes afrovenators, sinraptors, yangchuanosaurs, allosaurs, and carcharodontosaurs, some of which were enormous beasts. Smaller theropods of the same times and grade include compsognathids and probably ornitholestids.

Finally, there is a consensus that dromaeosaurs, troodonts, and oviraptorosaurs are the dino-avepods closest to birds. These form the core of the first group of avepectorans. We have already stressed that many have concluded that dromaeosaurs and archaeopterygiforms are especially close to one another.

So much for the broad agreements. The disagreements over the details are numerous, and the relationships of some groups are especially enigmatic. Consider *Elaphrosaurus,* for which we have no skull material. Barsbold and Osmólska (1990) continued to support the long-standing hypothesis that it was an ancestral ornithomimosaur. Holtz (1994a), Sereno (1997a), Rauhut (1998), and I (Paul 1988a) found that the connection between this baso-avepod and the advanced ostrich mimics is very weak, and Sereno and Rauhut ally it with ceratosaurs and abelisaurs. If the last view is correct, then elaphrosaurs were ostrich-mimic mimics. Sereno (1997a) suggested that dilophosaurids are less derived than coelophysids. This suggestion is interesting in that it obviates the need for a reversal in the size of the ilium that I proposed in 1988 (Paul 1988a). In recent years, coelophysoids have been joined with ceratosaurs and abelisaurs in a supposedly separate clade called Ceratosauria, but many of the attributes of this group appear to be merely primitive for avepods. The presence of auxiliary maxillary openings in ceratosaurs and abelisaurs indicates that they were averostrans, and their longer ilia and bladelike lesser trochanters of the femora were derived features. Therefore, I agree with Rauhut (1998), Carrano and Sampson (1999), and Britt et al. (2000) that Ceratosauria is probably paraphyletic.

I previously (Paul 1988a) considered the unusual spinosaurs to be advanced coelophysoids, mainly because of the presence of a subnarial gap, but Holtz (1994a), Charig and Milner (1997), Sereno (1997a), and Sereno et al. (1998) presented evidence that they were basal tetanurans. The tentative suggestion by Taquet and Russell (1998) that *Pelecanimimus* was a spinosaur rather than an ornithomimosaur is interesting because, if true, then this avepod would be another ostrich-mimic mimic. However, the suggestion remains to be verified. Elzanowski and Wellnhofer (1992, 1993) came to the even more surprising conclusion that spinosaurs, tiny *Archaeornithoides,* and troodonts form a monophyletic clade with birds. *Archaeornithoides* is known only from the snout and is very probably a juvenile (Elzanowski [1999] claims that it is a troodont), so its relationships remain obscure. Its teeth are more similar to those of *Archaeopteryx* than those of other avepod dinosaurs are, but its palate is not especially like that of the urvogel (Elzanowski and Wellnhofer 1995). As for the large spinosaurs, they may or may not have shared a diet of fish with the urvogel. Otherwise, these massive, vaguely croclike, rather primitive forms are about as nonavian as avepods get, and they share no distinctive features with troodonts. Elzanowski and Wellnhofer's hypothesis has not been accepted by others (Charig and Milner 1997).

One of the most interesting things about most cladograms is that they largely retain the old division between the large theropods (Carnosauria) and small theropods (Coelurosauria) (Gauthier 1986, Holtz 1994a, Sereno 1997a, Padian et al. 1999). This division may reflect more the absence of an adequate fossil record for small Jurassic theropods than actual relationships. In addition, some of the characters (such as the keyhole orbit and gracile hands) that seem to ally the carnosaurs may be the result of large size rather than phylogeny. Some characters that have been assigned to coelurosaurs are invalid. The orbit of *Ornitholestes* is no larger or more circular than that of *Coelophysis.* The pretibial process of *Coelurus* is much shorter than the processes of many carnosaurs. The characters that seem to characterize Coelurosauria are rather few, and their presence status in various basal coelurosaurs is inconsistent or not yet thoroughly known. Similar problems afflict the Maniraptora. A bowed ulna is weakly developed in coelophysids and allosaurs and strongly developed in *Coelurus.* Jugal expression in the rim of the antorbital fenestra is seen in a number of nonmaniraptoran theropods.

Well-preserved *Compsognathus* seems to be closely related to the even better preserved and later *Sinosauropteryx.* The latter is itself a question mark. On the one hand, the largest specimen has a relatively much shorter tail—both in length and in caudal count—than the other specimens, a tail in which deep chevrons are much more proximally limited. On the other hand, as it were, both forms share an exceptionally robust finger I. In any case, *Compsognathus* and *Sinosauropteryx* are classic basal avetheropods usually placed at the base of the Coelurosauria, but they are so generalized that they currently float unattached to other specific avepod groups (and were ignored by Sereno [1997a]).

The juvenile status of wee *Scipionyx*—it may have just emerged from the egg—hinders analysis of its phylogenetic position: some of its attributes may be features of youth and may have later been modified during growth. In particular, the low position of the superior temporal bar may be due to the large size of the juvenile braincase. The long arms may be a juvenile arboreal adaptation. Conversely, it is possible that birdlike features developed only with maturity. Because of these ontogenetic problems, *Scipionyx* is not included in the character list (Appendix 1A). The absence of most of the tail and the feet is an additional problem. *Scipionyx* is certainly an avetheropod because it has the defining characters of the group. However, the conventional shoulder girdle, the absence of a well-developed semilunate carpal, and the lack of other avepectoran characters imply that it is not a member of that clade.

*Coelurus,* the namesake of the Coelurosauria, is yet another basal avetheropod whose actual placement is obscure, both because the material found so far is rather generalized in nature and is not yet well documented. The lozenge-shaped carpal block was at best only a little more derived than that of an allosaur (Fig. 10.11A,B). The proximal femur had a very large lesser trochanter. The surprisingly short pretibial process was intermediate in size and form to the processes of sinraptors and allosaurs.

Also ambiguous is the basal avetheropod *Ornitholestes,* whose well-ossified vertebrae suggest it was not a juvenile, despite its small size. In most of the recent cladograms, *Ornitholestes* ends up in the Coelurosauria somewhere at or near the base of the Maniraptora, with no definite position vis-à-vis any other group. *Ornitholestes* did have some of the clade's attributes. However, its downward-curved dorsoposterior ilial border is not consistently present in maniraptors (such as oviraptorosaurs) but is seen in yangchuanosaurs. Owing to the current condition of the skeleton, the status of the obturator notch is not entirely clear. In addition, the little theropod—which has an unusually heavily built skull for something so small—shares some skull and skeletal features with some larger theropods (Paul 1988a), including a robust squamosal-quadratojugal contact that cannot be explained away as an artifact of large size. *Ornitholestes* seems most similar to medium-sized (juvenile?) *Proceratosaurus* (which has no particular link with its namesake) and to larger *Monolophosaurus.* In particular, *Ornitholestes* and *Proceratosaurus* share fragments of dorsal nasal projections that may be homologous with the intensely pneumatic crest of *Monolophosaurus.* Also interesting is the fact that all three creatures have widely spaced, relatively small teeth that tend to lack well-developed anterior serrations. *Monolophosaurus* is intriguing in that the ischium seems to have been short and to have had the triangular obturator process expected in coelurosaurs. Yet *Monolophosaurus* had an ischial foot and a closed obturator notch. Ornitholestids, proceratosaurs, and monolophosaurs may have been close relatives. They, in turn, may have been allosaur relatives that—at least in the case of ornitholestids—independently evolved a dynamic stabilizing tail for leaping. If *Ornitholestes* proves to have the bird type shoulder girdle and folding arm, then it should be a basal member of Avepectora. Alternatively, ornitholestids and its crested relatives may be a link between avepectorans and carnosaurs. All that can be said at this time is that we do not have enough fossil data to tell whether or not Carnosauria and Coelurosauria, as well as the division between big and little, are real or artifacts of our limited knowledge. The possibility that the various "carnosaurs" evolved independently from small avepod dinosaurs remains open. Similarly uncertain is the status of Maniraptora.

We now come to the related question of whether the great tyrannosaurs should be classified as carnosaurs or as arctometatarsalians. The often gigantic superpredators of the Cretaceous were long presumed to be the ultimate expression of Carnosauria, although a few researchers, such as Von Huene (1926), thought otherwise. Recently, Thomas Holtz (1994a,c) has presented evidence that all or most dino-avepods that share an arctometatarsalian pes (in which the central metatarsal is extremely narrow at its proximal end; Fig. 10.14Eh,i,Fi) form a monophyletic clade of their own, Arctometatarsalia. The smaller members of this clade include the troodonts and ornithomimosaurs and, perhaps, caenagnathids and avimimids, as well as the big tyrannosaurs. In Holtz's view, tyrannosaurs—which were the most speed adapted of the gigantic avepods (Paul 1988a, 1998)—evolved from small gracile ancestors. In particular, the pelves, hindlimbs, and tails of ornithomimids and tyrannosaurs were very much alike, albeit not identical (Paul 1998/2000; it is intriguing that tyrannosaurs lacked the well-developed supra-acetabular shelf common to ornithomimids as well as to nonavepectoran avepods), which suggests a common ancestry. At the same time, ornithomimids shared some distinctive characters with troodonts, most notably a strongly bulbous basisphenoid projecting forward from the base of the braincase (Fig. 10.4D). The highly pneumatic skull elements of tyrannosaurs also hint at a connection with ornithomimosaurs and troodonts, as do tyrannosaurs' forward-facing eyes. That the first tyrannosaurs are modest in size favors their descent from nongiants. In this scenario, tyrannosaurs may be avepectorans.

There are, however, problems with the Arctometatarsalia. One is the arctometatarsalian foot itself. The extreme narrowing of the foot seen in this group is exactly the kind of speed adaptation that can be expected to have evolved repeatedly. Indeed, we know it evolved at least twice among avepods, once in birds and once in predatory dinosaurs. We know the development was independent because the majority of birds have a severely pinched central metatarsal, but the urvogels, including *Archaeopteryx,* do not. In addition, the details of the proximal pinching differed in avepod dinosaurs and birds. There is evidence that a number of avepod dinosaur groups evolved proximally slender central metatarsals on their own. The arctometatarsalian caenagnathids were oviraptorosaurs, but oviraptorids lacked the strongly pinched central metatarsal. Unless the thicker metatarsal of the oviraptorids represents a reversal, the arctometatarsalian foot must have evolved independently within oviraptorosaurs. Basal ornithomimosaurs such as *Harpymimus* lacked the strong central pinching present in metatarsal III of the advanced ornithomimids. This indicates that the hindlimbs of the ostrich mimic paralleled or converged with those of tyrannosaurs. Basal alvarezsaurs had a proximally robust central metatarsal, but the advanced examples were arctometatarsalian. In this respect, the birdlike avepods paralleled the avian avepods.

The foot is not the only problem associated with the concept of Arctometatarsalia. The concept requires either that the shoulder girdle, arms, and pelvis of troodonts paralleled or converged with those of other avepectorans or that ornithomimids and tyrannosaurs re-evolved conventional scapulocoracoids, hands, and hips. In addition, it is not clear how troodonts could have so many flight-related characters while its supposedly close ornithomimosaur and tyrannosaur relatives do not. Indeed, there is little in tyrannosaurs to suggest they are especially close bird relatives in the manner of troodonts or even ornithomimids (Clark et al. [1994] also expressed skepticism about the placement of tyrannosaurs with the birdlike dinosaurs). Also important is that tyrannosaurs show evidence of being related to the allosaur group. Some of the characters are unrelated to large size, including a deep parasphenoid plate that sloped up and forward from the front of the braincase toward the frontals on the skull roof, and a broadened squamosal-quadratojugal bar. Both features are absent in many giant avepods. Also shared by sinraptors, ornitholestids, allosaurs, and tyrannosaurs was a central metatarsal that was distinctively L-shaped in proximal view (Fig. 10.Ee-h; Paul 1988a,b). This feature has not been reported in other avepods (contra Holtz 1994c), including ornithomimosaurs. Unfortunately, the L-shaped metatarsal has received little attention and has never been scored in a cladogram, because the L shape persists regardless of the size or proximal compression of the metatarsal. The shape therefore seems to have had little functional importance and may represent a consistent expression of a bit of genetic coding that marks a clade of small and large averostrans. In this scenario, giant tyrannosaurs may not have evolved directly from giant allosaurs. More probably, as mentioned above, large forms within this clade repeatedly evolved from the small ones, the tyrannosaurs being the last and most sophisticated example. This clade mimicked the Avepectora in a few regards, especially the pneumatic posterior skull and forward vision of tyrannosaurs. Most of all, the hindquarters of tyrannosaurs and ornithomimids paralleled each other in an evolutionary race for speed, one in which the latter were literally trying to escape from the former! Only the unearthing of more transitional forms will verify or falsify these scenarios.

The relationships of the ornithomimosaurs remain obscure, because these creatures merge a seemingly paradoxical set of characters (Paul 1988a, Makovicky and Norell 1998). These ostrich mimics are birdlike, as their generic titles imply. Troodont-like features of their braincase and, perhaps, the posterior plate both reinforce the bird relationship and suggest a link with the similarly gracile sickle claws. Short cervical ribs and an elongated sacrum are additional birdlike features, as are the large sternal plates observed in basal ornithomimosaurs. The plates also hint that ornithomimosaurs may have been neoflightless. However, the conventional shoulder girdle, hands, and pelvis suggest these are not as close to birds as troodonts and avepectorans in general. Ergo, the placement of these long-beaked dinosaurs remains enigmatic (Paul 1988a); even earlier transitional forms need to be found.

Let us next consider the troodonts. Because these had semididactyl feet with a sickle-clawed, hyperextendible second toe, troodonts were, from the 1960s until the mid-1980s, widely considered to have been close relations to the dromaeosaurs. However, the details of dromaeosaur and troodont feet differed in many regards. In the latter, the two bones of the second toe were not subequal in length, as in dromaeosaurs; the distal one was much shorter. Nor were the metatarsals subequally developed in troodonts, as they are in dromaeosaurs. In the former, metatarsal II was rather weakly developed, III was pinched proximally, and IV was unusually robust. In addition, hyperextendible second toes or semididactyl feet or both are turning out to be fairly common in other dinobirds, early birds, and even abelisaurs (Bonaparte and Powell 1980). Dromaeosaurs and troodonts do share unusual inverted L-shaped posterior rib shafts in anteroposterior view; whether any other avepectoran ribs were like this is unclear. But as we learn more about the braincase and palate of troodonts, we are finding that they are different from those of dromaeosaurs and archaeopterygiforms. The same is true of the ankles of troodonts. Few researchers now consider troodonts to be especially close relatives of dromaeosaurs (Osmólska 1981, Barsbold 1983, Currie 1987, Paul 1988a, Osmólska and Barsbold 1990, Currie and Zhao, 1993b, Holtz 1994a; Sereno [1997a, 1999a] continues to hold that they are close relatives). However, at the Florida dinosaur-bird symposium, some paleontologists noted that *Sinornithosaurus* and *Bambiraptor* combine dromaeosaur and troodont characters in a way that suggests that the two groups may be closely linked after all and that the two dinosaur groups may be sister taxa to dromaeosaurs rather than true dromaeosaurs. We have already discussed the evidence concerning whether troodonts were members of the Arctometatarsalia or close relatives of ornithomimosaurs or both. The flight-related and other avepectoran attributes of troodonts strongly suggest they were members of Avepectora, probably neoflightless ones at that. Although the last suggestion requires some degree of parallelism with other avepods, it minimizes the overall degree of parallelism required, with birds especially.

Let us now turn from the arctometatarsaled avepods to the herbivorous therizinosaurs. Perhaps no other group of dinosaurs has posed such a phylogenetic conundrum. The giant arms and hands with sabrelike claw cores approaching half a meter in length were initially thought to be those of a turtle. The first complete vertebral series and broad pelvis were assigned to a sauropod. The beaked, tiny-toothed skull and tetradactyl feet show similarities to those of prosauropods and ornithischians. In particular, the contact between metatarsal I and the ankle at first seemed to exclude the group from the Avepoda (Paul 1984c), but new fossils and reanalysis of the best skull show that these were avepods after all and that the earliest examples still had a hallux (Russell and Dong 1993b, Clark et al. 1994, Xu, Tang, and Wang 1999). In this case, the tetradactyl foot represents an exceptional reversal to the basotheropod condition. On the one hand, the fact that therizinosaurs' skull features were remarkably like those of herbivorous dinosaurs indicates strong convergence. On the other hand, the supposedly Early Jurassic therizinosaur jaw (Zhao and Xu 1998) may hint that the group originated independently of and earlier than avepectorans, although the convergence required in such a scenario is extensive.

Where do therizinosaurs fit in? The accessory maxillary fenestrae that mark averostrans are at best weakly developed in derived therizinosaurs, but because we do not yet know what these structures were like in basal therizinosaurs, the weak development may not mean much. The three-fingered hand is an avetheropod feature. The subrectangular coracoid and well-formed semilunate carpal of basal *Beipiaosaurus* are strongly avepectoran features. The lateral otic depression containing the inner ear and the retroverted pubes seen in derived therizinosaurs (and probably present in basal examples) are also avepectoran in nature. Sereno (1997a, 1999a,b) placed therizinosaurs closest to ornithomimosaurs, but as noted later in the chapter, others have suggested oviraptorosaurs were closer relatives. It is concluded here that therizinosaurs are avepectorans gone herbivorous.

### Avepectora

It is now time for us to see what we can do with the unambiguous members of Avepectora. As discussed earlier, the complex set of basal dinosaurian and derived birdlike characters in dromaeosaurs, as well as the presence of troodont-like features in basal members, leaves it uncertain as to whether this group is monophyletic, as usually thought, or paraphyletic. If it is paraphyletic, then characters we have been thinking of as dromaeosaurian may actually represent a basal dromaeo-avemorph grade from which a number of avepectoran clades emerged. Because we have good evidence that *Protarchaeopteryx*, dromaeo-avemorphs, troodonts, caudipterygians, oviraptorosaurs, and therizinosaurs were secondarily flightless, the possibility that these avepectoran avepods were closer to birds than *Archaeopteryx* was is strong. Detailed similarities in the twisted paraoccipital process, the diamond-shaped supraoccipital, the inverted V-shaped iliopubic articulation, the pubic aprons, and the overall body form make it clear that dromaeo-avemorphs, including archaeopterygiforms, were close relations. It is, however, possible that dromaeosaurs and archaeopterygiforms appear closer to one another than they really were for the reason that they retain numerous basal characters lost in other avepectorans that lie between them. In any case, both the extreme closeness of the relationship and the parallelism observed between dromaeosaurs, *Unenlagia,* and archaeopterygiforms obscure the details of their interrelationships. As previously noted, the evidence for post-urvogel status is, in some respects, weakest for dromaeosaurs because they are so close to the urvogel that they do not have a strong set of more-avian characters. In other regards dromaeosaurs possess some of the strongest neoflightless, post-urvogel features, including very large ossified sternal plates and strongly bowed third metacarpals, which indicate that they evolved from fliers more advanced than archaeopterygiforms. Ironically, their relatively unmodified urvogel-like condition may explain why dromaeosaurs retained more potentially neoflightless features than other, more modified avepectoran dinosaurs. *Unenlagia* shares with *Archaeopteryx* avian features not seen in other dromaeosaurs, including the absence of an anterior pubic boot and the presence of auxiliary dorsal ischial processes. *Unenlagia* also has a post-urvogel feature in the well-developed lateral process at the top of the ilium. *Unenlagia* retained a brevis shelf on the lower edge of ilium behind the hip socket. The shelf was a nonavian feature lost in most dromaeosaurs and basal birds, but *Sinornithosaurus* also retained it. *Unenlagia* may have been closer to modern birds than dromaeosaurs and even *Archaeopteryx* were, or it may have independently evolved at least some of its avian attributes. The absence of most of the skeleton makes it impossible for us to tell.

The situation in some regards is clearer with *Rahonavis*. Although its vertical pubis was not derived, its advanced scapula and associated ulna suggest that it was much closer to avebrevicaudan birds than any other dromaeo-avemorph, including archaeopterygiforms, was. Even if we discount forequarter elements, the derived avian characters of the vertebrae, femoral head, and fibula also place the Malagasy dino-bird above the urvogel and other sickle-clawed forms (Forster et al. 1998). Indeed, the scapula blade, radius, and ulna were more like those of modern birds than of any other avepectoran dino-bird, and we might take them to imply that rahonaviforms were closer to avebrevicaudans than were all the rest. Alternately, the less birdy forequarters of other avepectoran dinosaurs were an illusion due to reversal toward the dinosaur condition, a reversal caused by flight loss. Especially pertinent is that *Rahonavis* retained a square-tipped scapula blade, and this feature sometimes reappeared in neoflightless birds. Therefore, the evolution of the dromaeosaur scapulocoracoid from the rahonaviform arrangement poses no special problems. *Rahonavis* may have been, like *Archaeopteryx*, a flying "dromaeosaur" more advanced than archaeopterygiforms. Dromaeosaurs may even have descended from fliers whose flight performance and adaptations approached those of rahonaviforms. Yet another alternative is that rahonaviforms independently developed their fairly advanced level of flight in parallel with the avian mainstream. We await better remains of these flying dino-birds from Madagascar and elsewhere.

Considering the highly avian nature of the scapulocoracoid-sternum apparatus in troodonts, their unossified sternum and sternal ribs may represent a reversal or perhaps a juvenile condition. The birdlike braincase, posterior palate, and ankle of these gracile sickle-clawed forms suggest they lay closer to birds than the urvogel and dromaeosaurs did. But if, as noted above, some or all dromaeosaurs form a clade with troodonts, then the more-avian features of troodonts may represent parallelism with birds. The birdlike snout, palate, and occiput of caudipterygians and oviraptorosaurs suggest that they too were closer to birds, perhaps even closer than troodonts were.

Another important point is that dromaeosaurs, troodonts, caudipterygians, and oviraptorosaurs shared ornithothoracine characters not seen in archaeopterygiforms. Aside from the potentially neoflightless characters, these shared characters included short, non-overlapping cervical ribs (except in *Caudipteryx*), hypapophyses under the cervicodorsals, and sacralization of the posteriormost dorsal vertebrae. The neoflightless and avian characters observed in some or all of these groups imply that they form a clade with birds above the level of the urvogel. Also possible is that two or more of the groups form a distinct clade that is a sister group to more-derived birds. The probability that the four groups form a clade is reinforced by some special characters shared among them, most notably a short ischium in which the obturator process is large and subtriangular (this feature is also seen in less well-developed form in avimimids).

The specialized ischia of derived therizinosaurs are very different from those of the four groups mentioned in the previous paragraph, but the nature of the ischia as well as some of the other avepectoran features noted immediately above have not yet been documented in basal therizinosaurs. Most recent studies prior to the publication of *Beipiaosaurus* have allied therizinosaurs with oviraptorosaurs (Sereno [1997a, 1999a], however, allied them with ornithomimosaurs), in part because of the nature of the articulation between the posterior palate and the braincase and quadrate (Russell and Dong 1993b, Clark et al. 1994, Sues 1997, Makovicky and Sues 1998). With *Beipiaosaurus* added to the mix, Xu, Tang, and Wang (1999) also placed them closest to oviraptorosaurs; and the *Beipiaosaurus* ilium they figured is reminiscent of those of oviraptorosaurs. As we have seen, the jaws, hands, pelvis, and abbreviated tail of *Caudipteryx* are so similar to those of oviraptorosaurs that these two groups are probably very close relatives. The degree to which short tails may ally the two groups is not yet clear because we do not know whether the first oviraptorosaurs had this feature. If therizinosaurs, caudipterygians, and oviraptorosaurs form a distinct clade, then herbivory may have evolved early and only once in the group.

Regarding the last point, the many land-bound avepectorans may have been the product of numerous losses of flight while descending from a set of basal flying avepectorans. This would help explain why they share many derived characters on one hand, yet are different from one another in ways that otherwise seem inconsistent and perplexing. Perhaps *Sinornithosaurus, Microraptor, Bambiraptor, Dromaeosaurus,* velociraptorines, and various troodonts each lost flight as they descended on their own from flying ancestors whose form was broadly similar to, but in various ways more birdlike, than that of *Archaeopteryx.* Likewise, caudipterygians, oviraptorosaurs, and therizinosaurs may have shared a common herbivorous ancestor that had already evolved many of the features shared by these dinosaurs, but the three groups' divergence from one another may have occurred before flight was lost. In both of these examples, reversals associated with flight loss and other factors would have destroyed much of the data cladists normally use to restore relationships. This cladist's nightmare may well have happened, or maybe it did not.

On the one hand, it is unclear whether dromaeo-avemorphs, troodonts, or the therizinosaur-caudipterygian-oviraptorosaur group was closest to modern birds. Troodonts were the most birdlike in the braincase and ankle, the oviraptorosaurs in the snout and palate. But therizinosaurs and caudipterygians did not have such a markedly birdlike palate as oviraptorosaurs, which hints that the palate of the latter paralleled to some degree that of birds. Placing caudipterygians between dromaeosaurs and archaeopterygiforms, as Ji et al. (1998) did, is definitely questionable. The Yixian dinosaur is not nearly as similar to the urvogel as are dromaeosaurs. Among other things, Caudipteryx lacked the dromaeo-avemorph paraoccipital process and the urvogel-type pelvis. Its short tail and especially its advanced hand suggest that caudipterygians evolved from fliers markedly more sophisticated than *Archaeopteryx.* Postulating that *Caudipteryx* was a post-urvogel with advanced flying ancestors and that it underwent reversals that transformed it back into a more conventional dinosaurian is a superior hypothesis. Unknown is whether the oviraptorosaur pygostyle merely paralleled that of birds or whether the pygostyle was lost in other oviraptorosaurs and *Caudipteryx.* Why the caudipterygians' hands were much more birdlike than those of their apparently close, and seemingly more derived, oviraptorosaur relatives is perplexing (Fig. 10.11Af,j). The latter may have re-evolved a long outer finger.

On the other hand, there is evidence that archaeopterygiforms and dromaeosaurs may have been closer to birds than other dino-avepectorans were because their pelves—which had parallelogram-shaped ilia articulating with retroverted pubes via large, retroverted pubic peduncles—were most similar to those of early birds. Not enough has been published about early avebrevicaudan bird braincases and palates for us to say whether they are more similar to those of archaeopterygiforms and dromaeosaurs or to those of troodonts, oviraptorosaurs, and alvarezsaurs.

We next consider avimimids and alvarezsaurs. If one tries to explain these birdlike forms as being less avian than archaeopterygiforms, the parallelism required is so massive as to make the hypothesis inferior, although not impossible. If, however, we assume that they are closer to birds than archaeopterygiforms are, then avimimids and alvarezsaurs are so strongly avian in the skeleton and especially the skull that they should be significantly closer to birds than other dino-avepectorans. This is true even if the lack of a complete postorbital bar in alvarezsaurs and avimimids only parallels the same condition in the Ornithothoraces. As usual, a lack of transitional types leaves the relationships of these distinctive forms obscure.

Some of the avian characters attributed to *Yandangornis* by Cai and Zhao (1999) are also seen in dino-avepectorans (including large premaxilla, broad sternum with posterolateral processes, asymmetrical distal humeral condyles, retroverted pubis lacking an anterior pubic boot, reduced fourth trochanter on the femur). But other characters (reduced caudal count, completely fused tarsometatarsus) indicate that it was a post-urvogel and probably closer to avebrevicaudans than other dino-avepectorans, with the possible exception of avimimids and alvarezsaurs. Otherwise, we can say only that the interrelationships among these birdlike forms remain obscure.

## *Many Uncertainties*

In the end, our detailed knowledge of the interrelationships among many of the major dinotheropod groups remains grossly inadequate. The fact that some researchers have recently placed therizinosaurs near oviraptorosaurs whereas others have placed them near ornithomimosaurs reveals the inadequacy of our knowledge. This inadequacy is also made obvious by the fact that some consider tyrannosaurs as maniraptorans, others as relatives of allosaurs. We do not even know how close dromaeosaurs and troodonts are to each other. The fact that little attention is paid to the possibility that some dinosaurs were neoflightless is also a problem. So is our poor understanding of the small avepods of the Jurassic. To this we must add the endless alternative scenarios proposed for multiple losses of flight and the potential post-urvogel status of certain dinosaurs. For all we know, there may have been a host of small "carnosaurs" that gave rise to the giants of the Jurassic and Cretaceous. Another group of smaller, more birdlike avepods may have given rise to the avepectorans at the same time.

I have plotted a very tentative phylogenetic time chart that represents my current ideas about how predatory dinosaurs may be related to one another (Fig. 1.1). This chart is not likely to please the strict cladists in that it does not present a fully "testable" set of characters for others to reanalyze. However, I believe it more honestly reflects our current state of knowledge, including our current lack of critical information, than does a cladogram, with its single arrangement of nodes mapping out the most parsimonious analysis of the chosen set of characters.

### *Sauriurae: Myth or an Independent Group of Birds?*

The enantiornithines of the Mesozoic and the more familiar ornithurines are widely thought to have diverged from one another very early in avian evolution (Kurochkin and Molnar 1997). L. Martin (1983a,b, 1991), Hou et al. (1995, 1996), Kurochkin (1995), Zhou (1995a), Feduccia (1996), Cai and Zhao (1999), and L. Martin and Czerkas (2000) have gone so far as to contend that archaeopterygiforms and enantiornithines together form an independent clade, the Sauriurae, distinct from Ornithurae. If their view is correct, then *Archaeopteryx* was not a truly basal bird but already a fairly specialized member of a now-extinct group. Even more remarkably, almost everything we consider avian above the level of the urvogel—from cranial kinesis to the presence of a pygostyle—evolved at least twice. The massive parallelism required by this hypothesis is one of its fundamental weaknesses.

The other serious problem with the hypothesis is that most of the supposed Sauriurae characters are at best dubious (Chapters 3 and 4). Some did not exist in either *Archaeopteryx* or enantiornithines (for example, neither group had a quadrate-squamosal articulation, a direct scapula-vertebra connection, or a "stop" on the lateral femoral condyle). The status of other characters remains uncertain (the fusion of anterior dorsal vertebrae and the nature of proximal furcula articulation); a few characters are found in some ornithurine birds (robust furcula, distal contact between ilium and ischium); and others are primitive in nature (the absence of ossified uncinates, the presence of a dorsal process of the ischium). Basal enantiornithines retained a nearly dromaeosaur-urvogel-like pelvis, but so too did confuciusornithids, which according to Chiappe et al. (1999) is not the enantiornithine it was originally believed to be. Now *Apsaravis* (Norell and Clarke 2001) has further muddied the phylogenetic waters by combining factors once thought present only in ornithurines or enantiornithines, but not both.

There is little or no evidence that *Archaeopteryx* was a member of Sauriurae. This is not surprising, considering how dinosaurian the urvogel was. Indeed, nothing in the little flying dromaeosaur bars it from being the ancestral type for all birds. *Confuciusornis*—which was so primitive in form that it retained a long dorsal series, a nearly dinosaurian hand, a long fibula, and even a fifth metatarsal—is also a representative of the general avian condition at a somewhat higher (avebrevicaudan) level, although this little, toothless climber was specialized in some regards. The amount of parallelism that seems to have evolved in the enantiornithine and ornithurine clades after the Sauriurae is dissolved remains substantial, but not excessively so. Basic avian skull form and kinesis, pygostyles, well-developed supracoracoideus complexes, and other features were probably present in the common ancestors of enantiornithines and ornithurines. If so, then enantiornithines and ornithurines did diverge from each other early in the evolution of birds, but not as extremely as some have thought.

While we are on the subject of Mesozoic birds, we should note that many of the Early Cretaceous examples had the same degree of toe asymmetry seen in *Archaeopteryx* (App. Fig. 18A). This similarity leads to the question of whether this semididactyl condition was retained from sickle-clawed dromaeo-avemorphs. Feduccia (1996) suggested that *Chaoyangia* was an ornithurine, which, if true, would be interesting because *Chaoyangia* retained an urvogel-type pelvis quite different from the much more derived pelves characteristic of ornithurines. However, ossified uncinates, which are present in this avepod, are not, as Feduccia thought, limited to ornithurines. Fortunately, the detailed phylogenetics of Cretaceous-Tertiary birds, which is an incredibly complex nightmare of similarity, convergence, parallelism, and reversal, is beyond the scope of this study. For additional information on the subject, see S. Olson (1985), Chiappe (1995), Elzanowski (1995), and Feduccia (1996).

## Phylogenetic Conclusions

In accord with the methodology outlined earlier in this study, the following phylogenetic and taxonomic conclusions, alternatives, and questions can be made, ranked, and posed. The results of these analyses show that firm conclusions can be made concerning some basic matters of bird origins and the relationships of *Archaeopteryx*. At the more detailed levels, plausible alternatives are so numerous that they soon branch into a large set of possibilities that cannot be rigorously tested against one another at this time (so many that dozens of plausible cladograms of protoavian-bird interrelationships can be generated).

### *Firm Conclusions*

The evidence for the following conclusions is so strong that alternatives must be rejected unless new contrary information arises. These conclusions are therefore about as well established as the therapsid ancestry of mammals. In more general terms, they equal in probability the hypotheses that the sun is a fusion reactor and that species evolve from one another.

- Basal archosauromorphs, basal archosaurs, and basal crocodilians shared few if any avian characters that are not also observed in dinosaurs, and birds are not descended directly from or closely related to any of these groups.
- The anatomical evidence indicating that all dinosaurs and birds form a monophyletic clade is so overwhelming and so poorly explained by convergence that the dinosaur-bird link is as well demonstrated as the therapsid-mammal link.
- *Archaeopteryx* was almost entirely dino-avepodian in organization, to the point that it had very few avian characters not observed in other avepod dinosaurs, and rostral kinesis was at best poorly developed. In addition, *Archaeopteryx* was so primitive that it is a suitable ancestral type for all birds.
- *Archaeopteryx,* dromaeosaurs, and other dromaeo-avemorphs are very similar in overall form; they share many detailed characters; and they lack critical differences; therefore, they are the closest known relatives of one another.
- Avepectoran dinosaurs had avian characters not observed in *Archaeopteryx*.
- At least some dino-avepectoran dinosaurs had pectoral and caudal adaptations otherwise observed only in flying or secondarily flightless archosaurs.
- In at least some dino-avepectoran groups (dromaeosaurs, therizinosaurs), the early, basal members were more birdlike than the latter, more-derived members.
- Constructing a phylogeny of avepods that does not contain a high degree of parallelism and substantial reversals is impossible.
- Avian flight was achieved when the skull and the majority of the avepod skeleton were still largely dinosaurian in organization and before rostral kinesis had developed to more than a minimal level.
- Dinosaur ancestry of birds is compatible with both the arboreal and the terrestrial theories regarding origins of bird flight.

### *Very High Probabilities*

The evidence for the following conclusions is very strong, but not so strong that alternatives can be completely ruled out.

- Avepoda is a monophyletic clade.
- A direct ancestor-descendent relationship between dino-avepods and birds is strongly favored over a sister-group relationship.
- The best dino-avepod candidates for the ancestors of birds are avepectoran dinosaurs.
- *Archaeopteryx* was not a member of the Sauriurae.
- Alvarezsaurs were the sister group of birds above the level of dinosaurs and *Archaeopteryx*.
- Bird flight evolved among climbing dinosaurs.

### *High Probabilities*

The evidence strongly supports the following hypotheses, but, again, alternatives cannot be ruled out.

- Many or all dino-avepectorans were closer to birds than *Archaeopteryx* was. In general, most dino-avepectorans can be divided into three grades from furthest to closest to birds as follows: dromaeo-avemorphs (with at least *Protarchaeopteryx* and *Rahonavis* above *Archaeopteryx*); troodonts, caudipterygians, and oviraptorosaurs; and avimimids and alvarezsaurs.
- The reason that dino-avepectorans have essentially avian shoulder girdles, dynamic tails, and nonflight avian characters not observed in *Archaeopteryx* is that most or all were secondarily flightless.
- The presence of ossified rods on the tails of two potentially neoflightless groups indicates that early avian flight included a stage involving fairly long, pterosaur-like, stiffened tails.

### *Probabilities*

The evidence favors the following hypotheses, but alternatives cannot be ruled out.

- Feathers with central shafts did not evolve prior to the Avepectora.
- Highly complex contour feathers did not evolve prior to aerial aerodynamic use (gliding, flight, etc.)

### *Plausible Alternatives*

The available evidence neither confirms nor rules out the following hypotheses.

- If some or all dino-avepectorans formed a series of sister groups with birds above the level of *Archaeopteryx,* then the flight capabilities of *Archaeopteryx* may have been shared with other birds or may have arisen independently. If the former is true, then all those avepods in the clade above *Archaeopteryx* were secondarily flightless. If the latter is true, then some or all of the dino-avepods may not have been secondarily flightless.
- If *Archaeopteryx* was a sister taxon to birds above the level of all avepectoran dinosaurs, then the birdlike characters of the latter represented structural parallelism resulting from functional parallelism. In this scenario, dino-avepods may or may not have evolved flight independently from birds, and dino-avepectorans may or may not have been secondarily flightless.
- Either *Archaeopteryx* or some (or all) dino-avepectorans (dromaeo-avemorphs at least), or both, evolved their avian forms independently of the true bird clade, and this pseudo-avian clade was a sister group to the true bird clade, which included none or perhaps some dino-avepectorans. *Archaeopteryx* may have been either a basal sister taxon to the other members of the pseudoavian clade or at a higher level in the latter clade. In these cases, the flight capabilities of *Archaeopteryx* may have been shared with other birds or may have arisen independently; and the terrestrial dino-avepectorans may or may not have been secondarily flightless. In the pseudo-avian clade, all the birdlike characters not present in the common ancestor of both clades represented a series of structural parallelisms with birds resulting from functional convergence.
- Members of avepectoran groups that are usually considered to have descended from a common flightless ancestor—such as the dromaeosaurs *Sinornithosaurus, Microraptor, Bambiraptor,* dromaeosaurs, and velociraptorines—may instead have each lost flight on their own as they descended from a diverse set of flying avepectorans.
- In some basal avepectoran clades, flight may have been developed, lost, redeveloped, lost again, and so forth.
- Isolated Mesozoic feathers may belong either to avepod dinosaurs or to birds.

### *Improbable but Not Impossible Conclusions*

The available evidence works against the following hypotheses, but they cannot be ruled out at this time.

- Birds arose from basal dinosaurs in the Triassic. If so, *Archaeopteryx* or dino-avepectorans or both may represent long-surviving offshoots of this early radiation.
- The Olshevsky hypothesis is correct, and all dinosaurs and birds evolved from arboreal, flying ancestors.
- All of the dino-avepectoran dinosaurs descended from flightless ancestors, and any dino-avepectoran features that are similar to the flight-related features seen in birds were the result of structural convergence that occurred in spite of functional divergence.

### *Major Questions*

In some cases, the evidence raises more questions than it answers.

- How many times did flight arise in avepods?
- If flight was lost in dino-avepectorans, how many times was it lost? If it was lost once, did all terrestrial protoavians evolve from the group that lost flight, paralleling birds in the process? Alternatively, as basal flying avepods evolved toward birds, did they spin off a series of secondarily flightless groups each more like birds? If flight evolved more than once, did more than one clade produce secondarily flightless examples?
- Do the synapomorphies shared by archaeopterygiforms and dromaeosaurs mean these groups represent a distinct clade (in which case they may form one family), or is one of the two types closer to birds than the other (in which case the applicable characters were lost in the more-derived forms)? If the latter is correct, which type was closer to birds. In a simple quantitative sense, *Archaeopteryx* has more avian characters than dromaeosaurs do. However, many of these characters are minor features, and others are flight related and thus easily lost along with flight.
- Exactly what are the phylogenetic placements of all Cretaceous avepectorans relative to *Archaeopteryx* to birds?
- Regardless of their position vis-à-vis birds, what are the detailed interrelationships of avepectorans?
- What is the phylogenetic and pre- or post-flying status of ornithomimids and tyrannosaurs?

## What Is a Bird and What Is Not?

Determining what is a bird and what is not used to be fairly simple. A bird was an archosaur with feathers. Because *Archaeopteryx* was an archosaur with feathers, it was a bird. Dinosaurs lacked feathers and were not closer to birds than the urvogel; therefore, they were not birds. In this view, the only major question was where to draw the line as forms intermediate between predatory dinosaurs and *Archaeopteryx* were found.

Now that we have learned more about the evolution of birds from dinosaurs, this simple system is breaking down (much as the definition of Mammalia was initially made more obscure when mammals' relationships to therapsids were detailed). At one extreme, R. Thulborn (1975) placed all of Theropoda (which at that time was roughly equivalent to Avepoda) within Aves. At the other extreme, Gauthier (1986) excluded *Archaeopteryx* from Aves on an arbitrary cladistic basis, even though he placed *Archaeopteryx* closer to birds than he did any other dinosaur. Neither of these arrangements has been accepted, although some workers use the Gauthier system. Few wish to either dramatically alter the definition of Aves either to include a large group of classic dinosaurs or to exclude the classic urvogel. The problem with Aves is that the name was coined long ago, and it is not based on an unambiguous clade node or an apomorphy+clade combination. As a result, it is a floating name at its base and, therefore, up for taxonomic grabs. Chatterjee (1997), Padian (1998a), Padian and Chiappe (1998c), Sereno (1998), and I (Paul 1998a) have discussed these problems with Aves.

Nowadays, these problems are even more serious. The avian status of *Archaeopteryx* is challenged by its largely nonavian skull and skeleton, as well as by the possibility that some "dinosaurs" were closer to birds than the "original bird" was. The latter possibility simultaneously questions the "nonavian" status of the same avepectorans, some of which may have been secondarily flightless. For example, the birdlike alvarezsaurs are placed in Aves by Chiappe (1995) and Padian and Chiappe (1998c). If someone demonstrates that the less avian dromaeosaurs, troodonts, or oviraptorosaurs are closer to birds than *Archaeopteryx,* should we then place the dinosaurs in Aves, or should we exclude *Archaeopteryx?* What if the even less birdy therizinosaurs and even tyrannosaurs are in this category?

What if we define Aves as including dinosaurs that flew or descended from fliers? If we use this definition, all neoflightless dino-avepods are candidates for inclusion in Aves. But the situation is complicated. If some dinosaurs independently evolved flight, should Aves include them, birds, and the common ancestor of all these groups? This unsatisfactory situation can be avoided by defining Aves as including only those fliers that belong to the clade that includes modern birds. But what if *Archaeopteryx* is part of a clade that developed flight separately from modern birds?

Further complicating the definition of Aves—perhaps shaking it down to its foundations—is the newly discovered fact that small dino-avepods were feathered. This discovery is especially significant because feathers may extend all the way back to basal avepods, if not farther into basotheropods and perhaps even lower. Since birds are feathered archosaurs, should we place all feathered archosaurs in Aves even if they are early dinosaurs? If the answer is yes, then Thulborn's proposal is valid after all. Ji and Ji (1996, 1997a) placed all feathered avepods, including the new *Sinosauropteryx* specimens they were describing, in Aves. If feathers evolved at a certain level in the Avepoda, should we use that level to demarcate Aves? A problem with doing so is that determining exactly when and where feathers evolved remains an inherently difficult task that depends upon the vagaries of the fossilization process. We would first need to document the presence of fossil scales in the archosaurs immediately preceding those with preserved feathers. As it is, we cannot determine whether large sections of Theropoda or even Dinosauria belong in Aves defined as feathered dinosaurs. Similar problems apply to another possibility, which is to define Aves as pertaining to the presence of contour feathers. Such feathers may have evolved only in fliers well after simple fibrous feathers appeared, but it is possible they first appeared in earlier, nonvolant theropods. As Padian (1998a) emphasized, almost everything that has been used to define Aves is now known to be present in dinosaurs. All we have left as exclusively avian are unimpressive minor attributes, such as a fully reversed hallux. Of course, even that may show up in a dinosaur someday.

The problem of defining Aves has inevitably become complex and will be difficult to solve. In this study, I do not attempt to firmly define the group or to delineate its basal members. Instead, I offer some suggestions and cautions. Until recently, feathers were a poor criterion for defining higher fossil taxa because their preservability is so low. As we find more feathers, their usefulness may improve, but until the condition of the integument of small dinosaurs is well understood from feathers, perhaps of the complex contour type, defining Aves will continue to be a chancy business. If we are to use anatomical characters to define Aves, we should stick to readily preservable skeletal characters. Using functional features also requires care. What makes bats different from other mammals and pterosaurs different from other archosaurs is that bats and pterosaurs can fly. It has been suggested by some researchers that the acquisition of flight is the characteristic that most distinguishes birds from their ancestors, and basic flight performance should be readily determined in the fossil record via the examination of the development of the pectoral girdle and the size of the arms. Alternately, we could define Aves as including only those forms with avian cranial kinesis, including a push-pull quadrate-pterygoid articulation; such a definition would exclude *Archaeopteryx* and most dino-avepectorans but would include alvarezsaurs. We could also use the presence of pulmonary air sacs as an alternative criterion. In this case, the presence of pneumatic bones in the clade that includes birds would delineate the members of Aves. But this definition would include all avepods. Perhaps most probable is that Aves will ultimately be defined on cladistic grounds, with the common ancestor of *Archaeopteryx* and birds forming the base of this great group (if this definition is adopted, then archaeopterygiforms should be defined as limited to Aves).

In the end, we may need to designate an official body of researchers to settle the issue.

A related issue is whether birds, assuming that they descended from dinosaurs, should be called dinosaurs. Cladists say yes. Gould (2000) said no, observing that we do not call mammals polycosaurs just because they evolved from the latter. We do, however, call flying bats mammals. The anatomical and other differences between polycosaurs and mammals are great; those between bats and their mammalian insectivorous ancestors are much less. The differences between birds and the most birdlike dinosaurs are similar to the insectivore-bat example. On both cladistic and anatomical grounds, we should call birds dinosaurs.

### *If Some Dinosaurs Are Birds, Are Birds Not Dinosaurs?*

Recently, I have experienced a disquieting sense of déjà vu that reminds me of the days when some argued that dinosaurs could not be feathered because feathers are found only on birds and not on dinosaurs. As more and more feathered dinosaurs are being found, a similar tendency to base phylogenetic arguments on arbitrary taxonomy is threatening to muddle the debate over loss of flight in dinosaurs. The basic observation is that if some dinosaurs were closer to modern birds than *Archaeopteryx* was and if *Archaeopteryx* were in Aves, then the neoflightless dinosaurs would be birds. This last observation leads to two opposing views.

Proponents of the first view hold that because the dinosaurs in question were dinosaurs, they cannot be birds. Dinosaurs cannot be closer to modern birds than *Archaeopteryx* was because that would make them birds, which they are not. Those who take this view to the extreme can dismiss any evidence that some dinosaurs were closer to avebrevicaudans than the urvogel as the result of parallelism. Perhaps worrying that such an absurd line of logic could take hold seems silly, but absurd logic has already taken hold in regard to the feather issue!

Indeed, such distorted logic is already emerging at the opposite extreme of the dinosaur versus antidinosaur debate. We have previously discussed how, in this view, if some feathered dinosaurs were in Aves, then they were birds, not dinosaurs. In other words, some or all avepectoran "dinosaurs" are not really dinosaurs. Instead, they form an arboreal avian clade with birds that parallels the true, terrestrial theropod dinosaur clade. This arrangement may appeal to those who have opposed the dinosaur-bird connection but find it increasingly difficult to deny that the most birdlike dinosaurs are very close to birds, especially as feathered "dinosaurs" continue to appear. This kind of thought process is already happening in a certain sense with *Protarchaeopteryx* and *Caudipteryx,* which are being called secondarily flightless birds despite the strongly dinosaurian nature of their skeletons.

This way lies phylogenetic madness. The problems and issues that surround the naming of various groups and grades cannot be allowed to influence the restoration of taxonomic categories. First, we must gather and analyze the data contained within the various groups being considered *without being biased by preconceptions driven by traditional views of what is what.* Only after the data analysis is complete can the problem of what to call certain birdlike dinosaurs be considered. If we agree that *Archaeopteryx* is a basal member of Aves and that certain dinosaurs are even closer to modern birds than *Archaeopteryx,* then the dinosaurs are in Aves and are then birds in a broad sense. But, because nothing in the skeletons of these dinosaurs separates them from dinosaurs proper, and because their anatomical grade remains more dinosaurian than avian, they are still dinosaurs as well as birds. We find ourselves in the messy, murky gray zone between major groups; what we need is a simple, informal term that combines aspects of both types. That term is "dino-bird."

PART 6

# A MODEST SCENARIO

It is now time to put together the observations, possibilities, and conclusions we have discussed to construct a tentative scenario for the evolution of birds (a scenario that is a modification of that presented in Paul [1988a]). (The evolution of preavian respiration and energetics is detailed in Appendices 3 and 4.)

CHAPTER 13

# The Mesozoic

## The Triassic and Jurassic: Laying the Foundations

Near the end of the Paleozoic, the large fliers were dragonfly-like insects whose wings sometimes spanned over two-thirds of a meter. These big arthropods did not live to see the Mesozoic. There is no evidence that avian evolution began in the Paleozoic, except that the first primitive members of the archosaur clade are found in the Permian. Diapsids, including archosaurs, flourished in the Triassic, the predominant type being thecodonts, climbed with divergent outer digits. An distinguishing feature of archosaurs is that, for reasons that remain obscure, they evolved extensive cranial sinus complexes and intensely pneumatic skulls, which can still be seen in crocodilians and birds. It is intriguing that some Triassic archosauriforms—longisquamids, drepanosaurs, shuvosaurs, lotosaurs—seem to have converged with later avepods, including birds, in certain aspects of their morphology, especially in the head and neck, and some even had a superficially feather-like integument, but these experiments were ultimately sterile. Real bird evolution also got its start in the Middle Triassic, with the appearance of the gracile, digitigrade protodinosaurs. The running adaptations of the narrow feet imply that terrestrial rather than arboreal movements were driving the evolution of the first members of Dinosauria. One thing that is interesting about these little archosaurs was that they were so little, weighing much less than a kilogram. The dinosaur clade may have started out as small-bodied forms, but size rapidly increased. By the Late Triassic, tetradactyl baso-theropods weighing up to a few hundred kilograms had appeared. The erect-limbed, mesotarsal-ankled protodinosaurs and baso-theropods possessed a primitive proavian body form and posture, and aerobic capacity began to rise above reptilian levels in these early forms. Less discernible is whether resting metabolic rates were rising above the reptilian maximum. Herrerasaurs paralleled later avepods by developing partly foldable arms, raptorial hands, and retroverted pubes. One perplexing question is why the shoulder joint of strongly bipedal herrerasaurs faced so strongly down and back that they could not rotate the arm more sideways. The longer-fingered Triassic theropods had significant climbing abilities, and they climbed in a preavian style using inner rather than outer digits to better grasp branches.

Meanwhile, the other group of flying diapsids, the pterosaurs, appeared in full form in the later Triassic. If flying vertebrates had not evolved, it is quite possible that the skies would have again been graced by superinsects. Therefore, the advent of fliers with endoskeletons has probably prevented fliers with exoskeletons from weighing more than a few grams. The limitations of the decentralized oxygen-dispersal system of insects explains why flying examples have not exceeded a couple of hundred grams, but it is not obvious why swift and agile dragonflies never became larger than hummingbirds in the Mesozoic and Cenozoic.

In the scenario being presented here, the most critical event in the Triassic was the appearance of the birdlike avepod dinosaurs, with their nearly avian feet. Climbing ability seems to have been rather poor in these short-armed, running forms. One exaptation for climbing and flight, a laterally facing shoulder joint, probably evolved in order to allow them to swing their arms out to the side in order to better grasp at prey. Another exaptation for flight, the baso-theropod combination of limited aerobic capacity and long, erect, digitigrade legs, was unstable because the former prevented full exploitation of the locomotory potential of the latter. It is therefore not surprising that long ilia and the presence of pulmonary air sacs, as well as long-stride trackways, suggest that aerobic exercise capacity continued to rise. Resting metabolic rates have not been determined clearly, in part because no one has yet established whether

*Figure 13.1. Euparkeria.* It was from creatures like this small, Early Triassic African flesh-eater that archosaurs, including dinosaurs and birds, evolved.

*Figure 13.2. Marasuchus.* From small things, big things develop. Diminutive, gracile-limbed, semibipedal *Marasuchus* represents the Middle Triassic protodinosaurs from which dinosaurs and birds great and small evolved.

simple feather insulation appeared at this time; no early avepod has yet been found in fine-grained sediments, and the hairlike early feathers would make isolated feathers as difficult to recognize as mammal fur. It is notable that although mammals evolved before the end of the Triassic, no Mesozoic fur has yet been found. Nor is it yet certain whether the insulation of birds converged, or was shared, with that of pterosaurs, although convergence seems more likely. A major meteoritic impact in eastern Canada failed to terminate the avepod clade as the Triassic came to a close (see Chapter 14).

The following portion of the scenario rests upon two important facts: the first is the lack of solid evidence for birds in the Triassic and most of the Jurassic, and the second is the blossoming of fossil evidence for birds at the end of the Jurassic and during the Cretaceous. Bochenski (1999) speculated that enantiornithines evolved before *Archaeopteryx,* but his scenario stems from inferences based on the supposed flight range of basal enantiornithines relative to mid-Mesozoic biogeography and the presumption that *"Protoavis"* was a bird. To date, not a single, articulated, clearly avian skeleton, skull, or, for that matter, tooth has been found in sediments prior to the Late Jurassic. Nor has a single classic contour feather been found from those times. Also absent are the high-density footprint sites that large flocks flying birds tend to create. The temptation is to dismiss all this as negative evidence. Bird skeletons are small and delicate, and their large wing feathers are even more so. If only we looked harder, we might find bird fossils in the early Mesozoic. The problem with this argument is that people have been looking hard for, and successfully finding, small flying

*Figure 13.3. Staurikosaurus.* Late Triassic bird relatives included tetradactyl-footed basal theropods such as South American *Staurikosaurus*. Its long legs promoted high walking speeds, which the still-slender thigh muscles were hard pressed to sustain.

*Figure 13.4. Syntarsus.* In the Late Triassic and Early Jurassic, the first tridactyl-footed avepods appeared. Among the best known is the southern African coelophysoid *Syntarsus*. These avepods had the large leg muscles needed to fully exploit the sustained speed their long, bird legs were capable of generating. When this lavishly feathered portrait appeared in 1988, it was derided as wild speculation. Although today it is subject to less derision, these early avepods probably did not have such a flamboyant, well-developed plumage.

creatures in Jurassic and Late Triassic sediments since the early 1800s. Numerous fine-grained lagerstätten beds formed at the bottom of lagoons and lakes have produced a diverse array of pterosaurs and flying insects (Barthel et al. 1990, Wellnhofer 1991, Labandeira and Sepkoski 1993 and refs. therein, Fraser et al. 1996). Hundreds of pterosaur specimens have been found so far, and the flying invertebrates are even more numerous. The lack of avian remains and traces from these sediments cannot be ascribed to a sampling bias against small nonmarine birds, because the insects were terrestrial, and similar aquatic beds in the Cretaceous contain large numbers of apparently terrestrial bird skeletons, some contour feathers, and multitudes of footprints (see below in this chapter). Although the richness of the Early Cretaceous avifauna was not well known until very recently, bones were discovered in the nineteenth century (Seeley 1876), and feathers have been known for nearly half a century (Feduccia 1996). Besides, the diversity of the Early Cretaceous pterosaur fauna also was poorly understood until recent discoveries (Wellnhofer 1991). The Early Cretaceous aerial fauna was poorly known because it was not as well sampled as that of the Jurassic.

The continuing lack of unambiguous avian remains and traces from well-studied fine-grained sediments laid down before the advent of archaeopterygiforms and, for that matter, the fact that archaic (as far as birds go) archaeopterygiforms are themselves known only from times immediately before and after the J/K boundary suggest that earlier flying birds were either very rare or absent. The latter option seems preferable because it is difficult to understand why birds that had evolved the powerful advantage of flight would have remained so limited in form, type, and distribution for 70 million years if they first evolved in the Triassic. The unearthing of unambiguous bird remains from the Triassic and Jurassic would falsify this hypothesis, but until such remains are found, the notion that there was a radiation of Triassic and Jurassic birds is speculative. Note that the best evidence for such early birds would be skeletal remains, and asymmetrical pennaceous feathers would also constitute evidence of flying avepods. Symmetrical contour feathers would be less definitive because they may have evolved first in dino-avepods. As discussed in Chapter 11, birdlike footprints are not convincing evidence for Jurassic birds, whether they be little or as big as emus. Such footprints do suggest that some dino-avepods were becoming markedly birdlike before the middle of the Mesozoic. However, unambiguous avepectoran dinosaurs in the Early Jurassic or earlier, whether they be therizinosaurs or something else, have not yet shown up; if such early avepectorans are ever found, they will upset the scenario presented here to at least some degree.

The increasingly birdlike Averostra and Avetheropoda had evolved by the middle of the Jurassic. These creatures had an expanded air sac complex and possible unidirectional airflow through at least part of the lungs. Aerobic capacity rose to or very near to the avian minimum; therefore, most of the avian pulmonary complex (see Appendix 3B) must first have evolved for terrestrial rather than aerial purposes (Perry 1983, 1989, 1992, Carrier 1987, Paul 1988a, Britt 1997, Bramble and Jenkins 1998). The same appears to be true of feathers, which had evolved in simple form by this stage, if not earlier. In any case, the discovery of early feathers is allowing scientists to replace speculation (Rautian 1978, Regal 1985, Paul 1988a, Brush 1993, 1996, 2000, Bock and Buhler 1995, Feduccia 1996) with real data (Ji and Ji 1996, 1997a,b, Chen et al. 1998, Ji et al. 1998, Padian and Chiappe 1998a,c, Xu, Tang, and Wang 1999, Xu, Wang, and Wu 1999, Maderson and Alibardi 2000, Tarsitano et al. 2000). Effectively falsified are suggestions that feathers started as complex flight-related structures (contra Rautian 1978, Regal 1985, Feduccia 1996, Maderson and Alibardi 2000, Tarsitano et al. 2000). That feathers are present on the likes of *Protarchaeopteryx, Caudipteryx, Sinornithosaurus, Microraptor,* and *Beipiaosaurus* does not tell us much about feather origins, because those nearly avian dinosaurs may have been secondarily flightless. More informative are the simple feathers of *Sinosauropteryx,* because it was an only moderately advanced avepod dinosaur that appears not to have possessed, or evolved from ancestors with, adaptations for climbing or flight. The implication is that feathers first evolved for a purpose or purposes unrelated to gliding or flight, but just what other purposes is unclear. Feathers may have initially evolved as facial tactile sensors (similar to kiwi whiskers) or as display structures (Quinn 1997). If the first feathers were simple bristles, they were unlikely to have evolved to serve as insulators: initially they would have been too short or sparse or both to have formed an effective body covering. If feathers first evolved as frayed edges of broad scales—which is improbable—then they could have served a thermoregulatory function from the start. In any case, that the feathers adorning small avetheropods were short, simple, hollow structures distributed uniformly over most of the body supports the hypothesis that a primary preflight function was insulation (Chen et al. 1998). If so, then they should have boosted resting as well as active metabolism. Chatterjee (1999b) suggested that the cool temperatures among the treetops had something to do with the presence of feathers on small dino-avepods, but there is no correlation between the presence or absence of insulation in climbers on the one hand and nonclimbers on the other. Bock and Buhler's (1995) suggestion that feathers and high metabolic rates evolved so that eggs could be incubated in tree nests is more logical, but insulation and endothermy had probably evolved well before protobirds nested in trees (Appendix 4).

The last point brings us to the probability that the absence of medium-sized arboreal nondinosaurian tetrapods left a major niche opening for dinosaurs that could climb up against the planetary gravity well. In particular, the elongation of the arms improved the climbing abilities of small avetheropod dinosaurs. Brachial elongation may have originally occurred in order to improve prey capture or arboreal capabilities or both. Initial adaptations for arboreality—and exaptations for flight—included narrowing of the scapula blade, development of a furcula, elongation of the hand, and a decrease in the finger count to three. The tendency to use divergent inner rather than outer digits, including a hallux, to better grasp branches while climbing was further emphasized in climbing avepods. Climbing may have appeared in small adults or in the juveniles of large species or in both. If it appeared in small adults, the giant avepods of the Jurassic and later may have evolved from these small nonavepectoran climbers (a shift from arboreal to more terrestrial habits has occurred in some chameleons and in primates, including baboons and humans); the idea that climbing started in small creatures is contrary to Sereno's suggestion (1999a) that birds ultimately evolved from multitonne basal averostrans. If arboreality evolved among the young of giants, then the development of small adult climbers may have occurred via neoteny (see R. Thulborn 1985, J. Long and McNamara 1997), whereby animals attain sexual maturity at younger ages and smaller sizes, which results in a dwarfing effect. Eventually adulthood was reached at under the 100-kilogram size limit for climbers, which left the dinosaurs semi-arboreal for their entire life cycle. Whatever happened, climbing may have had a surprising influence upon the design of even gigantic predatory dinosaurs. The increased oxygen demands imposed by the hard work of quickly climbing up against gravity may have promoted at least part of the increase in respiratory efficiency seen in avetheropod dinosaurs. The need to keep distal toe bones long enough for juveniles to grasp branches may even explain the small mystery of why no avepod dinosaur evolved the severely abbreviated distal toes common in flightless cursorial birds.

Some small dino-avetheropods became even smaller as they became equally terrestrial and arboreal, as did small felids. Additional neoteny may have again been involved in this shrinkage. Neoteny may also be responsible for the retention of conical, unserrated teeth (often found in juveniles) into adulthood. The forelimbs, hands, and hallux were further enlarged and strengthened in order to facilitate bipedal and quadrupedal climbing. Juvenile hoatzins make a good model for this stage, as do tree kangaroos, whose arms have progressively elongated as their arboreal capacity has increased (Grzimek 1990, Nowak 1999). Avepod foot asymmetry may have developed when the second toe became hyperextendible as an adaptation for climbing vertical tree trunks (note that in this view such toes were not vestigial in urvogels, as suggested by Forster et al. [1998], but incipient). At the same time, the ilium was shortened because the hindlimb musculature was reduced, as is often true in arboreal tetrapods. Interbranch leaping and pouncing upon prey may have been improved as the tail began its transformation into a slender and lightweight body-orientation tool. This transformation had the secondary effect of forcing initial, weak retroversion of the pubis, which shifted the mass of the abdomen caudally. Large brains and well-developed forward vision may have begun to develop in climbing dino-avepods as adaptations for life in a complex three-dimensional environment; this would have been a convergence with primates. These diminutive scansorial dinoproavians remain largely undiscovered because of the low preservability of scarce, small nonfliers, which rarely drop into the water (for example, terrestrial birds outnumber small dino-avepods hundreds to one in the Jehol lagerstätten). However, the very small dromaeosaur-type teeth and isolated bones known from the last half of the Jurassic may record the presence of these pygmy semi-arboreal dino-birds. Indeed, the long-armed, pigeon-toed dromaeosaur *Sinornithosaurus* best represents this type of small climbing predatory dinosaur, despite its later geological age, its rather larger size, and its potentially neoflightless condition.

With arm length approaching leg length and with improved agility and information-gathering and -processing abilities, these small and aerobically capable predatory climbers were exaptated for flight. Flight developed as a result of selective pressures to further improve travel between high places. For one thing, the inability of dino-proavians to easily climb down trunks put selective pressure on them to evolve a faster way of descending. Another requirement for improved travel was longer and better-controlled interbranch leaps, both horizontal and descending. In order to better control body orientation during branch landings after long leaps or falls, small dermal airfoils developed distally on the forelimb. These were in the form of contour feathers with central support, probably developed from brachial fibrous feathers. As these aerodynamic control surfaces enlarged, they became able to produce significant lift that could extend the length of interbranch leaps beyond that possible via ballistics alone. These longer subflights may have been gravity-powered glides both horizontal and descending, or they may have been weakly powered by forelimb flapping. The latter would have created selective pressure to increase the muscle power of the forelimbs and to provide the oxygen needed by the muscles. Coracoids began to elongate and rotate toward the avian position (which repositioned the scapula more horizontally). The change in coracoid position improved the flapping action of the ventroproximal muscles and changed the articulation of the coracoid with the expanding sternum so that a hinge joint was formed between the two bones. The resulting bellows arrangement helped ventilate the air sac complex during flight. If lung airflow was not already strongly unidirectional, it should have became so during this stage. Aerobic exercise capacity of these fliers approached or equaled the lowest levels observed in modern flying birds. The increased size of the still poorly ossified sternum also improved the base for growing wing muscles. The coracoid tuber enlarged and migrated dorsally to form an incipient acrocoracoid to allow the operation of early-stage wing-elevating and -controlling muscles. The pectoral crest and furcula enlarged to anchor expanded wing-depressing muscles. At the same time, the area and aerodynamic sophistication of the dermal airfoils increased, which promoted the evolution of a wing-folding mechanism for protecting the delicate structure. An evolutionary feedback loop developed in these dino-birds and quickly expanded the airfoil dimensions and its power system to the level seen in Late Jurassic *Archaeopteryx*. The Avepectora had evolved.

The avepectoran *Archaeopteryx* was probably a crude but effective powered flier able to sustain at least level flight and perhaps able to take off from level ground with a leap or running start, climb slowly, and land in branches quadrupedally. The dominant limbs for locomotion were now the forward pair, which were used for both flying and climbing and were stronger than the hindlimbs. The still long tail and deep-keeled body provided stability in level flight (in a manner similar to long tailed pterosaurs), and the tail acted as a landing brake when erected vertically. *Archaeopteryx* shows that flight evolved when the skull and skeleton were still essentially dinosaurian in organization and function; in particular, avian kinesis had not yet evolved. If Garner et al. (1999) are correct that the urvogel lacked inner wing feathers, then it is possible that complete wings had not evolved by Solnhofen times. The absence of any skeletons of birds substantially more advanced than *Archaeopteryx* from the Solnhofen lithographic slates that have produced hundreds of pterosaurs, numerous flying insects, and a number of urvogel specimens (Wellnhofer 1991) indicates that such advanced birds were either absent at that time, very rare, or restricted in habitat, perhaps to continental areas. Of these three alternatives, the first appears superior. Bird skeletons are also absent in the Karabastau beds of Kazakhstan, another Late Jurassic lagerstätten that has produced numerous well-preserved insects and pterosaurs. The only possible trace of a bird is the very ambiguous *Praeornis* "feather." Because these are continental lake deposits and because some of the pterosaurs are insectivores rather than semi-aquatic, the absence of birds is especially informative. A Late Cretaceous bird fossil has been claimed to be present in North Korea, but the date has not been verified (see Padian and Chiappe 1998c). The earliest possible archaeopterygiform remains are teeth from about 10 Myr before the end of the Jurassic (Schudack 1993); the flight abilities of these avepods are not certain. It is improbable that *Archaeopteryx* was the sole bird, or perhaps more correctly flying dino-bird, alive circa 150 Myr ago, but the presence of only one such creature in such a flying creature–rich deposit suggests that there were few other birds extant at that time. This, in turn, implies that birds had just got their flying start in the last stages of the Jurassic (see immediately below). Despite the possibility of an extremely large meteoritic impact near or at the J/K boundary (see Chapter 14), there is no evidence of a major extinction of avepectorans at the end of the period. Indeed, there is increasing evidence that the J/K

*Figure 13.5. Rhamphorhynchus.* Long-tailed pterosaurs ruled the skies during the Late Triassic and Jurassic. Late Jurassic *Rhamphorhynchus* is a well-known example of this sort of small, stable-flight pterosaur.

transition was a gradual affair that began well before the boundary and lasted long after (Manabe et al. 2000). Few rhamphorhynchoid pterosaurs survived past the Jurassic (Barthel et al. 1990, Wellnhofer 1991). Whether this was because of competition from the new, more aerially agile pterodactyloids or for other reasons is not known. There is no evidence that competition from the few early birds of the time was a problem.

## The Jehol Group's Yixian Formation: Of Uncertain Age

The lower and therefore older part of the Yixian Formation (controversially labeled by Ji, Chiappe, and Ji [1999] the Chaomidianzi) and its avian fauna deserve a separate section for two reasons. First, the age of this formation has been a subject of vigorous debate, with various researchers arguing that it either belongs at the very end of the Jurassic or somewhere in the Early Cretaceous, a range that covers 145 to 110 Myr ago (Chen et al. 1998, L. Martin et al. 1998, J. Smith 1998, Wang 1998, Luo 1999, Swisher et al. 1999, Barrett 2000). On the one hand, the presence of psittacosaurs—small, herbivorous bipeds related to the big, horned dinosaurs—seems to imply a relatively late age. On the other hand, the compsognathid *Sinosauropteryx,* the archaeopterygiform *Protarchaeopteryx,* the archaic nature of the confuciusornithids *Confuciusornis* and *Changchengornis,* archaic tritylodont mammals, and even a long-tailed pterosaur of the rhamphorhynchoid type otherwise limited to the Jurassic suggest the time of deposition may not have been long after the Solnhofen deposit formed. However, the best radiometric dating (Swisher et al. 1999) has produced an age of 125 Myr, 20 Myr years after the end of the Jurassic. This date is in accord with the presence of the possible ornithurine birds *Liaoningornis* and "Archaeoraptor." Luo (1999) suggested that the lower Yixian was a Cretaceous refugium in which a Jurassic-like fauna survived well past its time. Manabe et al. (2000) countered that eastern Asia as a whole appears to have remained something of a "Jurassic Park." In either case, we can conclude that the lower Yixian bird fauna is representative of that which was common fairly soon after the J/K transition.

In the Yixian, bird skeletons—almost always still clothed with feathers—are abundant in the extreme, hundreds of times more so than are the pterosaurs. The abundance of these skeletons indicates that birds swiftly became a major component of the global fauna not long after their initial appearance in the fossil record. The sheer number of *Confuciusornis* skeletons (a thousand or so skeletons excavated and distributed so far) implies that the avian phenomenon of mass flocks had come to pass (L. Martin et al. 1998). But equally important is that the many birds that have come out of the lower Yixian are neither diverse nor greatly advanced. Only one example of the archaeopterygiform *Protarchaeopteryx,* which was only a little more derived than *Archaeopteryx,* has been found. Only three skeletons of enantiornithines (*Protopteryx*) and ornithurines, and archaic ones at that, have been described. The rest of the birds have been toothless yet primitive confuciusornithids. Of these almost all are *Confuciusornis; Changchengornis* (Ji, Chiappe, and Ji 1999) is known from only one skeleton, although it is possible that some specimens have been mistaken for its close relative. The posterior skull of at least one of the two confuciusornithids is more archaic than that of *Archaeopteryx,* and the rest of their skeletons are the most dinosaur-like outside those of archaeopterygiforms, rahonavids, and yandangithiforms. That only one kind of confuciusornithid dominates the lower Yixian, which has produced only one primitive ornithurine, means that birds had still not gone that far in terms of either diversity or anatomy, which, if true, has important implications for the timing of avian evolution discussed later in the chapter. The lack of strong competition from more advanced birds may explain why *Confuciusornis* was so abundant.

## The Cretaceous: The Age of Archaic Birds

The Cretaceous can also be called the "first age of the Avebrevicauda" in that nearly all flying birds had an abbreviated bony tail. It is in Early Cretaceous lagerstätten that birds begin to leave their mark in a big way, via both skeletal remains and isolated feathers that they dropped as they winged their way over Mesozoic waters.

Birds advanced rapidly, as indicated by the bones of *Ambiortus, Boluochia, Cathayornis, Chaoyangia, Concornis, Enaliornis, Eoalulavis, Eoenantiornis, Eurolimnornis, Gansus, Horezmavis, Iberomesornis, Liaoxiornis, Nanantius, Noguerornis, Otogornis, Palaeocursornis,* and *Sinornis* (except for one, these names were not coined before the seventies), as well as by an increasing number of isolated feathers progressing into the first half of the Cretaceous. Most of these birds have been classified as enantiornithines, "opposite birds," but in 2001, Norell and Clarke questioned some of these identifications. A few ornithurines have been more positively identified (of which some may have been carinates and even basal palaeognathes and neornithines depending how the incomplete remains are identified). The existence in the later Cretaceous of long-tailed rahonaviforms and yandangithiforms means that nonavebrevicaudans of some sort must still have been out and about through the entire Early Cretaceous. An archaeopterygiform and a ratite have been reported from the earliest Cretaceous of eastern Europe on the basis of isolated bones; of these, the urvogel identification is far more plausible, but neither claim can been verified (Chiappe 1995, Feduccia 1996). Fara and Benton (2000) convincingly argued that the Early Cretaceous fossil bird record is good enough to refute claims that some modern bird groups evolved this early in Earth's history.

Mass flocking is verified by the very high density—as many as one hundred to three hundred tracks per square meter—trackways seen in some bird ichnofaunas from this time (Lockley et al. 1992). The trackways represent shorebirds wan-

*Figure 13.6.* Two Chinese early birds. Despite its lack of teeth, the most primitive short-tailed bird yet discovered is *Confuciusornis (top);* the feathers are based upon the countless preserved specimens that include them. One of the new "opposite birds" of the Early Cretaceous is the much smaller toothed enantiornithine *Cathayornis (bottom).* Known from a nearly complete skull and skeleton, the plumage is restored. Not drawn to same scale.

dering about and feeding on ancient mudflats sediments, but most Early Cretaceous bones were those of land dwellers. Fully marine flying birds are at best comparatively rare; the oceanic skies were still the domain of the pterosaurs. With avian diversification in form and habits came diversification in size, some of the volant early birds being quite small (Kurochkin and Molnar 1997). None, though, were very large; it was the short-tailed pterodactyloids that were starting to become big. The persistent scarcity of strongly terrestrial pterosaurs throughout their existence—there were some insectivorous examples—implies that few ever became particularly successful at living away from the water. A less agile gait may have hindered the quadrupedal pterodactyloids from being as adept on the ground and in the trees as bipedal birds (S. Bennett 1997).

It is worth pausing to review the pattern of early avian evolution so far recorded from the fossil record. Some researchers active in early bird research have contended that the diversity and advanced nature of early Cretaceous birds indicates a long avian history extending well back into the Jurassic and even the Triassic (Feduccia 1996, Hou et al. 1996, Chatterjee 1997). If this were true, we would expect to find a diverse array of avians in a fauna representative of the mid–Early Cretaceous. However, no unambiguous bird skeletons or contour feathers are known prior to the pterosaur- and insect-rich Solnhofen of 150 Myr ago, which contains one still essentially dinosaurian bird. No more than 25 Myr later than the Solnhofen deposits, the Yixian deposits contain numerous specimens of a few kinds of birds, and the forms of these are only moderately advanced beyond the level of the urvogel. Other Early Cretaceous sediments are the oldest that contain a fairly wide variety of somewhat more advanced—but in some ways still archaic—bird skeletons, as well as a substantial number of shed contour feathers. This progressive and rapid appearance of increasingly sophisticated and diverse bird fossils in the Early Cretaceous is to be expected if birds appeared and began to diversify in the middle of the Mesozoic rather than in the Triassic. This pattern has all the appearance of a classic evolutionary "breakout," in which the selective advantages of flight were so strong that an evolutionary feedback loop continued to improve the flight apparatus at a dramatic pace, and flying avepods rapidly radiated into the many niches made accessible by the development and improvement of aerial travel. This rapid pace of evolution is similar to that observed in the Cambrian Explosion; the initial vertebrate conquest of land; the fast-paced post-K/T radiation of mammals (including the development of fully marine whales in only 10 million years, a transformation of form that was, if anything, greater than that seen in the transformation of already birdlike dinosaurs into birds) and Cenozoic birds; the rapid development of bipedalism, brain expansion, and culture in humans; and most especially the apparently rapid achievement of flight in pterosaurs and bats (Carroll 1988, 1996, Kaufman and Johnsen 1991, Feduccia 1995, Paul and Cox 1996).

In the scenario presented here, the avian flight apparatus improved rapidly in the Early Cretaceous. For improved streamlining, uniformly distributed fibrous feathers were replaced by body contour feathers arranged in tracts. These formed a streamlined aeroshell around the irregular contours of the muscles and bones.

Even more fundamental changes were underway. Except in *Rahonavis*-like forms, tails were shorter, to the point that they were reduced to a pygostyle, albeit somewhat larger than that in modern birds. This tail reduction marked the shift from the stable flight associated with long bony tails to more dynamic unstable flight. Interestingly, pterosaurs had made the same shift only a little earlier, in the Late Jurassic, but they made the shift some 70 Myr after they first appear in the known fossil record. The birds took only a third of that time, or less, to do the same thing. It is not surprising that rapid evolution caused birds, and for that matter bats, to switch to dynamic flight so quickly. The unanswered question, then, is why pterosaurs took so long. As for birds, without a long tail to balance the body, the pubes retroverted more strongly in order to shift the abdomen more posteriorly and to keep the balance in trim. Increased retroversion of the pubes also reduced the ventral keel of the body, further reducing stability in favor of maneuverability. The change from quadrupedal to bipedal branch landings was slow because the manus of many Early Cretaceous birds retained clawed fingers, including a grasping thumb. However, at least some of these clawed thumbs bore a well-developed alular feather as well as the claw (Sanz et al. 1996, Zhang and Zhou 2000). The appearance of the alula records the transition of the divergent thumb from a manipulative organ into a leading-edge slot (not found in pterosaurs or bats) that helped improve control at low speeds. In some of these archaic birds, the reduction of the central claw meant that the main finger had shifted from a grasping appendage to a wholly aerodynamic one. In a few Early Cretaceous birds, the hand was reduced and fused into a rigid, flattened, clawless unit, making the outer wing into a purely aerodynamic structure without any grasping function. Climbing, therefore, was accomplished with two legs. The improved level of flight control both negated the need for, and prevented, quadrupedal landings in favor of better-controlled bipedal touchdowns. Bipedal touchdowns were also enhanced by a much better developed supracoracoideus complex wing control surface, anchored upon a shallow-keeled sternum. The pectoralis also expanded with enlargement of the sternum, although very

*Figure 13.7. (opposite) Pterodaustro (top)* and *Anhanguera (bottom)*. In the Late Jurassic and especially the Cretaceous, the pterodactyloid pterosaurs lost most of their tail and became more adept fliers. At first, these improvements led to a new diversity of types, among them these two South American examples. Peculiar *Pterodaustro* used its flamingo-like strainer beak to filter tiny organisms from the water. Other pterodactyloids became large; the soaring wings of the *Anhanguera* spanned 4 meters. In the long run, the new level of sophistication did not save pterosaurs from decline and extinction by the end of the Mesozoic, perhaps at the hands of avian competition. Not drawn to same scale.

large pectoral crests suggest the mass of this muscle often remained more distally placed in these mesosterned birds than in more modern longosterned examples. As with the coracoids of the latter, the coracoids of Early Cretaceous flying birds were narrow and strutlike, and the scapula blade was pointed. The rib cage was strengthened via ossified uncinate processes, but this feature has yet to be found among some basal avebrevicaudans or in any enantiornithine.

Enlargement of the sternum and expansion of pneumatic spaces into the limb elements suggest that the air sac complex was somewhat enlarged in basal birds beyond the level seen in archaeopterygiforms, especially the abdominal sacs. Aside from a boost in maximum aerobic flight capacity, there should not have been a major alteration in power generation and thermoregulation at this time. Because power production in flight seems to have still been modest by later standards, the range of flights may have been limited. Birds may not have undertaken long migrations; such great journeys may still have been the specialty of pterosaurs and insects, whose flight capacity was fully developed.

If nonavebrevicaudan birds lacked proximal wing feathers, the inner wing was filled in to form complete wings in short-tailed Early Cretaceous birds, with a corresponding boost in flight performance.

Various workers have noted that the development of the flight apparatus in the arm and tail of Early Cretaceous birds was "ahead" of that in the skull, urvogel-type pelvis, and hindlimbs, which remained not all that different from those of *Archaeopteryx*. This is not surprising. The selective pressures to upgrade the systems that got and kept early birds in the air should have been much greater than those that guided the evolution of the secondary locomotory system and feeding apparatus. What is surprising is that the ilium continued to be shorter than normal in avepods; this shortness was presumably an adaptation for climbing. The long-bodied, short-necked, long-fingered Yixian confuciusornithids seem to have been adapted for climbing in a somewhat mammal-like manner. Why confuciusornithids and such basal enantiornithines as *Protopteryx* had such extremely short tail fans is perplexing in two ways. First, it seems aerodynamically disadvantageous. Second, the presence of such short tail feathers in the two groups suggests it was widespread among early birds, which is most odd. It is interesting that early birds retained the asymmetrical, semididactyl foot earlier evolved in sickle-clawed dinosaurs. Mental powers may have risen modestly in Early Cretaceous avians. The postorbital bar became incomplete, and the basic avian cranial kinetic system evolved in the Early Cretaceous. The evolution of tooth loss in Cretaceous birds was very inconsistent. Some had a significant number of teeth, others a few, some none. The overall development of flight and other features in Cretaceous birds exhibited the inconsistency characteristic of mosaic evolution. Yixian *Confuciusornis* retained a nearly dinosaurian temporal region and supple, clawed fingers, but had toothless jaws. Early Cretaceous *Concornis* combined an advanced, clawless aerodynamic hand with a relatively small sternal plate and keel. Late Cretaceous *Ichthyornis* combined a near modern sternal and wing form with toothed jaws.

In the Late Cretaceous, the ornithurine birds we are more familiar with became more common, but most of the remains found have been placed among more-archaic groups. Long known have been *Hesperornis, Ichthyornis,* and *Apatornis,* and newly named birds of the age include *Alexornis, Apsaravis, Avisaurus, Baptornis, Canadaga, Coniornis, Enantiornis, Gargantuavis, Gobipteryx, Gobipipus, Gurilynia, Judinornis, Kizylkumavis, Lectavis, Neogaeornis, Neuquenornis, Parahesperornis, Parascaniornis, Patagopteryx, Potamornis, Sazavis, Soroavisaurus, Vorona, Yandangornis, Yungavolucris,* and *Zhyraornis*. To these can be added the volant dino-bird *Rahonavis,* which bucked the general trend in flying birds by retaining a dinosaurian subvertical pubis and a long bony tail. The once dominant basal birds were giving way to carinates of various sorts with their deep-keeled sterna. Norell and Clarke (2001) have questioned whether as many of these birds are enantiornithines as previously thought. Of the above birds, only the flightless marine divers, in particular *Hesperornis,* have yet been described from complete skulls and skeletons. Therefore, beware of skeletal restorations of famous *Ichthyornis,* which usually include many elements of *Apatornis* (which, according to Hope [1998] is a duck relative [compare Fig. 10.9L and O]).

This brings us to the issue of whether modern birds got their big start in the Cretaceous or later, in the Cenozoic. Before enantiornithines were recognized, most bird remains that could not be assigned to toothed birds were identified, often incorrectly, as early representatives of one or another group of modern birds. Nowadays, many Cretaceous remains are likewise being tossed into the enantiornithines. Feduccia (1995, 1996) has stressed his opinion that the fossil record shows that few neornithines had evolved by the end of the Cretaceous, an opinion at least partly supported by Bleiweiss (1998), Marshall (1999), and Tuinen et al. (2000). Feduccia considered *Apatornis* to have been an ichthyornithid, for example. Feduccia's extreme view has been challenged by Norell and Chiappe (1996). Another extreme alternative, the notion that most modern bird groups first evolved in the Cretaceous, is based in part on molecular evidence (Hedges et al. 1996, Cooper and Perry 1997). Hope (1998) has also cited fossil evidence that birds distantly related to ducks, pelicans, and loons, as well as early shorebirds, were becoming the most abundant avians—at least in habitats near the shore—as the Cretaceous came to a close. Stidham (1998, 1999) identified a lower jaw tip as belonging to a Mesozoic parrot, but Dyke and Mayr (1999) disputed the identification. Someday, somewhere, bird-filled lagerstätten will clarify the matter. It is already apparent that Late Cretaceous avians inhabited a wide variety of habitats, from fully terrestrial to fully marine. It is interesting that the ornithurines appear to have been strongly linked with water as shore birds (their trackways, some webbed, were abundant [Lockley et al. 1992]) or marine birds (the long-jawed, ternlike ichthyornithids were especially widespread [Feduccia 1996]), and enantiornithines supposedly included both land- and water-loving examples. However, Hope (1998) and Tuinen et al. (2000) ascribe the seeming lack of aquatic neornithines to a biased fossil record, and parrots are not especially fond of water.

*Figure 13.8.* The Late Cretaceous Niobrara seaway of Kansas. As the Cretaceous progressed, birds did something no other dinosaurs ever did: they invaded the oceans. Some were sophisticated forms like tern-sized *Ichthyornis (top),* which used its high-capacity respiratory system and flight musculature to fly far over the waters. Others, such as 1.2-meters-long *Hesperornis (bottom)*, followed the persistent avian strategy of losing flight and became highly aquatic divers that returned to land only to reproduce.

Most Late Cretaceous flying birds were not large, but the wings of *Enantiornis* spread a respectable 1 meter or larger. Large footprints—if they were made by a bird rather than by a bird-footed dinosaur—suggest that some birds were as long legged and big as cranes (Fig. 10.19). However, it was the marine, and especially the continental, pterosaurs that won the aerial size race by growing as large and heavy as light aircraft. Another interesting aspect of big pterosaurs was that they were prone to developing very big heads adorned by spectacular crests of sometimes incredible dimensions. The largest pterosaur heads may have been 2 meters long not including the crest, three or four times longer than that of the largest bird. Why small-brained pterosaurs evolved such great heads is not understood, and neither is the purpose of their supercrests.

As far as we know, it was in the Late Cretaceous that birds reached an essentially modern level of organization in terms of their flight apparatus. By that time, pygostyles had been further shortened, and the coracoid was even more strutlike. Most obviously, the sternum was very large and deeply keeled, which indicates that the final expansion of the musculature had occurred. However, the retention of a very large pectoral crest in *Ichthyornis* suggests that the proximal migration of the pectoralis was not complete. Distally, hand reduction and fusion were now the norm. Ossified uncinate processes both strengthened the rib cage and improved its ventilation capacity in ornithurines, but enantiornithines never got around to developing these processes. The hindquarters caught up to the forequarters as the urvogel-type pelvis was abandoned in favor of a longer ilium better able to anchor enlarged, pelvis-based hindlimb retractors. The latter made up for the final reduction of the tail-based hindlimb retractors. The final abbreviation of the tail also required a last posterior shift of the abdomen. This shift was accomplished via a lateral split in the pubes and even in the ischia, so that the posterior belly could be carried tucked back and between both of them. The avian foot reverted to being fully symmetrical and tridactyl. How often these adaptations evolved remains to be determined: Did they evolve just once in both enantiornithines and ornithurines, or multiple times within one or the other of these groups.

The combination of a strengthened rib cage and a sternum that reached so far back that it ended under the pelvis allowed in-flight respiration rate and capacity to reach the highest avian levels. The powerful and sophisticated fliers that exhibited this combination may have been the first birds to undertake long migrations. By this stage, reptilian indeterminate growth was abandoned in favor of a more tightly regulated pattern, one in which mass was sometimes lost with the onset of adulthood. *Rahonavis,* however, retained a mid-Mesozoic flight and respiratory apparatus as the great era neared its end.

We know a surprising amount about avian reproduction in the Mesozoic thanks to the studies of Elzanowski (1981, 1985), Chiappe (1995), and Chatterjee (1997). The remains of embryos and eggs indicate that at least some birds continued the dinosaurian habit of nesting on the ground. As per their oviraptorid relatives, these birds probably brooded eggs that were partly or entirely exposed, using the feathery wings that had evolved for flight to help cover the eggs. Elzanowski (1985) concluded that it was the male that brooded the eggs among early birds. However, only one in ten or twenty feathered *Confuciusornis* specimens bears the hyperlong tail feathers that may have marked the males. If these feathers did indeed mark the males, then only females may have been numerous enough to brood the eggs. The embryonic skeletons found so far are well ossified, and the wing bones are especially large and highly developed. These features indicate that the hatchlings were precocial and were probably able to fly soon after popping out of the shell, perhaps within a day. If so, then parental care may have been absent, as per megapode fowl whose freshly hatched chicks literally take off on their own within a day of emerging from the nest. Alternatively, parenting was at most modest: the grown-ups may have guided and guarded their brood as the chicks fed themselves. Flight allows birds to nest in trees, which reduces the threat of predation upon eggs, chicks, and parents. A wide variety of birds nest in trees, including some waterbirds, so it is possible that this habit first appeared in the Cretaceous. If it did, then some birds may have engaged in intense parenting of altricial nestlings high up in trees by the end of the Mesozoic, but there is no positive evidence for this.

## The Great Retreat from the Air: Strange Days in the Late Mesozoic

As quickly as birds reached for the skies, the ability to fly was probably lost—first once, then again, and then again, and so on. Early flight loss should be no more surprising than the fact that soon after the first tetrapods crawled onto land, some of them were readapting to a fully aquatic lifestyle (Carroll 1988, Gould 2000). Being not that far from fish, these tetrapods had no difficulty returning to the water. Being little different from terrestrial avepods, the long-tailed, longer-fingered dino-birds found the arduous demands of flight easy to relinquish. Note that only one of the many flying mid-Mesozoic avepectorans needed to lose flight in order for a diverse radiation of secondarily flightless forms to occur.

Secondarily flightless dino-birds probably started to appear in small numbers as early as the Jurassic. The early trend toward flight loss was so strong that even some of the archaeopterygiform urvogels lost flight. The first neoflightless dinosaurs may have evolved from even earlier forms that had achieved only the initial stages of leaping-gliding flight. The first neoflightless avepectorans may be represented by the little dromaeosaur- and troodont-type teeth and bones that have been found in Middle and especially Late Jurassic deposits. Even coelurids and ornitholestids may have been neoflightless, albeit via ancestors with very weak aerial capability. We do not yet know how common neoflightless avepods were in the Jurassic, but they may well have laid the foundations for a proliferation of neoflightless dino-birds in the Early Cretaceous. Some or all of the neoflightless dinosaurs we now know of may have descended from fliers more advanced than the famous urvogel, fliers with larger sternal plates, ossified sternal ribs, and shorter tails. The presence of tails stiffened by ossified rods in dromaeosaurs may reflect a pterosaur-like level of flight ancestry.

The degree of reversal from the flying condition varied among the neoflightless dinosaurs and appears to have been lower among the earlier examples, a situation that is in accord with the predictions of the neoflightless hypothesis. The basal dromaeosaur *Sinornithosaurus,* with its very long arms and urvogel-like shoulder girdle with a highly reflexed coracoid and partly developed arm-elevating complex, was one of the dinosaurs least readapted for land-bound life. The other was squirrel-sized *Microraptor,* with its foot better designed for grasping branches than for dashing around on the ground. Other dromaeosaurs, *Protarchaeopteryx,* troodonts, caudipterygians, oviraptorosaurs, avimimids, and therizinosaurs retained at least some degree of arm folding, perhaps in order to protect their large, sharp claws when they were not in use (equivalent to claw retraction in cats); others did not. If ornithomimosaurs were neoflightless, then their constant use of large claws to grasp and manipulate vegetation left them with no need to protect them when not in use. Nor was there a need to retain large, ossified sternal plates. Tyrannosaurs severely reduced the size and function of their shoulder girdles and arms as they saved weight in favor of increased head and hindlimb muscle mass (Paul 1988a,b). Caudipterygians did the same thing, but to a lesser degree. Alvarezsaurs altered their arms into stout digging organs. In doing so, they either retained and modified a keeled sternum or evolved the keel in order to anchor massive pectoralis muscles. While dromaeosaurs and most troodonts modified the hyperextendible toe into a weapon and retained flight-adapted tails as dynamic stabilizers during leap-and-slash tactics during predation, some other secondarily flightless avepectoran dinosaurs re-evolved conventional toes and tails and deretroverted the pubes. Therizinosaurs did not the detrovert the pubes, because their tails remained short; and alvarezsaurs could not, because they had already slung the belly between split pubes. As in many flightless birds, the dino-avepectoran hallux tended to be reduced both in size and in degree of reversal as its arboreal grasping function was lost.

Lacking examples of integument from most avepectoran dinosaurs, we cannot yet undertake a comparative analysis of how flight loss may have affected whatever feathers adorned their frames. The well-developed and symmetrical arm and tail feathers of protarchaeopterygians and caudipterygians could have easily been transformed from asymmetrical flight vanes into display arrays and high-speed maneuvering surfaces. These may have been also used to shield eggs, at least in brooding oviraptorids, assuming they had such large arm feathers.

Dino-avepectoran features often explained as adaptations for predation and as exaptations for flight (large brains; forward vision; large furcula; large breastplates; long, foldable arms with lunate carpal blocks and raptorial fingers; hyperextendible toes; dynamic tails) may have originally evolved for climbing or flight or both and then have become exaptations for a neoflightless condition when they were readapted for predation. Neoflightlessness also explains why the sense of smell seems to have been reduced in arch-predatory dromaeosaurs and the troodonts, among which the ability to sniff out prey would have been an advantage. Note that if dromaeosaurs, troodonts, and even tyrannosaurs were neoflightless, then they partly re-evolved predatory habits, as opposed to inheriting them directly from their dino-avepod predecessors.

Of the dino-avepectorans, only the majority of dromaeosaurs remained unambiguous arch-predators, probably able to kill very large prey with their large, blade-toothed jaws, raptorial hands, and killer sickle-toed claws. Their attack mode probably focused on leaping onto their victims, using the hooked finger claws to hold on and wounding with the teeth and sickle claws or both. Little *Microraptor* had more conical, less serrated, and urvogel-like teeth suited for smaller prey. The killing ways of troodonts are more ambiguous. They may have used a sickle claw to kill, but the claw and the arms were smaller than those of dromaeosaurs. Troodont skulls and skeletons were also lighter than those of dromaeosaurs, and troodonts may have been omnivores rather than pure predators (also see similar conclusions based on teeth made by Holtz et al. [1998]). Toothless oviraptorosaurs were so bizarre and so unlike anything alive today that many different diets have been postulated. These include shellfish, fish, nuts, plants, and small animals. The association of lizard and baby troodont (not dromaeosaur) bones with oviraptorid skeletons (Norell et al. 1994, Sues 1997, Norell and Makovicky 1999) implies that little creatures were part of their diet. The similarity between short, blunt-beaked oviraptorosaur heads and caudipterygian heads is significant. Because the latter's small, short skulls, delicate teeth, and gizzard-stone bundles indicate they were herbivores or omnivores, their oviraptorosaur relatives probably were too. Even odder were alvarezsaurs. Their tubular jaws lined with small teeth imply that they were insectivores that used their

*Figure 13.9.* Shuvuuia. This alvarezsaur was remarkable not only for having an exceptionally birdlike skull but also for the hollow fibers found with the skeleton, fibers that have the characteristics of simple feathers. We do not yet know whether this dinosaur was also a secondarily flightless bird. It was a strange dino-bird, apparently adapted for breaking into the nests of social insects with its short but massive arms.

digging arms to break into hard-surfaced insect mounds or through tree bark. As strange as the therizinosaurs were, their diet is obvious. Broad beaks, small teeth, and, in the case of derived examples, enormous bellies are all signs of full-blown herbivory. The large abdomen implies the presence of an efficient fermenting gut. Some researchers have claimed that because ornithomimosaurs descended from predators, they could not have been herbivores (Russell 1972). However, all herbivores descended from predators, and the example of caudipterygians and therizinosaurs renders that argument moot. The ratitelike ornithomimosaurs were probably omnivores or herbivores. Their narrow bellies indicate that the gut was modest in size and not highly efficient at digesting plants; these features are found in large modern ratites (Withers 1983, Herd and Dawson 1984, Farlow 1987).

The break from classic dino-theropod predation made by avepectorans is further evidence that they emerged from an evolutionary event that separated them from conventional dino-avepods. That they passed through a flying stage in which arch-predation was lost explains this pattern well. This tendency away from predation parallels the condition in birds in general: early birds were not strongly predaceous, and the same is true of most extant avians. Flight, therefore, was the break from the classic dino-theropod pattern of predation that allowed avepectoran dinosaurs and birds to diversify into many new ways of living.

Flight and its loss also help explain the great success of avepectoran dino-birds. In the neoflightlessness scenario, they inherited improved neural, muscular, and skeletal systems from their arboreal-flying ancestors. It is interesting that the small avepod dinosaurs without flight-related characters present in the Triassic and Jurassic are largely absent from the Cretaceous. The less birdlike dinosaurs may have been hard pressed to compete with the new forms descending from the branches and skies. At the same time, the presence of well-developed hands and tails probably made secondarily flightless dinosaurs competitively superior to secondarily flightless land birds, which were rarer in the Cretaceous than in the Cenozoic, during which they faced mammalian rather than dinosaurian competition.

The scenario presented here successfully addresses a number of major issues. The most birdlike of the dinosaurs are younger than *Archaeopteryx* because they are not the ancestors of birds but rather secondarily flightless relatives. The semi-arboreal Jurassic dino-avepodian ancestors

*Figure 13.10.* Mongolia near the end of the Cretaceous. As counterintuitive as it may seem, even dinosaurian giants such as the tyrannosaur *Tarbosaurus (left),* and especially herbivorous *Therizinosaurus (right),* may have been part of a radiation of secondarily flightless dinosaurs. Such an event would explain the folding arms of the pot-bellied therizinosaurs, whose sabre claws measured up to two-thirds of a meter in length. These gigantic avepods descended from smaller, feathered forms; in fact, we know that small early therizinosaurs were insulated by such simple feathers. Were the latter, bigger examples also feathery? Were they entirely naked like equally large mammals? Or were they adorned with display feathers on their arms great and small?

of birds are unknown because they were too small, and too limited to interior habitats, to be readily preserved. Only when they lost flight and became larger did the most birdlike dinosaurs more often become part of the Cretaceous fossil record. Flight loss explains why avepectoran dinosaurs tended to start out small and become larger with time. Loss of flight explains the flight-related adaptations observed in these Cretaceous avepod dinosaurs.

So far, a flightless Early Cretaceous bird with a long-fingered hand and a very short (shorter than in oviraptorosaurs), pygostyle-tipped tail has not been found. Were there really no such creatures, or is this absence an artifact of the incomplete fossil record? If the former, were early short-tailed birds unable to compete with the longer-tailed avepectorans that also had clawed hands? *Yandangornis* may be an example of a long-fingered, long-tailed ground bird on the way to losing flight. More advanced Cretaceous birds with reduced and fused hands lost flight so readily that the hesperornithiforms became as water adapted as penguins as early as the Early Cretaceous. Why birds were the only dinosaurs to go oceanic is poorly understood. The reason may be that birds' flight capability makes it easy for them to invade marine habitats, where flight is not a crucial escape mechanism. We are only now beginning to realize that at least one bird group spun off flightless terrestrial examples before the end of the Mesozoic. South American *Patagopteryx* and European *Gargantuavis* appear too primitive to have been ornithurines. They may be related to one another, and neither appears to be related to later flightless birds. We do not know how big their heads were. *Patagopteryx* was chicken sized, and *Gargantuavis* may have been as large as an ostrich (Buffetaut et al. 1995, Buffetaut and Loeuff 1998). Such big Mesozoic land birds have so far been found only in southern Europe. Considering their absence from contemporary faunas where similar sized dinosaurs are well represented, it is probable that large land birds were absent or rare elsewhere. The radiation of Mesozoic neoflightless birds was modest compared with that of the possibly neoflightless dinosaurs. Conversely, it is possible that *Gargantuavis* evolved in the absence of ostrich mimics. Indeed, it may have been part of an island fauna that included a set of isolated dinosaurs. The insular isolation may explain why the bird was not especially well adapted for running. At least some of the large eggs that have long been known from the *Gargantuavis*-bearing deposits may be of avian, rather than dinosaurian, origin.

The archaeopterygiform *Protarchaeopteryx*, the various avepectoran dinosaurs with flight-related characters, and the basal bird *Patagopteryx* combine to suggest that avian flight was being lost at every major stage of its development, including its very beginning. If so, then the last half of the Mesozoic seems to have seen two major radiations of birds and near-birds: one radiation in the air and one—equally interesting and definitely stranger than previously imagined—back on the ground. This scenario is interesting because, if it is true, the evolution of birds from dinosaurs was not a simple, linear, progressive affair in which flight ability steadily improved in ever more birdlike small forms. Instead, the radiation of the Avepectora was an intricate, branching process that involved high levels of parallelism and significant reversals in flight ability and morphology. This scenario is strange because of the characters that may have been involved: fat-bellied herbivorous therizinosaurs that stood 5 meters tall and whose foldable arms carried claws more than 2 feet long in length; giant ornithomimosaurs with enormous arms; perhaps even great tyrannosaurs with just two degenerate fingers. At the smaller end of the size spectrum, psychedelic alvarezsaurs combined birdlike heads and gracile hindlimbs with molelike arms specialized for digging. Oviraptorids sported oddly contorted heads, jaws, and crests. Protarchaeopterygians and caudipterygians had bobbed tails ending with feather fans. Again, this seemingly extraordinary pattern should not startle us. Evolution is a dramatic process of alteration that produces seemingly outlandish results, many associated with major reversals in habitat. Early amphibians that readapted to underwater life included wing-skulled diplocerapids. Ungulates that returned to the seas evolved into filter feeders with mouths large enough to engulf an elephant. Loss of avian flight resulted in dodos and kiwis. In this view, modern birds are just one of many extraordinary branches of the advanced Avepoda.

CHAPTER 14

# The Great Extinction

By the final stage of the Cretaceous, birds were a thriving and very diverse group of (contour) feathered dinosaurs. A time traveler to the period would have found the skies filled with many kinds of avians, some of which mimicked modern birds, others of which lacked modern equivalents. Some were heavy-jawed, toothed marine ornithurines. However, a number of beaked ornithurine shorebirds probably did not look all that different from their modern equivalents. There may have been a large number of terrestrial ornithurines, maybe even parrots. A few fairly large ground birds were running about, but they were overshadowed by dinosaurs small and large, which still ruled the terrestrial roost. In the air, the number of pterosaurs species seems to have dwindled sharply, but they remained prominent simply because of their sheer size.

At the end of the Cretaceous, something went wrong, and many groups went completely extinct, among them all of the pterosaurs and nonavian dinosaurs, including the nonavian avepectorans. It has been observed that the Dinosauria did not really go extinct, because Cenozoic birds *are* dinosaurs (Dingus and Rowe 1998). Although this is true in a narrow taxonomic sense, it is misleading in the broader sense. If, for example, all mammals except bats went belly up, then mammals would not be extinct, but such an extinction would nevertheless be an extreme event with profound implications for the global fauna. The loss of all the many terrestrial dinosaurs, from small to large, from herbivorous to predaceous, was just such an event. In terms of terrestrial fauna, the K/T extinction was an even bigger affair than the otherwise more massive Permo-Triassic (P/T) extinction that had occurred 180 Myr earlier. While the P/T extinction left a small but significant number of large land animals—mainly therapsids—roaming about, the extinction that occurred 64 Myr ago left not a single large fully terrestrial animal alive in the world. The biggest creatures remaining on continents were freshwater crocodilians. No large fliers seem to have made it through the K/T crisis, and certainly the superpterosaurs did not.

Birds too suffered. That little of the scientific attention lavished on the K/T crisis has been directed toward the fate of the birds of the time is unfortunate, since their extinction pattern has important information to tell us about the character of the crisis. However, the situation is changing with the publication of a flurry of new studies.

On the one hand, Feduccia (1995, 1996), Bleiweiss (1998, 1999), and Fara and Benton (2000) have concluded that almost all the birds of the Cretaceous, few of which were neornithines, failed to survive the K/T troubles. Birds, therefore, experienced a classic evolutionary bottleneck, whereby they were squeezed down to just a handful of species and types, in this case some neognathous shorebirds and palaeognathous forms. On the other hand, Chiappe (1995), Hedges et al. (1996), Cooper and Perry (1997), Hope (1998), and Stidham (1998, 1999) have concluded that a large array of neornithines did make it through the boundary.

What we do know is that there is no evidence that a single enantiornithine or toothed ornithurine lived into the Cenozoic. Only the beaked neornithines seem to have survived. We do not know exactly how close to the K/T boundary enantiornithines and toothed ornithurines made it; some or all of them may have been gone before the final crisis, but the fossil record is too poor for us to tell (Padian and Chiappe 1998c). At or near the end of the Mesozoic, birds apparently took a heavy hit, although it was perhaps not as massive as some have argued (Marshall 1999).

## Why?

What happened to dinosaurs avian and nonavian? The simple answer is that we do not yet know for certain, despite the current near-consensus that the impact of an exceptionally large meteorite in Central America (near Chicxulub, Mexico) was largely or entirely responsible for the mass extinction both marine and continental (L. Alvarez et al. 1980, Raup 1991, W. Alvarez 1997, Chatterjee 1997, Dingus and Rowe 1998). The evidence that a super impact occurred and adversely affected the global environment is substantial. Yet questions and problems remain, so many that the consensus may be premature to a certain extent.

### *Big Extinctions Do Not Require Big Impacts*

So far impacts been definitively correlated with most other mass extinctions (Courtillot 1999). For two decades, attempts defined an impact event associated with the biggest extinction ever, the P/T event, failed. Becker et al. (2000) have presented evidence for an impact at that time, but there has not been time to verify or challenge their conclusion. Perhaps a nonimpact extraterrestrial event such as a nearby supernova or an influx of meteoritic dust (Kortenkamp and Dermott 1998) triggered the P/T extinction, as well as others. If not, then the biosphere must have been capable of producing severe mass extinctions without outside help. However, the terrestrial mechanisms that may cause such events remain poorly understood.

### *Superimpacts That Did Not Kill*

The solar system is a junkyard. Although the gas giants long ago cleared out most of the billions of comets that originally formed in the inner solar system (Paul and Cox 1996), the orbits of thousands of mountain-sized asteroids and comets still have the potential to cross paths with Earth. Our planet is therefore in a cosmic shooting gallery, one in which hits from very big, very high velocity objects will occur. Just how often these hits occur is not yet certain, and the hit rate may fluctuate. At this time, it is believed that a Chicxulub-scale impact may occur on an average of every 100 Myr, which means two or three of them should have occurred over the past 250 Myr.

What is becoming clear is that the more geologists look for giant impact craters, the more craters they find. At this time, four super impact events are known from the Mesozoic before the K/T Chicxulub event, and one is known from the Cenozoic (Paul 1989, Bottomley et al. 1997, Koeberi et al. 1997, Spray et al. 1998; the Indian Ocean K/T impact claimed by Chatterjee [1997] has not yet been confirmed). Three of these impact events—a possible multiple impact before the end of the Triassic, another at or near the J/K boundary, and a linked pair of craters from the mid Cenozoic (of which one formed the Chesapeake Bay)—seem to have been about as energetic as the Chicxulub explosion.

Although many craters have been found, the Mesozoic-Cenozoic crater survey is still incomplete. Since the oceans cover much of the earth's surface, there are probably many craters lying undetected on the deep ocean floor, and it is likely that many craters have been destroyed by subduction. The past 250 Myr worth of sediments have not been carefully surveyed for the reentry debris associated with big impacts; the K/T layer was found only because many people were intensely interested in that particular zone. It is therefore possible that Meso-Cenozoic impacts we do not know about approached or even surpassed the scale of Chicxulub. (In any case, a recent survey of the K/T crater indicates that the impact may not have been as large as previously thought; see Morgan et al. 1997.)

In general, the hypothesis has been that the largest Meso-Cenozoic extinction was so bad because it was associated with the biggest impact of the age. However, the fact that this impact may not have been unique creates an obvious problem with this premise. Another problem is that the other super-impacts did not produce correspondingly extreme extinctions of dinosaurs or other large creatures. The Late Triassic Manicouagan impact (Quebec) may or may not have been associated with some minor extinctions, but there is no evidence that dinosaurs suffered any lasting effects (Paul 1989). The Late Jurassic Morokweng impact (South Africa)—which possibly exceeded the power of the Chicxulub event—may have been associated with J/K dinosaur and pterosaur extinctions, but these extinctions were at most modest (see Chapter 13). In particular, there is no evidence that any avetheropod group, whether small or gigantic, went extinct. The well-dated mid-Cenozoic Popigai-Chesapeake impact (Siberia and the United States) does not closely coincide with a major extinction. This leads to an obvious question. If dinosaurs and other large land animals survived a number of super-impacts with few or no losses, why did they fail so utterly when yet another piece of space debris hit planet Earth?

One possible explanation has to do with the angle of impact. The Chicxulub meteorite appears to have impacted at a shallow angle, directing the bulk of the impact debris toward North America. This asymmetry of the postimpact phenomena may have worsened some global effects. But this would have lessened the effects upon the Southern Hemisphere. Besides, most impacts occur at shallow angles. What most differentiated the K/T impact was its location, a sulfur-rich carbonate shelf; an impact upon such a shelf is a statistically rare event. The extremely high level of atmospheric acidification and the carbon dioxide boost that probably resulted may distinguish the Chicxulub event from impacts of similar power.

### *Impacts Are Not* That *Bad*

At one time, comet impacts were truly horrific. Prior to 3.8 billion years ago, during the initial bombardment of the planet, comets the size of whole states and provinces walloped the planet on a regular basis. Such monsters produced enough heat to sterilize our planet's entire surface. Although such an ultraimpact could occur if one of the biggest comets from the distant Oort Cloud visits us, such an event is highly improbable. These days we are occasionally whacked by a Mount Everest–sized meteor. This may sound like a large meteor, and indeed the startling effects of a comet of that size, comet Shoemaker-Levy, upon giant Jupiter have impressed the advocates of impacts as dinosaur killers. However, on a relative scale, such an impact is equivalent to a globe being hit by a speck of dust, albeit one traveling at many dozens of times the speed of sound.

If a comet of this size hit the earth, the plants and animals within the blast zone would, of course, be wiped out directly, as would those along any coastline flooded by any colossal tsunamis, if the impact was at sea. But the great majority of the biosphere would not be in such direct danger. However, a super-impact explosion would project a debris cloud around the entire planet at suborbital velocities within 40 minutes. As the debris reentered en masse, it would produce an incandescent high-altitude pyrosphere that would heat up the surface as hot as a kitchen oven for some minutes. This heating effect would not only adversely affect exposed animals but also initiate mass forest fires. The atmosphere would be massively polluted by the resulting smoke and ash, a thousand times worse than the harshest modern smog. Sunlight would be blocked out for many months, which would shut down plant growth and cause a global winter that would bring snow to the equator. If the impact released materials locked up in a sulfur-rich carbonate shelf, the resulting intense, corrosive acid rain and airborne toxic metals would be lethal to nonburrowing animals. Water would also be polluted, and the resulting collapse of the marine food chain would devastate big, flightless marine diving birds. As the skies cleared, high levels of carbon dioxide—again the result of disruption of a sulfur-rich carbonate shelf—would cause a greenhouse effect that would drive global temperatures far above even the Mesozoic norm. Major droughts would ensue. Terrestrial and flying dinosaurs would be poisoned, baked, burned, overheated or chilled, and starved (Raup 1991, W. Alvarez 1997, Chatterjee 1997, Dingus and Rowe 1998).

The fact that such conditions would crush animal life is the very problem with the scenario. The projected conditions are too severe. We know that the situation could not have been *that* bad (Paul 1989, Archibald 1996) because had such conditions been prevalent everywhere, virtually every tetrapod would have been wiped out. Yet viable populations of reptiles, some of them large bodied, as well as amphibians, mammals, and birds did survive around much of the planet. The survival of amphibians and birds is especially significant. The former are exceptionally sensitive to environmental toxins because their thin skins easily absorb whatever they come in contact with. As for birds, their high metabolic rates have two effects. First, birds must constantly breath large volumes of air and eat lots of food, so their intake of any environmental toxins is rapid and high. Second, they starve quickly when denied food. Birds and amphibians are therefore considered key indicators of environmental degradation, whether it be in mines or in the biosphere as a whole. That thin-skinned amphibians and hyperenergetic birds survived the K/T impact shows that the environmental toxin and acid load could not have been consistently intolerable and that there must have been food available.

There is additional evidence that the K/T crisis was not as awful as some have estimated. In North America, a K/T "fern spike" indicates that almost all of the shrubs and trees were wiped out and replaced for a period by colonizing ferns. This fern spike may have been the result of the continent's being downrange of the primary blast produced by an oblique impact coming up from the south. The evidence suggests that half of the world's forests burned. But if we look at the glass as half full, we observe that half of the world's flora did not burn, especially in the Southern

Hemisphere, where there is no evidence of significant floral extinctions at the time. The simple presence of heavy cloud cover would have provided an effective local thermal shield against the short-lived pyrosphere, and subsequent rains would have put many of the fires out.

The combined evidence shows that the post-impact environment was not so harsh as to be unsurvivable and that large numbers of tetrapods found refuge to survive in. How then could entire groups have been completely lost?

### *Dinosaurs Were Hard to Kill Off*

The complete disappearance of pterosaurs is not surprising. By the end of the Mesozoic, only a few species remained (Unwin 1987, Wellnhofer 1991), and they were large-bodied, specialized types. When a group dwindles to a few types with constrained lifestyles, it is primed for extinction.

It is the dinosaurs, feathered and otherwise, that present a problem. The most ardent advocates of an extraterrestrial collision as the sole dinosaur killer argue that the worldwide fauna was healthy at the time of the impact. This argument makes it more difficult to explain the loss of dinosaurs because a healthy global population should been have roughly comparable in size to that of contemporary continental mammals weighing more than 10 kilograms: dinosaurs probably numbered in the billions and were spread among dozens or even hundreds of species. The Mesozoic birds should have been even more populous; the modern bird population before human interference numbered in the hundreds of billions (Chatterjee 1997).

An important assumption about most K/T extinction scenarios is that nonvolant dinosaurs were relatively easy to kill off, mainly because they were big. In order to understand why they were so easy to kill, we must look at how animals reproduce. Organisms can be sorted into two basic reproductive types, K-strategists, and r-strategists. The latter produce large numbers of young, which experience high mortality rates. Many insects and most small mammals are classic r-strategists. These can be thought of as "weed" species, in that their high rates of reproduction allow them to achieve very high rates of population growth and dispersal when conditions are favorable enough to let a large percentage of juveniles survive. Because of this reproductive potential, r-strategists can quickly recover from population losses. The r-strategists are therefore very hard to kill off, as anyone who has targeted cockroaches and mice for extermination knows all too well.

K-strategists reproduce slowly and try to keep juvenile mortality to a minimum, often via intense parental care. Large mammals are classic K-strategists. Because they cannot churn out lots of young, their maximum population growth rates are rather low even under the best of circumstances. In addition, big animals are always relatively few in number because each individual eats so much. Another problem for K-strategist mammals is that their young cannot survive without parental care, especially during the nursing phase. This means that the adults can care for only a limited number of young.

Modern birds are interesting, because they reverse the mammalian pattern. On the one hand, most small birds are K-strategists that lay a few eggs each season and then lavish lots of care and attention on their young. Flight may be responsible for this pattern. weight-conscious fliers must keep the weight of the eggs they carry to a minimum, so they make as few as possible. At the same time, the ability to fly allows avian parents to range far in search of the food demanded by their ravenous charges. On the other hand, it is the big continental ratites that are r-strategists. They lay many eggs and then fail to bring food to their young, which experience high rates of mortality.

Dinosaurs were r-strategists par excellence. As far as we know, dinosaurs of all types and sizes (including small avepods) laid large numbers of eggs, a dozen or more per season (Paul 1994b, 1997c, Norell et al. 1995, Varricchio et al. 1997, Carpenter 1999, Clark et al. 1999). Therefore, rates of population recovery should have been very high. Although some may have fed their young, especially when they were small nestlings, dinosaurian parental care as a whole was less intense than it is in K-strategist mammals and birds. This implies that just a few hundred adults of any particular dinosaur species needed to survive in order to reestablish their population over a short period. What is extraordinary is that rapid reproduction was true of giant dinosaurs as well as small ones. Ergo, the biggest dinosaurs were weed species, whose survival and recovery potential was very different from, and probably superior to, that of giant mammals.

The r-strategy reproduction of dinosaurs helps explain why they were so successful for so long. It is notable that so few major dinosaur groups went entirely extinct before the end of the Mesozoic, exceptions being prosauropods and stegosaurs. Otherwise, dinosaur history was a story of accumulative increase in diversity, with older groups continuing to live alongside the new. At no time was there a major "size squeeze," in which most or all of the large dinosaurs went extinct at the same time, to be replaced by an entirely new set of large forms that re-evolved from small-bodied stock (see Paul [1990] for speculations on what might have happened if a few small K/T dinosaurs had survived into the Cenozoic). Sauropods were persistently enormous and diverse for 130 Myr. In contrast, K-strategist mammalian giants have not been so successful to date. Uintatheres, arsinotheres, titanotheres, indricotheres, and megatheres have all come and gone within brief spans. Even proboscideans have been extant for only 40 Myr.

As for the small, Cretaceous avepod dinosaurs, they appear to have been persistently diverse and common for the 80 Myr of the period. The reproduction of Mesozoic birds is not yet well documented enough to allow us to assess directly their r versus K reproductive strategy. In general, annual egg production was modest. In any case, once they got going, birds also became increasingly diverse over the 80 Myr of the Cretaceous (Unwin 1987, Feduccia 1996, Chatterjee 1997).

Another common, tacit assumption about dinosaurs is that they were more vulnerable to climatic disruption than mammals. This assumption is a holdover from the traditional view of dinosaurs as reptiles. The presence of dinosaurs in polar regions where reptiles were sometimes absent is especially important, because it implies that the archosaurs' ability to cope with a postimpact winter was better than often assumed (Paul 1988c, Clemens and Nelms 1993). In addition, there is no reason to believe that the thermoregulation and energetics of the feathered avepod dinosaurs of the end of the Mesozoic were grossly inferior to those of the mammals and birds that survived (Appendix 4). Because the energy intake of terrestrial dinosaurs was probably somewhat less than that of birds, the former should have been less vulnerable to environmental pollutants than the latter. The large brains and sophisticated sensory systems of advanced avepod dinosaurs offered them the mental agility to adjust to new and adverse conditions. Birds did enjoy an advantage over earthbound dinosaurs. Birds' ability to fly allowed them to move away from bad local and regional conditions in search of less odious environments.

The rapid reproduction rates and sophisticated thermoregulatory abilities of dinosaurs and Mesozoic birds may have been an important reason that they survived a number of Mesozoic superimpacts in good order. This possibility returns us to events at the K/T boundary. After the K/T impact, those dinosaurs and birds that happened to be shielded by heavy cloud cover should have survived the initial pyrosphere. Postimpact pollution levels that were unable to destroy all hypersensitive amphibians and birds should have harmed nonvolant dinosaurs even less. These dinosaurs had sophisticated thermometabolic systems that allowed them to cope with unusual climatic fluctuations. Enough floras apparently survived to support viable populations of avian and nonavian dinosaurs. Even if most or all r-strategist dinosaur species were nearly wiped out, only a few hundred individuals of a given species needed to survive in order to lay the foundations for rapid recovery. It is understandable that some or even most dinosaur and bird species succumbed to the after effects of the Chixculub impact, especially in the Northern Hemisphere, where the habitat degradation was most severe. What remains inexplicable is why every single species of terrestrial dinosaur and nonneornithine bird in the entire world failed to survive in a world where Southern Hemisphere forests lay reasonably undisturbed. Why did the little herbivorous hypsilophodonts, which should have found enough plants to get by on, fail to pull through? Their sophistication and intelligence honed by flight, neoflightless avepectorans should have been especially resistant to total destruction (Paul 1989). Most of all, the troodonts, which were small and probably had some dietary flexibility, could have fed on the little mammals, reptiles, amphibians, and plants that survived the catastrophe. Why were some ornithomimids unable to find enough sustenance to make it through? Had just a few dinosaurs managed to hang on into the early Cenozoic, they could have been the seeds for a new radiation of terrestrial dinosaurs (Paul 1990).

Dinosaurs were such a large, diverse, and reproductively potent group that their total extinction at a time when numerous other tetrapods survived is incredible. At this time, the extraterrestrial impact hypothesis is at best incomplete in that no one has yet proposed a viable mechanism by which the aftereffects of the impact could have destroyed the entire Dinosauria with the exception of one branch of Aves. Lacking such a mechanism for total extermination, and without confirmation that the Chixculub event was uniquely powerful, the impact hypothesis cannot be considered verified.

*The Alternatives*

Over the years, a host of alternative explanations for the K/T extinction has been offered. Among them are a nearby supernova explosion (or some other galactic event), supervulcanism (which has effects broadly similar to those of an impact, only more extended), gradual climatic change (becoming either hotter or cooler), a sea-level drop, a decline in atmospheric oxygen, disease vectors, changes in the global flora, competition from new animal groups, diversity fluctuations due to evolutionary complexity and chaos, failure to reproduce properly, or some combination of the above (see Archibald 1996, Courtillot 1999). Many of these explanations have one simple problem. Things are always changing. The weather never remains the same, oceans rise and recede, vulcanism fluctuates, the earth runs into comets, and so forth. Things certainly were changing when nonneornithine dinosaurs went extinct, but they were changing even before that. Blaming whatever changes happened to be occurring at the time of an extinction for the extinction is easy, but there may have been no actual cause and effect.

One way to assess a particular alternative is to determine whether what happened at the K/T boundary was exceptional. Another is to estimate whether the victims were vulnerable to the postulated causal agent. Supernovas and other remote extraterrestrial events are difficult to assess because we cannot yet detect evidence of them.

Supervulcanism has more potential for assessment. It is known that at the K/T boundary a series of huge volcanic eruptions laid down the vast Deccan lava traps in India (Paul 1989, Archibald 1996, Chatterjee 1997, Courtillot 1999). The amount of pollutants put into the atmosphere by the Deccan eruptions must have been incredible, perhaps enough to produce global winters alternating with hot periods caused by the greenhouse effect. Supervulcanism has also been proposed as the cause of the P/T extinctions. On the one hand, vulcanism is superior to an impact as an extinction agent in that its effects extend over time, causing an attrition effect. On the other hand, repeated extinction events are similar to repeated applications of pesticides in that the victims tend to develop resistance; species that survive the first event are less likely to succumb to the subsequent repeats (McKinney 1987; the same principle applies to a series of impacts closely spaced in time). In addition, great trap-laying vulcanism was not limited to the end of the Mesozoic; others periods of vulcanism occurred without having a marked effect on dinosaurs (Paul 1989). In addition, supervulcanism does not explain why only the neornithine birds survived.

Disease as an explanation for mass extinction of dozens of species suffers from the same flaw as repeated eruptions or impacts, the classic Darwinian phenomenon of resistance (Paul 1989). It is very difficult to kill off even a single species with disease; the resistant individuals that almost invariably survive are well positioned to stage a comeback. Over the past half millennium, mortality rates among various human populations (for example, the Amerindians) and animal populations (consider the rinderpest epidemic among wildebeests) have often exceeded 90 percent; but no species has yet gone extinct, and full recoveries have often occurred. Killing off even a fraction of the dozens or hundreds of dinosaur and birds species via this mode may well have been impossible. Besides, birds had been flying across and between the continents and spreading disease among themselves, and to other tetrapods, for tens of millions of years without disastrous results (dinosaurs would have been more susceptible to being infected by their avian relatives than are mammals). Why K/T microbes would have been extremely virulent has not been convincingly demonstrated.

Some have claimed that disease spread like wildfire at the end of the Cretaceous because a global drop in sea level allowed the mixing of previously separated faunas. The Mesozoic, especially the Cretaceous, was an era of unusually high sea levels, and the K/T sea-level regression was a strong one by the standards of the time. The severity of the drop has led some to propose it as the primary cause of the extinction. The problem here is that *increasing* the total area of land for dinosaurs and birds to live on would probably help their fortunes, not hurt them. Some local populations might be adversely affected, but arguments that terrestrial dinosaurs collapsed because of continental expansion are convoluted, based on limited analysis of a few lowland populations, and unconvincing. Why all nonneornithines would have vanished because of receding shorelines is even more mysterious.

Some researchers claim to have detected high rates of abnormalities in dinosaur eggshells from very late in the Cretaceous, abnormalities that the researchers link to the extinction of the group (Erben et al. 1979). This data is limited to Europe, and at least some of the eggs in question may belong to big, island birds rather than to dinosaurs (Buffetaut and Loeuff 1998). A general and global state of excessive eggshell pathology has not been demonstrated.

Another extinction-reproduction link is based on the observation that some reptiles, including crocodilians, have temperature-dependent sex determination. The temperature at which a particular egg is incubated determines the sex of the embryo. Paladino et al. (1989) concluded that dinosaur reproduction was also temperature sensitive. They suggested that fluctuating temperatures at the end of the Cretaceous skewed the sex ratios so badly that the dinosaurs went extinct. The problems with this hypothesis are legion. In many reptiles and birds, sex is genetically determined, and the same may have been true for some or all dinosaurs, the near avian avepods especially. Even if dinosaur sex ratios were temperature dependent enough to be disrupted, it is hard to see how this problem would suddenly wipe out every single dinosaur species, after they had been spawning successfully for 160 Myr. The fact that crocodilians and turtles with temperature-sensitive sex determination survived any temperature fluctuations that may have occurred at the K/T crisis makes this scenario even more unlikely. The improbability of this scenario brings us to the matter of climate.

Climatic change is the classic dinosaur killer and has been invoked by many a paleontologist since the 1800s. Despite the popularity of this notion, climate change has never been accepted as the premier killing agent. Why? There are several reasons. The climate was changing throughout the Mesozoic. For example, a sudden and sharp drop and a subsequent rebound in temperature appear to have occurred well before the end of the Cretaceous, at a time when dinosaur diversity was increasing (Kuypers et al. 1999); and the weather change at the end of the era was by no means extreme. There was no ice age and no long-term superheating that left even the poles hot in the winter. For that matter, Mesozoic climates may not have been as universally warm and balmy as is usually thought (Barron and Washington 1982, Paul 1991, Sellwood et al. 1994), and there may have even been modest continental glaciation at the South Pole (Stoll and Schrag 1996). Dinosaurs and birds had long been living and reproducing in climates ranging from polar to tropical, from wet forests to deserts. They appear to have had well-developed thermoregulatory systems (Appendix 4). Their sex ratios were stable enough for most of the Mesozoic. Dinosaurs, especially birds, had the option of moving if climate change in a particular location became a problem. Climate change appears ill suited for explaining the entire collapse of the nonneornithine Dinosauria, or the Pterosauria for that matter. In particular, the suggestion that the giant yet lightweight and delicate-boned pterosaurs were grounded by increasing winds at the end of the Mesozoic does not fly. As explained in Chapter 4, pterosaurs were heavier bodied, had stronger bones, and were more powerfully muscled than is usually appreciated. Strong winds should have made it easier for these massive azhdarchids to get off the ground! The wave-soaring marine pterosaurs may or may not have been more adversely affected by changing wind speeds. However, the climatic modeling suggests that Cretaceous breezes were not significantly weaker than the breezes of today (Barron and Washington 1982, Paul 1991).



Were dramatic floral changes responsible for the K/T debacle? In the Late Cretaceous, flowering angiosperms displaced conifers, cycad relatives, and ferns as the dominant land flora, but this change had been well underway for tens of millions of years before the end of the Mesozoic. If anything, the new plants were better food sources than the old plants. They reproduced more rapidly, grew faster, and produced more-palatable leaves, larger seeds, and more-nutritious fruits. A whole array of dinosaurs and birds evolved along with the new flora, and birds would continue to thrive in the new forests and grasslands of the Cenozoic.

Another atmospheric change—a sharp drop in oxygen levels from a Mesozoic high to the modern Cenozoic level—is hotly disputed because the primary evidence is elevated oxygen levels in Mesozoic amber air bubbles (Hengst et al. 1996). Even if a drop did occur, moderate oxygen depletion appears ill suited for explaining the loss of birds with the high-performance respiratory complexes apparently present in Late Cretaceous enantiornithines and toothed ornithurines (Appendices 3B, 4). Birds are famous for being able to breathe effectively at extremely high altitudes via their exceptionally efficient cardiorespiratory systems (Schmidt-Nielsen 1972, Scheid and Piiper 1989, Whiteman 2000). For that matter, bradyaerobic and tachyaerobic (Appendix 4) tetrapods alike managed to breathe their way across the K/T boundary.

Let us now turn to the more exotic, and perhaps crucial, implications of information-processing theory, complexity theory, and chaos theory. Computer simulations of evolutionary trends and processes suggest that chaos-driven instability causes complex species communities to periodically experience self-initiated mass extinctions (Kaufman and Johnsen 1991, Levy 1992, Paul and Cox 1996). Sole et al. (1997) has shown that a nonlinear response to environmental perturbation that is itself insufficient to directly cause a mass extinction can nonetheless initiate a runaway effect that ultimately does cause extinction.

A rare attempt to explain the extinction of one particular Mesozoic bird group is Elzanowski's (1983) suggestion that the diving hesperornithiforms might have suffered at the hands of the new and explosive radiation of advanced bony fishes in the Late Cretaceous.

**Did Birds Kill Off the Pterosaurs?**

Were Cretaceous birds simply victims of the Cretaceous extinction, or did they cause the extinction of other groups? As birds became more common and diverse in the Late Cretaceous, pterosaurs—especially small ones—declined (Unwin 1987, Wellnhofer 1991). This correlation may have been coincidence, but it was too extended over time for us to dismiss it. One possibility is that small pterosaurs declined on their own, and birds merely took advantage of the situation by filling the niches pterosaurs vacated as they became specialized aerial giants. We can "test" this view via a thought experiment. Assume that birds never evolved and therefore that the Mesozoic skies were the domain of insects and pterosaurs. If there were no birds, would pterosaurs have abandoned the small side of the size spectrum so completely, if at all? After all, they had been doing well as small fliers for over 100 Myr. The evolution of giant pterosaur should not have influenced the evolution of small examples any more than the evolution of avian giants has hindered the evolution of small birds. So, while it is possible that small pterosaurs would have disappeared without the influence of avian competitors, the proposition can be considered dubious, if not implausible.

If birds did win a head-to-head competition with pterosaurs for the small-bodied niches, what made the former superior to the latter, which were themselves sophisticated, energetic fliers (Wellnhofer 1991)? Because pterosaur brains were no larger than those of reptiles, Mesozoic birds may have literally outwitted them over the long run. The avian feathered wing may have had subtle but important aerodynamic and safety advantages over the pterosaur membranous wing. On the one hand, the latter was more vulnerable to tearing; many birds regularly shed flight feathers and keep on flying. Layers of feathers can be built up into a more streamlined, subsymmetrical wing, whereas a thin, cambered airfoil cannot (see Chapter 7). On the other hand, the membranous wing can produce more lift and is better suited for tight maneuvering. Although trackways show that the semiquadrupedal pterosaurs were more adept on the ground than some have thought (S. Bennett 1997), they may have been unable to keep up with the new bipedal fliers that were frenetically dashing back and forth across Cretaceous mudflats in search of little things to eat. In the scenario presented here, birds prevented pterosaurs from being successful as small fliers. Giant pterosaurs would have evolved alongside small pterosaurs in a bird-free Cretaceous, but as it was, only the superpterosaurs were left cruising the skies at the end of the Late Cretaceous. When, as was inevitable in the long term, the few giant pterosaurs left went extinct for one reason or another, there was not a large population of small pterosaurs to carry the clade through the crisis. If there had been, then pterosaurs might have survived into the Cenozoic, had it not been for that other group of archosaurian fliers.

## Synthesis

The superimpact hypothesis is the leading explanation for the K/T extinctions, but it is still marred by important problems. In fact, at this time, no single hypothesis fully explains why toothless ornithurine birds survived a crisis that destroyed the rest of the Dinosauria. Within this observation may reside the germ of a solution. It is very possible that the terminal Mesozoic extinctions had multiple causes (Archibald 1996, Zinsmeister 1997). The pterosaurs may have been suffering from competition with birds to the point the former were reduced to a few oversized species. The flightless marine divers may have been under pressure from advanced fish. Perhaps pandemics resulting from ocean regression were depleting a number of species' populations. Population depletion by itself was not fatal, but repeated volcanic explosions may have polluted the biosphere and disrupted the climate enough to knock out some of the depleted species and drive others to the brink. Contrary to the pure impact hypothesis, the global archosaur fauna was not healthy when a mountain-sized object's trajectory converged with that of Earth, wreaking yet further havoc on species' populations. That the piece of cosmic dirt happened to hit a sulfur-rich carbonate shelf may have made matters worse. Even this blow might not have been enough to destroy every single species of nonneornithine dinosaur. A nonlinear, chaotic response to these environmental perturbations may have exaggerated what should have been survivable events such that they resulted in the total collapse of major groups.

What is certain is that something extraordinary happened 65 Myr ago, because all those dinosaurs and birds did disappear totally and in short order.

CHAPTER 15

# The Cenozoic

## *The Age of Neornithines*

When the K/T crisis ended and the Paleocene began, only a limited number of toothless ornithurines were still flying, and flightless birds were probably completely absent. The populations of some of these neornithines, especially shorebirds, may have boomed very quickly in a largely predator- and competitor-free global environment. Without any competition from pterosaurs or bats, birds dominated the skies in a way they never would again. But potential competition was already on the way. Bats evolved very early in the Cenozoic Era; the first, already modern examples date from 50 Myr ago, so they must have started evolving from climbing insectivores almost immediately after the K/T boundary, if not sooner. Of course, the Cenozoic is labeled the Age of Mammals, and since then mammals have commanded the land, flourished in the oceans, and owned the night skies. Even so, avian dinosaurs have done quite well for themselves. They rule the daylight skies and have established a foothold in the night airs, in the lakes, rivers, and oceans, and on the continents and islands. This time is also the second age of Avebrevicauda, in that all Cenozoic birds have had very short, bony tails. (Much of the information cited in the following discussion is from S. Olson [1985] and Feduccia [1996], with some details from S. Olson [1999]).

What was and is the reason for the avian-mammalian division between the day and night skies? Has it been due to competition between birds and bats? We can get a better grip on this question by undertaking another thought experiment. What would have happened if no birds had survived into the Cenozoic and bats had evolved early in the new era. One can easily imagine that bats would have evolved numerous diurnal forms, forms that used eyes rather than ears as their prime sensors. Most would probably have been terrestrial, unlike the water-loving pterosaurs. Would there have been bat hawks and bat vultures, with catlike or doglike jaws for rending and killing? Perhaps. But there probably would not have been song bats or bat ducks. There may have been shore bats skittering on four legs across mudflats and beaches after food, like mammalian caricatures of the lost pterodactyls. It is even possible some may have evolved into large marine soarers—albeit about 100 Myr after bats first appeared; giant oceanic pterosaurs and birds took a similarly long time to evolve. What we can be fairly sure of is that, with their mediocre ground capabilities, bats would never have become flightless. There are no examples of flightless bats, even on islands.

What if bats had never evolved? Would there now be more nocturnal flying birds? There are already a few nocturnal, predatory, insectivorous, and even nut-eating birds—primarily owls, goatsuckers, nightjars, and oilbirds—some of which employ a relatively crude form of echolocation. It is conceivable that there would be many more birds making a living off the countless insects that crowd the night skies if bats were not already doing such a good job of it.

What have been the respective advantages enjoyed, and disadvantages suffered, by bats and flying birds? Unlike pterosaurs and birds, bats and birds exhibit little difference in mental performance: both are big brained and intelligent. Although bats and birds evolved dramatically different respiratory complexes to provide the large amounts of oxygen needed by powered fliers, the overall effectiveness of the two systems seems similar (Maina 2000). In many other respects, bats are Cenozoic versions of pterosaurs, with some of the same limitations, among them a quadrupedal gait. Bats' thin wing membranes are not as streamlined or damage resistant as avian wings, but bats have taken advantage of the high maneuverability inherent in such strongly cambered airfoils for hunting flying insects. Bats' tails may also contribute to their agility. Because bats evolved from small-tailed ancestors, they started off with the short tails conducive to dynamic flight. In this, they differ from pterosaurs and birds, which evolved from ancestors with long tails and, in the initial versions, retained long tails that made them stable fliers. This feature was not entirely eliminated by the eventual abbreviation of the archosaurian fliers' tails. Birds have continued to emphasize speed over agility. They have also benefited from their superb archosaurian vision. The bats' biggest nighttime advantage stems from their being mammals. Having inherited the uniquely sensitive mammalian auditory complex, they have been able to develop a highly sophisticated sonar system that could hardly be better suited for navigating and tracking prey in the dark, an adaptation birds cannot hope to match with their archosaurian hearing (although, as we have seen, a few insectivorous nocturnal birds use a crude form of sonar). Bat sonar is so effective that some nocturnal insects have evolved countermeasures, including preprogrammed evasive maneuvers that are automatically initiated when the target detects that it is being tracked by sound. Whatever the reason, selective evolution has resulted in an arrangement in which dinosaurs best adapted for operating in well-lit conditions mainly fly during the day, and mammals well-suited for flying in the dark take over after sunset.

In one sense, the flying birds of the Cenozoic are remarkably uniform, in that all have beaks; streamlined hands; large, deeply keeled sterna; and broad, highly modified pelves. The days of birds with toothed jaws, clawed fingers, shallow breastplates, and dinosaur hips are long gone. Birds exhibit much less design variation than either mammals (compare kangaroos, shrews, elephants, and whales) or reptiles (compare lizards, turtles, and snakes). However, within these limitations, the story of Cenozoic flying birds has been one of increasing diversity in species and form, of invasion of new niches, and of expansion of the size range to include the tiny and the gigantic.

Paradoxically, the diets of flying birds are extraordinarily diverse yet limited by flight. The most specialized plant-eating volant bird is the hoatzin, whose large, fermenting crop limits its ability to fly. No other flying bird has evolved such a large crop, since it is difficult for fliers to carry the big guts needed for efficient herbivory. Geese are among the few other avian herbivores, and because they lack sophisticated digestive tracts, they are very inefficient at eating and digesting plants. Most nonpredatory, nonaquatic flying birds are herbivores or omnivores that consume the vast array of seeds, nuts, and fruits produced by the flowering plants that dominate the global flora. The interrelationship between the plants and birds is so well established that some plants depend on certain birds for seed dispersal. Some birds feed on plant nectars, and the plants often rely on the flying dinosaurs for pollination. Many land birds also hunt invertebrates, either as a specialty or as part of an omnivorous diet; the robin is an example of the latter. Raptors—the real raptors—and owls have become highly successful predators of the sky, from which they hunt other birds, insects, ground- and tree-dwelling mammals and reptiles, and fish. Interestingly, bats are seldom the victims of hunting birds; even owls prefer terrestrial small mammals to flying mammals. Many birds retain the habits of their Cretaceous predecessors in being water lovers that feed on plants or animals, or both, in or near the water. Tube-nosed birds, gulls, and pelicans tend to prevail over the oceans, and herons and ducks in freshwaters, although there is a lot of crossover.

Again, within the constraints imposed by flight, birds are remarkably diverse in form, much more so than either pterosaurs and bats. At one extreme are hovering hummingbirds, which function as helicopters—a feat achieved by no other flying vertebrates. At the other extreme are the soaring albatross and vulture groups, which can stay aloft for extended periods with hardly a wing beat. Why has the avian form proven more plastic than that of pterosaurs and bats? The feathered wing is more adaptatively flexible than the membranes of the other flying vertebrates, and the decoupling of the fore and aft locomotory systems into two independent and highly capable modes of movement has probably facilitated the evolution of birds to fit a variety of roles.

It is in the Cenozoic that bird size has gone to extremes. Again, hummingbirds are at one end, with the smallest living example, the bee hummingbird, being no larger than its namesake, a mere 1.6 grams. Ergo, bee hummingbirds are the smallest dinosaurs! And they lay the smallest known dinosaur eggs, weighing a third of a gram. The avian thermoregulatory system has reached the level of sophistication needed to cope with the extreme rates of heat loss inherent to such small sizes, in part by allowing a state of nocturnal torpor when the air is cool and food cannot be gathered. At the same time, some flying birds have become big. These days the biggest bustards and swans tip the scales at 18 to 19 kilograms. Albatross, marabou storks, and condors are not as heavy, but their great wings span distances up to 3 to 3.6 meters. The biggest hunting raptors are the Harpy and Stellar eagles, whose 2-meter wings carry bodies weighing up to 9 kilograms.

*Figure 15.1.* Pseudodontorns. It was in the Cenozoic that some birds became enormous, rivaling the earlier pterosaurs. The ultimate avian soaring machines were the oceanic pseudodontorns, whose wings may have spanned as much as 7 meters.

However, it is in the fossil record that we find the really big birds of the air; their wingspans and masses are detailed in Appendix 2. The biggest known hunting raptor was *Harpagornis* of New Zealand's South Island. With 3-meter wings carrying about 13 kilograms, it was as big as a condor. Talon marks on moa bones (in some sites, 10 percent of the individual ratites have such marks) indicate that *Harpagornis* hunted moa—apparently even the big ones—diving on them from above (Anderson 1989, Holdaway 1989, Holdaway and Worthy 1991)! They used their toe claws, especially an enlarged talon on the hallux, to pierce the lungs and sever the nerves and blood vessels along the back. Did these eagles improve the odds in their favor by attacking their far more massive prey in small packs, as do Australian wedge-tailed eagles that mob and kill large kangaroos (P. Olson 1995)? One also wonders whether the supereagles thought that the immigrant Maori, as they first strode ashore more than seven hundred years ago, were yet another group of tall bipeds worth sampling. If so, the eagles would have found the well-armed and quick-thinking humans to be surprisingly dangerous opponents; it was a contest the avian predators were destined to lose.

The oceanic pseudodontorns of the Eocene to Miocene (maybe Pliocene) were albatross-like pelican relatives whose jaws were lined with tooth-like projections. Their wings appear to have approached a span almost twice that of any albatross, probably as much as 7 meters. The longest wing feathers should have been well over a meter long! These lightly built superbirds should have weighed 50 kilograms. These avian giants came close to matching the biggest marine pterosaurs in wingspan and especially in heft. Pseudodontorns further mimicked giant pterosaurs by retaining the pelican-like adaptations of long jaws and necks with which to snatch up sea life while on the wing. These seabirds' lifestyles were probably broadly albatross-like, but if anything pseudodontorns were more specialized for soaring flight than the big tube-nosed seabirds, and even frigate birds and marine pterosaurs. Why pseudodontorns went extinct is a mystery that has been little investigated; perhaps these supersoarers were so specialized that they became vulnerable to changing circumstances.

Even bigger than pseudodontorns were the New World's condorlike teratorns, which are known from Miocene to very recent times. The largest example, *Argentavis*, probably had a wingspan of 6 meters and weighed at least 70 kilograms. *Argentavis* did not come close to matching the biggest pterosaurs in size or wingspan, but, clearly, birds as big as humans and large ratites were getting off the ground. Researchers have debated about whether teratorns were scavengers in the manner of the vultures and condors they resemble (Feduccia 1996)—in which case they could dine on the numerous megamammals of the time—or whether they used their strong legs to run down small prey, which they snapped up with albatross-like jaws (Campbell and Tonni 1982, 1983). Perhaps they did both. Alternatively, teratorns may have been oversized shorebirds that hunted in the shallows for frogs, fish, and the like. Campbell and Tonni (1983), who subscribe to the view that teratorns were soarers rather than flappers, argued that the strong west winds that constantly blow across southern South America spurred the evolution of these superbirds. Conversely, the partial blockage of these winds by the rise of the Andes grounded early *Argentavis*. Both of these ideas are speculative, but plausible. Teratorns disappeared during the general Pleistocene extinction. It is very possible that these birds, probably few in number and slow to take off, were not able to escape a new bipedal hunter recently arrived from up north.

Most birds are not mini-helicopters, or the size of hang gliders. The majority are small and moderate-sized perching forms. In the Paleogene, the coraciiforms—which today include rollers, hornbills, hoopies, bee-eaters, kingfishers, and trogons—were predominant. Also doing well are pigeons, woodpeckers and their relatives, and parrots. Parrots are in many regards convergent with primates. The two groups are generally tropical, arboreal fruit and nut eaters with manipulative appendages (highly kinetic beaks plus toes in parrots, fingers in primates) and well-developed color vision. Members of both groups parent intensely, have long lives, and are highly social, strongly vocal, and very intelligent. Controversial work suggests that parrots not only mimic human speech but also actually understand and use one hundred words. The most obvious difference between the two groups, aside from flight abilities, is that many primates have well-developed stereo vision, whereas parrots do not.

Concerning avian reproduction, it is notable that no member of the group gives birth to live young, unlike a number of reptiles and mammals. Flight does not seem to be the deciding factor, since bats bear live babies. The hard, calcified shells of birds may have prevented them, as well as their dinosaur ancestors, from switching to viviparity (Packard 1977, Paul 1994b). Almost all birds brood at least their eggs and care for the chicks either in the nest or outside it. The exception to the rule is the small group of Australo-Asian megapode fowl, which mimic reptiles and some dinosaurs by burying their eggs in soil or large mounds of fermenting vegetation. They carefully regulate the mound temperature by adjusting the depth and composition of the mound soil. When the chicks hatch, they are as completely on their own as most reptiles and are even able to fly in short order. It appears that the mound nesters evolved their pseudoreptilian nesting habits independently; why such habits have proven selectively advantageous is not obvious. Megapodes and some other Cenozoic birds, especially ratites, are r-strategists that produce many young that suffer high rates of juvenile mortality. But unlike most little reptiles and mammals, the majority of birds are—despite their small size—K-strategists that produce only a few eggs each year and lavish them with body heat and food in the hope of minimizing juvenile mortality. This strategy is an energy-expensive one that runs even these highly energetic fliers ragged; yet it has proven very successful, not least of all for the greatest group of avian K-strategists, the passerines, the songbirds.

Indeed, the greatest avian success story of all time has been the radiation of the passerines. Passerines were present in the Paleogene, but they became *the* big bird group in the Neogene. Of some 9,700 bird species alive today, about 5,700 are passerines. Why have crows, jays, wrens, sparrows, thrushes, flycatchers, finches, warblers, and their ilk become the greatest bird group? Their small size is one reason: little creatures tend to be more numerous than their larger counterparts in terms of population and number of species. The exceptionally high metabolic rates of passerines suggest that they enjoy some sort of energetic advantage. Finally, passerines tend to be large brained and intelligent, which is reflected in their sophisticated songs.

As numerous as they are, passerines have not been a source of flightless offshoots. With the possible exception of one, now extinct island species, no passerine has lost flight, although a few species have come close (Raikow 1985). This is not surprising for a group of such small, arboreal, strong fliers. Other bird groups have been more susceptible to flight loss, to the point that the Cenozoic has been the age of neoflightless Ornithothoraces. It has long been recognized that the loss of large nonavian dinosaurs allowed the subsequent, large-scale radiation of big mammals, some of which fulfilled broadly similar roles. It is equally true that the loss of near-avian, possibly secondarily flightless dinosaurs opened the way to later, modest radiations of secondarily flightless continental birds that often mimicked Cretaceous avepod dinosaurs. The big-headed phorusrhacoids filled the same bipedal, avepod predatory role that the big-headed predatory dinosaurs did. Small-headed, omnivorous ratites are similar to small-headed ornithomimosaurs.

Among the first neoflightless Cenozoic land birds that we know about are the ones we are most familiar with, the ratites. In South America, fragmentary remains suggest the presence of ratites as far back as the middle Paleocene: late Paleocene *Diogenornis* is the oldest unambiguous example. The big debate about ratites centers on their origins. Did most or all descend from a single flightless ancestor? One scenario has ratites appearing on the southern supercontinent, Gondwana, and then diversifying as the land masses split and went their separate tectonic ways (Tuinen et al. 1998). Islands were colonized via rafting on vegetation mats. One problem with this hypothesis is that the continental split occurred back in the Cretaceous, and so far ratite fossils have failed to show up in the same southern or for that matter northern fossil deposits that produce similar-sized dinosaurs and non-ornithurine birds. In addition, although the ratites are similar in gross design, they differ greatly in detailed skull, pectoral, and pelvic anatomy. The pelvis of an ostrich is easily distinguished from that of an emu, which cannot be confused with that of a moa, which is not like that of a kiwi. A second scenario points to the flying palaeognathous birds of the Cenozoic, the Paleogene (arboreal) lithornithids and their close Neogene relatives, the (more terrestrial) tinamous (Feduccia 1996). In this scenario, the various ratite groups each lost flight independently from lithornithids and, perhaps, from tinamous. This polyphyletic hypothesis avoids the problems with the single-origin hypothesis and best explains the island ratites.

Able to outsee and outrun predators and, when vision and speed failed, to replace the losses with lots of chicks, the big-eyed, gracile-limbed ratites have been successful on most of the major continents—North America not among them—while stockier, slower-breeding variants have flourished on more-isolated lands. *Diogenornis* appears to have been an early rhea; rheas have predominated on South America in the late Paleocene. As explained further below, rheas evolved on a continent filled with similarly large and swift avian predators. This ancient need to flee may help explain why rather sluggish rheas have the high aerobic exercise capacity that mystified Bundle et al. (1999). Eocene *Palaeotis* of Europe is either a rhea or an ostrich relative. Definitive ostriches are known from as early as the Miocene; they have been the big birds of Africa and Eurasia. Emus and cassowaries dominate Australia and the big islands to the immediate north. Sadly, some dwarf island emus quickly went extinct upon the arrival of Europeans. The fossil record of these Australo-Asian ratites is poor.

The fossil record is also poor for moa and the aepyornid elephant birds, in that both are mainly known from remains that date from historical times. Indeed, moa and elephant birds were thriving until people showed up on their islands during the medieval period. There is little doubt that humans, via hunting and habitat alteration, were the agents of extinction for these birds, which had been doing quite well on their own for millions of years (Anderson 1989, 2000, Holdaway 1989, Holdaway and Worthy 1991, Holdaway and Jacomb 2000a,b). Although the heavy-limbed ratites were probably better runners than their massive legs might imply (Alexander 1983a,b), their modest top speeds, lack of fear of humans, and low rates of reproduction—all resulting from isolation—left the avian giants so vulnerable that total extinction may have occurred over decades rather than centuries (but see Anderson [2000] versus Holdaway and Jacomb [2000a,b]). New moa restorations tend to de-emphasize their height with more horizontal postures, but Maori rock art suggests that at least some moa adopted a more erect pose (Anderson 1989, Holdaway and Worthy 1991, Naish 1998). Before they became extinct, moa were the tallest known birds, at well more than 3 meters. At as much as 380 kilograms, they were also among the heaviest, being similar in this respect to the shorter but massively built *Aepyornis* (Appendix 2). The elephant bird laid the largest known "dinosaur" egg: at 9 to 12 kilograms, these eggs were as big as the smaller adult dino-avepods.

Remarkably like the gracile ratites but definitely not palaeognathous (although their skulls remain little known) were the largely flightless neognathous ergilornithids and relatives that lived in the Northern Hemisphere from the Oligocene to the Pliocene. Some of these developed two-toed feet so similar to those of ostriches that some researchers have thought that ergilornithids may be ancestral to the swift ratites.

So much for the small-headed birds; the rest of the big-bodied birds had large heads. Long thought to be small-headed ratites, the robust dromornithids, or mihirungs, of Australia were actually big-headed neognathes, probably related to the duck group (S. Olson 1985, Murray and Megirian 1998, Paine 2000). The dromornithids had exceptionally large heads for birds, up to nearly half a meter long in *Dromornis*. *Dromornis* was also about as heavy as the moa and elephant birds (Appendix 2), and other dromornithids were as small as emus. Footprints suggest that dromornithids were walking around in the Oligocene; bones date from the Miocene to just a few tens of millennia ago. That these big birds' big, broad, massively constructed heads with deep but small-hooked beaks (Fig. 15.3) were like those of diatrymids has led some researchers to suggest that dromornithids and diatrymids had similar, possibly flesh-oriented, food preferences (but see below, Wroe 1998). However, analysis of dromornithid eggshell isotopes indicates that they were browsers (Miller et al. 1999), and Murray and Megiran (1998) came to the same conclusion on the basis of their examination of skull structure. Although the herbivorous ratites have been marked by their small, delicate skulls, the plant-eating takahe has a large, strongly constructed skull rather similar to that of the dromornithids. The possibility that dromornithids were omnivores that ate small animals, carcasses, and plants cannot be ruled out. If thunder birds were vulnerable plant-eaters, they would have been hunted by an array of marsupial predators, as well as by supersize monitor lizards weighing as much as a tonne! New analysis of abundant dromornithid eggshell sites favors newly arrived humans as the agent of their extinction, circa 50,000 BP, probably via hunting and via habitat alteration caused by fires (Miller et al. 1999).

Additional big-headed and big-bodied ground birds, probably related to the duck group, appeared very early in the Age of Mammals, in the late Paleocene and early Eocene. The gastornithoids and the diatrymids may actually have been, according to a recent report, one group that spanned the Northern Hemisphere. The best-known genus, *Diatryma,* may have dwelled both in North America and in Europe (Fig. 15.2). Diatrymids have been portrayed as everything from placid herbivores to arch-predators (Witmer and Rose 1991, Andors 1992). One problem for researchers trying to determine their dietary habits is that, on the one hand, the beak of well-known *Diatryma* seems to lack the very well developed hook that is characteristic of birds of prey, including phorusrhacoids. The rather heavy limbs also differ from those of some terror birds; but not all predators are lightly built, and the potential mammalian prey of the time was not all that swift either. On the other hand, the head and jaws were broader than those of phorusrhacoids. I wonder if diatrymids were not oversized versions of shoebill storks, which also have big, broad bills that end in a modest hook. Diatrymids may have used their massive beaks to crush turtles, and their broad jaws to gulp down other small to medium-sized aquatic creatures whole, as well as anything else they could catch. Working in favor of this possibility is that shoebills inhabit wetlands and that the remains of diatrymids are found in sediments deposited in swampy areas.

Perhaps diatrymids were omnivores that ate small animals and scavenged carcasses and also consumed low-fiber plants and fruits. The true feeding style of diatrymids is difficult to assess, to the point that it may always remain obscure. Andors suggested that the extinction of diatrymids was linked to the general reduction of warm wetlands as North America became cooler and drier.

In South America dwell the semipredatory seriemas, living, flying members of Cariamae. The Cariamae appear to be the group that has given rise to the great radiation of flightless predatory birds. Among them are the bathornithids and idiornithids of the later Eocene to early Oligocene, which lived in the Northern Hemisphere. The fairly long skull of *Bathornis* was moderately adapted for predation. Some of these hunting birds were medium sized, and others were quite large. Found in Europe are remains of Eocene cariamians, such as *Aenigmavis,* that either are phorusrhacoids or paralleled them. These medium-sized forms were flightless or close to it. The same was true of the South American psilopterids, which were members of the most spectacular group of big-headed Cenozoic land birds, the phorusrhacoid "terror birds" of the Americas (especially South America). Traces of these birds first appear in the Paleogene, but they reached their height in the Neogene (especially the Miocene).

Up to 3 meters tall and sometimes heavier than ostriches, these eagle-beaked, nonvolant phorusrhacoids were undoubtedly classic predators. L. Marshall (1994) portrayed terror birds as swallowing their prey whole. They probably could indeed have eaten small prey whole, but the transversely narrow jaws of phorusrhacoids indicate they were not specialized for gulping down animals intact. Instead, they probably dispatched and then dismembered large prey. For one thing, the beak was tipped with a great raptorial hook. Elevated shoulder spines—most unusual for birds, and otherwise seen among dinosaurs only in some sauropods and horned ceratopsids—provided an anchor for what must have been exceptionally powerful neck muscles. At least some phorusrhacoids share with seriemas an enlarged, strongly recurved claw on the second toe (C. Zimmer 1997), which is shortened but not hyperextendible (Fig. 4.5C). Chandler (1994) argued that at least

*Figure 15.2. Diatryma.* At nearly 2 meters tall, *Diatryma* was among the first great flightless birds of the Cenozoic. The habits of this massive-beaked dweller of Eocene American wetlands are obscure; here a flock is facing down a pack of *Pachyhyaena.*

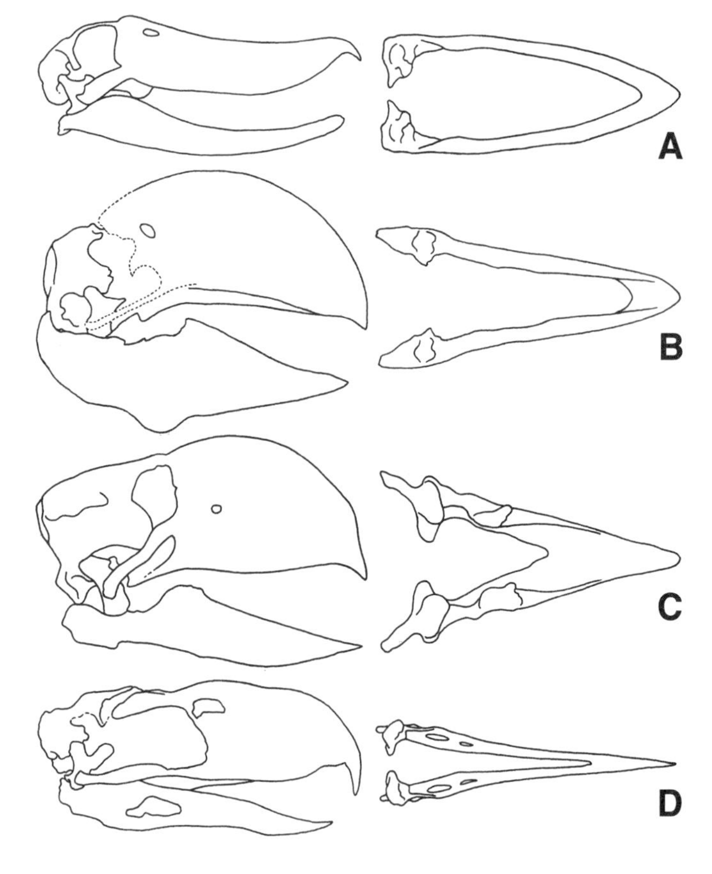


*Figure 15.3.* Skulls and lower jaws (shown in top view) of *(A)* shoebill stork, *(B) Bullockornis, (C) Diatryma,* and *(D) Phorusrhacos.* The skull of *Diatryma* appears similar to that of the terror bird *Phorusrhacos,* which hunted big game, but is even more like that of the shoebill stork, which crushed and swallowed small animals, and that of the possibly herbivorous dromornithid *Bullockornis* in terms of the great breadth and robustness of the jaws and the presence of a high-muscle-leverage coranoid process on the top of the lower jaws. Also note the posterior placement of the external nares in these big-headed, massive-beaked avians. Skulls and jaws drawn to same lengths for comparison.

some of the terror birds had spike-clawed hands, which could have been used to wound and kill big herbivores. By using their hands in this way, phorusrhacoids would have mimicked the hunting action of sickle-clawed dinosaurs, with the difference that the former lacked hook-clawed hands with which to hold onto the prey. When the victim was down, the phorusrhacoids could have used their great hooked beak to rip their meal into bite-sized pieces, as flying raptors do. Therefore, the hunting mode of terror birds appears to have combined aspects of predation both dinosaurian and avian.

Big terror birds came in two flavors: swift-running phorusrhacids with gracile limbs and heavier-legged brontornithids (Tonni 1980). It is obvious that the slender-legged phorusrhacids were pursuit predators. The big brontornithids are often thought to have been scavengers. Although scavenging is almost universal among predators, there is no such thing as a pure land scavenger (Paul 1988a). Locating carcasses from the ground is not as easy as locating them from the air, and walking is very energy inefficient. Besides, animals do not drop dead of their own accord often enough to supply a population of large predators with adequate food. As result, even hyenas hunt, and the only pure scavengers are soaring vultures and the like, which eat relatively little, expend little energy while traveling, and scan vast tracts of land for dead things (Houstan 1979). Stout brontornithids may have scavenged more often than gracile phorusrhacids, but both must have hunted for most of what they ate. The brontornithids probably went after the more massive, slower prey, whereas the phorusrhacids chased down the fleeter mammals and rheas.

The success and extinction of American terror birds, although by no means fully understood, are fascinating subjects, and researchers can call upon a fairly large set of fossil data to study these birds. Almost all phorusrhacoids were South American. For most of the time that they existed, their continent was isolated from the rest of the world and was inhabited by an unusual fauna: the herbivores were placentals—as in most of the rest of the world—but the nonavian predators were marsupials and reptiles—as in Australia. The predaceous reptiles were modified, highly terrestrial crocodilians whose aerobic power system and brainpower were probably limited. The marsupials tended to be heavy-limbed forms whose hunting abilities never equaled those of their placental counterparts on other land masses. These failings of the marsupials seem to have left the door open for birds to fill the role of fast, powerful hunters. There may have been some segregation between the groups.

On the one hand, the crocs and marsupials may have been ambush predators that lived in forests, whereas the tall birds tended to prefer more-open plains and savannas, where they could run down prey over long distances. In this view, the South American flesh-eaters tensely coexisted, occasionally dining on one another. On the other hand, L. Marshall (1994) has noted that the marsupials tended to decline as the great birds thrived, suggesting that the latter proved competitively superior. Why? Brain size was modest in both groups, and both were probably less intelligent than much of the ungulate prey; therefore, brain size does not appear to have been the deciding factor (also contradicted is the common opinion that predators, by necessity, must be smarter than their victims). Both were well armed. Most of the marsupials were vaguely doglike forms with good hunting teeth, and a few had enormous dental sabres. The birds made up for their lack of teeth with their hooked beaks and powerful claws. The height of the birds may have given them a scouting advantage over the low-slung marsupials, but any such advantage was probably not crucial. If anything, the marsupials, with their more-flexible bodies, should have been more agile than the stiff-legged birds. However, the marsupials evolved from little opossum-like mammals with short limbs. For some reason, the modestly aerobic marsupials have never evolved fast-running forms; the only really fast pouched mammals are kangaroos that use an unusually energy-efficient hopping gait. The phorusrhacoids were derived from cariamians with long striding legs, whose speed potential could be fully powered by means of the high-capacity avian cardiorespiratory system. The resulting speed and endurance advantage probably gave the birds the edge.

Explaining the extinction of the phorusrhacoids is a trickier problem, one punctuated by a surprising recent discovery. One of the most important factors in the extinction was the completion of the Panamanian land bridge that linked the Americas about 2.5 Myr ago. The completion of this bridge resulted in the Great Interchange, in which many North American creatures headed south and some South American animals went *al norte*. For example, proboscideans were soon roaming the pampas, and giant ground sloths eventually reached as far north as Alaska. More pertinent to our discussion was the sudden invasion of South America by an array of new pla-

*Figure 15.4. Phorusrhacos.* Avepods came full circle when, after having evolved from terrestrial flesh-eating dinosaurs, the large, flightless South American terror birds of the later Cenozoic readapted to the same hunting role. Among them was 1.5-meters-tall *Phorusrhacos.*

centals, including modern big cats and dogs that had evolved their sophisticated hunting abilities in the harsh arena inhabited by fauna from around the globe. The placental carnivores were big brained and correspondingly intelligent, and their superior neural controls and mammalian agility should have served them well. The basic theory holds that terror birds—as terrible as they were—were unable to compete with these new arrivals, either because the birds were unable to catch as much prey as the cats and dogs or the birds themselves (especially their eggs and chicks) became the prey of the newcomers, or both. Although it is certainly true that South American phorusrhacoids bought the farm at this time, there are problems with explanations based on simple competition.

For starters, we do not know whether phorusrhacoids increased in diversity up until the time of the Great Interchange (L. Marshall 1994) or whether they began to decline before then (Feduccia 1996, Tambussi et al. 1999). If they were already in decline, then the debut of placental carnivores on the scene was just the lethal blow for a group already in trouble.

Another problem is that we are not sure why terror birds, especially the larger examples, would have been so vulnerable to the new competition. Imagine how capably a terror bird that was more than 2 meters tall and had an eagle beak and slicing claws could have defended its nest and clutch! Some losses would undoubtedly have occurred, but ratites, and their eggs and babies, have managed to survive in a world full of mammalian carnivores (it is not clear whether terror birds reproduced rapidly like ratites and mammalian carnivores or whether they were K-strategists vulnerable to even modest rates of juvenile mortality). Savanna cats and dogs such as lions, big cheetahs, and wolves never made it through the dense forests of the Isthmus of Panama. The placental carnivores that did make it through were shorter-limbed jungle cats and wild dogs, which appear to have been more serious competitors for the forest-dwelling marsupials than were the savanna birds.

In fact, we know that a least one phorusrhacoid was able to hold its own against the best mammalian competitors. One of the big surprises of paleoornithology was the recent discovery of one of the biggest phorusrhacoids of all time. This phorusrhacoid, the last known survivor of the group, lived in what is now the southeastern United States at least 1.5 Myr ago (L. Marshall 1994) and perhaps as recently as 15,000 BP (C. Zimmer 1997)! This terror bird was able not only to survive among the most sophisticated open-country mammalian carnivores but also to move north against the competition. Why did all the rest of the great birds of prey disappear?

Size may provide an explanation for the predatory birds' extinction, but unfortunately the size pattern is a confusing one. In South America, most of the large phorusrhacoids disappeared in the Miocene, and the smaller psilopterids did best in the Pliocene prior to the Great Interchange (Tambussi et al. 1999). But that the last species were among the largest suggests that only the big birds were able to handle the mammalian carnivores. Perhaps the smaller species were not able to defend their nests against the new placental carnivores. With diversity suppressed, vulnerability to total extinction induced by adverse fluctuations or nonlinear chaos would have been elevated until the last survivor also disappeared, despite its aggressive move north.

Schultz et al. (1998) have proposed that an extraterrestrial impact just off the coast of South America 3.3 Myr ago was involved in the substantial extinctions that occurred around that time. The explosion was not as large as the K/T impact but may have been sufficient to either directly annihilate much of the South American fauna or to disrupt the continent's environment to the point that the phorusrhacoids were unable to fully recover. This scenario may help explain why the southern terror birds went bust before those that had moved north.

Big ground birds have done well in the Cenozoic continents, but their success has been sharply limited in some respects. On one hand, they were present in substantial numbers in both the Paleogene and Neogene, and large Neogene ratites radiated in the face of stiff mammalian competition on all but one continent. On the other hand, big birds have never been very numerous. These days there are just a handful of giant bird species, and they are entirely absent from some continents. None are flesh eaters. Meanwhile hundreds of species of mammals similar in size are running about on all of the continents except Antarctica. In addition, big birds have never dwelled in arctic regions despite their excellent insulation and high metabolic rates.

Ratites and other herbivorous ground birds lack the batteries of large grinding teeth that characterize most mammalian herbivores. Ratites have tended to compensate by using stone-rolling gizzard mills that are also lined with hard keratin-like surfaces (Withers 1983, Herd and Dawson 1984, Farlow 1987, Anderson 1989). Whether or not this substitution has been fully effective is not clear, since the digestive efficiency of ratites tends to be lower than that of ungulates of similar size (Withers 1983, Herd and Dawson 1984, Farlow 1987). However, this relative inefficiency may reflect the birds' simpler digestive tracts.

The inability to find an important use for the arms does not appear to have been a critical problem for flightless herbivorous birds because most mammalian herbivores use their forelimbs for the same locomotory purposes as the hindlimbs.

Large herbivorous birds have been able to survive among sophisticated mammalian predators for a number of reasons. Their avepod legs not only make them superb runners but also can be used as kicking weapons: the second toe of cassowaries is even modified into a lethal sabrelike weapon (Fig. 4.5C; Kofron 1999). The combination of tall head carriage and excellent vision constitutes an effective warning system. Another important reason for the persistence of big birds may be their high rates of reproduction. Because they lay large numbers of eggs, the large continental ground birds are "weed" species better able to recover from high mortality than slower breeding ungulates (Paul 1994b). Although r-strategy reproduction has not been the ticket to spectacular success, it has probably helped prevent ratites from going extinct.

The last issue brings us to another point. Although ostriches, terror birds, and mihirungs are quite large, there have never been any really gigantic birds. None has weighed more than a hefty, half-tonne horse, and even the largest birds are pygmies when compared with megamammals and dinosaurs. Birds have remained relatively small despite the twin facts that they can produce the large numbers of young (which gives animals an advantage in maintaining viable populations of giants; see Chapter 14) and that they certainly grow fast enough to become gigantic (Paul 1994b). That birds are bipeds should not be a limiting factor, since some two-legged dinosaurs weighed many tonnes (Paul 1988a, 1997a). The high-capacity, high-pressure avian cardiorespiratory complex should have been able to handle massive bulk and towering height. The lack of a large number of gigantic herbivores in the Cenozoic probably discouraged predatory birds from becoming as big as allosaurs and tyrannosaurs. But being truly enormous would have served to increase the digestive efficiency of herbivorous birds. Why none has scaled up to multitonne size remains a mystery. Perhaps there just have not yet been enough species for the generation of a true giant.

Flightless diving birds also failed to become truly gigantic: none has reached whale bulk. The

*Figure 15.5.* The big three of the Late Ice Age, the largest of known birds at 370 to 380 kilograms (see Fig. 8.4). *Left,* from New Zealand, the tallest bird, *Dinornis;* the feathers are preserved. *Upper right,* hailing from Madagascar, the elephant bird *Aepyornis,* nestling down on its 20-gallon eggs. *Lower right,* from Down Under, the nonratite thunder bird *Dromornis.* All of these giants were probably driven extinct by invading humans, the two slow-breeding ratites within surprisingly recent historical times. Drawn to same approximate scale.

reason for this particular size limitation is obvious. Because they cannot give birth to live young, flightless marine birds must remain small enough to venture ashore to lay their hard-shelled eggs. This necessity makes the size reached by some marine birds all the more impressive. The mid-Cenozoic plotopterids of the northern Pacific were pelican relatives whose design strongly converged with that of penguins. The biggest may have weighed as much as 200 kilograms. Perhaps even larger were some of the earliest known penguins. The Eocene giants were five to perhaps ten times heavier than the largest living penguins, and their bulk may have rivaled that of elephant birds (Appendix 2). Supersize penguins and plotopterids disappeared in the mid-Miocene. They may not have been able to compete with the bigger-brained, sonar-equipped porpoises and seals that were appearing at that time. Always limited to the Southern Hemisphere, penguins—as well as plotopterids and flightless auks—use their flipper-wings to literally fly underwater. Penguins seem to be close relatives of loons, the early examples of the former having had longer, more daggerlike spearing bills than today's penguins. The other large, flightless marine bird of the Cenozoic was the great auk, the last of which was killed in 1844. A few ducks and cormorants have also lost the ability to soar through the sky.

## Modern Times and Beyond

Avian extinctions have reached a new high with the advent of the genus *Homo,* especially its latest expression *H. sapiens.* Particularly vulnerable have been the flightless island birds, not just the giant ratites but also the rails, dodos, and other pigeons (S. Olson and James 1982, Holdaway 1989, Worthy 1999). The notion of the noble savages living in eco-harmony with their environment is more myth than reality. When the Polynesians first got to New Zealand in the late 1200s, they must have thought they had hit the culinary jackpot when they discovered great islands packed full of wonderfully pacific big birds offering super drumsticks and eggs so large that one or two could provide breakfast for a whole village. The only evil to mar this Maori paradise was the supereagle that descended from the skies with its piercing talons and was an awesome danger to be eliminated. The concept that these moas and eagles were avian marvels to be preserved probably did not occur to the Maori. It is possible, and hotly debated, that humans were—just before prehistoric times came to a close—responsible for the great wave of extinctions that hit the large animals of the polar regions and the Americas some ten millennia ago. Among the victims were the last teratorns, and the California condor went into a sharp decline as the big mammal carcasses that constituted the majority of its diet disappeared.

Modern civilization has been especially hard upon many birds. Hunting has been one factor in driving down bird populations, and neither pollution nor logging has helped. But the most disruptive force has been agriculture, which ties up vast tracts of land in order to feed the teeming billions of bipedal primates. Agricultural fields are able to support some birds, and a handful of bird species have directly benefited from agriculture, at least in terms of reproductive success, by being domesticated; but farmlands are highly artificial complexes that radically alter natural habitats. Although residential, commercial, and industrial areas are even more modified than farmland per square kilometer, these regions take up a comparatively small fraction of the land. Residential areas also support large populations of songbirds, often more so than agricultural areas, where bird feeders are lacking and pesticides are persistent.

The ultimate symbol of avian extinction, the passenger pigeon, is an ironic example of how modern agriculture and civilization can doom a bird. At one time, billions of these doves dwelled in the north-central United States. Passing flocks darkened the skies for days. By the early twentieth century, the species was extinct. Why? Although they were hunted extensively for meat, even an early ban on commercial hunting probably would not have saved the bird. As Bucher (1992) explained, the problem was more subtle and insidious. Passenger pigeons were mast specialists; they depended upon the countless nuts and acorns of beechnuts and other hardwood trees to breed successfully. Because nut production is very irregular, the doves could find enough food only by being nomads that moved in superflocks, ready to descend upon whatever nut-rich forest they encountered amid the vast pre-Columbian hardwood forests. So they could survive only by being extremely numerous, which required an enormous foodbase. By the mid-1800s, the hardwood forests had been heavily logged, and most had been converted into farmlands even more extensive than we see today. With their foodbase fragmented into small, scattered locales, the slow-breeding doves' population quickly crashed. Soon the flocks were no longer large enough to successfully find large amounts of mast in the depleted woodlands. The

*Figure 15.6.* Birds of the modern world. Despite the extreme impact of humans, birds continue to flourish in the modern world. Among them are a number of highly or entirely terrestrial examples. The hawklike secretary bird *(left)* is shown pinning down one of its favorite prey items, a snake. Also from Africa is the large-beaked ground hornbill or thunderbird *(lower right)*. The little roadrunner *(upper right)* inhabits the North American Southwest.

inherently gregarious doves were doomed. Passenger pigeons were therefore a hopeless case, in that no measure short of settling the country could have saved them. Viewed realistically, their extinction may have been for the best. As awe inspiring as the preindustrial superflocks may seem to us living in these hi-tech times, few would actually want to live with millions of birds in the neighborhood, covering the ground with their droppings and filling the air with their calls. Had the passenger pigeon survived, it could have done so only in small flocks, as a diminished version of what it once was.

A better example of an unnecessary avian extinction involves a flightless bird, the great auk. These penguinlike birds of the North Atlantic were very abundant until they were slaughtered for their meat and eggs to the point of extermination before the American Civil War. Had these birds been properly conserved, they would still be flourishing today, even with civilizations to the east and west.

Looking at the bigger picture, we must understand that not only is extinction normal—all but a few percent of the species that have yet lived are long extinct—but also even mass extinctions are a normal part of evolution. Consider that storms are brief but normal interruptions to the usual weather, disruptions that, although destructive, are useful because they alter the landscape and redistribute animals in ways that would not otherwise occur. Mass extinctions are likewise crucial to evolution because they shake things up and get life moving along new paths. Mass extinctions are not pretty, but evolution is not moral, and it does not have a conscious goal. Its most distinguishing feature is change (Paul and Cox 1996).

Some environmentalists and ecologists have claimed that we are in the midst of the greatest mass extinction yet seen. This is not true. The K/T extinction left not a single large, fully terrestrial animal alive. These days, elephants still number in the hundreds of thousands, great herds of big ungulates are still hunted by packs of carnivores, and deer are actually experiencing a population explosion. Even some free-living birds have benefited from humans and their civilization, sometimes to our annoyance. Rock pigeon, starling, and recently crow populations have soared to the point that these birds are often considered pests. Gulls thrive in our harbors. The global fauna is resilient. There are so many birds that the great majority have survived the human revolution in fairly good order. Even the big continental ratites have done surprisingly well. Although some rheas and cassowaries are threatened, none is currently endangered; ostriches are numerous and widespread, and emus are thriving despite long-standing efforts to reduce their numbers (del Hoyo et al. 1992–99). The big birds' ability to rapidly replace their losses to human hunters has certainly been helpful.

In the end, ground birds have an evolutionary ace up their sleeves. Even if they are all killed off, they can come back as long as there is a source of suitable ancestors, that is, flying birds. As for the latter, they are likely to survive in great numbers no matter what humans do to them. Even a full-scale nuclear war would probably not wipe them *all* out. In the very long term, the prognosis is grim. The sun is slowly heating up, and eventually, in a few hundred million years or more, the planet will become too hot to sustain vertebrate life of any kind. But by then space travelers may have carried birds to distant lands the way that sailors carried parrots on the old sailing ships. Until then, only a natural supercatastrophe greater than the K/T crisis could wipe out all of the feathered dinosaurs. Their age continues.

# APPENDICES

APPENDIX 1

# Character Lists

## A: General Character List

In the general character list given below, only potential bird relatives are analyzed, consisting of arboreal archosauromorphs, pseudosuchians, sphenosuchian protocrocodilians, and predatory dinosaurs. Drepanosaurs, also considered potential bird relatives, include *Megalancosaurus* and appropriate parts of *"Protoavis." Scipionyx* is not included here because it is a very small juvenile. Tyrannosaurs and poorly preserved and described *Yandangornis* are not analyzed. Only characters pertinent to avian origins are listed, so those pertinent only to the intrarelationships of nonavebrevicaudans are omitted. Characters for a group are based on the basal-most known member(s), but if a character is known only from the derived members of a group, it is usually assumed to exist in basal members. Among archaeopterygiforms, it is assumed that a character present in *Archaeopteryx* is present in *Protarchaeopteryx* unless known otherwise. Except for the critical characters detailed in Chapter 2, characters listed under new apomorphy+clade–based names are not part of the diagnosis for that taxon. Only skeletal characters are listed. In Chapter 10, character counts do not include broad category characters, such as "avian respiratory complex developing with . . . ," which are themselves formed by characters listed immediately after them.

The following symbols are used in the character lists:

- ! = Character potentially associated with flight, and therefore potentially subject to reversal if flight is lost.
- CU = condition unknown, undescribed, or ambiguous in . . .
- P = probably in . . .
- PN = probably not in . . .

*Drepanosaurs, caudipterygians, oviraptorosaurs, most ornithomimosaurs, avimimids, alvarezsaurs, avebrevicaudans*
Teeth very reduced or absent

*Drepanosaurs, theropods (including Aves)*
Neck long and slender (except giant dino-theropods)

*Drepanosaurs, herrerasaurs, avetheropods, archaeopterygiforms, alvarezsaurs, avebrevicaudans*
!Scapula blade is a thin strap

*Drepanosaurs, avepectorans*
Frontals posteriorly broad and anteriorly narrow, so eyes have overlapping fields of vision

*Drepanosaurs, archaeopterygiforms, dromaeosaurs*
Diamond-shaped dorsal supraoccipital
Opening to pneumatic sinus on the paraoccipital

*Drepanosaurs, archaeopterygiforms, dromaeosaurs, caudipterygians, oviraptorosaurs, avimimids, avebrevicaudans*
Premaxilla enlarged
External nares enlarged (P avimimids)

*Drepanosaurs, archaeopterygiforms, oviraptorosaurs, avimimids, avebrevicaudans*
External nares extend more posteriorly (P avimimids)

*Drepanosaurs, archaeopterygiforms, some dromaeosaurs, derived alvarezsaurs, avebrevicaudans (except some basal examples)*
Postorbital and posttemporal bars at least almost incomplete (CU drepanosaurs)

*Drepanosaurs, alvarezsaurs, avebrevicaudans (except some basal examples)*
Superior temporal bar incomplete (CU drepanosaurs)

*Drepanosaurs, avimimids, alvarezsaurs, avebrevicaudans (except some basal avebrevicaudans)*
Postorbital bar incomplete (CU drepanosaurs, archaeopterygiforms)
Posterior temporal bar incomplete (CU drepanosaurs, archaeopterygiforms)

*Drepanosaurs, dromaeosaurs, ornithomimosaurs, avebrevicaudans*
Cervical centra articulations at least incipiently

saddle shaped (PN archaeopterygiforms, CU various avepectorans)

*Drepanosaurs, dromaeosaurs, troodonts, oviraptorosaurs, ornithomimosaurs, avimimids, avebrevicaudans*
Cervical ribs too short to overlap

*Longisquamids, avepods (including Aves)*
!Furcula-like structure present

*Sphenosuchians, eoraptors, dino-avepods, archaeopterygiforms, alvarezsaurs,* Cathayornis
Antorbital fossa and fenestra expanded anteroventrally

*Sphenosuchians, avepods (including Aves)*
Braincase elements pneumatic

*Sphenosuchians, various dino-avepods, carinates*
Quadrate pneumatic

*Sphenosuchians, troodonts, oviraptorosaurs, avebrevicaudans*
Quadrate at least incipiently double headed

*Sphenosuchians, troodonts, oviraptorids, avimimids, alvarezsaurs, avebrevicaudans*
Quadrate contacts braincase

*Sphenosuchians, various dino-avepods, archaeopterygiforms, avebrevicaudans*
Teeth have constricted waists
Tooth replacement pits on medial surface of root

*Sphenosuchians, archaeopterygiforms, dromaeosaurs, troodonts, caudipterygians, avimimids, avebrevicaudans*
!Coracoid elongated

*Sphenosuchians, some dromaeosaurs, archaeopterygiforms, avebrevicaudans*
!Coracoid narrow (CU rahonaviforms, avimimids)

*Some sphenosuchians, dromaeo-avemorphs (including* Archaeopteryx*)*
Ischium at most one-half the length of pubis

*Sphenosuchians, lagosuchians, theropods (including Aves)*
Pubis elongated
Gait erect
Hindlimbs gracile
Femoral shaft bowed convex anteriorly
Toe V reduced and no longer divergent

*Sphenosuchians, dinosaurs (including Aves)*
Frontal participates in supratemporal fossa (except some avepectorans, CU in lagosuchians)
Postfrontal absent (CU lagosuchians)
Acetabulum perforated

*Sphenosuchians,* Pseudolagosuchus, *dinosaurs (including Aves)*
Pubis strongly elongated

*Lagosuchians, dinosaurs (including Aves)*
Neck S-curved because cervicals are parallelogram shaped
Clavicle-interclavicle brace no longer connected
!Coracoid tuber on coracoid enlarged
Pectoral crest elongated distally
Pectoral crest subrectangular
Femoral head inturned
Fossa trochanterica on femoral head
Lesser trochanter on femur (may be co-joined with greater trochanter)
Trochanter shelf on posterior femur (often reduced)
Cnemial crest present (except *Iberomesornis*)
Tibia quadrangular in distal view
Distal fibula fairly slender
Pretibial process formed by backing of astragalus ascending process by tibia
Calcaneum reduced
Ankle is a mesotarsal hinge joint
Metatarsal bundle laterally compressed and tightly bound together

Pseudolagosuchus, *theropods (including Aves)*
Metatarsal IV sigmoidally curved in anterior aspect

*Dinosaurs (including Aves)*
Skull transversely narrow (CU protodinosaurs)
Epipophyses on cervicals
At least three sacrals, at expense of dorsals
Sternal plates ossified (some exceptions)
Shoulder glenoids faces ventroposteriorly (except most avepods)
Pectoral crest further elongated distally
Finger I medially divergent (CU protodinosaurs)
Fingers IV and V vestigial or absent
Distal ischium elongated, leaving obturator process sited proximally (except alvarezsaurs, avebrevicaudans)
Femoral head more strongly inturned
Femoral lesser trochanter better developed
Pretibial process of astragalus taller (CU eoraptors)

*"Saurischians"*
Dorsal series stiffened with extra hyposphene-hypantrum articulations
At least some degree of superficial postcranial pneumaticity present

*Theropods (including Aves)*
Nasal elongated at expense of frontal (except therizinosaurs, CU lagosuchians, staurikosaurs)
Posterior temporal fenestra closed (CU lagosuchians, staurikosaurs)
Vomers posteriorly elongated (CU lagosuchians, staurikosaurs)
Internal nares posteriorly elongated (CU lagosuchians, staurikosaurs)
Postcranial elements thin walled
Avian respiratory complex developing with:
dorsal column further stiffened by partly ossified interspinal ligaments

dorsal series shortened
number of double headed ribs increased posteriorly
Fingers I–III long and slender
Fibula braced at midshaft by fibular process of tibia (CU eoraptors)
Obligatorily bipedal when on the ground
*Dino-theropods (including archaeopterygiforms)*
Distal tail stiffened by elongated prezygapophyses that overlap preceding vertebrae (CU eoraptors)
Medially divergent thumb is a large, clawed weapon
Manual and pedal claws large; former strongly recurved and sharp
*Herrerasaurs, avepods*
!Scapula blade more slender
Hand more raptorial (with some exceptions)
Hand at least one-half the combined humerus-radius length (with some exceptions)
Pubis narrow (CU *Pseudolagosuchus*)
*Herrerasaurs, dino-avepods, some avebrevicaudans*
Incipient intramandibular joint (CU lagosuchians, nonavetheropods, archaeopterygiforms)
*Herrerasaurs, dino-tetanurans, archaeopterygiforms, basal avebrevicaudans*
Pubic boot present (with some exceptions)
*Herrerasaurs, dino-avetheropods, archaeopterygiforms, basal avebrevicaudans*
Pubic boot well developed (with some exceptions)
*Herrerasaurs, archaeopterygiforms, dromaeosaurs, therizinosaurs, alvarezsaurs, avebrevicaudans*
Pubis retroverted
*Herrerasaurs, most avepectorans*
!Arm-folding mechanism developing with:
!at least moderately asymmetrical distal humeral condyles
!ecte/entepicondyles on distal humerus
*Avepods*
Jugal at most three pronged, because antorbital process is very small or absent (except some large forms)
Vomers co-fused (CU eoraptors, herrerasaurs)
Pterygoids somewhat slender to very slender
Neck more strongly S-curved (some exceptions)
At least five sacrals, reducing dorsals to thirteen
Gastralia articulate via interlocking zigzag pattern (CU more-basal theropods; condition may be partly developed in coelophysoids)
!Single-element furcula present (except some examples)
Avian respiratory complex further developing with:
at least cervicals and cervical ribs pneumatic
all but last posterior ribs double headed
posterior ribs elongated
Shoulder glenoid faces laterally (except some dino-avepods)
Finger II is the longest
Ilium elongated anteroposteriorly
Ilium is a subrectangular plate
Pubes very narrow
Vertical supra-acetabular ridge present on ilium
Femoral head more strongly inturned
Cnemial crest further enlarged (except dromaeo-avemorphs, *Iberomesornis*)
Cnemial crest subrectangular
Metatarsal V at most a thin splint
Metatarsal I does not contact tarsals
Foot tridactyl
*Dino-avepods, archaeopterygiforms*
*Interpterygoid-palatine fenestra present (closed in some large forms, CU alvarezsaurs, baso-avebrevicaudans)*
Lateral pro-otic depression (absent in some dino-avepods, CU alvarezsaurs, baso-avebrevicaudans)
Anteromedial process of palatine has a semi-hook shape (CU alvarezsaurs, baso-avebrevicaudans)
At most twenty-three presacrals; ten are functional cervicals
*Dino-avepods, archaeopterygiforms, some enantiornithines*
Ectopterygoid jugal process strongly hooked (CU alvarezsaurs, baso-avebrevicaudans)
*Cerato-saurans, ornithothoracines*
Pelvic elements co-ossified
*Coelophysids, ceratosaurs, some oviraptorosaurs, archaeopterygiforms, avimimids, most avebrevicaudans*
Tarsometatarsus elements at least incipiently co-ossified
*Averostrans including ceratosaurs, archaeopterygiforms, basal avebrevicaudans*
Interpterygoid-palatine fenestra larger (CU ceratosaurs)
Bar connecting anterior and posterior pterygoid more slender
Accessory antorbital fossa (or fossae) lead(s) to maxillary sinus(es)
Nasal airway L-shaped (except oviraptorids)
Lacrimal lateral pneumatic excavation
Ilium strongly elongated anteroposteriorly (except dromaeo-avemorphs, basal avebrevicaudans)
Lesser trochanter of femur bladelike
Femoral head more transversely elongated
*Elaphrosaurs, yangchuanosaurs, various dromaeo-avemorphs and avebrevicaudans*
Dorsal process(es) on ischium

*Yangchuanosaurs, ornitholestids, avepectorans*
*Posterodorsal edge of ilium curves ventrally (CU monolophosaurs)*
*Some nonavepectoran avepods, dromaeosaurs, rahonaviforms, troodonts, caudipterygians, oviraptorosaurs, avimimids, most alvarezsaurs, avebrevicaudans*
Well-developed hypapophyses under cervicodorsals (PN *Archaeopteryx*)
*Tetanurans*
Bar connecting anterior and posterior pterygoid even more slender
All maxillary teeth anterior to antorbital bar
Cervical epipophysis better developed
Avian respiratory complex further developing in basal tetanurans with:
at least some dorsals pneumatic
pneumatic spaces camellate
length of dorsal series further shortened
rib cage deepened
rib cage ventilation shifted posteriorly with:
first dorsal ribs shortened
posterior ribs more elongated
better-developed double heads on posterior ribs
Proximal half of metacarpal I closely appressed to metacarpal II
Metacarpal I short (one-half or less the length of metacarpal II) and stout
Iliopubic larger than ilioischial articulation (also abelisaurs)
Fibular process of tibia elongated proximodistally
Metatarsal III at least somewhat pinched proximally (except baso-avebrevicaudans, enantiornithines)
*Dino-tetanurans, archaeopterygiforms*
Chevron bases with paired anterior and posterior processes
Lesser femoral trochanter expanded and separated from femoral head by a cleft
Lesser femoral trochanter proximal (except caudipterygians)
*Dino-tetanurans, archaeopterygiforms, baso-avebrevicaudans*
Pectoral crest has a broad, curved-hatchet shape
*Avetheropods*
Jugal participates in antorbital fenestra rim (also dilophosaurs)
Connection between anterior and quadrate processes of pterygoid is slender
Surangular does not interdigitate with dentary, so intramandibular joint is well developed (latter absent in some dino-avetheropods and most avians)
Anteromedial process of palatine shortened (CU nonavepectorans)
Vomers further elongated
!Transition point of tail located more proximally
Ectopterygoid pneumatic
Avian respiratory complex further developing with:
first dorsal ribs further shortened
Humerus head and distal condyles transversely narrower
!Arm folding partly developing with:
!Incipient semilunate carpal (CU in some nonavetheropods)
At least a small extensor process on metacarpal I (a few exceptions)
Hand at most tridactyl, with metacarpal IV severely reduced or absent
Obturator foramen absent
Ilial preacetabular fossa present (except *Patagopteryx*, ornithurines)
Pretibial process taller
Pretibial process plate shaped
Pretibial process separated from condyles by prominent horizontal groove
*Allosaurs, compsognathians, avepectorans*
Pretibial process taller (CU ornitholestians)
*Coelurosaurs? (coelurids, compsognathians, ornitholestians [tentatively including monolophosaurs, proceratosaurs], avepectorans)*
Fifteen or fewer caudals with transverse processes (P compsognathians, CU monolophosaurs)
Ischium at most two-thirds the length of pubis (P monolophosaurs, CU ornitholestids)
Obturator notch on ischium present (except monolophosaurs, CU ornitholestids)
Obturator process of ischium subtriangular
Ischial foot absent (except monolophosaurs, at least some ornithomimosaurs)
*Maniraptors? (ornitholestians, avepectorans)*
Cervical zygapophysis flexed (except and CU some dino-avepectorans)
!Fewer proximal caudals lack elongated prezygapophyses
!Majority of distal chevrons very shallow
Fourth trochanter of femur very reduced or absent
*Avepectorans (dromaeo-avemorphs [including archaeopterygiforms], troodonts, caudipterygians, oviraptorosaurs, therizinosaurs, ornithomimosaurs, alvarezsaurs, avebrevicaudans)*
Beak at least incipient (indicated by hyperdevelopment of foramina on premaxilla and/or tip of dentary that may have fed keratin sheath)

Two accessory antorbital fenestrae (except therizinosaurs, oviraptorosaurs, avebrevicaudans)
Frontals subtriangular, so eyes have overlapping fields of vision (except oviraptorids, some alvarezsaurs, avebrevicaudans, CU therizinosaurs)
Elongated frontal process of postorbital directed vertically (except caudipterygians, oviraptorids)
Upper temporal bar set low, so upper temporal fenestra faces laterodorsally (except oviraptorids; CU caudipterygians, oviraptorosaurs)
Palatine nearly or completely triradiate
Palatine very close to or contacts ectopterygoid (if latter is present)
Ectopterygoid process of pterygoid very reduced or absent
Pterygoid flange of ectopterygoid reduced to the point that it is not visible in lateral view
Epipterygoid reduced or absent
!Brain/body mass ratio in avian range
 Teeth modified in some manner from standard dino-theropod pattern including:
  serrations strongly asymmetrical, reduced in size or number, or absent
 teeth reduced in size or absent (except dromaeosaurs)
Sacrum narrows posteriorly (except ornithomimids, oviraptorosaurs)
!Tail modified in some manner from standard dino-theropod pattern including:
 !caudal count of forty or less (P rahonaviforms, avimimids)
!Avian respiration further developed with:
 !increased pneumatization of dorsal, sacral and caudal series (with exceptions)
 !transversely broad hinge joint between coracoids and broad sternal plate (except ornithomimids, alvarezsaurs, P therizinosaurs)
 !ossified sternal plates elongated posteriorly (except archaeopterygiforms, ornithomimids; CU adult troodonts, therizinosaurs)
!Sternocostal articulations form hinge joints
!Arm-folding mechanism well developed with (except ornithomimosaurs):
 !strongly asymmetrical distal humeral condyles (CU therizinosaurs)
 !Well-developed semilunate carpal block
Metacarpal I at most one-third the length of II (except some oviraptorosaurs, ornithomimosaurs, alvarezsaurs)
Supra-acetabular shelf narrow (except ornithomimosaurs, alvarezsaurs)
Pubic apron reduced so pelvic canal is deepened (except *Unenlagia;* CU troodonts, therizinosaurs)
Long-distal ischial rod absent (except ornithomimosaurs)
Tibia subrectangular in distal view (rather than with narrow lateral apex, as in other more basal dino-avepods)
Distal fibula very slender or absent

*Archaeopterygiforms, dromaeosaurs, troodonts, caudipterygians, oviraptorosaurs, avebrevicaudans*
 !Scapula horizontal
 !Coracoid strongly reflexed relative to scapula
 !Superficial surface of coracoid faces strongly anteriorly
 !Coracoid tuber at least beginning to be enlarged into acrocoracoid process

*At least some archaeopterygiforms, dromaeosaurs*
 Nasal depressed (except some derived dromaeosaurs)
 Maxillary dorsal process posteriorly elongated, and maxillary portion of anterior lacrimal process short, so that maxilla almost reaches preorbital bar
 Preorbital bar straight and narrow in lateral view
 Posterior facing step on postorbital process of frontal (P archaeopterygiforms)
 Paraoccipital process so strongly twisted that:
  anterior surface faces partly ventrally and helps form an external auditory meatus
  posterior surface has a pronounced bulge at midlength (most visible in dorsoventral view)
 Shallow subrectangular ventral expansion of paraoccipital process
 Pneumatopore on anterior surface of paraoccipital process opens into pneumatic sinus
 Ectopterygoid set more posteriorly
 !Finger II at least as robust as I
 Iliopubic articulation inverted V-shape
 Ilial ischial peduncle much shorter than pubic peduncle
 Pubic midshafts are flat plates, angled about 140° to one another

Archaeopteryx, *some dromaeosaurs,* Unenlagia
 Distal pubis deflected posteriorly (inconsistent in *Archaeopteryx;* CU *Protarchaeopteryx,* rahonaviforms)

Archaeopteryx, Unenlagia, *some avebrevicaudans*
 Medial wall of acetabulum better developed (CU *Protarchaeopteryx,* basal avebrevicaudans)

*Archaeopterygiforms, dromaeosaurs, some basal avebrevicaudans*

Small dorsal ectopterygoid depression (except *Dromaeosaurus*)
Ilium short relative to most other tetanurans
Cnemial crest small compared with most other avepods

*At least some archaeopterygiforms and dromaeosaurs, most avebrevicaudans*
!Arms at least approximately 80% of leg length (except *Protarchaeopteryx*, derived dromaeosaurs)
!At least some broadening of the proximal phalanx of finger II (CU archaeopterygiforms)

*Dromaeo-avemorphs, avebrevicaudans*
!Metacarpal III bowed posterolaterally (CU rahonaviforms)
Ilium parallelogram shaped (CU troodonts, alvarezsaurs)
Vertical supra-acetabular ridge set posteriorly
Partial medial wall of acetabulum present (and a few avebrevicaudans)
Pubic peduncle retroverted

*Archaeopterygiforms, some dromaeosaurs, derived alvarezsaurs, avebrevicaudans*
Postorbital and posterior temporal bars nearly incomplete or incomplete
Pubis strongly retroverted

*Dromaeo-avemorphs, troodonts*
Anterior pterygoid very slender (CU rahonaviforms)
!Tail base dorsally hyperflexible (P troodonts)
!Transverse processes and neural spines restricted to proximal tail (except *Protarchaeopteryx*)
!Distal caudal neural spines reduced to fine ridges (CU *Protarchaeopteryx*, most other dino-avepectorans)
!Distal three-quarters of caudal centra elongated and slender tail (except *Protarchaeopteryx*)
!Distal three-quarters of caudal chevrons elongated and shallow tail (except *Protarchaeopteryx*)
!Large notch in dorsoanterior edge of coracoid (CU rahonaviforms, oviraptorosaurs)
Posteromedial flange of metatarsal IV backs III, but II does not back III (PN *Protarchaeopteryx;* CU rahonaviforms)
Second toe hyperextendible (PN *Protarchaeopteryx*)

Archaeopteryx, *dromaeosaurs, rahonaviforms, troodonts, therizinosaurs, some basal avebrevicaudans*
Toe IV nearly as long as III and much longer than II

*Archaeopterygiforms, dromaeosaurs, troodonts, avebrevicaudans*
Finger II at least as robust as I (CU dromaeosaurs)

*Dromaeo-avemorphs, troodonts, caudipterygians, oviraptorosaurs, avimimids*
Obturator process of ischium large and mid- or distally placed

*Dromaeo-avemorphs, some oviraptorosaurs, alvarezsaurs, avebrevicaudans*
Ilium at least as long anteriorly as posteriorly to acetabulum (CU alvarezsaurs)
Posterior ilium subtriangular
Apex of posterior ilium directed posteroventrally

*Archaeopterygiforms, basal dromaeosaurs, oviraptorosaurs, avimimids, alvarezsaurs, avebrevicaudans*
Quadratojugal and jugal at least fairly slender (condition marginal in confuciusornithids)

*Archaeopterygiforms, avimimids, alvarezsaurs, avebrevicaudans*
Quadratojugal anteroposteriorly short
Quadratojugal and squamosal do not contact one another (PN archaeopterygiforms, avimimids)

*Archaeopterygiforms, oviraptorosaurs, therizinosaurs, alvarezsaurs, avebrevicaudans*
Coranoid absent (CU avimimids)

*Archaeopterygiforms, ornithomimids, some avebrevicaudans*
Quadrate strongly procumbent

*Archaeopterygiforms, ornithomimids, some avebrevicaudans*
Jugal postorbital process very short

*Archaeopterygiforms, ornithomimosaurs, troodonts, therizinosaurs, at least some avebrevicaudans*
Palatine triradiate

*Archaeopterygiforms, ornithomimosaurs, troodonts, avimimids, avebrevicaudans*
Frontal rim of orbit forms a semicircular arc in lateral view (CU caudipterygians)

*Archaeopterygiforms, troodonts, ornithomimosaurs, oviraptorosaurs, therizinosaurs, avimimids, avebrevicaudans*
Frontal largely or entirely excluded from supratemporal fossa (CU caudipterygians)

*Archaeopterygiforms, ornithomimosaurs, oviraptorosaurs, therizinosaurs, most or all avebrevicaudans*
Anterolateral process of palatine at least twice length of anteromedial process
Intraramural articulation absent
Quadrate head bent posteriorly
Coranoid process absent

*Archaeopterygiforms, troodonts, ornithomimosaurs, oviraptorosaurs, most avebrevicaudans*
Volume of rostrum reduced

*Archaeopterygiforms, ornithomimosaurs, caudipterygians, avimimids, most alvarezsaurs, avebrevicaudans*
Tarsometatarsus and/or tibioastragalus longer relative to femur than in other avepods
*Archaeopterygiforms, caudipterygians, most oviraptorosaurs, alvarezsaurs, avebrevicaudans*
Caudal count thirty or less (CU therizinosaurs)
*Archaeopterygiforms, caudipterygians, avebrevicaudans*
Caudal count twenty-three or less
*Archaeopterygiforms, some troodonts, alvarezsaurs, avebrevicaudans*
Teeth combine unserrated conical crowns with constricted waist (CU serrations *Protarchaeopteryx*)
*Archaeopterygiforms,* Bambiraptor, *rahonaviforms, most alvarezsaurs, avebrevicaudans*
Posterior ilium shallow
*Archaeopterygiforms, rahonaviforms, most alvarezsaurs, avebrevicaudans*
Quadratojugal process of jugal short (CU basal troodonts)
Ilium shallow (CU alvarezsaurs)
*Archaeopterygiforms, rahonaviforms, at least some oviraptorosaurs, alvarezsaurs, avebrevicaudans*
Pubic apron very reduced or absent, so that pelvic canal is further deepened
*Archaeopterygiforms, basal dromaeosaurs, rahonaviforms, most avebrevicaudans*
!Supracoracoideus forelimb elevating system at least incipient with:
!acromion process well developed (except basal dromaeosaurs)
!acrocoracoid at least partly developed (CU rahonaviforms)
!coracoid strongly retroverted relative to scapula (CU rahonaviforms)
*Archaeopterygiforms, rahonaviforms, caudipterygians, most avebrevicaudans*
Hallux reversed
*Archaeopterygiforms, rahonaviforms, most avebrevicaudans*
Hallux enlarged
Hallux more distally placed
*Archaeopterygiforms, alvarezsaurs, some basal avebrevicaudans*
Quadrate about one-quarter the length of skull
*Archaeopterygiforms, caudipterygians, avebrevicaudans*
!Caudal/dorsosacral length ratio $1^2/_3$ or less (CU avimimids)
*Archaeopterygiforms, rahonaviforms, caudipterygians, avebrevicaudans*
!Nine or fewer caudal transverse processes (CU avimimids)
*Archaeopterygiforms, caudipterygians, avebrevicaudans*
!Eight or fewer transverse processes (CU *Protarchaeopteryx*, avimimids)
*Archaeopterygiforms, caudipterygians, probably therizinosaurs, probably oviraptorosaurs, avebrevicaudans*
!Caudal/dorsosacral length ratio approximately 1.7 or less (CU rahonaviforms, CU avimimids)
*Archaeopterygiforms, avimimids, avebrevicaudans*
!Carpal block articulates more with metacarpal I than with II
!Finger II markedly more robust than I
*Archaeopterygiforms, at least some avebrevicaudans*
Anteromedial process of the palatine is very small
Posteromedial process of palatine points straight posteriorly, and aligns with anterolateral process
!Very short proximal chevrons (except *Protarchaeopteryx*)
!Forelimbs hyperenlarged
!Metacarpal I narrower distally than proximally (CU avimimids)
!Proximal phalanx of finger II has posterior expansion (CU avimimids)
!Sharp dorsal ridges on phalanges of finger II (CU avimimids)
Protarchaeopteryx, *caudipterygians, probably some oviraptorosaurs, avebrevicaudans*
!Caudal/dorsosacral length ratio approximately 1.0 or less (CU avimimids)
Protarchaeopteryx, *caudipterygians, avebrevicaudans*
!Caudal/dorsosacral length ratio approximately 0.7 or less (CU avimimids)
*Dromaeosaurs, caudipterygians, oviraptorosaurs, most avebrevicaudans*
!At least three fully ossified sternal ribs (CU troodonts, therizinosaurs, avimimids, alvarezsaurs, some basal avebrevicaudans)
!Uncinate processes ossified (CU troodonts, therizinosaurs, avimimids, alvarezsaurs, some basal avebrevicaudans)
*Dromaeosaurs, troodonts, oviraptorosaurs, avimimids, most avebrevicaudans*
Quadrate short
*Derived dromaeosaurs, rahonaviforms, troodonts, caudipterygians, oviraptorosaurs, therizinosaurs, avimimids, alvarezsaurs, avebrevicaudans*
Sacrals increased to six or more at expense of dorsals, which do not number more than twelve (sacral count CU in caudipterygians)

*Dromaeo-avemorphs, troodonts, oviraptorosaurs, therizinosaurs, avimimids, avebrevicaudans*
Overhanging supra-acetabular shelf very reduced or absent (CU caudipterygians)
*Dromaeosaurs, troodonts, ornithomimosaurs, therizinosaurs, avimimids, alvarezsaurs, avebrevicaudans*
Calcaneum further reduced
*Dromaeosaurs, rahonaviforms, troodonts*
Hyperextendible second toe and didactyl well developed (except some troodonts)
*Some dromaeosaurs, avebrevicaudans*
!Scapula tip pointed (P *Bambiraptor*)
!Coracoid proximally narrowed (P rahonaviforms)
!Four or more sternocostal articulations (P *Sinornithosaurus*)
!Well-developed posterolateral flange on proximal phalanx of finger II
*Some dromaeosaurs,* Protarchaeopteryx, *most avebrevicaudans*
!Metacarpal III bowed more strongly posterolaterally (CU rahonaviforms)
Unenlagia, *Rahonaviforms, avebrevicaudans*
Iliolateral process well developed (either absent or very small in *Archaeopteryx;* CU in basal avebrevicaudans)
*Dromaeosaurs, rahonaviforms, avebrevicaudans*
!Scapular acromion process better developed
*Rahonaviforms, alvarezsaurs, avebrevicaudans*
Ischia do not contact one another distally (CU archaeopterygiforms)
*Rahonaviforms, ornithothoraces*
!Scapula tapers posteriorly
!Scapula and coracoid not fused in adult fliers
*Most dromaeo-avemorphs, troodonts, derived alvarezsaurs, avebrevicaudans*
Anterior pubic boot absent
*Ornithomimosaurs, troodonts, avimimids, avebrevicaudans*
Cervical ribs fused to vertebrae
*Troodonts, derived alvarezsaurs, most avebrevicaudans*
Pubic boot absent
*Troodonts, oviraptorosaurs, avimimids, avebrevicaudans*
Basipterygoid processes subhorizontal or horizontal (CU caudipterygians)
Occiput semilunate in posterior profile (CU caudipterygians)
*Troodonts, oviraptorosaurs, avimimids, alvarezsaurs, avebrevicaudans*
Quadrate articulates with braincase (PN archaeopterygiforms)
*Troodonts, ornithomimosaurs, alvarezsaurs*
Premaxilla and external nares reduced
*Troodonts, ornithomimosaurs, avebrevicaudans*
Parasphenoid so pneumatic that it is bulbous (CU archaeopterygiforms, alvarezsaurs)
Basitubera reduced and occipital condyle directed ventrally, so that head is strongly flexed on neck (except some avebrevicaudans)
Posterior braincase and brainstem flexed ventrally
*Troodonts, avimimids, derived alvarezsaurs, avebrevicaudans*
Fibula does not contact calcaneum (except some avebrevicaudans)
*Troodonts, avimimids, alvarezsaurs, avebrevicaudans*
Astragalocalcaneums co-fused
Astragalocalcaneums laterally compressed
Astragalocalcaneal condyles project strongly anteriorly
*Troodonts, ornithomimosaurs, oviraptorosaurs, avebrevicaudans*
Olfactory lobes reduced (CU therizinosaurs)
*Troodonts, therizinosaurs, at least some oviraptorosaurs, avebrevicaudans*
Fenestra ovalis and associated structures set in subcircular lateral otic depression (CU caudipterygians, oviraptorids)
*Derived troodonts, derived alvarezsaurs, ornithothoraces*
Lesser and greater femoral trochanters confluent
*Troodonts, caudipterygians, oviraptorosaurs, therizinosaurs, ornithomimosaurs, alvarezsaurs, avebrevicaudans*
Antorbital fossa and fenestra reduced, compared with dino-avepod standard
*Therizinosaurs, oviraptorosaurs, troodonts, avimimids, alvarezsaurs, avebrevicaudans*
Squamosal is an integral part of the occiput (CU caudipterygians, alvarezsaurs; appears to be at least partly developed in therizinosaurs)
Ectopterygoid shifted anteriorly, if present, and no longer articulates with ectopterygoid process of pterygoid (CU avimimids, alvarezsaurs; appears to be at least partly developed in therizinosaurs)
Anterior and quadrate processes of pterygoid continuous and aligned (CU caudipterygians, alvarezsaurs; appears to be at least partly developed in therizinosaurs)
Articular and surangular co-ossified (except oviraptorids)
Lateral process on articular
Medial process on articular
*Caudipterygians, probably therizinosaurs, probably oviraptorosaurs, avebrevicaudans*
!Caudal/dorsosacral length ratio approximately 1.2 or less (CU rahonaviforms, CU avimimids)

*Caudipterygians, avebrevicaudans*
!Six or fewer caudal transverse processes (except some avebrevicaudans; CU avimimids)
*Caudipterygians, oviraptorosaurs, avebrevicaudans*
Maxilla and antorbital fossa, fenestra and maxillary sinuses strongly reduced (except *Cathayornis*)
Deep bifurcation of the posterior dentary
*Oviraptorosaurs, avimimids, alvarezsaurs, avebrevicaudans*
Quadratojugal and jugal very slender
*Oviraptorosaurs, avimimids, avebrevicaudans*
Premaxilla greatly expanded
Basal process on pterygoid present
*Oviraptorosaurs, avimimids,* Patagopteryx
Quadrate fused to pterygoid
*Oviraptorosaurs, avebrevicaudans*
Ectopterygoid contacts lacrimal
Mandibles co-fused
Semilunate carpal continues laterally onto metacarpal III (CU basal avebrevicaudans)
*Some oviraptorosaurs, avebrevicaudans*
!Distal caudals fused into pygostyle
*Derived oviraptorosaurs, most avebrevicaudans*
Nasal airway not L-shaped (except confuciusornithids, *Cathayornis*)
*At least some oviraptorosaurs, caudipterygians, alvarezsaurs, avebrevicaudans*
Palatal elements almost aligned in lateral view (P caudipterygians)
Ectopterygoid placed very anteriorly if not absent
Cervical neural spines severely reduced
*At least some oviraptorosaurs, alvarezsaurs, avebrevicaudans*
Sacrals increased to seven or more at expense of dorsal series (CU caudipterygians, alvarezsaurs)
Carpometacarpus present
*Therizinosaurs, derived alvarezsaurs, avebrevicaudans*
Pubis articulates with ischium distally
*Ornithomimids, oviraptorosaurs,* Confuciusornis, *some enantiornithines, neornithines*
Well-developed beak present
*Ornithomimids, caenagnathids, avimimids, derived alvarezsaurs, ornithurines*
Metatarsal III very narrow proximally
*Ornithomimosaurs, oviraptorosaurs, therizinosaurs, most or all avebrevicaudans*
Palatine shelf of premaxilla and maxilla well developed (CU some basal avebrevicaudans)
*Ornithomimosaurs, alvarezsaurs, ornithothoraces*
Metatarsal V absent

*Caudipterygians, alvarezsaurs, avebrevicaudans*
Medial ball of femoral head reduced
*Avimimids, most avebrevicaudans*
Ilial surfaces face partly dorsally, and posterior pelvis is broad (CU alvarezsaurs)
!Ulna twice as thick as radius
*Avimimids, alvarezsaurs, avebrevicaudans (except some basal avebrevicaudans)*
Postorbital bar incomplete (CU archaeopterygiforms)
Quadratojugal and/or squamosal lack quadrate processes (P avimimids)
Mobile quadratoquadratojugal joint
Bending zone in skull roof (PN avimimids)
Acetabular antitrochanter well developed
*Alvarezsaurs, avebrevicaudans*
Foramen magnum enlarged relative to occipital condyle
Neural canals tall relative to centra
Seven or more sacrals
Ischial obturator process extremely reduced or absent
!Long axis of humerus head and distal carpals in same plane (some exceptions)
Pubes contact one another at extreme distal ends, if at all
Iliofibularis tuber well developed
*Derived alvarezsaurs, most or all avebrevicaudans*
Pubic apron absent
Pubes do not contact one another distally
Lateral cnemial crest well developed
Fibula short
Pretibial process laterally placed (CU archaeopterygiforms)
*Caudipterygians, alvarezsaurs, ornithothoraces*
!Number of finger elements reduced
*Caudipterygians, ornithothoraces*
Eleven or more cervicals
Ten or fewer dorsals
Base of finger II braced by finger III
*Alvarezsaurs, derived enantiornithines, carinates*
!Sternum deeply keeled (with exceptions)
*Alvarezsaurs, ornithurines*
Two cnemial crests present
*Derived alvarezsaurs, ornithurines*
Pubis and ischium subparallel one another
*Avebrevicaudans*
External nares sited more posteriorly than in archaeopterygiforms
Premaxillae fused at least anteriorly
Dentaries fused anteriorly (CU alvarezsaurs)
Neck more strongly S-curved (CU some basal examples)
Hyposphene-hypantrum articulations absent
Dorsal vertebral foramina large (some exceptions)

!Post-synsacrum caudal count fifteen or fewer
!Tail severely reduced in length relative to dorsosacral series
!Coracoid very narrow in fliers
!Coracoid strongly retroverted in fliers
!Supracoracoideus wing elevating complex well developed in fliers
!Shoulder glenoid faces more dorsally in fliers
!Radius much more slender than ulna in fliers
!Wing-folding mechanism improved with:
!V-shaped ulnare
!more posteriorly directed carpal pulley
!Posterolateral flange of proximal phalanx of finger II enlarged

*Ornithothoraces*
Fully developed, saddle-shaped cervical centra articulations
Eight or more sacrals at expense of dorsals
Gastralia absent
!Flexion at scapula-coracoid juncture (except some flightless examples)
!Furcula more slender in most fliers
!Furcula's hypocleidium well developed in most fliers
!Sternal keel at least weakly developed in fliers
!Distal humeral condyles face anteriorly
!Ulna longer than humerus in fliers
Ischia do not contact one another distally (CU *Iberomesornis;* a few exceptions)
Tibia, astragalus, and calcaneum completely fused
Distal vascular foramen of tarsometatarsus present (except some enantiornithines)

Patagopteryx, *ornithurines*
!Bladelike extensor process on metacarpal I
Ilium elongated
Ilial preacetabular fossa absent
Posterior trochanter of femur absent
Extensor canal present on tibia
Tarsometatarsus completely fused

*Ornithurines*
Maxillary process of premaxilla forms a significant part of side of face
Posterior maxillary sinus has a cup-shaped depression
Quadrate has an orbital process
Humerus head subspherical in proximal view
Pubic shaft flattened along entire length
Acetabulum small relative to rest of pelvis
Posterior ilium and ischium subparallel one another
Femur has a deep patellar groove
Two cnemial crests present
M. iliofibularis tuber of fibula directed posteriorly
Well-developed anterior intercondylar process on tarsometatarsus
Proximal end of metatarsal III posterior to II and IV

*Carinates*
Distal humerus has a brachial depression

*Ichthyornithids, neognathes*
Quadrate head bifurcated by penetration of dorsal tympanic recess

*Neornithines*
Most skull elements strongly fused
Ascending process of maxilla is reduced
Lacrimal-jugal contact absent
Contralateral dorsal tympanic diverticula communicate within dermal skull roof
Contralateral posterior tympanic diverticula communicate within cranium
Communication of ipsilateral dorsal and posterior tympanic diverticula
Long bones pneumatic (possibly in some enantiornithines also)

*Neognathes*
Anterior maxillary sinus small or absent
Vomers very reduced or absent
Basipterygoid processes absent
Flexible joint between anterior and posterior sections of pterygoid

## B: Character List for *Archaeopteryx* and Dromaeosaurs

- * = Character not always observed in basal avebrevicaudans, and so may represent parallelism.
- Other symbols same as those used in the general character list.

*Characters found in at least some dromaeosaurs and archaeopterygiforms, but not other dino-avepectorans (except as noted)*
Nasal depressed
Preorbital bar straight and narrow in lateral view
Posterior facing step on postorbital process of frontal (P *Archaeopteryx*)
Diamond-shaped dorsal supraoccipital
Paraoccipital process so strongly twisted that:
Anterior surface faces partly ventrally, and helps form an external auditory meatus
Posterior surface has a pronounced bulge at midlength (most visible in dorsoventral view)
Shallow subrectangular ventral expansion of paraoccipital process
Pneumatopore on anterior surface of paraoccipital process opens into pneumatic sinus

Ectopterygoid set more posteriorly
Dorsal ectopterygoid depression (except *Dromaeosaurus*)
!Metacarpal III bowed posterolaterally
Ilium parallelogram shaped (also found in other dromaeo-avemorphs)
Apex of posterior ilium directed posteroventrally
Ilium shorter
Vertical supra-acetabular ridge set posteriorly
Partial medial wall of acetabulum present (also found in *Unenlagia*)
Ilial ischial peduncle much shorter than pubic peduncle
Iliopubic articulation inverted V-shape (also found in other dromaeo-avemorphs)
Pubic midshafts are flat plates, angled about 140° to one another
Distal pubis deflected posteriorly (inconsistent in *Archaeopteryx;* also found in *Unenlagia*)
Ischial peduncle much shorter than pubic peduncle
Pubic peduncle retroverted
Cnemial crest small relative to most other avepods

*Characters found in at least some dromaeosaurs and avebrevicaudans, but not* Archaeopteryx
Ectopterygoid process of pterygoid absent
Quadrate shorter
Sacrals increased to six at expense of dorsals (except basal dromaeosaurs)
Cervical ribs short and do not overlap
Cervical centra articulations incipient saddle shaped (PN *Archaeopteryx*)
*Well-developed hypapophyses under cervicodorsals (PN *Archaeopteryx*)
Ossified uncinate processes
Ossified sternal ribs
Four or more sternocostal articulations
Long ossified sternal plate
*Scapula tip pointed (P *Bambiraptor*)
Scapular acromion process better developed
Coracoid proximally narrowed
Metacarpal III bowed more strongly posterolaterally
Well-developed posterolateral flange on proximal phalanx of finger II
Pubis more retroverted
Supraacetabular shelf narrower
Incipient antitrochanter present
Lesser and greater trochanters more confluent
Calcaneum reduced

*Characters found in* Archaeopteryx *and avebrevicaudans but not dromaeosaurs*
External nares enlarged
External nares sited more posteriorly
Frontals excluded from supratemporal fossa
Frontal rim of orbit forms a semicircular arc in lateral view
*Upper and lower temporal, postorbital, and jugal bars slender
Quadratojugal and squamosal do not contact one another (CU *Archaeopteryx*)
Quadratojugal anteroposteriorly short
Anteromedial process of the palatine is very small
Palatine fully triradiate
Posteromedial process of palatine points directly backward, and is in same line as anterolateral process
Anterolateral process of palatine at least twice the length of anteromedial process
Intraramural articulation absent
Quadrate head bent posteriorly
Volume of rostrum reduced
Epipterygoid smaller
Coranoid absent
Teeth combine unserrated conical crowns with constricted waist
!Caudal/dorsosacral length ratio lower
!Caudal count lower
!Fewer caudal transverse processes
!Proximal chevrons shallower
!Forelimbs hyperenlarged
!Carpal block articulates more with metacarpal I than with II
!Metacarpal I more narrow distally than proximally (CU avimimids)
!Finger II markedly more robust than I
!Sharp dorsal ridges on phalanges of finger II
!Proximal phalanx of finger II has posterior expansion
Ilium shallower
Pubic apron very reduced or absent, so that pelvic canal is deeper
Hallux more distally placed
Hallux fully reversed
Hallux enlarged

APPENDIX 2

# *Musculature, Body Mass, and Wing Dimensions*

The mass of the biggest modern birds is difficult to determine precisely. Data sources often differ in the details; for example, Matthews and McWhirter (1993) and del Hoyo et al. (1992–99) reported that wild Andean condors reach 11 or 15 kg, respectively. The mass problem is even greater regarding extinct birds, whose weight can only be restored, either by extrapolating from modern birds of similar form, or by volumetric models. More difficult still are pterosaurs, whose form diverged from all modern fliers. Methods for restoring the musculature and the body mass of extinct tetrapods via volumetric restorations are detailed by Paul (1988a, 1997a) and Paul and Chase (1989). Those studies, and McGowan (1982), noted that it is often not possible to restore the detailed musculature; more often only gross patterns can be discerned. The statement by Bramwell and Whitfield (1974) that 20 percent of total volume may consist of lungs in birds is incorrect: avian lungs are actually unusually small (Appendix 3). It is the air sacs that take up considerable space in bird bodies, 20% according to Dorst (1974) and del Hoyo et al. (1992–99). Comparing a volumetric restoration of a frigate bird with the observed mass produced a specific gravity of 0.7. In the only comprehensive survey of avian specific gravities, Hazelhurst and Rayner (1992) observed an average of 0.73. In arriving at this figure, they correctly reinflated the air sacs, but it is possible that volumes obtained in this manner are a little high, because bird air sacs are not constantly maximally inflated in living birds. If so, the average avian specific gravity should be somewhat higher. Suggestions that soaring birds have much lower specific gravities (Bramwell and Whitfield 1974) have not been confirmed. In flying birds, approximately 6% of the total mass is feathers (Turcek 1966). Mass distribution within the body cannot be as reliably restored as total mass because the internal distribution of air spaces and bone mass will never be sufficiently known in extinct forms. In this regard, the centers of gravity calculated for dino-avepods in Jones et al. (2000b) are unavoidably unreliable (also, the *Caudipteryx* restoration is not accurate). In birds other than hummingbirds, mass scales to span$^{2.56}$, according to Alexander (1971). Additional mass and dimensional data are available from Amadon (1947), Hartman (1961), Greenwalt (1962), J. L. Long (1965), Prange et al. (1979), and Dunning (1993). Selected span and mass data is tabulated in App. Table 1, and plotted in App. Fig. 18G,H.

## *Archaeopteryx*

Because *Archaeopteryx* probably had a moderately well-developed air sac complex (Appendix 3), it is assigned a specific gravity of 0.85 (Paul 1988a), a value that if anything may be too high.

Body mass estimates for *Archaeopteryx* range from 200 to 500 g for HMN 1880 and BMNH 37001 (Yalden 1984). Yalden used a multiview skeletal restoration of *Archaeopteryx* HMN 1880 to produce a mass estimate of 271 g, which may be somewhat too high because the chest is too deep.

A key reason for the variation in mass estimates centers on the uncertain mass of the flight muscles. Paul (1988a) assumed that the pectoralis made up 8% of total mass in the small JM 2257, and 15% in other specimens. Restoring the mass of the pectoralis is difficult because its extent is not always well defined by skeletal features. In tetrapods, the posterior fibers of the pectoralis muscles often spread beyond the limits of the sternum onto the rib cage, and even onto the abdomen (horse and lion in Ellenberger et al. 1956). Bats support a very large wing depressor complex without a keeled sternal plate. In big-armed theropods such as *Allosaurus,* very large pectoral crests probably supported pectoralis muscles that were too large to be contained on their very small sternal plates, so the posterior fibers of the pectoralis probably spread out onto the abdominal gastralia. The sternal plates of all known Early Cretaceous birds were too short to anchor all of the posterior fibers of pectoralis muscles if these muscles made up 5% or more of body mass (Figs. 10.1Bt,u, 10.9H–K), so it is possible that the muscle spread more posteriorly (onto the gastralia, if pres-

*Appendix Table 1. Wingspan and Total Body Mass*

| | Span (mm or m) | Mass (g or kg) |
|---|---|---|
| Dragonflies and protodonates | | |
| *Anax* | 109 mm | 1.2 g |
| *Megaloprepus* | 191 | ~4 |
| *Meganeura* | ~700 | ~60 |
| Birds | | |
| *Archaeopteryx* | | |
| JM 2257 | 360 | 67 |
| BSP 1999 | 506 | 140 |
| HMN 1880 | 560 | 234 |
| BMNH 37001 | ~700 | 333 |
| S6 | | 520 |
| *Confuciusornis* | 780 | 325 |
| Great black-backed gull | 1.7 m | 1.9 kg |
| Frigate bird | 2 | 1.6 |
| Wandering albatross | 4 | 13 |
| Wild turkey | 1.8 | 10 |
| Harpy eagle | 2 | 9 |
| Andean condor | 3.2 | 14 |
| Marabou stork | 3.3 | 10 |
| Mute swan | 2.8 | 18 |
| Kori bustard | 2.6 | 19 |
| *Harpagornis* | 3 | 13 |
| *Teratornis* composite | 3.3 | 15 |
| *Argentavis* MLP 65-VIII-29-49 | 6 | 70–100+ |
| Pseudodontorn ChM PV4768 | 7+ | 50 |
| Pterosaurs | | |
| *Pterodaustro* PVL 3860 | 1.3 | 1.0 |
| *Tupuxura* | 2.2 | 1.6 |
| *Anhanguera* AMNH 22555 | 3.8 | 2.3 |
| *Nyctosaurus* FHSM VP 2148 | 2.7 | 1.1 |
| *Pteranodon* | | |
| FHSM VP 184 | 5.9 | 20–25 |
| FHSM VP 339 | 8 | ~50 |
| *Quetzalcoatlus* TMM 41961-1 | 4.7 | 25–30 |
| *Q. northropi* TMM 41450-3 | 10–11 | 200–250 |

NOTE: Values for living examples are often approximate owing to divergent data in various sources. Spans for extinct birds were measured from complete wings.

ent). The more expanded, deeper-keeled sternum of modern birds may be a means to anchor the large supracoracoideus complex (S. Olson and Feduccia 1979, Rayner 1991). Therefore, size of the pectoralis in early birds may not have been as closely linked to the size of the keeled sternum as Ruben (1991) assumed. Besides, the sternum of the urvogel may have been twice as large as indicated by the bony element (Chapter 4). Nor is the depth of any cartilaginous keel known, although it was probably shallow. S. Olson and Feduccia (1979) observed that the pectoral crest and furcula of *Archaeopteryx* were larger than in most tetrapods and most birds (a point confirmed in App. Fig. 18F), indicating that they anchored large anterior and distal pectoralis muscle fibers. Modeling of the pectoralis in clay suggests that there was sufficient area on the chest and upper arm of *Archaeopteryx* easily to support pectoralis fibers making up 5–7% of body mass; 10% could be accommodated with some difficulty; and 15% is implausible (Fig. II.1). There is room for supracoracoideus muscles making up 1–2% of total mass, and because the forelimb was so large, they could support more distal muscles making up 4–5% of body mass. Total flight muscle mass may have been 8–14% of total mass.

Mass estimates for complete *Archaeopteryx* specimens from Paul (1988a) are revised, and the new specimens added, with the assumption that the pectoralis made up 6% of total mass. The estimate for HMN 1880 is in good agreement with Yalden's (1984) estimate if the difference in chest dimensions is taken into account. Those who wish to disagree with the flight muscle masses used here may adjust the values correspondingly. Wing dimensions are calculated in Chapter 5.



## Giant Birds

The marine pseudodontorns appear to have had albatross-like body/wing proportions and shape (Fig. 8.2, App. Fig. 1). S. Olson (1985) estimated the wingspan of the biggest pseudodontorns as having been as great as 6 m. Newer and fairly complete remains suggest that this is an underestimate. Assuming that the ratio of the span of the wing bones to the total wingspan was similar to that of albatross, the largest known pseudodontorn wings would have spanned approximately 7 m. The span/mass ratio was higher than in albatross, and using a 7-m span, the mass is estimated to be nearly 50 kg.

The continental teratorns appear to have had condorlike body/wing proportions (Fig. 8.2, App. Fig. 1). The biggest teratorns, which are poorly known, have been estimated to span 6–8.5 m, and weigh 65–120 kg (Campbell and Tonni 1980, 1983, Paul 1991, Campbell and Marcus 1992, Matthews and McWhirter 1993). In condors, total span is consistently about ten times humerus length (Campbell and Tonni 1983). Applying this value to well-known *Teratornis* (H. Fisher 1945) gives a wingspan of 3.15 (average) to 3.3 m. Although the humerus of fragmentary *Argentavis* is incomplete, its length is preserved at about 600 mm, so span should not have exceeded 6 m. Extrapolating from condor mass/span values and using a volumetric estimate, an average *Teratornis* is estimated to have a mass of 15 kg, *Argentavis* approximately 70 kg. The latter value is close to that estimated by Campbell and Marcus (1992) using hindlimb cross section/mass relationships. However, the few existing body elements of the superteratorn appear to be larger relative to the humerus than their counterparts in its smaller relative: perhaps too large to fit inside a 70-kg body. It therefore remains possible that *Argentavis* reached or exceeded 100 kg.

Turning to flightless avian giants, the big question is, which was the very largest? For many years, elephant birds were ranked the champions at over 400 kg. More recently, thunder birds have often been given the edge at nearly half a tonne, and moa were considered relative lightweights at just 250 kg. But as Figure 8.4 shows, the largest specimens of the three types have nearly equal-sized bodies (tibia lengths of specimens: *Dinornis maximus* 990 mm [Archey 1941], *Aepyornis* 810 mm [Monnier 1913], *Dromornis* 780 mm). Volumetric models for the ratites produce practically equivalent mass estimates of 370–380 kg, and the less complete *Dromornis* does not appear distinguishable in this regard. (The moa estimates agree with Amadon [1947] and Alexander [1983a,b]. The higher elephant bird estimates by Amadon and by Paul [1997a] are excessive.)

We next consider big flightless ocean birds. The second largest sphenisciform, the long beaked king penguin, has a 185-mm-long head and a total length of 0.95 m and weighs 15 kg (del Hoyo et al. 1992–99). The shorter-beaked, 1.15-m-long emperor penguin weighs 45 kg. Anterior skull elements assigned to *Palaeoeudyptes* show it had a long beak. S. Olson (1985) restored an approximately 570-mm-long skull; approximately 435 mm is the minimum possible. Postcranial remains include a 240-mm-long sacrum and a 150-mm-long femur. The postcranial elements suggest total lengths of approximately 1.25 m. The skull appears to come from a much larger individual: if king penguinlike proportions are assumed, then the total length was 2.2–2.9 m. Mass could have been 200–400 kg. The largest plotopterid femur is 225 mm long (S. Olson 1985). If the femur-length/total-length ratio was similar to that of penguins and great auks, then total length was approximately 2 m, and mass should have been approximately 200 kg.

## Giant Pterosaurs

Researchers tend to underestimate the mass of giant pterosaurs (Bramwell and Whitfield 1974 and references therein, Stein 1975, Brower and Veinus 1981, Langston 1981, Wellnhofer 1991), to the point that absurdly low specific gravities have been assumed (Paul 1991). For example, Bramwell and Whitfield restored the volume of a 7-m span *Pteranodon* as having been over 40 liters, but estimated a mass of less than 17 kg. It is improbable that the insides of a vertebrate can be 60% air! Many pterosaurs had extremely thin-walled long bones, and their large-volume beaks were probably little denser than Styrofoam, but they lacked the pneumatic dorsals common to birds. Therefore, the extent of any pterosaur air sac complex was probably not greater than that seen in birds, and the specific gravity of the two groups should be similar. The restored volume of a 6-m span *Pteranodon* is 35 liters (Fig. 8.2). Mass is restored at 20–25 kg. The largest specimen is tentatively restored at 8 m and approaching 50 kg. The new mass values place giant pterosaur wingspan/mass ratios at the upper limit of the avian range.

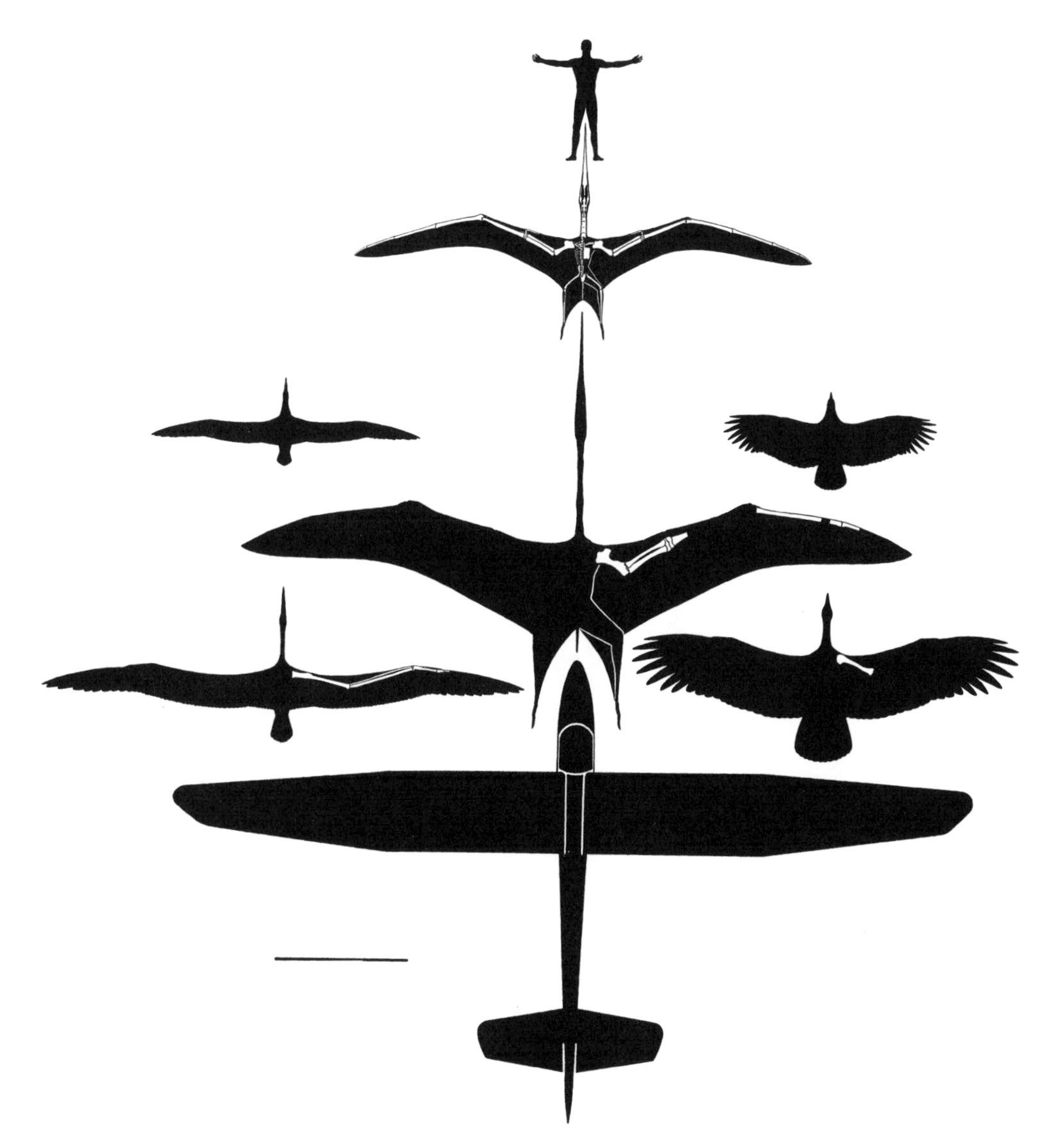

*Appendix Figure 1.* Size comparisons. *Top to bottom: center column:* human male; *Pteranodon* scaled to 7-m wingspan; *Quetzalcoatlus northropi* (the configuration of the attachment of the pterosaurs' wing membrane to the hindlimbs is provisional); 12-m sailplane; *left column:* wandering albatross; 7-m-wingspan pseudodontorn; *right column:* Andean condor; *Argentavis.* Scale bar equals 2 m.

Half-sized *Quetzalcoatlus* had a larger body and legs, more robust wing bones, and a larger pectoral crest relative to span than *Pteranodon* (Fig 8.2, App. Fig. 1). Therefore, an azhdarchid with a 4.7-m wingspan probably weighed 25–30 kg. An individual with a span more than two times as great should have weighed about ten times as much. Yet past mass estimates for the largest known *Quetzalcoatlus* have ranged from only 54 to 127 kg, in the human to ostrich range (Brower and Veinus 1981, Langston 1981, Wellnhofer 1991, Shipman 1998). However, the arms spanned six times those of a man (App. Fig. 1), and the head was as long as a person is tall. Considering the enormous size of the pectoral crest, and the tremendous power needed to fly, the flight muscles should have been a quarter of total mass. Therefore, if total mass was no greater than that of a human, the leg muscles would have been smaller than those of a human, even though the pterosaur's legs were about twice as long. Attempting to apply only 50 kg of muscle to the projected skeletal framework of giant *Quetzalcoatlus* results in an implausible, atrophied appearance. Conversely, scaling down the pterosaur's body until it represents only 70 kg renders it much too small for the great wings. In either case, the approximately 15 kg of flight muscles are not large enough to fit the space available on the inner arm bones. Extrapolating the ratio of wing-muscle mass to pectoral-crest length restored in smaller pterosaurs suggests that the flight muscles of *Quetzalcoatlus* alone should have weighed as much as an entire human. The giant azhdarchid was also much larger than the biggest teratorn, with a head five times longer, a humerus much more massive, and a span nearly twice as great. It is improbable that *Quetzalcoatlus* weighed as little as a human, *Argentavis*, or even an ostrich (Fig. 8.2). Scaling up the mass of the smaller *Quetzalcoatlus* individuals to the large span results in a mass of 200–250 kg.

## Big Insects

While on the subject of estimating the mass of extinct fliers, it is worth examining the largest aerial insects. Nowadays the heaviest flying insects are goliath beetles, which reach 70–100 g: heavy enough that these hard-shelled bugs can crack window glass (Linsenmaier 1972, Matthews and McWhirter 1993). The insects with the greatest known wingspans were the protodonates, dragonfly-like cruisers of Paleozoic skies. Their wings spanned as much as 700 mm, almost half a dozen times greater than that of the largest modern dragonfly (May 1982, Matthews and McWhirter 1993) and equaling pigeons and medium-sized falcons. Estimating the mass of protodonates is hindered by the lack of completely preserved bodies. The latter were more slender relative to span, and probably had higher length/width ratios, than those of dragonflies. Protodonate bodies seem to have been lower in overall volume relative to wingspan than observed in dragonflies. Using body-width/total-mass ratios observed in dragonflies (May 1981) and extrapolated to the extinct giants, May (1982) estimated the mass of a 650-mm-span protodonate to be 17.8 g. This is an extremely low value, only some four times greater than that for the heaviest modern dragonfly, and 20–25% that predicted for a dragonfly or for a bird with the same wingspan and a low wingspan/mass ratio (Greenwalt 1962). Reanalysis of the body-width/total-mass data published by May (1981, 1982) boosts the mass of a 650-mm-span protodonate is approximately 30 g, but even this is probably too low in view of their elongated bodies. Approximately 50–60 g is a more plausible mass range for the biggest winged insects (App. Fig. 18H).

# *Respiratory Tract Anatomy and Function*

This appendix addresses a set of technical questions and problems concerning the origins and nature of the dino-avepod respiratory complex. This subject has become important in recent years largely as a result of a series of papers arguing that theropod and avian respiration are not compatible with one another.

## A: Nasal Airway

### *Anatomy*

The first problem centers on the structure of the nasal airway (NA) in avepod dinosaur skulls, including urvogels. The NA is that part of the nasal passage between the external and internal nares through which the main airstream flows, and is therefore potentially suitable for containing respiratory conchae; the olfactory portion of the nasal passage is excluded from the NA. Detailed examinations of the anatomy of archosaur nasal passages and sinuses are to be found in Witmer (1990, 1995a, 1997a,c).

Ruben (1996) and Ruben et al. (1996, 1997b) asserted that the NA of tyrannosaurs and ornithomimids consisted only of a narrow tube. In the same studies, they restored *Archaeopteryx* and/or *Dromaeosaurus* with short, simple, lizardlike NA (App. Fig. 2Ab). In particular, the internal nares were set just below and behind the nostrils. These conclusions are either incomplete or incorrect, because the actual situation is more complicated—and more interesting.

The dromaeo-avemorph NA restorations by Ruben and company were entirely speculative; indeed, in one case they restored the internal nares too far forward to exit through the bony internal opening (App. Fig. 2Ab). The anterior NA is not exposed in any *Archaeopteryx* specimen. Nor is the NA preserved in the only known *Dromaeosaurus* skull, in which only fragments of the premaxilla, maxilla, vomers, and the palatines—which define the posterior border of the internal nares—are present, and the nasals are entirely missing (that these elements were missing was obscured in the Ruben papers by using an artistically complete restoration of the skull by Currie [1995], rather than using the same study's figure guide showing which elements were missing [App. Fig. 2Ac]). What was the actual path of the NA in these and other predatory dinosaurs and early birds?

Coelophysoids lack bony evidence of well-developed accessory maxillary sinuses forward of the antorbital fenestra (Fig. 10.2E, App. Fig. 2Ba). The NA could run in a straight, simple path, exiting via internal nares that were set more posteriorly than in most reptiles—posteriorly enough to be visible in the anteroventral corner of the antorbital fenestra in lateral view in most theropod dinosaurs and some basal birds (Fig. 10.2D–H,J–L).

The averostran nasal complex was more complex. Accessory maxillary foramina and fenestrae led into the well-developed mediorostral maxillary sinuses (Fig. 10.2G–H,J–N,P, App. Fig. 2Ab–d,Bb, c,D), which were composed of the medial portion of the combined premaxillary and maxillary structures (Witmer 1997a). The ventral shelf of this sinus complex projected so far medially that it contacted or nearly contacted the anterior palate (App. Fig. 2C). This had the effect of closing off the anterior end of the bony internal nares. Although the result was to ensure that the internal nares were set fairly posteriorly in the oral cavity, it is questionable whether this arrangement qualifies as a true secondary palate, because the anterior closure was at the same level as the palate—even in *Tyrannosaurus,* in which the ossification of the closure was unusually extensive. Secondary palates have been reported in spinosaurs (Sereno et al. 1998), ornithomimosaurs (Osmólska et al. 1972, Barsbold 1983), oviraptorosaurs (Barsbold et al. 1990, Sues 1997), dromaeosaurs (Norell and Makovicky 1999), and troodonts (Norell et al. 2000), and they may be widespread among dino-avetheropods. The large sinus complex essentially closed off the ventral portion of the NA, thereby barring a straight course for the main flow of air between the external and internal nares. As a result, the NA followed a long, L-shaped path. Disarticulated maxillas, complete snouts, and the results of

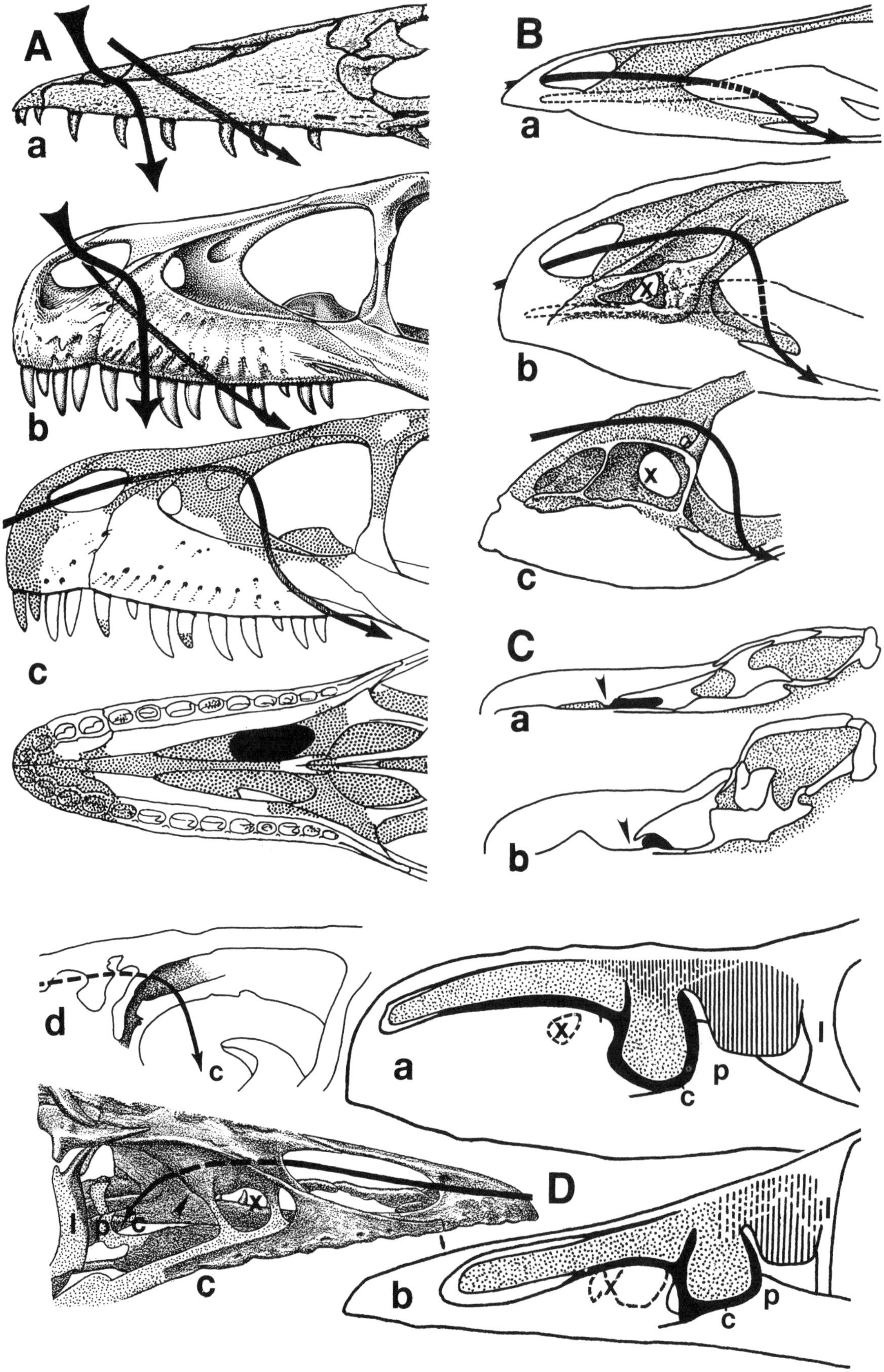
A
a
b
c
B
a
b
x
c
x
C
a
b
d
c
a
x
l
p
c
D
l
p
c
x
c
b
x
c
p

*Appendix Figure 2. (opposite)* Dino-avepod nasal passages and conchae. Main airflow through the nasal airway (NA) is indicated by arrows. View *A* illustrates the lizard-dromaeosaur problem. *Aa,* NA in varanid lizard, as variably indicated in Ruben et al. (1996, 1997b). *Ab,* short course of NA, as variably and inaccurately restored as lizardlike in the same references using the restored snout of *Dromaeosaurus* AMNH 5356 rather than the bones as actually preserved. *Ac,* bones as actually preserved (white) in side and bottom views: In the bottom view, the snout is restored as too broad because the ventral shelves of the maxillary sinuses should nearly contact each other on the midline (see *Ca*); the probably posterior position of the internal nares is indicated with solid black (see *C*). In the side view, the arrow indicates the probable course of the main NA above the accessory maxillary sinuses following other dino-averostrans, including *Velociraptor. d, Velociraptor* as preserved, showing exit of anterior NA tube (Pl. 11B), shaded, above the internal nares; note that anterior NA is broadest above the accessory maxillary sinuses and narrower between them. *B,* the maxillary sinuses and NA (both shaded) in inner view of dino-avepods: *a,* snout of cerato-sauran *Syntarsus-Coelophysis* without well-developed maxillary sinuses; averostrans showing how maxillary sinuses cause the main airflow to follow an L-shaped curve: *b,* snout of basal *Sinraptor* with moderately developed sinuses, palatal elements uniformly stippled; *c,* maxilla of *Tyrannosaurus* CM 973 with very well-developed sinuses, which are not completely preserved. *C,* functional internal nares (solid black) in *(a) Allosaurus* and *(b) Tyrannosaurus* AMNH 5027, in which strongly medially projecting ventral shelves of the maxillary sinuses restrict the internal nares to a posterior position. *D,* hypothetical reconstructions of nasal conchae in *(a)* juvenile *Tyrannosaurus* and *(b) Archaeopteryx;* nasal airway black, vertical dashed lines indicate possible preserved conchae, postulated anterior and posterior respiratory conchae stippled, postulated olfactory conchae vertical solid lines, position of accessory maxillary fenestrae indicated by dashed lines (see App. Fig. 3 for additional examples of avepod NA). *Dc,* study of the snout of *Archaeopteryx* JM 2257 (from Pl. 7), shaded areas on main slab, dotted areas from counterslab, long arrow indicates probable path of NA, short arrow points to probable ventral border of anterior nasal passage. Abbreviations: c = internal nares, l = lacrimal, p = palatines, x = mediorostral maxillary sinuses and/or accessory maxillary fenestrae. *Aa* and *b* from Currie (1995). Not drawn to same scale, except *Db* and *c.*

CT scans (of ornithomimids and tyrannosaurs by Ruben et al. 1996, 1997b) show that the anterior portion of the NA was a tube—sometimes short and fairly deep, sometimes long and slender—that ran under the nasals along the dorsal section of the maxilla. The tube ended close to the anterodorsal rim of the antorbital fenestra, where the postantral struts (Witmer 1997a) converged ventrally to form a V- or U-shaped posteroventral rim to the anterior NA tube (Pl. 11B; App. Figs. 2b,c, 3A). This sharpened the inside corner of the L-shaped NA. The posterior NA was a subvertical passage that connected the anterior tube with the internal nares.

The presence of the anterior tube/vertical posterior passage in dromaeosaurs is indicated by the presence of typical averostran accessory maxillary openings and verified by complete examples of the anterior tube in articulated *Velociraptor* skulls (Pl. 11B, App. Fig. 2Ac). When the *Dromaeosaurus* snout is correctly restored (Chapter 3), the ventral shelf of the maxillary sinuses nearly contacts the anterior palate, closing off the seemingly anteriorly elongated bony internal nares. The accessory maxillary fenestrae and narrow snouts of basal birds such as *Archaeopteryx* and *Cathayornis* (Fig. 10.2K,P) strongly suggest that they also shared the basic averostran L-shaped, long NA. The short and simple airways restored by Ruben and company are therefore in error and seem to have been abandoned by Ruben and Jones (2000). Oviraptorids differed from other averostran dinosaurs (Fig. 10.2M). The NA was still somewhat L-shaped. But the mediorostral maxillary sinuses were sharply reduced, and strong posterodorsal migration of the external nares set them almost directly dorsal to, and well above, the internal nares. The NA was therefore a straighter and more continuous subvertical tube from the external to the internal nares (App. Fig. 3N,O). This arrangement resembles that of some big-beaked birds in which the external nares are posteriorly placed (App. Fig. 3g,s, Fig. 15.3).

### *Measuring Airway Dimensions*

Ruben et al. (1996) proposed that the most reliable method of gathering the data for living forms is by sampling either living animals or fresh or spirit-preserved carcasses, whose NA dimensions can be measured by sectioning or by CT scanning, and whose bodies can be directly weighed. However, there are pitfalls to this method. Carcasses may be underweight because of extended terminal illness or malnutrition before death—a particular problem with zoo specimens—or as a result of autopsy and dissection. Estimating masses by correlating skeletal dimensions with masses of healthy individuals is a well-established technique that can provide

L
X
x
T
X
V
X
x
O
s
e
G
X
c
d
o
k
g
R
D
A
X
P

superior results, and most of the bird and reptile specimens in this study are part of a large database constructed by using this method on—preferably—large adult specimens. The body mass of the larger kiwi plotted here was measured well before death, but even so may be somewhat low, because the individual was never completely healthy and died young. (All the wet-preserved kiwi specimens at the USNM were captives that were not weighed when alive and may have lost weight before death; they were autopsied and partly gutted after death.)

Aside from the data taken from Ruben et al. (1996), NA measurements tabulated in App. Table 2 were taken from uncrushed skulls and guided by whole tissue cross sections (in Bang 1966, 1971). This is most effective in those birds in which most or all of the NA walls are well ossified, including kiwis. Nasal wall tissues and septums are usually too thin in birds to alter the values significantly. Nasal dimensions of the echidna were measured from a sagittally sectioned skull.

The difficulties posed by fossils are even more severe. It is not possible to weigh exactly fossil forms even when complete skulls and skeletons are available (Appendix 2), and the level of inaccuracy rises with the incompleteness of an individual. Body masses in this study are based on carefully restored volumetric models and dimensional scaling by Paul (1988a, 1997a), but all dinosaur estimates are subject to error. Extinct ratites were so similar in form to their modern relatives that the estimates are more reliable (and are in good agreement with estimates by Alexander [1983a,b]).

Although CT scanning can reliably measure NA cross sections as preserved, the data can still be in error. The often-subtle crushing that mars the great majority of fossil skulls invariably reduces NA cross sections from their true values. For example, the snout of the juvenile *Tyrannosaurus* scanned by Ruben et al. (1996) is at least 30% narrower than other skulls (Paul 1988a), so both preserved and restored NA cross sections need to be plotted (App. Figs. 3T, 4). Conversely, nasal wall tissues are thick in some tetrapods (Parsons 1970, Witmer 1995a), so the fossil NA cross-sectional measurements tabulated in App. Table 2 represent maximum possible values that may be far above the true figure. As with living tetrapods, important sections of the NA are contained by little bone in some fossils. For example, the external nares of large ceratopsid dinosaurs are so large and extend so posteriorly that the cross section of the preserved NA at its broadest can be measured on any sufficiently complete and reasonably uncrushed, prepared skull. It is even easier to restore the NA dimensions of extinct large ratites because their external nares are also very large and extend far posteriorly; are similar to those of modern ratites; and many complete, uncrushed skulls and skeletons are known (R. Owen 1879, Monnier 1913, Archey 1941, Oliver 1949). In dinosaurs, well-preserved, disarticulated snout elements can be used to estimate the dimensions of the NA, after CT scans have determined the basic characteristics of similar passages in related forms. For additional anatomical details, and an examination of the metabolic implications of nasal airways, see Appendix 4. Many NA are diagrammed in Appendix Figures 3 and 5.

*Appendix Figure 3. (opposite)* Same-scale comparison of nasal airways of avepod dinosaurs and birds (identifying symbols and body masses listed in App. Table 2). *L:* left figure and rest of figures NA (solid black) are sagittally sectioned in left lateral view; nostrils are at the left end, and internal nares are at the lower right end. Transverse sections of the snout and/or the anterior or middle NA are at the upper right of each sagittal section, and the location of the sections made in this study are indicated by short vertical bars. Soft nasal tissues are not indicated, and uncertain borders of nasal passages not defined by bone edged with lines; x = mediorostral maxillary sinuses and/or accessory maxillary fenestrae; position of latter indicated by dashed lines. *L* includes detail of inner view of *Allosaurus* snout showing how maxillary sinuses influence primary NA (both structures are shaded) to follow the L-shaped path common to most averostran dinosaurs; *inset* shows that NA is sagittally sectioned at the point that the process of maxillary sinuses nearly contacts the vomer, effectively closing off the airway from taking a more ventral course; the probable posterior extent of NA is indicated by a dashed line; outline of anterior palatal elements is indicated by a heavy dashed line (also see App. Fig. 2). *T,* juvenile *Tyrannosaurus* cross section as preserved and restored on left and right sides, respectively, of left section; possible extent of posterior airway above internal nares shown on the right section; vertical dashed lines indicate possible conchae. *d,* NA cross-sectional area required to reach modern large ratite level indicated by dashed line on *Dinornis.* In part drawn after Ruben et al. (1996) and Witmer (1997a). Scale bar equals 100 mm.

## B: Lungs and Related Structures

In a paper published in 1997, Ruben et al. made an exceptional claim. They not only contended that theropod dinosaurs possessed a crocodilian-like lung ventilation system dominated by a pelvis-based diaphragmatic muscle pump, they believed an example

*Appendix Table 2. Dimensions of Nasal Airways Relative to Body Mass*

| | Mass (kg) | Nasal airway dimensions: Cross-sectional area ($cm^2$) Broadest | Anterior | Length (mm) |
|---|---|---|---|---|
| Birds | | | | |
| e—Elephant bird | 370 | 23 | — | 105 |
| d—*Dinornis* | | | | |
| CM1.12.5 | 260 | 13 | — | 70 |
| AM 64 | 150 | 8 | — | 50 |
| o—Ostrich USNM 429070 | 125 | 11.6 | — | 70 |
| p—*Pachyornis* CM XXB | 125 | 8 | — | 50 |
| m—*Emeus* CM viiiC | 84 | 7.8 | — | 40 |
| u—Emu | 40 | 5 | — | — |
| r—Rhea | 20 | 4 | — | — |
| b—Bustard USNM 289732 | 14.5 | 2 | — | 50 |
| n—Swan USNM 489348 | 12.6 | 2.6 | — | 45 |
| a—Albatross USNM 488376 | 12 | 2.1 | 0.7 | 38 |
| c—Condor USNM 492447 | 9.8 | 2.3 | — | 25 |
| t—Turkey USNM 488189 | 7.4 | 1.8 | — | 28 |
| s—Shoebill stork USNM 345070 | 6.5 | 2 | — | 55 |
| g—Ground hornbill USNM 321839 | 4.5 | 1.1 | — | 70 |
| h—Heron USNM 553852 | 4.1 | 1.15 | — | 25 |
| z—Goose USNM 320117 | 3.3 | 1.2 | — | 38 |
| k—Kiwis | | | | |
| USNM 614807 | 2 | 0.5 | 0.08 | 140 |
| USNM 18279 | 0.8 | 0.3 | 0.04 | 85 |
| v—Turkey vulture USNM 346785 | 1.3 | 1.6 | — | 15 |
| l—Gull USNM 347936 | 1.1 | 0.7 | — | 65 |
| w—Hawk USNM 614338 | 1.1 | 0.6 | — | 14 |
| d—Duck USNM 432298 | 1.05 | 0.7 | — | 26 |
| i—Prairie chicken USNM 289376 | 1 | 0.45 | — | 15 |
| p—Pheasant USNM 322386 | 0.95 | 1.0 | — | 17 |
| f—Falcon USNM 291186 | 0.9 | 0.4 | — | 10 |
| Mammals | | | | |
| ––Echidna USNM 236714 | 3 | — | 0.25 | 0.75 |
| +—Giant anteater | 15 | — | 0.8 | 1.9 |
| X—African elephant | 6,000 | — | 25 | — |
| Reptiles | | | | |
| 1—Alligator | 110 | — | — | 175 |
| 2—Crocodile | 50 | — | — | 125 |
| 3—Monitors | | | | |
| USNM 228163 | 45 | — | — | 40 |
| USNM 220283 | 5 | — | — | 20 |
| 4—Iguanas | | | | |
| USNM 35633 | 3 | — | — | 17 |
| USNM 220221 | 3.1 | — | — | 16 |
| USNM 220217 | 0.7 | — | — | 11 |
| 5—Gila monster USNM 220205 | 4 | — | — | 8 |
| Dinosaurs | | | | |
| L—*Allosaurus* | 1,300 | ~66 | 31 | 275 |
| T—*Tyrannosaurus* | 5,700 | ~160 | 82 | 470 |
| Juvenile | 500 | ~36 | 12–18 | 210 |
| G—*Gallimimus* | 440 | ~11 | ~2.5 | 160 |
| U—*Ornithomimus* | 110 | ~37 | — | 80 |

Appendix Table 2. *(continued)*

| | Mass (kg) | Nasal airway dimensions: Cross-sectional area ($cm^2$) Broadest | Anterior | Length (mm) |
|---|---|---|---|---|
| Dinosaurs *(continued)* | | | | |
| V—*Velociraptor* | 6.5 | ~1.8 | 0.6 | 75 |
| O—*Oviraptor* | 50 | ~9 | — | 58 |
| N—*Ingenia* | 5 | ~3 | — | 33 |
| P—*Parasaurolophus* | 2,600 | — | 32 | 2,300 |
| H—*Hypacrosaurus* | 2,800 | — | — | 670 |
| Juvenile | 300 | — | 13 | — |
| P—*Panoplosaurus* juvenile | 900 | 37–47 | — | 170 |
| E—*Euplocephalus* | 2,000 | ~100 | 38 | 340 |
| M—*Monoclonius* | 1,400 | 230 | — | 220 |
| R—*Triceratops* | 5,500 | 560 | — | 550 |
| A—*Apatosaurus* | 11,000–17,500 | — | 88 | — |
| D—*Diplodocus* | 9,500 | — | 46 | — |

NOTE: Kiwis are large birds that weigh up to 2–4 kg (McNab 1996). The larger specimen (USNM 614807) is an ex-NZP specimen with a measured peak weight, and 614807 and other specimens were used to estimate the weight of the smaller specimen (USNM 18279). Crocodilian masses were calculated from skull/mass relationship data. *Allosaurus* dimensions are based on disarticulated elements, including those described in Madsen (1976). Adult *Tyrannosaurus* dimensions primarily after type maxilla (Osborn 1912). *Gallimimus* anterior NA dimensions were estimated from the external dimensions of the complete skull (Osmólska, Roniewicz, and Barsbold 1972). *Velociraptor* dimensions were measured from the exposed posterior end of anterior NA tube in an articulated skull. Measurements for oviraptors are based on internally prepared complete skulls. Data for hadrosaurs come from Weishampel (1981), for ankylosaurs from Coombs (1978) and Witmer (1997a), and for sauropods from Berman and McIntosh (1978). Dinosaur mass data includes data from Paul (1997a). Two masses are included for *Apatosaurus* because we do not know which of two skeletons the measured skull belonged to. Data are plotted in Appendix Figure 2.



was preserved as soft tissues in the new *Sinosauropteryx* type specimen. Ruben et al. (1999) further argued that the preserved soft tissues of *Scipionyx* confirm the conclusions of their earlier study. If correct, this hypothesis would overturn the broad consensus that avepod dinosaurs evolved a preavian, air sac–ventilated lung complex (Perry 1983, 1989, 1992, Paul 1988a, Claessens 1996, Reid 1996, 1997, Britt 1997, Currie 1997, Larson 1997, Bramble and Jenkins 1998, Britt et al. 1998, Claessens and Perry 1998, Xu, Wang, and Wu 1999, Burnham et al. 2000, Christiansen and Bonde 2000, Martill et al. 2000). To verify the presence of a pelvovisceral muscle pump, some requirements must be met. It must be shown that it is possible to discriminate between a crocodilian-like versus a birdlike set of central organs when the latter are preserved. This has become an important issue following the preservation, in one example in three dimensions, of organs in avepod dinosaurs. In such cases it must be shown that the fossil soft tissues in question really have been preserved and correctly identified. Finally, it must be demonstrated that the skeletal anatomy is compatible with the operation of a pelvovisceral muscle pump.

## *Soft Tissue Analysis*

The last point brings us to a problematic aspect of the conclusions of Ruben et al. (1997a) regarding the preserved soft tissues. Until recently, Yixian specimens were illegally collected and prepared by local residents rather than by trained personnel. Slabs were often shattered as they were removed from the ground piece by piece, so the type *Sinosauropteryx* was symmetrically broken into almost two dozen pieces (Pls.16–18). After each piece was split to reveal all the preserved parts of the specimen, the fragments were then glued or plastered back together by the amateur collector. Edges were often not properly lined up in any dimension, and cement was often used to fill in broad cracks and in some cases was colored to match the sediment, thereby obscuring the damage.

A dark area is present in the abdominal region on both slabs of the type specimen (Pls. 16 and 17, App. Fig. 6Aa). Because the material lies medial to the ribs, it appears to be carbonized internal tissues, but separate organs are not distinguishable. The critical point centers on the opinion of Ruben et al. that the anterior border of the carbonized tissues form a smooth, anteriorly convex arc spanning the middle

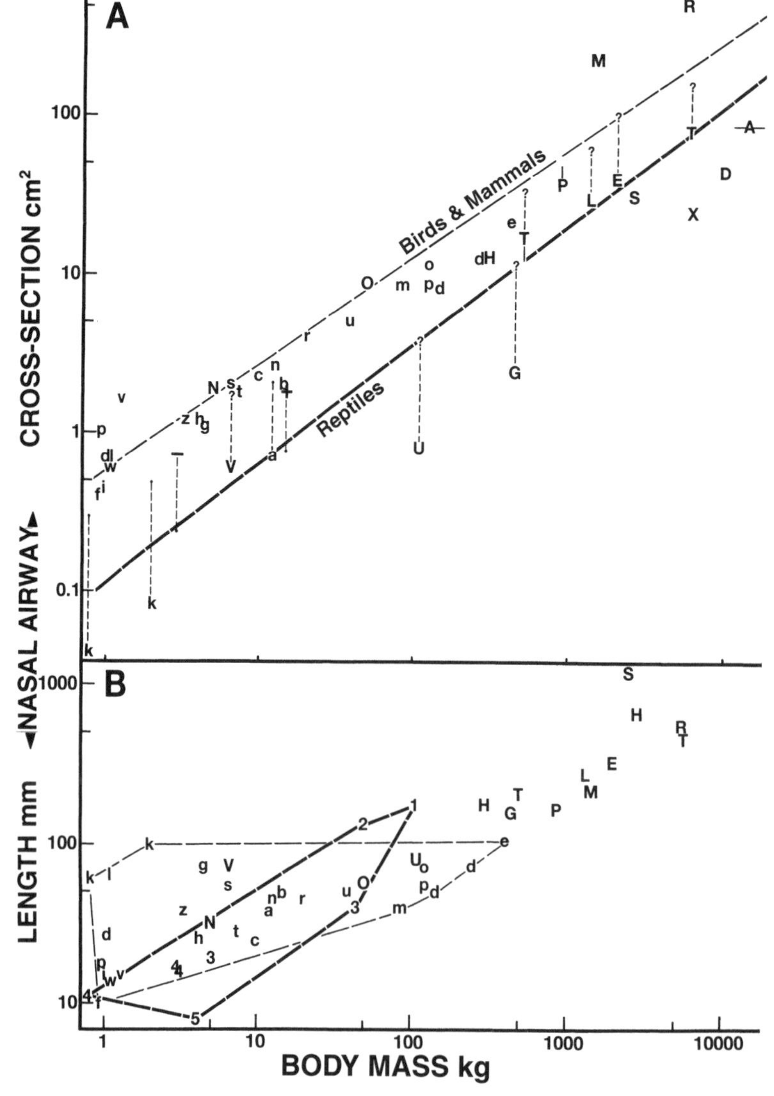


*Appendix Figure 4.* Plots of *(A)* nostril (elephant only) or nasal airway (all others) cross-sectional area as a function of body mass and *(B) nasal airway* length relative to body mass. Identifying symbols and data are listed in App. Table 2. The slender anterior and broader posterior (vertical dashed lines) passage cross sections are plotted for kiwis, albatross, and most theropods (question marks indicate approximations for posterior sections). The regression slopes are taken from Ruben et al. (1996). Note that owing to the steepness of the slopes, even modest errors in body mass significantly alter the results.

of the body cavity from top to bottom, similar to the postpulmonary septum of crocodilians that partitions the body into fore-and-aft compartments (the characteristics of various tetrapod groups' respiratory complexes are detailed below). Within the anterior crocodilian compartment are the lungs, and the posterior compartment begins with a liver that is often very large (App. Figs. 6Bb, 7B). Before directly examining the specimen, Ruben et al. published a low-quality photograph of the main slab, with three arrows pointing to the edge of the alleged septal boundary (Fig. 5A in Ruben et al. 1997a). In 1999, Ruben et al.—after examining the specimen—continued to assert that the supposed septal boundary and liver are "distinct" and "remarkably crocodilian-like." No mention was made of the breakage or other possible problems in either of these studies.

Direct examination of the main slab shows extensive breakage at the dorsoanterior edge of the carbonized material (Pl. 16). The ribs are also broken at this location. This break occurred when a thin layer of sediment dislodged from the main slab and remained attached to its counterpart. The dorsal arrow in Ruben et al. (1997a) points directly to the edge of this break. A large, complex set of breakage astride and running perpendicular to the anterocentral edge of the dark tissue is present on both slabs. It is partially filled with cement, which appears to have been colored to better match the dark material. A narrow, irregular zone of dark material appears to

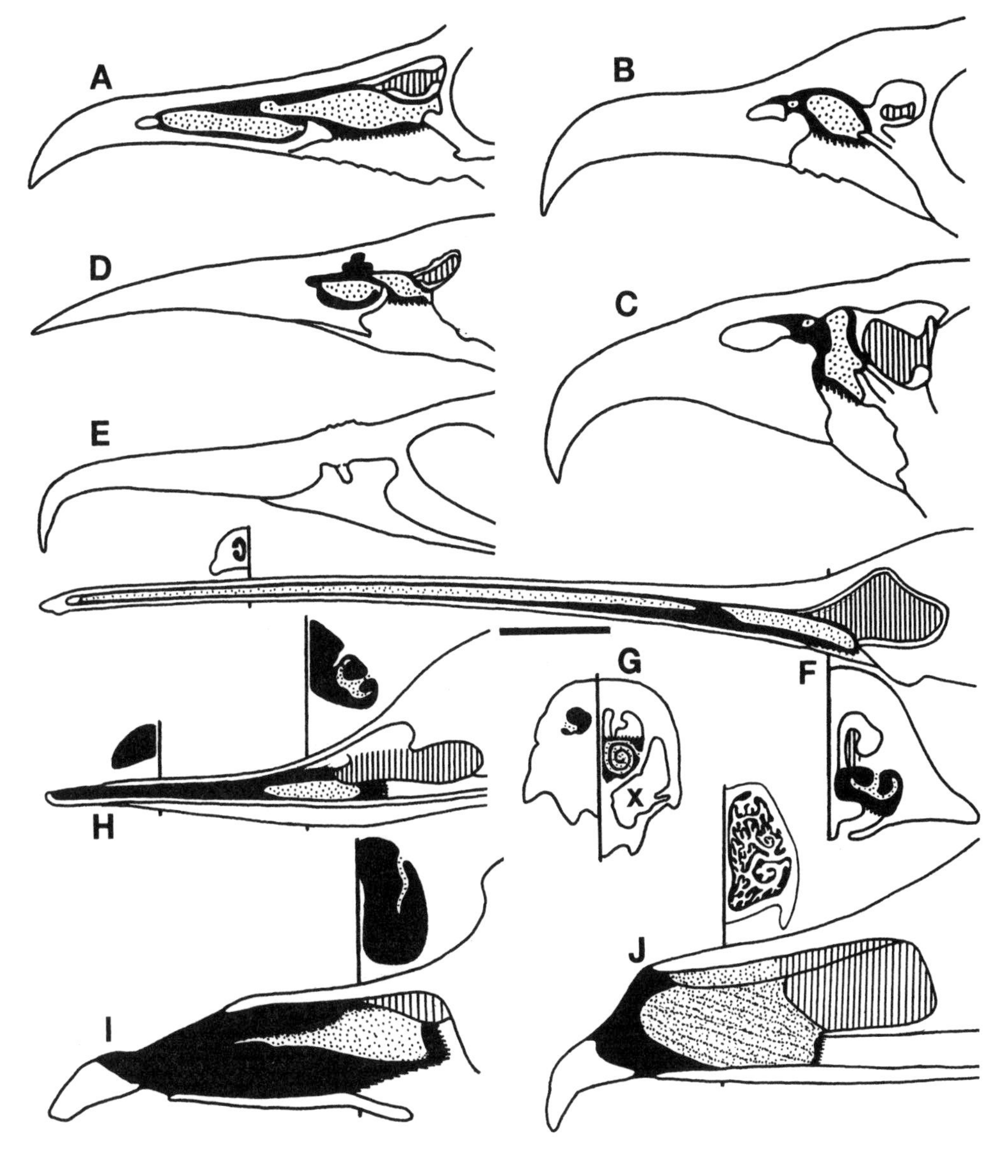


*Appendix Figure 5.* Variation in size and configuration of nasal airway (NA, solid black) and respiratory concha (RC, stippled, divided into anterior and middle conchae in birds; primarily maxilloturbinal in mammals) in birds and mammals (olfactory conchae vertical solid lines), x = pre-orbital sinus. *A,* gull with large RC; RC is smaller in the following birds: *B,* black vulture; *C,* kite; *D,* tropic bird; and *E,* cormorant, in which nasal passage is blocked and functional conchae are absent; the NA is exceptionally narrow in the following birds: *F,* large kiwi; *G,* albatross, comparison of small anterior and larger posterior sections; *H,* echidna with small RC in narrow NA; *I,* baboon with poorly developed RC in large NA; *J,* dog with well-developed RC in large NA. Drawn, in part, after Bang (1971) and Baker (1979). Nasal soft tissues is included in birds but not in mammals. Not to same scale, except for *F* and *H,* in which scale bar equals 10 and 20 mm for transverse and sagittal sections, respectively.

lie immediately forward of the crack. The material's anterior extent is further obscured by the presence of a rib, but it appears to extend too far anterodorsally to conform to the smooth convex arc described by Ruben et al. On both slabs, the ventroanterior border of the dark material is another illusion created by an irregular zone of flakage of the superficial layer of sediment: the ventral arrows in Ruben et al. (1997a, 1999) point toward this pseudoborder. Posterior to the central and ventral damage there is a very irregular zone of darker material. Because more than 60% of the anterior edge of the dark material consists of breakage and the preserved edge is irregularly formed, there is no well-formed, semicircular structure present on either slab. Instead, the smooth septum is a mirage created by cracks, breaks, and cement filler shared by the symmetrically damaged slabs and accentuated by poor-quality photography. Neither of the other two *Sinosauropteryx* specimens shows any trace of a septal boundary (Pl. 18; pp. 78–79 in Ackerman

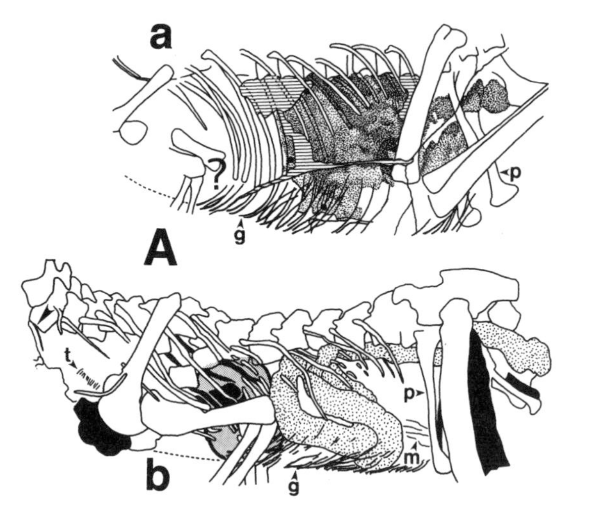


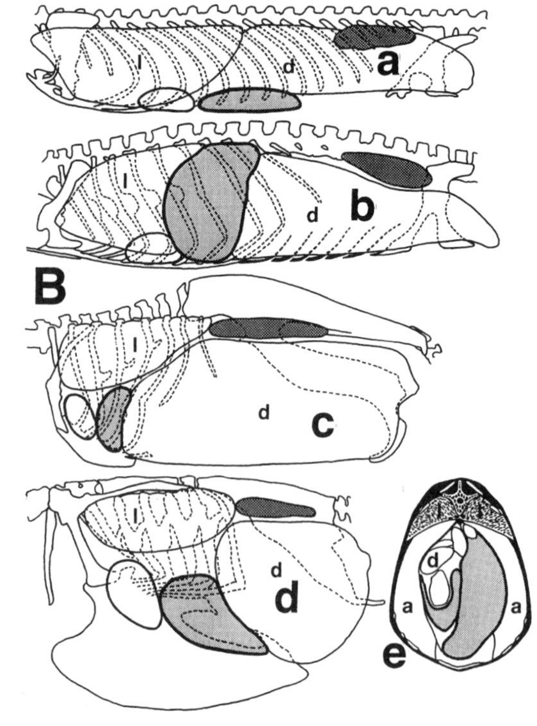


*Appendix Figure 6.* Comparative anatomy of the thorax in archosaurs, with emphasis on the respiratory complex. *A,* compressed dino-avepod specimens with preserved soft tissues: *a, Sinosauropteryx* type (after Pl. 16), in which the ill-defined, badly damaged dark material probably represents the digestive tract; dots mark points of arrow tips in Figure 5A in Ruben et al. (1997a); *b, Scipionyx* type (Sasso and Signore 1998, Ruben et al. 1999), with exquisitely preserved intestines and probable liver and trachea; ventrally displaced cervical vertebra indicated by arrow. In both *Aa* and *Ab,* the probable posterior cartilaginous sternum is indicated by the dashed line; note that there is no organ preserved in *Aa* in the position (indicated by question mark) in which the liver is preserved in *Ab. B,* contents of thorax in diapsids: *a,* varanid lizard with conventional liver; *b,* modern crocodilian with fore-and-aft sliding pelvovisceral pump and strong fore-and-aft partitioning of body cavity; *c,* flightless ostrich with small, anteriorly placed liver (in part after Fowler 1991); *d,* flying pigeon with normal-sized liver (after Dorst 1974); *e,* cross section at midthorax of flying carrion crow with enlarged liver that extends dorsally between lungs. Also note the air sacs lateral to the guts (after Fig. 2.18d in Duncker 1979); livers are heavy bordered and lightly stippled, the heart is heavy bordered, the kidneys are heavy stippled, and the muscles are solid black. *C,* rib cages in dorsolateral view showing ribhead and vertebrae articulations, ventral head of ribs stippled: *a,* varanid lizard with long trunk containing simple dead-end lungs, transverse processes very short, ribheads simple, lumbar region absent; *b,* crocodilian with long trunk containing pelvis-based diaphragmatic muscle pump, most ribs articulate only with distal ends of hyperelongated transverse processes, lumbar region well developed; avepods with shorter, deeper trunks in which even posterior ribs are long and double headed in order to form well developed, angled hinge joints with moderately elongated transverse processes: *c,* basal dino-avepod *Coelophysis* based in part on Figures 100 and 102 in Colbert (1989); *d,* kiwi with air sac ventilation. *D,* articulation between ribheads (solid black) and dorsal series (rib articulations solid back) in side view, elements shown depend upon completeness of specimens: *a, Euparkeria; b,* modern crocodilian; basal theropods: *c, Staurikosaurus; d, Herrerasaurus;* avepods: *e, Coelophysis;* avetheropods: *f, Yangchuanosaurus; g, Allosaurus;* avepectorans: *h,* dromaeosaur; *i,* duck. *E,* relationship between the mid–rib cage ceiling and respiratory anatomy and function shown via transverse cross section and sagittal sections along transverse processes in side view: *a,* modern crocodilian with liver solid (black), liver-supporting ligaments (lined), liver sac (stippled); arrows indicate anteroposterior motion of liver; *b,* generalized avetheropod-bird with lung and bone diverticula sections on right side. *F,* variation of ossification of sternum and sternal ribs (white zones) and absence of uncinate processes: *a,* ostrich chick; *b,* very large emu; *c,* precocial king rail chick at about two and a half weeks; *d,* adult screamer; *e,* fully grown but immature six-week-old chicken in side and bottom views. Abbreviations: a = air sac, l = lung, d = digestive tract, r = rib, t = transverse process; other abbreviations are listed in Figure 4.1. Not drawn to same scale.

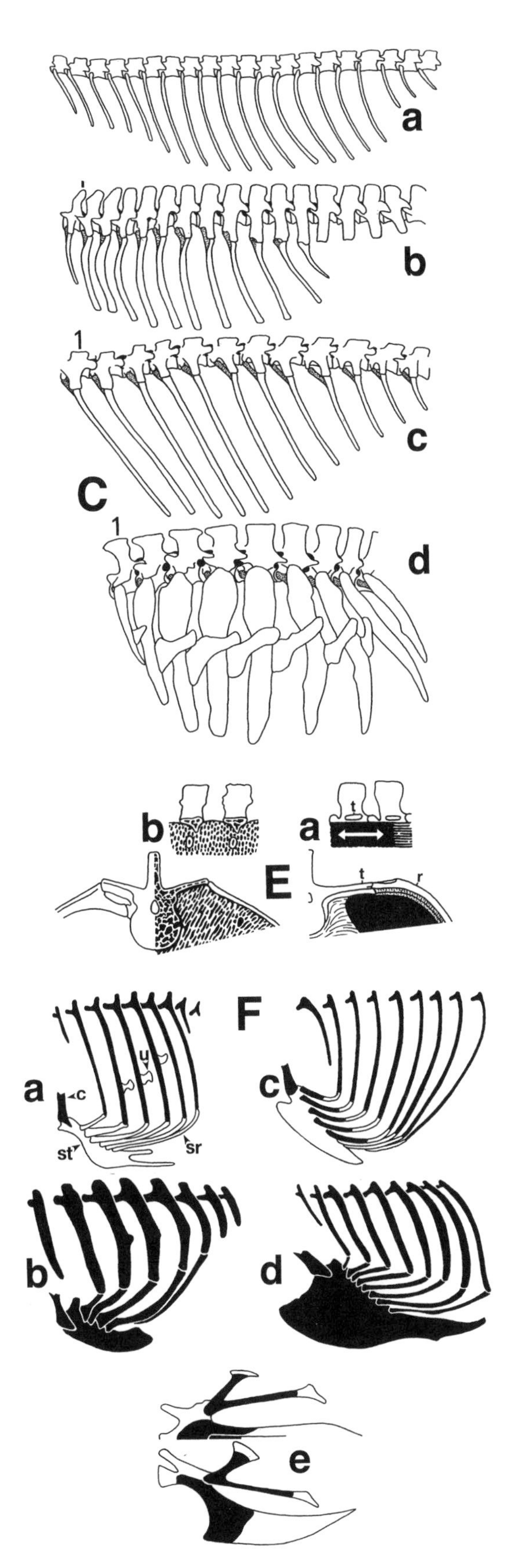
a
b
1
c
C
1
d
b
a
t
E
t
r
F
u
a
c
c
st
sr
b
d
e

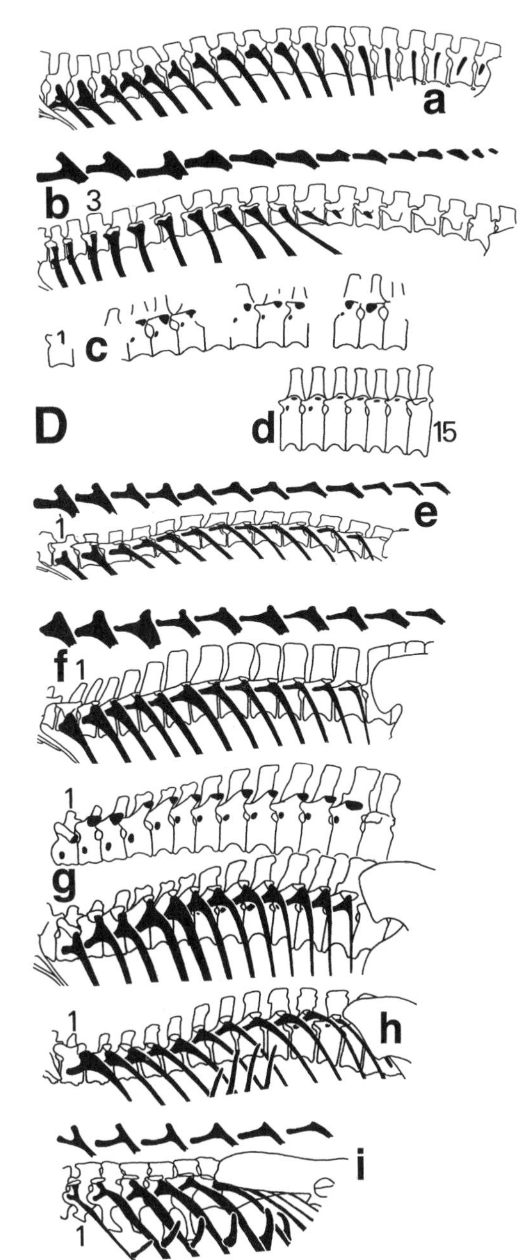
a
b 3
1
c
D
d
15
e
1
f 1
1
g
1
h
i
1

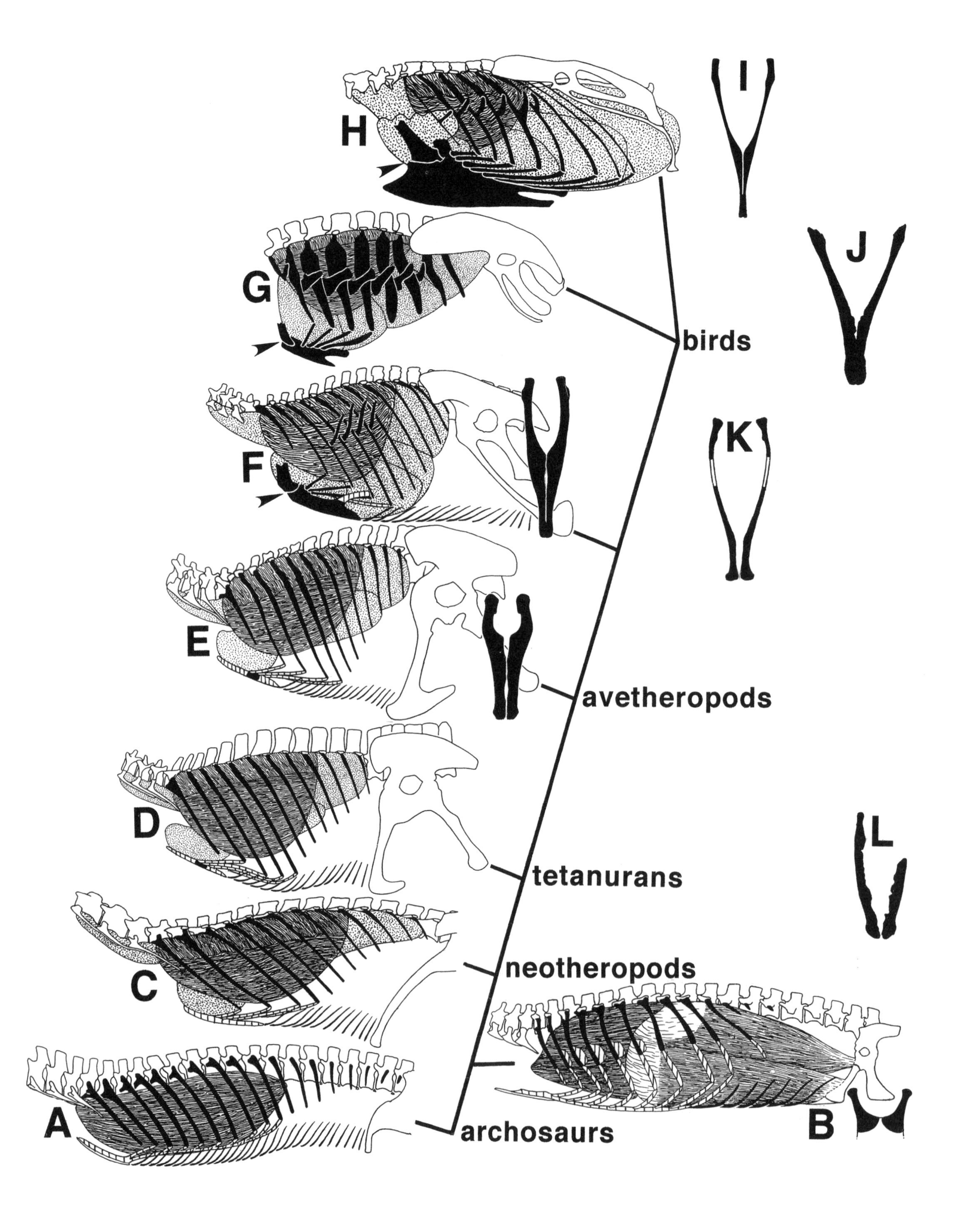
H
I
J
G
birds
F
K
E
avetheropods
D
tetanurans
L
C
neotheropods
A
archosaurs
B

[1998], Fig. 4a in Chen et al. [1998] in which eggs contribute to the anterior portion of an irregular abdominal dark area). What is the origin of the dark material, and why does it lie in only the posterior half of the body cavity?

Answering the above question brings us to the problem of telling whether the soft tissues preserved in a fossil are more like those of crocodilians or of birds. This is a major problem because the differences between the two groups are not as dramatic as might be expected. Livers, for example, are multipurpose organs with both physiological and anatomical functions, and whose size, shape, and location are highly variable. In particular, the size of a liver can vary substantially between individuals and even within them over time. Livers tend to be larger in growing juveniles and in eaters of flesh than in plant eaters (Siwe 1937, Secor and Diamond 1995). In modern crocodilians, the enlarged liver is usually so tall that it almost spans the body cavity from the top to the bottom of the rib cage (App. Fig. 6Bb,Ea; Duncker 1979; Ruben et al. 1997a). In many birds, the liver is much less enlarged and is set low in the body cavity, often well forward in the chest region (App. Fig. 6Bc,d). But this is not always the case. In some birds, the liver is so large and tall that it almost spans the distance from the sternum to the vertebrae, and even extends up between the high-set lungs (App. Fig. 6Be; Duncker 1979; Fig. 1 in Brackenbury 1987). The anteriorly convex arch that characterizes the anterior border of the liver and the postpulmonary septum in crocodilians is also seen in birds (App. Fig. 6Bc,d; Fig. 1 in Brackenbury 1987). For that matter, major organs are typically rounded in profile and may give the impression of arched septums. The presence of a fossil liver that spans the entire body cavity and is anteriorly convexly arched is therefore compatible with either a crocodilian- or birdlike arrangement. That the fossil liver is likely to have been flattened by sedimentary pressure only further complicates its interpretation. To put it another way, if a large-livered fossil bird's skeleton and internal organs were preserved on their sides, the deep liver would give the illusion of being crocodilian-like.

Thus, the presence of a tall, anteriorly convex liver is compatible with either a crocodilian- or avian-like arrangement. Therefore, the finding of a tall liver tells us little about whether the liver was mobile and divided the body cavity into distinct fore-and-aft compartments. To determine liver mobility via fossilized soft tissues would require the discovery of unambiguous diaphragmatic muscles, and/or lungs deep enough and properly positioned to be ventilated by such muscles. However, lungs are mainly air and therefore poorly suited for preservation. Because lungs may inconsistently lie partly lateral to more easily preserved central organs, especially the heart and/or liver, their original extent may be obscured if the specimen is flattened on its side. Also bear in mind that in crocodilians, lobes of the lungs are lateral to heart but not the liver, whereas in some birds the opposite is true (App. Fig. 6Bb–e). Because the walls of air sacs are very thin (McLelland 1989b), fossilizing these bags of air is as improbable as fossilizing a balloon. Air sac preservation is especially unlikely if the specimen is flattened. The majority of the thoracic sacs lie lateral to the internal organs (App. Figs. 6Be, 7G,H), so organ-free areas will not necessarily mark the sites of air sacs if the specimen is flattened and lying on its side.

In the extraordinary type *Scipionyx*, some internal organs are very well preserved in two dimensions (App. Fig. 6Ab), but the crushed condition casts doubt on the exact position of the organs (see below). The object that Sasso and Signore (1998) and Ruben et al. (1999) identified as the liver is set

*Appendix Figure 7. (opposite)* Phylogenetic chart of thoraces, pelves, and respiratory tracts restored and known in archosaurs. *A,* basal archosaur *Euparkeria,* which probably had a little-modified, basal tetrapod system; *B,* modern crocodilian with divergent pelvovisceral pump; avepods with increasingly well-developed air sac complexes; *C, Coelophysis* with long chest ribs; tetanurans with shorter chest ribs; *D,* basal tetanuran *Yangchuanosaurus* with moderately shortened chest rib (see Fig. 51 in Dong et al. 1983), avetheropods with short anterior chest rib; *E, Allosaurus,* avepectorans with bellows-action ossified sterna, sternal ribs, and uncinate processes; *F,* dromaeosaur composite, sternum even longer than shown in some examples (Fig. 1.7, App. Fig. 10B); *G,* flightless kiwi with short sternum and free posterior ribs; *H,* flying duck with extremely large sternum attached to all ribs. The articulated pubes are shown in anterior view in *B, E, F,* and *I, Archaeopteryx; J, Avimimus; K,* basal alvarezsaur *Patagonykus; L,* basal crocodilian *Hesperosuchus* (medially incomplete and possibly broader than shown). Lungs are indicated by irregular lines; pulmonary diverticula are stippled; pelvodiaphragmatic muscles are indicated by fine lines; ossified gastralia, uncinate processes, ribs, and sterna are solid black; restored cartilaginous sternal elements and sternal ribs are indicated by heavy lines. Sternocostal articulations in *D* and *E* are taken from Currie and Zhao (1993a). In *F–H,* the hinge articulation of the sternum with the coracoid (only partly included) is indicated by arrows; all rib cages based on articulated examples. The air sac that may have been present immediately posterior to the dino-avepod pubis (Martill et al. 2000) is not included. Not drawn to same scale.

in the proper position for this organ. It is less likely that the organ is part of the digestive tract, or that the heart should have been placed farther forward. The intestines are exquisitely preserved, the liver less well preserved, and the heart entirely lost. The preferential preservation of the intestines implies that their chemistry favored their fossilization. The liver sits directly above the juncture between the gastralia and the probable posterior edge of the cartilage sternum. The liver also sits immediately behind and below the posterior-most end of the scapula blade, and between the distal ends of the tucked-up humeri. The digestive tract is set in the posterior half of the body cavity, with the anterior edge of the intestines below dorsal vertebra 9. In the type *Sinosauropteryx,* there is an empty space above the anterior end of the gastralia, just behind the posterior-most extent of the scapula and between the tucked-up humeri (App. Fig. 6Aa,b), so the liver was probably not preserved. The anterior-most dark material is below dorsal 8, and fecal material is present. Thus, we conclude that the dark material is all that remains of the digestive tract, barely preserved because of its distinctive chemistry. If so, then both the liver and heart rotted away.

Even more remarkable is the preservation in three dimensions of organs involved with the pelvis in a small, basal averostran dinosaur from the Santana Formation (Martill et al. 2000). This specimen has a well-preserved colon. Perhaps even more striking is a vacuity immediately posterior to the entire length of the pubic shafts, which Martill et al. interpreted as an air sac. Because air sacs consist of air contained by very thin walls, their preservation as fossils is astonishing, and may be limited to such uncrushed specimens.

The liver appears to have been deep in *Scipionyx* (Ruben et al. 1999), filling the entire body cavity from top to bottom. This configuration is compatible with either the crocodilian or with avian conditions, especially since flesh eaters are likely to have large livers. The septum is not preserved in *Scipionyx,* or in *Sinosauropteryx*. Therefore, fore-and-aft partitioning of the body cavity has not been demonstrated in either dinosaur.

Because bird lungs are shallow and sit high in the body cavity, it is common for the trachea to run just under the cervicodorsal vertebrae (Ruben et al. 1999), but there are birds in which the trachea runs more ventrally (Figs. 3 and 6b in Duncker 1971; Fig. 69 in Nickel et al. 1977, Minnaar and Minnaar 1997). In deep-lunged reptiles, the trachea sits low in the neck base, especially in crocodilians (Ruben et al. 1999). In *Scipionyx,* the probable trachea appears to be set low (App. Fig. 6Ab). However, cervical 10 is displaced ventrally relative to its neighboring vertebrae. Breakage of the neck may have pushed the trachea from a more dorsal life position: it is not possible to determine the true position of the trachea without additional, less damaged specimens. For that matter, the position of the trachea may have been variable in the living dinosaur. In some birds, especially those with long necks, the trachea is remarkably mobile, and the air passage may drop well below the cervicodorsals just anterior to the shoulder girdle (Fig. 2.8a in McLelland 1989a). The position of the trachea in *Scipionyx* does not offer definitive evidence regarding lung depth.

The dorsal position of the colon as preserved in *Scipionyx* is more like that of crocodilians and mammals than that of birds, in which the last section of the alimentary canal is more ventrally placed. Ruben et al. (1999) suggest that the avian arrangement is tightly linked to function of the posterior air sacs, although exactly why this should be so is not clear. Nor has it been demonstrated that dorsally placed colons are not found in any tetrapods without pelvovisceral breathing (see further discussion below). In any case, the hypothesis that nonavepectoran dinosaurs possessed *pre*avian respiration does not necessarily require fully avian posterior air sacs. Direct refutation of the Ruben et al. argument appears to come from the Santana avepod (Martill et al. 2000). Its colon was set well below the dorsal column, immediately prior to the pelvic canal. Because preservation is three-dimensional, this birdlike colon position is more reliable than that of the crushed *Scipionyx*. As explained below, there is skeletal evidence that other dino-avepods had low-set colons.

To establish the presence of pelvovisceral mobility via fossilized soft tissues would require the discovery of unambiguous pelvis-based diaphragmatic muscles. Ruben et al. (1999) noted the presence of possible muscle traces immediately anterior to the distal pubes. They proposed that these are oriented in the manner expected of crocodilian-like diaphragmatic muscles. The small patch of possible tissues does not reach anteriorly to the liver, is poorly preserved, and lacks the fine fibers that characterize other muscles preserved with the specimen. Therefore, the actual length and orientation of the muscle fibers is not certain. The longitudinally oriented undulations observed by Ruben et al. appear to be superficial. The configuration of these tissues more closely resembles that of the outer layer of pubis-based abdominal wall muscles of birds, especially the posterior-most sections of the *M. obliquus* and *M. rectus* components of the abdomoni (compare Fig. 3 in Ruben et al. 1999 with Fig. 1 in Fedde 1987). There is no definitive evidence for pelvo-diaphragmatic muscles in *Scipionyx*.

To sum up the soft-tissue evidence, the type *Sinosauropteryx* can only be said to include very poorly preserved and badly damaged amorphous dark material that lies between the ribs in the posterior half of the thoracic cavity, and is most compatible with representing traces of the stomach and/or intestines. Despite its extraordinary preservation, the *Scipionyx* is a single, somewhat damaged, very small juvenile whose internal structure—most especially the depth of the liver—may have changed with ontogeny, and whose preserved anatomy is compatible with either a crocodilian or preavian respiratory tract. To date, neither a diaphragm nor fore-and-aft partitioning of the thoracic cavity has been directly observed in a dino-avepod, and evidence for pelvodiaphragmatic muscles is questionable. Therefore, the soft-tissue evidence that any dinosaur possessed a crocodilian-like respiratory complex is at best weak. Instead, the best-preserved avepod dinosaur soft tissues—the low-set colon and possible air sac of the Santana specimen—are most compatible with a birdlike system.

## *Osteological Analysis*

Because the soft-tissue evidence is not yet definitive, we must turn to osteological data to better restore the respiratory complex of dinosaurs. To do so, we must first cross-correlate the skeletal and pulmonary adaptations of modern diapsids.

**Tuataras, lizards, snakes (App. Fig. 6Ba,Ca)**

The trunk is long, so shallow that it is at least as broad as it is deep, and flexible. The flexible lungs are not tightly attached to the vertebrae or ribs. The lungs are large, and internal complexity ranges from very simple to moderately complex (Bellairs 1970, Lawiewski 1972, Duncker 1978, Perry 1983, 1989, 1992). As in all amniotes, a thin sheet of tissue forms a septum between the lung and abdominal cavities. The liver is normal in size and position, and is not actively moved during respiration. The lungs are primarily ventilated by expansion of the chest ribs. To accomplish this, all of the anterior ribs, including the first, are long. However, ventilation rates in reptiles are low (Appendix 4), so the rib head-vertebrae articulations tend to be simple. Because the posterior ribs are not involved in respiration, they are not strongly elongated, and a well-developed lumbar region is absent. The pubes, which have no particular attachment to the respiratory apparatus, are short, about as broad as they are long, and procumbent. The pelvic canal is set immediately below the sacrum. Gastralia may be present or absent.

Most reptiles' lungs are dead-end, but nonvascular diverticula of the lungs are present in geckos, monitor lizards, and especially in chameleons and certain snakes (Lawiewski 1972, Duncker 1978, Ruben et al. 1997a). Their function is little studied, and there are no apparent skeletal adaptations to strongly ventilate the relatively extensive diverticula of chameleons and snakes.

**Crocodilians (App. Figs. 6Bb,Cb,Db,Ea, 7B,L)**

These archosaurs retain a long, shallow, flexible trunk, long chest ribs, simple vertebrae-rib articulations along most of the dorsal series, and procumbent pubes. The lungs are fairly tightly attached to the ribs (Perry 1988). The dead-end lungs are highly flexible and deep, but are smaller than in other diapsid reptiles, perhaps because they are internally more complex. A subvertical septum separates the liver from the rest of the abdominal cavity, forming a fore-and-aft separation of the body cavity. The liver is so deepened that it fills the middle of the body cavity from top to bottom. The pelvic canal is set immediately below the sacrum.

What is really unique and remarkable about crocodilians is their pelvis-based diaphragmatic muscle pump system for ventilating the lungs. The lungs are highly expandable, and are deeply indented by only the first two dorsal ribs (Perry pers. comm.). Just the first three ribs are double-headed and articulate with the vertebrae in the usual archosaur manner. The mid-dorsal ribs articulate only with the lateral ends of hyperelongated, dorsoventrally flattened transverse processes (Mook 1921). This very unusual feature forms an exceptionally smooth bony ceiling to the rib cage (App. Fig. 6Ea). The gastralia, which do not articulate along the midline (Fig. 10.7E), are set in a continuous cartilage sheet. The combination of a smooth rib cage ceiling and cartilage-embedded gastralia forms a well-braced, subcylindrical body tube, through which the viscera can readily slide as the lungs are ventilated (Gans and Clark 1976, Perry 1988, 1990). The diaphragmatic muscles that operate this pump attach to the capsule that encloses the liver, rather than directly to that organ (Duncker 1978). The pulmonary septum is airtight and maintains a pressure differential between the fore and aft thoracic compartments. The large pelvodiaphragmatic muscles are anchored to the anterior edges of the ventral pelvis and posterior gastralia. Because of the involvement of the pelvis, and because organs other than the liver participate in the pump action, the term *pelvovisceral* rather than *hepatic* more fully describes this piston/pump. Modern crocodilian pubes are about as broad as they are long, so the pubes are not "elongate" as Ruben et al. (1997a) claimed. Even in gracile basal crocodilians with apparently mobile pubes, the length/distal breadth ratio is not under 4, and the length/midshaft ratio is over 2. Nor is there a pubic boot (Hutchinson 2001). Instead, crocodilian pubes are distally transversely broad enough that they form shovel-like plates. The abdominal surface of this expansion faces strongly forward and somewhat medially, thereby helping to support the broad abdomen. Crocodilian pubes are unique in being able to swing back and forth. Their function was previously little studied, but new work (Farmer and Carrier 2000b) indicates that visceral respiration is markedly enhanced by pubic mobility, which helps create space for posterior displacement of the viscera (whether pubic immobility is compatible with some degree of crocodilian-like breathing has not been tested by disabling this complex). A rib-free lumbar region in front of the pelvis also allows the volume of the abdomen to change dramatically as the pelvovisceral muscle pump operates. This feature is also seen in mammals with diaphragm lung ventilation. The respiratory function of the unusual unossified cartilaginous double sternal ribs is obscure. Cartilaginous or poorly ossified uncinate processes are present on crocodilian ribs; their function is also not well documented.

### Modern birds
### (App. Figs. 6Bc–e,Cd,Di,Eb,F, 7G,H)

Birds have also evolved an unusual respiratory system. The small lungs are internally very complex (Huxley 1882, King 1966, Duncker 1971, Lawiewski 1972, Schmidt-Nielsen 1972, Fedde 1987, McLelland 1989b, Maina 1989, 2000). Their lungs are semirigid, so the dorsal column is correspondingly inflexible. The series is also very short, the number of dorsal vertebrae being very reduced. Also short are the anterior chest ribs, which must only hold rather than ventilate the shallow, stiff lungs. The mid-dorsal rib heads' ventral processes are long and set well below moderately long, T-cross-sectioned transverse processes. These structures create a series of deep ridges along the entire ceiling of the rib cage. Dorsal lobes of the lung pack the intervening recesses. The lungs are further immobilized by invasion of pulmonary diverticula into pneumatic vertebrae and ribs in most birds. In those birds with pneumatic bones, the excavation is consistently present. However, the vertebrae and ribs of many small fliers, most divers, and kiwis are reported to be weakly or not pneumatic, despite the presence of well-developed air sacs (Bellairs and Jenkin 1960, King 1966, McLelland 1989b). The various posterior air sacs breach the pulmonary septum, which does not maintain a strong pressure differential. The pubes are elongated, slender, and extremely retroverted. The pelvic canal is very deep, an adaptation that facilitates a ventrally placed colon. The pubic boot and gastralia are absent. The liver and pubes are not highly mobile.

How is the semirigid avian lung ventilated? The lungs do not dead-end, because diverticula are very well developed. In particular, large ventroposterior air sacs (clavicular, thoracic, and abdominal) fill much of the body cavity. These act as bellows to ventilate the lungs, via a complex set of aeroplumbing (K. Zimmer 1935, King 1966, Duncker 1971, Lawiewski 1972, Schmidt-Nielsen 1972, Brackenbury 1987, Fedde 1987, McLelland 1989b, Scheid and Piiper 1989, Brackenbury and Amaku 1990). Air flow through the lungs is predominantly unidirectional, moving from the back to the front. The bronchi connecting the lungs and posterior air sacs penetrate the arched horizontal septum that separate the lungs from the rest of the body cavity. Another septum separates the liver from the abdomen (Schmidt-Nielsen 1972, Duncker 1979). The posterior ribs of all birds are elongated, so there is no lumbar region. The posterior ribs also have well-developed double heads, which are also angled relative to the main axis of the body. This unusual adaptation serves to tightly control the rotation of these ribs (K. Zimmer 1935, King 1966, Duncker 1971, 1978, McLelland 1989b). They are prevented from rotating too far medially, thus avoiding collapse of the lateroposterior air sacs. Instead, the hinge joint adaptation allows the ribs to move freely in a single plane. Because the double heads are angled, the rotation of the ribs is forward and outward, or backward and inward, which significantly changes the volume of the posterior body cavity in transverse as well as fore-and-aft directions. The rotation of the posterior ribs is accomplished at least in part via intercostal and other axial muscles. The resultant rotation changes the volume of the thoracic and abdominal air sacs enclosed by the ribs. Anteriorly, the sternum is connected to the ribs via ossified sternal ribs, which are not doubled. The hinged joint formed by the transversely broad articulation between the coracoids and sternum, as well as the hinged joints between the sternum and sternal ribs, allows the sternal plate—which can be raised or depressed—to help ventilate the anterior and ventral air sacs. The mobile ribs, sternal ribs, and sternum form an integrated unit that works in concert to ventilate the air sacs (K. Zimmer 1935, King 1966, Duncker 1971, Lawiewski 1972, Schmidt-Nielsen 1972, Fedde 1987, McLelland 1989b, Scheid and Piiper 1989, Ruben et al. 1997a).

The ribs of almost all birds sport well-developed ossified uncinate processes. These processes pose some interesting questions. It is generally thought that they assist respiration by improving the action of the intercostal muscles (K. Zimmer 1935, Duncker 1971, Fedde 1987, Ruben in Feduccia 1996, Hou et al. 1996). It has also been suggested that the processes help strengthen the rib cage (Bellairs and Jenkin 1960). However, Heilmann (1926) and K. Zimmer (1935) noted that not all birds have uncinate processes (contra Ruben in Feduccia 1996). Strong-flying screamers lack any trace of the processes (App. Fig. 6Fd); the projections are rather small in herons (Fig. 23 in K. Zimmer 1935). In weak-flying hoatzins, the processes are so anteroposteriorly short and deep that they represent expansions, rather than projections, of the ribs (Fig. 22 in K. Zimmer 1935). In flightless emus, uncinate processes are generally absent, except vestigially in very large individuals (App. Fig. 6Fb). Flightless relatives of screamers, the diatrymids and dromornithids, lacked ossified uncinates (Matthew and Granger 1917, Andors 1992, Murray and Megirian 1998); it is not clear whether phorusrhacoids and elephant birds had them (Monnier 1913). Ossified uncinate processes are also absent, or small and poorly ossified, in some active, precocial bird chicks (App. Fig. 6Fa,c). What is clear is that the absence or very small size of uncinate processes in various birds shows that their development is not critical to air sac–driven respiration on the ground, or even in the air.

The above analysis applies to birds in general. Almost all studies of avian respiration have focused on modern flying birds—not surprising in view of the intense interest in their ability to fly so far and high. This has had the unfortunate effect of neglecting those birds whose flight abilities are reduced or absent (Schmidt-Nielsen's [1972] limited work on the ostrich is a rare exception). This is unfortunate because ground birds are probably better models for the nonaerial ancestors of birds. After all, the thorax of an airborne bird must do a number of things at once: anchor the large, highly aerobic flight musculature; operate the respiratory complex that oxygenates these muscles; and resist collapse caused by the stresses induced by the flapping action of all those powerful flight muscles. The posterior air sacs are very large, in most cases extending along the sides of the abdominal cavity to the posterior end of the body cavity. The posterior ribs are very elongated to help ventilate these sacs. All but the anterior-most ribs are connected to the sternum by ossified sternal ribs. The sternum is often as long as the entire rib cage, and its posterior end is beneath the anterior end of the pelvis. This fully developed longosterna condition is limited to flying birds. The resultant geometry amplifies dorsoventral movement at the posterior end, so that the sternum helps ventilate the enormous posteroventral air sacs. The sternum may be a broad plate, or transversely narrow, as it often is in short-range fliers.

Some flightless birds retain a thorax and air sac complex similar to that of flying birds, but others do not (App. Figs. 6Bc,Fa–c, 7G). Flightless birds need not oxygenate highly aerobic flight muscles for hours at a time, nor does their rib cage have to be strong enough to support powerful flight muscles in action. Schmidt-Nielsen's (1972) observation that ratite respiration is similar to that of flying birds was true only in the broadest sense: there are major differences in thoracic and air sac development between ratites and flying birds. In all ratites, the abdominal air sacs are smaller than those of most other birds (Huxley 1882, King 1966, McLelland 1989b). Kiwis and cassowaries have the smallest abdominal air sacs among modern birds (the only laterally illustrated description of ratite diverticula is of the kiwi by Huxley [1882]; Beale [1985] stated that an unpublished radiograph confirmed the earlier kiwi study). The sacs do not extend posteriorly into the abdominal cavity, but they remain large and extend to the pelvis. The reduced air sacs of female kiwis may be due to the extreme size of their eggs, but this does not explain the condition in male kiwis or in cassowaries. The posterior ribs of ratites are shorter than in other birds, but are still long, and have well-developed, offset, mobile double heads. Because the posterior ribs are not attached to the sternum, they must work independently of, albeit in concert with, the sternal complex, which is relatively weakly attached to the rib cage. The hinged coracoid-sternal joint is retained, but ratite sterna—which are always broad plates—are also shorter than in flying birds. The shortness of these mesosterna geometrically limits their dorsoventral rotation at the posterior end, limiting their ability to alter thoracic volume. Mesosterna do not extend posteriorly to under the pelvis, even with the aid of a short cartilaginous extension. Emu sterna are shorter than those in other large living ratites. In kiwis and elephant birds (Figs. 8.3, 8.4A), the ossified sternal plates are set well forward on the chest and are very short, a third or less the length of the rib cage. These ratites' sterna are so short that they appear unable to influence the posterior air sacs, which should therefore be ventilated primarily by the free posterior ribs via the muscles that operate them. Short sterna therefore appear to be limited to ventilating the anteroventral air sacs. Ratites demonstrate that large sterna are not necessary to operate an effective air sac complex. The number of ossified sternal ribs may be only two or three.

The precocial young of some birds also show that large ossified sternal plates are not critical for air sac ventilation in nonfliers. Rail and ratite chicks quickly become very active outside the nest (del Hoyo et al. 1992–99). For weeks after hatching, their sterna remain short and cartilaginous, and the sternal ribs are only partly ossified (App. Fig. 6Fc; S. Olson 1973).

The sternum-sternal rib complex of rails enlarges and ossifies only when they approach the flight stage. For an example closer to home, fully grown domestic chickens that are not yet fully mature have poorly ossified sterna (App. Fig. 6Fe). The respiration of precocial juvenile birds with poorly ossified sternal and rib breathing apparatus has not been examined—a situation that will hopefully change—but their example shows that even a cartilaginous, weakly connected sternum can help ventilate anteroventral air sacs.

Also informative are adult birds with reduced flight performance. Roadrunners, for example, have sterna that appear too short to ventilate posterior air sacs (Fig. 8.3). Only three ribs are directly connected to the sternum via ossified sternal ribs, and a forth indirectly by a partial sternal rib. This reduced respiratory apparatus is able to oxygenate sustained high-speed ground movement as well as short flights. Short sterna are also observed in other poor fliers, such as the ash-colored tapaculo (Feduccia 1996, p. 264).

Because flightless birds descended from, or are the juveniles of, flying birds, their respiratory tracts probably reflect a heritage of flight. These pulmonary complexes are therefore not ideal representations of the preavian flightless condition. In particular, the air sacs and associated skeletal apparatus (extreme shortening of the trunk, ossification and size of sternum, sternal ribs, uncinate processes, and so on) may be better developed than they were in preavians. Even so, the initial development of air sacs must have included a stage approximately similar to that of flightless birds, so the respiratory anatomy and function of flightless birds is the best comparative model we have for restoring the respiration of the bird ancestors, certainly one far superior to that of flying birds. Thus there is a pressing need for more study of the respiratory anatomy and function of these earthbound birds. Emus may be the best research subject because they are readily available, do not grow extremely large, begin life as active precocial chicks with poorly ossified pectoral girdles, and never develop large ossified sterna or uncinate processes.

**Avepod dinosaurs and basal birds (Figs. 4.1, 10.1Bd–u, App. Figs. 6Ce,Dc–h,Eb, 7C–F,I–K)**

The thoraxes of these dinosaurs and early noncarinate birds (those lacking a deep-keeled sternum) were so similar that they can be discussed together. Their trunks were shortened by at least three vertebrae relative to crocodilians, and were stiffened by extra intervertebral articulations and/or partly ossified interspinal ligaments (Ostrom 1969, Paul 1988a). The trunk was transversely narrow, especially in the abdomen. The ceiling of the body cavity was deeply corrugated by deep vertebrae-rib articulations of the avian type along most or all of the dorsal series, and there were no hyperelongated transverse processes. Therefore, there was no smooth bony ceiling of the rib cage. In most of these avepods, at least some vertebrae and ribs were consistently pneumatic, in some cases as far posteriorly as the sacrum and tail (Russell and Dong 1993b, Britt 1997). The posterior ribs had mobile double heads and were elongated, so a lumbar region was absent. The pubes were always transversely narrow, especially distally. Length/distal width ratios were not lower than 4.5, and reached as high as 20, and length/midwidth ratios were 3 to 6. Development of the distal boot was highly variable. Boots were absent in early dino-avepods and birdlike troodonts, and were probably absent in a few derived alvarezsaurs (Fig. 10.12Be,n,p). Otherwise, boots were fairly large to very large in most predatory dinosaurs and basal birds. The main surface of the boot faced laterally and the abdominal surface was very narrow. Pubic orientation was highly variable. In the dinosaurs, it ranged from strongly propubic to strongly retropubic. Among basal birds it was minimally to strongly retropubic. Even among close relatives, retroversion was often highly variable, ranging from minimal to strong in dromaeo-avemorph dinosaurs and alvarezsaurs, respectively (Fig. 10.12Bo,p,r–v). Only in those avepods with strongly retroverted pubes did the boot project posteriorly to the distal tip of the ischia. This was true for dromaeosaurs and many basal birds (Fig. 10.12Br–t,Cc–g), but probably not for *Archaeopteryx* (Fig. 10.12Da). In all nonavepectorans and some avepectorans, the pelvic canal was immediately below the sacrum, and was set above exceptionally deep distal pubes (Fig. 10.12B). In dromaeosaurs, oviraptorosaurs, rahonavids, avimimids, alvarezsaurs, archaeopterygiforms, and all other birds, the pelvic canal was deepened to a lesser or greater extent compared with more basal theropods. In avimimids, alvarezsaurs, and ornithothoraces, the pelvic canal was extremely deep; in derived alvarezsaurs and most ornithurines, the pubes do not contact one another distally (Fig. 10.12Bo,q,C; Fig. 10 in Sues 1997).

Contrary to the opinions of Ruben et al. (1997b, 1998) and Hengst (1998), the trunks of avepod dinosaurs and crocodilians could hardly have been more different. The double heading of the entire avepod dorsal rib series is in sharp contrast to the crocodilians' winglike transverse processes (contra Hengst 1998). The avepods' long, gracile pubes differ dramatically from the broader, stouter crocodilian structure. A triradiate pelvis is not only a common diapsid feature, but is absent in many avepectoran dinosaurs. Appendix Table 3 shows that the two groups do not share a single osteological adaptation

*Appendix Table 3. Skeletal Adaptations Associated with Crocodilian- and Avian-Type Respiratory Systems*

Adaptations associated with the operation of a pelvovisceral pump
- Exclusive to system
  - Bony rib cage ceiling smooth (formed by elongated, flattened transverse processes articulating with ribs only with tips of processes)
  - Gastralia do not meet on midline and are set in cartilage sheet
  - Pubes mobile
- Probably critical to system but may be found in other tetrapods
  - No consistent excavation of bones, especially posterior to pulmonary septum
  - Well-developed lumbar region
  - Sternal ribs doubled
  - Pubes at least fairly broad
- Associated with but not exclusive to system
  - Uncinate processes are at most poorly ossified
- Number of these adaptations observed in theropods: none

Adaptations associated with air sac ventilation
- Exclusive to system
  - Various postcranial bones consistently pneumatic (owing to invasion by pulmonary diverticula), sometimes posterior to pulmonary septum
  - Bony rib cage ceiling strongly corrugated (because medial rib heads are set well below transverse processes)
  - Posterior ribs elongated and highly mobile in stereotypical manner (because of well-developed, angled double heads)
  - Large sternum articulates with coracoids via hinge joint
  - Sternocostal articulations are hinge joints
- Probably critical to system but may be found in other tetrapods
  - Dorsal series short and stiff
  - Sternal ribs single
- Often associated with but not exclusive to system
  - Short anterior ribs
  - Ossified uncinate processes
  - Sternal ribs ossified
  - Pubes very narrow
  - Pubes retroverted
  - Pelvic canal deep
- Present in basal birds
  - Gastralia articulate on midline via overlapping zigzag pattern
- Number of these adaptations observed in all, most, or some theropod dinosaurs: all



that supports the presence of a pelvodiaphragmatic muscle pump in any predatory dinosaur. Quite the contrary, the operation of a pelvovisceral pump appears to have been impossible in the latter. The absence of a smooth, bony rib cage ceiling means that the ceiling was not specialized to accommodate, and may have hindered, the strong back-and-forth movement of an expandable lung, even if intracostal muscles and other tissues lined the ceiling of the thorax. The lack of either a lumbar region or mobile pubes would have hindered the abdominal volume changes inherent in the operation of a pelvodiaphragmatic muscle pump. Because the abdominal surfaces of dino-avepod pubes were usually narrow, especially the most gracile examples, they appear too transversely narrow to anchor large diaphragmatic muscles (see Hutchinson [2001] for additional comments). Strongly retroverted pubes were especially poorly suited for supporting these respiratory muscles. Of the derived examples, the pubes of alvarezsaurs were in particular too slender and weak to anchor and resist the pull of diaphragmatic muscles. For that matter, it is difficult to see how the pubis could have been an important part of the respiratory complex, when its orientation and boot development was so variable within the group as a whole, and even among close relatives.

The Santana avepod suggests that the colon was placed ventrally in at least some avepod dinosaurs, despite the shallowness of the pelvic canal. Whether this condition was limited to averostrans or appeared earlier is not known. The deep pelvic canals of some dino-avepods are indicative of ventrally placed colons. Ruben et al. (1999) associated low placed colons with the presence of posterior air sacs, but such air sacs are absent in some birds. It is not yet certain whether the depth of the pelvic canal and colon is tightly linked to respiratory anatomy and function, but if it is, then the anatomy of avepod dinosaurs supports the presence of well-developed thoracic air sacs.

Claessens and Perry (1998) argued that the unusual interlocking, zigzag midline articulations of theropod gastralia (Figs. 4.1A, 10.7) allowed them to help ventilate abdominal air sacs. Whether theropod abdominal air sacs extended so far posteroventrally that they could be directly operated by the gastralia is uncertain. But it is significant that basal bird gastralia had the same midline articulation—and presumably the same function—as seen in theropods and was markedly different from that of crocodilians and other archosaurs (Fig. 10.7E). Furthermore, the complex articulations suggest that theropod gastralia were not set in a cartilage sheet, as they are in crocodilians.

Carrier and Farmer (2000) also suggested that theropod (including basal bird) gastralia participated in respiration, and they added a novel mode of operation in which muscles based on the tail base and/or distal ischium looped via tendons under the pubic boot to pull on the gastralia. They did so in part because both crocodilians and birds have forms of pelvic respiration, suggesting a shared ancestor. However, the function of this system is markedly different in the two groups, leaving open the possibility of independent origin. If such a respiratory tendon was present in dino-theropods, the ventral surface of the pubic boot should have been broad enough to include a deep, smooth-surfaced groove that would have ensured the stability of the tendon, and prevented it from being pinched, immobilized, or injured when the dinosaur rested on the boot. The always narrow, sometimes flat, and usually entirely rugose undersurfaces of theropod pubic boots favor the presence of a cartilaginous resting pad over a respiratory tendon. It is difficult to imagine a tendon loop operating when the boot was bearing hundreds or thousands of kilograms, and the extremely narrow pubic boots of some avepods (Fig. 10.12Be,j,u,Ca,c) were especially unsuited for supporting a tendon loop.

The key to understanding the respiratory anatomy and function of avepod dinosaurs is that their vertebrae, rib cages, and pubes, especially those of the advanced examples, were strikingly similar to those of flightless birds. That the thorax of *Archaeopteryx* was still basically dinosaurian in organization is also key. Appendix Table 3 shows that many skeletal features associated with the avian air sac complex are also found in dino-theropods, especially dino-avepods. It is not logical to conclude that dino-avepod respiration was similar to that of crocodilians when the dinosaur's apparatus was birdlike, and that avian respiration is fundamentally different when the thoraxes of birds—*Archaeopteryx* especially—are dinosaur-like. Applying crocodilian respiratory muscles to the retropubic pelvis of dromaeosaurs and alvarezsaurs, whose forms were entirely noncrocodilian and essentially avian, is particularly illogical. The thoraxes of avepod dinosaurs had the attributes expected in bird ancestors that were evolving a preavian air sac complex. The possible presence of an air sac in the Santana fossil supports this conclusion. A tentative outline of the evolution of the theropod-bird respiratory complex in preavian archosaurs is given below (App. Fig. 7; for a corresponding analysis of the respiratory capacity of these systems, see Appendix 4).

### Basal archosaurs and protodinosaurs (App. Fig. 7A)

These animals retained long, shallow, flexible trunks, long chest ribs, short posterior ribs without double heads, short procumbent pubes, nonpneumatic postcranial bones, and other reptilian features that suggest deep and flexible lungs and the absence of true air sacs. Nor were basal archosaur gastralia like those of basal birds (Fig. 10.7E). However, an increase in dorsal ribs with double heads (App. Fig. 6Da) compared with more-basal diapsids (Gow 1975) suggests that ventilatory capacity had increased. Although a well-developed lumbar region was absent, the posterior ribs were fairly short. So it is possible that a weakly developed abdominal pump was present, operated by diaphragmatic muscles anchored on the short pubes. Carrier and Farmer (2000) suggested that tail base and ischial-anchored muscles pulled on the gastralia to assist in respiration. Because the pubis was distally broad, and too short to bear a heavy load when the animal laid down, this hypothesis is plausible. The presence of uncinate processes—whether cartilaginous or ossified—in crocodilians, some ornithischians (Zou 1983, P. Fisher et al. 2000), and various avepods including basal avebrevicaudans (Derstler and Hembree 1999) suggests but does not prove that uncinate processes were a general archosaur character whose ossification was irregular.

### Baso-theropods

Absence of pneumatic postcranial bones or elongated posterior ribs with double heads indicate that diverticula were still absent or poorly developed. Long anterior chest ribs suggest that the lungs remained deeper and more flexible than in birds. However, the dorsal series was shortened and correspondingly more rigid than in protodinosaurs. The ceiling of the rib cage cavity was more strongly corrugated, because the number of double-headed ribs increased to 9–11 ribs, whereas the number of posterior dorsals that lacked mobile ribs declined to 2–5 (App. Fig. 6Dc,d). The rigidity and corrugation of the rib cage suggest that the lungs were less flexible than those in reptiles. The dominance of double-headed ribs indicates that the ventilation capacity of the rib cage had increased. The combination of these adaptations suggests that the dorsal portion of the lung was rigid, and flexible, nonvascular ventroposterior diverticula were beginning to be ventilated by the highly mobile ribs that encompassed them. The incipient development of posterior diverticula ventilation implies that any abdominal pump was being or already had been lost. Development of very long pubes, some with narrow distal boots, may have eliminated gastralial retraction via ischial and tail-based muscles.

### Cerato-saurans (App. Fig 7C)

Complete rib cages (Paul 1988a, Colbert 1989, Bonaparte et al. 1990) show that the chest ribs were still long, and corrugation of the body-cavity ceiling remained modest in that the depth of the vertebrae-rib articulations was not great. Therefore, central lung anatomy may not have changed much from the baso-theropod condition. However, the dorsal series was shorter, and pneumatic cervicals suggest that well-developed lung diverticula were present. Most importantly, all but the last of the posterior ribs had well-developed, strongly angled double heads—for a total of 12 (App. Fig. 6Cc,De)—and the posterior ribs were more elongated than in more basal archosaurs (compare App. Fig. 7C with A). The resulting loss of the lumbar region precluded the operation of an abdominal pump. It is difficult to explain bird-like posterior ribs that were able to alter dramatically the volume of the upper abdominal cavity as anything other than a means to ventilate the initial backward extension of the respiratory tract. This extension probably took the form of a saclike organ. As in many rails, the fairly large, cartilaginous sternum may have ventilated anteroventral air sacs.

### Tetanurans and avetheropods (App. Fig. 7D–F)

The trunk was deeper relative to its length. The most basal tetanurans, such as abelisaurs, retained a long first dorsal rib (Bonaparte et al. 1990), but the anterior chest ribs of the more derived yangchuanosaurs were beginning to shorten (App. Fig. 7D), and in compsognathians, allosaurs, and other avetheropods, the anterior ribs were as short as those of birds (App. Figs. 6A, 7E–H). The shortening of the anterior ribs cannot be related to bipedalism or forelimb reduction. Nontetanuran theropods were small-armed bipeds, yet they had long anterior dorsal ribs. The same ribs were persistently short in avetheropods regardless of arm size. The ceiling of the body cavity was strongly corrugated, because the middle vertebrae-rib articulations were deep. The anterior dorsals and ribs were pneumatic. All of these vertebral, rib, and liver adaptations suggest that the lungs were becoming locked firmly into the rib cage ceiling, semirigid, and as shallow or nearly as shallow as those in birds, while diverticula were further enlarged (Claessens and Perry 1998). At the same time, more elongated posterior ribs—equaling in length those of ratites—imply a further posterior shift in the volume expansion capacity of the rib cage (compare App. Fig. 7D,E with C). The Santana specimen indicates that low-set colons and pelvic air sacs were present by this stage, if not earlier.

Therefore, primary ventilation of the rigid lungs should have been via large thoracic and abdominal air sacs. Considering the condition in ratites, kiwis, and cassowaries in particular, the abdominal sacs may not have extended far behind the rib cage. In this view, any involvement of the gastralia in respiration (Perry 1989, Claessens 1996, Claessens and Perry 1998) in theropods and the basal birds that retained gastralia may have been indirect, acting to resist collapse of the abdominal walls during air sac ventilation. Alternately, the gastralia may have actively pushed on the abdominal mass, which in turn helped ventilate the posterior air sacs. That the abdominal air sacs extended sufficiently posteriorly to be directly ventilated by the gastralia cannot be ruled out.

The very long-booted pubes common in tetanurans, and especially in avetheropods, were a most unusual feature. It does not appear to have been primarily an adaptation for limb function, since such long-booted pubes are absent in other erect, bipedal dinosaurs. The long pubes would have allowed the air sac system to operate unhindered while the dinosaur rested on the cartilage- and skin-padded boot, with the chest clear of the ground. Especially useful for the giant avepods, this would have eliminated the need for the caudopelvic system for air sac ventilation used by resting birds (see below). Because the broad pelvic canal remained shallow, the colon could still be placed dorsally, but this may not have had a major influence on respiratory anatomy and function.

Unidirectional pulmonary airflow may have been developing in nonavepectoran tetanurans. Perry (1992) has outlined how avian pulmonary air flow could have evolved from the archosaur pattern. Again note that the shortening of the chest ribs and further increase in posterior rib length and mobility is contrary to that of crocodilians, and entirely inconsistent with the presence of a dynamic diaphragm (contra Ruben et al. 1997b, 1998, Hengst 1998). Hengst's (1998) data indicating an expansion in rib cage ventilation capacity in tetanurans over the more basal cerato-saurans is more compatible with the presence of an increasingly avian system.

### Avepectora, including basal birds (App. Fig. 7F, I–K)

In mesosterna troodonts, caudipterygians, oviraptorosaurs, dromaeo-avemorphs, and basal birds, the hinged sternum—whether cartilaginous, ossified, or both—should have helped ventilate the anteroventral air sacs in a manner approximating that of kiwis and other adult and juvenile birds with similarly proportioned sterna. The sternal plates of dromaeosaurs were especially large. Being almost 40% as long as the rib cage (the ossified sterna are about half as long as the ilia and femora in *Velociraptor* MIG 100/25 and 100/985 [Fig. 4.1A,B; Norell and Makovicky 1997]), the plates were relatively larger than those of some island ratites (compare App. Fig. 7F and G). The ossified sterna of *Sinornithosaurus* and *Bambiraptor* were even more impressive, being about half the length of the rib cage (App. Fig. 10B). At the other extreme, the ossified sternum of *Archaeopteryx* was only about 12% of rib cage length (Figs. 4.2Ad, 10.9D). The fairly long but narrow ossified sternum of alvarezsaurs may also have been useful in respiration.

It is interesting that vertebral aeration was as variable in avepectoran dinosaurs as it is in birds themselves (McLelland 1989b, Russell and Dong 1993b). In some it was absent; in others posterior dorsals, sacrals, and even anterior caudals were pneumatic (Britt 1997). The tendency to posterior pneumatization of the vertebrae suggests that the posterior air sacs were fairly well developed (see above). The pneumatic pelvis of *Archaeopteryx* is also in line with the presence of well-developed posterior air sacs in this and other dino-avepectorans (Christiansen and Bonde 2000). The depth of the pelvic canal tended to be deeper in dino-avepectorans than in more basal avepods (Fig. 10.12B,C, App. Fig. 7C–F, I–K; Xu et al. 2000), suggesting that the colon was placed more ventrally in the former. This may be related to an expansion of the posterior air sacs. If unidirectional air flow had not previously appeared, it probably developed at this stage (Claessens and Perry 1998), and indeed may have been fully developed.

Caudipterygians, oviraptorosaurs, and dromaeosaurs were especially avian in having ossified uncinate processes, as well as up to five ossified sternal ribs that connected the sternum to the true ribs via hinged sternocostal joints (Xu, Wang, and Wu 1999). These systems were better developed than those of urvogels, and these dinosaurs' air sacs probably approached or equaled the ratite condition. Only in possessing a deep pelvic canal was the archaeopterygiform respiratory apparatus more birdlike than that of dromaeosaurs and caudipterygians, but it was not as avian as that of oviraptorosaurs. Very slender, rodlike, retroverted, closely applied to the ischia, and with a very deep pelvic canal, the pubes of derived alvarezsaurs were remarkably like those of derived birds, much more so than those of archaeopterygiforms (Fig. 10.12Bo,p).

Ruben et al. (1997a) suggested that *Archaeopteryx* was capable of the caudopelvic air sac ventilation practiced by modern birds. The latter can breathe in this manner because their pubic shafts—which are anteriorly fused to the rest of the pelvis—are so slender that they are flexible, and project posterior to the ischium. Tail-based muscles pull on the pubis to operate this system (also see Carrier and Farmer 2000). Since the pubes of the urvogel probably did not extend strongly posteriorly, and its pubic shafts were fairly robust, it is improbable that it could use caudal muscles to help breathe in this manner. Nor does it seem that *Archaeopteryx* had the hypopubic cup Ruben et al. cited as evidence for caudopelvic respiration (Chapter 4), although modern birds also lack this feature. Pubic retroversion in *Protarchaeopteryx* cannot yet be assessed, but its pubic shafts were even more robust than those of its close relative and were correspondingly less suited for avian-type caudopelvic respiration. Strong pubic shafts probably inhibited dromaeosaurs from practicing avian-type caudopelvic breathing despite the projection of their pubes posterior to the ischia. Because the pubes of derived alvarezsaurs were slender, and may have projected more posteriorly than the ischia, they may have been able to practice caudopelvic respiration of the avian type.

Ruben et al. (1997a) concluded that even in basal birds, the air sac complex was only weakly developed, because they generally lacked ossified uncinate processes or sternal ribs, and their ossified sternal plates were short. However, the latter were as large as in ratites and precocial chicks, and the lack of ossified sternal ribs and uncinate processes does not preclude the presence of well-developed air sac ventilation in modern birds that lack these features. What is true is that respiratory capacity was almost certainly not as extremely high as in flying birds with hyperelongated sterna, and numerous ossified sternal ribs. This condition was present in derived enantiornithines (Chiappe and Calvo 1994) and in ornithurines.

### Summary

One could hardly ask for a better pattern of incremental evolution progressing to the avian skeletal features needed to operate respiratory air sacs. This fact reinforces the case for preavian pulmonary air sac ventilation in predatory dinosaurs. No evidence for progressive evolution of a pelvis-based diaphragmatic muscle pump in dinosaurs has been presented. Nor did Ruben et al. (1997a, 1999) properly correlate the anatomical features associated with crocodilian versus avian respiratory complexes. They did not demonstrate that crocodilians and theropods share any unique respiratory adaptations, and they mistook damage and repair work by an untrained collector for soft-tissue anatomy. There is no unambiguous soft-tissue or skeletal evidence that theropods had a pelvis-based diaphragmatic muscle pump, and there is abundant osteological evidence that they did not have such a pump. It is therefore probable that the fully developed, pelvovisceral pump is limited to crocodilians.

The last probability is not surprising. The unusual crocodilian pelvovisceral pump is elaborate compared with other reptiles whose aerobic capacity is equally low, and it does not increase respiratory capacity, so it may have evolved for purposes other than respiration. Gans (1976) presented data indicating that the ability to extend the air-filled lungs strongly backward via the pelvovisceral pump acts to regulate precisely the center of buoyancy, allowing strict control of the orientation of the submerged body. The critical features associated with this respiratory pump—elongated transverse processes and short, broad, mobile pubes—were absent among the most basal sphenosuchian protocrocodilians (Crush 1984, Carrier and Farmer 2000). A lumbar region was present, so some form of abdominal pump was probably operational (also see Carrier and Farmer 2000). Mobile pubes indicative of a well-developed pelvovisceral pump appear in more derived sphenosuchians, and are further developed in other early crocodilians (Carrier and Farmer 2000). If even gracile-limbed crocodilians were frequently aquatic (see the protocrocodilian section in Chapter 10), the strong-action pelvovisceral pump may have evolved in part or primarily for center of buoyancy control, with increased aerobic capacity being incidental (contra Carrier and Farmer 2000, Farmer and Carrier 2000a). This casts doubt on whether a similar center-of-buoyancy–shifting pump would evolve among such strongly terrestrial animals as theropods.

Ironically, the main effect of the dinosaurian pelvovisceral pump hypothesis has been to inspire various researchers to investigate further and improve our understanding of the saurischian air sac–driven respiratory complex (saurischian, because this system may not have been limited to the avepod clade). Prosauropods and especially sauropods had pneumatic vertebrae, a strongly corrugated rib cage ceiling, mobile elongated posterior ribs, and other adaptations that suggest they, too, used pulmonary diverticula to ventilate their lungs (Perry 1983, 1989, 1992, Paul and Leahy 1994, Reid 1996, 1997, Britt 1997, Larson 1997). Pneumatic postcrania and large sterna suggest that pterosaurs also evolved some form of pulmonary air sacs (Perry 1983, Wellnhofer 1991, Larson 1997). However, the dorsal ribs had little or no mobility and were unable to help ventilate the sacs. Large, mobile prepubes may have served this purpose. If so, then pterosaur respiration may have combined features similar to those seen in crocodilians and birds (also see Carrier and Farmer 2000).

### Ornithischians

Briefly turning to the "bird-hipped" dinosaurs, ornithischians lack evidence of the pneumatic and other adaptations associated with air sacs. Lacking any living analogs, it is difficult to reconstruct the nature and operation of ornithischian lungs. As part of their hypothesis that all archosaurs inherited some form of pelvic respiration, Carrier and Farmer (2000) suggested that ornithischian pelves were kinetic in a manner that made them an integral part of the respiratory complex. However, in ceratopsids, the anterior rim of the prepubis directly articulated with the posterior edge of one of the posterior dorsal ribs, which in turn were tightly bundled to one another distally (Paul 1987a). The resultant rigidity (probably for protection against physical impacts) would have severely limited movement within the pelvis, and even within the rib cage. Perry (1983, 1989) suggested that the retroverted pubes supported abdominal muscles, which acted as a pseudo-diaphragm to ventilate the lungs. This is most applicable to non-ornithopods, which lacked a lumbar region. Ornithopods are interesting in that they did have a well-developed lumbar region, and its topography was more like that of mammals than the lumbar regions seen in some other archosaurs. They further mimicked mammals in lacking gastralia and a long procumbent pubis. It is therefore possible that a muscular diaphragm was supported by the steeply sloped ribs that defined the anterior border of the lumbar region (Paul and Leahy 1994, Paul 1998).

The metabolic implications of avepod respiratory function are examined in Appendix 4.

APPENDIX 4

# *Evolving the Power to Walk, Run, and Fly*

Parallel to the debate over bird origins is an equally vigorous dispute over the energetics that powered the development of birds and their flight. The evolution of avian energy systems, or paleometabolics, is important to understanding the ancestry of birds, and the origin and loss of their flight. Assume, for instance, that it has been shown that basal birds were flying reptiles in terms of power production, and that predatory dinosaurs had high aerobic exercise capacity (AEC). It would seem improbable that a reptilian system would evolve from a more energetic one at the same time that locomotory power requirements were increasing and feather insulation was evolving. The problem is also important for understanding how avian energetics evolved. If, for instance, avepod dinosaurs had high aerobic capacity, then birds merely took a power system that was already operating and boosted it for the needs of sustained, powered flight. In this case, nonavian avepods were primed for the development of flight, so that the evolution of flight was made easier than it would have been if the direct ancestors of birds had had a reptilian power plant. The paleometabolic problem is also tied to the question of flight loss in dinosaurs. If preavian dinosaurs were still reptiles in thermoenergetic terms, and urvogels were more like birds, then what was the metabolic status of secondarily flightless dinosaurs that descended from urvogels?

A book on bird origins would, therefore, be incomplete without an analysis of paleometabolics. This appendix is also a response to a section on the same subject by Feduccia (1996), but it is much more extensive and technical, to avoid the pitfalls of his superficial discussion. To reconstruct the energetics of fossil forms, one must consider in depth as many aspects of their condition as possible (see below). Lack of space prevents this from being a thorough review of all the issues. And because this work focuses on the origins of birds and their flight, and birds are never truly gigantic, this analysis will not delve deeply into the problems raised by giant dinosaurs (for the latter, see Paul and Leahy 1994, Paul 1998).

As for the history of the science of preavian paleometabolics, the pre-1960s "modernist consensus" (Gould 1993) on the evolution of avian energetics outlined a rather simple story. The ancestors of dinosaurs were scaly reptiles with low metabolic rates, low AEC, and widely fluctuating body temperatures. Traditionally, feathered *Archaeopteryx* was reconstructed as a flying bird that used high metabolic rates to generate the majority of its body heat internally, with a stable body temperature, and a high AEC. In this view, avian thermoenergetics evolved in the context of the development of flight. If so, this would be very different from mammalian flight, in which bats inherited their basic power system long after it had evolved in flightless ancestors.

The postmodern situation is much more complicated, with a number of hypotheses vying with one another. At the same time that dinosaurs were being revived as bird ancestors, the largely unsubstantiated modern consensus was challenged by Ostrom (1970, 1980b) and Robert Bakker (1968, 1975, 1980), who suggested that dinosaurs were "warm-blooded" to a greater or lesser extent. In this view, birds inherited much or most of their aerobic power system from dinosaurs that had developed it for more terrestrial purposes, although the extremely high sustainable AEC of birds is certainly a flight-related adaptation (Paul 1988a, Paul and Leahy 1994). In this case, the dinosaur-bird pattern is broadly similar to the mammal-bat pattern. At the opposite extreme, Ruben (1991) suggested that even urvogels may have been physiologically reptilian, and that they powered their flight with a small set of highly anaerobic flight muscles. Ruben et al. (1996, 1997a,b) further offered that dinosaur and early bird energetics were at least as reptilian as avian, a view also supported by Regal and Gans (1980), Sereno and Rao (1992), Randolph (1994), Chinsamy et al. (1995), Chinsamy, Chiappe, and Dodson (1995), Feduccia (1996), Paladino et al. (1997), and Ruben and Jones (2000). This view is radical in that it proposes that high rates of sustainable aerobic energetics did not evolve until carinate birds had evolved essentially modern flight. Nor could it be more different from the situation with the always highly energetic origin of bats. More recently, Ruben et al. (1999) proposed that predatory dinosaurs combined a reptilian resting metabolism with an AEC elevated well above that observed in living reptiles (early bird energetics were not discussed). This view also is radical in that no living vertebrate combines low resting and boosted exercise metabolisms. Many researchers have concluded that the energetics of many or all nonavian dinosaurs were in some significant way different from those of both reptiles and birds (Hotton 1963, 1980, Ostrom 1980b, Ricqles 1980, Reid 1987, 1990, 1993, 1996, 1997, Dunham et al. 1989, Russell 1989, Lambert 1991, Chinsamy 1993, 1994, Paul and Leahy 1994, Farlow et al. 1995, Barrick et al. 1997, 1998, Chinsamy, Rich, and Vickers-Rich 1998, Fricke and Rogers 2000).

## Terminology and Modern Metabolic Types

### *The Modern Metabolic Gap: How Big Is It?*

Much of the literature gives the impression that there is a dramatic difference between living bradymetabolic, ectothermic reptiles on the one hand and tachymetabolic, endothermic mammals on the other, so it may have been possible for extinct animals to have had minimal metabolic rates that are both "well below those of modern endotherms, although still higher than in any modern reptiles" (Reid 1990). The situation is not so clear cut (Paul 1998, Horner et al. 1999). The Kleiber curve that describes the typical resting metobolic rates (RMRs) of mammals is about 70 kcal/kg$^{0.75}$/day. RMRs are as low as 16.5–24 kcal/kg$^{0.75}$/day in echnidas, tenrecs, pangolins, and golden moles when body temperature is 28–31°C (Crompton et al. 1978, McNab 1988, Seymour et al. 1998). Manatee RMR can be just 10 kcal/kg$^{0.75}$/day at a body temperature of 36.5°C; Irvine 1983). RMR can be as high as approximately 16.5 kcal/kg$^{0.75}$/day in small lizards when operating at 37°C, and are sometimes only a little lower at 30°C (A. Bennett and Dawson 1976, R. Andrews and Pough 1985). RMR of leatherback sea turtles is 7.2 kcal/kg$^{0.75}$ at 29°C. Manatees are sluggish and cannot thermoregulate in cold water (Irvine 1983); naked mole-rats are poikilothermic (Buffenstein and Yahav 1991).

The reptile-mammal gap is barely present. If a dinosaur is postulated to have limited thermoregulatory ability and an RMR of 20 kcal/kg$^{0.75}$/day—only one-half to one-sixth that of typical eutherians—it is not intermediate to reptiles and mammals: instead, it is within the mammalian range! The separation of reptiles and birds is greater, with kiwi RMR being an exceptionally low 40 kcal/kg$^{0.75}$/day (P. Bennett and Harvey 1987: the RMRs of other potentially low-energy birds such as hoatzins and kakapo parrots have not yet been measured).

On a related matter, Ruben (1995) stated that warm-bodied tuna do not possess high metabolic rate endothermy. The RMR of muscle immobilized skipjack tuna is a strongly elevated 47 kcal/kg$^{0.75}$/day (Brill 1986, Brill and Bushnell 1991).

### *Terminology*

The terminology currently used to describe the thermoenergetics of animals is inadequate to discuss and understand the issues (Horner and Dobb 1997, Horner et al. 2000). This is particularly true when considering extinct groups, whose metabolic complexes may have been much more diverse and were often quite different from those of the few groups we are left with today. Modern reptiles, birds, and mammals are usually segregated into "ectotherms" and "endotherms," but this division is simplistic and misleading when considering other organisms. For example, skunk cabbage, large flying insects, certain fish, incubating pythons, and giant reptiles can generate and conserve enough heat to maintain high body temperatures without external heat sources, even though their minimal MRs are low (McNab 1983, Heinrich 1993). Such organisms are endothermic, but not in the manner of birds and mammals. So it is not entirely clear what Farlow et al. (1995) meant when they stated that "no evidence convincingly shows that dinosaurs were endotherms."

Let us focus on the term *endothermic.* "Endo" simply means internal, "therm" means heat, so the term *endotherm* means "is internally heated." Strictly applied to animals, endothermic should mean merely that a creature generates most of its body heat internally. This is the logical opposite of ectothermic, which means that an animal acquires most of its heat from the external environment. However, endothermic as it is usually used has become a loaded term, one that applies to an animal that uses a sophisticated, controlled, and high rate of thermogenesis to generate internal heat, and then uses the head to maintain a constant body temperature, at least when not torpid. There are multiple problems with this definition of the term. What, for instance, do we call an animal that relies mainly on internal heat, but does not fully meet the thermogenesis and homeothermy requirements? By no means is it an ectotherm that depends upon outside heat sources. Nor can we call it a mesotherm, which properly defined, refers to an animal that is dependent upon internal and external heat in roughly equal amounts. It is telling that large flying insects with high aerobic capacity are called "hot-blooded" endotherms (Heinrich 1993), even though their RMRs are very low. The only sensible decision is to define these words according to their etymology. Prefixes and word combinations can then be used to characterize a given thermal physiology. Because animal thermoenergetics is complex, the necessary descriptive terms are not as simple as one might wish, but this is the norm in scientific discourse. Although RMRs are a little higher than the standard and basal rates that mark minimal normal energy flow, for the purposes of this study, they are considered roughly equivalent, and RMR is used to cover all three. Terms as used here are defined as follows.

**Body heat**

- Ectothermic—Most is acquired from the environment.
- Mesothermic—Approximately equal portions are generated internally and acquired from the environment.
- Endothermic—Most is generated internally. Includes the following:
  - Alphaendothermic—Sophisticated and elevated thermogenesis is used to actively control body temperature.
  - Betaendothermic—Internal thermogenesis is elevated above reptilian level, but is not actively used to help adjust body temperature.
  - Inertioendothermic—Internal thermogenesis is not elevated above reptilian level, but large bulk results in endothermy.

**Resting metabolic rate**

These rates apply at the level of cells, muscles, organs, or whole bodies.

- Bradymetabolic—Low, does not exceed reptilian maximum.
- Tachymetabolic—Moderate to extremely high, exceeds reptilian maximum. Includes the following:
  - Mesometabolic—Moderate, between reptilian maximum and avian minimum.
  - Supra-aerobic—High, at or above avian minimum.
  - Hyperaerobic—Extremely high, observed in shrews and passerines.

**Aerobic capacity**

These terms apply at the level of cells, muscles, organs, or whole bodies.

- Bradyaerobic—Low, does not exceed reptilian maximum.
- Tachyaerobic—Moderate to extremely high, exceeds reptilian maximum. Includes the following:
  - Mesoaerobic—Moderate, between reptilian maximum and avian minimum.
  - Supraaerobic—High, at or above avian minimum.
  - Hyperaerobic—Very high, at or above minimum observed in birds capable of sustained powered flight.

**Anaerobic exercise capacity**

These terms apply at the level of cells and muscles.

- Bradyanaerobic—Muscle-mass–specific anaerobic power generation is low or absent, as in some insect muscles.
- Tachyanaerobic—Power generation is modest to high. Includes the following:
  - Mesoanaerobic—Power generation is modest, as in bird and mammal red fiber–dominated muscles.
  - Supraanaerobic—Power generation is moderately high, as in bird and mammal white fiber–dominated muscles.
  - Hyperanaerobic—Power generation exceeds avian-mammalian maximum, observed in reptile muscles.

**Body temperature**

- Heterothermic—Fluctuates strongly on either a daily or seasonal basis.
- Poikilothermic—Fluctuates in accordance with ambient temperature.
- Semihomeothermic—Often constant, but subject to modest periodic fluctuations.
- Homeothermic—Maintained at a precise or fairly constant level on at least a daily basis. Includes the following:
  - Alphahomeothermic—Constancy achieved via active response of internal thermogenesis.
  - Inertiohomeothermic—Constancy achieved via bulk insulation.

These terms can be combined to describe the thermal physiology of a given animal. For example, a typical reptile is a bradyaerobic, heterothermic ectotherm. A typical bird is a supraaerobic, homeothermic alphaendotherm.

**Abbreviations**

It is convenient in a discussion of this type to abbreviate some key terms.

- AEC—Aerobic exercise capacity. Includes the following:
  - LoAEC—Low, does not exceed reptilian maximum.
  - ElvAEC—Elevated above reptilian maximum.
  - IntAEC—Intermediate, between reptilian maximum and avian minimum.
  - HiAEC—At or above avian minimum.

- MaxAEC—The maximum attainable in a given animal.
- ISAEC—Indefinitely sustainable for many hours, always a fraction of MaxAEC.

- MR—Metabolic rate.
- RMR—Resting metabolic rate. Includes the following:
  - LoRMR—Low, does not exceed reptilian maximum.
  - ElvRMR—Elevated above reptilian maximum.
  - IntRMR—Intermediate, between reptilian maximum and avian minimum.
  - HiRMR—At or above avian minimum.

## *Metabolic Types*

### Modern amniotes

Before we can examine the metabolic systems of extinct creatures, we must set a baseline by outlining those of living forms. Sources for energetic, thermoregulatory, and activity data include Schmidt-Nielsen (1964, 1972, 1984, 1990), Jansky (1965), Bellairs (1970), Whittow (1970–1973, 2000), Calder and King (1974), Wigglesworth (1974), Peaker (1975), A. Bennett and Dawson (1976), A. Bennett (1983, 1994), Grubb et al. (1983), McNab (1983, 1988, 1996), Barnes and Rautenburg (1987), P. Bennett and Harvey (1987), Buttemer and Dawson (1989), Hill and Wyse (1989), Groscolas (1990), Butler (1991), del Hoyo et al. (1992–99), Heinrich (1993), Jones and Lindstedt (1993), Saunders and Fedde (1994), Gill (1995), Jansky (1995), T. Johnson (1996), Eldershaw et al. (1997), Paladino et al. (1997) Thompson and Withers (1997), Pough et al. (1998), Seymour et al. (1998), Bundle et al. (1999), Rose et al. (1999), and Kortner et al. (2000).

### Reptiles

These are all bradymetabolic, bradyaerobic, ectothermic, and heterothermic. RMR is 2–12 kcal/kg$^{0.75}$/day at preferred body temperature, the high range includes some varanids. AEC 10–250 kcal/kg$^{0.75}$/day, the high range is observed in some varanids. Highest MaxAEC/RMR ratio is approximately 40, again observed in varanids. Long-term energy budgets are 20–30 times lower than in birds and mammals. Muscles are often hyperanaerobic, producing about twice as much anaerobic power per unit mass as those of birds and mammals (Ruben's [1993] defense of greater anaerobic power output in reptile muscles is tentatively accepted over Speakman's [1993] arguments to the contrary). Some examples are poikilothermic, whereas others actively control body temperature when active by seeking out favorable thermal conditions. Tolerable internal temperature range is always high, and reaches 46.5°C in some examples. Preferred body temperatures range from 12 to 41°C.

### Birds

These are all suprametabolic or hypermetabolic, supra-aerobic or hypermetabolic, and alphaendothermic. RMR is 40–200 kcal/kg$^{0.75}$/day at normal operating body temperature (kakapo and/or hoatzins may have lower values). In ratites, some birds of prey, and some water birds, the RMR is well below the avian "standard" of 75 kcal/kg$^{0.75}$/day; at the other extreme are passerines and some other examples. MaxAEC is supraaerobic to hyperaerobic in some fliers (kakapos may be mesoaerobic). MaxAEC/RMR ratios have never been measured in flying birds (Saunders and Fedde 1994), but are probably high, as per the value of 36 just observed in the rhea (Bundle et al. 1999). Respiration rates are two to three times slower than in mammals of similar size, tidal volumes much higher than in the latter, and breathing and wing flapping rates are not co-linked in most flying birds. No bird has hyperanaerobic muscles: flight muscles are either mesoanaerobic or supraanaerobic. Although mass-specific anaerobic power is lower in bird muscles than in those of reptiles, overall muscle mass relative to total mass is higher in birds than in reptiles, so total anaerobic burst power is roughly equivalent at a similar total mass. In general, birds are alphahomeothermic with a strong metabolic response to cold. This response and arousal from torpor are achieved primarily via shivering. Thermogenic brown fat is absent. Whether heat is generated by nonshivering muscle is a subject of debate (Eldershaw et al. 1997). A few birds are baskoendothermic, in that they use direct solar radiation to raise body temperatures a few degrees. Some birds are daily or subweekly heterotherms in that they can allow body temperature to drop a few or even dozens of degrees during nocturnal and/or diurnal torpor. Some can allow metabolism to drop when fasting, and some examples hibernate; these thermal declines occur for all sizes, including ratites, but tend to be most dramatic in smaller examples. Altricial bird chicks are often poikilothermic, but are suprametabolic at high body temperatures. Some birds are hyperthermic, allowing body temperature to rise to 45°C when external temperatures are very high. Evaporative cooling is via panting and gular fluttering (it has not been demonstrated that air sacs are involved in cooling [Britt 1997], and in birds such as rheas, exercise cooling is not well developed [Schmidt-Nielsen 1972, Bundle et al. 1999]). Normal body temperatures are usually 38–42°C, 38–40°C in flightless examples, and up to 44°C in a few small fliers.

### Mammals

The energetics and thermoregulation of mammals is extraordinarily diverse, ranging from energy levels not strongly elevated above reptilian maximums to extremely high aerobic capacities. Among terrestrial examples, mesometabolic monotreme, marsupial, edentate, tenrec, hedgehog, and mole-rat RMRs are 18–40 kcal/kg$^{0.75}$/day at normal operating body temperature. Some canids and other carnivores and camels have RMRs markedly below the suprametabolic placental "standard" of 70 kcal/kg$^{0.75}$/day. AEC also ranges from not far above the reptilian maximum to as high as seen in flying birds. Highest MaxAEC/RMR ratio is 100. No mammal has hyperanaerobic muscles, but large muscle volume means that total anaerobic burst power is usually similar to that of reptiles. Most mammals are alphahomeothermic and alphaendothermic. Aside from shivering cold response, some eutherians also employ nonshivering brown-fat-generated heat, and nonshivering-muscle-generated heat may also be a factor (Rose et al. 1999). A significant number are heterothermic, allowing body temperatures to decline during torpor on a daily to seasonal basis. Some mole-rats are poikilothermic and mesothermic. Some mammals are hyperthermic, allowing body temperature to rise to 46.5°C when external temperatures are very high. Evaporative cooling is via panting or sweating. Normal body temperatures are around 30°C in monotremes, 34–36°C in marsupials and edentates, and 36–38°C in most eutherians.

### Extinct systems

As diverse as modern amniote energy systems are, and despite the lack of a dramatic RMR gap between reptiles and mammals, there remains a basic division between reptiles on the one hand and birds and mammals on the other. As a result, there must have been "intermediate" thermal physiologies that are now extinct. However, the narrowness of the gap between reptiles and mammals, and the extensive variation in the two groups, leave the term *avian-mammalian* highly ambiguous when it comes to describing thermal physiology and thermoregulation. Even if an extinct tetrapod is postulated to have a RMR, AEC, and thermal stability not far above the reptilian level, it may approach or be within the lower mammalian zone.

## Procedures and Methods

The question of preavian energetics has been afflicted by false presumptions concerning the burden of evidence. Because dinosaur energy systems were long thought to have been reptilian in nature, it has often been presumed that the burden of evidence still lies upon the hypothesis that dinosaur energetics were avian or near avian. This is not so. Because dinosaurs were an exotic group whose metabolic processes are unknown—unlike, for example, fossil land turtles—all hypotheses are equal until demonstrated otherwise. Also irrelevant is whether a given hypothesis is "extreme" or "radical" versus "moderate." That the view that all dinosaurs had avian energetics is at one extreme of the hypothesis spectrum, and that even early birds had reptilian energetics is at the other end, does not make either less viable than "compromise" views of dinosaur energetics as having been intermediate. All hypotheses need to be considered as seriously as the evidence supporting them warrants.

Another critical issue in paleometabolics is single versus multiple character analysis. Some studies on avepod energetics have focused on a single system as being a key to understanding a major aspect of the energetics of extinct animals. A recent example is the presence of nasal respiratory conchae (RC), which Ruben et al. (1997b) cited as a "bellwether indicator" of RMRs in "virtually all terrestrial taxa, living or extinct." The problem with relying on isolated features is obvious. Animals are complex machines that are the sum of their many parts. Anatomical features evolve under complex factors, and the preservation of characters is often irregular. To understand how the whole animal works, one must examine as many parts as possible: the more the better. A single character that may appear critical may prove otherwise when put in the context of the whole animal. For example, it may be possible for animals to use an alternative to RC to achieve the same purpose—in this case water conservation. As appealing as it may be for there to be a single "Rosetta Stone" character that can be used to restore the energetics of any given fossil tetrapod, it is improbable that such a simple solution exists. This study therefore examines a wide range of characters that may prove useful in diagnosing and restoring the energetics of extinct avepods.

For a character to be a useful and reliable means for restoring a given metabolic rate, it must meet the following criteria. It must be readily preserved, and easily observed and measured in fossils. If the relevant attributes of a character cannot be demonstrated, then its diagnostic value is nil. There must be a consistent correlation between the status of the character and MR, and preferably the correlation should be demonstrated in a quantitative manner. The correlation must be at least empirical, and preferably there should be a strong causal link. Some characters are prerequisite characters that are necessary to have an elevated AEC, but do not necessarily mean that ElvAEC is present (for example, posterior placement of internal nares). Others are diagnostic characters found only in forms with ElvAEC (for example, erect legs). In general, the most sensitive diagnostic characters are those associated directly with the generation of aerobic or anaerobic power. The more prerequisite and diagnostic characters indicative of a given metabolic level present in a given form, the more confident the diagnosis.

Another paleometabolic problem centers on the known versus the speculative. Although there must have been exotic thermal physiologies that are now extinct, their nature will always remain murky and speculative. This study is conservative, in that systems known to work in living vertebrates are preferred over more speculative alternatives. Only when the evidence compels it is the application of an exotic system restored, and even then its speculative aspects are kept to a minimum.

This study emphasizes the power systems of tetrapods over the manner in which they regulate body temperatures, or their RMRs. The many attempts to estimate the resting metabolism of extinct forms, for example, are somewhat misplaced in that they focus on when the animal was withdrawn from its habitat, rather than actively participating in it. In land tetrapods, the most important energy requirement is to generate power to do work and be active. Other metabolic and thermoregulatory adaptations are secondary results of, or requirements for, operating at a given power level in a given habitat. Raising the activity level that can be sustained on an hour-by-hour basis in order to walk, run, and fly requires a higher hour-by-hour AEC, which leads us to the next section.

## The Aerobic Exercise Capacity Hypothesis

It is a primary thesis of this analysis that tetrapods of the land and air that practice differing levels and types of athletic activity will also differ significantly in hard structures that are preserved in the fossil record. A tetrapod that can aerially migrate at 50 km/h can be expected to differ in skeletal anatomy from one that migrates while walking 5 km/h, and the latter can be expected to differ from a creature that can walk just a few kilometers at only 0.5 km/h. This contrasts with the view of Ruben (1995) and Ruben and Jones (2000) that major metabolic differences are largely limited to soft tissues that do not preserve as fossils.

My assertion is valid because animals are machines that must work to acquire, produce, and expend energy in order to produce the power they need to achieve different levels of performance. In particular, animals are not heat pumps that can use body or external temperature to power activity. Instead, animals are internal "combustion" machines that use muscle-generated power to do work, which—even when initially generated anaerobically—is ultimately sustained aerobically. Therefore, a high and stable body temperature may extend the daily period of activity, but will not by itself raise the basic activity level that can be sustained over one hour. This is a crucial but often ignored fact.

Aerobiosis is energy efficient and moderately powerful. Its main advantage is that an animal can take in oxygen constantly as it uses the molecules to power its activity, so even high-level aerobiosis can be sustained indefinitely. Oxidative metabolism is therefore ideal for power-sustained activity: its ability to power bursts of activity is modest but useful. Because anaerobic power works differently, it has a different advantage. Since it can quickly utilize molecules normally held in reserve in muscle cells, the initial level of power can greatly exceed that achievable via the maximum level of aerobiosis. The drawbacks of anaerobiosis are that this power quickly drains down, and that high levels of toxic by-products are generated. Therefore, anaerobiosis is suited for producing quick bursts of intense power but is ill-suited for sustained activity. Anaerobiosis and aerobiosis are inherently antagonistic systems that cannot be simultaneously maximized in the same cell, so animals can be optimized to be either highly aerobic or highly anaerobic, but not both (Hill and Wyse 1989).

One can think of anaerobic power as the rough equivalent of non-oxygen rockets, briefly producing extreme power before quickly being expended. Aerobic power is more akin to a jet engine that never can produce as much burst power as a similar-sized rocket, but can sustain high levels of energy for long periods.

Because all activity, especially sustained activity, is critically dependent upon the aerobic power system, it follows that the form and function of an animal should also be dependent upon and correlated with its AEC. This is most true of animals that live on land or fly, because terrestrial and aerial movements consume very large amounts of energy. Depending on exactly how streamlined they are, hydrodynamically streamlined fish and reptiles swim a given distance five to twelve times more energy efficiently than a land animal can walk or run the same distance (Seymour 1982, Hill and Wyse 1989, A. Bennett 1991, Eckert 1992). Swimmers can, therefore, cruise at high speeds for long distances with little work, so they do not need a high aerobic capacity. In comparison, even walking at a modest pace of 3–5 km/h is surprisingly hard work and doing so for any substantial distance requires an elevated AEC. Although level, powered flying (as opposed to gliding or soaring) a given distance is about three times more energy efficient than walking or running, it is always very energy demanding on a unit-time basis, because the flier is constantly working to support itself against the planetary gravity field (this does not apply to fliers weighing less than 1 g, for which flight is very energy efficient [Heinrich 1993]). Thus, flapping flight costs about ten times as much per unit time as does walking at a modest pace. The sustainable AEC of a long-distance powered flying bird must be as high as that of a fast running bird.

The emergence of animals onto land, and later their climb into the air, presented serious power problems. The same LoAEC that powers sustained high-speed travel under water can sustain only low walking speeds in 1 *G*. Short-term, toxic anaerobic power generation cannot be used to escape this constraint. As shown below, configuration of the limbs cannot solve the problem either. It is sometimes stated that bradyaerobic animals can be quite active, but this claim can be misleading. In general, bradyaerobic land animals can be fairly active in that they can engage in short or long periods of low-level aerobic activity, punctuated by brief anaerobically powered bursts of high-level activity (A. Bennett 1983, 1991, 1994). An ora can dash at 14 km/h for a couple of minutes (Auffenberg 1981). However, low sustainable power severely limits sustained walking speed, which in turn precludes migration, movement in organized herds, and sustained powered flight. So an ora cannot walk at 5 km/h for hours on end, as can similar-sized birds and mammals. Aerobic power production and food intake must be boosted to achieve the higher walking speeds that underpin high levels of sustained activity, migration, living in social groups on land, and ultimately sustained powered flight. So, although bradyaerobic and tachyaerobic land animals can be both active and athletic, they do so in very different ways, with bradyaerobes being much more limited in what they can do. Bradyaerobes can be sprinters, but not marathoners. Tachyaerobes can be either.

The AEC hypothesis states that, on land, AEC must be elevated to sustain activity levels higher than those practiced by classic reptiles (which excludes extinct exotic forms such as ancient archosaurs and therapsids). This hypothesis encompasses the aerobic-capacity hypothesis, which goes on to argue that minimal and maximal aerobic MR are intimately linked to one another (see below; A. Bennett and Ruben 1979, A. Bennett 1991, Hayes and Garland 1995), an issue the AEC hypothesis does not itself address.

## Locomotion on Land and in the Air

The AEC hypothesis implies that the greater the evidence for high levels of sustained activity in an extinct tetrapod, the higher its aerobic capacity probably was. This is a powerful concept for restoring the energetics of extinct tetrapods.

### *Cruising and Migrating in Organized Groups*

Cruising is sustained movement over many minutes, hours, or days. It excludes unusually athletic locomotory performances, such as when animals are being actively chased, are in hot pursuit, or are displaying or playing. For several reasons, cruising speeds offer an excellent means for helping to determine the general athletic types and activity levels of land tetrapods. First, cruising represents the activities of animals when they are up and about and expending large amounts of locomotory energy. Second, land travel is so energy expensive that walking at even a moderate pace requires an elevated AEC.

This brings us to crucial point not always understood by critics: sustained walking speeds, *not burst running speeds,* are central to the hypothesis. The ability of an ora to run briefly as fast as a human is not as significant as the reptile's inability to keep pace with a walking person and her dog for more than a few minutes. It is also important to understand that the use of anaerobiosis to power cruising is probably untenable (A. Bennett 1983, 1991, Hertz et al. 1988). Aside from the rapid drop-off inherent to anaerobic power, either constant use of anaerobic power for periods exceeding many minutes, or repetitive bouts of anaerobiosis during sustained cruising, will quickly induce fatigue and the potentially debilitating effects associated with excessive anaerobiosis, thereby degrading aerobic performance. Anaerobic power is therefore only used for emergencies that briefly require high speeds, such as attack or escape. Thus, it is not feasible for a bradyaerobic animal to use anaerobiosis to regularly power sustained high walking speeds for a short distance, rest, and then continue for another brief period of fast walking. The animal would suffer chronic fatigue and toxic illness. Therefore, the AEC is the critical factor in setting the cruising speed. In particular, indefinitely sustainable AEC is the power source behind cruising for hours each day, especially during migrations. ISAEC is a major fraction of MaxAEC, which cannot be sustained for many minutes or hours because of energy depletion within muscle cells (Hill and Wyse 1989).

Another crucial point in favor of using cruising speeds centers on the surprising fact that moving a given distance costs about the same regardless of speed, limb design, or posture (Fedak and Seeherman 1979, Hill and Wyse 1989, Alexander 1992, Jones and Lindstedt 1993, Langman et al. 1995). Instead, the cost of locomotion is largely a function of size. Therefore, a slow-walking, flat-footed, sprawling monitor burns about as much oxygen to walk a mile as does a trotting, erect-limbed, digitigrade wolf. It is not possible, therefore, to use an optimal limb design to achieve high sustained walking speeds with LoAEC. This means that a sprawling tetrapod can sustain a high walking speed only if it has ElvAEC, and an animal with a reptilian AEC will not be able to sustain high walking speeds even if its legs are long and erect. In particular, it is improbable that a combination of long, erect legs and anaerobic power can be used consistently to sustain walking speeds above those observed in reptiles, contrary to speculations to that effect by Farlow (1990).

The final advantage of cruising speeds is that they can be recorded via trackways, which are preserved in large numbers as trace fossils. At a given limb length—which can often be estimated from footprint length—speed is largely a function of stride length. Alexander (1989) has presented an equation for calculating the speed at which a given trackway was made. Basic similarities of limb dynamics shared by sprawling and erect legs (Farley and Ko 1997) suggests that the equation applies to both types, a view confirmed by the results obtained with the trackways of sprawling fossil reptiles (see below). The resulting values are not precise. Anton (1998) concluded that the formula overestimated the speed of bipeds by one-third, but it underestimated the observed speeds of an ostrich (in Muybridge 1957) by a similar amount. A few trackway speed estimates are not definitive, so the larger the sample set, the better. Cruising speeds as recorded by trackways therefore meet all the paleometabolic criteria of observability, correlation, and causal linkage with aerobic capacity.

Migrations on land are such arduous journeys that they demand "the total physiological attention" of tetrapods (Meier and Fivizzari 1980). Migrating animals must minimize travel time and exposure to danger and so must walk rapidly (Pennycuick 1979, Meier and Fivizzani 1980). Even hiking without a burden and on level, graded ground nonstop for tens of kilometers is hard work—a fact the author can confirm from hiking 80 km in 20 h! A HiAEC is therefore required to migrate on land. These facts are contrary to Spotila et al. (1991), who calculated that an energy-efficient bradyaerobic land animal can migrate thousands of kilometers at 50 km/day, and outrange less efficient tachyaerobes. The pace required (3–4 km/h, assuming a few hours rest each day) is above that sustainable by reptilian AEC. Spotila et al. failed to appreciate that aerobic power is more important than energy efficiency when migrating on land.

Interacting and especially moving in organized groups is an another form of activity that requires sustained effort at high levels of performance. On land, this requires a HiAEC, because an animal must have enough energy to spare for extended activity in addition to foraging and breeding. Bradyaerobic tetrapods are therefore not expected, and are probably unable, to form and move in organized groups (A. Bennett 1983), unlike very efficient swimmers, which can form schools even if they are bradyaerobic. Birds and mammals are noted for their tendency to form flocks, herds, and packs.

The ISAEC of reptiles is so low that they are limited to sustained cruising speeds of only 0.1–2.0 km/h (App. Fig. 8A; A. Bennett 1983, 1991, 1994, Paul and Leahy 1994, Farmer and Carrier 2000a). Even the most aerobically capable monitors using a gular pump (see below) to enhance breathing during exercise are not able to sustain higher speeds (Thompson and Withers 1997, Owerkowicz et al. 1999). The 4- to 5-km/h speeds cited by Auffenberg (1981), Farlow (1990), and Suzuki and Hamada (1992) would represent an extraordinary performance for a reptile, if this represents the normal cruising behavior of oras. No speed frequency data was published and requests for same were not met. Auffenberg (1994) cites a more plausible 0.3-km/h normal walking and foraging speed for the large Bengal monitor. An extensive data set indicates that oras normally progress at a typically reptilian pace of less than 2 km/h; the few cases with speeds over this value appear to represent individuals occasionally resorting to bursts of anaerobic power. The story was much the same with archaic amphibians and reptiles: 83% of the fossil trackways examined produce estimated speeds of 0.1–2.0 km/h, and only one is over 3 km/h. The low speeds of archaic reptiles were not just the result of their "primitive" limbs, because they moved about as fast as their modern counterparts. Some reptiles walk for many hours each day, but they do so slowly (A. Bennett and Gorman 1979, A. Bennett 1983). Faster walking and running speeds are always limited in duration.

No land reptile migrates. That bradyaerobic sea turtles migrate long distances at high cruising speeds (Spotila et al. 1991) is irrelevant, because they swim so much more efficiently than similar-sized land animals (Paul 1991, 1994a, Paul and Leahy 1994). Leatherback metabolics can indefinitely sustain a swimming speed of 3–5 km/h (Standora et al. 1984), but on land the sixfold decrease in locomotory efficiency would result to a land speed of only 0.5–0.8 km/h. No land reptile regularly moves in organized groups. Researchers often do not understand that migration and high degrees of socialization are evi-

*Appendix Figure 8. A,* cruising speeds in extinct and living tetrapods (all capable of running gait). The speeds for the extinct tetrapods were estimated from trackways, as per text, and those of the living tetrapods from motion pictures (including Pennycuick 1979) or direct observation. Sample sizes are shown on the right. Dinosaur data include trackways identified as belonging to avepods (solid black), ornithischians, and unidentified forms; the triangle on the lower left represents a short trackway in which decreasing stride length indicates that the animal slowed down and perhaps came to a stop. *B,* hip height/foot length ratios in theropods. Only complete hindlimbs are figured, all as lefts. Leg posture was restored after Paul (1987a, 1988a); the metatarsals may have been less vertical, in which case hip height would have been lower. Because the degree of knee flexion is unlikely to have been less, more-vertical limb segments would exaggerate hip height. For restoration of the entire foot length, complete toe III with its claw sheath, the posterior portion of toe IV, and restored toes pads are included. *Left to right: Herrerasaurus* PVSJ 373; *Coelophysis* AMNH 7223; *Dilophosaurus* UCMP 37302; *Allosaurus* USNM 4734; *Gorgosaurus* AMNH 5458; *Struthiomimus* AMNH 5339; emu. The top row of lines under the feet measures foot length as restored by Anton (1998); the second row indicates foot length as restored by T. Thulborn (1990).

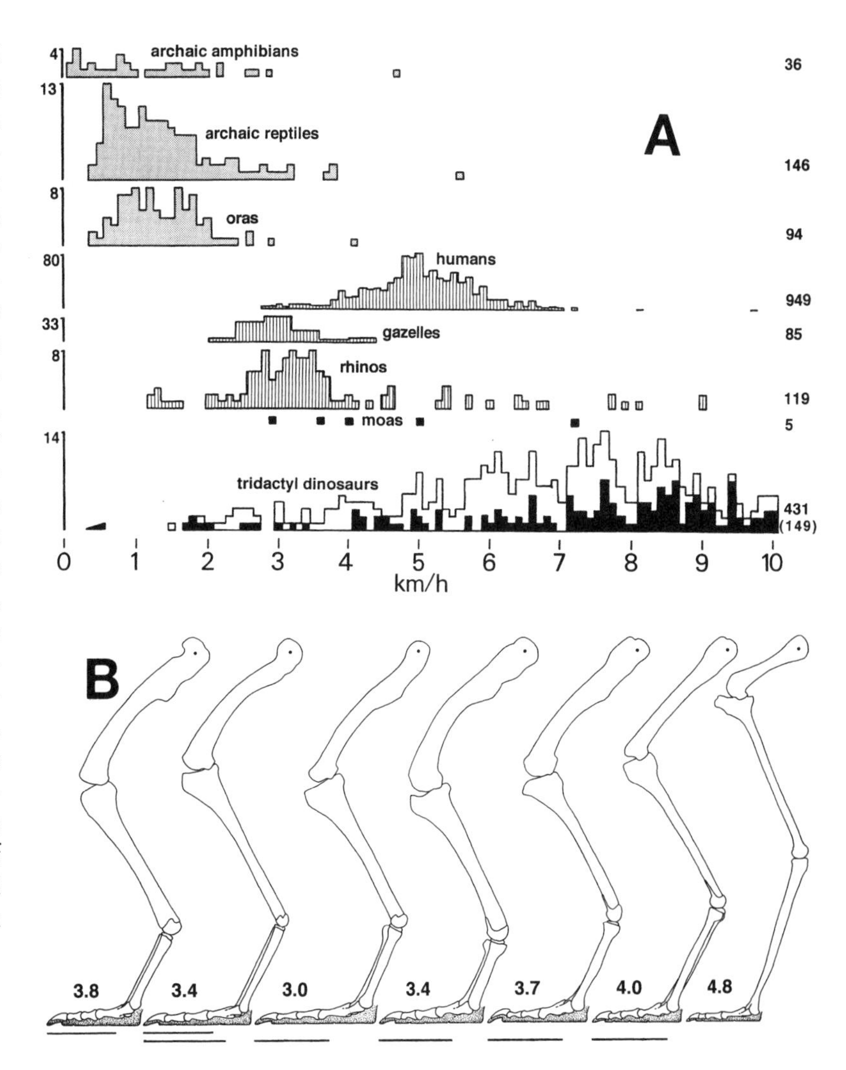


dence for high AEC and high energy budgets not because of the implied level of sophistication of the animals—after all, schools of bradyaerobic fish swim across entire oceans—but because of the high energy cost of these activities on land.

Some land mammals migrate long distances, caribou and gnu being the premiere examples. Many mammals move in organized herds or packs. This is not surprising. With their high ISAEC, tachyaerobic mammals easily sustain walking speeds of more than 2–3 km/h, and some can sustain more than 10 km/h, for many hours or even days (App. Fig. 8). It is also important that tachyaerobes rarely walk more slowly than 2–3 km/h (see below). Unfortunately, little direct data is available for birds, but a few moa trackways record high walking speeds (App. Fig. 8A). Tachyaerobes enjoy marked cruising speed superiority over bradyaerobes, and do so at all sizes. Thus squirrels regularly dash about much more swiftly than little plodding lizards, and humans cruise at 5 km/h while big monitors can manage only 1 or 2 km/h. Note that many mammals cruise at moderate speeds that do not fully exploit their ISAEC. This means that cruising speeds may set a gross minimum for the AEC, rather than its maximum.

The above results will be falsified if it can be shown that reptiles can consistently walk at high speeds. As it is, a large set of trackways in which the great majority of estimated speeds are above 2–3 km/h, plus evidence for migration and/or movement in organized groups, are all diagnostic characters that indicate that the subjects' AEC was elevated above the reptilian level. Because these characters are directly linked to or record aerobic power production, their diagnostic value is further enhanced. Because of the imprecision of the data, and the failure of some animals to exploit fully their ISAEC, it cannot be established or disproved that the MaxAEC was in the supra-aerobic zone (unless the estimated trackway speeds are consistently very high). Evidence for migration and/or social organization favors HiAEC. What about animals whose aerobic exercise capacity is just above the reptilian level? Lacking living examples, it can be speculated that their walking speeds should straddle the reptilian-mammalian boundary of 2–3 km/h.

Speeds have been estimated for some 400 tridactyl dinosaur trackways made by individuals weighing from 1 kg to 5 tonnes, and spanning all three periods of the Mesozoic (App. Fig. 8A). Of these individuals, about one-third have been identified as probable avepods, many are ornithopods and other ornithischians, and the rest are ambiguous. T. Thulborn (1990) and Anton (1998) have estimated that hip height/foot length ratios—used to estimate the size of the track makers—were 4.5–5.2 for the dinoavepods, but neither presented accurate diagrams confirming these high values, which are falsified by careful restorations of legs and feet in realistic postures (App. Fig. 8B). A general ratio of 4 has been used here. Even this value may be somewhat too high for most tridactyl dinosaurs: boosting hip height has the effect of somewhat reducing the speed estimates. The great majority of trackways in the sample show long, avian-mammalian–like stride lengths. Ninety-eight percent were made at speeds exceeding approximately 2 km/h, and 95% at speeds greater than approximately 3 km/h. Among probable theropods, the respective figures are 96 and 93%. Only if abnormally low stride frequencies and/or inaccurately high hip height/foot length ratios are assumed can the estimated speeds of the long-stride trackways be driven down to the level of shorter-stride reptiles. The 0.8- to 9-km/h speeds estimated by Anton (1998) for a set of Triassic tridactyl trackways is probably too low by as much as 50%, because he used excessive hip/foot ratios and an overly conservative speed formula. Instead, even small avepod dinosaurs tended to walk much faster than monitors of similar size. The peak speed for avepod dinosaurs is approximately 7.5 km/h. The power needed to sustain such a speed is some four times the MaxAEC that could be generated by reptilian aerobiosis. The large data set shows that dinosaur cruising speeds were entirely nonreptilian, show no evidence for an intermediate pattern, are well within the avian-mammalian speed range, and required sustained aerobic power output beyond that which could be generated by reptiles.

There is widespread agreement, based in part on trackway evidence as well as single species bone beds, that a number of dinosaurs migrated, and/or moved in organized groups (Hotton 1963, 1980, Currie and Dodson 1984, Paul 1988c, 1994a, 1997b, Dunham et al. 1989, Russell 1989, T. Thulborn 1990, Spotila et al. 1991, Horner and Dobb 1997).

If the estimated speeds recorded by dinosaur trackways were in the reptilian range, and there was no evidence that they moved in herds or migrated, then it could only be concluded that their AEC was similarly low. As it is, there is no justification for applying a reptilian AEC to dinosaurs based on the locomotory performance they left on Mesozoic mudflats: reptilian energetics are entirely contradicted. The assertion by Ruben et al. (1997b) that "if dinosaurs possessed aerobic metabolic capacities and predatory habits equivalent to those of modern tropical-latitude lizards (for example, *Varanus komodoensis*), they may well have maintained large home ranges" neglects certain facts. If dinosaurs were slow-walking reptiles, then they could have moved only a small fraction of the distance that terrestrial birds and mammals cover in a given day. Of course, the trackways show that dinosaurs walked as fast as ground birds and mammals. In this case, dinosaurs could have consistently walked fast and had reptilian AEC and energy budgets only if they powered almost every step partly anaerobically, and if they moved no farther, and for much shorter periods of time, than reptiles. Bradyaerobic dinosaurs that cruised fast but did so only briefly, and then spent the rest of the day recovering from their chronic use of anaerobiosis, would be less active per total time each day than many modern reptiles! Boosted AEC easily and fully explains the high cruising speeds, long range, and social interactions observed in dinosaurs. Indeed, for avepods and other dinosaurs to have walked so fast so far, and done so in herds and packs, their ISAEC must have been well above the reptilian level. The trackway data alone does not tell us whether the MaxAEC of these dinosaurs was in the mesoaerobic or supraaerobic level, but the later is favored by the probability of migration and herding. The above conclusions can be contradicted if a large set of short-stride, low-speed predatory trackways are discovered, or if it can be shown that they made long-stride trackways at low speeds.

A large sample of Cretaceous bird trackways is available (Lockley et al. 1992), but a dimensional data set has not been published. In addition, it is difficult to estimate hip height/foot length ratios in birds because they are extremely variable. It is therefore not possible to generate a set of estimated speeds at this time. It has recently been suggested that some brevisternan birds of the Cretaceous probably lived in large flocks, judging from the great abundance of their remains in some locales (Chapter 13). The level of activity implied by this data is more compatible with elevated than with low AEC. The ability of toothed Mesozoic marine birds to swim long distances is not, as believed by Feduccia (1996) and Chinsamy, Martin, and Dodson (1998), evidence for high aerobic capacity, because their cost of locomotion was low.

## *The Causal Link between Limb Posture and Aerobic Exercise*

The cruising speed differential between bradyaerobes and tachyaerobes helps reveal the long-sought causal link between limb posture and AEC (such a relationship was not apparent to Hayes and Garland [1995], Feduccia [1996], or Ruben et al. [1997b]). The link starts with the observation that all modern, terrestrial tetrapods with sprawling limbs are bradyaerobic, and all those with limbs that are habitually and highly erect are tachyaerobic. In addition, all obligatory bipeds are tachyaerobic.

Sprawling and quadrupedalism are well suited for slow-cruising bradyaerobes. The wide gauge and multiple ground contacts provide the stable platform needed by animals taking short, low-frequency strides (Alexander 1992). Sprawling legs can also run the high-speed bursts powered by brief anaerobiosis by assuming a more dynamic, partly erect gait (Fieler and Jayne 1998), and then setting the body down for the much-needed recovery. Belly crawling is also adaptive for chilled and therefore weak reptiles that need to reach a sunny site to bask.

The higher aerobic speeds of walking tachyaerobes leave them without the need for the slow speed and belly-resting potential of sprawling limbs. Instead, they require erect legs, whose strong fore-and-aft pendulum effect promotes the longer strides and higher step frequencies that they are capable of powering (note that this is a physical action effect, not an energetic one because locomotory energy efficiency remains the same). Indeed, erect legs favor long strides to the point that walking below approximately 2 km/h is physically awkward, and the tall body carriage inhibits the belly resting that is so advantageous for animals that regularly use anaerobic power to move more rapidly. The instability inherent to erect carriage and a narrow gauge further discourage slow speeds, in favor of a faster pace in which dynamic balance keeps the animal from falling (Alexander 1992). Obligatory bipedalism is inherently unstable too, so high dynamic walking speeds are favored over a slower, more stable pace. A fully bipedal and/or erect-limbed reptile would find itself tending to walk too fast to avoid the excesses of anaerobic power, yet would not be able to exploit fully the potential of the legs for a fast sustained walking pace. In this view, the evolution of erect legs probably forces aerobic capacity to be elevated above, and anaerobic capacity to be reduced below, reptilian levels.

The above arguments do not apply to arboreal chameleons, which combine erect limbs (Paul 1987a) with bradyaerobic power. The erect limbs of chameleons evolved to permit very slow and stealthy motion along slender branches, and so no boost in AEC was required.

Carrier (1987, 1998) links erect and bipedal gaits with ElvAEC via another argument. In salamanders and lizards, the laterally flexible dorsal columns associated with their non-erect legs prevent breathing when running. An alternative means of lung ventilation is used by some running lizards (see below), but it remains possible that lateral flexion of the trunk during locomotion hinders lung ventilation to the point that suprareptilian rates of oxygen uptake are prevented. Birds and mammals have escaped this constraint by adopting bipedal and/or erect gaits that do not involve lateral spinal flexion. The latter is also absent in sprawling turtles, semi-erect crocodilians, and erect chameleons, but their lack of elongated limbs suited for sustained high speeds renders their example moot.



Limb and body posture meet the criteria of observability, because they are usually readily restorable in fossil tetrapods via a combination of skeletal anatomy and trackways. The consistent correlation between leg posture and MR observed in modern tetrapods, and the causal link between the two factors, mean that sprawling legs are diagnostic for bradyaerobiosis, whereas strong bipedalism and highly erect legs—especially but not necessarily when they are long—are diagnostic for tachyaerobiosis. The close connection with power production strengthens the diagnostic value of these characters. The situation with semi-erect forms is ambiguous, because this group includes both crocodilians and some mammals (Court 1994, Reilly and Elias 1998). Berman et al. (2000) described a small basal reptile with erect limbs, but it appears that the animal was able to adopt a less erect gait when not running.

Limb bone morphology and trackways show that the hindlimbs of protodinosaurs, dinosaurs, and early birds were erect (Chapter 4), and that theropods and early birds were obligatory bipeds. Feduccia's (1996) argument that dinosaurs evolved erect limbs to support their enormous bodies is false because the dinosaurian erect gait first evolved in small forms: the erect gait may have promoted gigantism, not the reverse. Neither erect legs or bipedalism is compatible with reptilian energetics. Both fully support the idea that dinosaurs and early birds were tachyaerobic.

### *Muscle Mass and Pelvic Dimensions*

The profoundly different aerobic performance of bradyaerobes and tachyaerobes is expressed in their musculature. The low-capacity cardiorespiratory system of reptiles cannot oxygenate large, tachyaerobic muscles, whereas the avian-mammalian system can. Nor do reptiles need large leg muscles to run fast, because their bradyaerobic, hyperanaerobic muscles can produce about twice as much anaerobic burst power per unit mass as those of mammals and birds (see above). The latter have to compensate for their less anaerobically capable muscles by having larger muscles. Note that if an animal's leg muscles were as large as a mammal's, and had the mass specific anaerobic power typical of reptiles, then its sprint speed would exceed that of any mammal, and would be too high to be practical. Therefore, for a given mass, the skeletal and leg muscles of reptiles are significantly smaller than those of birds and mammals of similar mass. This disparity has been measured in small tetrapods (Ruben 1991, 1993). Quantitative data is not available for larger forms. However, a visual comparison of large monitors and crocodilians with those of equally large mammalian predators indicates that the limbs and muscles of the latter are markedly larger than those of the former. The muscle-mass hypothesis would be falsified by the discovery that the locomotory muscle/body mass ratios of big reptiles are similar to those of large birds and mammals.

Because the musculature is supported by the skeleton, the dimensions of the latter should reflect the volume of the former. This study concentrates on the proximal muscles of the hindleg, because they are involved in ground locomotion in both quadrupeds and bipeds, and are less affected by modifications of the distal segments and feet. The dorsopelvis (ilium in reptiles and birds, ilium plus ischium in mammals) supports the large set of proximal locomotory muscles. In reptiles, the ilium is short, so the muscle attachment area is narrow, this being all that is needed to support a low volume of bradyaerobic thigh muscles (App. Fig. 17). In birds and mammals, the dorsopelvis is an elongated plate that anchors a large set of tachyaerobic thigh muscles. A comparison of dorsopelvis length/mass ratios confirms that this structure is consistently longer, by an average of 2, in longoschian mammals and birds than in brevischian diapsid reptiles (App. Fig. 18I). There is a progressive correlation between dorsopelvis length and aerobic capacity in that mesoschian monotremes and insectivores with more modest MR tend to have shorter pelves than more energetic longoschian mammals. These comparisons somewhat understate the size difference, because they do not take into account the increase in the depth of the ilial plate in mammals and birds (it would be preferable to compare the area rather than length, but the former proved impractical to measure in a large sample).

Within given groups—quadrupedal lizards, for example—hip dimensions tend to remain isometric, so there is not a strong size effect (App. Fig. 18I). Even large and gigantic monitors have much shorter ilia than birds and mammals of similar size. Ilial expansion is not simply a means of strengthening the sacral-pelvic connection, because the latter remains short in longoschian mammals, even for bipedal examples. Nor is dorsopelvic expansion necessary for an erect or bipedal gait, because neither gait directly requires an expanded musculature. This is confirmed by the observation that the ilia of the most bipedal modern reptiles are—and some bipedal, erect-limbed dinosaurs were—much shorter than those of birds and bipedal mammals. Although the dorsopelvis of bipedal birds and saltorial mammals tend to be longer than those of humans and quadrupedal mammals, most of the latter are also well above the reptilian maximum (see below). Dorsopelvic expansion is not a function of severe tail reduction, because the structure is enlarged in most big-tailed dinosaurs. The dorsopelvis is hyperelongated in many armored forms (paraeisaurs, stegosaurs, ankylosaurs, armored edentates), and apparently the anterior pelvis is enlarged to assist in supporting the armor. The hyperelongated anterior ilia of frogs function as part of a pelvovertebral joint that is critical to their specialized leaping gait (Emerson 1979). Such very long and slender anterior ilial processes do not support thigh muscles.

These results show that, like any character subject to the influence of multiple factors, the enlargement of the pelvis is partly guided by limb and body posture, but only in a minor way. The major factors determining elongation of the dorsopelvis are increased thigh-muscle volume, followed by spring leaping and support of armor. Aside from armored or spring-leaping forms, it is not logical for tetrapods to have a dorsal pelvic plate larger than needed to anchor the thigh muscles. Nor can a small pelvis support a large volume of muscles. It is therefore concluded that dorsopelvis length is largely a function of, and a good indicator of, gross thigh muscle volume in nonarmored, nonspringing tetrapods. Because locomotory muscle volume is correlated with aerobic capacity, short reptilian ilia are diagnostic for LoAEC, platelike dorsopelves as large as those of birds and mammals are a diagnostic character that indicates ElvAEC, and elongation of the dorsopelvis above the monotreme-insectivore level is indicative of HiAEC. The direct connection between the size of the dorsopelvis and muscle power ranks the diagnostic value as high. Elongation of the legs can also increase the volume of the limb musculature, other factors being equal.

Protodinosaurs and basal dinosaurs (basotheropods and prosauropods) are interesting because their reptile-shaped ilia were only modestly elongated compared with those of reptiles (Fig. 5.1Ba–c, App. Fig. 18I). The pelves were therefore mesoschian, and most strongly resembled those of monotremes and insectivores in this regard. Their short pelves were not able to support a thigh musculature as broad and large as that of birds. However, the modest expansion of the ilia combined with elongation of the limbs suggests that limb muscle volume was higher than in semibipedal reptiles. If so, then it is improbable that the leg muscles were either hyperanaerobic or mesoanaerobic, or they may have been supraanaerobic. Equally important is that the boost in aerobic capacity should have limited anaerobic power generation, leading to the loss of reptilian hyperanaerobiosis. An avian-mammalian aerobic capacity is falsified, but reptilian AEC is not supported either. Some form of marginal tachyaerobiosis, in the lower mesoaerobic range, is diagnosed.

The great majority of dinosaurs had long, nonreptilian ilia (Fig. 5.1Bd–s) whose length falls in the avian-mammalian range (App. Fig. 18I). In particular, the ilia of all known dino-avepods were birdlike plates in size and gross form. Those of coelophysoids tended to be shorter than those of later avepod dinosaurs. The ilia of averostrans small enough to be within the avian mass range falls in or barely below the lower avian zone. Ornithomimid and tyrannosaurid ilia were especially large. The shorter ilia of dino-avepods relative to avebrevicaudan birds may reflect latter's elongation of pelvis to compensate for absence of tail-based limb retractors. It would have been illogical for the large ilia of longoschian dinosaurs to have been largely bare of muscles, and to have anchored only the narrow thigh muscles restored by Knight (in Paul 1996). Romer (1923, 1927), Galton (1974), and Paul (1987a, 1988b) restored longoschian dinosaurs with broad, birdlike thighs. The long length of most dinosaur legs also indicates that their leg muscles were larger than in reptiles. It is very improbable that these large muscles favored anaerobic over aerobic power generation, because the resulting sprint speeds would have been as fantastical as they were impractical. Boosted AEC easily and fully explains the long ilia of dinosaurs. Bradyaerobic, hyperanaerobic reptilian muscle energetics are therefore falsified in favor of strongly elevated AEC, probably at the higher mesoaerobic and/or supra-aerobic levels. There is some evidence of a substantial but modest increase in aerobic capacity as pelvis size increased in avepod dinosaurs.

It is interesting that the ilia of basal birds tended to be rather short. Those of dromaeosaurs were on the short side for avetheropod dinosaurs, but they still qualified as longoschian. The ilia of archaeopterygiforms and confuciusornithids were shorter than those of other avetheropods avian and nonavian (Fig. 10.1Bm,t, App. Fig. 8I). This probably reflected a shift from locomotory power generation from the hindlimbs during terrestrial locomotion to the forelimbs during quadrupedal climbing and especially during flight (Chapters 7 and 9). Even so, basal bird ilia were in the upper mesoschian ilia—much larger than those of reptiles and significantly longer than those of protodinosaurs and baso-theropods—and in combination with their long legs were sufficient to support a moderately broad thigh musculature. A higher mesoaerobic/lower supraaerobic AEC is therefore diagnosed. The ilia of other early birds with urvogel-type ilia, such as *Sinornis* and *Cathayornis*, were in the lower general avian range (Fig. 10.1Bt,u, App. Fig. 18I).

## Synthesis of Terrestrial Locomotion

The results of the above analysis of locomotory adaptations and performance in protodinosaurs, predatory dinosaurs, and basal birds are conclusive. In longoschian predatory dinosaurs and early birds, the combination of high estimated walking speeds, erect limbs, obligatory bipedalism, and enlarged pelves anchoring expanded muscles are fully and solely diagnostic of a strongly elevated AEC. There may have been a gradual increase in AEC from the upper mesoaerobic into the lower supraaerobic range as the size of the ilium and the leg muscles increased. There is absolutely no evidence that longoschian dinosaurs or basal birds retained a reptilian, bradyaerobic, hyperanaerobic locomotory apparatus.

The situation with protodinosaurs and basotheropods is also definitive, albeit somewhat more obscure. Their estimated walking speeds, erect legs, bipedal gait, and mesoschian ilia are not compatible with fully reptilian locomotory energetics. At the same time, their short ilia falsify the presence of highly aerobic leg muscles. It can only be concluded that the power system of these archosaurs was neither reptilian nor birdlike, but represented some form of a marginally tachyaerobic condition that is now extinct. Note that this result suggests that fully erect, digitigrade legs can evolve in animals with only modest ElvAEC.

## Flight and Forelimb Muscle Mass

Gliding is observed in both brady- and tachyaerobic tetrapods and is not diagnostic for any particular AEC. It is usually presumed that more sophisticated flight requires high AEC in all but the tiniest fliers, whose flight is very energy efficient. In part, this is because supra-aerobiosis can readily power the high per-unit-time demands of powered flight for creatures that exceed 1 g in mass. In addition, soaring and powered flight are observed only in supra-aerobic birds, bats, and flying insects of mass greater than 1 g (Heinrich 1993). Even short-ranged, burst-flying birds are supra-aerobic, and most birds and bats have exceptionally high ISAEC (Yalden and Morris 1975, Pennycuick 1989, Maina 2000).

Ruben (1991, 1993, 1996) argued that soaring and powered flight can evolve among bradyaerobic, ectothermic reptiles. In this scenario, soaring flight is so energy efficient (see Chapter 7) that it does not require tachyaerobiosis, and short bursts of flapping flight are powered by small but highly anaerobic muscles. In principle, this may be possible, but in view of the absence of any living fliers with such energetics, the idea is speculative. For one thing, even soarers must occasionally power fly for short distances, and bursts of anaerobic-powered flight would quickly fatigue the flight muscles, which is not a suitable condition for a creature that must use these muscles to do the substantial work needed for lengthy, controlled flight with wings outstretched (Pennycuick 1989). Thermal and wave soaring are very rapid, complex, and potentially dangerous activities that require the brain to work constantly at peak performance. The fluctuating and low body temperatures that would accompany a combination of a low MR and air cooling (especially on cool marine nights) might interfere with neural activity. The flight performance and postflight recovery of the anaerobic-powered flier postulated by Ruben (1991) is inferior to that observed in even short-range burst flying birds. Even very short flights of a few dozen meters incur recovery periods of about a minute. Maximum flight range is a relatively short 1–2 km, followed by long-lasting anaerobic-induced fatigue, and thus sustained aerial excursions are not possible. It is therefore questionable whether selective pressures have ever been strong enough to produce complex, dynamic airfoils in aerodynamic bradyaerobes.

The last point is contrary to an implication of Ruben's hypothesis, which is that flight should evolve in bradyaerobes because flight must first evolve in forms with arm muscles too small to power flight. This is false because bats retain the high AEC of their mammalian ancestors (Yalden and Morris 1975), so they must have solved the muscle mass problem during the initial stages of flight. There is no compelling reason to think birds could not have solved the same problem. In fact, the example of bats proves that flight can evolve in tachyaerobic, homeothermic endotherms. Thus, much as climbing bats hinder arguments that vertebrates evolved flight from the ground up, tachyaerobic bats work against the bradyaerobic hypothesis of the development of vertebrate flight. Ruben's (1991, 1993) arguments for anaerobic-powered flight in basal birds are comparatively speculative and depend upon questionable assumptions, among them that animals with metabolics like those of lizards can soar and power fly. It is therefore probable that evolving and exercising powered flight is practical only in tachyaerobes, or more specifically, supraaerobes. To demonstrate otherwise it must be shown that a powered or soaring flier has characters strongly indicative of a bradyaerobic, hyperanaerobic condition.

The latter has not yet been done. Ruben (1991) tried to use the size of pectoral elements, particularly the sternal plate, and the mass of flight muscles to discern between bradyaerobic and tachyaerobic fliers. However, bats lack enlarged pectoral elements. Nor is there a general correlation between the size of the flight musculature and basic metabolic type, because most tachyaerobes—including the insectivore ancestors of bats—do not have large flight muscles. As discussed in Chapter 7, aerobically powered flyers do not require the extreme expansion of flight musculature that Ruben asserts is necessary. Ruben (1991) at one point acknowledges that 15% of total mass needs to be flight muscles in typical birds and bats, but then seems to set a minimum of 15% for the pectoralis alone for a flier to be tachyaerobic. In other modern birds with well-developed powered flight and reasonable climb rates, the flight muscles make up only 11–14% of the total mass (Hartman 1961, Greenwalt 1962); in weak flying birds, the value can be as low as about 6% (Raikow 1985). In quadrupedal mammals, the forelimb muscles range from less than 4% of body mass to close to 9% (Grand 1977). The lower values tend to be found in arboreal forms, but the values for small cats and small primates can be almost as high as in terrestrial forms. Developing minimal-powered flight in tachyaerobes does not, therefore, necessarily require a dramatic increase in muscle mass or the size of supporting structures, or a massive alteration of muscle-cell energetics. However, the development of aerobically powered flight does require that the arm muscles be much larger than observed in reptiles. If the pectoral girdle and/or arm of a winged tetrapod is larger than that of reptiles, a tachyaerobic flight musculature is indicated. Therefore, the presence of wings suited for either soaring or powered flight is considered a diagnostic character for tachyaerobiosis, all the more so because the link with power production is direct.

Ruben (1991) suggested that *Archaeopteryx* powered its flight anaerobically rather than aerobically. Paladino et al. (1997) and Easley (1999) agreed. This hypothesis assumes that the "Eichstätt scenario" (Chapter 7) is correct (that *Archaeopteryx* was a powered flier), but that it could also take off from level ground. It also assumes that *Archaeopteryx* lacked flight muscles large enough to power-fly aerobically. The first assumption is reasonable but not sufficiently definitive. Lacking direct observation of its flight performance, it remains possible that *Archaeopteryx* could only glide or flap only weakly, and may not have been able to take off from the ground, so these are weak links upon which to construct a hypothesis. The second assumption is also not definitive. Because the actual mass of the urvogel's flight muscles is unknowable, it cannot be shown that it lacked enough aerobic power to fly: thus the Ruben and Easley hypotheses cannot be confirmed. What can be demonstrated is that *Archaeopteryx* probably had a near-avian rather than a reptilian set of arm muscles. Although a lack of a large flight musculature is not indicative of bradyaerobiosis or hyperanaerobiosis—to conclude otherwise would mean restoring the insectivore ancestors of bats as having had reptilian energetics—the presence of a brachial musculature larger than observed in reptiles should be correlated with an aerobic capacity above the reptilian maximum. The large, stout furcula, substantial ossified sternal plate, large pectoral crest of the humerus, and very large arms of *Archaeopteryx* indicate that its forelimb muscles were greatly enlarged compared with those of any reptile (Chapter 7). The presence of hyperanaerobic, bradyaerobic flight muscles is therefore contradicted, and a set of such large muscles suggests that AEC easily surpassed that of any reptile. As detailed in Appendix 2, such an expanded skeletal framework could probably support pectoralis and combined flight muscles making up 5–7% and 8–14% to of total mass, respectively (Fig. II.1). This would have approached or entered the lower flying-avian range, and is in the range expected in initial, tachyaerobically powered fliers. Such modest-sized, tachyaerobic flight muscles should have been sufficient for minimally powered flight using anaerobic power for takeoff followed by aerobic power during horizontal flight (Marden 1994). Ruben's (1991) assumption that the combined flight muscles of the urvogel made up only 7% of total mass is not impossible, but may be too low. Therefore, the argument that it needed highly anaerobic reptilian flight muscles is unfounded. Also refuted is Speakman's (1993) assumption that *Archaeopteryx* could not support an adequate flight musculature no matter what its power system.

The urvogel's flight muscles were probably supraanaerobic to compensate for the modest volume of the flight musculature, and to emphasize burst power for short-range flight. More advanced basal birds also had large arms, often extremely large pectoral crests, and larger sternal plates than the urvogel. Thus, flight-muscle mass should have been higher than in the urvogel and in the lower avian range. In this case, flight muscles of such basal birds would have been highly aerobic and probably supraanaerobic, not bradyaerobic and hyperanaerobic. If the early avian system was highly aerobic, postflight fatigue should have been much less than postulated in the reptilian scenario. The flight apparatus was not well enough developed to power long-distance flight, but this does not necessarily contradict HiAEC, considering that many modern birds are similarly short ranged. As with the terrestrial locomotory apparatus, the aerial locomotory apparatus of basal birds conclusively favors avian over reptilian exercise energetics. The ability of the small, toothed ornithurine *Ichthyornis* to power-fly long distances over oceans is strong evidence of sustained high aerobic capacity (Feduccia 1996, Chinsamy, Martin, and Dodson 1998).

### *The Energetics of Evolving Powered Flight from the Ground Up*

While on the subject of locomotion and paleometabolics, it is worthwhile to look at the energetic problems associated with the hypothesis that avian flight evolved among ground-running insectivores. Aside from large forms that feed largely on insect colonies, most insectivorous tetrapods weigh only 3–100 g (Nowak 1999). Most flying insects are small and each provides a modest number of calories when eaten: being a small insectivore limits the number of insects that must be caught each day. A 50-g mammalian insectivore, for example, must eat about sixteen 1-g insects per day.

Protobirds such as *Archaeopteryx* were rather large for nonsocial insectivory, weighing up to 500 g (Appendix 2). A typical mass for members of the flying insect orders that were prevalent in the Jurassic (Orthoptera, Hempitera, Neuroptera, and Coleoptera) is about 0.5–1.0 g (Greenwalt 1962). With a water content of about 80% and a caloric digestibility of 90%, each insect is worth 0.45–0.9 kcal (not 5 kcal/g, as suggested by Caple et al. [1983]; see Cummins and Wuycheck [1971], Avery [1971]). With an active, marginally supraaerobic MR of approximately 50 kcal/day, a leaping protobird would need to consume some 55–110 insects each day. Small mammals about the size of the protobird have daily foraging ranges of 0.5–4.0 km, which cost about 1–5% of their daily energy budgets (Garland 1983). If the protobird cruises and pursues aerial insects at the rather modest speed of 10 km/h suggested by Caple et al., then it would exhaust a normal foraging range in only 6–24 min. This would require catching an insect every 2–16 sec, which is not a realistic capture rate except on the rare occasions when single-species concentration of winged insects is available. Assume a still-optimistic capture rate of an insect every minute or two. In this case, foraging distances would be 15–70 km, and the locomotory energy costs would range from one-third to more than the entire daily energy budget: values so high that their viability is questionable.

If the protobird is modeled as bradyaerobic, then the number of insects that must be captured drops to 9–18 per day. However, pursuing insects at 10 km/h would require an intense level of anaerobic power that could last at most for only a few seconds if recovery times are to be kept to about a minute, and at most 4 min with a recovery period of over an hour (Ruben 1991). Just a few 1- or 2-min-long chases of aerial targets covering a few hundred meters would result in down times of many hours. Only by limiting runs to a few dozen meters each would profound exhaustion be avoided. The daily anaerobic activities of real reptilian hyperanaerobes are much more limited than postulated for the leaping insectivore (Hertz et al. 1988). In addition, with an insect caught every 1 or 2 min, required daily foraging ranges of 1.5–3.0 km are typical of mammals, and are well above those observed in reptiles (1.0 km in an 80-g teiid [A. Bennett and Gorman 1979], and an average of 1.8 km in large oras [Auffenberg 1981]). These energetic problems may explain why no living tetrapod obtains the bulk of its sustenance by chasing after flying insects from the ground, and why protobirds probably did not do so either.

## Respiration

Respiration is divided into two main subjects: the characteristics of the nasal passage and the structures they contain; and the form and function of the pulmonary complex.

### *Respiratory Conchae*

RC are thin, sheetlike conchae covered with respiratory epithelia. In birds, the RC are the anterior and middle conchae (Bang 1971); in mammals, they are the maxilloturbinals and in some cases the nasoturbinals (Hillenius 1992, 1994). RC are always set in the nasal airway (NA), through which the main airflow passes. The NA includes the nonolfactory portion of the nasal cavity proper, and in many cases the anterior vestibule. It excludes the nasopharyngeal passage (App. Figs. 2D, 4, 5; Witmer 1995a, 1997a,c; Ruben et al. 1996).

Ruben (1996), Ruben et al. (1996, 1997b, 1999), and Ruben and Jones (2000) argued that the dimensions of NA and RC offer a means to reconstruct the energetics of extinct amniotes, one so powerful they may constitute a "Rosetta Stone" or "bellwether indicator" for diagnosing paleometabolics. They observed that RC are absent in bradymetabolic reptiles, and are usually well developed in tachymetabolic birds and mammals (App. Fig. 5A,J). It was further observed that RC minimize the loss of water and/or body heat that occurs during the extensive breathing associated with high RMR (Geist 2000). The combination of causal links and empirical correlation led to the formation of the hypothesis that extinct animals lacking RC were probably bradymetabolic, whereas those possessing these structures were probably tachymetabolic. Ruben and company further presented data that suggested that the cross-sectional area of the NA is larger in birds and mammals than in reptiles. They explained this pattern as resulting from the requirement that the NA be broad to accommodate the high rates of airflow associated with HiRMR. Because they do not breathe very hard, reptiles can get by with slender NA. However, there are reasons to doubt whether nasal passage dimensions and RC development meet the paleometabolic criteria of preservability or consistency of correlation. The anatomy of avepod dinosaur NA is detailed in Appendix 3.

#### Correlation and causality

There are birds and mammals that lack well-developed RC. In some terrestrial birds, the anterior conchae are much less well developed than the middle conchae, and in many cases neither is very large (App. Fig. 5B–G, Bang 1966, 1971). The simple anterior conchae of kiwis and albatross are especially notable (App. Fig. 5F,G). Note that in some birds, the middle conchae sit immediately above the internal nares (App. Fig. 5A–D,F). RC are poorly developed in some raptorial birds (App. Fig. 5B,C). They are also poorly developed or even absent in various marine birds, whose external nares are nearly or entirely closed (App. Fig. 5D,E). Most of the diving marine birds that have blocked-off external nares may be able to drink copious amounts of water to compensate for high rates of respiratory water loss (Ruben et al. 1996, Geist 2000), but this is not always true of emperor penguins. They continuously brood their egg up to four months in a habitat where all water is frozen and thus thermally expensive to ingest (Groscolas 1990). The penguins' fasting MR is one-half the normal level, but remains well above reptilian levels. The respiratory water and heat loss experienced by obligatory oral breathers is not known, but the simple RC of penguins with open nostrils keep respiratory water loss low (Murrish 1973). Water conservation is as effective in pigeons (Geist 2000) as it is in birds with better-developed RC (Schmidt-Nielsen et al. 1970, Withers et al. 1981, Withers and Williams 1990). Savanna ground hornbills are interesting in that the beak is not able to close fully (App. Fig. 3g). The lack of a secondary palate may force hornbills to be habitual oral breathers and regularly bypass the RC, unless the glottis connects directly with the internal nares.

RC are absent in whales, and are not well developed in spiny anteaters, bats, elephants, saiga antelope, and most primates (App. Fig. 5H,I; Matthes 1934, J. Scott 1954, Negus 1958, Coulombe et al. 1965, Dorst 1973, Bhatnagar and Kallen 1975). Note that unlike tube-nosed birds, no tube-nosed mammal has RC in its anterior airway. In whales, a specialized respiratory water loss reduction system that may not be applicable to land animals is operative. Little data on respiratory water and heat loss is available for terrestrial mammals lacking well-developed RC.

It has been suggested that the long, narrow nasal passages in elephant proboscises help compensate for their poorly developed RC (Sikes 1971, McFadden et al. 1985). RC reduction in primates cannot be ascribed solely to snout reduction. Even in short-faced primates and snoutless humans, the anterior nasal cavity is large enough to accommodate RC as large and complex as those of other mammals. Male baboon snouts are as long and large as those of large canids and small ungulates, and their anterior nasal passages are very capacious, yet their RC are simple and do not project fully into the main airflow (App. Fig. 5I). The presence of baboons in semi-arid habitats where humidity is low and water and fruits are scarce (especially in the dry season, when 90% of baboons' diets may be dry grasses [Nowak 1999]) shows that poorly developed RC are compatible with high MR even in dry climates. Lizards without RC recover one-third of exhaled water (Murrish and Schmidt-Nielsen 1970), so even poorly developed RC may increase water savings to levels acceptable for more energetic animals.

The poor development or absence of RC in some birds and mammals lends support to the suggestion by Nagy (in Fischman 1995) that RC may be a useful but noncritical adaptation for tachymetabolic animals. The correlation between NA dimensions and MR is even less credible. Ruben et al. (1996, 1997b) plotted the cross-sectional area of NA as a function of body mass in a limited number of tetrapods. This included just four birds, of which three were large ratites and only one was a neognathous. This inadequate sample did not capture the full range of NA dimensions present in Aves. In addition, they sampled only one section of each nasal passage, thereby failing to capture any variation in NA dimensions within species. An expanded sample of birds (Appendix 3) includes the intra-individual variation in NA cross section when present (App. Fig. 4A). The double data points plotted for some specimens inhibits regression analysis. Instead, zones of separation and overlap are the primary basis for conclusions in this examination. The results show that most birds fall within the tachymetabolic zone observed by Ruben et al. (1996). This verifies both their results in part, and the methodology for calculating NA dimensions and body masses used here (see Appendix 4A).

However, because the avian sample in this study is six times larger than that in Ruben et al. (1996, 1997b), and because the anterior NA is much narrower than the posterior section in some birds, much greater variation is observed in this study. Cross-sectional areas of airways vary up to twentyfold within birds of a given body mass. Of particular interest are ratites, which tend to have narrower NA than other birds, particularly recent and living island ratites. The broadest section of moa and kiwi NA plot between those of other birds and of reptiles. The narrower NA of moa result from their beaks being little larger than those of ostriches, even when their bodies are up to twice as massive (App. Fig. 3d,o). The NA values for moa can be brought up to the modern ratite level only by ballooning the nasal capsules far outside of the external nares (App. Fig. 3d), an improbable condition not observed in a completely preserved moa head, or in Maori sketches of moa (Anderson 1989, Naish 1998). In kiwis, the posterior NA is constricted by hyperenlargement of the nasal capsule (App. Fig. 3F). Because kiwis have such extremely slender beaks, their anterior NAs fall *well below the reptile line* (App. Fig. 4A). The maximum-minimum NA dimensions of kiwis therefore straddle the reptile mean. Tube-nosed seabirds also have very narrow anterior NAs (App. Figs. 4x, 5G, Bang 1966). Another terrestrial bird with a severely constricted nasal passage is the roadrunner. Although its RC are very well developed and highly ossified, the internal nares are very narrow slits as little as 0.7-mm wide. The plot cannot include those marine diving birds, particularly adult penguins and cormorants, in which the external nares are entirely closed (App. Fig. 5E). The anterior nasal passages of long-nosed insectivorous spiny and giant anteaters mimic those of kiwis in being so slender they are in the reptilian range. The posterior NAs of spiny and giant anteaters are the narrowest so far observed among mammals. The anterior nasal passages of elephants are remarkably narrow, being only 40 mm across in a large bull (App. Fig. 4A). The hypothesis that tachyaerobic tetrapods must breathe through broad NAs is falsified.

The connection between RC and MR is further complicated because the development of the former may be influenced by nonmetabolic factors. Among these is the thermoregulatory function of cooling large brains on hot days or after exercise. Although brain cooling is not needed constantly, it is critical because brain overheating is soon lethal (Schmidt-Nielsen 1964, Taylor and Lyman 1972, Baker and Chapman 1977, Baker 1979, 1982, Bernstein et al. 1979, 1984, Kilgore et al. 1981, H. Johnson et al. 1987, Schroter and Watkins 1989, Witmer and Sampson 1999). Note that the large size of mammal brains may require a large nasal apparatus for adequate cooling. A need to empty the nasal cavity for use as a resonating chamber in the vocal primates may help explain the poorly developed RC in such primates. RC may indirectly improve olfaction by warming, humidifying, and slowing air before it reaches the olfactory region (Negus 1958, Schneider and Wolf 1960, Williams et al. 1989). There may be a correlation between RC development and olfactory performance (Gross et al. 1982, Bang and Wenzel

1985), but the correlation is not always consistent (Negus 1958, Bang 1971). RC may serve to protect the deeper respiratory tract from noxious gases, particles, and microbes (Bang and Bang 1959, Schreider and Rabbe 1981, Harkema et al. 1987, Schroter and Watkins 1989, Williams et al. 1989).

**Preservability**

The other major problem with utilizing nasal structures and dimensions is their preservability, or lack of same, in the fossil record. RC have never been reported in fossil birds. Taken at face value, this would indicate that birds were bradyaerobic until just before modern times!

RC are always supported on thin sheets of cartilage and/or bone that are sometimes very variable in form, and may be small (Matthes 1934, J. Scott 1954, Negus 1958, Bang 1966, 1971, Dorst 1973, Bhatnagar and Kallen 1975). As a result, they are rarely preserved in fossils, and even when preserved may go undetected. Therefore, the presence of RC is difficult to demonstrate or exclude, and must often be inferred from the characteristics of the NA and the bones that make up the passage. This is a serious problem, because these characteristics are highly variable and inconsistent in tetrapods with and without RC (Negus 1958, Bang 1971, Hillenius 1992, 1994, Witmer 1995a, 1997a,c). For example, the vestibule is very short and empty in crocodilians, whereas in many birds it is long, involves many bones including the maxilla, and contains a concha.

In synapsids, RC are often supported by distinctive ridges associated with the nasolacrimal duct, and the presence of such ridges can be used to infer the presence of RC (Hillenius 1994). The problem with detecting RC in fossil archosaurs is more acute. RC are usually made entirely of soft tissues in birds (Bang 1971; Fig. 2B in Ruben et al. 1996). Birds usually lack supporting ridges for RC. The latter may be anchored on smooth, flat surfaces of bone, or even on the cartilage spanning the external nares. Ruben (1996) listed a number of features of bird nasal passages often associated with RC, including the presence of an osseous ventromedial "schwele" associated with the lateral nasal ducts, and the shape of the maxillary palatine process. However, these features are either not consistently present in birds with RC, or are not preservable in the fossil record. Nor has it yet proven possible to know whether any or all of these characters were associated with RC in non-avian archosaurs. This is true because the avian nasal passage is strongly modified from the general archosaur pattern. In particular, expansion of the premaxilla and orbit at the expense of the maxilla has strongly telescoped and rotated the avian nasal passage (Buhler 1981, Witmer 1995a, 1997a).

Because of the above problems, Ruben et al. (1996, 1997b) turned to using the breadth and length of archosaur NAs to try and restore the presence or absence of RC. This poses a series of additional problems. Because the slopes that characterize the relationship between NA cross-sectional area and mass are steep, the data for both functions must be accurate. Yet this requirement is inherently difficult to fulfill even in living tetrapods, much less fossils (see Appendices 2 and 3A). That NA walls are not always fully ossified is an additional problem; that they are almost invariably crushed to some degree in fossils is yet another problem.

The problems continue. While most RC are contained in broad, long NA, this is not always the case. The anterior NAs of albatross are as slender as those of reptiles, yet they contain a simple anterior conchae (App. Fig. 5G). The situation with kiwis is even more remarkable. Although their anterior NAs are narrower than those of reptiles, they contain a simple but very long anterior conchae (App. Fig. 5F; Bang 1971)! Therefore, even extremely slender NAs can contain RC. Appendix Figure 3B shows that reptile NAs, in particular those of crocodilians, are not always shorter than those of birds, so this is not a reliable means of assessing the presence of RC. The correlation between NA dimensions and RC presence is inconsistent and unreliable. Thus, there is currently no means by which to establish or refute the presence of RC in fossil tetrapods.

The NAs of baso-theropods and coelophysoids have not been measured. The claim by Ruben et al. (1996, 1997b) and Ruben (1996) that the NAs of dromaeosaurs and *Archaeopteryx* were short and lizardlike was refuted in Appendix 3. Instead, most averostran dinosaurs and basal birds had a long, L-shaped NA, in which most of the anterior section was a horizontal tube, and the poorly ossified posterior section was vertical and potentially broader. Ruben et al. plotted the cross-sectional area of the NA as a function of body mass for a limited number of dinosaurs, including two averostrans. Their results implied that the NA was as narrow as that of reptiles, and could not contain RC. However, they plotted only the anterior NA. An expanded sample (App. Figs. 3A, 4) includes both the anterior and posterior sections of the NA as measured in Appendix 3. Note that the values for the posterior section are inherently tentative.

The new anterior NA data is in broad agreement with that of Ruben et al. (1996), which again verifies both their results in part and this study's methodology. In deep-snouted dino-averostrans, the bony anterior nasal tubes were relatively broader than those of kiwis. Those of the tube-snouted ornithomimids were like those of kiwis in being well below the reptile line. The posterior NA is important in that the space available to contain the vertical passage was potentially as much as two to four times broader than the anterior section, because it was in a deeper and broader part of the skull. The posterior NAs of small-headed ornithomimids were small, but may have been significantly broader than the anterior NAs of kiwis.

Although the bones involved in forming the sections of dinosaur and bird NAs are different—to the point that it is not clear whether the anterior tube was the vestibule and/or part of the nasal passage proper—the configuration of a slender anterior tube and a potentially much broader posterior NA in dino-averostrans was broadly similar to the condition observed in kiwis and tube-nosed seabirds. The dinosaurs' narrow anterior tubes may have contained simple RC similar to those seen in these birds. The potentially broad posterior NAs of most averostran dinosaurs were a potential site for complex middle conchae set directly above the internal nares in a birdlike manner. It is not possible at this time to assess whether selective forces had worked to favor the available space being filled largely by RC or by sinuses. The possible arrangement of RC, if they were present in dino-averostrans including *Archaeopteryx,* is presented in Appendix Figure 2A,B.

More birdlike than other dino-averostrans, including even *Archaeopteryx,* are deep-beaked oviraptorids (App. Fig. 3O). A vertical and potentially capacious NA is similar to that of that of big-beaked birds with big, deep beaks (diatrymids, phorusrachids, toucans, shoebills, and hornbills), whose external nares are set so posteriorly that the NA is subvertical (App. Fig. 3g,s). As per deep-beaked birds, the NAs of other deep-beaked oviraptorids may have contained well-developed RC.

Briefly considering some other dinosaurs, the extremely narrow external nares of diplodocid sauropods are no more indicative of a limited aerobic metabolism than are the similarly constricted nostrils of elephants (contra Hengst et al. 1996). Instead, the enlarged NA characteristic of sauropods was compatible with the presence of well-developed RC (Witmer and Sampson 1999). The very long anterior NA of lambeosaurine hadrosaurs is proportionally as slender as that of kiwis, and may have contained simple RC. The greatly enlarged anterior NA of hadrosaurines is compatible with the presence of well-developed RC (Witmer and Sampson 1999). The latter researchers also observed the presence of enlarged NAs and bony evidence for RC in ankylosaurids. A sectioned, slightly crushed NA of a nodosaurid is long and capacious, and appears to be rather mammal-like (App. Fig. 2A). The lateral boundaries of the NA are uncertain (Witmer 1997a), so two possible values are plotted. Because both values are well above reptilian levels, there was room for well-developed RC if the nasal lining was thin. The preserved NAs of ceratopsids are extremely capacious anteriorly and at midsection, all the more so because ceratopsid skulls were oversized relative to the body (App. Fig. 2A, Witmer and Sampson 1999). The posterior NA was narrower, to what extent is not yet documented. Very large anterior conchae could have been set on the cartilage spanning the enormous external nares, and the well-developed and complex rostral septum may have helped divert airflow to them. Taken as a whole, preserved dinosaur NA cross-sectional area varies up to fortyfold for a given mass.

### Summary

The problems with using RC to diagnose paleometabolics are illustrated by the isolated juvenile tyrannosaur skull emphasized by Ruben et al. (1996, 1997b). Its body mass could have been from 350–650 kg, the crushed anterior NA plots on the reptile line, the estimated posterior cross section falls on the bird line, and the preserved nasal passage is on the whole larger than that of a similarly sized bird (App. Figs. 3A, 4e,T, 9). Does this mean that the dinosaur had a higher RMR too? Or are nasal dimensions and structures less diagnostic than Ruben et al. (1996, 1997b) concluded? Also consider a hypothetical archosaur, whose entire NA is as slender as the anterior NA of the kiwi. Can the presence of simple RC be excluded in the fossil, given that RC are present in the kiwi? If a simple RC were present in the fossil's slender NA, was the animal's RMR as high as in a kiwi, penguin, or baboon, as low as in a reptile, or somewhere in between?

Or consider the situation with fossil birds. Should the current absence of preserved RC in extinct carinates be taken as reason to hypothesize that *Ichthyornis,* teratorns, and ancient ducks had LoRMR, or should it be ascribed to the often cartilaginous nature of avian RC? If the latter is postulated, why cannot the same be considered a possible reason for the failure to find RC in predatory dinosaurs?

The data presented in Ruben (1996), Ruben et al. (1996, 1997b), and Ruben and Jones (2000) was in some cases inaccurate, and in other cases inadequate because the database was too small. The results of the present study effectively falsify a number of conclusions in the two previous studies. There is not a clear-cut separation of cross-sectional areas or length of RC containing nasal passages between reptiles on one hand and birds on the other; in particular, kiwis and elephants prove that narrow nasal passages can accommodate high rates of oxygen intake, even when the passages contain conchae. Nasal passages need

*Appendix Figure 9.* Because *Aepyornis (bottom)* had a very small head, the potential space for respiratory conchae (solid white, see App. Fig. 3e,T) was comparable in size to that of a big-headed juvenile *Tyrannosaurus* (*top;* body size estimated). Drawn to same scale; scale bar equals 1 m.

not be broad to accommodate functioning RC, or to permit high rates of airflow. The statement by Ruben and Jones that "the presence of respiratory turbinates in extant endotherms is *inevitably* associated with marked expansion of the proportionate cross-sectional area of the nasal cavity proper" (italics added) is therefore incorrect. The nasal passages of predatory dinosaurs and urvogels were not short, simple, or distinctly reptilelike. Reptile, dinosaur, and early bird NAs are not consistently shorter than those of birds and mammals.

Other conclusions by Ruben (1996), Ruben et al. (1996, 1997b), and Ruben and Jones (2000) cannot be verified or falsified. It remains uncertain how well developed RC must be to have an elevated RMR. No reliable anatomical feature for inferring the presence of RC in archosaurs has been identified. In particular, it cannot be demonstrated that the entire NA of most predatory and other dinosaurs was consistently different in width from those of birds and mammals. Therefore, RC cannot be firmly excluded from the extinct archosaurs on this basis. Although their conclusion that *Archaeopteryx* had nasal passages similar to those of other averostrans, it is not possible to quantitatively assess the actual dimensions of its NA.

Broad NAs and RC have so far been observed only in tachymetabolic animals, and there may be a causal link between RC and respiratory water savings and/or heat retention. Therefore, the demonstrated presence of these features in a fossil form may be diagnostic for an elevated RMR. It is interesting that the NA cross sections of less energetic ratites, monotremes, and edentates plot lower than more energetic relatives. This suggests that there may be graded correlation between NA breadth and RMR. However, other evidence suggests that the relationship between RC development, NA dimensions, and RMR is complicated, inconsistent, and possibly weak. For example, the RC of marsupials and tenrecs are often better developed than those of other mammals with higher RMR (J. Scott 1954, Negus 1958, Hillenius 1994). RC are simple, small, or even absent in other birds and mammals, including some examples that live in habitats where water is scarce. Some birds and mammals breathe through extremely narrow NAs, or have none at all. NA dimensions and RC are not related to exercise energetics because exercising animals can breathe through the mouth (Niinimaa 1983). The dimensions of the NA are fairly preservable and observable, at least in terms of determining maximum dimensions, but it is possible that the NA of the living animal was much narrower than the bones presently indicate. The observability of RC in fossils ranges form poor to nil. Neither RC or NA is directly linked to power generation; they are merely secondary adaptations to the consequences of having a certain metabolism. This further lessens their diagnostic value.

It is concluded that RC and NA are not "bellwether" or "Rosetta Stone" indicators, because they do not meet most of the criteria of good paleometabolic characters. Specifically, the absence of preserved RC is not necessarily diagnostic of a low RMR, because of the difficulty in determining the status of RC in many fossils, because RC are not always well developed in animals with high RMR, because of nonmetabolic influences on nasal passage form and function (such as relative head size, slender proportions of the snout, large sinuses and olfactory apparatus, and use of the nasal cavity for vocalization), because the alternative functions of RC are not yet well understood, and because extinct tetrapods may have evolved alternative systems. Short, narrow nasal airways are also not necessarily diagnostic of low RMR, because they are present in some birds and mammals.

It is therefore not surprising that RC and NA dimensions have not provided definitive information on the thermoenergetics of dinosaurs and early birds. Our current knowledge of the status of RC in all dinosaurs remains so poor that all conclusions pro and contra are speculative. Continued failure to demonstrate the presence of RC and broad NAs in these archosaurs will probably remain nondefinitive, because of the preservability problem. Finding these characteristics would constitute evidence for tachymetabolism. Witmer and Sampson (1999) proposed that the RC of dinosaurs were involved more with brain cooling rather than with elevated metabolism, but it is questionable whether the small brains of these dinosaurs would have required especially well-developed cooling surfaces. It is possible that at least some dinosaurs, especially the earlier forms with less complex NAs, had no or poorly developed RC, and were correspondingly mesometabolic. It is also possible that at least some dinosaurs had RC as well developed as ratites, and similar RMR. It is interesting that predatory dino-avepods tended to have much smaller NAs than herbivorous dinosaurs of similar size, and that the giant extinct predatory ground birds had abbreviated NAs (Fig. 15.3). The functional and energetic implications of this parallel development are not yet clear. The expansion of the maxillary sinuses may have taken precedence over NA volume and RC development in dino-avepectorans in the same manner that RC have been reduced in some birds and mammals in order to better fulfill other respiratory needs. The narrow NAs of ornithomimids may have resulted from an evolutionary tradeoff in which reduction of head size and beak width for feeding purposes was selected over optimal RC function.

That dinosaur nasal passages were different from those of birds does not constitute evidence that they lacked RC; it only makes the question of whether dinosaurs had the conchae more difficult to answer and therefore reduces the diagnostic value of RC. This is all the more germane, because the radical alterations to the avian snout were primarily associated with the development a specialized kinetic feeding system (Buhler 1981). Then the major alterations to the nasal passage observed in Cretaceous birds were therefore a secondary effect, rather than a primary adaptation for the development of RC (contra Ruben 1996). Instead, the RC may have had to be reconfigured to fit into the modified NA. The beginnings of this transition may be seen in the still very dinosaurian skull of *Archaeopteryx,* in which posterior elongation of the external nares may have forced any anterior concha to be anchored in part on the cartilage spanning the nostrils (App. Fig. 2Db). The great diversity of NAs and RC in modern birds shows that their co-evolution has been a long process.

## *Mouth Roof*

Tachyaerobes need an oral adaptation that allows breathing to continue even when food is present in the mouth, whether this be a vaulted palate, or internal nares that are set more posteriorly than in most reptiles. These characters are prerequisites for ElvMR, but they are not diagnostic for the same, because crocodilians and teiids have posteriorly placed internal nares. With a few exceptions, such as oviraptorosaurs, dinosaurs consistently had vaulted palates, and there is evidence for a secondary palate in a few dino-avepods (Appendix 3).



## *Pulmonary Cardiac Complex*

The anatomy and function of living and fossil diapsid respiratory complexes are detailed in Appendix 3.

### Modern amniotes

Reptile cardiorespiratory systems are in principle able to oxygenate HiRMR, but can achieve a MaxAEC only moderately higher than the RMR of birds and mammals (Ruben 1995, Thompson and Withers 1997). Reptile lungs are often very large, but this is because they are inefficient. Their inability to exchange large amounts of gas rapidly is a result of low blood-flow rates combined with relatively simple lung structure having low surface-area/volume ratios (Bellairs and Jenkin 1960, Duncker 1978, Perry 1983, 1989; Ruben et al. [1996] labeled such lungs "septate"). Most reptiles lungs are dead-end structures. The pulmonary diverticula seen in a few reptiles appear unable to boost the gas exchange capacity of the lungs; instead, they serve to inflate the body for display and other purposes. The same is true of the diaphragm-like septums present in teiids and crocodilians, or the pelvis-based diaphragmatic muscle pump of the latter. The inability of reptiles with laterally flexible dorsal columns (see above) to breathe when running has been at least partially overcome via a gular pump (Owerkowicz et al. 1999). Even the semi-four-chambered crocodilian heart has a limited capacity. It is concluded that the presence of a reptilelike respiratory and cardiac complex is diagnostic of nonavian-mammalian energetics.

Increasing respiratory capacity to higher levels requires upgrading internal lung complexity, boosting lung ventilation values, and increasing cardiac output. The high-capacity hearts of birds and mammals have only one systemic aorta and are completely four chambered, which prevents blood shunting between the two ventricles, so this configuration is considered diagnostic of nonreptilian energetics.

In mammals, elevated lung capacity is accomplished (despite dead-ending of their large lungs) via extreme internal complexity that greatly increases surface area/volume ratios (alveolar condition), an effective muscular diaphragm, vertical flexion of the dorsal trunk, and a heart that is not only completely four chambered, but so powerful that it has a high capacity. Bats have further boosted respiratory capacity by having unusually large lungs and hearts, and by having blood whose oxygen-carrying capacity is higher than that of other mammals and even of birds (Yalden and Morris 1975, Maina 1989, 2000). Air sacs and associated skeletal specializations are therefore not prerequisites for aerobically powered flight.

The lungs of birds remain "septate," and although less internally intricate than those of mammals, they are parabronchial organs that are much more complex than those of reptiles (King 1966, Duncker 1971, Schmidt-Nielsen 1972, 1984, McLelland 1989b, Scheid and Piiper 1989). Breathing during locomotion is unhindered, because the trunk and lungs are both semirigid. The small lungs are very efficient (gas exchange efficiency is approximately 25% superior to the mammalian lungs [Schmidt-Nielsen 1984]), because they are ventilated mainly by large air sacs that are operated by the ribcage, sternum, and pelvic apparatus.

Ruben (1996) and Ruben et al. (1996, 1997b) appeared to conclude that only a fully developed avian system of the type seen in flying birds—including a sternum so long that it extends posteriorly to under the pelvis and is connected via ossified sternal ribs to true ribs adorned by ossified uncinate processes, plus an internally complex, parabronchial, unidirectional airflow lung—can oxygenate avian-level energetics. This view is implausible on theoretical grounds. The extremely large sternum probably evolved primarily to support extremely large flight muscles (Olson and Feduccia 1979). The ossified sternal ribs firmly attach this major muscle support to the ribcage, and ossified uncinate processes further strengthen the ribcage, producing a strong skeletal box that can resist the stresses of flapping flight while operating the respiratory tract. The presence of such an enormous sternum and associated features inevitably affected the respiratory system. Not only did the sternum and associated structure have to be incorporated into the respiratory system, it is possible that as the longosterna-rib complex developed, flying birds took advantage of the large lever it provided to increase air sac ventilation and thereby boost gas exchange capacity to very high levels. These, however, are flight adaptations, not requirements for achieving a basic mesoaerobic or supra-aerobic AEC in land-bound tetrapods.

In empirical terms, the assumption by Ruben and company that a flying-bird–level respiratory complex is needed to oxygenate a HiAEC when lungs are "septate" is falsified in three ways. First, Brackenbury and Amaku (1990) demonstrated that blocking up to 70% of the air sacs of domestic fowl reduced MaxAEC (and altered blood gas composition), but AEC could still be sustained at mesoaerobic levels sufficient to power walking speeds of more than 3 km/h. In addition, the abdominal air sacs are regularly disabled by hyperenlargement of the gonads in breeding females, yet flight performance is not critically impaired (Brackenbury pers. comm.). Even more important is the absence of long sterna, ossified sternal ribs, ossified uncinate processes, and well-developed posterior air sacs in various supra-aerobic birds. Most informative are the flightless examples, especially ratites. The very short sterna of kiwis, emus, and elephant birds have not kept them from being highly energetic. Neither have the reduced abdominal air sacs of kiwis and cassowaries: the oxygen consumption of a running cassowary matches that of a flying bird (Grubb et al. 1983). Maina (2000) emphasized the redundancy and overcapacity of the avian respiratory tract. Nonflying tachyaerobes would not need such a well-developed system.

Nor is it obvious that fully developed parabronchial unidirectional ventilated lungs are critical to boosting respiratory capacity significantly above reptilian levels. Up to 20% of some birds' lungs have bidirectional airflow (Duncker 1971, Scheid and Piiper 1989). This suggests that increasing the parenchymal density of a crocodilian-like lung, expanding its nonvascularized terminal passages, and narrowing the tubular chamber walls could increase the effectiveness of the cross-current gas exchange system, and significantly raise the system's gas-exchange capacity (Perry 1992, Hicks and Farmer 1998a,b).

Flightless birds are, therefore, the best living archosaurs for reconstructing the correlation between respiratory tract form and function and thermoenergetics in the ancestors of birds. However, because flightless-bird pulmonary complexes evolved from those developed earlier for an advanced level of flight, even their systems are probably overdeveloped for what would be needed in a supra-aerobic avepod that neither flew nor descended from early flying ancestors. For example, the extreme reduction of the trunk vertebrae and ribs, and the large size and ossification of the sternum and its connecting ribs even in ratites, are probably retained flight features rather than critical adaptations for oxygenating a HiMR.

The existence of some or all of the skeletal adaptations indicative of the presence of a nonreptilian air sac complex are diagnostic of ElvAEC. Because the connection to aerobic power production is direct, the diagnostic value is ranked as high. In general, the more associated features that were present, the better developed the complex probably was, and the higher the AEC was likely to have been.

### Dinosaurs and basal birds

Turning to dinosaurs, there has long been a broad consensus that dinosaur hearts were four-chambered double pumps, although it has not always been clear whether the ventricles were thought to be completely separate (Hotton 1963, 1980, Bakker 1968–1980, Ostrom 1970, 1980b, Regal and Gans 1980, Ricqles 1980, Paul 1988a, Paul and Leahy 1994, Ruben 1995, 1996, Reid 1997). This is in accord with phylogenetic bracketing following the presence of four-chambered hearts in crocodilians and birds, but this analysis does not establish whether the heart was a completely four-chambered double pump. Such a heart has reportedly been preserved as a three-dimensional fossil in an ornithischian dinosaur (P. Fisher et al. 2000), but this claim is controversial.

Phylogenetic bracketing indicates that predatory dinosaurs must have had nonaveolar ("septate," Ruben et al. 1997a) lungs as per their reptilian ancestors and avian descendants (App. Fig. 10A). Randolph (1994) argued that the development of an air sac complex in birds indicates a simultaneous rise in MR, but she did not discuss the presence or absence of preavian respiration in dino-avepods. As shown in Appendix 3, the skeletal anatomy of theropods conclusively refutes speculations that they had a crocodilian-like pelvovisceral pump (contra Ruben et al. 1997a). Baso-theropods probably had a respiratory complex that differed from those of modern reptiles, but was nonavian. Dino-avepods developed an increasingly sophisticated, preavian pulmonary complex that included well-developed diverticula, of which the largest were probably air sacs that ventilated increasingly unidirectional flow lungs.

A critical question is why the preavian system evolved in avepod dinosaurs. Were the pneumatic bones that resulted from their invasion by pulmonary diverticula a means of saving weight for predation, large size, and/or flight? This is a poor explanation for multiple reasons. Aerated bones first appeared, and then became highly developed, in small flightless avepod dinosaurs. Many giant animals and flying bats lack pneumatic skeletons. Weight reduction for flight and giant size are therefore not the reasons thin-walled bones appeared. At most, thin-walled bones were adaptations for these purposes, but even this view is challenged, because the percentage of total body mass made up by the mass of the skeleton is similar in birds and in mammals (Prange et al. 1979). If anything, it is maladaptive for predators that regularly struggle with their prey to evolve thin-walled bones that are more vulnerable to impact injury, and such bones have not appeared in any reptilian or mammalian hunters. The conclusion is that the aerated skeleton of avepod dinosaurs is an extraordinary adaptation that evolved despite its structural implications for small flightless predators, and that there must have been a compelling selective reason for the invasion of bones by diverticula in nonflying predators. The reason appears to have been largely related to increased respiratory capacity during exercise. The energetic implications of the evolving preavian system of theropods is as follows (see Appendix 3 for corresponding assessment of skeletal respiratory adaptations).

### Basal archosaurs, protodinosaurs, and baso-theropods

Aside from the evidence for a modest increase in the ventilation capacity of the ribcage, and the lack of any evidence for air sacs, it is difficult to reconstruct the exact lung form, function, and capacity of these archaic dinosaurs. The almost certainly bidirectional lungs were probably flexible or semiflexible in baso-theropods. They may have been internally simple, as in reptiles, or intermediate between reptiles and birds in complexity. Diverticula may have been absent or weakly developed in baso-theropods. Although it is probable that the heart was more completely four chambered than in fully bradymetabolic/aerobic crocodilians, it is possible that it was not entirely so, especially in protodinosaurs. Maximum respiratory capacity may have been at the high reptilian level, or mesoaerobic. Supraaerobic AEC can be ruled out.

### Cerato-saurans

The probable presence of weakly developed air sacs suggests that internal lung complexity at least matched and probably exceeded that of the most intricate reptilian examples. Airflow of the still large, partially flexible lungs was probably bidirectional: any unidirectional flow was probably incipient. The heart was probably completely four chambered at this point. Maximum respiratory capacity very probably exceeded the highest reptilian level. It may have been in the lower to higher mesoaerobic range, less possibly the lowest supraaerobic level.

### Tetanurans and avetheropods

The probable presence of well-developed air sacs that ventilated small, rigid lungs indicates that the latter were complex parabronchial organs with at least partial unidirectional airflow. Respiratory capacity should have been well above reptilian levels, and was probably in the higher mesoaerobic (especially the most basal members) and/or lower supraaerobic levels. This is in accord with the conclusion by Bramble and Jenkins (1998) that the evolution of birdlike respiration in predatory dinosaurs should have resulted in a corresponding enhancement of their oxygen consumption.

### Avepectorans (including basal birds)

The bellows action of the hinge-jointed sterna may have further increased respiratory capacity, particularly in dromaeosaurs with long plates (App. Figs. 7F, 10B). So may have the ossified sternal ribs and

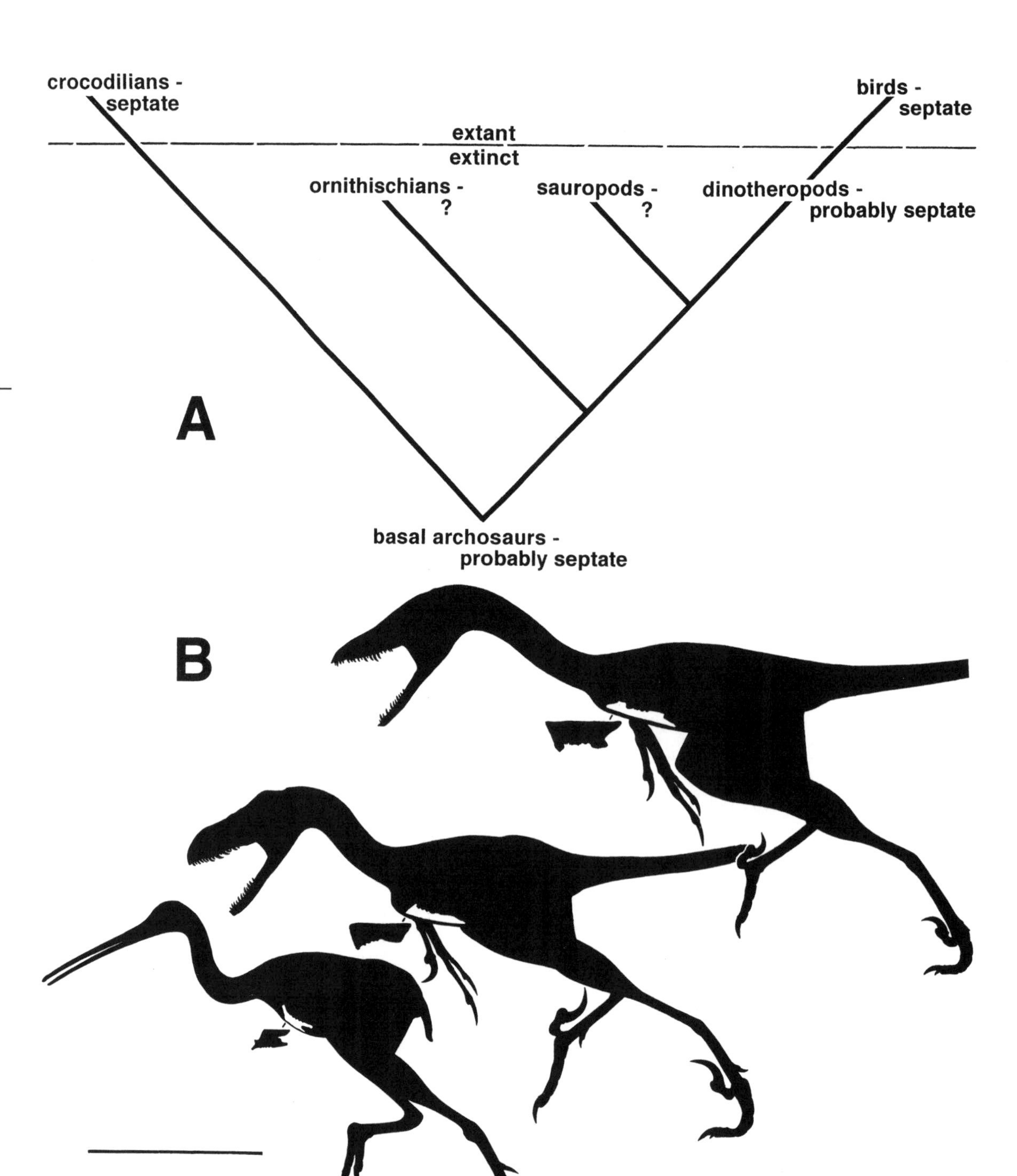


*Appendix Figure 10. A,* phylogenetic bracketing of internal lung anatomy in dinosaurs. The presence of "septate" lungs in living crocodilians and birds makes it nearly certain that theropod dinosaurs had the same. The lack of descendants of herbivorous dinosaurs precludes restoration of their lung configuration via this methodology. *B,* same-scale comparison of kiwi (*top*, Fig. 8.3) and basal dromaeosaurs *Sinornithosaurus* (*center,* Fig. 1.6) and juvenile *Bambiraptor* (*bottom,* Fig. 1.7), showing that the sternal plates (shown in dorsal profile and in place) of the dinosaurs were about twice as long as, and therefore reached more posteriorly than, that of the similar-sized bird. Scale bar equals 200 mm.

uncinate processes seen in many examples. All these features may reflect metabolic adaptations related to powered flight and its loss. Articulated hyoids are scarce in this group, but the slender, anteriorly placed set in *Archaeopteryx* (Pl. 7) suggest a gular pump was absent. AEC was probably supra-aerobic, albeit not as extremely high as seen in modern flying birds.

**Dino-avepod gular pump**

The basic avepod dinosaur breathing system may have been augmented by a gular pump in at least some examples. Robust hyoids at midthroat in an articulated juvenile specimen (Figs. 1–3 in Sasso and Signore 1998) suggest a throat-breathing apparatus; perhaps this was a condition retained from more basal diapsids. However, such hyoids have not yet been reported in adult theropods, whose bony hyoids tend to be more gracile and anteriorly located (Colbert 1989). It is possible that the gular pump was a retained juvenile feature lost with maturity.

**Other dinosaurs**

Because of the lack of living descendants, it is not possible to use simple phylogenetic bracketing to discern the internal structure of nontheropod dinosaur lungs (App. Fig. 10A), whose skeletal respiratory adaptations were briefly discussed in Appendix 3. Prosauropods may have had a poorly developed air sac system, so their respiratory capacity may have been mesoaerobic. The sauropod air sac complex appears to have been well developed. If so, a boosted AEC is probable, but it was probably modest in the manner of similarly slow elephants. The presence of air sacs implies that prosauropod and sauropod lungs were "septate," but were probably more internally complex than those of reptiles. The moderately sophisticated respiratory complexes that have been attributed to non-ornithopod ornithischians may have boosted respiratory capacity well into the mesoaerobic range, especially if the lungs were alveolar to some extent. The probability that ornithopods had a well-developed and possibly mammal-like diaphragm strongly implies that the lungs were alveolar. In this case MaxAEC was probably supraaerobic.

## *Summary of Cardiorespiratory Adaptations*

At this time, RC and NA dimensions do not provide definitive evidence on the resting and exercise metabolics of dinosaurs and early birds. Oral cavity adaptations imply elevated breathing capacity. More definitive results are provided by the restored condition of the pulmonary complexes in early birds and their ancestors. The respiratory capacity of protodinosaurs and baso-theropods was too limited to oxygenate an avian-level AEC. The preavian air sac complex that evolved in avepod dinosaurs indicates that their MaxAEC approached or reached the lower avian level.

## Feather Insulation

Both ectotherms and endotherms may lack insulation in the form of feathers or fur. This is true of all reptiles, and some mammals. Among the former are multitonne equatorial examples, in which bulk insulation is operative. Suids and humans are more modest-sized endotherms that lack integumentary insulation in warm climates. The nearly hairless southern Asian naked bat (Nowak 1999) shows that even flying endotherms weighing less than 200 g do not require insulation to compensate for their high surface-area/mass ratios.

Ruben (1991) and Ruben and Jones (2000) argued that feather insulation is compatible with LoMR ectothermy and perhaps even homeothermy, a view supported by Regal and Gans (1980), Randolph (1994), Chinsamy et al. (1995), Feduccia (1996), and Martin and Czerkas (2000). The example of basking birds using direct solar radiation to raise body temperatures (also see Kortner et al. 2000) does not constitute evidence that ectotherms can be insulated. For one thing, the birds are tachymetabolic endotherms, so their basking only shows that insulated HiMR endotherms can use basking to raise body temperatures a few degrees. Specifically, the modest temperature rises observed in basking birds, less than 10°C (Ruben 1991, Kortner et al. 2000), are not sufficient for bradymetabolic ectotherms, which need to raise body temperatures by 10–30°C or more on a daily basis to be active and emergent (Bellairs 1970). Ectotherms cannot be insulated because they are then too decoupled from the environmental heat they critically depend upon. Nor does insulation confer homeothermy when the MR is low (A. Bennett 1991). The only potential benefit that reptiles can gain from a pelage or plumage is a thermal shield from high external heat loads. This is apparently not worth the loss of heat-gain potential, because no hot climate reptile has featherlike insulation. This is probably why considerably more than ten thousand insulated living tetrapod species are ElvMR endotherms, and none are LoMR ectotherms (some immobile altricial bird chicks that cannot thermoregulate are insulated, but this situation is not applicable to active juveniles and adults). Some ectothermic, bradyaerobic insects are insulated with furlike structures, but these insects are so small that gaining heat from the environment is not significantly hindered.

Absence of superficial insulation is not diagnostic of any thermoregulatory or energetic condition in land and flying tetrapods weighing more than approximately 150 g, as long as they live in warm climes. Below that size, or in colder climates, absence of insulation is probably indicative of ectothermy, or at best mesothermy. I conservatively conclude that extensive body feathers are strictly limited to, and therefore diagnostic of, endotherms with intermediate or high rates of internal heat generation. In this regard, the presence of extensive feather insulation may come as close as any single character to being a paleometabolic "Rosetta Stone" that—unless strongly contradicted by other data—marks its owner as a certain tachymetabolic endotherm. This is true even though feathers are not directly connected to energy production. Until recently, the most serious problem with feathers as a paleometabolic character has been the paucity of samples, but this is changing.



In view of the above facts, Ruben's (1991) assertion that "*Archaeopteryx* feathers probably reveal little about its thermoregulatory physiology" rejects the crucial effect insulation must have upon the thermoregulation and metabolism of an organism. Furthermore, to conclude that the absence of potentially unpreservable RC (which are weakly developed in some tachymetabolic birds and mammals) is a bellwether indicator, while dismissing the value of preserved insulation (which is consistently and logically correlated with endothermy), is a reversal of sound logic. Therefore, Ruben's (1991) statement that "a fully feathered *Archaeopteryx* could have been either ectothermic or endothermic" is a speculative, radical, and unsubstantiated conclusion. Instead, ElvMR endothermy is fully compatible with and easily explains the presence of insulation in avepods. Ergo, that at least some small predatory dinosaurs and early birds were fully insulated on the head, neck, body, tail, arms, and legs (see Chapter 5) is compelling evidence that they were not ectotherms (Ji et al. 1998, Xu, Wang, and Wu 1999), but were instead endotherms with high rates of internal thermogenesis and relatively stable body temperatures (Paul 1988a, Bock and Buhler 1995, Griffiths 1996, 1998/2000, Chen et al. 1998, Schweitzer et al. 1999, Xu, Wang, and Wu 1999, Rensberger and Watabe 2000). That the insulation appears to have been hollow fibers, well suited for using trapped air as an insulating medium, reinforces the conclusion that insulated avepods were not ectothermic. Griffiths (1996, 1998/2000) noted that the plumulous base of the isolated feather often assigned to *Archaeopteryx* is evidence of insulation and endothermy. We do not yet known when insulation first appeared in small dinosaurs, or exactly how widely distributed it was. The possible absence of insulation in various small dinosaurs is not definitive, because they weighed over a kilogram and lived in relatively warm locales. Griffiths (1998/2000) suggested that prior to the evolution of feathers, theropods relied on fat for insulation, but this may have not been necessary if feathers were present early in the group.

## Reproduction and Growth

I discuss the manner in which living amniotes and avepectoran dinosaurs produce eggs and nest and brood in Chapter 10.

### *Eggs*

Randolph (1994) observed that birds avoid the weight problems associated with mammalian viviparity and lactation by quickly producing large, yolk-rich eggs, and then feeding the chicks with unprocessed food. She goes on to cite this as evidence that avian endothermy evolved along with flight, not before. Why the difference between avian and mammalian reproduction is indicative of dramatically different metabolic histories is not clear. After all, diapsid and synapsid reproductive systems have been markedly divergent since the Paleozoic, and it is probably not possible for bats to break away from the mammalian standard. Nor did Randolph analyze egg type and parental feeding in avepod dinosaurs. The latter also produced large, presumably yolk-rich eggs, and did not nurse their young. *If* such a combination is indicative of elevated metabolic rates, then this applies to the dinosaurs as well as their descendants.

It is interesting that avepod dinosaurs were like their avian relatives in being able to produce eggs that were dozens of times larger than those laid by reptiles (Chapter 10). This implies a difference between avepods and reptiles in the energetics of the mother and/or embryos, but this subject has not been investigated. In 1998, Norton argued that if "dinosaur eggs were buried in sand or vegetation, the highly effective avian gas exchange apparatus could not have developed." This is disproven by megapode fowl that deeply bury their eggs in mounds. And not all dino-avepods buried their eggs.

### *Incubating*

Eggs that are not incubated with body heat are deposited in warm places that offer adequate protection from the elements. If the eggs are not to be incubated with body heat, then it is maladaptive to leave them exposed to extreme fluctuating temperatures. Eggs that are exposed to the weather therefore constitute evidence of incubation via brooding (Norell and Makovicky [1999] do not offer a viable nonincubating explanation for the birdlike situation with oviraptorid nests). Note that brooding to protect exposed eggs from excessive heat or rain, or from predators, is a secondary action that occurs only because the eggs had to be exposed to allow body heating in the first place. A few reptiles and many birds body brood eggs, but do so in very different ways. A snake broods by coiling its sinuous body around a pile of eggs and using low-frequency contractions of the long thoracic muscles to generate heat that is transferred to the eggs. Reptilian brooding is therefore a form of aerobic exercise that raises heat production to a level similar to the RMR of tachymetabolic animals. Bird eggs are not laid in a pile, and the bird sits among or atop the egg(s), covering it or them with its body, wings, and tail. Body heat generated by an ElvRMR, plus elevated thermogenesis when conditions are especially cold, is transferred to the eggs, often but not always via a bare brooding patch. Avian brooding behavior is an extension of resting energy production augmented by metabolic cold response (in this case shivering), and is assessed as being diagnostic of an ElvRMR.

The brooding posture of at least some avepectoran dinosaurs—body set amid a ring of partially exposed eggs, and the arms and tail draped over the eggs—appears to have been a preavian arrangement that allowed both soil and body heat to be used for incubation. In the best-preserved oviraptorid nest, all eggs were in at least partial contact with the body of the brooding adult (Pl. 15A, Fig. 10.16; Clark et al. 1999, Horner 2000, contra Carpenter 1999). The probability that the body and appendages sported feathers may have allowed for complete coverage of the eggs—feathers complete the coverage of eggs in large ratite nests—and the feathers would have trapped body heat. Bare brooding patches that enhanced the release of body heat would have been advantageous, but they are absent in ratites. The incubating dinosaur could regularly rotate its body around the nest to more evenly heat the eggs (but eggs in large ratite nests do not necessarily receive equal heating; del Hoyo et al. 1992–99). The short, rigid, lightly muscled (Paul 1987a, 1988a, 2000) trunks of dinosaurs were poorly suited for using low-frequency muscle contractions to generate body heat in the manner of brooding snakes. It is therefore probable that dinosaurs warmed their eggs with general body heat via an ElvRMR, augmented when needed by shivering cold response (Horner and Dobb 1997, Horner 2000, Varricchio and Jackson 2000). The nesting behavior of birdlike dinosaurs is not compatible with classic reptilian energetics, and favors the presence of a preavian reproductive thermal physiology.

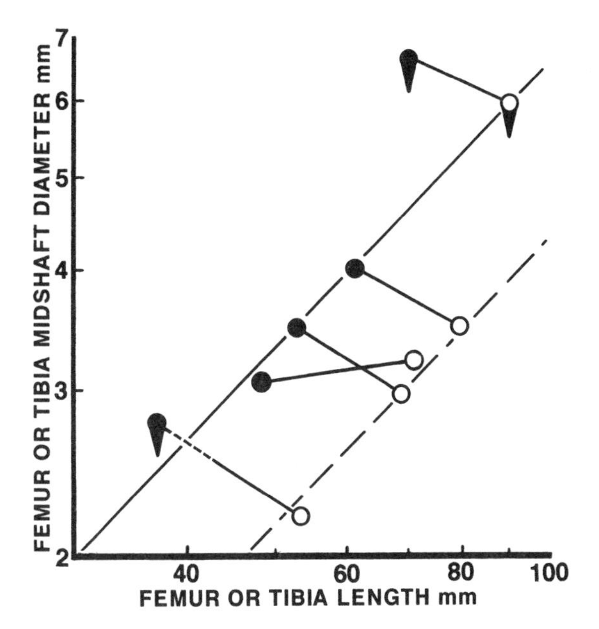


*Appendix Figure 11.* Comparison of femur *(solid circles)* and tibia *(open circles)* midshaft diameters as a function of total femur or tibia length in *Archaeopteryx*. Femur and tibia measurements of each specimen are linked by a bar, and the arrows indicate elements that have been overbroadened by crushing, as indicated by extensive longitudinal cracking and flattening of the bone. Note that minimum shaft diameters of the right and left elements often differ in the same specimen because of crushing and that the most robust bones are crushed regardless of overall size. Crushing and differential rotation of the bones around the long axis hinder comparison of bone proportions in the specimens. The solid and dashed lines indicate isometric scaling for femora and tibiae, respectively, relative to HMN 1880. The available evidence suggests that *Archaeopteryx* hindlimb elements remained essentially isometric with growth. Femur lengths and diameters and tibia lengths and diameters, respectively, are as follows (in mm): JM 2257—37, <2.8, 52.5, 2.2; BSP 1999—48, 3, 71, 3.2; HMN 1880—52.5, 3.5, 68.5, 3: BMNH 37001—60.5, 4, 80.5, 3.5; S6—approximately 70, <6.5, 89.5, <6.0.

## *Proportional Bone Ontogeny*

Houck et al. (1990) suggested that the energetics of *Archaeopteryx* were birdlike, because its long-bone growth was negatively allometric, as in birds. However, within given hindlimb bones, growth was actually isometric in the known specimens (App. Fig. 11), as in reptiles. The energetic implications of long-bone growth in *Archaeopteryx* are not clear. We do not know whether metabolic status and the proportions of long-bone growth influence each other. Bone allometry may be more closely related to the presence of determinate versus indeterminate growth.

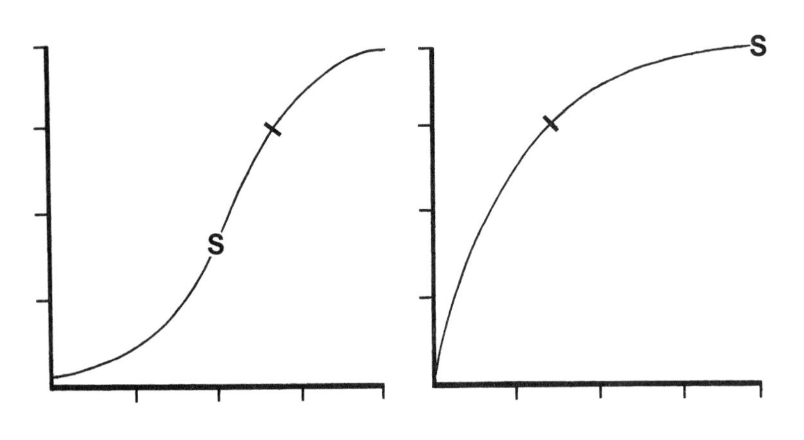


*Appendix Figure 12.* Growth curves, S-shaped versus sigmoid, for two hypothetical animals that start growth at the same mass and cease growth at the same mass and age. In both curves, a simple linear calculation of the growth rate (GR) between the start and cessation of growth gives the same result, even though the peak juvenile GR (JGR) of the sigmoid example is twice that of the S-shaped example. When one example matures while still growing rapidly and the other matures after completing growth, measuring growth as a function of age of sexual maturity (S) also gives a misleadingly low result. The best way to compare juvenile growth is to measure growth in both examples as a function of the majority of growth *(heavy bar)* before growth has begun the subadult plateau. When a full growth curve is available, peak JGR is estimated (for App. Fig. 14) by calculating values along the curve and choosing the highest value. When estimated JGR is calculated from age of maturity alone, a plausible growth curve is drawn, and a maximum value is included to give a maximum-minimum range.

## *Growth Rates and Interruptions*

### Estimating and comparing growth data

It is important to compare quantitatively the growth rates (GRs) of living and extinct tetrapods in a manner that is both equivalent for all groups and observable in fossils. In general, juvenile growth rates (JGRs) are the highest, subadult growth rates are much lower, and indeterminate adult growth (if it occurs) is very slow (Case 1978a). Quantitatively meaningful comparisons should only made between the individuals in the same fast juvenile growth stage, plotted as a function of average adult mass. Because various animals' growth curves are different, there may be no precise way to compare them (Ricklefs 1968, 1973, Case 1978a). In general, I follow Case (1978a) in plotting daily JGR as the highest value of rapid juvenile growth as a function of adult mass. Calculating JGR by simply dividing the difference between adult and initial juvenile mass by the total number of days of juvenile growth often understates the true value, because the value for the last, slower stage of subadult growth (and possibly even slower, indeterminate adult growth) is included. This pertains when the growth curve is S-shaped, as in most reptiles, and is even more significant when the curve is sigmoidal, as in most birds and mammals (App. Figs. 12, 13). When the growth curve is known, JGR is taken to be the main part of the growth curve, excluding slower subadult and adult growth rates. When the curve is not known in fossils, a range of estimates is made, assuming plausible growth curves. Although the results are not exact, they are adequate for broad comparisons between tetrapod types. Per-day JGR must be compared between animals of similar size: for example, a JGR of 250 g/day is in the high avian-mammalian range for animals of about 100 kg, but it is in the range predicted for multitonne reptiles (App. Fig. 14).

Farlow et al. (1995), Ruben (1995), and Ruben and Jones (2000) suggest that using mass at sexual maturity when comparing GRs. They do so because sexual maturity is a roughly equivalent stage of the life cycle in tetrapods, but it is not equivalent in terms of the growth period. At one extreme, body mass can be lost before sexual maturity (some birds); at the other extreme, many animals are still in rapid juvenile growth and at their minimum adult body mass when they begin reproducing. Between these extremes, many animals experience a period of slow, subadult growth before fecundity (Ricklefs 1968, 1973, Case 1978a, R. Andrews 1982). Therefore, using sexual maturity as the critical end-of-growth point arbitrarily lowers the GR values of animals that become sexually mature late in the growth period. Nor is it always possible to determine the onset of sexual maturity in extinct forms. For example, Ruben's (1995) comparison of GRs of large (more than 50-kg) *Troodon* specimens (that were probably adults that had completed most growth) to 30-kg subadult crocodilians (that are still growing rapidly) probably understated the overall GR of the dinosaur compared with the crocodilians, whose adult mass is hundreds of kilograms (App. Fig. 13A).

The method of using lines of arrested growth (LAG) set deep in the cortex of bones to age and construct growth curves for tetrapods fossil is based on the presumption that LAG were annual (Ferguson 1984, Reid 1987–1997, Chinsamy 1993, 1994, Chinsamy et al. 1995; see Horner et al. [1999] for an overview of the problems associated with LAG). If this is true, then ring counts give a minimum age of death, because the central rings are often lost to central bone loss or remodeling. The spacing of lost central rings must be estimated as a correction factor. The more regular the spacing between LAG, the

*Appendix Figure 13.* Body mass as a function of age. *A,* medium-sized amniotes; *B,* small amniotes. Wild alligators (both sexes prior to age 5 *[vertical lines* in *A, squares* in *B],* after age 5 females *[open ovals],* males *[solid squares]),* farm-raised alligators *(A);* estuarine crocodiles, both sexes *(vertical lines with horizontal tips);* oras *(K);* Aldabran tortoise *(T);* humans *(H);* kiwi (*small solid circles,* male and female, neither growth series complete); *Patagopteryx (p); Syntarsus (s).* M indicates male, F female for various species. Data sources include Roth (1965), Ricklefs (1968), Frith and Calaby (1969), Eisenberg and Gould (1970), Robinette and Archer (1971), Brody (1974), Bourn and Coe (1978), Webb et al. (1978), Fairall (1980), Pomeroy (1980), Widdowson (1981), Degen et al. (1991), and Scheidler and Sell (1997). Kiwi data were obtained courtesy of Rimlinger, ora data courtesy of Walsh and Birchard.

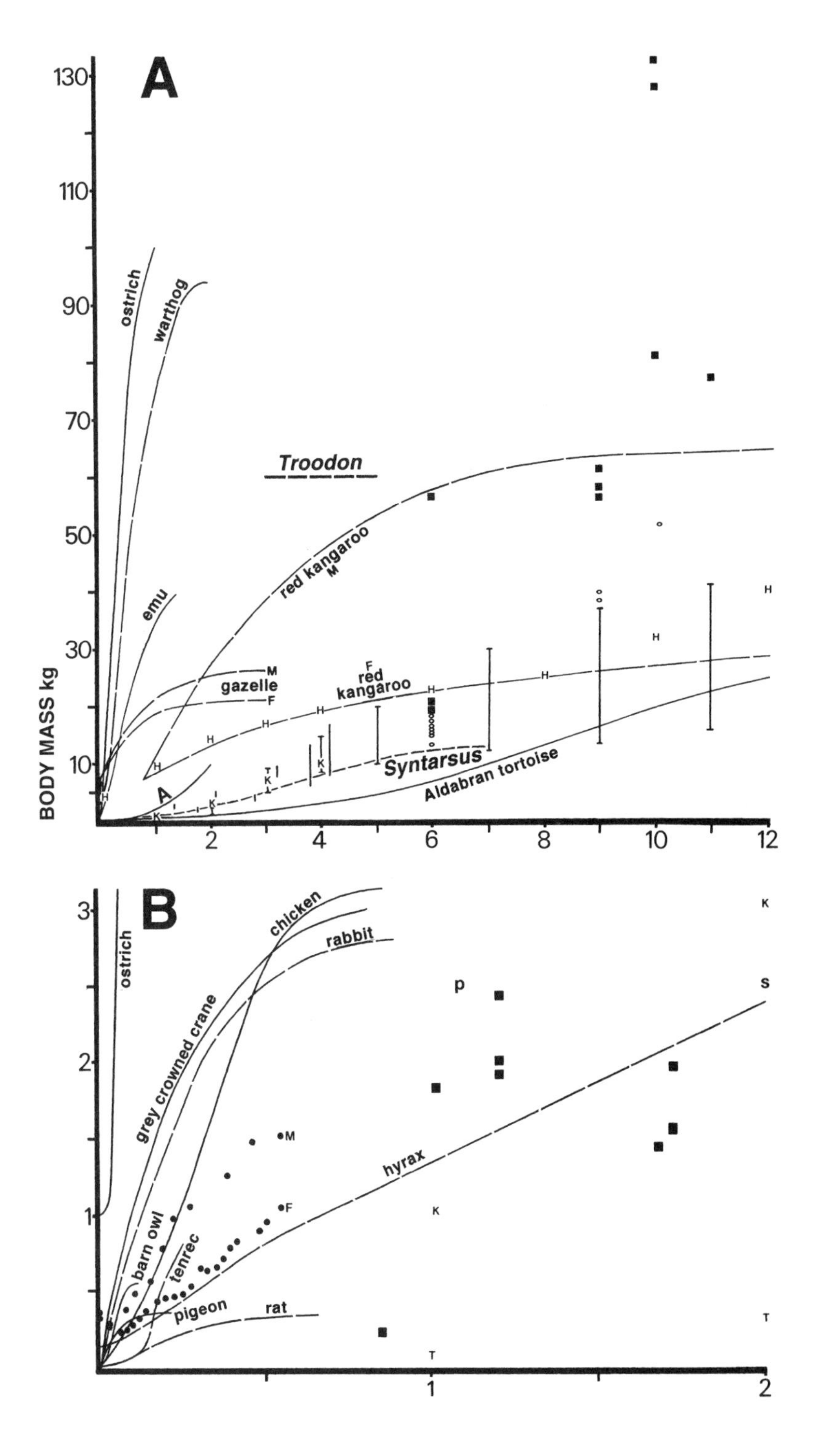


more likely they are to have been deposited on an annual basis. In a few reptiles as well as some birds and mammals, peripheral and deep-set LAG are, or probably are, multiannual or irregular annuli that cannot be used to accurately estimate the age of an individual (Lewis 1979, Nelson and Bookhout 1980, Zug et al. 1986). It has also been shown that in some animals, the number of deep LAG varies within the bones in a single individual (Reid 1996, 1997, Horner et al. 1999, 2000), which casts doubt on the assumption that the dinosaurs' rings are reliable age markers. The fact is that deep LAG deposition in mammals remains too poorly sampled and documented to know how to interpret their presence in fossils with confidence. I am not aware of any cases of LAG being regularly deposited on a less-than-annual basis, and no mechanism for such timing had been proposed. It is therefore concluded that—unless one can show that the rings are annual—a set of bone rings may not record the information needed to restore a growth curve, and actually gives only a maximum possible age for the subject at death. Again, the

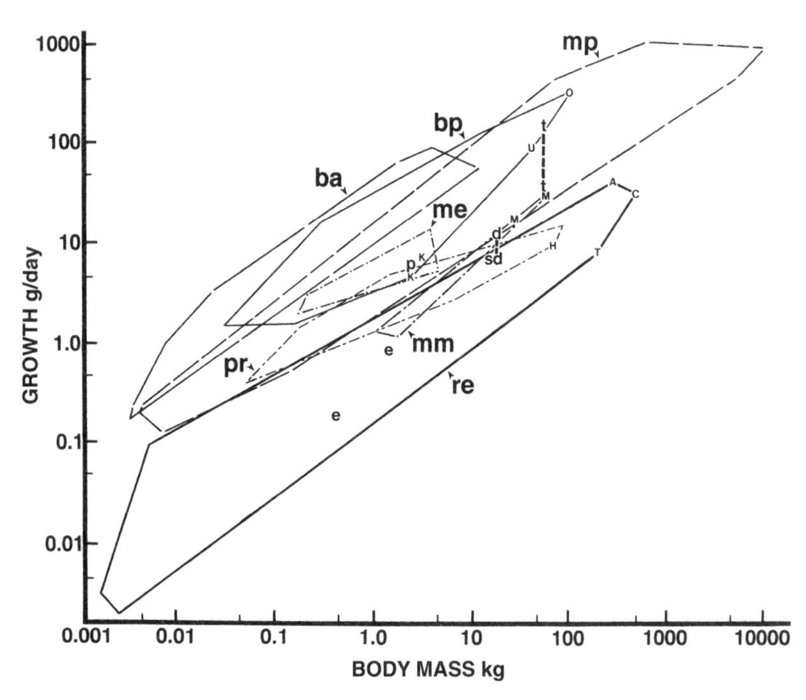


*Appendix Figure 14.* Daily growth rate as a function of adult body mass in nonmarine amniotes. Least-area polygons are shown for the following: birds, altricial *(ba);* birds, precocial *(bp);* marsupials *(mm);* eutherians *(mp),* excluding primates *(pr),* tenrecs, monotremes, and edentates *(me);* reptiles *(re).* Reptiles: alligator *(A),* estuarine crocodile *(C),* tortoise *(T);* mammals: human *(H),* red kangaroo (*M,* female and male); birds: brown kiwi (*K,* female and male); ostrich *(O);* emu *(U);* extinct birds: enantiornithines *(e); Patagopteryx (p);* dino-avepods: dromaeosaur *(d); Syntarsus (s); Troodon (t).* Data were obtained either from wild animals or from captives whose growth is similar to that of their wild counterparts; range of growth rate estimates are indicated for some dinosaurs. Growth rates calculated as per Appendix Figure 12. Data sources include Ricklefs (1968, 1973), Case (1978a), Andrews (1982), Appendix Figure 13, and text. This figure is a revised version of Figure 15.5 in Paul (1994b).

number of innermost rings lost must be estimated. This may be impractical if the unaltered cortical bone is limited to less than approximately 50% of the bone radius. This problem probably precludes aging fossils with very thin-walled bones.

### Modern amniotes

Arendt (1997) observed that it is usually to the advantage of animals to grow as rapidly as possible. Yet, as Case (1978a) observed, all continental reptiles grow slowly in the wild, more slowly than most mammals and birds. An updated data set confirms this conclusion (App. Fig. 14). At any given mass, growth rates vary by a few hundredfold. Horses grow many hundreds of kilograms in just 2 years, whereas wild crocodilians reach only a few kilograms in the same period. The fastest-growing large, wild, nonmarine reptiles yet documented are male alligators living on Avery Island (McIlhenny 1934). They reached 15 kg in about 4 years, 50 kg in about 5.5–8.5 years, and 70 kg in about 6.5–10 years (App. Fig. 13A). The fastest-growing individual reached 56 kg in about 6 years; another reached 160 kg in about 11 years. The JGR is calculated to have been 40 g/day. Published data on farm-raised crocodilians terminates at rather low juvenile masses, implying that captive JGRs can be more than twice the natural mean (Coulson et al. 1973). Inadequate data is available on the JGRs of wild oras (Auffenberg 1981). New data on captives (App. Fig. 13A) shows that growth is slow when food intake is kept modest in imitation of natural conditions.

Among wild birds, altricial chicks are the fastest growing. Ostriches also grow with remarkable rapidity for juveniles that feed themselves, emus less so. New data shows that the slowest-growing bird yet documented is the kiwi (App. Fig. 13B). Among wild birds and mammals, those with the lowest metabolics rates (kiwis, monotremes, marsupials, edentates, and tenrecs) often grow more slowly than more energetic examples. There are exceptions to these patterns. For example, precocial large ratite chicks are less energetic than nursing savanna mammals and semi-altricial children, yet the birds grow as rapidly as the savanna mammals and much more rapidly than humans.

In the wild, most birds and mammals grow faster than most nonmarine reptiles. Although the growth rate for *Alligator* is 50% higher than calculated by Case (1978a), crocodilians and giant lizards grow significantly less rapidly than all nonprimate, terrestrial placentals and birds of similar size (App. Fig. 14). There is a zone of overlap between the fastest-growing wild reptiles (especially crocodilians) and slowest-growing mammals (some primates, monotremes, marsupials, edentates, and tenrecs). For example, female red kangaroos do not grow more rapidly than alligators (App. Fig. 13A). However, the claim by Ruben (1995) that JGR of alligators exceeds that of marsupials manyfold is exaggerated: male red kangaroos grow faster than wild male alligators (App. Fig. 13A).

In this analysis, rapid growth at any given body mass is empirically defined as that above the maximum rate observed in wild terrestrial and freshwater reptiles. According to the data plotted in Appendix Figure 14, this scales as growth in g/day = approximately 1.6 $(kg)^{\sim 0.57}$. Rapid growth is divided into moderately rapid for those not far above the reptilian maximum, and very rapid for the fastest-growing animals. Slow growth is defined as that observed below the minimum observed in mammals, which appears to be 45% of the noncaptive reptilian maximum. Moderate growth covers the overlap between mammals and wild nonmarine reptiles.

What prevents reptiles from enjoying the advantages of rapid growth? The above pattern shows that rapid growth only occurs when high rates of food consumption and relatively stable body temperatures are present in juveniles (Case 1978a, Arendt 1997), and this requires the presence of tachyaerobic homeothermy in either the juveniles themselves or their caretakers. Tachyaerobic, homeothermic, self-feeding juveniles have the high activity levels needed to seek out and consume large amounts of food and keep themselves warm. Poikilothermic bird-chick parents are alphaendotherms that use body heat to keep their charges warm, and bring the latter enormous quantities of food at no energy cost to the juveniles. Much the same is true of farm-raised reptiles. Their human caretakers are alphaendotherms that also use solar and fossil energy to keep their charges warm, and supply the reptiles with amounts of food that bradyaerobes could not hope to gather on their own at so little cost in the wild. The ability of tachyaerobes to sustain the high activity levels needed to gather large amounts of food apparently more than compensates for their inefficiency in utilizing food for growth. Conversely, although their energy efficiency allows reptiles to dedicate a large portion of their energy intake to growth, they are apparently unable to sustain the high activity levels needed to gather enough food to grow rapidly. This also explains why adult land reptiles are unable to gather and carry food for their young.

Reid (1996) argued that the primary basis for rapid growth is "not high metabolic rates, but high circulatory efficiency." In this view, a circulatory system with a capacity higher than observed in reptiles may be necessary to transport the large amounts of nutrients and energy needed by fast-growing tissues. However, the example of poikilothermic bird chicks discussed above casts doubt on this hypothesis. Even if the hypothesis is true, the cardiac tissues must expand and do more aerobic work if the capacity of the circulatory system is to increase. The same is true of the respiratory and digestive systems that support the heart, and thus food consumption must rise correspondingly (see below). Therefore, increasing the MR is critical for powering the expanded circulatory system central to Reid's hypothesis: it is not the uncertain and incidental adaptation he implies. Reid (1987, 1990, 1996) further argued that fast-growing animals are not necessarily endotherms, but the HiMR associated with rapid growth should heat the body to the point that it is no longer critically dependent upon environmental heat.

As for attempts to compare the JGR of dinosaurs to those of farm-raised reptiles, no extinct tetrapod lived on a farm or a laboratory, and to compare the growth of a precocial baby dinosaur to that of a farm-raised alligator is no more logical than comparing the dinosaur to a well-slopped farm hog.

I conclude that the inability of all continental reptiles to grow rapidly under natural conditions indicates that bradyaerobic ectotherms are unable to do so. Therefore, rapid juvenile growth is diagnostic of ElvMR and some degree of homeothermy in juveniles and/or parents. Conversely, slow juvenile growth appears to be diagnostic of LoMR and strong heterothermy. Moderate JGRs are compatible with either energy level.

If growth at reduced but substantial rates continues well into adulthood, then the growth is termed indeterminate. If growth slows down to extremely low levels or ceases entirely with maturity, then it is determinate. Growth is indeterminate in most but not all reptiles, and determinate in most but not all mammals and birds (R. Andrews 1982, Hayflick 1995). Some male kangaroos never stop growing (Jarman 1989), and cessation of growth and typical age of death are essentially coincident in bull African elephants (Lindeque and van Jaarsveld 1993). Adult growth is therefore not diagnostic of any thermoenergetic status.

Interrupted growth that results in the deposition of well-developed LAG in the deep cortex of postcranial bones is common in reptiles (Ricqles 1980, Ferguson, 1984, Zug et al. 1986, Reid 1987–1997, Chinsamy 1993, 1994). Deep postcranial LAG in living birds have not been published yet, although the fact that most birds—including those that regularly enter torpor—complete growth in less than one year usually precludes the deposition of deep LAG. LAG is often absent in mammals, but has been reported in marsupials (Leahy 1991) and polar placentals regardless of whether they hibernate (Chinsamy, Rich, and Vickers-Rich 1998). Based on additional examples, Padian (1997b) and Horner et al. (1999, 2000) disputed a tight correlation between LAG and energetics in tetrapods. It is concluded that the presence of consistent, well-developed reptilian postcranial growth rings is suggestive of, but not proof of, heterothermy. The absence of growth rings is suggestive of homeothermy. The metabolic implications of the irregular presence of postcranial LAG, especially weakly developed deep rings, may be suggestive of a degree of thermal stability intermediate to reptiles and birds associated with betaendothermy, but may also be compatible with homeothermic alphaendothermy, especially in habitats with a pronounced cold or dry season.

**Dinosaurs and early birds**

Growth rates of nonavian dinosaurs and mesosternan birds have been estimated to have been as rapid as that of their bird relatives and mammals, as low as that of reptiles, or between the two groups (Case 1978b, Ostrom 1980b, Ricqles 1980, Reid 1987–1997, Paul 1988a, 1994b, Dunham et al. 1989, Chinsamy 1993, 1994, Varricchio 1993, 1997, Chinsamy et al. 1995, Farlow et al. 1995, Ruben 1995, Chinsamy, Rich, and Vickers-Rich 1998, Horner et al. 2000, Rensberger and Watabe 2000). The data available on predatory dinosaur GRs is much more extensive than it was just a few years ago, but remains limited and tentative. The growth patterns of protodinosaurs or baso-theropods have not yet been examined. The available data are plotted in Appendix Figures 13 and 14. To standardize results, a day is defined as 24 h, regardless of changes in the spin rate of the planet.

LAG in a growth series of femora have been used to calculate that the coelophysoid *Syntarsus* grew to adult mass (approximately 13 kg) in 7 years (Chinsamy 1993, 1994). The growth zones become somewhat thinner progressing outward, and a statistical analysis is compatible with annual LAG. The data plots on a reptilian S-curve, and the maximum estimated JGR is 6.0 g/day, a value that straddles the upper reptilian and lower mammalian ranges. The rather irregular spacing of the rings and their position between zones of highly vascular fibrolamellar bone indicate that the rings might have been multiannual, and growth more rapid. The presence of determinate growth may favor this possibility.

Reid (1990) observed 14 LAG in a small-animal bone, tentatively identified as the femur of a small dino-avepod (approximately 15 kg). The growth zones tend to become thinner progressing from the inside out. Because the cortex is thin, either a large number (30 or more) of deeper rings were lost during bone reformation, or initial growth to nearly adult size was rapid and the innermost preserved rings represent slower, late-juvenile growth. If the first option is correct, then growth was very slow, even for a reptile, and this appears to be contrary to the moderate vascularization of the growth zone bone. The irregularly spaced rings are therefore good candidates for being multiannual. It is not possible to choose between the options, and the growth curve and rate cannot be reliably estimated.

Reid (1993, 1997) used LAG in a tibia, probably assignable to a dromaeosaur (approximately 15 kg), to assign an age of maturation of 7 years, but 5 years is equally possible, assuming the rings were annual. In this case, roughly linear growth was an average of 6–8 g/day. A more realistic JGR might be 10 g/day. These values are in the upper reptilian and lower mammalian zone. A more complete growth curve cannot be constructed. The combination of growth-zone matrix that was more vascularized than typical in reptiles, but was not fibrolamellar, is compatible with a modest pace of growth. Considering the single sample and the irregular spacing of the LAG, it is possible that the rings were multiannular and growth was more rapid than calculated, a view that may be supported by the presence of determinate growth.

Varricchio (1993, 1997) tentatively assigned a maximum age of death of 3–5 years based on LAG in a small sample of bones from *Troodon*. By modern standards, *Troodon* was a large predator with long bones and a skeletal volume similar to that of a human male. The mass at death is modeled at 60 kg (corrected from the 50-kg estimate by Varricchio [1993]). Insufficient data has been published to plot a growth curve. Assuming a linear growth curve, mass increased an average of 33–54 g/day; assuming more realistic curves, the JGR values are approximately 50–150 g/day (compared with 27–47 g/day calculated by Ruben [1995]). *Troodon* appears to have outgrown wild, semi-aquatic alligators by factors of 2–6. Ruben's (1995) assertion that the dinosaur grew no more rapidly than modern reptiles was unfounded. Indeed, the estimated JGR equals or exceeds that of small ungulates, carnivores, and male red kangaroos, and may have approached that of large birds (note that lowering the estimated adult mass to 50 kg does not drop the value into the reptile range). If the irregularly spaced, rather weakly developed LAG set among well vascularized, fibrolamellar growth zones were multiannular, as Varricchio (1993) cautions they may be and as determinate growth suggests, then the JGR may have equaled that of large birds.

All of the above dinosaurs lived in warm climates. Chinsamy, Rich, and Vickers-Rich (1998) described nine LAG in the femur of a polar ornithomimosaur. Only the thin outermost zone of the bone is preserved, so it is not possible to restore a juvenile growth curve or rate. The peripheral placement of the zonal bone, and the rapid thinning of the growth zones progressing distally, suggest that only the last stage of juvenile growth and/or adult growth have been preserved.

Rensberger and Watabe (2000) concluded that the fine structure of ornithomimid bones is most similar to that of birds, and indicates similarly rapid growth.

Chiappe (1995) suggested that the linear size trajectory of *Archaeopteryx* specimens (up to approximately 500 g) hints that it grew slowly, because it is improbable that all the individuals died differing in age by just a few months. An alternative is that *Archaeopteryx* grew rapidly, became emergent while still only a fraction of adult mass, and that inexperienced subadults suffered high rates of mortality. Another alternative is that *Archaeopteryx* grew rapidly in the nest, and then grew slowly as a subadult. Or perhaps the specimens represent two or more species of different adult size rather than a growth series. The growth curves and rates of the urvogel cannot be restored at this time.

Chinsamy et al. (1995) used growth rings to estimate that some Cretaceous birds took many years to mature. Fibrolamellar bone was well developed and growth was indeterminate in *Patagopteryx* (approximately 2,500 g), which may have died at the age of 1 year or more. In this case, JGR was 6 g/day or less, which places it close to that of the kiwi, or perhaps in the upper reptile range. Two enantiornithines (approximately 500 and 1,200 g) lacked fibrolamellar bone, and were estimated to have died after at least 5 and 4 years of growth, respectively, in which case linear JGR were approximately 0.2 and 0.73 g/day or less, respectively. These values are entirely in the reptile range. Indeed, they are implausibly low, even for ancient birds. The GR estimates are not, in fact, reliable because all the bones are so thin walled that bone deposited during the most of the juvenile growth period is missing. That the birds are fairly large suggests they were adults or subadults, and 60–70% of adult mass may have been reached when the first preserved ring was deposited. It is therefore equally possible that growth to nearly adult size was slow or rapid, but the scarcity of juvenile remains is more compatible with the latter. It is therefore possible that the preserved rings and nonfibrolamellar bone record slow, indeterminate subadult and adult growth, rather than slow juvenile growth. Alternately the rings may have been multiannual. The last two alternatives are compatible with the irregular spacing of the rings. If the birds achieved the most of their final body mass in a year or less, then they were growing faster than continental reptiles of similar adult size (Horner et al. 1999). The bone microstructure of toothed ornithurines matches that of their modern relations (Chinsamy, Martin, and Dodson 1998). This indicates that juvenile growth was rapid, and adult growth was minimal or absent.

Growth rates of various prosauropods, sauropods, and ornithischians have been tentatively estimated to have been moderate to rapid, based on ring counts and size distribution patterns in large bone beds (Currie and Dodson 1984, Reid 1987–1997, Russell 1989, Chinsamy 1993, 1994, Horner and Dobb 1997, Padian 1999, Horner 2000, Horner et al. 2000). Growth was variably determinate and indeterminate in these dinosaurs. Rensberger and Watabe (2000) observed mammal-like fine structure in ornithischian bones, which they interpret as evidence of rapid growth. It has yet to be demonstrated that any dinosaur combined fully reptilian bone growth microstructure with slow JGR, and the growth rings that are present are usually less well developed than is typical in reptiles. Some dinosaurs do exhibit typically avian-mammalian growth microstructure (Chinsamy 1994, Reid 1997).

Because the calculated growth rates of dinosaurs and brevisternan birds are tentative, all the following conclusions are similarly tentative. The data suggests that some dinosaurs grew more rapidly than modern wild reptiles, and that others may have grown at moderate rates observed in both reptiles and mammals. Although a few examples of LAG may record slow annual reptilian growth, in each case it is possible that the rings were multiannual. At this time, therefore, there is no unambiguous example of a dinosaur that grew in a manner exactly paralleling that of reptiles in terms of JGR, growth curve, bone matrix, and LAG development. Some early prosauropod and theropod dinosaurs *may* have had reptile-like growth curves and rates, but even the lowest estimated JGRs fall within the minimum mammalian range. This is in accord with the dinosaurs' highly vascularized growth zones with extensive fibrolamellar deposition. Other advanced dinosaurs, including avepectorans, appear to have grown much more rapidly than reptiles. Some also deposited LAG, but they appear to have done so in a manner more like those observed in some mammals (LAG infrequently present and not as well developed as in reptiles, growth zones highly vascularized and fibrolamellar). Other dinosaurs appear to have grown in a very avian-mammalian–like manner in terms of growth curves and rates, and in having LAG-free bone made of extensive fibrolamellar structure.

The thermoenergetic interpretation of the data is as follows. Because the growth of no single known dinosaur exactly followed the reptilian pattern, it is concluded that the growth energetics and thermoregulation of wild reptiles do not adequately or fully explain those of any known dinosaur (Padian 1997b, Horner et al. 1999, 2000, Padian et al. 1999). However, the growth pattern of a number of dinosaurs—early forms from Triassic-Jurassic boundary times in particular—also do not appear to match those of most birds and mammals, so that their growth metabolics and/or thermoregulation may not have matched the avian-mammalian pattern. Presumably, energetics and thermoregulation intermediate to the reptilian and avian conditions were involved. Reptilian growth energetics and thermoregulation are still less adequate to explain the growth of more advanced dinosaurs, including the birdlike avepectorans (Horner et al. 1999, 2000, Padian et al. 1999).

Minimal requirements for rapid growth in precocial juvenile dinosaurs probably included AEC higher than those of reptiles and the ability to thermoregulate well enough to keep warmer when inactive than can young reptiles. Both conditions should have required that resting metabolic rates and energy budgets be higher than in reptiles, probably substantially so. The indeterminate growth observed in some dinosaur bones does not, as Feduccia (1996) claimed, imply that these animals were ectotherms. Feduccia also ignored the presence of determinate growth in small avepod dinosaurs (but this evidence is not definitive).

The evidence is sometimes ambiguous, but it is compatible with non-ornithurine birds growing about as rapidly as modern examples, and therefore having nearly or fully avian energetics (Horner et al. 1999). The determinate growth in Mesozoic ornithurines does not necessarily constitute evidence for a dramatic metabolic upgrade.

The moderate and inconsistent frequency of LAG deposition in various dinosaurs, including avepods, implies that RMR and/or metabolic cold response may not have not been fully developed in some examples. Such betaendothermy would have allowed body temperatures to drop sharply during torpor in dry or cool seasons, and stopped body and bone growth more often than in birds and mammals. But as Horner, Ricqles, and Padian (1999, 2000) emphasized, dinosaurian LAG are probably not inconsistent with alphaendothermy. Juvenile dinosaurs may have dedicated the available energy to maintaining thermal stability, rather than to growth, during harsh seasons, leading to LAG deposition while they were homeothermic. In other dinosaurs, the absence of LAG and avian-mammal–like growth curves are compatible with either marsupial- or eutherian-like growth energetics and thermoregulation. It is possible that low-latitude dinosaurs tended toward the former, and polar dinosaurs thermoregulated more like eutherians (Paul 1988c, 1994a). Chinsamy, Rich, and Vickers-Rich (1998) considered the presence of LAG in a polar avepod dinosaur to be evidence that it grew in a less birdlike manner than a LAG-free ornithopod from the same sediments. This may be correct, but the poor preservation of the juvenile-growth bone in the avepod hinders analysis. Altricial baby dinosaurs may have been poikilothermic, but this does not necessarily mean they were bradyaerobic, and their apparently rapid growth probably required that their parents be tachyaerobes able to forage for the large amounts of food for their charges.

Modest declines in resting metabolic rates (as much as 30%) are observed in birds and mammals as they mature (Brody 1974). It has been suggested that dinosaurs achieved rapid juvenile growth with high metabolic rates, and then experienced a dramatic decline to reptilian or nearly reptilian energy levels upon maturation (Ricqles 1980, Farlow 1990, Lambert 1991). Such extreme shifts in metabolism may not be possible. They would require a massive and implausible reconfiguration of cellular structure and physiology (Ruben 1995), and leave adult dinosaurs with an aerobic exercise capacity grossly inferior to that of their young. The latter would contradict the evidence that some juvenile dinosaurs moved in herds with the adults (see above).

## Teeth and Beaks, Lazy Flesh-Eating Dinosaurs, and Food Passage Rates

Randolph (1994) contrasted the loss of teeth in birds with their retention in bats. She suggested that this difference means that increased rates of food intake evolved along with the development of flight in the former, rather than before, as in mammals. Predatory birds replaced teeth with hooked bills and talons for dispatching and dismembering prey. Herbivorous birds were able to lose their teeth because an alternate food grinding system, grit or stone-rolling gizzards, were a common archosaur feature. The loss of teeth and masticatory jaw muscles was adaptive for avian flight because it reduced the mass of the head, thereby shifting the center of gravity aft in compensation for loss of the tail. Mammals lack beaks or thoracic grinding mechanisms, so bats were compelled to retain teeth. This basic difference between the two tetrapod types precludes knowing whether already-tachyaerobic mammals would, if they could, lose their teeth as they lost their tails while evolving flight. Thus, Randolph's hypothesis cannot be verified. Feduccia (1996) implied that, because predatory dinosaurs lacked the precisely occluding, heterodont teeth of mammalian carnivores, their food consumption should have been more like that of reptiles. This view is falsified by the example of predatory birds, which establishes that occluding, heterodont teeth are not necessary to eat enough flesh to stoke a high-energy budget.

Feduccia then went on to imply that predatory dinosaurs were ambush predators that spent so little time hunting that they had low food budgets. Big cats are also ambush predators that spend most of their time sleeping and resting (Paul 1988a), so this criteria does not meet the qualifications of a good paleometabolic character. Besides, various avepod dinosaurs may have been pursuit predators (Paul 1988a,d, 1998/2000, Holtz 1994c).

Feduccia (1996) further argued that predatory dinosaurs digested their food slowly enough that ingested bone was completely destroyed and thus is absent in their fossil coprolites. If so, then they were more like reptiles than birds and mammals. The high energy demands of birds and mammals compel food intake, passage, and assimilation to be rapid, so that intact bone fragments are evacuated in the feces. The three-decades-old source on bone composition in dinosaur coprolites cited by Feduccia was obsolete: dino-avepod scat had not yet been positively identified when it was published. A recently discovered megacoprolite attributable to *Tyrannosaurus* (Chin et al. 1998) contains numerous, sometimes almost pristine, bone fragments, indicating that food passage was rapid in the tachymetabolic manner (P. Andrews and Fernandez-Jalvo 1998, Bartlett et al. 1998). Even more extraordinary evidence for very rapid rates of food passage derives from the presence of undigested muscle cells (probably those of a herbivorous dinosaur) in another large coprolite assignable to a tyrannosaur (Chin et al. 1999)! At the other end of the avepod dinosaur size spectrum, the intestines of *Scipionyx* are short and deep in section, suggesting to Sasso and Signore (1998) that rates of food absorption were high. The fecal and soft-tissue evidence is fully compatible with high rates of food passage, but cannot be considered definitive.

## Bone Isotopes

Barrick et al. (1997) and Fricke and Rogers (2000) have used oxygen isotope ratios in fossil bones to discriminate between ectothermic heterotherms and endothermic homeotherms. Their success in diagnosing fossil reptiles as heterothermic, while other skeletons in the same sediments exhibited homeothermic patterns, supports the viability of their methodology. This especially applies to the Fricke and Rogers work, which sampled Late Cretaceous dinosaurs and crocodilians over a range of latitudes. The presence of homeothermy is diagnostic of ElvMR in animals too small to be inertiohomeothermic, especially when the pattern does not change with latitude. It is theoretically possible that a homeotherm gigantotherm had a LoMR in a region that did not experience a cool season, but this is not true in more strongly seasonal high latitudes. The presence of marked heterothermy indicates either a LoMR or strong torpor on a daily or seasonal basis. Barrick et al. (1998) used oxygen isotope ratios in fossil bones to estimate mass-specific metabolic rates. So far these studies have not investigated the thermodynamics of marsupials and other mammals with moderate energy budgets.

To date, almost all dinosaurs—including some small adults and juveniles—have tested as homeothermic by Barrick et al. (1997) and Fricke and Rogers (2000). The degree of homeothermy appears to have been too high to be fully compatible with LoMR in even the larger examples, especially as some of the large dino-avepods sampled dwelled at high latitudes with a pronounced cool season. Because the Mesozoic dinosaurs' degree of homeothermy was similar to that of birds and mammals living in more seasonal Cenozoic climates, Barrick et al. concluded that their level of homeothermy was somewhat less than the mammalian standard. Fricke and Rogers further concluded that their data was most compatible with avepod dinosaurs having body temperatures and MRs somewhat lower than modern birds and mammals (by which they appear to mean eutherians). However, they noted that differences in isotope ratios in the Cretaceous versus the present is partly responsible for the disparity in the data. These views are compatible with either a MR somewhat lower than in placentals, less developed metabolic response to cold, or both. One large armored dinosaur tested as strongly heterothermic (Barrick et al. 1997). It is not clear whether this reflects a LoMR, or hibernation during a high-latitude winter (see below). Barrick et al. (1998) concluded that the energy budgets of the dinosaurs they examined were intermediate to those of reptiles and suprametabolic eutherians. However, Barrick et al. (1998) underestimated the mass-specific MR of dinosaurs, in that they overstated their body mass by a factor of about 2 (Paul 1997a). Even so, bone isotopes provide the best evidence to date that the dinosaurs sampled may not have been as homeothermic as are alphaendothermic birds and eutherians.

## Size Squeezes

McNab (1978) hypothesized that HiMR endothermy evolves during a size squeeze. In this scenario, a degree of homeothermy develops at large body sizes due to inertiohomeothermy. As size decreases, total MR remains about the same, so mass-specific MR increases. Insulation appears to retain the abundant body heat despite a high surface/mass ratio. Voilà, endothermy has evolved! The size decrease that marked the evolution of mammals from therapsids is cited as an example of this effect. Because many dinosaurs were large, it has been argued that they did not experience the initial size squeeze necessary to induce endothermy (Reid 1987). The size decrease that accompanied the appearance of the first birds is, conversely, cited as evidence for the evolution of elevated metabolism and endothermy. More recently, in accordance with the view that even *Archaeopteryx* and other brevisternan birds were not endotherms, it has

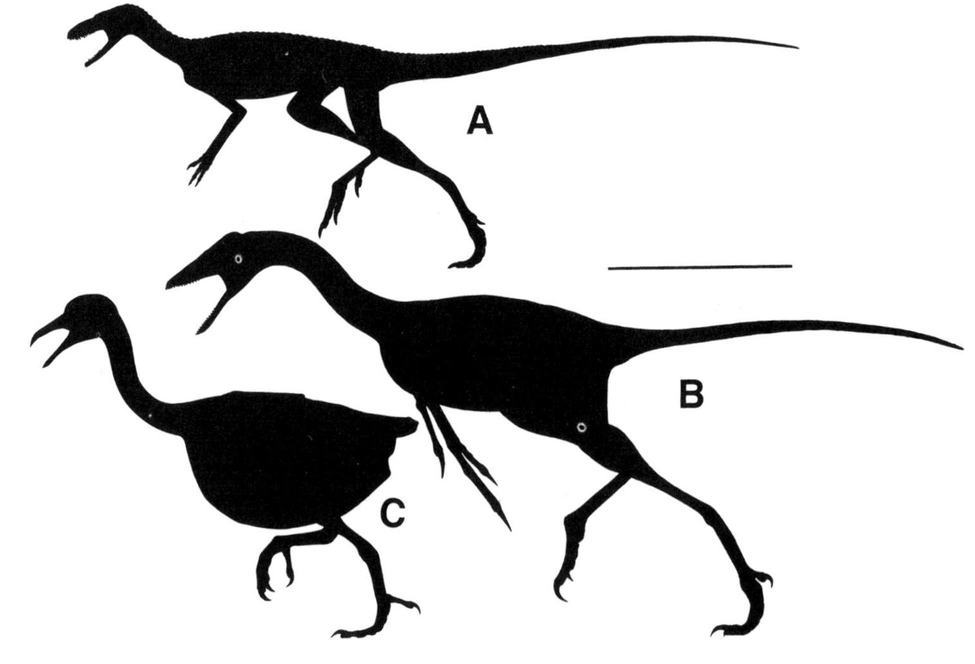


*Appendix Figure 15.* Archosaur size squeezes. *A,* protodinosaur *Marasuchus,* 180 g; *B,* early bird *Archaeopteryx,* 234 g; *C,* pigeon. Drawn to same scale; scale bar equals 100 mm.

been suggested that the continuing size squeeze observed among Cretaceous birds represents the fundamental size squeeze that resulted in avian endothermy (Feduccia 1996, Paladino et al. 1997).

The size-squeeze hypothesis is open to question on theoretical grounds. First, it is just a hypothesis rather than documented fact, so the absence of a size squeeze cannot be used to falsify ElvMR endothermy in a group that never experienced a size squeeze. In addition, the size-squeeze hypothesis assumes that the primary motive for evolving ElvMR and related traits is improved thermoregulation. The aerobic-capacity hypotheses do not require a size squeeze, because the advantages of increased aerobic capacity accrue at any size (Hayes and Garland 1995). Various reptiles groups have undergone size reduction without evolving higher MR or endothermy.

As for dinosaurs, it is an often-neglected fact that they *did* experience an initial size squeeze! The protodinosaurs from which the group evolved were quite small at only 200 g (App. Fig. 15). Whether this decrease in size from earlier archosaurs in of itself initiated the development of ElvMR endothermy is not clear, but the possibility is certainly open. The same can be said for the second dinosaur size squeeze during the origin of birds. In this view, the progressive size decline—from multikilogram dino-avepods, to kilogram-or-more dino-proavians, to subkilogram minidromaeosaurs and urvogels, to low-gram–class avebrevicaudans—was driven primarily by ecological requirements and diversification into new niches: thermoregulatory factors were, at best, secondary selective agents. Griffiths (1998/2000) suggested that a switch from inefficient fat to efficient feathers as insulation allowed and promoted the dinosaur-to-bird size reduction without a dramatic increase in an already ElvMR. This view will be supported if basal avepods prove to be unfeathered and falsified if they were feathered. In any case, the protodinosaur-protobird size squeezes discredit claims that dinosaurs and basal birds could not have been endotherms, and are compatible with their having been endothermic, but do not necessarily constitute firm evidence that they raised their MR.

## Brains

Bradyaerobic reptile brains are small and simple, whereas tachyaerobic birds and mammals have large, complex brains. This correlation has inspired some researchers to suggest that the high sustained-activity level associated with a high energy budget requires that the brain be large and sophisticated, and so small, simple-brained dinosaurs could not have been tachyaerobic (Hopson 1980, Feduccia 1996). Hopson logically extended this argument to conclude that those dinosaurs—in particular certain small avepods—which have brains as large and developed as those of birds had grossly similar activity and energy levels. Feduccia, however, denied that the bird-brained dinosaurs were as energetic as birds.

Relative brain size data presented in Appendix Figure 16 contradicts the brain-metabolism correlation.

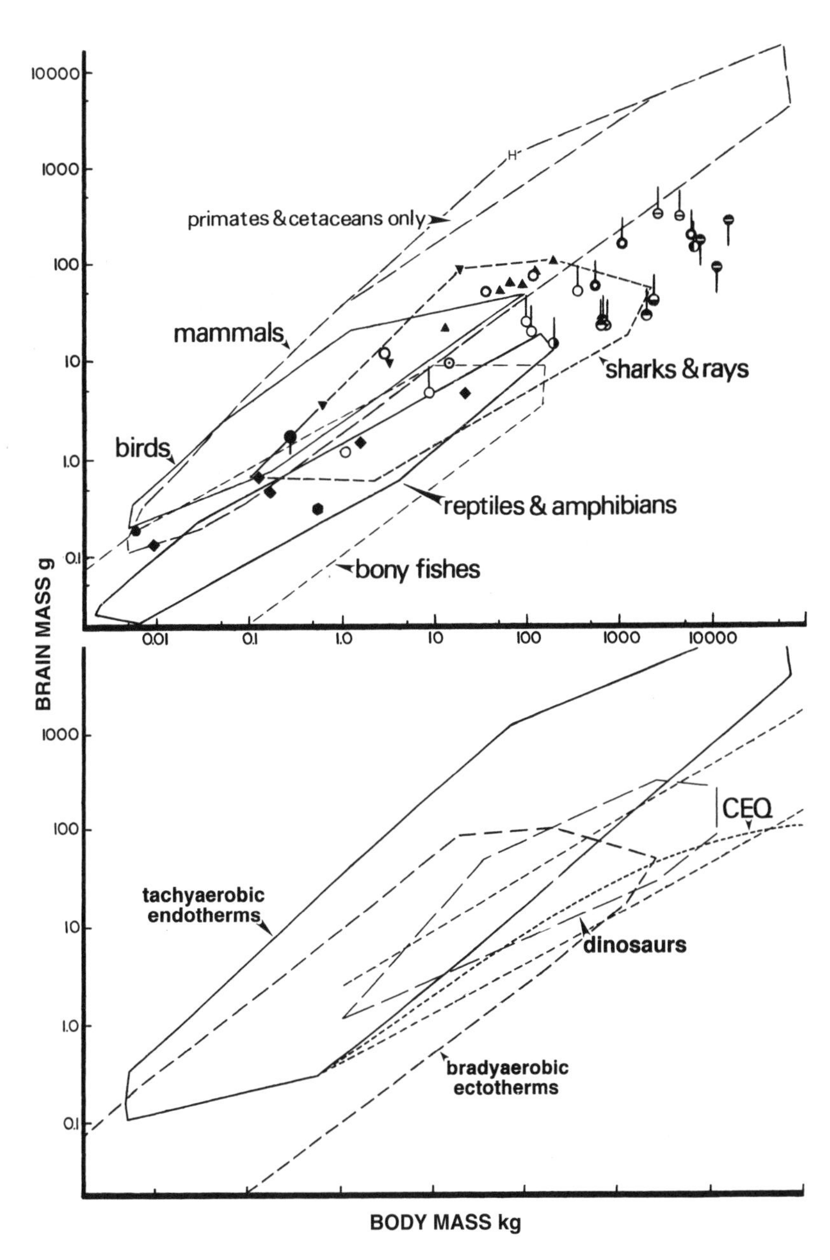


*Appendix Figure 16. Upper plot,* brain mass as a function of body mass in vertebrates. Humans (H); bradyaerobic myliobatiform rays *(inverted triangles);* reef sharks *(triangles);* elephant nose electric fish *(small hexagon);* tachyaerobic tuna *(large hexagon);* dinosaurs *(circles):* sauropodomorphs *(solid circles with horizontal white lines): Brachiosaurus* (body mass 20,000 kg, brain mass 140–280 g), *Diplodocus* (11,000, 45–90), *Camarasaurus* juvenile (6,500, 75–150), *Plateosaurus* (700, 25–50); ceratopsians: *Triceratops* (6,000, 150–300, *circles with solid right halves*), *Protoceratops* (164, 15–30, *circles with solid left halves*), *Euplocephalus* (2,300, 40–80, *circles with solid lower halves*); stegosaurs *(circles with solid upper halves): Stegosaurus* (2,100, 28–56), *Kentrosaurus* (640, 24–48); large ornithopods *(open circles with horizontal lines): Edmontosaurus* (4,100, 295–590), *Kritosaurus* (2,300, 300–600); small ornithopods *(very thin rings): Camptosaurus* (800, 24–48), *Tenontosaurus* (350, 45–90), *Thescelosaurus* (110, 12–24), *Dryosaurus* (100, 19–38), *Hypsilophodon* (9, 4–8), *Leaellynasaura* juvenile (1.1, 1.3); large theropods *(thick rings): Tyrannosaurus* (5,700, 200–400), *Allosaurus* (1.0, 168–336), *Ceratosaurus* (524, 62–124); small cerato-saurans *(circle with central dot): Syntarsus* (11, 9); avepectorans *(thin rings):* ornithomimid *Dromiceiomimus* (144, 70), *Troodon* (35, 49), *Sinornithoides* juvenile? (2.5, 13); *Archaeopteryx (solid circles),* (0.3, 1.1–1.7); pterosaurs *(diamonds): Pteranodon* (20 kg, 4.8), *Scaphognathus* (1.3, 1.7), *Rhamphorhynchus* (0.12, 0.7), *Pterodactylus* (0.18, 0.42), and juvenile (0.09, 0.14). For cases in which the brain may have filled only half the available volume, vertical lines indicate the potential range. *Lower plot,* demonstration of the extensive overlap of vertebrates with high and low metabolic rates. Note that extent of overlap would be even greater if larger tachyaerobic tuna and bradyaerobic myliobatiform rays were included. Note also that dinosaurs are either close to or above the corrected encephalization quotient (CEQ) value *(very short-dashed curve),* and scaling values observed in egg-laying animals (*short-dashed slopes* following 0.58 power, upper line is birds), when CEQ or 0.58 slope is set at the level of the lowest EQ tachymetabolic animals. Data are taken in part from Jerison (1973), Northcutt (1977), Hopson (1980), Martin (1981), Nilsson (1996), Van Dongen (1998), Ito et al. (1999), and Larsson et al. (2000).

The brains of the probably highly active and energetic pterosaurs were in the upper reptilian range. Even though some tuna are tachymetabolic and tachyaerobic (see above), their brains are very small. Some bradyaerobic sharks and rays have large brains, remarkably large and complex in the case of some relatively inactive bottom-dwelling rays (Northcutt 1977). Large ratites and large primates have grossly similar metabolic and activity levels, yet the brains of these birds are only a fraction the size and complexity of those of the mammals. There is no significant correlation between brain development and metabolic rate in mammals (McNab and Eisenberg 1989), or in vertebrates in general, and most researchers reject the brain-activity link (Hayes and Garland 1995). It appears, therefore, that small-brained animals can achieve the mental activity levels needed to be highly physically active simply by increasing the daily frequency of brain utilization needed to carry out a given food-gathering action. It is concluded that neither the reptilian brains characteristic of most dinosaurs, nor the more birdlike brains of a few of them, are good indicators of their metabolic condition.

## Mesozoic Climates

Ruben and Jones (2000) repeated a common belief of those who support ectothermic dinosaurs and early birds (such as Feduccia 1996) when they stated that Mesozoic "climates were so mild and equable" that ectotherms "were probably fully capable of relatively precise behavioral thermoregulation." Even in tropical regions, small or medium-sized reptiles are unable to maintain stable body temperatures on a daily or longer basis. This is true because daily temperature changes are substantial and because rain, especially when combined with wind, produces a powerful cooling effect. In the Mesozoic, a carbon dioxide–induced greenhouse effect and high, temperature-stabilizing sea levels did make Mesozoic climates much warmer and less seasonal than they are in today's interglacial period. This should not, however, be exaggerated into the unnaturally extreme warmth and lack of climatic variability that would have allowed bradyaerobes, especially small dino-birds, to be homeothermic. Sellwood et al. (1994) showed that the Mesozoic may have been somewhat cooler than is often thought. Daily and seasonal temperature fluctuations were strong enough to require tachyaerobic endothermy in order to maintain stable body temperatures (Barron and Washington 1982, Paul 1991, Barrick et al. 1997), especially away from the coastlines and at higher altitudes, where nightly and winter temperatures may have been quite low. Harsh winter conditions probably applied to higher latitudes as well, which brings us to the next subject.

## Polar Dinosaurs

Dinosaurs, including small and moderately large avepods, have been found at very high paleolatitudes both northern and southern (Paul 1988c, 1994a, Clemens and Nelms 1993, Chinsamy, Rich, and Vickers-Rich 1998, Constantine et al. 1998, Tarduno et al. 1998). Winter conditions included a chronic lack of the solar and ambient heat needed by ectotherms to operate normally. The severity of the cold has been a subject of controversy, and probably varied with time and place. Some polar dinosaur habitats appear to have been relatively mild, with reptiles present. Some researchers have cited mild Mesozoic polar conditions as being compatible with reptilian dinosaur energetics. However, there is evidence that some of the polar habitats were markedly harsher, with winters marked by frosts and even hard freezes as well as modest glaciation (Paul 1988c, 1994a and references therein, Stoll and Schrag 1996). It is significant that reptiles appear to be entirely absent from some of the polar sediments that produce a substantial number of dinosaur remains (Clemens and Nelms 1993): the indication is that the winters were too harsh for bradymetabolic ectotherms in these locales. The possibility that dinosaurs migrated away from the poles to reach warmer latitudes is implausible, because the great distances required are impossible for terrestrial bradyaerobes to cover, and are excessive even for tachyaerobes (Paul 1988a,c, 1994a, 1997b, Clemens and Nelms 1993, Chinsamy, Rich, and Vickers-Rich 1998, Constantine et al. 1998). This suggests that the thermal physiology of dinosaurs was in one or more respects superior to that of reptiles in coping with the chilly, dark arctic winters. Insulation and/or ElvMR were probably the critical adaptations. It is not certain whether small avepod dinosaurs remained active, or denned and hibernated during the winter. Chinsamy, Rich, and Vickers-Rich (1998) and Constantine et al. (1998) suggested that the presence of LAG in a polar ornithomimosaur is compatible with its having hibernated during the winter, in contrast to the continuous activity implied by the absence of LAG in a small ornithopod. However, polar bears do not hibernate, and yet their bones record periods of arrested growth. Larger dinosaurs were ill suited for torpor because they were too large to seek shelter from predators and the elements, frostbite being a particular danger. The exception was armored forms, whose thick covering would have provided protection against both the weather and predation.

## Mesozoic Air

Hengst et al. (1996) and Dudley (1998) have argued that atmospheric oxygen levels were elevated well above the current value in the Mesozoic. If correct, this would have profound implications for the evolution of AEC in ancient archosaurs. They would have been able to achieve high levels of sustained aerobic exercise capacity even if their respiratory, pulmonary, and muscular apparatus were not as well developed as those of modern birds and mammals. In particular, Dudley proposed that the ease of acquiring oxygen was fundamental to the initiation of pterosaur and bird flight. However, there are serious problems in reconstructing atmospheric oxygen content in the remote past, so that the issue is highly controversial and the results uncertain. Therefore, I assume in this study that the Mesozoic oxygen level was little different from what it is now.

Contradicting the idea of high Mesozoic oxygen levels are the well-developed adaptations for taking in oxygen at high rates, seen in avepods and many other dinosaurs of that era. For example, such adaptation is seen in Late Cretaceous birds, whose respiratory apparatus appears to have been as capable as that of their modern relatives. If dinosaurs avian and otherwise had to work as hard then as their modern counterparts do now to breath in large volumes of air, the implication is that the oxygen content was not markedly higher in the Mesozoic than in the Cenozoic.

## Power versus Efficiency

A common theme of those who doubt that dinosaurs—and these days, basal birds—were highly energetic is that these animals should not have consumed large amounts of energy because doing so is inefficient, and energy inefficiency is not a good thing (Regal and Gans 1980, Spotila et al. 1991, Feduccia 1996). This is in line with a view that is almost axiomatic in modern zoology, namely, that evolution constantly works to increase the efficiency of organisms. Applied to dinosaurs, the view is that dinosaurs would have been better off if they were as energy efficient and bradyaerobic as possible, rather than energy inefficient like birds and mammals. This is especially true because dinosaurs lived in a warm world, where high rates of internal thermogenesis were not necessary to maintain body warmth.

The problems with the energy efficiency hypothesis are obvious. Priede (1985) notes that if energy efficiency really has been the primary goal of natural selection, then "there should be a progressive reduction in energy expenditure by animals together with a general increase in food intake. This is clearly absurd since, taken to its logical conclusion, it would be suggest an evolutionary progression toward sessile animals with minimal locomotory energy expenditure. Evolution has in fact proceeded from sessile forms toward more active forms of life." Indeed, some 10,000 species of birds and mammals are very energy inefficient. Even in the warm tropics, mammals are the dominant tetrapods large and small: Feduccia's (1996) statement that one "need only look at the highly active, large tropical reptiles, such as the Nile crocodiles and Amazonian anacondas, to see how ectotherms have surpassed endotherms in the warm climates of the tropical zones" does not accord with reality. Crocodilians and giant snakes are largely freshwater forms, and are inactive much of the time. It is the giraffes, rhinos, and elephants that are the big and vigorous rulers of tropical lands.



The fundamental reason energy inefficiency can be advantageous is simple. Greater levels of power production increase the ability to sustain high levels of activity, which increases speed of sustained movement and range, which in turn increases the ability to acquire more resources that can be dedicated to reproduction. Greater levels of power production are inherently energy inefficient, but so are aircraft compared with horses as means of human transport. Indeed, the primary factor behind the performance of leading-edge life forms has been increased power, and this is the basis of the aerobic-exercise-capacity hypothesis. Priede was discussing fish, but the ultimate expression of boosted aerobic capacity increasing activity levels and range are migrating birds. It is ironic that although Feduccia (1996) thinks energy-efficient dinosaurs are logical, he accepted that the exceptionally high energy consumption of passerines is one reason for the great success of the group: this is a double standard.

The advantages of high aerobic power do not mean that all animals should be energy inefficient, as the existence of multitudes of reptiles and other LoMR creatures shows. What it does mean is that assumptions as to what animals *should* be like are of little use in reconstructing paleometabolics. Instead, character analysis must be used to diagnose the metabolic and thermoregulatory status of fossil organisms.

## The Link between AEC and RMR: Or Why Dinosaurs Were Not Insects

In modern vertebrates, ElvAEC is always associated with ElvRMR. This is energy inefficient, because even a resting bird or mammal uses as much oxygen and food as an active reptile. Larger flying insects are very different. They can achieve very high AEC, but the RMR drops to reptilian levels. This is much more efficient, because energy is not being expending when the insects are inactive. Such tachyaerobic/bradymetabolic insects can achieve extremely high maximal/minimal aerobic ratios because of their dispersed respiratory system, in which short trachea directly oxygenate the muscles systems (Heinrich 1993). The high aerobic ratios of insects partly compensates for their lack of significant anaerobic burst power. Why vertebrates appear to be hard pressed to achieve high aerobic capacity during exercise without raising the resting metabolism a corresponding amount is not entirely understood, but is probably linked to properties inherent in a centralized cardiorespiratory system and tachyaerobic cells (Else and Hulbert 1985, 1987; Porter and Brand 1993). Expanding the aerobic capacity of the musculature requires expanding the size and/or energy capacity of the heart and other central organs that meet the oxygen, nutrient, and waste-removal needs of the muscles. The enlarged organs consume more energy when at rest. This constraint appears to be so severe that even the most aerobically capable monitor lizards, using a gular pump to overcome the breathing limitations imposed by strong lateral trunk flexion during rapid locomotion, are unable to achieve ElvAEC (Owerkowicz et al. 1999).

Some vertebrates have unusually high maximal/minimal aerobic ratios (Jones and Lindstedt 1993, Bundle et al. 1999), but these are still below the hundredfold ratios seen in insects. The absence of any modern tetrapod that combines the energy efficiency of a resting reptile with the sustained power of a bird makes such a combination, if not impossible, highly speculative. Ruben et al. (1999) suggested that dinosaur maximal/minimal aerobic ratios were high enough to allow them to have an AEC elevated above that seen in modern reptiles, while retaining a reptilian RMR. They did not explain whether this would require a higher ratio than seen in modern tetrapods. However, a combination of features that would suggest dinosaurs had an ElvAEC and a LoRMR has not been demonstrated. While the case for ElvAEC is solid, the evidence for LoRMR (in particular the absence of RC) remains at best weak. Because all dinosaurs and Mesozoic birds appear to have been tachyaerobic, it is probable that they were all tachymetabolic. It remains possible that aerobic ratios were somewhat higher in dinosaurs than is typical in modern mammals and birds, so they may have had lower (but still suprareptilian) RMRs than do mammals and birds with similar aerobic capacity. It is also possible that some of the less aerobically capable dinosaurs, such as baso-theropods, were mesoaerobic, yet had an aerobic ratio high enough to be bradymetabolic, but this is the inferior hypothesis.

## The Big Picture

This analysis does not support either extreme possibility (entirely reptilian or entirely avian-mammalian) of where dinosaurs and basal birds are located in the spectrum of thermoenergetics. If all dinosaurs walked slowly, had small pelves, grew slowly, avoided polar regions, and so forth, then one must conclude that they retained a reptilian power system. Instead, the evidence strongly contradicts the possibility that all dinosaurs and even mesosternan birds had reptilian exercise energetics in which anaerobic burst power was emphasized over sustained aerobic power. In fact, no dinosaur or bird appears to fall into the reptilian metabolic camp. The absence of diagnostic characters indicative of LoMR, LoAEC, and hyperanaerobic power generation is sufficiently well established to falsify the hypothesis that dinosaur exercise energetics were largely reptilian. Recent attempts to revive this hypothesis are based on arguments that are either not subject to verification (absence of RC), or demonstrably false (presence of a pelvis-based diaphragmatic muscle pump), and are directly contradicted by other evidence (high estimated trackway speeds). Indeed, with the revised view of Ruben et al. (1999), there currently is no active advocacy of reptilian exercise energetics in predatory dinosaurs. In this sense, a postmodernist consensus of dinosaurs as nonreptilian has eclipsed the classic modernist view. At the same time, the evidence strongly contradicts the hypothesis that all dinosaurs had avian level aerobic and thermoregulatory capacity.

Much as the bird-origins debate is no longer about whether they descended from dinosaurs but how, the debate on dinosaur energetics is no longer about whether they were tachyaerobic, but how much so, how early, and how well they thermoregulated.

### *Mesoschian Protodinosaurs and Dinosaurs*

The skeletons of protodinosaurs, baso-theropods, and prosauropods exhibited a mix of characters no longer seen in tetrapods. These characters included an unsophisticated thorax and short hips and with long, erect, digitigrade legs. This extinct and unusual combination suggests that these dinosaurs had a correspondingly unusual thermal physiology not found in modern animals. In particular, these mesoschian archosaurs could not have had the large limb muscles associated with HiAEC, and their respiratory and cardiovascular capacity appears to have been accordingly limited. At the same time, the erect legs and bipedal gait of the predatory examples should have forced normal walking speeds to be above those sustainable by reptiles. The exercise energetics of basal predatory dinosaurs appears to have been intermediate to those of brevischian reptiles and longoschian avepods. This implies that aerobic capacity had been boosted above the maximum observed in reptiles. If so, then the incompatibility of elevated aerobic capacity with hyperanaerobiosis implies that the latter was lost. Mesoschian dinosaurs appear to have been mesoaerobic, which in turn implies that they were mesometabolic, both probably at levels not greatly above the reptilian maximum. With so many uncertainties, it is difficult to evaluate AEC/RMR ratios. The early dinosaurs may have been mesoendotherms—equally dependent upon environmental and internal heat—or marginal endotherms. In this case, basking was important, but not always as critical as in reptiles. Thermoregulation may have been poorly developed, so that body temperatures were subject to strong fluctuations. It will be interesting to see whether these dinosaurs were or were not insulated with feathers: the second condition suggests less thermal stability than the first. Because these dinosaurs generally lived in warm climates, body temperatures should have been similar to those of diurnal tropical mammals with modest metabolic rates (such as kangaroos) at around 36°C. Energy budgets should have been higher than in reptiles, but only by a factors of 2 or 3.

Mesoschian dinosaur energetics cannot have been closer to that of birds than reptiles, but lacking the bradyaerobiosis and hyperanaerobiosis typical of reptiles, it was not reptilian either. Instead it seems to have been a truly intermediate metabolic system, one with too many contradictory traits to long remain in place. The contradictions arise from the initial combination of erect legs and a low-volume leg musculature. Such a combination would have been unstable, because the limited aerobic system was overtaxed, while the sustained speed capacity of the legs was not fully exploited.

### *Longoschian Dinosaurs and Early Birds*

In the non-mesoschian dinosaurs as well as the first birds, such diagnostic paleometabolic characters as sophisticated thoraxes, enlarged hips, and erect digitigrade limbs come together in an anatomical form that is grossly similar to that observed in modern flightless birds and mammals, and is entirely outside the reptilian pattern. The assertion by Ruben et al. (1997b) that "the dynamic skeletal structure of many dinosaurs strongly suggests that they (even if they were fully ectothermic) possessed a bird- or mammal-like capacity for at least burst activity" is contradicted by skeletal features that indicate most dinosaurs possessed a bird- or mammal-like capacity for burst *and sustained* activity. In particular, increasingly mammal-like (ornithischians) or birdlike (saurischians) respiratory complexes, large leg muscles, and erect legs combine with fast-walking trackways to demonstrate that AEC was approaching or reaching the level of ground birds. The rapid growth tentatively observed in some dinosaurs supports this proposition. If this interpretation is correct, then longoschian dinosaur hearts should have been large, complete double pumps. The low aerobic and high anaerobic exercise capacity of reptiles are not present in any living animals with these features, and such power systems are inappropriate for animals with such features, so attempts to apply reptilian energetics to longoschian dinosaurs and basal birds are contrived and implausible. Nor does the respiratory and locomotory anatomy and performance of longoschian avepod dinosaurs favor some form of intermediate energetics over an avian level rate of energy use, although it does not entirely rule out the former condition.

Because most body heat was probably generated by internal aerobiosis, these dinosaurs were probably endotherms that normally operated at high body temperatures. To maintain advantageous temperature gradients in warm climates, body temperatures should have been at least as high as they are in marsupials living in similarly hot climates. Conversely, the presence of some dinosaurs at high-latitude locales from which all reptiles are absent indicates a colder-climate endothermy. Not only do the egg incubation and feather insulation characteristic of at least some small avepod dinosaurs confirm this conclusion, but the feathers come closer than anything else to being a "Rosetta Stone" for ElvMR endothermy.

However, unlike analyses that rely primarily on one or two "Rosetta Stones," this study emphasizes the totality of the evidence. It is the combination of large hips *and* erect legs *and* long-stride trackways *and* sophisticated respiratory adaptations *and* egg incubation *and* rapid growth *and* feather insulation that builds up a case for ElvAEC in longoschian dinosaurs and basal birds sufficiently strong that the presence of tachyaerobic endothermy is no longer a central question.

What remains less certain is how these dinosaurs and birds heated themselves, and how well they controlled their body temperatures: were they alphaendotherms or betaendotherms? The feathery insulation and bone-isotope data suggest that homeothermy was generally well developed, but if further work verifies recent analyses on isotopes, then homeothermy may not have been as well developed as in

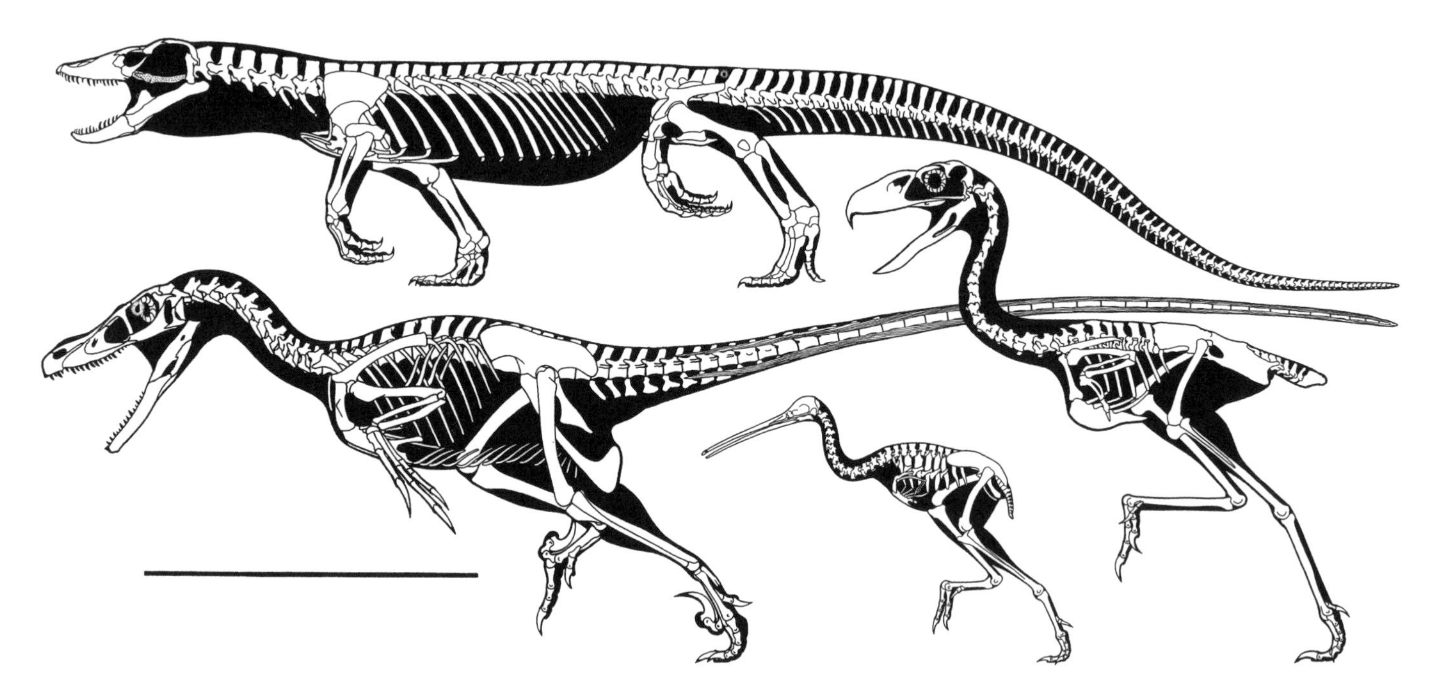

*Appendix Figure 17.* Which is the better model for the energetics of the predatory dinosaur *Velociraptor,* with its birdlike erect legs and respiratory apparatus: the phorusrhacoid terror bird *Peleycornis,* or the sprawling komodo monitor? With its exceptionally modest energy needs, the kiwi may be the best of all. Drawn to same scale; scale bar equals 0.5 m.

most birds and mammals. The moderate frequency of growth rings may be compatible with the thermal stability being somewhat variable. It is possible that dinosaurs' and basal birds' high rates of internal thermogenesis were primarily the result of their tachyaerobic power system, both in terms of the muscular effort and supporting organ work that directly powered high levels of sustained activity, and indirectly from the ElvRMR that resulted from having an ElvAEC. Body temperatures were defended by dumping and retaining heat when necessary, but when such heat-management techniques were insufficient to maintain thermal stability, the metabolic response to falling body temperature was weak or absent. In this view, temperature stability was maintained most of the time, but was lost when the metabolic system was strongly challenged. The latter event may have been rare in milder Mesozoic climes, in which case these dinosaurs and birds were semihomeothermic and betaendothermic. Such a condition may apply in particular to basal avepods if they were not insulated by feathers. If this semihomeothermic, betaendothermic system was true of longoschian dinosaurs and basal birds, then they were less capable thermoregulators, and consequently less homeothermic, than most living birds (note that these metabolic shifts are not as dramatic as those suggested by Farlow [1990]). Casual basking may have been a common means of raising body temperature a few degrees in smaller examples, but was never required, as it is for most reptiles. However, the new analysis by Horner et al. (1999) may show that the presence of deep LAG is not evidence for nonavian-mammalian thermal stability, making it feasible that some or all longoschian dinosaurs were alphaendothermic. Alphaendothermy is especially logical for polar examples that needed a strong metabolic response to winter chill. The metabolic response to cold was almost certainly limited to muscular thermogenesis in dinosaurs, as it still is in living avepods and nonplacental mammals; whether a nonshivering mode was also used is unclear. Evaporative cooling was probably limited to the respiratory tract (but probably did not involve the air sacs), except for the use of regurgitated water and urine spread on body parts. When water was not readily available, some species may have allowed body temperatures to rise in order to save water, and to maintain a favorable temperature gradient to facilitate heat loss (Paul 1991, 1998). If AEC was very high while RMR was less so, then AEC/RMR ratios may have been approached, equaled, or less probably exceeded the maximum observed in modern tetrapods. If so, then maximal/minimal aerobic ratios may have tended to be higher on average than in birds or mammals. With strongly elevated exercise MRs (and, probably, elevated RMRs), total energy budgets should have been many times above reptilian levels, and may have reached marsupial standards.

As envisioned here, the tachyaerobic condition of longoschian dinosaurs was not merely an "intermediate" condition, nor were these animals "failed endotherms" (Reid 1987–1997). Rather, theirs was a sophisticated nonreptilian system that could power high levels of sustained activity in a manner most resembling, but not necessarily identical to, that of birds and mammals.

Turning to avepods specifically, cerato-saurans probably had an incipient air sac system, and pelvic expansion was modest by later avepod standards. These factors suggest that these basal avepods were mesometabolic and mesoaerobic, but well above the reptilian and even baso-theropod maximums in both regards. The awkward, slothlike body and limbs of therizinosaurs suggest they too were mesoaerobic, in the manner of similarly heavy-footed edentates, but this probably represents a reversal from the more energetic, general averostran condition. This is because the large ilia of averostrans imply a further boost in aerobic capacity. So does the increasingly well-developed air sac system of tetanuran and especially avetheropod averostrans. Averostrans may have reached the lower suprametabolic and supra-aerobic levels. The development of flight in avepectorans, and their essentially avian sternal-ribcage apparatus, are indicative of HiAEC. The action of such a high tidal-flow system probably resulted in a low frequency of inhalation.

Low-level suprametabolic-resting and supra-aerobic-exercise capacities were probably retained in mesosternan birds, flying and otherwise. This brings us to the kiwi (see also below). Although this bird probably evolved from more highly suprametabolic flying ancestors, its marginally suprametabolic condition may be similar to the maximum reached by even the advanced predatory dinosaurs. The big island ratites are candidates for a modest AEC. The swift phorusrhacoids probably had the HiAEC observed in the similarly speedy continental ratites.

## Dinosaurs, Basal Birds, and the Aerobic-Capacity Hypotheses

Dinosaurs fall within the parameters of the aerobic-exercise-capacity hypothesis, in that they broke through the sustained activity barrier, which so severely limits land reptiles, by boosting their AEC to increasingly high levels. In doing so, they probably also had to meet the requirement of the aerobic-capacity hypothesis that RMR must be raised above the reptilian maximum to support the boost in exercise capacity. This leads to an interesting situation. Bennett and Ruben (1979) and A. Bennett (1991) have proposed and advocated the aerobic-capacity hypothesis, in which high aerobic capacity first evolves to power high walking speeds in the harsh energy habitat of 1 *G*. Yet these researchers have failed to recognize its expression in dinosaurs, whose anatomy was clearly nonreptilian and became increasingly birdlike. Indeed, the evidence for high-level sustained locomotory performance is better in the long-legged, digitigrade-footed dinosaurs than it is among advanced therapsids that retained short, plantigrade limbs.

Considering that powered flight is the ultimate expression of the aerobic-capacity hypothesis, it is even more perplexing that Ruben (1991) limited its expression to advanced birds. He does this even though tachyaerobiosis first appeared long before powered flight was developed in mammals, and despite lack of evidence that it was otherwise in avepods. There is no evidence that powered flight has ever evolved in bradyaerobic tetrapods. It is much more likely that tachyaerobiosis encouraged the evolution of avian powered flight than the reverse. It follows that the development of an advanced flight apparatus in avebrevicaudan birds had the same influence upon energetics as did the development of advanced flight in bats. Following this vein, the aerobic energetics of neoflightless avepectoran dinosaurs was probably not dramatically different from preavian averostrans, or from urvogels.

Finally, Ruben et al. (1999) effectively abandoned the aerobic-capacity hypothesis by postulating that predatory dinosaurs combined a LoRMR with an ElvAEC. Although not impossible, this combination is not based on strong evidence, and so is highly speculative. It also leaves unexplained why birds and mammals persist in being energy-inefficient tachymetabolic tachyaerobes, if they do not have to be the former in order to be the latter.

## Our Furry-Feathered Friend, the Kiwi

Among modern tetrapods, which comes closest to being a living match for the anatomy and thermal physiology of avepod dinosaurs? (Here I have in mind especially the simple fiber-feathered averostrans and neoflightless avepectorans, with their preavian air sac complexes [App. Fig. 17].) Certainly not oras and other reptiles, nor mammals with their diaphragm-ventilated alveolar lungs. Not surprisingly, it is to the modern avepods we must turn, and among these we can dismiss flying birds with their enormous sterna. The large flightless birds are tempting, but even their MaxAEC rates are exceptionally high.

There is a neoflightless bird that is highly active, has a high, stable body temperature, has a modest resting metabolic rate, breathes through narrow nasal passages, has a moderately developed set of air sacs ventilated by a short sternum and free posterior ribs, is insulated with furlike feathers, has a complete four-chambered heart, grows at a modest pace, has two functioning ovaries, is an obligatory biped with degenerate arms, and even has a hallux that is only partly reversed (Calder 1978, del Hoyo 1992–99, T. Johnson 1996. McNab 1996): the kiwi. Not that this peculiar bird is a perfect model for Mesozoic dinosaurs, kiwis being highly specialized, nocturnal insectivores that have small eyes and lay oversized eggs. Indeed, the furlike insulation, modest MR, well-developed sense of smell, nocturnal habits, burrow habitat, twin functioning ovaries, and extended embryonic and juvenile growth periods inspired Calder to label the kiwi an honorary mammal. But this may not be the honor it was intended to be. Kiwis are avepods, not therians, so the kiwi is more fittingly recognized with the status of a living dinosaur that, better than any other modern animal—and much better than any reptile—gives an impression of what Mesozoic avepods were like.

APPENDIX 5

# Locomotory Data Plots

Most of the dinosaur body masses and skeletal figures used in the plots in Appendix Figure 18 are presented in Paul (1988a, 1997a) and in Figure 10.1. Data for birds and pterosaurs were taken in part from Hartman (1961), Greenwalt (1962), Alexander (1971), Hazlehurst and Rayner (1992), Livezey (1992), and Appendix 2.

Appendix Figure 18A plots the lengths of toes II and IV as a function of central toe length in avepods, measured from toe base to claw tip. The slope indicates that when toe lengths are equal, the differential between the lengths of toes II and IV marks the degree of asymmetry; when IV is much longer than II and nearly as long as III, the foot is at least partly didactyl. In most dino-avepods and birds, including *Protarchaeopteryx,* the feet are subsymmetrical and the foot is tridactyl. In dromaeosaurs, most troodonts, *Archaeopteryx* (pathological feet of S6 not included), baso-avebrevicaudans, enantiornithines, and the seriema, the foot is asymmetrical and is often potentially didactyl.

Coracoid length as a function of scapula length in archosaurs is plotted in Appendix Figure 18B, as measured from the center of the glenoid. The slope indicates when the scapula and coracoid lengths are equal. Birds, especially those that fly, have elongated coracoids. Archaeopterygiforms and a dromaeosaur are in the lower avian range. Most other archosaurs have shorter coracoids, crocodilians being an exception.

Appendix Figure 18C shows forelimb length as a function of hindlimb length in archosaurs, as measured from the proximal end of the humerus or femoral head to the end of the longest finger or toe. The slope indicates when the fore- and hindlimb lengths are equal. Dromaeosaurs and oviraptorids have unusually long arms for dino-avepods, as long as those of *Protarchaeopteryx*. *Archaeopteryx* have the elongated arms characteristic of flight-capable birds. Note that the juvenile JM 2257 is shorter armed than more-mature specimens, suggesting relatively limited flight performance.

Appendix Figure 18D graphs humerus circumference as a function of femur circumference in archosaurs. When the shafts were not completely exposed for direct measurement, circumference was estimated by multiplying the minimum shaft diameter by 3.34 (also applies to Appendix Figure 18E). The slope indicates when the humerus and femur circumferences are equal. In flightless archosaurs (including *Protarchaeopteryx*), the arm bones were much weaker than those of the load-carrying legs. The unequally strong humeri and femora of *Archaeopteryx* indicate that it had the strong arm bones needed by flight-capable birds.

The plot in Appendix Figure 18E shows humerus circumference as a function of body mass in volant birds. The humeri of *Archaeopteryx* appear as robust as those of flight-capable birds and are more robust than those of flightless dino-avepods.

Appendix Figure 18F shows furcula area as a function of body mass in volant birds. The furcula of *Archaeopteryx* appears to have been able to support large flight muscles as ably as those of flight-capable birds.

Wing area as a function of body mass in modern birds and *Archaeopteryx* is plotted in Appendix Figure 18G. A slope of 2/3 indicates isometry. *Archaeopteryx* wing area was estimated as per Chapter 5 and is in the mid-avian range: note that the juvenile JM 2257 is smaller winged than more-mature specimens, which suggests a relatively limited flight performance.

Appendix Figure 18H plots wingspan as a function of body mass in dragonfly-type insects, bats, birds, pterosaurs, and light aircraft. Data are plotted for *Harpagornis* (E); pseudodontorn (P); *Teratornis, Argentavis* (T); wandering albatross (a); kori bustard (b); Andean condor (c); harpy eagle (e); frigate bird (f); semiflightless kakapo male and female (k); marabou stork (m); mute swan (s); and wild turkey (t). The dashed-line least-area polygon encompasses bats; the solid-lined least-area polygon encompasses flying birds. The 2.56-power slope marks avian regression (Alexander 1971). *Archaeopteryx* falls into the mid-avian range in terms of wingspan.

Appendix Figure 18I shows dorsopelvis length as a function of body mass in amniotes. Crurotarsal-

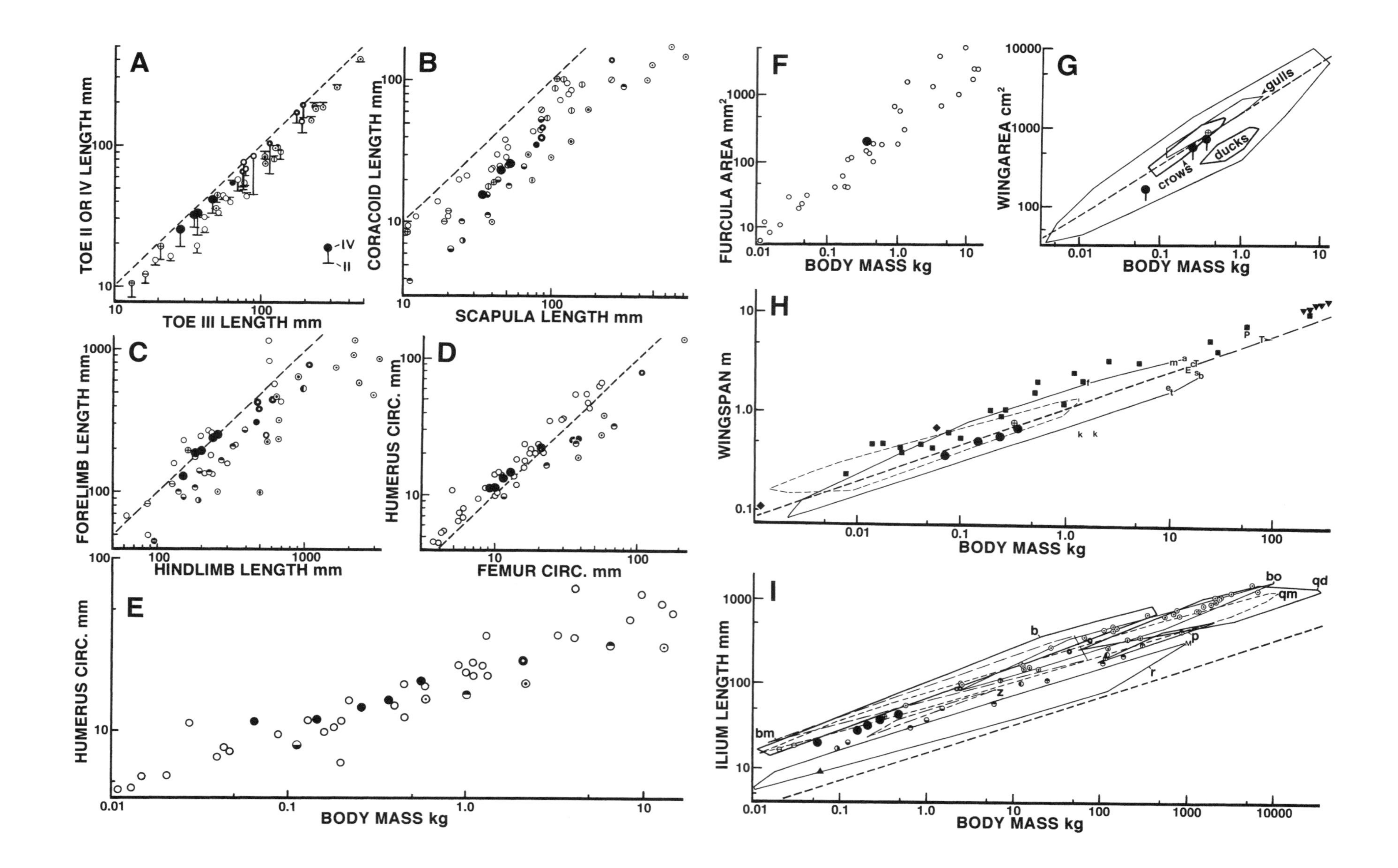
A
TOE II OR IV LENGTH mm
TOE III LENGTH mm
IV
II
B
CORACOID LENGTH mm
SCAPULA LENGTH mm
C
FORELIMB LENGTH mm
HINDLIMB LENGTH mm
D
HUMERUS CIRC. mm
FEMUR CIRC. mm
E
HUMERUS CIRC. mm
BODY MASS kg
F
FURCULA AREA mm²
BODY MASS kg
G
WINGAREA cm²
gulls
crows
ducks
BODY MASS kg
H
WINGSPAN m
BODY MASS kg
I
ILIUM LENGTH mm
bm
b
z
r
p
bo
qd
qm
BODY MASS kg

ankled noncrocodilian archosaurs, protocrocodilians, coelophysoids, and elaphrosaur are indicated with a vertical line in either lower or upper half of circles. The smallest dino-avepod was a juvenile. Also represented are modern crocodilians and lizards (r), including *Megalania* (M); neornithine birds, most modern, but extinct ratites are included (b); bipedal ornithischians (bo); prosauropod dinosaurs (p); quadrupedal dinosaurs, excluding prosauropods and armored ornithischians (qd); bipedal mammals (bm); quadrupedal mammals (qm), excluding monotremes and tenrec (z). A slope power 1/3 indicates isometry. Note that due to the shallowness of the slopes, major variations in body mass are required to alter the results dramatically. For additional information, see Figure 10 in Paul (1998).



*Appendix Figure 18. (opposite)* Locomotory data plots.

▲ Semi-bipedal lizard

■ Pterosaurs

◓ Basal archosauriforms and thecodonts

◒ Crocodilians

⊖ *Scleromochlus*

◑ *Marasuchlus*

◐ Baso-theropods

⊙ Dino-avepods

◉ *Caudipteryx* and oviraptorids

○ Troodonts

○ Dromaeosaurs

● Archaeopterygiforms

⦶ Alvarezsaurs

⊕ Baso-avebrevicaudans

⊖ Enantiornithines

⦶ Ratites

⊘ Flightless neognathus birds

○ Flying neognathus birds

◆ *Megaloprepus* and dragonfly

▼ Light aircraft and sailplanes

# A New Yixian Dromaeosaur, and the Flight-Adapted Finger of a Flightless Basal Sickle-Clawed Dinosaur

When this book was too far along in production to allow significant changes to the main text, Ji et al. (2001) described the articulated skeleton of a small dromaeosaur from the Yixian Formation (Pl. 16), which I was subsequently able to examine. They observed that the unusually large-headed specimen has the proportions expected in a subadult sinornithosaur, so the features of *Sinornithosaurus* discussed in the main text generally apply to the new specimen and vice versa.

The preservation of the subadult skeleton in a dorsoventrally flattened posture with both sets of limbs splayed out to the sides is unusual for a predatory dinosaur, but it does parallel the condition in most post-urvogel bird fossils. Although the new sinornithosaur skeleton is complete, much of it was badly damaged when the slabs were split. The vertebrae in particular are so ruined that a complete and accurate skeletal restoration is not yet practical. The tail was long, but a caudal count is not possible. Ossified uncinate processes demonstrate their existence in these dromaeosaurs, ossified sternal ribs may be present, and the absence of an ossified sternum is probably a juvenile attribute. The articulation of the furcula and scapulocoracoids follows and verifies the avepectoran pattern. The arms are tightly folded at the elbows with no signs of disarticulation. The observation that both wrists are nevertheless not tightly folded is in line with the avian arm folding mechanism being only partly developed at this stage of avepectoran evolution. Possible skin impressions are preserved in association with the feet, which were correspondingly naked, as in most birds.

The specimen's most spectacular attribute is its complete set of feathers still in their proper place. The spread-eagled pose of the remains facilitated its preservation in this condition. The feathers are preserved within multiple layers of the surrounding sediment. The subadult's feathers follow and add to the pattern described for the adult *Sinornithosaurus* (Chapter 5), and they are included in the restorations presented in Figure 6.2. Feathers extend even farther onto the snout than I have previously restored on dino-avepectorans, but it is not clear that the entire muzzle was feathered (also see frontispiece, Figs. 9.1, 9.2); the extent of snout feathers in more basal avepods also remains uncertain. There is a modest tuftlike expansion of the feathers at the tip of the tail. Especially notable are the short, ragged-edged, but apparently pennaceous feathers on the distal forelimbs.

The most interesting and important skeletal feature of adult and subadult sinornithosaurs is found in the central finger (App. Fig. 19). At first glance minor and subtle in nature, this feature has such crucial implications for bird origins and the early loss of flight—as well as the limitations of cladistics—that it required the addition of this appendix. To understand why this is so, consider that in almost all other known dino-avepods the proximal finger bone (first phalanx) of digit II was a long, slender, distally tapering rod (except those examples with extremely reduced central fingers, but they are not relevant to this discussion). In *Archaeopteryx* the same bone was still basically a rod, but it had a slightly expanded distal end according to Zhou and Martin (1999). They may have somewhat exaggerated the expansion: its extent seems inconsistent between specimens, and may have been increased in some cases by crushing. The condition in *Protarchaeopteryx* is not entirely clear, but if any expansion was present it was at most very small. In avebrevicaudan birds whose hands are not extremely reduced, the first phalanx continues to be long, but it is flattened owing to the presence of a broad posterolateral flange that is somewhat wider distally than proximally. Well-developed posterolateral flanges are often retained in the central fingers of neoflightless birds, but they are often reduced compared with those of their flying relatives.

Outer flight feathers are subjected to high bending forces at their bases during flight. Posterolateral flanges on central finger bones evolved to greatly improve the anchoring of these large feathers, whose bases are set on the dorsal surface of the flange. It is therefore surprising that *Archaeopteryx* had at most only a trace of an expansion, since a well-developed set of outer flight feathers was anchored on finger II (see Chapters 2 and 3). It is difficult to imagine an alternative selective force that would favor the development of feather-anchoring flanges on the central finger. A posterolateral flange on a finger bone would serve no apparent purpose in the context of improving hand function for grasping prey, branches, or other items, especially since other avepod dinosaurs failed to evolve such flanges for these purposes. Support of distal display feathers is not a viable explanation for the flanges, since *Archaeopteryx* was able to anchor very large feathers on a still-slender proximal finger bone. Because even flying *Archaeopteryx* lacked well-developed finger flanges supporting flight feathers, it must be concluded that the presence of this feature is both a strong post-urvogel character and as close as one can get to proof of a level of flight more advanced than that seen in the urvogel.

The last conclusion is reinforced if an avepod with posterolateral finger flanges has other post-urvogel flight characters. Even though the hand-borne feathers are strongly reduced and no longer subjected to the loads associated with flight, neoflightless avebrevicaudan birds tend to retain posterolateral flanges on the central fingers if the hand is not extremely reduced. The flanges are often reduced compared with those of fliers. If an avepod with posterolateral central finger flanges lacks the basic features needed to fly (mainly fully developed wings) and possesses other advanced flight features, then the only sound conclusion is that the flightless subject evolved from fliers whose aerial performance exceeded that of *Archaeopteryx*. There may be no better evidence of a neoflightless condition.

The proximal bone of the finger II of the adult *Sinornithosaurus* is markedly expanded posterolaterally (App. Fig. 19, Fig. 4d in Xu, Wang, and Wu 1999). My direct examination of the specimen could not discern whether or not the expansion resulted from asymmetrical postmortem crushing—although it was difficult to see how such an extensive flange could be pressed into existence when the rest of the finger bones were only slightly distorted—so the potentially neoflightless feature was set aside until further information became available. The new information is provided by the subadult, which shows exactly the same degree and form of posterolateral expansion at the bases of both central fingers, while the rest of the finger bones remain slender rods (Pl. 16). The presence of a well-developed posterolateral flange in sinornithosaur fingers is therefore well established, even if some pressure damage occurred in the adult. The flange is slightly broader distally than proximally, as in birds. The expansion is much better developed than that of *Archaeopteryx,* but is somewhat less than that in basal avebrevicaudans such as *Confuciusornis*. It is difficult to compare the

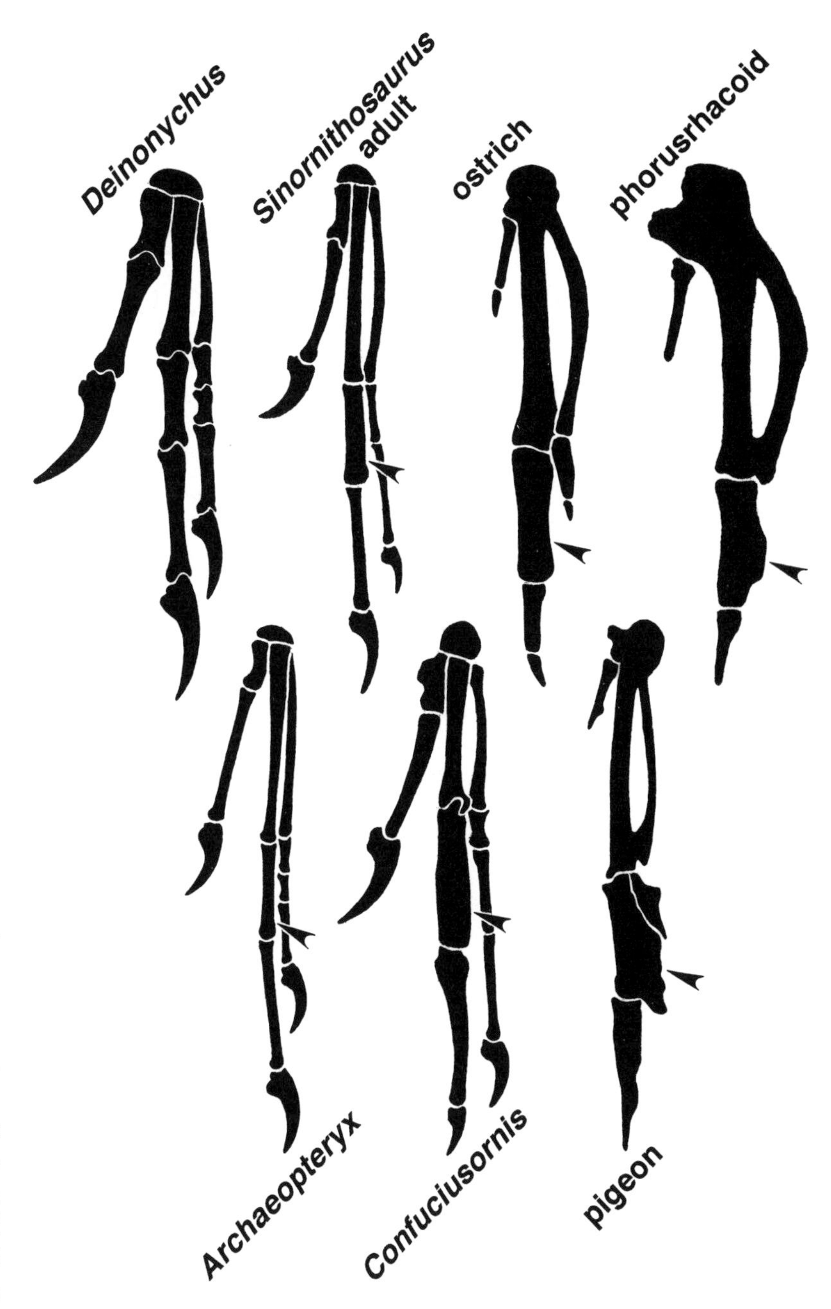


*Appendix Figure 19.* Hands of avepectorans, with a focus on the development of a posterolateral flange on the proximal bone of the central finger (indicated by arrows). Top row, flightless examples. Bottom row, flying examples. In both rows hands are least modified from the dinosaur condition on the left, and become increasingly avian progressing to the right. Note that the combination of a well-developed posterolateral flange and a strongly bowed metacarpal III made the hand of flightless *Sinornithosaurus* better suited for supporting primary feathers than was the hand of flying *Archaeopteryx*. Also note that other finger bones are also posterolaterally expanded in avebrevicaudans.

detailed morphology of the sinornithosaur central finger's first phalanx to the same bone in *Archaeopteryx* and other birds. In the adult *Sinornithosaurus* the bone is exposed ventrally, rather than dorsally as in the urvogel specimens. Crushing may have altered the surface topography of the bone in the adult sinornithosaur, and the element's surfaces are badly damaged in the subadult. The dorsal surface of the flange almost certainly supported the pennaceous feathers anchored on the base of the central finger. What is interesting is that the degree of support provided by the flange was excessive, considering the shortness of the feathers and the low bending loads that their bases experienced.

The situation observed in sinornithosaurs is summarized as follows. A posterolateral flange on the proximal bone of finger II was considerably better developed in all three articulated hands than in any *Archaeopteryx* hand, was a little less prominent than that in basal avebrevicaudans, and anchored a set of quite short, apparently symmetrical and ragged outer forelimb feathers. This arrangement was combined with arms that (although very long) were too short for flight, as well as potential neoflightless characters (as detailed in Chapter 11), including a strongly bowed metacarpal III, ossified uncinates and sternal ribs, and a large sternal plate. Indeed, the combination of strong bowing of the outer metacarpal with the posterolateral flange on the central finger and a stiffened central finger made the small feathered sinornithosaur hand much better adapted for supporting large feathers than was the large feather-bearing hand of the urvogel (App. Fig. 19), exactly the paradoxical condition predicted in neoflightless post-urvogels. (The variable degree of bowing of the outer metacarpal in sinornithosaurs is also consistent with this feature no longer having an important flight function.) The dromaeosaurs' numerical superiority over other dino-avepectorans in the number of potentially neoflightless characters is further boosted. Because the flanges were better developed than those in *Archaeopteryx* and were not supporting large primary feathers, a protoflight status is contradicted. Since the posterolateral flange on the central finger is an extraordinary flight adaptation that is probably explicable only as a post-urvogel aerodynamic character, since numerous other strong potentially neoflightless characters were present in sinornithosaurs, and since dromaeosaurs clearly could not fly, it is concluded that these dinosaurs were neoflightless, with an ancestral flight performance that exceeded that achieved by *Archaeopteryx*. That posterolateral finger bone flanges are present only in basal dromaeosaurs adds to the evidence that neoflightless dromaeosaurs experienced a set of reversals as they evolved away from their flight heritage during the Cretaceous (Fig. 11.1). The posterolaterally expanded first phalanx of finger II also adds to the evidence for the dromaeosaur's posturvogel phylogenetic status.

Norell et al. (2001) briefly noted the unusual robustness of the proximal phalanx of the central finger of the subadult, but their conventional, cladistically based conclusion that sinornithosaurs lay outside the clade of flying avians precluded them from putting the peculiar finger base in a logical functional or phylogenetic context. To refute a neoflightless condition in dromaeosaurs now requires the following: either falsify the presence of posterolateral finger bone flanges in sinornithosaurs, or propose a convincing explanation for why basal dromaeosaurs evolved an advanced flight feature without passing through a correspondingly advanced stage of flight, and why dromaeosaurs then lost the posterolateral finger bone flange and reverted to the normal dinosaur condition as the group continued to evolve. Unless one of these improbable events occurs, the conventional hypothesis that dromaeosaurs were preflight pre-urvogels is inferior to the hypothesis that they were neoflightless post-urvogels (see the discussion of this point in Chapter 11). The presumption by Norell et al. (2001) that pennaceous feathers first evolved in nonvolant dinosaurs is also challenged, but not refuted.

The implications of the posterolateral flange of sinornithosaur central fingers for cladistics are significant. It is probable that, even when all the flight-related and other characters discussed in this book are properly taken into account, dromaeosaurs will still fall outside the *Archaeopteryx*-Avebrevicauda clade in standard cladistic studies. What is to be done with such results? The feather-anchoring finger flange by itself directly challenges standard cladistic analyses of dinosaur-bird relationships. When it is added to all the other evidence for a post-urvogel neoflightlessness in dromaeosaurs, then cladistic studies that persist in placing dromaeosaurs outside the *Archaeopteryx*-Avebrevicauda clade are contradicted to the point that they are effectively falsified. Scientists should never be slaves to a particular methodology when, as useful as it may be, it is overwhelmed by another set of data and analysis. Science requires practical flexibility. Researchers must therefore no longer automatically adhere to the phylogenetic conclusions derived from anatomically based cladograms when working to restore detailed interrelationships, and must consider them as only one component within more comprehensive phylogenetic analyses.

Those who oppose direct descent of birds from dinosaurs should not take comfort from this new exposure of the limitations of cladistics. The sinornithosaurs' feather-anchoring finger flange delivers yet more blows to all nondinosaurian hypotheses of bird origins. A once exclusively avian character has been added to the many others present in predatory dinosaurs, and the already close similarity between the hands of the two groups is dramatically reinforced. The convergence hypothesis is further weakened. Why would flightless dinosaurs have evolved such a distinctive avian flight character? The presence of posterolateral flanges on a central finger II in dinosaurs, but on a central finger III in birds, also pushes evolutionary convergence far beyond what is plausible. The nonhomology of the dinosaur and bird fingers argument is further undermined by the presence of the same kind of flight adaptation on the same central finger in both groups. As discussed in Chapter 10, the probability that sinornithosaurs and other dromaeosaurs were post-urvogels does make them into secondarily flightless birds, but in no way divorces them from other dinosaurs. They were neoflightless dino-birds.

The finger flanges of sinornithosaurs verify the presence of feathers in dinosaurs; affirm the descent of birds from dinosaurs; add to the near certainty that some, if not all, dromaeosaurs descended from ancestors that flew better than *Archaeopteryx;* bolster the already high probability that some or all dromaeosaurs were post-urvogels, at least some of which experienced reversal away from flight as they evolved in the Cretaceous; demonstrate that cladistics alone cannot reliably restore relationships at such a fine level; and help challenge cladistically based depictions of the evolution of flight, feathers, and other aspects of early avian functional anatomy. Remarkable what a little process on the finger bone of a small, often arboreal, predatory dinosaur can tell us about the origin of birds and their flight!

# *Figure Abbreviations for Skull Elements*

| | |
|---|---|
| a | Angular |
| ar | Articular |
| bs | Basisphenoid |
| c | Nasal concha |
| d | Dentary |
| e | Epiotic |
| ec | Ectopterygoid |
| ep | Epipterygoid |
| et | Ectopterygoid process of pterygoid |
| f | Frontal |
| h | Hyoid |
| if | Intramandibular fenestra |
| j | Jugal |
| l | Lacrimal |
| ls | Laterosphenoid |
| m | Maxilla |
| n | Nasal |
| o | Preorbital |
| oc | Occipital condyle |
| p | Prootic |
| pa | Parietal |
| pb | Basipterygoid process |
| pf | Postfrontal |
| pm | Premaxilla |
| po | Postorbital |
| pp | Paraoccipital process |
| pr | Prearticular |
| ps | Parasphenoid |
| pt | Pterygoid |
| q | Quadrate |
| qj | Quadratojugal |
| s | Splenial |
| sa | Surangular |
| sc | Sclerotic ring |
| so | Supraoccipital |
| sq | Squamosal |
| tb | Basitubera |
| v | Vomer |
| z | Palatine |
| zc | Posteromedial process of palatine |
| zr | Anterolateral process of palatine |

# Bibliography


Abel, O. 1911. Die Vorfahren der Vögel und ihre Lebensweise. *Verhandlungen, Zoologisch-Botanische Gesellschaft in Wien* 61:144–91.

Ackerman, J. 1998. Dinosaurs take wing. *National Geographic* 194(1):74–99.

Alexander, R. McN. 1971. *Size and Shape.* Edward Arnold, London.

———. 1983a. Allometry of the leg bones of moas and other birds. *Journal of Zoology* (London) 200:215–31.

———. 1983b. On the massive legs of a moa. *Journal of Zoology* (London) 201:363–76.

———. 1989. *Dynamics of Dinosaurs and Other Extinct Giants.* Columbia University Press, New York.

———. 1992. *Exploring Biomechanics.* Scientific American Library, New York.

Alvarez, L., et al. 1980. Extraterrestrial cause for the Cretaceous-Tertiary extinction. *Science* 208:1095–108.

Alvarez, W. 1997. T. rex *and the Crater of Doom.* Princeton University Press, Princeton, N.J.

Amadon, D. 1947. An estimated weight of the largest known bird. *Condor* 49:159–64.

Anderson, A. 1989. *Prodigious Birds: Moas and Moa-Hunting in New Zealand.* Cambridge University Press, Cambridge.

———. 2000. Less is moa. *Science* 289:1472–73.

Anderson, D., and Eberhardt, S. 2001. *Understanding Flight.* McGraw-Hill, New York.

Andors, V. A. 1992. Reappraisal of the Eocene groundbird *Diatryma. Science Series of the Los Angeles County Museum of Natural History* 36:109–26.

Andrews, P., and Fernandez-Jalvo, Y. 1998. 101 uses for fossilized faeces. *Nature* 393:629–30.

Andrews, R. M. 1982. Patterns of growth in reptiles. In *Biology of the Reptilia,* ed. C. Gans and H. Pough, 13:273–320. Academic Press, New York.

Andrews, R. M., and Pough, F. H. 1985. Metabolism of squamate reptiles: Allometric and ecological relationships. *Physiological Zoology* 58:214–31.

Anton, J. A. 1998. Velocity estimates of selected dinosaurian ichnotaxa preserved in Triassic Age mudstones. In *The Dinofest Symposium,* ed. D. L. Wolberg et al., 1–2. Academy of Natural Sciences, Philadelphia.

Archey, G. 1941. The moa: A study of the Dinornithiformes. *Bulletin of the Auckland Institute and Museum* 1:1–145.

Archibald, J. D. 1996. *Dinosaur Extinction and the End of an Era: What the Fossils Say.* Columbia University Press, New York.

Arcucci, A. B. 1986. Nuevos materiales reinterpretacion de *Lagerpeton chanarensis* Romer del Triasico medio de La Rioja, Argentina. *Ameghiniana* 23:233–42.

———. 1987. Un nuevo Lagosuchidae de la fauna de Los Chanares, La Rioja, Argentina. *Ameghiniana* 24:89–94.

Arendt, J. D. 1997. Adaptative intrinsic growth rates: An integration across taxa. *Quarterly Review of Biology* 72:149–77.

Attenborough, D. 1998. *The Life of Birds.* Princeton University Press, Princeton, N.J.

Auffenberg, W. 1981. *The Behavioral Ecology of the Komodo Monitor.* University Presses of Florida, Gainesville.

———. 1994. *The Bengal Monitor.* University Presses of Florida, Gainesville.

Austin, J. J., et al. 1997. Problems of reproducibility—Does geologically ancient DNA survive in amber preserved insects? *Proceedings of the Royal Society of London B* 264:467–74.

Avery, R. A. 1971. Estimates of food consumption by the lizard *Lacerta vivipara. Journal of Animal Ecology* 40:351–65.

Baker, M. A. 1979. A brain-cooling system in mammals. *Scientific American* 79(5):130–39.

———. 1982. Brain cooling in endotherms in heat and exercise. *Annual Review of Physiology* 44:85–96.

Baker, M. A., and Chapman, L. W. 1977. Rapid brain cooling in exercising dogs. *Science* 195:781–83.

Bakhurina, N. N., and Unwin, D. M. 1995a. A preliminary report on the evidence for "hair" in *Sordes pilosus,* an Upper Jurassic pterosaur from Middle Asia. In *Sixth Symposium on Mesozoic Terrestrial Ecosystems and Biota, Short Papers,* ed. A. Sun and Y. Wang, 79–82. China Ocean Press, Beijing.

———. 1995b. The evidence for "hair" in *Sordes* and other pterosaurs. *Journal of Vertebrate Paleontology* 15:17A.

Bakker, R. T. 1968. The superiority of dinosaurs. *Discovery* 3:11–22.

———. 1975. Dinosaur renaissance. *Scientific American* 232(4):58–78.
———. 1980. Dinosaur heresy—Dinosaur renaissance: Why we need endothermic archosaurs for a comprehensive theory of bioenergetic evolution. In *A Cold Look at the Warm-Blooded Dinosaurs,* ed. R.D.K. Thomas and E. C. Olson, 351–462. AAAS, Washington, D.C.
Bakker, R. T., and Galton, P. M. 1974. Dinosaur monophyly and a new class of vertebrates. *Nature* 248: 168–72.
Balda, R. P., Caple, G., and Willis, W. R. 1985. Comparison of the gliding to flapping sequence with the flapping to gliding sequence. In Hecht et al. 1985, 267–77.
Balter, M. 1997. Morphologists learn to live with molecular upstarts. *Science* 276:1032–34.
Bang, B. G. 1966. The olfactory apparatus of tubenosed birds (Procellariiformes). *Acta Anatomica* 65:391–415.
———. 1971. Functional anatomy of the olfactory system in 23 orders of birds. *Acta Anatomica* 79 (suppl. 58):1–76.
Bang, B. G., and Bang, F. B. 1959. A comparative study of the vertebrate nasal chamber in relation to upper respiratory infections. *Bulletin of the Johns Hopkins Hospital* 104:107–49.
Bang, B. G., and Wenzel, B. M. 1985. Nasal cavity and olfactory system. In *Form and Function in Birds,* ed. A. S. King and J. McLelland, 3:195–225. Academic Press, London.
Barnes, G. M., and Rautenburg, W. 1987. Temperature control. In *Bird Respiration,* ed. T. Seller, 1:131–53. CRC Press, Boca Raton, Fla.
Barrett, P. M. 2000. Evolutionary consequences of dating the Yixian Formation. *Trends in Ecology and Evolution* 15(3):99–103.
Barrick, R. E., Stoskopf, M. K., and Showers, W. J. 1997. In *The Complete Dinosaur,* ed. J. O. Farlow and M. Brett-Surman, 474–90. Indiana University Press, Bloomington.
Barrick, R. E., Russell, D. A., and Showers, W. 1998. How much did dinosaurs eat: Metabolic evidence from oxygen isotopes. *Journal of Vertebrate Paleontology* 18:26A.
Barron, E. J., and Washington, W. M. 1982. Cretaceous climate: A comparison of atmospheric simulations with the geological record. *Palaeogeography, Palaeoclimatology, Palaeoecology* 40:103–33.
Barsbold, R. 1974. Saurornithoididae, a new family of small theropod dinosaurs from Central Asia and North America. *Palaeontologica Polonica* 30: 5–22.
———. 1976. New data on *Therizinosaurus. Joint Soviet-Mongolian Palaeontological Expedition Transactions* 3:68–75.
———. 1981. Toothless carnivorous dinosaurs of Mongolia. *Joint Soviet-Mongolian Palaeontological Expedition Transactions* 15:28–39.
———. 1983. Carnivorous dinosaurs from the Cretaceous of Mongolia. *Joint Soviet-Mongolian Palaeontological Expedition Transactions* 19:1–117.
Barsbold, R., and Perle, A. 1984. The first record of a primitive ornithomimosaur from the Cretaceous of Mongolia. *Paleontological Journal* 2:118–20.
Barsbold, R., Osmólska, H., and Kurzanov, S. M. 1987. On a new troodontid from the Early Cretaceous of Mongolia. *Acta Palaeontologica Polonica* 32:121–32.
Barsbold, R., and Maryanska, T. 1990. Segnosauria. In *The Dinosauria,* ed. D. B. Weishampel, P. Dodson, and H. Osmólska, 408–15. University of California Press, Berkeley.
Barsbold, R., Maryanska, T., and Osmólska, H. 1990. Oviraptorosauria. In *The Dinosauria,* ed. D. B. Weishampel, P. Dodson, and H. Osmólska, 249–58. University of California Press, Berkeley.
Barsbold, R., and Osmólska, H. 1990. Ornithomimosauria. In *The Dinosauria,* ed. D. B. Weishampel, P. Dodson, P., and H. Osmólska, 225–44. University of California Press, Berkeley.
———. 1999. The skull of *Velociraptor* from the Late Cretaceous of Mongolia. *Acta Palaeontologica Polonica* 44:189–219.
Barsbold, R., et al. 2000. A pygostyle from a non-avian theropod. *Nature* 403:155–56.
Barthel, K. W., Swinburne, N.H.M., and Morris, S. C. 1990. *Solnhofen: A Study in Mesozoic Paleontology.* Cambridge University Press, Cambridge.
Bartlett, J., et al. 1998. Digestive evidence for metabolism from carnivore coprolites. *Journal of Vertebrate Paleontology* 18:26A.
Beale, G. 1985. A radiological study of the kiwi (*Apteryx australis mantelli*). *Journal of the Royal Society of New Zealand* 15:187–200.
Becker, L., Poreda, R. J., Hunt, A. G., Bunch, T. E., and Rampino, M. 2001. Impact event at the Permian-Triassic boundary: Evidence from extraterrestrial noble gases in fullerenes. *Science* 291: 1530–1533.
Bellairs, A. 1970. *The Life of Reptiles.* Universe Books, New York.
Bellairs, A., and Jenkin, C. R. 1960. The skeleton of birds. In *Biology and Comparative Physiology of Birds,* ed. A. Marshall, 1:241–300. Academic Press, New York.
Bennett, A. F. 1983. Ecological consequences of activity metabolism. In *Lizard Ecology,* ed. R. B. Huey, E. R. Painka, and T. W. Schoener, 11–23. Harvard University Press, Cambridge, Mass.
———. 1991. The evolution of aerobic capacity. *Journal of Experimental Biology* 160:1–23.
———. 1994. Exercise performance of reptiles. *Advances in Veterinary Science and Comparative Medicine* 38B:113–38.
Bennett, A. F., and Dawson, W. 1976. Metabolism. In *Biology of the Reptilia,* ed. C. Gans and W. Dawson, 127–223. Academic Press, New York.

Bennett, A. F., and Gorman, G. C. 1979. Population density and energetics of lizards on a tropical island. *Oecologia* 42:339–58.

Bennett, A. F., and Ruben, J. 1979. Endothermy and activity in vertebrates. *Science* 206:649–54.

Bennett, P. M., and Harvey, P. H. 1987. Active and resting metabolism in birds: Allometry, phylogeny, and ecology. *Journal of Zoology* (London) 213:327–63.

Bennett, S. C. 1996. The phylogenetic position of the Pterosauria within the Archosauromorpha. *Zoological Journal of the Linnean Society* (London) 118: 261–308.

———. 1997. Terrestrial locomotion of pterosaurs: A reconstruction based on *Pteraichnus* trackways. *Journal of Vertebrate Paleontology* 17:104–13.

Benton, M. J. 1985. Classification and phylogeny of the diapsid reptiles. *Zoological Journal of the Linnean Society* (London) 84:97–164.

———. 1999. *Scleromochlus taylori* and the origin of dinosaurs and pterosaurs. *Philosophical Transactions of the Royal Society of London B* 354:1423–46.

Berman, D. S., and McIntosh, J. S. 1978. Skull and relationships of the Upper Jurassic sauropod *Apatosaurus*. *Bulletin of the Carnegie Museum of Natural History* 8:1–35.

Berman, D. S., et al. 2000. Early Permian bipedal reptile. *Science* 290:969–72.

Bernstein, K. P., Duran, H. L., and Pinshow, B. 1984. Extrapulmonary gas exchange enhances brain oxygen in pigeons. *Science* 226:564–66.

Bernstein, M. H., Sandoval, I., Curtis, M. B., and Hudson, D. M. 1979. Brain temperature in pigeons: Effects of anterior respiratory bypass. *Journal of Comparative Physiology B* 129:115–18.

Bhatnagar, K. P., and Kallen, F. C. 1975. Quantitative observations on the nasal epithelia and olfactory innervation in bats. *Acta Anatomica* 91:272–82.

Bidar, A., Demay, L., and Thomel, G. 1972 *Compsognathus corallestris* nouvelle espèce de dinosaurien théropode du Portlandien de Canjuers. *Extrait des annales du Museum d'Histoire Naturelle de Nice* 1:1–33.

Bleiweiss, R. 1998. Fossil gap analysis supports early Tertiary origin of trophically diverse avian orders. *Geology* 26:323–26.

———. 1999. Fossil gap analysis supports early Tertiary origin of trophically diverse avian orders: Comment and reply. *Geology* 27:95–96.

Bochenski, Z. 1999. Enantiornithes: Earlier birds than *Archaeopteryx?* In Olson 1999, 285–88.

Bock, W. J. 1986. The arboreal origin of avian flight. *Memoirs of the California Academy of Sciences* 8: 57–72.

Bock, W. J., and Buhler, P. 1995. Origin of birds: Feathers, flight, and homoiothermy. *Archaeopteryx* 13:5–13.

Bonaparte, J. F. 1975. Nuevos materiales de *Lagosuchus talampayensis* Romer y si significado en el origen de los Saurischia. *Acta Geologica Lilloana* 13:1–90.

———. 1984. Locomotion in rauisuchid thecodonts. *Journal of Vertebrate Paleontology* 3:210–18.

Bonaparte, J. F., and Powell, J. E. 1980. A continental assemblage of tetrapods from the Upper Cretaceous beds of El Brete, northwest Argentina. *Mémoires de la Société Géologique de France* 139:19–28.

Bonaparte, J. F., Novas, F. E., and Coria, R. A. 1990. *Carnotaurus sastrei,* the horned, lightly built carnosaur from the Middle Cretaceous of Patagonia. *Contributions in Science* 416:1–41.

Bottomley, R., et al. 1997. The age of the Popigai impact event and its relation to the events at the Eocene/Oligocene boundary. *Nature* 388:365–68.

Bourn, D., and Coe, M. 1978. The size, structure, and distribution of the giant tortoise population of Aldabra. *Philosophical Transactions of the Royal Society of London B* 282:139–75.

Bower, B. 2000. Out on a limb: The science of body development may make kindling out of evolutionary trees. *Science News* 18:346–47.

Brackenbury, J. 1987. Ventilation of the lung–air sac system. In *Bird Respiration,* ed. T. Seller, 1:39–69. CRC Press, Boca Raton, Fla.

Brackenbury, J., and Amaku, J. 1990. Effects of combined abdominal and thoracic air sac occlusion on respiration in domestic fowl. *Journal of Experimental Biology* 152:93–100.

Bramble, D. M., and Jenkins, F. A. 1998. Locomotor-respiratory integration: Implications for mammalian and avian divergence. *Journal of Vertebrate Paleontology* 18:28A.

Bramwell, C. D. 1971. Flying ability of *Archaeopteryx*. *Nature* 231:128.

Bramwell, C. D., and Whitfield, G. R. 1974. Biomechanics of Pteranodon. *Philosophical Transactions of the Royal Society of London B* 267:503–81.

Brett-Surman, M. K., and Paul, G. S. 1985. A new family of bird-like dinosaurs linking Laurasia and Gondwanaland. *Journal of Vertebrate Paleontology* 5:133–38.

Briggs, D.E.G., et al. 1997. The mineralization of soft tissue in the Lower Cretaceous of Las Hoyas, Spain. *Journal of the Geological Society* (London) 154: 587–88.

Brill, R. W. 1986. On the standard metabolic rates of tropical tunas, including the effect of body size and acute temperature change. *Fishery Bulletin* 85:25–35.

Brill, R. W., and Bushnell, P. G. 1991. Metabolic and cardiac scope of high energy demand teleosts, the tunas. *Canadian Journal of Zoology* 69:2002–9.

Brinkman, D. L., Cifelli, R. L., and Czaplewski, N. J. 1998. First occurrence of *Deinonychus antirrhopus* from the Antlers Formation of Oklahoma. *Oklahoma Geological Survey Bulletin* 146:1–27.

Britt, B. B. 1995. The nature and distribution of pneumatic vertebrae in Theropoda. *Journal of Vertebrate Paleontology* 15:20A.

———. 1997. Postcranial pneumaticity. In Currie and Padian 1997, 590–93.

Britt, B. B., Makovicky, P. J., Gauthier, J., and Bonde, N. 1998. Postcranial pneumatization in *Archaeopteryx*. *Nature* 395:374–75.

Britt, B. B., et al. 2000. A reanalysis of the phylogenetic affinities of *Ceratosaurus* based on new specimens from Utah, Colorado, and Wyoming. *Journal of Vertebrate Paleontology* 20:32A.

Brochu, C. A., and Norell, M. A. 2000. Temporal congruence and the origin of birds. *Journal of Vertebrate Paleontology* 20:197–200.

Brody, S. 1974. *Bioenergetics and Growth*. Hafner Press, New York.

Brower, J. C., and Veinus, J. 1981. Allometry in pterosaurs. *University of Kansas Paleontological Contributions* 105:1–33.

Brush, A. H. 1993. The origin of feathers: A novel approach. In *Avian Biology*, ed. D. Farner, J. A. King, and K. C. Parkes, 9:121–62. Academic Press, London.

———. 1996. On the origin of feathers. *Journal of Evolutionary Biology* 9:131–42.

———. 1998. Protofeathers: What are we looking for? In *The Dinofest Symposium*, ed. D. L. Wolberg et al., 3. Academy of Natural Sciences, Philadelphia.

———. 2000. Evolving a protofeather and feather diversity. *American Zoologist* 40(4):631–39.

Bucher, E. 1992. The causes of extinction of the passenger pigeon. In *Current Ornithology*, ed. D. M. Power, 9:1–36. Plenum Press, New York.

Buffenstein, R., and Yahav, S. 1991. Is the naked molerat *Heterocephalus glaber* an endothermic yet poikilothermic mammal? *Journal of Thermal Biology* 16: 227–32.

Buffetaut, E., Loeuff, J. L., Mechin, P., and Mechin-Salessy, A. 1995. A large French Cretaceous bird. *Nature* 377:110.

Buffetaut, E., and Loeuff, J. L. 1998. A new giant ground bird from the Upper Cretaceous of southern France. *Journal of the Geological Society* (London) 155:1–4.

Buhler P. 1981. Functional anatomy of the avian jaw apparatus. In *Form and Function in Birds*, ed. A. S. King and J. McLelland, 2:439–68. Academic Press, London.

———. 1985. On the morphology of the skull of *Archaeopteryx*. In Hecht et al. 1985, 135–40.

———. 1992. Light bones in birds. *Science Series of the Los Angeles County Museum of Natural History* 36:385–94.

Buisonje, P. H. de. 1985. Climatological conditions during deposition of the Solnhofen Limestones. In Hecht et al. 1985, 45–65.

Bundle, M. W., et al. 1999. High metabolic rates in running birds. *Nature* 397:31–32.

Burgers, P., and Chiappe, L. M. 1999. The wing of *Archaeopteryx* as a primary thrust generator. *Nature* 399:60–62.

Burke, A. C., and Feduccia, A. 1997. Developmental patterns and the identification of homologies in the avian hand. *Science* 278:666–68.

Burke, A. C., Feduccia, A., and Hinchliffe, R. 1998. Counting the fingers of birds and dinosaurs: Reply. *Science* 280:355.

Burnham, D. A., and Zhou Z. 1999. Comparing the furcula in birds and dinosaurs. *Journal of Vertebrate Paleontology* 19:34A.

Burnham, D. A., et al. 2000. Remarkable new birdlike dinosaur from the Upper Cretaceous of Montana. *University of Kansas Paleontological Contributions*, n.s., 13:1–14.

Bustard, H. R. 1967. Reproduction in the Australian gekkonid genus *Oedura*. *Herpetologica* 23:276–84.

Butler, P. J. 1991. Exercise in birds. *Journal of Experimental Biology* 160:233–62.

Buttemer, W. A., and Dawson, T. J. 1989. Body temperature, water flux, and estimated energy expenditure of incubating emus. *Comparative Biochemistry and Physiology A* 94:21–24.

Cai, Z., and Zhao, L. 1999. A long tailed bird from the Late Cretaceous of Zhejiang. *Science in China* 42:434–41.

Calder, W. A. 1978. The kiwi. *Scientific American* 239 (1):132–42.

Calder, W. A., and King, J. R. 1974. Thermal and caloric relationships of birds. In *Avian Biology*, ed. D. S. Farner, J. King, and K. C. Parkes, 4:260–413. Academic Press, New York.

Calzavara, M. G., Muscio, G., and Wild, R. 1980. *Megalancosaurus preonensis* n. g. n. sp., a new reptile from the Norian of Friuli. *Gortania-Atti del Museo Friulano di Storia Naturale* 2:49–54.

Camp, C. L. 1936. A new type of small bipedal dinosaur from the Navajo sandstone of Arizona. *University of California Publications, Bulletin of the Department of Geological Sciences* 24:36–56.

Campbell, K. E., and Tonni, E. P. 1980. A new genus of teratorn from the Huayquerian of Argentina. *Los Angeles County Museum of Natural History, Contributions in Science* 330:59–68.

———. 1982. Preliminary observations of the paleobiology and evolution of teratorns. *Journal of Vertebrate Paleontology* 1:265–72.

———. 1983. Size and locomotion in teratorns. *Auk* 100:390–403.

Campbell, K. E., and Marcus, L. 1992. The relationship of hindlimb bone dimensions to body weight in birds. *Science Series of the Los Angeles County Museum of Natural History* 36:395–412.

Caple, G., Balda, R. P., and Willis, W. R. 1983. The physics of leaping animals and the evolution of preflight. *American Naturalist* 121:455–76.

Carpenter, K. 1999. *Eggs, Nests, and Baby Dinosaurs*. Indiana University Press, Bloomington.

Carrano, M. T., and Sampson, S. D. 1999. Evidence for paraphyletic "Ceratosauria" and its implications for theropod dinosaur evolution. *Journal of Vertebrate Paleontology* 19:36A.

Carrier, D. R. 1987. The evolution of locomotor stamina in tetrapods: Circumventing a mechanical restraint. *Paleobiology* 13:326–41.

———. 1998. The evolution of locomotor stamina in tetrapods. *Journal of Vertebrate Paleontology* 18:31A.

Carrier, D. R., and Farmer, C. G. 2000. The evolution of pelvic aspiration in archosaurs. *Paleobiology* 26: 271–93.

Carroll, R. L. 1988. *Vertebrate Paleontology and Evolution.* W. H. Freeman, New York.

———. 1996. Revealing the patterns of macroevolution. *Nature* 381:19–20.

Case, T. J. 1978a. On the evolution and adaptive significance of postnatal growth in the terrestrial vertebrates. *Quarterly Review of Biology* 53:243–82.

———. 1978b. Speculations on the growth rate and reproduction of some dinosaurs. *Paleobiology* 4: 320–28.

Chandler, R. M. 1994. The wing of *Titanis walleri* from the late Blancan of Florida. *Bulletin of the Florida State Museum of Natural History, Biology Series* 36 (6):175–80.

Charig, A. J. 1976. Dinosaur monophyly and a new class of vertebrates: A critical review. In *Morphology and Biology of Reptiles,* ed. A. A. Bellairs and C. B. Cox. Academic Press, New York.

Charig, A. J., and Milner, A. C. 1997. *Baryonyx walkeri,* a fish eating dinosaur from the Wealden of Surrey. *Bulletin of the Natural History Museum, Geology Series* 53:11–70.

Chatterjee, S. 1985. *Postosuchus,* a new thecodont reptile from the Triassic of Texas and the origin of tyrannosaurs. *Philosophical Transactions of the Royal Society of London B* 309:395–460.

———. 1991. Cranial anatomy and relationships of a New Triassic bird from Texas. *Philosophical Transactions of the Royal Society of London B* 332:277–346.

———. 1993. *Shuvosaurus,* a new theropod. *National Geographic Research and Exploration* 9:274–85.

———. 1995. The Triassic bird *Protoavis. Archaeopteryx* 13:15–31.

———. 1997. *The Rise of Birds.* Johns Hopkins University Press, Baltimore.

———. 1998a. The avian status of *Protoavis. Archaeopteryx* 16:99–122.

———. 1998b. Counting the fingers of birds and dinosaurs. *Science* 280:355.

———. 1999a. *Protoavis* and the early evolution of birds. *Palaeontographica A* 254:1–100.

———. 1999b. Feathered coelurosaurs and the early evolution of avian flight. *Journal of Vertebrate Paleontology* 19:37A.

Chen, P., Dong, Z., and Zhen, S. 1998. An exceptionally well-preserved theropod dinosaur from the Yixian Formation of China. *Nature* 391:147–52.

Chiappe, L. M. 1995. The first 85 million years of avian evolution. *Nature* 378:349–55.

———. 1996. Late Cretaceous birds of southern South America: Anatomy and systematics of Enantiornithes and *Patagopteryx deferrariisi. Münchner Geowissenschaftliche Abhandlungen* 30:203–44.

Chiappe, L. M., and Calvo, J. O. 1994. *Neuquenornis volans,* a new Late Cretaceous bird from Patagonia, Argentina. *Journal of Vertebrate Paleontology* 14: 230–46.

Chiappe, L. M., Norell, M., and Clark, J. 1996. Phylogenetic position of *Mononykus* from the Late Cretaceous of the Gobi Desert. *Memoirs of the Queensland Museum* 39:557–82.

———. 1997. *Mononykus* and birds: Methods and evidence. *Auk* 114:300–302.

———. 1998. The skull of a relative of the stem-group bird *Mononykus. Nature* 392:275–78.

Chiappe, L. M., Ji S., Ji Q., and Norell, M. A. 1999. Anatomy and systematics of the Confuciusornithidae from the late Mesozoic of northeastern China. *Bulletin of the American Museum of Natural History* 242:1–89.

Chin, K., Tokaryk, T. T., Erickson, G. M., and Calk, L. C. 1998. A king-sized theropod coprolite. *Nature* 393:680–82.

Chin, K., Eberth, D. A., and Sloboda, W. J. 1999. Exceptional soft-tissue preservation in a theropod coprolite from the Upper Cretaceous Dinosaur Park Formation of Alberta. *Journal of Vertebrate Paleontology* 19:37A.

Chinsamy, A. 1993. Bone histology and growth trajectory of the prosauropod dinosaur *Massospondylus carinatus* Owen. *Modern Geology* 18:319–29.

———. 1994. Dinosaur bone histology: Implications and inferences. *Paleontological Society Special Publication* 7:213–28.

Chinsamy, A., Chiappe, L. M., and Dodson, P. 1995. Mesozoic avian bone microstructure: Physiological implications. *Paleobiology* 21:561–74.

Chinsamy, A., Martin, L. D., and Dodson, P. 1998. Bone microstructure of the diving *Hesperornis* and the volant *Ichthyornis* from the Niobrara Chalk of western Kansas. *Cretaceous Research* 19:225–35.

Chinsamy, A., Rich, T., and Vickers-Rich, P. 1998. Polar dinosaur bone histology. *Journal of Vertebrate Paleontology* 18:385–90.

Christiansen, P. 1999. Long bone scaling and limb posture in non-avian theropods: Evidence for differential allometry. *Journal of Vertebrate Paleontology* 19:666–80.

Christiansen, P., and Bonde, N. 2000. Axial and appendicular pneumaticity in *Archaeopteryx. Proceedings of the Royal Society of London* 267:2501–5.

Chure, D. J. 1994. *Koparion douglassi,* a new dinosaur from the Morrison Formation of Dinosaur National Monument; The oldest Troodontid. *Brigham Young University Geology Studies* 40:11–15.

———. 1995. The teeth of small theropods from the Morrison Formation, UT. *Journal of Vertebrate Paleontology* 15:23A.

———. 1999. The wrist of *Allosaurus* and the evolution of the semilunate carpal. *Journal of Vertebrate Paleontology* 19:38A.

Chure, D. J., and Madsen, J. H. 1996. On the presence of furculae in some non-maniraptoran theropods. *Journal of Paleontology* 16:573–77.

Claessens, L. 1996. Dinosaur gastralia and their function in respiration. *Journal of Vertebrate Paleontology* 16:28A.

Claessens, L., and Perry, S. F. 1998. Using comparative anatomy to reconstruct theropod respiration. *Journal of Vertebrate Paleontology* 18:34A.

Clark, J. M., Perle, A., and Norell, M. 1994. The skull of *Erlicosaurus andrewsi,* a Late Cretaceous "segnosaur" from Mongolia. *American Museum Novitates* 3115:1–39.

Clark, J. M., Norell, M., and Chiappe, L. 1999. An oviraptorid skeleton from the Late Cretaceous of Ukhaa Tolgod, Mongolia, preserved in an avian-like brooding position over an oviraptorid nest. *American Museum Novitates* 3265:1–36.

Clemens, W. A., and Nelms, L. G. 1993. Paleoecological implications of Alaskan terrestrial vertebrate fauna in latest Cretaceous time at high paleolatitudes. *Geology* 21:503–6.

Colbert, E. H. 1989. The Triassic dinosaur *Coelophysis. Museum of Northern Arizona Bulletin* 57: 1–160.

Colbert, E. H., and Russell, D. A. 1969. The small dinosaur *Dromaeosaurus. American Museum Novitates* 2380:1–49.

Constantine, A., Chinsamy, A., Vickers-Rich, P., and Rich, T. H. 1998. Periglacial environments and polar dinosaurs. *South African Journal of Science* 94: 137–41.

Coombs, W. P. 1978. The families of the ornithischian dinosaur order Ankylosauria. *Journal of Paleontology* 21:143–70.

———. 1989. Modern analogs for dinosaur nesting and parental behavior. *Geological Society of America, Special Paper* 238:21–53.

Cooper, A., and Perry, D. 1997. Mass survival of birds across the Cretaceous-Tertiary boundary. *Science* 275:1109–13.

Cott, H. 1961. Scientific results of an inquiry into the ecology and economic status of the Nile crocodile in Uganda and northern Rhodesia. *Transactions of the Zoological Society of London* 29:215–337.

Coulombe, H. N., Ridgeway, S. H., and Evans, W. E. 1965. Respiratory water exchange in two species of porpoise. *Science* 149:86–88.

Coulson, T. D., Coulson, R. A., and Hernandez, T. 1973. Some observations on the growth of captive alligators. *Zoologica* 58:47–52.

Court, N. 1994. Limb posture and gait in *Numidotherium koholensis,* a primitive proboscidean from the Eocene of Algeria. *Zoological Journal of the Linnean Society* (London) 111:297–338.

Courtillot, V. 1999. *Evolutionary Catastrophes: The Science of Mass Extinctions.* Cambridge University Press, Cambridge.

Cowen, R., and Lipps, J. H. 1982. An adaptive scenario for the origin of birds and their flight. *Proceedings of the Third North American Paleontological Convention* 1:109–12.

Cracraft, J. 1971. Caenagnathiformes: Cretaceous birds convergent in jaw mechanism with dicynodont reptiles. *Journal of Paleontology* 45:805–9.

Crompton, A. W., Taylor, C. R., and Jagger, A. 1978. Evolution of homeothermy in mammals. *Nature* 272:333–36.

Crush, P. J. 1984. A late Upper Triassic sphenosuchid crocodilian from Wales. *Palaeontology* 27:131–57.

Culick, F.E.C. 1979. The origins of the first powered, man-carrying airplane. *Scientific American* 241(1): 86–100.

Cummins, K. W., and Wuycheck, J. C. 1971. Caloric equivalents for investigations in ecological energetics. *Mitteilung International Veröffentlichung Limnologische* 18:1–157.

Currie, P. J. 1985. Cranial anatomy of *Stenonychosaurus inequalis* and its bearing on the origin of birds. *Canadian Journal of Earth Sciences* 22:1643–58.

———. 1987. Bird-like characteristics of the jaws and teeth of troodontid theropods. *Journal of Vertebrate Paleontology* 7:72–81.

———. 1995. New information on the anatomy and relationships of *Dromaeosaurus albertensis. Journal of Vertebrate Paleontology* 15:576–91.

———. 1997. Theropoda. In Currie and Padian 1997, 731–37.

Currie, P. J., and Dodson, P. 1984. Mass death of a herd of ceratopsian dinosaurs. *Third Symposium on Mesozoic Terrestrial Ecosystems,* ed. W.-E. Reif and F. Westphal, 61–66. ATTEMPTO Verlag, Tübingen.

Currie, P. J., and Zhao, X.-J. 1993a. A new carnosaur from the Jurassic of Xinjiang, People's Republic of China. *Canadian Journal of Earth Sciences* 30: 2037–81.

———. 1993b. A new troodontid braincase from the Dinosaur Park Formation of Alberta. *Canadian Journal of Earth Sciences* 30:2231–47.

Currie, P. J., and Padian, K. 1997. *Encyclopedia of Dinosaurs.* Academic Press, San Diego.

Czerkas, S. 1997. Skin. In Currie and Padian 1997, 669–75.

Dahn, R. D., and Fallon, J. F. 2000. Interdigital regulation of digit identity and homeotic transformation of modulated BMP signaling. *Science* 289:438–41.

Dalton, R. 2000. Feathers fly in Beijing. *Nature* 405:992.

Darwin, C. 1859. *On the Origin of Species by Means of Natural Selection, or the Preservation of Favoured Races in the Struggle for Life.* Murray, London.

Davis, P. G. 1996. The taphonomy of *Archaeopteryx. Bulletin of the National Science Museum, Tokyo, Series C* 21:1–25.

Davis, P. G., and Briggs, D.E.G. 1995. Fossilization of feathers. *Geology* 23:783–86.

———.1998. The impact of decay and disarticulation on the preservation of fossil birds. *Palaios* 13:3–13.

Dawson, A., et al. 1994. Ratite-like neoteny induced by neonatal thyroidectomy of European starlings. *Journal of Zoology* (London) 232:633–39.

De Beer, G. 1954. Archaeopteryx lithographica: *A Study Based on the British Museum Specimen.* British Museum of Natural History, London.

Degen, A. A., Kam, M., Rosenstrauch, A., and Plavnik, I. 1991. Growth rate, total body water volume, dry-matter intake, and water consumption of domesticated ostriches. *Animal Productivity* 52:225–32.

del Hoyo, J., Elliot, A., and Sargatal, J., eds. 1992–99. *Handbook of the Birds of the World.* Vols. 1–5. Lynx Edicions, Barcelona.

Demment, M. W., and Soest, P. J. 1985. A nutritional explanation for body-size patterns of ruminant and nonruminant herbivores. *American Naturalist* 125: 641–72.

Derstler, K., and Hembree, D. 1999. Observations of *Confuciusornis sanctus* and other fossil Aves. *Journal of Vertebrate Paleontology* 19:42A.

Dilkes, D. W. 1998. The Early Triassic rhynchosaur *Mesosuchus browni* and the interrelationships of basal archosauromorph reptiles. *Philosophical Transactions of the Royal Society of London B* 353: 501–41.

Dingus, L., and Rowe, T. 1998. *The Mistaken Extinction: Dinosaur Evolution and the Origin of Birds.* W. H. Freeman, New York.

Dong, Z. 1984. A new theropod dinosaur from the Middle Jurassic of Sichuan Basin. *Vertebrata PalAsiatica* 22:213–19.

Dong, Z., Zhou, S., and Zhang, Y. 1983. The dinosaurian remains from Sichuan Basin, China. *Paleontologica Sinica* 162:1–145.

Dong, Z., and Currie, P. J. 1996. On the discovery of an oviraptorid skeleton on a nest of eggs at Bayan Mandahu, Inner Mongolia, People's Republic of China. *Canadian Journal of Earth Sciences* 33: 631–36.

Dorst, J. 1973. Appareil respiratoire. In *Traité de Zoologic.* Vol. 16, *Mammifères: Splanchnologie.* L'Académie Saint-Germain de Médecine, Paris.

———. 1974. *The Life of Birds.* Columbia University Press, New York.

Downs, A. 2000. *Coelophysis bauri* and *Syntarsus rhodesiensis* compared, with comments on the preparation and preservation of fossils from the Ghost Ranch *Coelophysis* quarry. *New Mexico Museum of Natural History and Science Bulletin* 17:33–38.

Dudley, R. 1998. Atmospheric oxygen, giant Paleozoic insects, and the evolution of aerial locomotor performance. *Journal of Experimental Biology* 201: 1043–50.

Duncker, H.-R. 1971. The lung air sac system of birds. *Advances in Anatomy, Embryology, and Cell Biology* 45:1–171.

———. 1978. General morphological principles of amniote lungs. In *Respiratory Function in Birds, Adult and Embryonic,* ed. J. Piiper, 2–15. Springer-Verlag, Heidelberg.

———. 1979. Coelomic cavities. In *Form and Function in Birds,* ed. A. King and J. McLelland, 1:39–67. Academic Press, New York.

Dunham, A. E., Overall, K. L., Porter, W. P., and Forster, C. A. 1989. Implications of ecological energetics and biophysical and developmental constraints for life-history variation in dinosaurs. *Geological Society of America, Special Paper* 238:1–19.

Dunning, J. B. 1993. *CRC Handbook of Avian Body Masses.* CRC Press, Boca Raton, Fla.

Dyke, G. J., and Mayr, G. 1999. Did parrots exist in the Cretaceous period? *Nature* 399:317–18.

Earls, K. D. 2000. Kinematics and mechanics of ground take-off in the starling *Sturnis vulgaris* and the quail *Coturnix coturnix. Journal of Experimental Biology* 203:725–39.

Easley, R. R. 1999. Ground effect and the beginnings of avian flight. *Journal of Vertebrate Paleontology* 19:43A.

Eckert, S. A. 1992. Bound for deep water. *Natural History* 101(3):28–35.

Eisenberg, J. F., and Gould, E. 1970. The tenrecs: A study in mammalian behavior and evolution. *Smithsonian Contributions to Zoology* 27:1–97.

Eldershaw, T.P.D., et al. 1997. Potential for nonshivering thermogenesis in perfused chicken. *Comparative Biochemistry and Physiology A* 117:5455–554.

Ellenberger, P. 1974. Contribution à la classification des Pistes de Vertébrés du Trias: Les types du Stormberg d'Afrique du Sud. *Paleovertebrata, Mémoire Éxtraordinaire* 1974:1–141.

———. 1977. Quelques précisions sur 1'anatomie et la place systématique très spéciale de *Cosesaurus aviceps. Cuadernos Geologia Ibérica* 4:169–88.

Ellenberger, P., and de Villalta, J. F. 1974. Sur la présence d'un ancêtre probable des oiseaux dans le Muschelkalk supérieur de Catalogne (Espagne): Note préliminaire. *Acta Geológica Hispánica* 9: 162–68.

Ellenberger, W., Dittrich, H., and Baum, H. 1956. *An Atlas of Animal Anatomy for Artists.* Dover Publications, New York.

Else, P. L., and Hulbert, A. J. 1985. An allometric comparison of mammalian and reptilian tissues: The implications for the evolution of endothermy. *Journal of Comparative Physiology B* 156:3–11.

———. 1987. Evolution of mammalian endothermic metabolism: Leaky membranes as a source of heat. *American Journal of Physiology* 253:R1–7.

Elzanowski, A. 1981. Embryonic bird skeletons from the Late Cretaceous of Mongolia. *Palaeontologica Polonica* 42:147–79.

———. 1983. Birds in Cretaceous ecosystems. *Acta Palaeontologica Polonica* 28:75–92.

———. 1985. The evolution of parental care in birds with reference to fossil embryos. In *Acta: XVIII Congressus Internationalis Ornithologici,* ed. V. D. Ilyichev and V. M. Gavrilov, 1:178–83. Nauka, Moscow.

———. 1988. Ontogeny and evolution of the ratites. In *Acta: XIX Congressus Internationalis Ornithologici,* ed. H. Ouellett, 2:2037–46. University of Ottawa Press, Ottawa.

———. 1995. Cretaceous birds and avian phylogeny. *Courier Forschungsinstut Senckenberg* 181:37–53.

———. 1999. A comparison of the jaw skeleton in theropods and birds, with a description of the palate in the Oviraptoridae. In Olson 1999, 311–23.

Elzanowski, A., and Wellnhofer, P. 1992. A new link between theropods and birds from the Cretaceous of Mongolia. *Nature* 359:821–23.

———. 1993. Skull of *Archaeornithoides* from the Upper Cretaceous of Mongolia. *American Journal of Science* 293A:235–52.

———. 1995. The skull of *Archaeopteryx* and the origin of birds. *Archaeopteryx* 13:41–46.

———. 1996. Cranial morphology of *Archaeopteryx:* Evidence from the seventh skeleton. *Journal of Vertebrate Paleontology* 16:81–94.

Elzanowski, A., and Pasko, L. 1999. A skeletal reconstruction of *Archaeopteryx. Acta Ornithologica* 34: 123–29.

Elzanowski, A., Hope, S., Paul, G. S., and Stidham, T. A. 2000. The quadrate of an odontognathus bird from the Late Cretaceous Lance Formation of Wyoming. *Journal of Vertebrate Paleontology* 20:712–19.

Emerson, S. P. 1979. The ilio-sacral articulation in frogs: Form and function. *Biological Journal of the Linnean Society* 11:153–68.

Erben, H. K., Hoefs, J., and Wedepohl, K. H. 1979. Paleobiological and isotopic studies of eggshells from a declining dinosaur species. *Paleobiology* 5:380–414.

Evans, S. E., and Milner, A. R. 1994. Middle Jurassic microvertebrate assemblages from the British Isles. In *In the Shadow of the Dinosaurs: Early Mesozoic Tetrapods,* ed. N. Fraser and H.-D. Sues, 303–21. Cambridge University Press, Cambridge.

Ewer, R. F. 1965. The anatomy of the thecodont reptile *Euparkeria capensis* Broom. *Philosophical Transactions of the Royal Society of London B* 248:379–435.

Fairall, N. 1980. Growth and age determination in the hyrax. *South African Journal of Zoology* 15:16–21.

Fara, E., and Benton, M. J. 2000. The fossil record of Cretaceous tetrapods. *Palaios* 15:161–65.

Farley, C. T., and Ko, T. C. 1997. Mechanics of locomotion in lizards. *Journal of Experimental Biology* 200:2177–88.

Farlow, J. O. 1987. Speculations about the diet and digestive physiology of herbivorous dinosaurs. *Paleobiology* 13:60–72.

———. 1990. Dinosaur energetics and thermal biology. In *The Dinosauria,* ed. D. B. Weishampel, P. Dodson, and H. Osmólska, 43–62. University of California Press, Berkeley.

Farlow, J. O., Dodson, P., and Chinsamy, A. 1995. Dinosaur biology. *Annual Review of Ecological Systematics* 26:445–71.

Farlow, J. O., Gatsey, S. M., Holtz, T. R., Hutchinson, J. R., and Robinson, J. M. 2000. Theropod locomotion. *American Zoologist* 40(4):640–63.

Farmer, C. G., and Carrier, D. R. 2000a. Ventilation and gas exchange during treadmill locomotion in the American alligator. *Journal of Experimental Zoology* 203:1671–78.

———. 2000b. Pelvic aspiration in the American alligator. *Journal of Experimental Zoology* 203:1679–87.

Fedak, M. A., and Seeherman, H. J. 1979. Reappraisal of energetics of locomotion shows identical cost in bipeds and quadrupeds including ostrich and horse. *Nature* 282:713–16.

Fedde, M. R. 1987. Respiratory muscles. In *Bird Respiration,* ed. T. Seller, 1:3–37. CRC Press, Boca Raton, Fla.

Feduccia, A. 1980. *The Age of Birds.* Harvard University Press, Cambridge, Mass.

———. 1985. On why the dinosaur lacked feathers. In Hecht et al. 1985, 75–79.

———. 1986. The scapulocoracoid of flightless birds: A primitive avian character similar to that of theropods. *Ibis* 128:128–32.

———. 1993a. Evidence from claw geometry indicating arboreal habits of *Archaeopteryx. Science* 259: 790–93.

———. 1993b. Birdlike characters in the Triassic archosaur *Megalancosaurus. Naturwissenschaften* 80: 564–66.

———. 1995. Explosive evolution in Tertiary birds and mammals. *Science* 267:637–38.

———. 1996. *The Origin and Evolution of Birds.* Yale University Press, New Haven.

———. 1999. 1,2,3 = 2,3,4: Accommodating the cladogram. *Proceedings of the National Academy of Sciences* 96:4740–42.

Feduccia, A., and Tordoff, H. B. 1979. Feathers of *Archaeopteryx:* Asymmetric vanes indicate aerodynamic function. *Science* 203:1021–22.

Feduccia, A., and Wild, R. 1993. Birdlike characters in the Triassic archosaur *Megalancosaurus. Naturwissenschaften* 80:564–66.

Feduccia, A., Martin, L. D., and Simmons, J. E. 1996. Nesting dinosaur. *Science* 272:1571.

Feduccia, A., and Martin, L. D. 1998. Theropod-bird link reconsidered. *Nature* 391:754.

Feduccia, A., et al. 1998. Birds of a feather. *Scientific American* 278(6):8.

Fell, H. B. 1939. The origin and developmental mechanics of the avian sternum. *Philosophical Transactions of the Royal Society of London B* 229:407–64.

Ferguson, M.W.J. 1984. Craniofacial development in *Alligator mississippiensis. Symposia of the Zoological Society of London* 52:223–73.

Fieler, C. L., and Jayne, B. C. 1998. Effects of speed on the hindlimb kinematics of the lizard *Dipsosaurus dorsalis. Journal of Experimental Biology* 201:609–22.

Fischman, J. 1995. Were dinos cold-blooded after all? The nose knows. *Science* 270:735–36.

———. 1999. Feathers don't make the bird. *Discover* 20(1):48–49.

Fisher, H. 1945. Locomotion in the fossil vulture *Teratornis. American Midland Naturalist* 33:725–42.

Fisher, P., et al. 2000. Cardiovascular evidence for an intermediate or higher metabolic rate in an ornithischian dinosaur. *Science* 288:503–5.
Forster, A., Sampson, S. D., Chiappe, L. M., and Krause, D. W. 1998. The theropod ancestry of birds: New evidence from the Late Cretaceous of Madagascar. *Science* 279:1915–19.
Fowler, M. E. 1991. Comparative clinical anatomy of ratites. *Journal of Zoo and Wildlife Medicine* 22: 204–27.
Fox, D. L., Fisher, D. C., and Leighton, L. R. 1999. Reconstructing phylogeny with and without temporal data. *Science* 284:1816–19.
Fraser, N. C., et. al. 1996. A Triassic lagerstätten from eastern North America. *Nature* 380:615–19.
Frey, E., and Martill, D. M. 1999. Soft tissue preservation in a specimen of *Pterodactylus kochi* from the Upper Jurassic of Germany. *Neues Jahrbuch für Geologie und Paläontologia Abhandlungen* 210: 421–41.
Fricke, H. C., and Rogers, R. R. 2000. Multiple taxon–multiple locality approach to providing evidence for warm-blooded theropod dinosaurs. *Geology* 28:799–802.
Frith, H. J., and Calaby, J. H. 1969. *Kangaroos.* C. Hurst, London.
Galton, P. M. 1970a. Ornithischian dinosaurs and the origin of birds. *Evolution* 24:448–62.
———. 1970b. The posture of hadrosaurian dinosaurs. *Journal of Paleontology* 44:464–73.
———. 1974. The ornithischian dinosaur *Hypsilophodon* from the Wealden of the Isle of Wight. *Bulletin of the British Museum (Natural History)* 25:1–152.
———. 1977. On *Staurikosaurus pricei,* an early saurischian dinosaur from the Triassic of Brazil. *Paläontologische Zeitschrift* 51:234–45.
Gans, C. 1976. Ventilatory mechanisms and problems in some amphibious aspiration breathers. In *Respiration of Amphibious Vertebrates,* ed. G. M. Hughes, 357–74. Academic Press, London.
Gans, C., and Clark, B. 1976. Studies on ventilation of *Caiman crocodilus. Respiration Physiology* 26:285–301.
Gardiner, B. G. 1982. Tetrapod classification. *Zoological Journal of the Linnean Society* (London) 74: 207–32.
Garland, T. G. 1983. Scaling the ecological cost of transport to body mass in terrestrial mammals. *American Naturalist* 121:571–87.
Garner, J. P., and Thomas, A.L.R. 1998. Counting the fingers of birds and dinosaurs. *Science* 280:355.
Garner, J. P., Taylor, G., and Thomas, A.L.R. 1999. On the origin of birds. *Proceedings of the Royal Society of London B* 266:1259–66.
Gatesy, S. M. 1990. Caudofemoralis musculature and the evolution of theropod locomotion. *Paleobiology* 16:170–86.
———. 1991. Bipedal locomotion: Effects of speed, size, and limb posture in birds and humans. *Journal of Zoology* (London) 224:127–47.
Gatesy, S. M., and Dial, K. P. 1996. From frond to fan: *Archaeopteryx* and the evolution of short-tailed birds. *Evolution* 50:2037–48.
Gatesy, S. M., et al. 1999. Three-dimensional preservation of foot movements in Triassic theropod dinosaurs. *Nature* 399:141–44.
Gauthier, J. A. 1986. Saurischian monophyly and the origin of birds. *Memoirs of the California Academy of Sciences* 8:1–55.
———. In press. *New Perspectives on the Origins of Early Evolution of Birds: Proceedings of the International Symposium in Honor of John H. Ostrom.* Yale Peabody Museum, New Haven.
Gauthier, J. A., and Padian, K. 1985. Phylogenetic, functional, and aerodynamic analyses of the origin of birds and their flight. In Hecht et al. 1985, 185–97.
Gauthier, J., Kluge, A. G., and Rowe, T. 1988. Amniote phylogeny and the importance of fossils. *Cladistics* 4:105–209.
Gauthier, J. A., and Gishlick, A. D. 2000. Re-examination of the manus of *Compsognathus* and its relevance to the original morphology of the coelurosaur manus. *Journal of Vertebrate Paleontology* 20:43A.
Geist, N. R. 2000. Nasal respiratory turbinate function in birds. *Physiological and Biochemical Zoology* 73: 581–89.
Geist, N. R., and Jones, T. D. 1996. Juvenile skeletal structure and the reproductive habits of dinosaurs. *Science* 272:712–14.
Geist, N. R., Jones, T. D., and Ruben, J. A. 1997. Implications of soft tissue preservation in the compsognathid dinosaur, *Sinosauropteryx. Journal of Vertebrate Paleontology* 17:48A.
Geist, N. R., and Feduccia, A. 2000. Gravity-defying behaviors: Identifying models for protoaves. *American Zoologist* 40(4):664–75.
Gibbons, A. 1998a. Missing link ties birds, dinosaurs. *Science* 279:1851–52.
———. 1998b. Dinosaur fossils, in fine feather, show link to birds. *Science* 280:2051.
Gierlinski, G. 1996. Feather-like impressions in a theropod resting trace from the Lower Jurassic of Massachusetts. *Museum of Northern Arizona Bulletin* 60: 179–84.
———. 1997. What type of feathers could nonavian dinosaurs have, according to an Early Jurassic ichnological evidence from Massachusetts? *Przeglad Geologiczny* 45:419–22.
Gierlinski, G., and Sabath, K. 1998. Protoavian affinity of the *Plesiornis* trackmaker. *Journal of Vertebrate Paleontology* 18:46A.
Giffin, E. B. 1995. Postcranial paleoneurology of the Diapsida. *Journal of Zoology* (London) 235:389–410.
Gill, F. B. 1995. *Ornithology.* W. H. Freeman, New York.
Gingerich, P. D. 1973. Skull of *Hesperornis* and early evolution of birds. *Nature* 243:70–73.

Gishlick, A. D. 2000. An evaluation of the climbing abilities of basal Maniraptora. *Journal of Vertebrate Paleontology* 20:44A.

Glut, D. F. 1997. *Dinosaurs, the Encyclopedia.* McFarland and Company, Jefferson, N.C.

———. 1999. *Dinosaurs, the Encyclopedia.* Suppl. 1. McFarland and Company, Jefferson, N.C.

Gordon, J. E. 1978. *Structures, or Why Things Don't Fall Down.* Penguin Books, New York.

Gould, S. J. 1993. *The Book of Life: An Illustrated History of Life on Earth.* W. W. Norton, New York.

———. 2000. Tales of a feathered tail. *Natural History* 110(11):32–42.

Gow, C. E. 1975. The morphology and relationships of *Youngina capensis* Broom and *Prolacerta broomi. Palaeontologica Africana* 18:89–131.

Gower, D. J., and Weber, E. 1998. The braincase of *Euparkeria,* and the evolutionary relationships of birds and crocodilians. *Biological Reviews* 73:367–411.

Grand, T. I. 1977. Body weight: Its relation to tissue composition, segment distribution, and motor function. *American Journal of Physical Anthropology* 47:211–40.

Greenwalt, C. H. 1962. Dimensional relationships for flying animals. *Smithsonian Miscellaneous Collections* 144:1–46.

Grenard, S. 1991. *Handbook of Alligators and Crocodiles.* Krieger Publishing, Malabar, Fla.

Griffiths, P. J. 1993. The claws and digits of *Archaeopteryx lithographica. Geobios* 16:1010–106.

———. 1996. The isolated *Archaeopteryx* feather. *Archaeopteryx* 14:1–26.

———. 1998/2000. The evolution of feathers from dinosaur hair. *Gaia* 15:399–403.

Groscolas, R. 1990. Metabolic adaptations to fasting in emperor and king penguins. In *Penguin Biology,* ed. L. S. Davis and J. T. Darby, 269–96. Academic Press, London.

Gross, E. A., Swenberg, J. A., Field, S., and Popp, J. A. 1982. Comparative morphometry of the nasal cavity in rats and mice. *Journal of Anatomy* 135:83–88.

Grubb, B., Jorgensen, D. D., and Conner, M. 1983. Cardiovascular changes in the exercising emu. *Journal of Experimental Biology* 104:193–201.

Grzimek, B., ed. 1973. *Grzimek's Animal Life Encyclopedia.* Vol. 9, *Birds III.* Van Nostrand Reinhold, New York.

———. 1990. *Grzimek's Encyclopedia of Mammals.* Vol. 1. McGraw Hill Publishing, New York.

Hai, T. 1993. *Some New Discoveries about the Groups of Palaeoecological Geography in Xinjiang and the Study of Them* (in Chinese). Xinjiang Science and Technology and Hygiene Publishing House, China.

Harkema, J. R., et al. 1987. Nonolfactory surface epithelium of the nasal cavity of the bonnet monkey: A morphologic and morphometric study of the transitional and respiratory epithelium. *American Journal of Anatomy* 180:266–79.

Harrison, C.J.O., and Walker, C. A. 1973. *Wyleyia:* A new bird humerus from the Lower Cretaceous of England. *Paleontology* 16:721–28.

———. 1975. The Bradycnemidae, a new family of giant owls from the Cretaceous of Romania. *Paleontology* 18:563–70.

Hartman, F. A. 1961. Locomotor mechanisms of birds. *Smithsonian Miscellaneous Collections* 143:1–91.

Hassenpflug, W., and Kopp, G. 1997. The scientist's bookshelf. *American Scientist* 85:479–81.

Haubitz, B. M., Prokop, W., Dohring, W., Ostrom, J. H., and Wellnhofer, P. 1988. Computed tomography of *Archaeopteryx. Paleobiology* 14:206–13.

Haubold, H., and Buffetaut, E. 1987. Une nouvelle interprétation de *Longisquama insignis,* reptile énigmatique du Trias supérieur d'Asie centrale (in English). *Comptes rendus de l'Acadçmie des Sciences,* ser. 2A, 305:65–70.

Hayes, J. P., and Garland, T. 1995. The evolution of endothermy: Testing the aerobic capacity model. *Evolution* 49:836–47.

Hayflick, L. 1995. *How and Why We Age.* Ballantine, New York.

Hazelhurst, G. A., and Rayner, J. M. 1992. Flight characteristics of Triassic and Jurassic pterosaurs. *Paleobiology* 18:447–63.

Hecht, M. K. 1985. The biological significance of *Archaeopteryx.* In Hecht et al. 1985, 149–60.

Hecht, M. K., and Tarsitano, S. 1982. The paleobiology and phylogenetic position of *Archaeopteryx. Geobios mémoires spéciales* 6:141–49.

———. 1984. Paleontological myopia. *Nature* 309:588.

Hecht, M. K., Ostrom, J. H., Viohl, G., and Wellnhofer, P., eds. 1985. *The Beginnings of Birds: Proceedings of the International Archaeopteryx Conference, Eichstätt, 1984.* Freunde des Jura-Museums Eichstätt, Eichstätt.

Hecht, M. K., and Hecht, B. M. 1994. Conflicting developmental and paleontological data: The case of the bird manus. *Acta Palaeontologica Polonica* 38:329–38.

Hedges, S. B., and Schweitzer, M. H. 1995. Detecting dinosaur DNA. *Science* 268:1191–94.

Hedges, S. B., et al. 1996. Continental breakup and the ordinal diversification of birds and mammals. *Nature* 381:226–29.

Heilmann, G. 1926. *The Origin of Birds.* Witherby, London.

Heinrich, B. 1993. *The Hot Blooded Insects: Strategies and Mechanisms of Thermoregulation.* Harvard University Press, Cambridge, Mass.

Hembree, D. 1999. Re-evaluation of the posture and claws of *Confuciusornis. Journal of Vertebrate Paleontology* 19:50A.

Hengst, R. 1998. Lung ventilation and gas exchange in theropod dinosaurs. *Science* 281:47.

Hengst, R., et al. 1996. Biological consequences of Mesozoic atmospheres: Respiratory adaptations and functional range of *Apatosaurus.* In *Cretaceous-Tertiary Mass Extinctions: Biotic and Environmen-*

*tal Changes,* ed. N. Macleod and G. Keller, 327–47. W. W. Norton, New York.

Hennig, W. 1966. *Phylogenetic Systematics.* Trans. D. D. Davis and R. Zangerl. University of Illinois Press, Urbana.

———. 1981. *The Phylogeny of Insects.* Pitman Press, Bath, England.

Heptonstall, W. B. 1971a. Quantitative assessment of the flight of *Archaeopteryx. Nature* 228:185–86.

———. 1971b. Flying ability of *Archaeopteryx. Nature* 231:128.

Herd, R. M., and Dawson, T. J. 1984. Fiber digestion in the emu, *Dromaius novaehollandiae,* a large bird with a simple gut and high rates of food passage. *Physiological Zoology* 57:70–84.

Hertz, P. E., Huey, R. B., and Garland, T. 1988. Time budgets, thermoregulation, and maximal locomotor performance: Are reptiles Olympians or Boy Scouts? *American Zoologist* 28:927–38.

Hicks, J. W., and Farmer, C. G. 1998a. Lung ventilation and gas exchange in theropod dinosaurs. *Science* 281:45–46.

———.1998b. Gas exchange in vertebrate lungs: The dinosaur-avian. *Journal of Vertebrate Paleontology* 183:50A.

Hill, R. W., and Wyse, G. A. 1989. *Animal Physiology.* Harper and Row, New York.

Hillenius, W. J. 1992. The evolution of nasal turbinates and mammalian endothermy. *Paleobiology* 18: 17–29.

———. 1994. Turbinates in therapsids: Evidence for Late Permian origins of mammalian endothermy. *Evolution* 48:207–29.

Hinchliffe, R. 1977. The chondrogenic pattern in chick limb morphogenesis: A problem of development and evolution. In *Vertebrate Limb and Somite Morphogenesis,* ed. D. A. Ede, R. Hinchliffe, and M. Balls, 293–309. Cambridge University Press, Cambridge.

———. 1985. "One, two, three" or "two, three, four": An embryologist's view of the homologies of the digits and carpus of modern birds. In Hecht et al. 1985, 141–47.

———. 1997. The forward march of the bird-dinosaurs inhibited? *Science* 278:596–97.

Hoffman, H. 2000. Messel, window on an ancient world. *National Geographic* 197(2):34–51.

Holdaway, R. N. 1989. New Zealand's pre-human avifauna and its vulnerability. *New Zealand Journal of Ecology* 12:11–25.

Holdaway, R. N., and Worthy, T. 1991. Lost in time. *New Zealand Geographic* 4:51–68.

Holdaway, R. N., and Jacomb, C. 2000a. Rapid extinction of the moas: Model, test, and implications. *Science* 287:2250–54.

———. 2000b. Less is moa: Response. *Science* 289: 1473–74.

Holtz, T. R. 1994a. The phylogenetic position of the Tyrannosauridae: Implications for theropod systematics. *Journal of Paleontology* 68:1100–17.

———. 1994b. Predatory adaptations of the skull and unguals of modern and extinct carnivorous amniotes. *Journal of Vertebrate Paleontology* 14:29A–30A.

———. 1994c. The arctometatarsalian pes, an unusual structure of the metatarsus of Cretaceous Theropoda. *Journal of Vertebrate Paleontology* 14:480–519.

———. 1996. Phylogenetic taxonomy of the Coelurosauria. *Journal of Paleontology* 70:536–38.

———. 1998/2000. A new phylogeny of the carnivorous dinosaurs. *Gaia* 15:5–61.

Holtz, T. R., Brinkman, D. L., and Chandler, C. L. 1998. Denticle morphometrics and a possibly omnivorous feeding habit for the theropod dinosaur *Troodon. Gaia* 1:159–66.

Homberger, D. G., and de Silva, K. N. 2000. Functional microanatomy of the feather-bearing integument: Implications for the evolution of birds and avian flight. *American Zoologist* 40(4):553–74.

Hope, S. 1998. The Mesozoic record of the neornithes. *Journal of Vertebrate Paleontology* 18:51A.

Hopp, T., and Orsen, M. 1998. Dinosaur brooding behavior and the origin of flight feathers. In *The Dinofest Symposium,* ed. D. L. Wolberg et al., 27. Academy of Natural Sciences, Philadelphia.

Hopson, J. A. 1980. Relative brain size in dinosaurs: Implications for dinosaurian endothermy. In *A Cold Look at the Warm-Blooded Dinosaurs,* ed. R.D.K Thomas and E. C. Olson, 287–310. AAAS, Washington, D.C.

Hopson, J. A., and Chiappe, L. M. 1998. Pedal proportions of living and fossil birds indicate arboreal or terrestrial specialization. *Journal of Vertebrate Paleontology* 183:52A.

Horner, J. R. 2000. Dinosaur reproduction and parenting. *Annual Review of Earth and Planetary Sciences* 28:19–45.

Horner, J. R., and Dobb, E. 1997. *Dinosaur Lives: Unearthing an Evolutionary Saga.* Harper Collins, New York.

Horner, J. R., Ricqles, A., and Padian, K. 1999. Variation in dinosaur skeletochronology indicators: Implications for age assessment and physiology. *Paleobiology* 25:295–304.

———. 2000. Long bone histology of the hadrosaurid dinosaur *Maiasaura peeblesorum:* Growth dynamics and physiology based on an ontogenetic series of skeletal remains. *Journal of Vertebrate Paleontology* 20:115–29.

Hotton, N. 1955. A survey of adaptive relationships of dentition to diet in North American Iguanidae. *American Midland Naturalist* 53:88–114.

———. 1963. *Dinosaurs.* Pyramid Press, New York.

———. 1980. An alternative to dinosaur endothermy: The happy wanderers. In *A Cold Look at the Warm-Blooded Dinosaurs,* ed. R.D.K Thomas and E. C. Olson, 311–50. AAAS, Washington, D.C.

Hou, L. 1995. Morphological comparisons between *Confuciusornis* and *Archaeopteryx.* In *Sixth Symposium on Mesozoic Terrestrial Ecosystems and Biota,*

*Short Papers,* ed. A. Sun and Y. Wang, 193–201. China Ocean Press, Beijing.

Hou, L., and Chen, P. 1999. *Liaoxiornis delicatus* gen. et. sp. nov., the smallest Mesozoic bird. *Chinese Science Bulletin* 44:834–38.

Hou, L., and Zhang J. 1993. A new fossil bird from Lower Cretaceous of Liaoning. *Vertebrata PalAsiatica* 31:217–24.

Hou, L., Zhou Z., Martin, L. D., and Feduccia, A. 1995. A beaked bird from the Jurassic of China. *Nature* 377:616–18.

Hou, L., Martin, L. D., Zhou Z., and Feduccia, A. 1996. Early adaptative radiation of birds: Evidence from fossils from northeastern China. *Science* 274: 1164–67.

———. 1999. *Archaeopteryx* to opposite birds—Missing link from the Mesozoic of China. *Vertebrata PalAsiatica* 37:88–95.

Houck, M. A., Gauthier, J. A., and Strauss, R. E. 1990. Allometric scaling in the earliest fossil bird, *Archaeopteryx lithographica. Science* 247:195–247.

Houstan, D. C. 1979. The adaptations of scavengers. In *Serengeti: Dynamics of an Ecosystem,* ed. A.R.E. Sinclair and M. Norton-Griffiths, 263–86. University of Chicago Press, Chicago.

Howgate, M. 1983. *Archaeopteryx*—No new finds after all. *Nature* 306:644–45.

———. 1984a. On the supposed difference between the teeth of the London and Berlin specimens of *Archaeopteryx lithographica. Neues Jahrbuch für Geologie und Paläontologie Monatshefte* 1984:654–60.

———. 1984b. The teeth of *Archaeopteryx* and a reinterpretation of the Eichstätt specimen. *Zoological Journal of the Linnean Society* (London) 82:159–75.

———. 1985. Problems in the osteology of *Archaeopteryx:* Is the Eichstätt specimen a distinct genus? In Hecht et al. 1985, 105–12.

Huelsenbeck, J. P., and Crandall, A. 1997. Phylogeny estimation and hypothesis testing using maximum likelihood. *Annual Review of Ecology and Systematics* 28:437–66.

Hutchinson, J. R. 2001. The evolution of pelvic osteology and soft tissues on the line to extinct birds. *Zoological Journal of the Linnean Society* 131:123–68.

Huxley, T. 1882. On the respiratory organs of *Apteryx. Proceedings of the Zoological Society of London* 64:560–69.

Irvine, A. B. 1983. Manatee metabolism and its influence on distribution in Florida. *Biological Conservation* 25:315–34.

Ito, H. M., Yoshimoto, M., and Somiya, H. 1999. External brain form and cranial nerves of the megamouth shark *Megachasma pelagios. Copeia* 1999: 210–13.

Janensch, W. 1925. Die Coelurosauria und Theropoden der Tendaguru-Schichten Deutsch-Ostafrikas. *Palaeontographica Suppl.* 7(1):1–99.

Jansky, L. 1965. Adaptability of heat production mechanisms in homeotherms. *Acta Universitatis Carolinae Biologica* 1:1–91.

———. 1995. Hormonal thermogenesis and its role in maintaining energy balance. *Physiology Review* 75: 237–59.

Janvier, P. 1983. Le divorce de 1'oiseau et du crocodile. *La Recherche* 14:1430–32.

Jarman, P. J. 1989. Sexual dimorphism in Macropodoidea. In *Kangaroos, Wallabies, and Rat-Kangaroos,* ed. G. Grigg, P. Jarmen, and I. Hume, 1:433–47. Surrey Beatty and Sons, Chipping Norton, New South Wales.

Jenkins, F. A. 1993. The evolution of the avian shoulder joint. *American Journal of Science* 293A:253–67.

Jensen, J. A., and Padian, K. 1989. Small pterosaurs and dinosaurs from the Uncompahgre, Late Jurassic, western Colorado. *Journal of Paleontology* 63: 364–73.

Jepsen, G. L. 1970. Bat origins and evolution. In *Biology of Bats,* ed. W. A. Wimsatt, 1–64. Academic Press, New York.

Jerison, H. J. 1973. *Evolution of the Brain and Intelligence.* Academic Press, New York.

Ji Q., and Ji S. A. 1996. On discovery of the earliest bird fossil in China and the origin of birds. *Chinese Geology* 233:1164–67.

———. 1997a. Advance in *Sinosauropteryx* research. *Chinese Geology* 237:30–32.

———. 1997b. Protarchaeopteryid bird (*Protarchaeopteryx* gen. nov.)—Fossil remains of archaeopterygids from China. *Chinese Geology* 238:38–41.

Ji Q., Currie, P. J., Norell, M. A., and Ji S. 1998. Two feathered dinosaurs from northeastern China. *Nature* 393:753–61.

Ji Q., Chiappe, L. M., and Ji S. 1999. A new late Mesozoic confuciusornithid bird from China. *Journal of Vertebrate Paleontology* 19:1–7.

Ji Q., Lou, Z., and Ji S. 1999. A Chinese triconodont mammal and mosaic evolution of the mammalian skeleton. *Nature* 398:326–30.

Ji Q., Norell, M., Gao K.-Q., Ji S.-A. and Ren, D. 2001. The distribution of integumentary structures in a feathered dinosaur. *Nature* 410:1084–88.

Ji S. 1998. New pterosaurs from northeastern China and the geological age problem of *Confuciusornis. Journal of Vertebrate Paleontology* 18:54A.

Johnson, H. K., Blix, A. S., Mercer, J. B., and Bolz, K. D. 1987. Selective cooling of the brain in reindeer. *American Journal of Physiology* 253:R848–53.

Johnson, T. 1996. *Husbandry Manual for North Island Brown Kiwi.* Kiwi Stud Book Project, Rotorua, New Zealand.

Jollie, M. 1977. A contribution to the morphology and physiology of the Falconiformes. Part 3. *Evolutionary Theory* 2:209–300.

Jones, J. H., and Lindstedt, S. L. 1993. Limits to maximal performance. *Annual Review of Physiology* 55: 547–69.

Jones, T. D., et al. 2000a. Nonavian feathers in a Late Triassic archosaur. *Science* 288:2202–5.

———. 2000b. Cursoriality in bipedal archosaurs. *Nature* 406:716–18.

Jones, T. D., Ruben, J. A., Maderson, P. F. A., and Martin, L. D. 2001. *Longisquama* fossil and feather morphology: Response. *Science* 291:1900–1902.

Karhu, A. A., and Rautian, A. S. 1996. A new family of Maniraptora (Dinosauria: Saurischia) from the Late Cretaceous of Mongolia. *Paleontological Journal* 30:583–92.

Kaufman, S., and Johnsen, J. 1991. Coevolution on the edge of chaos: Couple fitness landscapes, poised states, and coevolutionary avalanches. *Journal of Theoretical Biology* 149:467–505.

Kellner, A.W.A. 1996a. Fossilized theropod soft tissue. *Nature* 379:32.

———. 1996b. Remarks on Brazilian dinosaurs. *Memoirs of the Queensland Museum* 39:611–26.

Kemp, R. A., and Unwin, D. M. 1997. The skeletal taphonomy of *Archaeopteryx:* A quantitative approach. *Lethaia* 30:229–38.

Kilgore, D. L., Birchard, G. F., and Boggs, D. F. 1981. Brain temperature in running quail. *Journal of Applied Physiology* 50:1277–81.

King, A. 1966. Structural and functional aspects of the avian lung and air sacs. *International Review of General Experimental Zoology* 2:171–267.

Kobayashi, Y., et al. 1999. Herbivorous diet in an ornithomimid dinosaur. *Nature* 402:480–81.

Koeberi, C., Armstrong, R. A., and Reimond, W. U. 1997. Morokweng, South Africa: A large impact structure of Jurassic-Cretaceous boundary age. *Geology* 25:731–34.

Kofron, C. P. 1999. Attacks to humans and domestic animals by the southern cassowary on Queensland, Australia. *Journal of Zoology* (London) 249:375–81.

Kollar, J., and Fisher, C. 1980. Tooth induction in chick epithelium: Expression of quiescent genes. *Science* 207:993–95.

Kortenkamp, S. J., and Dermott, S. F. 1998. A 100,000-year periodicity in the accretion rate of interplanetary dust. *Science* 280:874–76.

Kortner, G., Brigham, R. M., and Geiser, F. 2000. Winter torpor in a large bird. *Nature* 407:318.

Kranz, P. M. 1998. Mostly dinosaurs: A review of the vertebrates of the Potomac Group, USA. *New Mexico Museum of Natural History and Science Bulletin* 14:235–38.

Kundrat, M. 1998. Comments on the significance of integumentary impressions of the early Jurassic theropod *Eubrontes minusculus. Journal of Vertebrate Paleontology* 18:57A.

Kurochkin, E. N. 1995. Synopsis of Mesozoic birds and early evolution of class Aves. *Archaeopteryx* 13:47–66.

Kurochkin, E. N., and Molnar, R. 1997. New material of enantiornithine birds from the Early Cretaceous of Australia. *Alcheringa* 21:291–97.

Kurzanov, S. M. 1981. *Avimimus* and the problem of the origin of birds. *Joint Soviet-Mongolian Palaeontological Expedition Transactions* 24:104–9.

———. 1982. Structural characteristics of the fore limbs of *Avimimus. Paleontological Journal* 16:108–12.

———. 1983. New data on the pelvic structure of *Avimimus. Paleontological Journal* 17:110–11.

———. 1985. The skull structure of the dinosaur *Avimimus. Paleontological Journal* 19:92–99.

———. 1987. Avimimidae and the problem of the origin of birds. *Joint Soviet-Mongolian Paleontological Expedition Transactions* 31:1–92.

Kuypers, M.M.M., et al. 1999. A large and abrupt fall in atmospheric $CO_2$ concentration during Cretaceous times. *Nature* 399:342–45.

Labandeira, C. C., and Sepkoski, J. J. 1993. Insect diversity in the fossil record. *Science* 261:310–15.

Lambe, L. M. 1917. The Cretaceous theropodous dinosaur *Gorgosaurus. Geological Survey of Canada, Memoirs* 100:1–84.

Lambert, W. D. 1991. Altriciality and its implications for dinosaur thermoenergetic physiology. *Neues Jahrbuch für Geologie und Paläontologie Abhandlungen* 182:73–84.

Lamin, T. 2000. Wild gliders. *National Geographic* 198 (4): 68–85.

Langman, V. A., et al. 1995. Moving cheaply: Energetics of walking in the African elephant. *Journal of Experimental Biology* 198:629–32.

Langston, W. 1981. Pterosaurs. *Scientific American* 244 (2): 122–36.

Larson, P. L. 1997. Do dinosaurs have class? Implications of the avian respiratory system. In *The Dinofest International,* ed. D. L. Wolberg, E. Stump, and G. D. Rosenberg, 105–11. Academy of Natural Sciences, Philadelphia.

Larsson, H.C.E., Sereno, P. C., and Wilson, J. A. 2000. Forebrain enlargement among nonavian theropod dinosaurs. *Journal of Vertebrate Paleontology* 20: 615–18.

Lawiewski, R. C. 1972. Respiratory function in birds. In *Avian Biology II,* ed. D. S. Farner, J. A. King, and K. C. Parkes, 2:287–342. Academic Press, New York.

Leahy, G. D. 1991. Lamellar-zonal bone in fossil mammals: Implications for dinosaur and therapsid paleophysiology. *Journal of Vertebrate Paleontology* 11: 42A.

Levy, S. 1992. *Artificial Life.* Random House, New York.

Lewis, J. C. 1979. Periosteal layers do not indicate ages of sandhill cranes. *Journal of Wildlife Management* 43:269–71.

Lindeque, M., and van Jaarsveld, A. S. 1993. Postnatal growth of elephants *Loxodonta africana* in Etosha National Park, Namibia. *Journal of Zoology* (London) 229:319–30.

Linsenmaier, W. 1972. *Insects of the World.* McGraw Hill, New York.

Livezey, B. C. 1992. Morphological corollaries and ecological implications of flightlessness in the kakapo. *Journal of Morphology* 213:105–45.

Lockley, M. G., et al. 1992. The track record of Mesozoic birds: Evidence and implications. *Philosophical Transactions of the Royal Society of London B* 336: 113–34.

Long, J. A., and McNamara, K. J. 1997. Heterochrony: The key to dinosaur evolution. In *The Dinofest International,* ed. D. L. Wolberg, E. Stump, and G. D. Rosenberg, 113–23. Academy of Natural Sciences, Philadelphia.

Long, J. L. 1965. Weights, measurements, and food of the emu in the northern wheatbelt of western Australia. *Emu* 64:214–19.

Longrich, N. 1999. On the semilunate carpal and trochanteric crest of maniraptoran theropods. *Journal of Vertebrate Paleontology* 19:60A.

———. 2000. Myrmecophagus maniraptora? Alvarezsaurs as aardraptors. *Journal of Vertebrate Paleontology* 20:54A.

Lovtrup, S. 1985. On the classification of the taxon Tetrapoda. *Systematic Zoology* 34:463–70.

Luo, Z. 1999. A refugium for relicts. *Nature* 400: 23–24.

MacCready, P. 1985. The great pterodactyl project. *Engineering and Science* 49(11):18–24.

McDonald, K. 1996. A dispute over the evolution of birds. *Chronicle of Higher Education* 43(9):A14–A15.

McFadden, E. R., et al. 1985. Thermal mapping of the airways in humans. *Journal of Applied Physiology* 58:564–70.

McGowan, C. 1982. The wing musculature of the brown kiwi *Apteryx australis mantelli* and its bearing on ratite affinities. *Journal of Zoology* (London) 197:179–219.

———. 1984. Evolutionary relationships of ratites and carinates: Evidence form ontogeny of the tarsus. *Nature* 307:733–35.

———. 1985. Homologies of the avian tarsus. *Nature* 315:160.

McIlhenny, E. A. 1934. Notes on incubation and growth of alligators. *Copeia* 1934:80–88.

McKinney, M. L. 1987. Taxonomic selectivity and continuous variation in mass and background extinctions of marine taxa. *Nature* 325:143–45.

McLelland, J. M. 1989a. Larynx and trachea. In *Form and Function in Birds,* ed. A. King and J. M. McLelland, 4:69–103. Academic Press, New York.

———. 1989b. Anatomy of the lungs and air sacs. In *Form and Function in Birds,* ed. A. King and J. M. McLelland, 4:221–79. Academic Press, New York.

McNab, B. K. 1978. The evolution of endothermy in the phylogeny of mammals. *American Naturalist* 112: 1–21.

———. 1983. Energetics, body size, and the limits to endothermy. *Journal of Zoology* (London) 199:1–29.

———. 1988. Complications inherent in scaling the basal metabolism in mammals. *Quarterly Review of Biology* 63:25–54.

———. 1994. Energy conservation and the evolution of flightlessness in birds. *American Naturalist* 144: 628–42.

———. 1996. Metabolism and temperature regulation of kiwis. *Auk* 113:687–92.

McNab, B. K., and Eisenberg, J. F. 1989. Brain size and its relation to the rate of metabolism in dinosaurs. *American Naturalist* 133:157–67.

Maderson, P. F. A., and Alibardi, L. 2000. The development of the sauropsid integument: A contribution to the problem of the origin and evolution of feathers. *American Zoologist* 40(4):513–29.

Maderson, P. F. A., and Homberger, D. G. 2000. Evolutionary origin of feathers. *American Zoologist* 40(4).

Madsen, J. H. 1976. *Allosaurus fragilis:* A revised osteology. *Bulletin Utah Geological Survey* 109:1–163.

Madsen, J. H., and Welles, S. P. 2000. *Ceratosaurus,* a revised osteology. *Utah Geological Survey Miscellaneous Publication* 00-2:1–80.

Maina, J. N. 1989. The morphometry of the avian lung. In *Form and Function in Birds,* ed. A. King and J. M. McLelland, 4:307–68. Academic Press, New York.

———. 2000. What it takes to fly: The structural and functional respiratory refinements in birds and bats. *Journal of Experimental Biology* 203:3045–64.

Makovicky, P. J., and Currie, P. J. 1998. The presence of a furcula in tyrannosaurid theropods, and its phylogenetic and functional implications. *Journal of Vertebrate Paleontology* 18:143–49.

Makovicky, P. J., and Norell, M. A. 1998. A partial ornithomimid braincase from Ukhaa Tolgod (Upper Cretaceous, Mongolia). *American Museum Novitates* 3247:1–16.

Makovicky, P. J., and Sues, H.-D. 1998. Anatomy and phylogenetic relationships of the theropod dinosaur *Microvenator celer* from the Lower Cretaceous of Montana. *American Museum Novitates* 3240:1–27.

Manabe, M. 1999. The early evolution of the Tyrannosauridae in Asia. *Journal of Paleontology* 73: 1176–78.

Manabe, M., Barrett, P. M., and Isaji, S. 2000. A refugium for relicts? *Nature* 404:953.

Marden, J. H. 1987. Maximum lift production during takeoff in flying animals. *Journal of Experimental Biology* 130:235–58.

———. 1994. From damselflies to pterosaurs: How burst and sustainable performance scale with size. *American Journal of Physiology* 266:R1077–84.

Marden, J. H., O'Donnell, B. C., Thomas, M. A., and Bye, J. Y. 2000. Surface-skimming stoneflies and mayflies: The taxonomic and mechanical diversity of two-dimensional aerodynamic locomotion. *Physical and Biochemical Zoology* 73:751–64.

Marsh, O. C. 1880. *Odontornithes: A monograph on the extinct toothed birds of North America. Report of the U.S. Geological Exploration of the Fortieth Parallel* 7.

Marshall, C. R. 1999. Fossil gap analysis supports early Tertiary origin of trophically diverse avian orders: Comment and reply. *Geology* 27:95.

Marshall, L. G. 1994. The terror birds of South American. *Scientific American* 270(2):90–95.

Martill, D. M., et al. 2000. Skeletal remains of small theropod dinosaur with associated soft structures from the Lower Cretaceous Santana Formation of northeastern Brazil. *Canadian Journal of Earth Sciences* 37:891–900.

Martin, J., Martin-Rolland, V., and Frey, E. 1998. Not cranes or masts, but beams: The biomechanics of sauropod necks. *Oryctos* 1:113–20.

Martin, L. D. 1983a. The origin of birds and of avian flight. *Current Ornithology* 1:105–29.

———. 1983b. The origin and early radiation of birds. In *Perspectives in Ornithology*, ed. A. H. Brush and G. A. Clark, 291–338. Cambridge University Press, Cambridge.

———. 1991. Mesozoic birds and the origin of birds. In *Origins of the Higher Groups of Tetrapods*, ed. H.-P. Schultze and L. Trueb, 466–77. Cornell University Press, Ithaca, N.Y.

———. 1995. A new skeletal model of *Archaeopteryx*. *Archaeopteryx* 13:33–40.

———. 1997. The difference between dinosaurs and birds as applied to *Mononykus*. In *The Dinofest International*, ed. D. L. Wolberg, E. Stump, and G. D. Rosenberg, 337–42. Academy of Natural Sciences, Philadelphia.

Martin, L. D., Stewart, J. D., and Whetstone, K. N. 1980. The origin of birds: Structure of the tarsus and teeth. *Auk* 97:86–93.

Martin, L. D., and Stewart, J. D. 1985. Homologies of the avian tarsus. *Nature* 315:159–60.

———. 1999. Implantation and replacement of bird teeth. In Olson 1999, 295–300.

Martin, L. D., and Miao, D. 1995. Evolutionary patterns in the early history of birds and mammals. In *Sixth Symposium on Mesozoic Terrestrial Ecosystems and Biota, Short Papers*, ed. A. Sun and Y. Wang, 217–19. China Ocean Press, Beijing.

Martin, L. D., and Zhou Z. 1997. *Archaeopteryx*-like skull in enantiornithine bird. *Nature* 389:556.

Martin, L. D., and Simmons, J. 1998. Theropod dinosaur nesting behavior. In *The Dinofest Symposium*, ed. D. L. Wolberg et al., 39. Academy of Natural Sciences, Philadelphia.

Martin, L. D., Zhou Z., Hou, L., and Feduccia, A. 1998. *Confuciusornis sanctus* compared to *Archaeopteryx lithographica*. *Naturwissenschaften* 85:286–89.

Martin, L. D., and Czerkas, S. A. 2000. The fossil record of feather evolution in the Mesozoic. *American Zoologist* 40(4):687–94.

Martin, R. D. 1981. Relative brain size and basal metabolic rate in terrestrial vertebrates. *Nature* 293: 57–60.

Maryanska, T. 1997. Segnosaurs (Therizinosaurs). In *The Complete Dinosaur*, ed. J. O. Farlow and M. Brett-Surman, 234–41.

Maryanska, T., and Osmólska, H. 1997. The quadrate of oviraptorid dinosaurs. *Acta Palaeontologica Polonica* 42:361–71.

Matthes, E. 1934. Geruchorgan. In *Handbuch der Vergleichenden Anatomie der Wirbeltiere*, ed. L. Bolk et al., 879–948. Urban and Schwarzenberg, Berlin.

Matthew, W. D., and Granger, W. 1917. The skeleton of *Diatryma*, a gigantic bird from the Lower Eocene of Wyoming. *Bulletin of the American Museum of Natural History* 37:307–26.

Matthews, P., and McWhirter, N. D. 1993. *The Guinness Book of Records*. Guinness Publishing, Middlesex, England.

May, M. L. 1981. Allometric analysis of body and wing dimensions of male Anisoptera. *Odonatologica* 10: 279–91.

———. 1982. Heat exchange and endothermy in Protodonata. *Evolution* 36:1051–58.

Mayr, G., and Mourer-Chauvire, C. 2000. Rollers from the Middle Eocene of Messel (Germany) and the Upper Eocene of the Quercy (France). *Journal of Vertebrate Paleontology* 20:533–46.

Meier, A. H., and Fivizzari, A. O. 1980. Physiology of migration. *Animal Migration, Orientation, and Navigation*, ed. S. A. Gauthreaux, 225–82. Academic Press, London.

Metcalf, S. J., and Walker, R. J. 1994. A new Bathonian microvertebrate locality in the English Midlands. In *In the Shadow of the Dinosaurs: Early Mesozoic Tetrapods*, ed. N. Fraser and H.-D. Sues, 322–31. Cambridge University Press, Cambridge.

Middleton, K. M. 1999. Morphological basis for hallucal orientation in fossil birds. *Journal of Vertebrate Paleontology* 19:64A.

Middleton, K. M., and Gatesy, S. M. 2000. Theropod forelimb design and evolution. *Zoological Journal of the Linnean Society* (London) 128:149–87.

Mikulic, D. G. 1997. Footprints in the sands of time: The early misidentification of dinosaur tracks. In *The Dinofest International*, ed. D. L. Wolberg, E. Stump, and G. D. Rosenberg, 13–18. Academy of Natural Sciences, Philadelphia.

Miles, C. A., Carpenter, K., and Cloward, K. 1998. A new skeleton of *Coelurus fragilis* from the Morrison Formation of Wyoming. *Journal of Vertebrate Paleontology* 18:64A.

Miller, G. H., et al. 1999. Pleistocene extinction of *Genyornis newtoni:* Human impact on Australian megafauna. *Science* 283:205–8.

Mindell, D. P. 1997. *Avian Molecular Evolution and Systematics*. Academic Press, San Diego.

Minnaar, P., and Minnaar, M. 1997. *The Emu Farmer's Handbook*. 7th ed. Nyoni Publishing Co., Groveton, Tex.

Molnar, R. E. 1985. Alternatives to *Archaeopteryx:* A survey of proposed early or ancestral birds. In Hecht et al. 1985, 209–17.

Monnier, L. 1913. Paléontologie de Madagascar. Part 7, Les *Aepyornis*. *Annales de paléontologie* 8: 125–72.

Mook, C. C. 1921. Notes on the postcranial skeleton in the Crocodilia. *Bulletin of the American Museum of Natural History* 44:67–99.

Moran, B., and Van Dam, L. 1996. Robots on all twos. *Technology Review* 99:10–11.
Morell, V. 1997. The origin of birds: The dinosaur debate. *Audubon* 99:36–45.
Morgan, J., et al. 1997. Size and morphology of the Chicxulub impact crater. *Nature* 390:472–76.
Morton, E. S. 1978. Avian arboreal folivores: Why not? In *The Ecology of Arboreal Folivores,* ed. G. G. Montgomery. Smithsonian Institution Press, Washington, D.C.
Murray, P. F., and Megirian, D. 1998. The skull of dromornithine birds: Anatomical evidence for their relationship to Anseriformes. *Records of the South Australian Museum* 31:51–97.
Murrish, D. E. 1973. Respiratory heat and water exchange in penguins. *Respiration Physiology* 19: 262–70.
Murrish, D. E., and Schmidt-Nielsen, K. 1970. Exhaled air temperature and water conservation in lizards. *Respiration Physiology* 10:151–58.
Muybridge, E. 1957. *Animals in Motion.* Dover Publications, New York.
Naish, D. 1998. Cryptozoology of the moa: A review. Part 1. *Cryptozoology Review* 2(3):15–24.
Negus, V. 1958. *The Comparative Anatomy and Physiology of the Nose and Paranasal Sinuses.* E. and S. Livingston, Edinburgh.
Nelson, R. C., and Bookhout, A. T. 1980. Counts of periosteal layers invalid for aging Canada geese. *Journal of Wildlife Management* 44:518–21.
Nicholls, E. L., and Russell, A. P. 1985. Structure and function of the pectoral girdle and forelimb of *Struthiomimus altus. Palaeontology* 28:643–77.
Nickel, R., Schummer, A., Seiferle, E., Siller, W. G., and Wight, P.A.L. 1977. *Anatomy of the Domestic Birds.* Springer-Verlag, Berlin.
Nieuwland, I. 1999. Gerhard Heilmann and the artist's eye in paleontology. *Journal of Vertebrate Paleontology* 19:66A.
Niinimaa, V. 1983. Oronasal airway choice during running. *Respiration Physiology* 53:129–33.
Nikbakht, N., and McLachlan, J. C. 1999. Restoring avian wing digits. *Proceedings of the Royal Society of London B* 266:1101–4.
Nilsson, G. E. 1996. Brain and body oxygen requirements of *Gnathonemus petersii,* a fish with an exceptionally large brain. *Journal of Experimental Biology* 199:603–7.
Norberg, R. 1985. Function of vane asymmetry and shaft curvature in bird flight feathers. In Hecht et al. 1985, 303–18.
———. 1995. Feather asymmetry in *Archaeopteryx. Nature* 374:221.
Norberg, U. M. 1985a. Evolution of flight in birds: Aerodynamic, mechanical, and ecological aspects. In Hecht et al. 1985, 293–302.
———. 1985b. Evolution of vertebrate flight: An aerodynamic model for the transition from gliding to active flight. *American Naturalist* 126:303–27.
Norell, M. A., et al. 1994. A theropod dinosaur embryo and the affinities of the Flaming Cliffs dinosaur eggs. *Science* 266:779–82.
Norell, M. A., et al. 1995. A nesting dinosaur. *Nature* 378:774–76.
Norell, M. A., and Chiappe, L. M. 1996. Flight from reason. *Nature* 384:230.
Norell, M. A., and Makovicky, P. J. 1997. Important features of the dromaeosaur skeleton: Information from a new specimen. *American Museum Novitates* 3215:1–28.
———. 1999. Important features of the dromaeosaurid skeleton. Part 2, Information from newly collected specimens of *Velociraptor mongoliensis. American Museum Novitates* 3282:1–45.
Norell, M. A., Makovicky, P. J., and Clark, J. M. 1997. A *Velociraptor* wishbone. *Nature* 389:447.
———. 1998. Theropod-bird link reconsidered—Reply. *Nature* 391:754.
———. 2000. A new troodontid theropod from Ukhaa, Mongolia. *Journal of Vertebrate Paleontology* 20: 7–11.
Norell, M. A., and Clarke, J. A. 2001. Fossil that fills a critical gap in avian evolution. *Nature* 409:181–84.
Norman, D. B. 1990. Problematic Theropoda: "Coelurosaurs." In *The Dinosauria,* ed. D. B. Weishampel, P. Dodson, and H. Osmólska, 280–305. University of California Press, Berkeley.
Northcutt, R. G. 1977. Elasmobranch central nervous system organization and its possible evolutionary significance. *American Zoologist* 17:411–29.
Norton, J. M. 1998. Eggs and nest place constraints on dinosaur lung physiology and function. *Journal of Vertebrate Paleontology* 18:67A.
Novas, F. E. 1991. Los evolución los dinosaurios carnivoros. In *Los Dinosaurios y su entorno biotico.* Vol. 2, *Curso de Paleontologia en Cuenca.* Instituto "Juan de Valdes," Ayuntamiento de Cuenca.
———. 1994. New information on the systematics and postcranial skeleton of *Herrerasaurus ischigualastensis* from the Ischigualastro Formation of Argentina. *Journal of Vertebrate Paleontology* 13:400–423.
———. 1996. Dinosaur monophyly. *Journal of Vertebrate Paleontology* 16:723–41.
———. 1997. Anatomy of *Patagonykus puertai* (Theropoda, Avialae, Alvaerzsauridae), from the Late Cretaceous of Patagonia. *Journal of Vertebrate Paleontology* 17:137–66.
Novas, F. E., and Puerta, P. F. 1997. New evidence concerning avian origins from the Late Cretaceous of Patagonia. *Nature* 387:390–92.
Nowak, R. M. 1999. *Walker's Mammals of the World.* 6th ed. Johns Hopkins University Press, Baltimore.
Oliver, W.R.B. 1949. The moas of New Zealand and Australia. *Dominion Museum Bulletin* 15:1–206.
Olshevsky, G. 1994. A revision of the Parainfraclass Archosauria Cope, 1869, excluding the advanced Crocodylia. *Mesozoic Meanderings* 2:1–196.

Olson, P. 1995. *Australian Birds of Prey.* Johns Hopkins University Press, Baltimore.

Olson, S. L. 1973. Evolution of the rails of the South Atlantic islands. *Smithsonian Contributions to Zoology* 152:1–53.

———. 1985. The fossil record of birds. In *Avian Biology,* ed. D. S. Farner, J. A. King, and K. C. Parkes, 8:80–256. Academic Press, New York.

———, ed. 1999. Avian paleontology at the close of the 20th century: Proceedings of the 4th International Meeting of the Society of Avian Paleontology and Evolution. *Smithsonian Contributions to Paleontology* 89.

Olson, S. L., and Feduccia, A. 1979. Flight capability and the pectoral girdle of *Archaeopteryx. Nature* 278:247–48.

Olson, S. L., and James, H. F. 1982. Fossil birds from the Hawaiian Islands: Evidence for wholesale extinction by man before western contact. *Science* 21: 633–35.

Osborn, H. F. 1903. *Ornitholestes hermanni,* a new compsognathoid dinosaur from the Upper Jurassic. *Bulletin of the American Museum of Natural History* 19:459–64.

———. 1912. Crania of Tyrannosaurus and Allosaurus. *American Museum of Natural History, Memoirs,* n.s., 1:1–30. *Bulletin of the American Museum of Natural History* 35:733–71.

———. 1916. Skeletal adaptations of Ornitholestes, Struthiomimus, Tyrannosaurus. *Bulletin of the American Museum of Natural History* 35:733–71.

———. 1924. Three new Theropoda, *Protoceratops* zone, Central Mongolia. *American Museum Novitates,* no. 144:1–12.

Osmólska, H. 1976. New light on the skull anatomy and systematic position of *Oviraptor. Nature* 262: 683–84.

———. 1981. Coossified tarsometatarsi in theropod dinosaurs and their bearing on the problem of bird origins. *Palaeontologica Polonica* 42:79–95.

———. 1987. *Borogovia gracilicrus* Gen. Et. Sp. N., A new troodontid dinosaur from the Late Cretaceous of Mongolia. *Acta Palaeontologica Polonica* 32:133–50.

Osmólska, H., Roniewicz, E., and Barsbold, R. 1972. A new dinosaur, *Gallimimus bullatus* N. Gen., N. Sp., from the Upper Cretaceous of Mongolia. *Palaeontologica Polonica* 27:103–43.

Osmólska, H., and Barsbold, R. 1990. Troodontidae. In *The Dinosauria,* ed. D. B. Weishampel, P. Dodson, and H. Osmólska, 259–68. University of California Press, Berkeley.

Ostrom, J. H. 1969. Osteology of *Deinonychus antirrhopus,* an unusual theropod from the Lower Cretaceous of Montana. *Peabody Museum of Natural History Bulletin* 30:1–165.

———. 1970. Terrestrial vertebrates as indicators of Mesozoic climates. In *North American Paleontological Convention, Chicago, 1969, Proceedings,* ed. E. L. Yochelson, D:347–76. Allen Press, Lawrence, Kans.

———. 1972. Dinosaur. In *McGraw-Hill Yearbook of Science and Technology,* 176–79. McGraw-Hill, New York.

———. 1973. The ancestry of birds. *Nature* 242:136.

———. 1974a. *Archaeopteryx* and the origin of flight. *Quarterly Review of Biology* 49:27–47.

———. 1974b. The pectoral girdle and forelimb function of *Deinonychus:* A correction. *Postilla* 165:1–11.

———. 1976a. *Archaeopteryx* and the origin of birds. *Biological Journal of the Linnean Society* 8:91–182.

———. 1976b. Some hypothetical anatomical stages in the evolution of avian flight. *Smithsonian Contributions to Paleobiology* 27:1–27.

———. 1978. The Osteology of *Compsognathus longipes. Zitteliana* 4:73–118.

———. 1980a. *Coelurus* and *Ornitholestes:* Are they the same? In *Aspects of Vertebrate History,* ed. L. Jacobs, 245–56. Museum of Northern Arizona Press, Flagstaff.

———. 1980b. The evidence for endothermy in dinosaurs. In *A Cold Look at the Warm-Blooded Dinosaurs,* ed. R.D.K. Thomas and E. C. Olson, 15–54. AAAS, Washington, D.C.

———. 1985. The meaning of *Archaeopteryx.* In Hecht et al. 1985, 161–76.

———. 1986. The cursorial origin of avian flight. *Memoirs of the California Academy of Sciences* 8:73–81.

———. 1990. Dromaeosauridae. In *The Dinosauria,* ed. D. B. Weishampel, P. Dodson, and H. Osmólska, 269–79. University of California Press, Berkeley.

———. 1991. The question of the origin of birds. In *Origins of the Higher Groups of Tetrapods,* ed. H.-P. Schultze and L. Trueb, 467–84. Cornell University Press, Ithaca, N.Y.

———. 1994. On the origin of birds and of avian flight. In *Major Features of Vertebrate Evolution,* ed. D. R. Prothero and R. M. Schoch, 7:160–77. Paleontological Society, Knoxville, Tenn.

———. 1995. Wing biomechanics and the origin of bird flight. *Neues Jahrbuch für Geologie und Paläontologie Abhandlungen* 195:253–66.

———. 1996. The questionable validity of *Protoavis. Archaeopteryx* 14:39–42.

Ostrom, J. H., Poore, S. O., and Goslow, G. E. 1999. Humeral rotation and wrist supination: Important functional complex for the evolution of powered flight in birds? In Olson 1999, 301–9.

Otschev, V. G. 1975. The palate of the Proterosuchia. *Paleontological Journal* 4:515–21.

Owen, M. 1972. Some factors affecting food intake and selection in white-fronted geese. *Journal of Animal Ecology* 41:79–92.

Owen, R. 1879. *Memoirs on the Extinct Wingless Birds of New Zealand.* J. Van Voorst, London.

Owerkowicz, T., et al. 1999. Contribution of gular pumping to lung ventilation in monitor lizards. *Science* 284:1661–62.

Packard, G. C. 1977. The physiological ecology of reptilian eggs and embryos, and the evolution of viviparity within the class Reptilia. *Biological Review* 52:71–105.

Padian, K. 1982. Macroevolution and the origin of major adaptations: Vertebrate flight as a paradigm for the analysis of patterns. *Proceedings of the Third North American Paleontological Convention* 2:387–92.

———. 1984. The origin of pterosaurs. In *Third Symposium on Mesozoic Terrestrial EcoSystems*, ed. W. E. Reif and F. Westphal, 163–68. ATTEMPTO Verlag, Tübingen.

———. 1985. The origins and aerodynamics of flight in extinct vertebrates. *Paleontology* 28:413–33.

———. 1986. A comparative phylogenetic and functional approach to the origin of vertebrate flight. In *Recent Advances in the Study of Bats*, ed. M. B. Fenton, P. Racey, and J. M. Rayner, 3–23. Cambridge University Press, Cambridge.

———. 1997a. The continuing debate over avian origins. *American Scientist* 85:178–80.

———. 1997b. Growth lines. In Currie and Padian 1997, 288–91.

———. 1998a. When is a bird not a bird? *Nature* 393: 729–30.

———. 1998b. Bird, dinosaur link. *Science* 280:986–87.

———. 1999. Dinosaurian growth rates and the evolution of life history strategies. *Journal of Vertebrate Paleontology* 19:67A.

Padian, K., and Chiappe, L. M. 1998a. The origin of birds and their flight. *Scientific American* 278(2): 38–47.

———. 1998b. Birds of a feather: Reply. *Scientific American* 278(6):8–8A.

———. 1998c. The origin and evolution of birds. *Biological Reviews* 73:1–42.

Padian, K., Hutchinson, J. R., and Holtz, T. R. 1999. Phylogenetic definitions and nomenclature of the major taxonomic categories of the carnivorous Dinosauria. *Journal of Vertebrate Paleontology* 19: 69–80.

Paine, S. 2000. The demon duck of doom. *New Scientist* 166(2240):36–39.

Paladino, F. V., Dodson, P., Hammond, J. K., and Spotila, J. R. 1989. Temperature-dependent sex determination in dinosaurs? Implications for population dynamics and extinction. *Geological Society of America* 238:63–70.

Paladino, F. V., Spotila, J. R., and Dodson, P. 1997. A blueprint for giants: Modeling the physiology of large dinosaurs. In *The Complete Dinosaur*, ed. J. O. Farlow and M. Brett-Surman, 491–504. Indiana University Press, Bloomington.

Parrish, J. M. 1986. Locomotor adaptations in the hindlimb and pelvis of the Thecodontia. *Hunteria* 1(2): 1–35.

———. 1987. The origin of Crocodilia. *Paleobiology* 13:396–414.

Parsons T. S. 1970. The nose and Jacobson's organ. In *Biology of the Reptilia*, ed. C. Gans and T. S. Parsons, 191–299. Academic Press, London.

Paul, G. S. 1984a. The archosaurs, a phylogenetic study. In *Third Symposium on Mesozoic Terrestrial Ecosystems*, ed. W. E. Reif and F. Westphal, 175–80. ATTEMPTO Verlag, Tübingen.

———. 1984b. The segnosaurian dinosaurs: Relics of the prosauropod-ornithischian transition? *Journal of Vertebrate Paleontology* 4:507–15.

———. 1984c. The hand of *Archaeopteryx*. *Nature* 310:732.

———. 1987a. The science and art of restoring the life appearance of dinosaurs and their relatives. In *Dinosaurs Past and Present*, ed. S. J. Czerkas and E. C. Olson, 2:4–49. Natural History Museum of Los Angeles County, Los Angeles.

———. 1987b. Pterodactyl habits—Real and radio controlled. *Nature* 328:481.

———. 1988a. *Predatory Dinosaurs of the World*. Simon and Schuster, New York.

———. 1988b. The horned theropods of the Morrison and Great Oolite, and the Cloverly, Djadokhta and Judith River. *Hunteria* 2:1–9.

———. 1988c. Physiological, migratorial, climatological, geophysical, survival, and evolutionary implications of Cretaceous polar dinosaurs. *Journal of Paleontology* 62:640–52.

———. 1988d. Predation in the meat eating dinosaurs. *Occasional Papers of the Tyrrell Museum of Paleontology* 3:173–78.

———. 1989. Giant meteor impacts and great eruptions: Dinosaur killers? *Bioscience* 39:16272.

———. 1990. An improbable view of Tertiary dinosaurs. *Evolutionary Theory* 9:309–15.

———. 1991. The many myths, some old, some new, of dinosaurology. *Modern Geology* 16:69–99.

———. 1993. Are *Syntarsus* and the Whitaker Quarry theropod the same genus? *New Mexico Museum of Natural History and Science Bulletin* 3:397–402.

———. 1994a. Physiology and migration of North Slope dinosaurs. In *1992 Proceedings of the International Conference on Arctic Margins*, ed. D. K. Thurston and K. Fujita, 405–8. U.S. Department of the Interior, Anchorage, Alaska.

———. 1994b. Dinosaur reproduction in the fast lane: Implications for size, success, and extinction. In *Dinosaur Eggs and Babies*, ed. K. Carpenter, K. F. Hirsch, and J. R. Horner, 244–55. Cambridge University Press, Cambridge.

———. 1996. The art of Charles R. Knight. *Scientific American* 274(6):74–81.

———. 1997a. Dinosaur models: The good, the bad, and using them to estimate the mass of dinosaurs. In *The Dinofest International*, ed. D. L. Wolberg, E. Stump, and G. D. Rosenberg, 129–54. Academy of Natural Sciences, Philadelphia.

———. 1997b. Migration. In Currie and Padian 1997, 444–46.

———. 1997c. Reproductive behavior and rates. In Currie and Padian 1997, 630–37.

———. 1998. Terramegathermy and Cope's Rule in the land of titans. *Modern Geology* 23:179–217.

———. 1998/2000. Limb design, function, and running performance in ostrich-mimics and tyrannosaurs. *Gaia* 15:257–70.

Paul, G. S., and Chase, T. L. 1989. Reconstructing extinct vertebrates. In *The Guild Handbook of Scientific Illustration,* ed. E.R.S. Hodges, 239–56. Van Nostrand Reinhold, New York.

Paul, G. S., and Leahy, G. D. 1994. Terramegathermy in the time of the titans: Restoring the metabolics of colossal dinosaurs. *Paleontological Society Special Publication* 7:177–98.

Paul, G. S., and Cox, E. 1996. *Beyond Humanity.* Charles River Media, Rockland, Mass.

Peaker, M. 1975. *Avian Physiology.* Academic Press, London.

Pennycuick, C. J. 1979. Energy costs of locomotion and the concept of "foraging radius." In *Serengeti: Dynamics of an Ecosystem,* ed. A.R.E. Sinclair and M. Norton-Griffiths, 164–84. University of Chicago Press, Chicago.

———. 1986. Mechanical constraints on the evolution of flight. *Memoirs of the California Academy of Sciences* 8:83–98.

———. 1989. *Bird Flight Performance: A Practical Calculation Manual.* Oxford Science Publications, Oxford.

Pennycuick, C. J., Einarsson, O., Bradbury, T.A.M., and Owen, M. 1996. Migrating whooper swans *Cygnus cygnus:* Satellite tracks and flight performance calculations. *Journal of Avian Biology* 27:118–34.

Perez-Moreno, B., et al. 1994. A unique multitoothed ornithomimosaur dinosaur from the Lower Cretaceous of Spain. *Nature* 370:363–67.

Perez-Moreno, B. P., Holtz, T., Sanz, J. L., and Moratalla, J. 1998/2000. Aspects of theropod paleobiology. *Gaia* 15, entire volume.

Perle, A. 1979. Segnosauridae—A new family of theropods from the Late Cretaceous of Mongolia. *Sovmestnaya Sovetsko-Mongol'skaya Paleontologicheskaya Ekspiditsiya, Trudy* 8:45–55.

———. 1985. Comparative myology of the pelvic femoral region in the bipedal dinosaurs. *Paleontological Journal* 19:105–9.

Perle, A., et al. 1981. A new segnosaurid from the Upper Cretaceous of Mongolia. *Trudy Soviet-Mongolian Expeditions* 15:50–59.

Perle, A., Norell, M. A., Chiappe, L. M., and Clark, J. 1993. Flightless bird from the Cretaceous of Mongolia. *Nature* 362:623–26.

Perle, A., Chiappe, L. M., Barsbold, R., Clark, J. M., and Norell, M. 1994. Skeletal morphology of *Mononykus olecranus* from the Late Cretaceous of Mongolia. *American Museum Novitates* 3105:1–29.

Perle, A., Chiappe, L. M., and Clark, J. 1999. A new maniraptoran theropod—*Achillobator giganticus*—from the Upper Cretaceous of Burkhant, Mongolia. *Mongolian-American Museum Paleontological Project* 101:1–106.

Perry, S. F. 1983. Reptilian lungs. In *Functional Anatomy and Evolution.* Springer-Verlag, Berlin.

———. 1988. Functional morphology of the lungs of the Nile crocodile, *Crocodylus niloticus:* Nonrespiratory parameters. *Journal of Experimental Biology* 134:99–117.

———. 1989. Mainstreams in the evolution of vertebrate respiratory structures. In *Form and Function in Birds,* ed. A. S. King and J. McLelland, 4:1–67. Academic Press, London.

———. 1990. Gas exchange strategy in the Nile crocodile: A morphometric study. *Journal of Comparative Physiology B* 159:761–69.

———. 1992. Gas exchange strategies in reptiles and the origin of the avian lung. In *Physiological Adaptations in Vertebrates,* ed. S. Wood et al., 149–67. Marcel Dekker, New York.

Peters, D. S. 1985. Functional and constructive limitations in the early evolution of birds. In Hecht et al. 1985, 243–49.

Peters, D. S., and Gorgner, E. 1992. A comparative study on the claws of *Archaeopteryx. Science Series of the Los Angeles County Museum of Natural History* 36:29–37.

Peterson, A. 1985. The locomotor adaptations of *Archaeopteryx:* Glider or cursor? In Hecht et al. 1985, 99–103.

Pettigrew, J. D. 1995. Flying primates: Crashed, or crashed through? *Symposium of the Zoological Society of London* 67:3–26.

Poinar, H. N., et al. 1996. Amino acid racemization and the preservation of ancient DNA. *Science* 272: 864–66.

Pomeroy, D. E. 1980. Growth and plumage changes of the grey crowned crane. *Bulletin of the British Ornithological Conference* 100:219–23.

Poore, S. O., Sanchez-Haiman, A., and Goslow, G. E. 1997. Wing upstroke and the evolution of flapping flight. *Nature* 387:799–802.

Porter, R. K., and Brand, M. D. 1993. Body mass dependence of $H^+$ leak in mitochondria and its relevance to metabolic rate. *Nature* 362:628–30.

Pough, F. H., et al. 1998. *Herpetology.* Prentice Hall, Upper Saddle River, N.J.

Prange, H. D., Anderson, J. F., and Rahn, H. 1979. Scaling of skeletal mass to body mass in birds and mammals. *American Naturalist* 113:103–11.

Priede, I. G. 1985. Metabolic scope in fishes. In *Fish Energetics: New Perspectives,* ed. P. Tyler and P. Calow, 33–64. Johns Hopkins University Press, Baltimore.

Proctor, N. S., and Lynch, P. J. 1993. *Manual of Ornithology: Avian Structure and Function.* Yale University Press, New Haven.

Prum, R. O. 2001. *Longisquama* fossil and feather morphology. *Science* 291:1899–900.

Quinn, T. 1997. An alternative hypothesis on the information origin of feathers and their original function. *Evolutionary Theory* 11:273.

Raath, M. A. 1977. The anatomy of the Triassic theropod *Syntarsus rhodesiensis* and a consideration of its biology. Ph.D. diss., Rhodes University, Salisbury.

———. 1985. The theropod *Syntarsus* and its bearing on the origin of birds. In Hecht et al. 1985, 219–27.

Raikow, R. J. 1985. Locomotor system. In *Form and Function in Birds,* ed. A. S. King and J. McLelland, 3:57–147. Academic Press, London.

Randolph, S. E. 1994. The relative timing of the origin of flight and endothermy: Evidence from the comparative biology of birds and mammals. *Zoological Journal of the Linnean Society* (London) 112: 389–97.

Rauhut, O.W.M. 1998. *Elaphrosaurus bambergi* and the early evolution of theropod dinosaurs. *Journal of Vertebrate Paleontology* 18:71A.

Raup, D. M. 1991. *Extinction: Bad Genes or Bad Luck.* W. W. Norton, New York.

Rautian, A. S. 1978. A unique bird feather from Jurassic lake deposits in the Karatau. *Paleontological Journal* 12:520–28.

Rayner, J.M.V. 1985a. Mechanical and ecological constraints on flight evolution. In Hecht et al. 1985, 279–88.

———. 1985b. Cursorial gliding in proto-birds: An expanded version of a discussion contribution. In Hecht et al. 1985, 289–92.

———. 1991. Avian flight evolution and the problem of *Archaeopteryx. Biomechanics in Evolution,* ed. J.M.V. Rayner and R. J. Wootton, 183–212. Cambridge University Press, Cambridge.

Regal, P. J. 1985. Common sense and reconstructions of the biology of fossils: *Archaeopteryx* and feathers. In Hecht et al. 1985, 67–74.

Regal, P. J., and Gans, C. 1980. The revolution in thermal physiology: Implications for dinosaurs. In *A Cold Look at the Warm-Blooded Dinosaurs,* ed. R.D.K. Thomas and E. C. Olson, 167–88. AAAS, Washington, D.C.

Reid, R.E.H. 1987. Bone and dinosaurian "endothermy." *Modern Geology* 11:133–54.

———. 1990. Zonal "growth rings" in dinosaurs. *Modern Geology* 15:19–48.

———. 1993. Apparent zonation and slowed late growth in a small Cretaceous theropod. *Modern Geology* 18:391–406.

———. 1996. Bone histology of the Cleveland-Lloyd dinosaurs and of dinosaurs in general. Part 1, Introduction to bone tissues. *Brigham Young University Geology Studies* 41:25–71.

———. 1997. Histology of bones and teeth. *Brigham Young University Geology Studies* 42:329–39.

Reilly, S. M., and Elias, J. A. 1998. Locomotion in *Alligator mississipiensis:* Kinematic effects of speed and posture and their relevance to the sprawling-to-erect paradigm. *Journal of Experimental Biology* 201: 2559–74.

Reisz, R. R., and Sues, H.-D. 2000. The "feathers" of *Longisquama. Nature* 408:428.

Renesto, S. 1994. *Megalancosaurus,* a possibly arboreal Archosauromorph from the Upper Triassic of Italy. *Journal of Vertebrate Paleontology* 14:38–52.

———. 2000. Bird-like head on a chameleon body: New specimens of the enigmatic diapsid reptile *Megalancosaurus* from the Late Triassic of northern Italy. *Rivista Italiana di Paleontologia e Stratigrafia* 106: 157–80.

Rensberger, J. M., and Watabe, M. 2000. Fine structure of bone in dinosaurs, birds, and mammals. *Nature* 406:619–22.

Rich, P. V. 1979. The Dromornithidae. *Bureau of Mineral Resources, Geology, and Geophysics, Australia, Bulletin* 184.

Ricklefs, R. E. 1968. Patterns of growth in birds. *Ibis* 110:419–51.

———. 1973. Patterns of growth in birds. Part 2, Growth rate and mode of development. *Ibis* 115: 177–210.

Ricqles, A. J. de. 1980. Tissue structures of dinosaur bone: Functional significance and possible relation to dinosaur physiology. In *A Cold Look at the Warm-Blooded Dinosaurs,* ed. R.D.K. Thomas and E. C. Olson, 103–40. AAAS, Washington, D.C.

Ridley, M. 1996. *Evolution.* Blackwell Science, Cambridge, Mass.

Rietschel, S. 1985. Feathers and wings of *Archaeopteryx,* and the question of her flight ability. In Hecht et al. 1985, 251–60.

Robinette, W. L., and Archer, A. L. 1971. Notes on aging criteria and reproduction of Thomson's gazelle. *East African Wildlife Journal* 9:83–98.

Romer, A. S. 1923. The pelvic musculature of saurischian dinosaurs. *Bulletin of the American Museum of Natural History* 48:605–17.

———. 1927. The pelvic musculature of ornithischian dinosaurs. *Acta Zoologica* 8:225–75.

———. 1971. Two new but incompletely known long limbed pseudosuchians. *Breviora* 378:113.

———. 1972. Further remains of the thecodonts *Lagerpeton* and *Lagosuchus. Breviora* 394:1–7.

Rose, W. R., et al. 1999. Nonshivering thermogenesis in a marsupial is not attributable to brown adipose tissue. *Physiological and Biochemical Zoology* 72: 699–704.

Roth, H. H. 1965. Observations on the growth and aging of warthog (*Phacochoerus aethiopicus*). *Zeitschrift Saügetiere* 30:367–80.

Ruben, J. 1991. Reptilian physiology and the flight capacity of *Archaeopteryx. Evolution* 45:1–17.

———. 1993. Powered flight in *Archaeopteryx:* Response to Speakman. *Evolution* 47:935–38.

———. 1995. The evolution of endothermy in mammals and birds. *Annual Review of Physiology* 57:69–95.

———. 1996. Evolution of endothermy in mammals, birds and their ancestors. In *Animals and Tempera-*

*ture,* ed. I. A. Johnston and A. F. Bennett, 347–76. Cambridge University Press, Cambridge.
———. 1998. Gliding adaptations in the Triassic archosaur *Megalancosaurus. Journal of Vertebrate Paleontology* 18:73A.
Ruben, J., et al. 1996. The metabolic status of some Late Cretaceous dinosaurs. *Science* 273:120–47.
———. 1997a. Lung structure and ventilation in theropod dinosaurs and early birds. *Science* 278:1267–70.
———. 1997b. New insights into the metabolic strategy physiology of dinosaurs. In *The Complete Dinosaur,* ed. J. O. Farlow and M. Brett-Surman, 505–18. Indiana University Press, Bloomington.
———. 1998. Lung ventilation and gas exchange in theropod dinosaurs. *Science* 281:4748.
———. 1999. Pulmonary function and metabolic physiology of theropod dinosaurs. *Science* 283: 514–16.
Ruben, J. A., and Jones, T. D. 2000. Selective factors associated with the origin of fur and feathers. *American Zoologist* 40(4):585–96.
Ruppell, G. 1975. *Bird Flight.* Van Nostrand Reinhold, New York.
Russell, D. A. 1969. A new specimen of *Stenonychosaurus* from the Oldman Formation of Alberta. *Canadian Journal of Earth Sciences* 6:595–612.
———. 1972. Ostrich dinosaurs from the Late Cretaceous of western Canada. *Canadian Journal of Earth Sciences* 9:375–402.
———. 1989. *An Odyssey in Time: The Dinosaurs of North America.* University of Toronto Press, Toronto.
Russell, D. A., and Dong, Z.-M. 1993a. A nearly complete skeleton of a new troodontid dinosaur from the Early Cretaceous of the Ordos Basin, Inner Mongolia, People's Republic of China. *Canadian Journal of Earth Sciences* 30:2163–73.
———. 1993b. The affinities of a new theropod from the Alaxa Desert, Inner Mongolia, People's Republic of China. *Canadian Journal of Earth Sciences* 30: 2107–27.
Sampson, S. D., Krause, D. W., and Foster, C. A. 1997. Madagascar's buried treasure. *Natural History* 106 (3):24–27.
Sanz, J. L., and Lopez-Martinez, N. 1984. The prolacertid lepidosaurian *Cosesaurus aviceps,* a claimed "protoavian" from the Middle Triassic of Spain. *Geobios* 17:747–53.
Sanz, J. L., and Bonaparte, J. 1992. A new order of birds from the Lower Cretaceous of Spain. *Science Series of the Natural History Museum of Los Angeles County* 36:39–49.
Sanz, J. L., et al. 1996. An Early Cretaceous bird from Spain and its implications for the evolution of avian flight. *Nature* 382:442–45.
———. 1997. A nestling bird from the Lower Cretaceous of Spain: Implications for avian skull and neck evolution. *Science* 276:1543–46.
Sanz, J. L., Perez-Moreno, B. P., and Poyato-Ariza, F. J. 1998. Living with dinosaurs. *Nature* 393:32–33.
Sasso, C. D., and Signore, M. 1998. Exceptional soft-tissue preservation in a theropod dinosaur from Italy. *Nature* 392:383–87.
Saunders, D., and Fedde, M. R. 1994. Exercise performance in birds. In *Comparative Vertebrate Exercise Physiology: Phyletic Adaptations,* ed. J. H. Jones, 139–90. Academic Press, San Diego.
Saunders, F. 2001. Pushing the envelope on robots. *Discover* 22(3):50–55.
Scheid, P., and Piiper, J. 1989. Respiratory mechanisms and air flow in birds. In *Form and Function in Birds,* ed. A. S. King and J. M. McLelland, 436–91. Academic Press, New York.
Scheidler, S. E., and Sell, J. L. 1997. Designing a successful feeding program for your birds. *Emu Today and Tomorrow* 6(7):43–47.
Schlein, M. 1996. *The Puzzle of the Dinosaur-Bird.* Dial Books, New York.
Schmidt-Nielsen, K. 1964. *Desert Animals: Physiological Problems of Heat and Water.* Clarendon Press, Oxford.
———. 1972. *How Animals Work.* Cambridge University Press, Cambridge.
———. 1984. *Scaling: Why Is Animal Size So Important?* Cambridge University Press, Cambridge.
———. 1990. *Animal Physiology: Adaptation and Environment.* Cambridge University Press, Cambridge.
Schmidt-Nielsen, K., Hainsworth, F. R., and Murrish, D. E. 1970. Counter-current heat exchange in the respiratory passages: Effect on water loss and heat balance. *Respiration Physiology* 9:263–76.
Schneider, R. A., and Wolf, S. 1960. Relation of olfactory acuity to nasal membrane function. *Journal of Applied Physiology* 15:914–20.
Schreider, J. P., and Rabbe, O. G. 1981. Anatomy of the nasal-pharyngeal airway of experimental animals. *Anatomical Record* 200:195–295.
Schroter, R. C., Robertshaw, D., and Zino Filali, R. 1989. Brain cooling and respiratory heat exchange in camels during rest and exercise. *Respiration Physiology* 78:95–105.
Schroter, R. C., and Watkins, N. V. 1989. Respiratory heat exchange in mammals. *Respiration Physiology* 78:357–68.
Schudack, M. E. 1993. Charophyten aus dem Kimmeridgium der Kohlengrube Guimarota: Mit einer eingehenden Diskussion zur Datierung der Fundstelle. *Berliner geowissenschaftliche Abhandlungen E* 9:211–31.
Schultz, P. H., et al. 1998. A 3.3-Ma impact in Argentina and possible consequences. *Science* 282: 2061–63.
Schweitzer, M. H., et al. 1999. Beta-keratin specific immunological reactivity in feather-like structures of the Cretaceous alvarezsaurid, *Shuvuuia deserti. Journal of Experimental Zoology* 285:146–57.
Scott, J. H. 1954. Heat regulating function of the nasal mucous membrane. *Journal of Laryngology Otology* 68:308–17.

Scott, S. P., ed. 1974. *The World Atlas of Birds.* Crescent Books, New York.
Secor, S. M., and Diamond, J. 1995. Adaptive responses to feeding in Burmese pythons: Pay before pumping. *Journal of Experimental Biology* 198: 1313–25.
Seeley, H. G. 1876. On the British fossil Cretaceous birds. *Quarterly Journal of the Geological Society of London* 32:496–512.
Sellwood, B. W., Price, G. D., and Valdes, P. J. 1994. Cooler estimates of Cretaceous temperatures. *Nature* 370:453–55.
Sereno, P. C. 1991. Basal archosaurs: Phylogenetic relationships and functional implications. *Society of Vertebrate Paleontology, Memoir* 2:1–53.
———. 1993. The pectoral girdle and forelimb of the basal theropod *Herrerasaurus ischigualastensis. Journal of Vertebrate Paleontology* 13:425–50.
———. 1997a. The origin and evolution of dinosaurs. *Annual Review of Earth and Planetary Sciences* 25: 435–89.
———. 1997b. Ancient aviary, featherweight phylogeny. *Evolution* 51:1689–90.
———. 1998. A rationale for phylogenetic definitions, with application to the higher-level taxonomy of Dinosauria. *Neues Jahrbuch für Geologie und Paläontologie Abhandlungen* 210:41–83.
———. 1999a. The evolution of dinosaurs. *Science* 284: 2137–47.
———. 1999b. Alvarezsaurids: Birds or ornithomimosaurs? *Journal of Vertebrate Paleontology* 19:75A.
———. 1999c. A rationale for dinosaurian taxonomy. *Journal of Vertebrate Paleontology* 19:788–90.
Sereno, P. C., and Novas, F. 1992. The complete skull and skeleton of an early dinosaur. *Science* 258: 1137–40.
Sereno, P. C., and Rao, C. 1992. Early evolution of avian flight and perching: New evidence from the Lower Cretaceous of China. *Science* 255:845–48.
———. 1993. The skull and neck of the basal theropod *Herrerasaurus ischigualastensis. Journal of Vertebrate Paleontology* 13:451–76.
Sereno, P. C., and Arcucci, A. B. 1993. Dinosaurian precursors from the Middle Triassic of Argentina: *Lagerpeton chanaernsis. Journal of Vertebrate Paleontology* 13:385–99.
———. 1994. Dinosaurian precursors from the Middle Triassic of Argentina: *Marasuchus lilloensis,* gen. nov. *Journal of Vertebrate Paleontology* 14: 53–73.
Sereno, P. C., Forster, C. A., Rogers, R. R., and Monetta, A. M. 1993. Primitive dinosaur skeleton from Argentina and the early evolution of Dinosauria. *Nature* 361:64–66.
Sereno, P. C., Wilson, J. A., Larsson, H.C.E., Dutheil, D. B., and Sues, H.-D. 1994. Early Cretaceous dinosaurs from the Sahara. *Science* 266:267–70.
Sereno, P. C., et al. 1998. A long snouted predatory dinosaur from Africa and the evolution of spinosaurids. *Science* 282:1298–302.
Seymour, R. S., and Ackerman, R. A. 1980. Adaptations to underground nesting in birds and reptiles. *American Zoologist* 20:437–47.
———. 1982. Physiological adaptations to aquatic life. In *Biology of the Reptilia,* ed. C. Gans and R. S. Seymour, 1–51. Academic Press, New York.
Seymour, R. S., Withers, P. C., and Weathers, W. W. 1998. Energetics of burrowing, running, and free-living in the Namib Desert golden mole. *Journal of Zoology* (London) 244:107–17.
Sharov, A. G. 1970. An unusual reptile from the Lower Triassic of Fergana. *Paleontological Journal* 1970: 127–30.
Shipman, P. 1998. *Taking Wing:* Archaeopteryx *and the Evolution of Bird Flight.* Simon and Schuster, New York.
Shores, C. 1983. *Air Aces.* Bison Books, Greenwich, Conn.
Shrine, R. 1988. Parental care in reptiles. In *Biology of the Reptilia,* ed. C. Gans and K. A. Nagy, 16:275–329. Wiley, New York.
Shubin, N. 1994. History, ontogeny, and evolution of the archetype. In *Homology: The Hierarchical Basis of Comparative Biology,* ed. B. K. Hall, 248–71. Academic Press, New York.
———. 1998. Evolutionary cut and paste. *Nature* 394: 12–13.
Sikes, S. K. 1971. *Natural History of the African Elephant.* American Elsevier, New York.
Simmons, N. B. 1995. Bat relationships and the origin of flight. *Symposia of the Zoological Society of London* 67:27–43.
Simons, L. M. 2000. Fossil trail. *National Geographic* 198(4):128–32.
Siwe, S. A. 1937. Die grossen Druesen des Darmkanals—A. Die Leber. In *Darmsystem, Atmungssystem, Coelom,* ed. L. Bolk et al., 3:725–74.
Sloan, C. P. 1999. Feathers for *T. rex*? *National Geographic* 196(5):98–107.
Smith, D. 1992. The type specimen of *Oviraptor philoceratops,* a theropod dinosaur from the Upper Cretaceous of Mongolia. *Neues Jahrbuch für Geologie und Paläontologie Abhandlungen* 186:365–88.
Smith, J., You H., and Dodson, P. 1998. The age of the *Sinosauropteryx* quarry, northeastern China. *Journal of Vertebrate Paleontology* 18:78A.
Sober, E. 1988. *Reconstructing the Past: Parsimony, Evolution, and Inference.* MIT Press, Cambridge, Mass.
Sokoloff, A., Gray, J., and Harry, J. 1994. Supracoracoideus is not necessary for take-off in the starling. *American Zoologist* 311:64A.
Sole, R. V., et al. 1997. Self-similarity of extinction statistics in the fossil record. *Nature* 388:764–67.
Speakman, J. R. 1993. Flight capabilities in *Archaeopteryx. Evolution* 47:336–40.
Speakman, J. R., and Thomson, S. C. 1994. Flight capabilities of *Archaeopteryx. Nature* 370:514.
Spearman, R.I.C., and Hardy, J. A. 1985. Integument. In *Form and Function in Birds,* ed. A. S. King and J. McLelland, 1–56. Academic Press, London.

Spinar, Z. V., and Burian, Z. 1972. *Life before Man.* American Heritage Press, New York.

Spotila, J. R., O'Connor, M. P., Dodson, P., and Paladino, F. V. 1991. Hot and cold running dinosaurs: Body size, metabolism and migration. *Modern Geology* 16:203–27.

Spray, J. G., Kelley, S. P., and Rowley, D. B. 1998. Evidence for a Late Triassic multiple impact event on Earth. *Nature* 392:171–73.

Standora, E. A., Spotila, J. R., Keinath, J. A., and Shoop, C. R. 1984. Body temperatures, diving cycles, and movement of a subadult leatherback turtle, *Dermochelys coriacea. Herpetologica* 40:169–76.

Stein, R. S. 1975. Dynamic analysis of *Pteranodon ingens:* A reptilian adaptation to flight. *Journal of Paleontology* 49:534–48.

Stephen, B. 1974. *Urvogel Archaeopterygiformes.* A. Ziemsen Verlag, Wittenberg.

Sternberg, C. M. 1940. A toothless bird from the Cretaceous of Alberta. *Journal of Paleontology* 14: 81–85.

Stettenheim, P. 1974. The bristles of birds. *Living Bird* 12:201–34.

———. 2000. The integumentary morphology of modern birds—An overview. *American Zoologist* 40(4): 461–77.

Stidham, T. A. 1998. A lower jaw from a Cretaceous parrot. *Nature* 396:29–30.

———. 1999. Did parrots exist in the Cretaceous period? *Nature* 399:318.

Stokstad, E. 2000. Feathers, flight or fancy? *Science* 288:2124–25.

———. 2001. Exquisite Chinese fossils add new pages to book of life. *Science* 291:232–36.

Stoll, H. M., and Schrag, D. P. 1996. Evidence for glacial control of rapid sea level changes in the Early Cretaceous. *Science* 272:1771–74.

Sues, H.-D. 1977. Dentaries of small theropods from the Judith River Formation of Alberta, Canada. *Canadian Journal of Earth Sciences* 14:587–92.

———. 1978. A new small theropod dinosaur from the Judith River Formation of Alberta Campanian. *Zoological Journal of the Linnean Society* (London) 62:381–400.

Sumida, S. S., and Brochu, C. A. 1997. On *Chirostenotes,* a Late Cretaceous oviraptorosaur from western North America. *Journal of Vertebrate Paleontology* 17:698–716.

———. 2000. Phylogenetic context for the origin of feathers. *American Zoologist* 40(4):486–503.

Sutter, E., and Cornaz, N. 1965. Zum Wachstum der Grossfusshühner (*Alectura* und *Megapodius*). *Ornithologischer Beobachter* 62:43–60.

Suzuki, N., and Hamada, T. 1992. Quadrupedal function of the Komodo dragon in a wild habitat. *Science Papers, College of Arts and Sciences, University of Tokyo* 41:65–105.

Swartz, S. M., Bennett, M. B., and Carrier, D. R. 1992. Wing bone stresses in free flying bats and the evolution of skeletal design for flight. *Nature* 359:726–29.

Swisher, C. C., et al. 1999. Cretaceous age for the feathered dinosaurs of Liaoning, China. *Nature* 400: 58–61.

Sy, M. 1936. Funktionell-anatomische Untersuchungen am Vogelflügel. *Journal für Ornithologie* 84:253–67.

Tambussi, C., Ubilla, M., and Perea, D. 1999. The youngest carnassial bird from South America. *Journal of Vertebrate Paleontology* 19:404–6.

Taquet, P., and Russell, D. A. 1998. New data on spinosaurid dinosaurs from the Early Cretaceous of the Sahara. Comptes rendus de l'Académie des Sciences, ser. 2, *Sciences de la terre et des planètes* 327: 347–53.

Tarduno, J. A., et al. 1998. Evidence for extreme climatic warmth from Late Cretaceous arctic vertebrates. *Science* 282:2241–45.

Tarsitano, S. F. 1985. The morphological and aerodynamic constraints on the origin of avian flight. In Hecht et al. 1985.

———. 1991. *Archaeopteryx:* Quo Vadis? In *Origins of the Higher Groups of Tetrapods,* ed. H.-P. Schultze and L. Trueb, 485–540. Cornell University Press, Ithaca, N.Y.

Tarsitano, S., and Hecht, M. K. 1980. A reconsideration of the reptilian relationships of *Archaeopteryx. Zoological Journal of the Linnean Society* (London) 69:149–82.

Tarsitano, S. F., Russell, A. P., Horne, F., Plummer, C., and Millerchip, K. 2000. On the evolution of feathers from an aerodynamic and constructional view point. *American Zoologist* 40(4):676–86.

Taylor, C. R., and Lyman, C. P.. 1972. Heat storage in running antelopes: Independence of brain and body temperatures. *American Journal of Physiology* 222: 114–17.

Teeling, E. C., et al. 2000. Molecular evidence regarding the origin of echolocation and flight in bats. *Nature* 403:188–92.

Thewissen, J.G.M., and Babcock, S. K. 1991. Distinctive cranial and cervical innervation of wing muscles: New evidence for bat monophyly. *Science* 251: 934–36.

Thompson, G. G., and Withers, P. C. 1997. Standard and maximal metabolic rates of goannas. *Physiological Zoology* 70:307–23.

Thulborn, R. A. 1975. Dinosaur polyphyly and the classification of archosaurs and birds. *Australian Journal of Zoology* 30:611–34.

———. 1984. The avian relationships of *Archaeopteryx,* and the origin of birds. *Zoological Journal of the Linnean Society of London* 82:119–58.

———. 1985. Birds as neotenous dinosaurs. *Records of the New Zealand Geological Survey* 9:90–92.

———. 1993. A tale of three fingers: Ichnological evidence revealing the homologies of manual digits in theropod dinosaurs. *New Mexico Museum of Natural History and Science Bulletin* 3:461–63.

Thulborn, R. A., and Hamley, T. L. 1984. On the hand of *Archaeopteryx. Nature* 311:218.

———. 1985. A new palaeoecological role for *Archaeopteryx*. In Hecht et al. 1985, 81–89.

Thulborn, T. 1990. *Dinosaur Tracks*. Chapman and Hall, London.

———. 1992. The demise of the dancing dinosaur. *The Beagle, Records of the Northern Territory Museum of Arts and Sciences* 9:29–34.

Tomida, Y., and Sato, T. 1995. *Dinosaurs: Their Life and Their Behavior*. Vol. 2, *Deinonychus* (in Japanese). Kaisei-sha Publishing, Tokyo.

Tonni, E. P. 1980. The present state of knowledge of the Cenozoic birds of Argentina. *Los Angeles County Museum of Natural History, Contributions in Science* 330:105–14.

Tuinen, M., Sibley, C. G., and Hedges, S. B. 1998. Phylogeny and biogeography of ratite birds inferred from DNA sequences of the mitochondrial ribosomal genes. *Molecular and Biological Evolution* 15: 370–76.

Tuinen, M., et al. 2000. The early history of modern birds inferred from DNA sequences of nuclear and mitochondrial ribosomal genes. *Molecular and Biological Evolution* 17:451–57.

Turcek, F. J. 1966. On plumage quantity in birds. *Ekologia Polska* 14:617–33.

Unwin, D. M. 1987. Pterosaur extinction: Nature and causes. *Mémoires de la Société Géologique de France* 150:105–11.

Unwin, D. M., and Benton, M. J. 2001. *Longisquama* fossil and feather morphology. *Science* 291:1900–1901.

Van Dongen, P.A.M. 1998. Brain size in vertebrates. In *The Central Nervous System of Vertebrates*, ed. R. Nieuwenhuys, H. J. Ten Donkelaar, and C. Nicholson, 3:2099–134. Springer, Berlin.

Van Tyne, J., and Berger, A. J. 1976. *Fundamentals of Ornithology*. Wiley, New York.

Varricchio, D. J. 1993. Bone microstructure of the Upper Cretaceous theropod dinosaur *Troodon formosus*. *Journal of Vertebrate Paleontology* 13:99–104.

———. 1997. Growth and embryology. In Currie and Padian 1997, 282–88.

Varricchio, D. J., Jackson, F., Borkowski, J. J., and Horner, J. R. 1997. Nest and egg clutches of the dinosaur *Troodon formosus* and the evolution of avian reproductive tracts. *Nature* 385:247–50.

Varricchio, D. J., Jackson, F., and Trueman, C. N. 1999. A nesting trace with eggs for the Cretaceous theropod dinosaur *Troodon formosus*. *Journal of Vertebrate Paleontology* 19:91–100.

Varricchio, D. J., and Jackson, F. 2000. Physiological implications of reproductive behavior in the dinosaur *Troodon formosus*. *Journal of Vertebrate Paleontology* 20:75A.

Vaughan, T. A., and Bateman, M. M. 1980. The molossid wing: Some adaptations for rapid flight. In *Proceedings of the Fifth International Bat Research Conference*, ed. D. E. Wilson and A. L. Gardner, 69–78. Texas Tech Press, Lubbock.

Vazquez, R. J. 1992. Functional osteology of the avian wrist and the evolution of flapping flight. *Journal of Morphology* 211:2–268.

Vermeij, G. J. 1999. A serious matter with character-taxon matrices. *Paleobiology* 25:431–33.

Viohl, G. 1985. Geology of the Solnhofen lithographic limestones and the habitat of *Archaeopteryx*. In Hecht et al. 1985, 31–44.

Von Huene, F. R.. 1926. The carnivorous saurischia in the Jura and Cretaceous formations principally in Europe. *Revista Museo La Plata* 29:35–167.

Wagner, G. P., and Gauthier, J. A. 1999. 1,2,3 = 2,3,4: A solution to the problem of the homology of the digits in the avian hand. *Proceedings of the National Academy of Sciences* 96:5111–16.

Walker, A. D. 1964. Triassic reptiles from the Elgin area: *Ornithosuchus* and the origin of carnosaurs. *Philosophical Transactions of the Royal Society of London B* 248:53–134.

———. 1972. New light on the origin of birds and crocodiles. *Nature* 237:257–63.

———. 1977. Evolution of the pelvis in birds and dinosaurs. In *Problems in Vertebrate Evolution*, ed. S. M. Andrews, R. S. Miles, and A. D. Walker, 319–57. Academic Press, London.

———. 1985. The braincase of *Archaeopteryx*. In Hecht et al. 1985, 123–34.

Walker, C. A. 1981. New subclass of birds from the Cretaceous of South America. *Nature* 292:51–53.

Wang, J. 1998. Scientists flock to explore China's "site of the century." *Science* 279:1626–27.

Watabe, M., et al. 2000. New nearly complete skeleton of the bird-like theropod, *Avimimus*, from the Upper Cretaceous of the Gobi Desert, Mongolia. *Journal of Vertebrate Paleontology* 20:77A.

Webb, G.J.W., Messel, H., Crawford, J., and Yerbury, M. J. 1978. Growth rates of *Crocodylus porosus* from Arnhem Land, northern Australia. *Australian Wildlife Research* 5:385–99.

Weems, R. E. 1987. A Late Triassic footprint fauna from the Culpeper Basin, northern Virginia. *Transactions of the American Philosophical Society* 77 (1):1–79.

Wegener, P. P. 1997. *What Makes Airplanes Fly?* Springer-Verlag, New York.

Weigert, A. 1995. Isolierte Zähne von cf. *Archaeopteryx* sp. aus dem Oberen Jura der Kohlengrube Guimarota (Portugal). *Neues Jahrbuch für Geologie und Paläontologie Monatshefte* 1995:562–76.

Weis-Fogh, T. 1975. Unusual mechanisms for the generation of lift in flying animals. *Scientific American* 233(5):81–87.

Weishampel, D. B. 1981. The nasal cavity of lambeosaurine hadrosaurids: Comparative anatomy and homologies. *Journal of Paleontology* 55:1046–57.

Weishampel, D. B., Dodson, P., and Osmólska, H., eds. 1990. *The Dinosauria*. University of California Press, Berkeley.

Welles, S. P. 1984. *Dilophosaurus wetherilli*, osteology and comparisons. *Palaeontographica A* 185:85–180.

Wellnhofer, P. 1974. Das Fünfte Skelettexemplar von *Archaeopteryx. Palaeontographica A* 147:169–216.
———. 1985. Remarks on the digit and pubis problems of *Archaeopteryx*. In Hecht et al. 1985, 113–22.
———. 1988. Ein neues Exemplar von *Archaeopteryx. Archaeopteryx* 6:1–30.
———. 1991. *The Illustrated Encyclopedia of Pterosaurs.* Crescent Books, New York.
———. 1993. Das siebte Exemplar von *Archaeopteryx* aus den Solnhofener Schichten. *Archaeopteryx* 11: 1–48.
———. 1994. New data on the origin and early evolution of birds. *Comptes rendus de l'Académie des Sciences* 319:299–308.
Welman, J. 1995. *Euparkeria* and the origin of birds. *South African Journal of Science* 91:533–37.
Whetstone, K. N. 1983. Braincase of Mesozoic birds. Part 1, New preparation of the "London" *Archaeopteryx. Journal of Vertebrate Paleontology* 2:439–52.
Whetstone, K. N., and Martin, L. D. 1979. New look at the origin of birds and crocodiles. *Nature* 279: 234–36.
Whiteman, L. 2000. The high life. *Audubon* 102(6): 104–8.
Whittow, G. C., ed. 1970–73. *Comparative Physiology of Thermoregulation.* Vols. 1–3. Academic Press, New York.
———, ed. 2000. *Sturkie's Avian Physiology.* 5th ed. Academic Press, San Diego.
Widdowson, E. M. 1981. Growth of the body and its components and the influence of nutrition. In *The Biology of Normal Human Growth,* ed. M. Ritzen et al., 253–63. Raven Press, New York.
Wigglesworth, V. B. 1974. *Insect Physiology.* Chapman and Hall, Cambridge.
Wild, R. 1984. A new pterosaur from the Upper Triassic of Friuli, Italy. *Gortania-Atti del Museo Friulano di Storia Naturale* 5:45–62.
Wiley, E. O. 1981. *Phylogenetics: The Theory and Practice of Phylogenetic Systems.* Wiley, New York.
Williams, P. L., Warwick, R., Dyson, M., and Bannister, L. H., eds. 1989. *Gray's Anatomy.* Churchill Livingston, London.
Withers, P. C. 1983. Energy, water, and solute balance of the ostrich *Struthio camelus. Physiological Zoology* 56:568–79.
Withers, P. C., Siegfried, W. R., and Louw, G. N. 1981. Desert ostrich exhales unsaturated air. *South African Journal of Science* 77:569–70.
Withers, P. C., and Williams, J. B. 1990. Metabolic and respiratory physiology of an arid-adapted Australian bird, the spinifex pigeon. *Condor* 82:99–100.
Witmer, L. M. 1990. The craniofacial air sac system of Mesozoic birds. *Zoological Journal of the Linnean Society* (London) 100:327–78.
———. 1991. Perspectives on avian origins. In *Origins of the Higher Groups of Tetrapods,* ed. H.-P. Schultze and L. Trueb, 427–66. Cornell University Press, Ithaca, N.Y.
———. 1995a. Homology of facial structures in extant archosaurs (birds and crocodilians), with special references to paranasal pneumaticity and nasal conchae. *Journal of Morphology* 225:269–77.
———. 1995b. The extant phylogenetic bracket and the importance of reconstructing soft tissues in fossils. In *Functional Morphology in Vertebrate Paleontology,* ed. J. Thomason, 19–33. Cambridge University Press, Cambridge.
———. 1997a. The evolution of the antorbital cavity of archosaurs: A study in soft-tissue reconstruction in the fossil record with an analysis of the function of pneumaticity. *Journal of Vertebrate Paleontology* 17 (suppl. to 1): 1–73.
———. 1997b. Flying feathers. *Science* 276:1209–10.
———. 1997c. Craniofacial air sinus systems. In Currie and Padian 1997, 151–59.
Witmer, L. M., and Rose, K. D. 1991. Biomechanics of the jaw apparatus of the gigantic Eocene bird *Diatryma:* Implications for diet and mode of life. *Paleobiology* 17:95–120.
———. 1999. Nasal conchae and blood supply in some dinosaurs: Physiological implications. *Journal of Vertebrate Paleontology* 19:85A.
Witmer, L. M., and Sampson, S. D. 1999. Nasal conchae and blood supply in some dinosaurs: Physiological implications. *Journal of Vertebrate Paleontology* 19: 72A–73A.
Woodward, A. S. 1907. On a new dinosaurian reptile from the Trias of Lossiemouth, Elgin. *Quarterly Journal of the Geological Society of London* 63:140–45.
Wootton, R. 2000. From insects to microvehicles. *Nature* 403:144–45.
Worthy, T. H. 1999. The role of climate change versus human impacts—Avian extinction on South Island, New Zealand. In Olson 1999, 111–23.
Wroe, S. 1998. Bills, bones, and bias: Did thunder birds eat meat? *Riversleigh Notes* 40:24.
Xu, X., Tang, Z., and Wang, X. 1999. A therizinosauroid dinosaur with integumentary structures from China. *Nature* 399:350–54.
Xu, X., Wang, X., and Wu, X. 1999. A dromaeosaurid dinosaur with a filamentous integument from the Yixian Formation of China. *Nature* 401:262–66.
Xu, X., Zhou Z., and Wang, X. 2000. The smallest known non-avian theropod dinosaur. *Nature* 408: 705–8.
Xu, X., Zhou Z., and Prum, R. O. 2001. Branched integumental structures in *Sinornithosaurus* and the origin of feathers. *Nature* 410:200–203.
Yalden, D. W. 1971a. Flying ability of *Archaeopteryx. Nature* 231:127–28.
———. 1971b. Flying ability of *Archaeopteryx. Ibis* 113:349–56.
———. 1984. What size was *Archaeopteryx*? *Zoological Journal of the Linnean Society* (London) 82: 17788.
———. 1985. Forelimb function in *Archaeopteryx*. In Hecht et al. 1985, 91–97.

Yalden, D. W., and Morris, P. A. 1975. *The Lives of Bats.* The New York Times Book Co., New York.

Zelenitsky, D. K., and Hirsch, K. F. 1997. Fossil eggs: Identification and classification. In *The Dinofest International,* ed. D. L. Wolberg, E. Stump, and G. D. Rosenberg, 279–86. Academy of Natural Sciences, Philadelphia.

Zhang, F., and Zhou Z. 2000. A primitive enantiornithine bird and the origin of feathers. *Science* 290: 1955–59.

Zhao, X.-J., and Currie, P. J. 1993. A large crested theropod from the Jurassic of Xinjiang, People's Republic of China. *Canadian Journal of Earth Sciences* 30:2027–36.

Zhao, X.-J., and Xu, X. 1998. The oldest coelurosaurian. *Nature* 394:235–36.

Zhou Z. 1995a. The discovery of Early Cretaceous birds in China. *Courier Forschungsinstitut Senckenberg* 181:9–22.

———. 1995b. Is *Mononykus* a bird? *Auk* 112:958–63.

———. 1998. Origin of avian flight: Evidence from fossil and modern birds. *Journal of Vertebrate Paleontology* 18:88A.

Zhou Z., and Martin, L. D. 1999. Feathered dinosaur or bird? A new look at the hand of *Archaeopteryx.* In Olson 1999, 289–93.

Zhou Z., and Wang, X. 2000. A new species of *Caudipteryx* from the Yixian Formation of Liaoning, northwest China. *Vertebrata PalAsiatica* 38:111–27.

Zhou Z., Wang, X., Zhang, F.-C., and Xu, X. 2000. Important features of *Caudipteryx*—Evidence from two nearly complete new specimens. *Vertebrata PalAsiatica* 38:241–54.

Zimmer, C. 1997. Terror, take two. *Discover* 18 (6): 68–74.

Zimmer, K. 1935. Beiträge zur Mechanik der Atmung bei den Vögeln in Stand und Flug. *Zoologica* 88:1–142.

Zinke, J. 1998. Small theropod teeth from the Upper Jurassic coal mine of Guimarota. *Paläontologische Zeitschrift* 72:179–89.

Zinsmeister, W. J. 1997. Did the world of the dinosaurs end with a bang or a pop? In *The Dinofest International,* ed. D. L. Wolberg, E. Stump, and G. D. Rosenberg, 541–57. Academy of Natural Sciences, Philadelphia.

Zou, H., and Niswander, L. 1996. Requirement for BMP signaling in interdigital apoptosis and scale formation. *Science* 272:738–41.

Zou, S. 1983. A nearly complete skeleton of stegosaur from Middle Jurassic of Dashanpu, Zigong, Sichuan. *Journal of Chengdu College of Geology Suppl.* 1: 15–26.

Zug, G. R., Wynn, A. H., and Ruckdeschel, C. 1986. Age determination of loggerhead sea turtles, *Caretta caretta,* by incremental growth marks in the skeleton. *Smithsonian Contributions to Zoology* 427: 1–34.

Zusi, R. L. 1984. A functional and evolutionary analysis of rhynchokinesis in birds. *Smithsonian Contributions to Zoology* 395:1–40.

Zusi, R. L., and Warheit, K. I. 1992. On the evolution of intraramal mandibular joints in pseudodontornes. *Science Series of the Natural History Museum of Los Angeles County* 36:351–60.

# Index

Page numbers followed by the letters *f* and *t* indicate entries in figures and tables, respectively.